Stanley G. Prussin
Nuclear Physics for Applications

1807–2007 Knowledge for Generations

Each generation has its unique needs and aspirations. When Charles Wiley first opened his small printing shop in lower Manhattan in 1807, it was a generation of boundless potential searching for an identity. And we were there, helping to define a new American literary tradition. Over half a century later, in the midst of the Second Industrial Revolution, it was a generation focused on building the future. Once again, we were there, supplying the critical scientific, technical, and engineering knowledge that helped frame the world. Throughout the 20th Century, and into the new millennium, nations began to reach out beyond their own borders and a new international community was born. Wiley was there, expanding its operations around the world to enable a global exchange of ideas, opinions, and know-how.

For 200 years, Wiley has been an integral part of each generation's journey, enabling the flow of information and understanding necessary to meet their needs and fulfill their aspirations. Today, bold new technologies are changing the way we live and learn. Wiley will be there, providing you the must-have knowledge you need to imagine new worlds, new possibilities, and new opportunities.

Generations come and go, but you can always count on Wiley to provide you the knowledge you need, when and where you need it!

William J. Pesce
President and Chief Executive Officer

Peter Booth Wiley
Chairman of the Board

Stanley G. Prussin

Nuclear Physics for Applications

A Model Approach

WILEY-VCH Verlag GmbH & Co. KGaA

The Author

Prof. Stanley G. Prussin
Dept. of Nuclear Engineering
University of California
Berkeley, California 94720
USA

■ All books published by Wiley-VCH are carefully produced. Nevertheless, authors, editors, and publisher do not warrant the information contained in these books, including this book, to be free of errors. Readers are advised to keep in mind that statements, data, illustrations, procedural details or other items may inadvertently be inaccurate.

Library of Congress Card No.:
applied for

British Library Cataloguing-in-Publication Data
A catalogue record for this book is available from the British Library.

Bibliographic information published by the Deutsche Nationalbibliothek
The Deutsche Nationalbibliothek lists this publication in the Deutsche Nationalbibliografie; detailed biliographic data are available in the Internet at <http://dnb.d-nb.de>.

© 2007 WILEY-VCH Verlag GmbH & Co. KGaA, Weinheim

All rights reserved (including those of translation into other languages). No part of this book may be reproduced in any form – by photoprinting, microfilm, or any other means – nor transmitted or translated into a machine language without written permission from the publishers. Registered names, trademarks, etc. used in this book, even when not specifically marked as such, are not to be considered unprotected by law.

Typesetting Dörr + Schiller GmbH, Stuttgart
Printing betz-druck GmbH, Darmstadt
Binding Litges & Dopf Buchbinderei GmbH, Heppenheim
Wiley Bicentennial Logo Richard J. Pacifico

Printed in the Federal Republic of Germany
Printed on acid-free paper

ISBN: 978-3-527-40700-2

To Sadie, Abe and Joanne, Sophie, Camille, John, Avery A. Ashdown, Charles D. Coryell, Rengen, W.W. Meinke, Jack Hollander, Mike, Jerry, Hans Mark, Don Olander, Gunther Herrmann, Norbert and Jens, Adnan, Jorma, George and Fathi. But most of all to Traudel, Stephanie and Alexa, without whom it never would have been possible.

Table of Contents

Preface *XIII*

1 Introduction *1*
1.1 Low-Energy Nuclear Physics for Applications *1*
1.2 Some General Observations and Notations *3*
1.3 Overview of Radioactive Decay Processes and Nuclear Reactions *4*
1.3.1 Alpha Decay *4*
1.3.2 Beta Decay *5*
1.3.3 Spontaneous Fission *7*
1.3.4 Gamma Decay *8*
1.3.5 Nuclear Reactions *9*
1.4 The Model-based Character of this Text *10*
1.5 Sources of Nuclear Data *11*

2 Nuclear Masses and Energetics of Radioactive Decay and Nuclear Reactions *13*
2.1 Introduction *13*
2.2 Review of the Special Theory of Relativity *13*
2.3 Masses of Atoms and Particles *18*
2.4 Comments Concerning "Nuclear Stability" and Energetics *21*
2.4.1 Spontaneous Transformations and Nuclear Masses *21*
2.4.2 Nuclear Stability *24*
2.5 Bound and Unbound States and Their Energetics: Potential Wells *25*
2.6 Nuclear and Atomic Masses and Binding Energies *29*
2.6.1 β^- Decay *30*
2.6.2 β^+ Decay or Positron Emission *32*
2.6.3 Electron Capture Decay *34*
2.6.4 Competitive Decay Modes *35*
2.6.5 α Decay *36*
2.6.6 Spontaneous Fission *37*
2.7 Nuclear Reactions *38*

3 Phenomenology of Radioactive Decay and Nuclear Reactions 41

3.1 Introduction 41
3.1.1 The Phenomenology of Radioactive Decay 41
3.1.2 Units for Describing Radioactive Decay 45
3.1.3 Radioactive Growth and Decay 45
3.1.4 Simple Decay Schemes and Decay Chains 51
3.2 Statistical Considerations in Radioactive Decay 54
3.2.1 The Binomial Distribution 54
3.2.2 The Poisson Distribution 56
3.2.3 Application of Statistical Analysis to Common Experimental Conditions 61
3.2.4 Propagation of Errors 62
3.3 Reaction Cross Sections 66

4 Nuclear Binding Energies: Empirical Data and the Forces in Nuclei 75

4.1 Empirical Masses and Average Binding Energies of Nucleons 75
4.2 The Forces Acting Between Nucleons 77
4.3 The Average Nuclear Interaction Between Nucleons in the Nucleus and Nuclear Radii 83
4.4 Quantization of the Nucleus: Pairing of Identical Nucleons 92
4.5 Quantization of the Nucleus: Asymmetry Energy 100

5 The Semi-Empirical Mass Formula and Applications to Radioactive Decay 109

5.1 Introduction 109
5.2 The Semi-Empirical Mass Formula 110
5.3 The Nuclear Mass Surface 115
5.4 The Semi-Empirical Mass Formula and β Decay 117
5.5 The Semi-Empirical Mass Formula and α Decay 123
5.6 The Semi-Empirical Mass Formula and Nuclear Fission 125
5.7 Discrepancies Between Experimental Masses and those Predicted by the Semi-Empirical Mass Formula 128

6 Elements of Quantum Mechanics 133

6.1 Introduction 133
6.2 Elements of Quantum Mechanics 134
6.2.1 The Schrödinger Equation and Conservation Laws 134
6.2.2 Elementary Properties of Operators 135
6.2.3 Elementary Properties of Wave Functions 137
6.2.4 Operators, Eigenfunctions and Conservation Laws 138
6.2.5 Parity 146
6.3 Angular Momentum in Quantum Mechanics 147
6.3.1 Operators for Orbital Angular Momentum 148
6.3.2 Angular Momentum and Magnetic Moments 150

6.4	The Vector Model for Angular Momentum	*152*
6.5	The Wave Functions of Many-Particle Systems	*158*

7 Nuclear Structure: The Spherical Shell Model *161*
7.1 Introduction *161*
7.2 The Independent Particle Model *161*
7.2.1 The Angular Equations: Angular Momentum and Parity *164*
7.2.2 Some Properties of the Wave Functions *170*
7.2.3 The Radial Equation and the Centrifugal Potential *172*
7.2.4 Models for the Average Central Potential in the Independent Particle Approximation *175*
7.2.5 The Infinite Spherical Potential Well *178*
7.2.6 The Isotropic Harmonic Oscillator *185*
7.3 The Single-Particle Levels of Spherical Nuclei *189*
7.4 Comparison of the Predictions of the Single-Particle Model with Experiment *195*

8 Nuclear Shapes, Deformed Nuclei and Collective Effects *205*
8.1 Introduction *205*
8.2 Collective Excitations *212*
8.3 Rotational Excitations in (Even,Even) Nuclei *213*
8.4 Rotational Excitations in Odd-A Nuclei *219*
8.5 Vibrational Excitations in Nuclei *222*
8.6 Nuclear Structure in a Deformed Potential *229*
8.7 The Nilsson Model *234*

9 α Decay and Barrier Penetration *245*
9.1 Introduction *245*
9.2 Q_α and α Decay Half-Lives *248*
9.3 Binding of Valence Nucleons and the Potential for Interaction Between an α Particle and a Heavy Nucleus *253*
9.4 The Wave Functions for Particles in Finite Potential Wells and Barrier Penetration *258*
9.5 A Simple Model for α Decay *266*
9.6 Application of the Model to the Decay of Even-Even Nuclei *268*
9.7 Angular Momentum Effects in α Decay *272*
9.8 Decay of Odd-A Nuclides and Structure Effects *274*

10 β Decay *281*
10.1 Introduction *281*
10.2 β Decay and Conservation Laws: The Neutrino and the Weak Interaction *282*
10.3 The Fermi Golden Rule No. 2 *285*
10.4 The Fermi Theory of Allowed β Decay *287*
10.5 β Spectra *292*

10.6 Decay Probabilities for β^- and β^+ Decay *296*
10.7 Some Implications of the Simple Theory of Allowed β Decay *300*
10.7.1 Angular Momentum Effects *300*
10.7.2 Nuclear Matrix Elements: Fermi Transitions *302*
10.7.3 Nuclear Matrix Elements: Gamow–Teller Transitions *305*
10.8 Classification of β Transitions and Experimental Log_{10}ft *306*
10.9 Electron Capture Decay *307*
10.9.1 X-ray Emission *308*
10.9.2 Auger Electron Ejection *311*
10.10 Elementary Theory of Electron Capture *314*
10.11 Ratio of Electron Capture to Positron Emission *318*
10.12 β-Decay Schemes *320*
10.13 β-Delayed Particle Emission *325*
10.14 Comments on Fermi Transitions *327*

11 γ Decay and Internal Conversion *331*
11.1 Introduction *331*
11.2 The Angular Momentum of Photons and Conservation Laws *332*
11.3 Introduction to the Theory of Photon Emission *334*
11.3.1 The Radiation Field and Matrix Elements for Photon Emission *334*
11.3.2 Matrix Elements and Transition Rates *341*
11.4 Examples of Nuclear Isomerism *347*
11.5 Some General Observations *350*
11.5.1 E1 Transitions *350*
11.5.2 E2 and M1 Transitions *351*
11.5.3 Other Transitions *351*
11.6 Internal Conversion *351*
11.6.1 Elementary Theory of Internal Conversion *353*
11.7 Decay Schemes *357*

12 Nuclear Fission *373*
12.1 Introduction *373*
12.2 The Discovery of Nuclear Fission *374*
12.3 The Liquid-Drop Model and Nuclear Fission: The Nuclear Potential Energy Surface *375*
12.4 Empirical Data on Spontaneous and Neutron-Induced Fission *383*
12.5 Energy Release in Fission *392*
12.5.1 Fission Fragment Kinetic Energy *392*
12.5.2 Kinetic Energy of Prompt Neutrons *395*
12.5.3 The Spectrum of Prompt γ-Rays *399*
12.5.4 Summary of the Sources of Energy Release in Fission *399*
12.6 Fission Barriers and Fission Probabilities *401*

13	**Low-Energy Nuclear Reactions** *405*
13.1	Introduction *405*
13.2	Kinematics of Nonrelativistic Reactions *407*
13.2.1	Kinematics of Elastic Scattering in the Laboratory Coordinate System *408*
13.2.2	Kinematics of Elastic Scattering in the Center of Mass Coordinate System *412*
13.2.3	Kinematics of General Nonrelativistic Nuclear Reactions *418*
13.3	Cross Sections for Nuclear Reactions from First-Order Perturbation Theory *424*
13.4	The Reciprocity Theorem *435*
13.5	Qualitative Considerations of the Mechanisms of Low-Energy Nuclear Reactions *440*
13.5.1	Potential Scattering *440*
13.5.2	The Compound Nucleus *441*
13.5.3	Direct Reactions *446*
13.6	The Properties of Time-Dependent States *447*
13.7	A Physical Approach to the Form of Cross Sections for Compound Nucleus Reactions: The Breit–Wigner Single-Level Formula *451*
13.8	Scattering in Quantum Mechanics: Partial Wave Analysis *457*
13.9	Extension of the Partial Wave Analysis to Nuclear Reactions *468*
13.10	S-Wave Scattering and Reactions in the Limit of the Spherical Potential Well Model *472*
13.11	The Breit–Wigner Single-Level Formula and Experimental Cross Sections *478*
13.12	About Fission Cross Sections *487*
14	**The Interaction of Ionizing Radiation with Matter** *493*
14.1	Introduction *493*
14.2	The Interaction of Photons with Matter *495*
14.2.1	Elastic Scattering of Photons on Unbound Electrons *496*
14.2.2	Compton Scattering *502*
14.2.3	The Photoelectric Effect *516*
14.2.4	Pair Production *519*
14.2.5	Total Cross Sections and Attenuation Coefficients *520*
14.3	The Interaction of Charged Particles with Matter *524*
14.3.1	The Stopping of Heavy Charged Particles in Matter *525*
14.3.2	The Stopping of Electrons and Positrons in Matter *538*

Appendix 1
Atomic Masses *545*

Appendix 2
Nuclide Table *565*

Appendix 3
Physical Constants *607*

Appendix 4
First-Order Time-Dependent Perturbation Theory *611*

Index *619*

Preface

Nuclear chemistry and low-energy nuclear physics have developed enormously since the time I was first introduced to the subjects formally by Charles D. Coryell. They were absolutely fascinating then and they remain so to this day. These areas have not only contributed greatly to our understanding of the nature of matter and the universe but they have had profound effects on almost all fields of science and technology and on many areas in the humanities as well.

I have taught upper division and graduate courses in low-energy nuclear physics for applications for many years. The students with whom I have had the pleasure of interacting have come from almost all areas of engineering, as well as from chemistry, physics and the geosciences. For the most part, these students have not had the extensive training in mathematics and physics found in most physics curricula and the majority of them have had but a fleeting introduction to quantum mechanics. Most have been interested in applications of nuclear physics and the interaction of radiation with matter but some have been keenly interested in nuclear physics and chemistry research.

The present text is an outgrowth of the lecture notes that I have developed over the years. It treats those aspects of low-energy nuclear physics that appear to me of greatest importance to applications and nuclear chemistry. As such, it is quite limited in its scope. It does not deal at all with many current research areas and it is almost devoid of reference to particle physics. It treats only, in some detail, the fundamentals of nuclear structure, radioactive decay and low-energy nuclear reactions, and provides an introduction to the interaction of ionizing radiation with matter.

The approach taken here is centered in the use of simple limiting models that emphasize the fundamental physics of the different topics considered. This approach is taken for a number of reasons. First and foremost, it is my belief that the simplest of models allow the student to grasp the central ideas of the physics involved in different phenomena without struggling simultaneously with the mathematical complexities needed for more sophisticated treatments. Second, it provides the student with a means of correlating information that is commonly treated either empirically or with reference to results of theoretical calculations that are too sophisticated to be presented and for which many students have no basis for interpreting physically. The use of simple models clearly has serious limita-

Nuclear Physics for Applications. Stanley G. Prussin
Copyright © 2007 WILEY-VCH Verlag GmbH & Co., Weinheim
ISBN: 978-3-527-40700-2

tions. We are not treating the physics of a problem in its entirety and thus all of the nuances of theory will not be understood and the agreement we can expect with experimental data will be limited. Those parameters that are defined solely by conservation laws will be given "exactly" and we endeavor to provide the background for these in as simple and understandable a form as possible.

For most of the models and theory discussed here, a solid background in lower division mathematics, physics and chemistry will prove sufficient. However, it is my opinion that it is impossible to treat low-energy nuclear physics in a reasonable manner without resort to the fundamentals of quantum mechanics, and for that purpose a summary of the necessary "facts" from quantum theory is presented with a discussion that I hope provides some insight into their meaning and the physical ramifications that stem from them.

The text is written for students at the advanced undergraduate and beginning graduate level. I have attempted to provide all of the essential mathematical details in the text so that derivations can be followed with relative ease. I have tried to keep the admonition "it can be shown that" to a minimum. It appears most often in relation to the properties of some mathematical functions and differential equations. I am not a great fan of directing students to an appendix for a derivation or more detailed discussion of a topic. Almost all of what is needed is found within the text along with, I hope, sufficient discussion to make the physical relevance of the mathematics clear. Finally, while it is efficient and has much pedagogical value, I also am not a fan of collecting all of the fundamental problems in quantum mechanics in an introductory chapter to be referred to as needed later in the text. Rather, I have incorporated many of the models in those sections where it makes most sense to the topic under study. I have also included discussions of some topics in classical mechanics that are treated in detail in upper division physics courses, but which will not have been part of the curricula of many students.

The text has been written to provide a solid working foundation for those students who have had little exposure to things quantized and nuclear. I have tried to make it approachable for both beginners and more advanced students. The first five chapters are designed to provide such students with sufficient information so that they can do some useful things without the need for a detailed introduction to the more theoretical aspects considered in the remainder of the text. Most of the energetics and phenomenology of radioactive decay and nuclear reactions are covered here, as well as an introduction to statistical considerations. More advanced students might only wish to go through this material lightly to refresh some of their understanding and to become familiar with the notations and styles found in the remainder of the text. In the remaining chapters I have tried to separate general discussions and simple physical considerations from the more formal developments that might not be of interest in the first course of study.

The majority of what is presented here is "classical" and has been treated previously in many excellent texts. Unfortunately, almost all of these are no longer in print. I have learned a great deal from them over the years and I am greatly indebted to the authors who took the time to produce them. At many points, those familiar with the wonderful texts by Evans, Marmier and Sheldon and von Buttlar may see their influence more or less strongly.

This book owes much to many people. There have been about two hundred and fifty captive students who have used various drafts of the text in their course work. Their comments and criticisms have provided the type of input needed to make the text as open and "friendly" as I can make it. I am indebted to Dr. Kexing Jing for a first and expert reading of the text. Brett Isselhardt and Bethany Lyles helped prepare the solutions to many problems and provided editorial assistance and much advice on issues with the text.

I am forever indebted to Brian Quiter who suffered through my lectures, carefully read the manuscript, provided continuous and expert comments on various topics and, of greatest importance, provided insight into what makes a text useful to students. sgp

Berkeley, July 2007   *Stanley G. Prussin*

1
Introduction

1.1
Low-Energy Nuclear Physics for Applications

During the past century the development of the physical sciences has been nothing short of astounding. We now take the existence of nuclear atoms and the molecules formed from them as the self-evident constituents of matter, without much consideration for the enormous energy, talent and genius that was needed to elucidate them and to define their fundamental properties. It is commonplace for high school students to see "pictures" of individual atoms on surfaces as revealed by atomic force microscopes. Three-dimensional structures of many of the fundamental molecules that define the function of our cells and tissues can be found in elementary texts on molecular biology. The understanding of atoms and molecules and their interactions has led to practical advances in almost every sphere of human endeavor. Fundamental studies in the physical sciences are continuing to provide the basis for practical applications at an ever-increasing rate.

The discovery of radioactivity during the latter part of the 19th century and the experimental and theoretical studies that took place early in the 20th century, led to our understanding of the nuclear atom and engendered intense interest in the nucleus itself. Advances came swiftly. The discovery of the neutron led to an understanding of the proton and neutron as the main "constituents" of the nucleus. The understanding that the radioactive decay process known as β^- decay led to an increase in the atomic number of an atom by one and the ability to create small neutron sources led to the search for elements beyond uranium and to the phenomenon of nuclear fission. Within a matter of a few years, nuclear weapons and the harnessing of fission for the production of electricity were both in development.

Today, applications of this type of physics, referred to as low-energy nuclear physics, are found almost everywhere. A sizeable fraction of the electric power generated in the industrialized nations is derived from nuclear fission reactors. Nuclear reactions are used to modify semiconductors to produce desirable properties and are used to study the properties of the surfaces of many of these materials. Radioactive decay is applied to determine the age of once-living objects, as well as the age of various minerals in the earth's crust. Indeed, radioactive decay is one of the fundamental means by which we know the age of the earth itself.

Nuclear Physics for Applications. Stanley G. Prussin
Copyright © 2007 WILEY-VCH Verlag GmbH & Co., Weinheim
ISBN: 978-3-527-40700-2

The radiations emitted in radioactive decay are one of the means by which we gain an understanding of the physics of nuclei and they also represent the basis of important practical applications. Three-dimensional images of organs within the human body are often based on the measurements of γ-rays emitted following radioactive decay of atoms that have been introduced in forms that concentrate on specific tissues.

The title of this book, Nuclear Physics for Applications: A Simple Approach, is meant to indicate that the subjects treated are those that are most closely connected to current and likely applications of low-energy nuclear physics. In the present context, the reference to low energy is meant to indicate that we are concerned only with nuclear particles, nuclei and their reactions under conditions that do not require us to delve into the structure of the particles themselves or to consider the creation and characteristics of nuclear particles that are more massive than the neutron or proton. The material presented is neither exhaustive nor is it limited to a statement of facts and formulas for use in applications. Rather, it is a textbook on low-energy nuclear physics that attempts to provide a sound theoretical basis for understanding the fundamentals of nuclear structure, radioactive decay and nuclear reactions. In so doing, it introduces and applies the simplest models that make sense and that are often the basis of the much more sophisticated models that provide quantitative predictions. We chose this approach because a purely empirical and phenomenological presentation does not provide the student with the means for correlating different data, nor does it provide the student with a reasonable basis for advanced study. We will demand that the student pay attention to some of the underlying physics that can be presented only with use of a significant amount of mathematics. But the mathematics is, for the most part, familiar, having been encountered in previous mathematics and physics courses, and can be made reasonably understandable by explicit reference to the physical significance of the expressions that are developed.

In this chapter, we will introduce some of the fundamental concepts, notations and nomenclature that will be used throughout the text. We will also introduce the reader to some of the sources of information on low-energy nuclear physics and the interaction of radiation with matter that are useful for both fundamental and applied purposes.

At the very start, it is appropriate to issue a word of warning concerning nomenclature. Over the years a number of terms have been introduced to increase the precision of what we say and write, and we will try to stress the correct usage of such terms. But, as is often the case in older texts and even in some current literature, we may lapse into "common" usage from time to time. We will try to point out where common usage can be particularly confusing to the novice.

1.2
Some General Observations and Notations

At the present time, we have evidence for the existence of some 117 chemical elements. Of these, 86 exist on earth in more than trace abundance. The chemical elements are defined by the number of protons, Z, in their nuclei and their chemistry is defined by the behavior of their electron distributions. All elements have *isotopes* that contain different numbers of neutrons, N. Isotopes are generally indicated by their atomic number Z and mass number A, where A = N + Z. The standard symbol for an isotope is $^A_Z\text{El}_N$, where El represents the symbol for the chemical element (e.g., H, Cl, U, etc.) This designation is clearly redundant and the atomic and neutron numbers are often omitted. In many cases, it is common to refer to neutrons and protons collectively as *nucleons* and we will follow this practice when appropriate. Further, the term *nuclide* is used when referring to an arbitrary nucleus of given Z and N.

As far as we know, the heaviest known elements do not possess any isotopes with sufficiently long half-lives that they could have existed on earth for a time comparable to its age of about 4.5 billion years. Curiously, this is also true of the elements technetium (Tc; Z = 43) and promethium (Pm; Z = 61). Technetium was first discovered with certainty through reactions of neutrons and *deuterons* (nuclei of the hydrogen isotope $^2_1\text{H}_1$) with isotopes of the element molybdenum (Z = 42) [1] and promethium was first identified as a product of the nuclear fission of uranium and the reaction of neutrons with the element neodymium [2].

As indicated above, the chemical properties of the elements are defined primarily by the properties of their electron distributions. Under normal conditions, the mass of an atom has little effect on its chemistry. With the exception of the isotopes of the lightest elements, and conditions in which a process is employed specifically to make use of the very small mass dependence of chemical properties, all isotopes of an element have essentially the same chemistry.

The same is definitely not true for the nuclear properties of different isotopes of an element. The nuclear properties are sensitively dependent on the number of neutrons and protons they contain and their reactions with other nuclei, their nuclear chemistry[1], is also sometimes exquisitely sensitive to these. It is quite common to find an isotope with certain properties more closely related to those of adjacent elements than to isotopes of the same element. For example, for some specific values of the neutron number N, the low-energy nuclear properties of isotopes of different elements can be remarkably similar. This is particularly true for the properties of isotopes with N = 51 and N = 83, respectively, which possess an even atomic number. Nuclei with the same number of neutrons but different number of protons are called *isotones*.

Remarkably, it is also found that the low-energy nuclear properties of isotopes with Z = 51 or Z = 83 are also similar if the neutron number is even, but these

---

1) Nuclear chemistry deals with changes in or transformations of the atomic nucleus. In many parts of the world, such topics are normally included in the definition of nuclear physics.

properties are quite distinct from those of the N = 51 and N = 83 isotones. Indeed, if one produces a *chart of the nuclides* by plotting the atomic number Z versus the neutron number N, and records the low-energy properties of nuclides at each point, the similarities between the properties of certain isotopes and isotones become evident. A study of the properties of *isobars* – nuclei with the same mass number A but different N and Z – shows that even here some remarkably similar properties exist with respect to their low-energy behavior.

All of these characteristics are a reflection of the structure of the nuclei themselves. Just as for atoms, the quantized structure of atomic nuclei leads to regularities in properties that are akin to the periodic structure demonstrated in the Mendelev periodic table of the elements. These regularities point to a type of shell structure that has a profound influence on the properties of nuclei and their reactions. Unlike the shell structure in atoms, however, shell structure in nuclei is complicated by the presence of two different nuclear constituents, the neutrons and protons, and the reference to nuclei of Z or N = 51 or 83 is suggestive that shell structure must exist separately for neutrons and protons. This idea is not altogether so strange if one considers the fundamental notion of the Pauli exclusion principle that defines the number of *identical* particles that can be placed into a given quantum state.

1.3
Overview of Radioactive Decay Processes and Nuclear Reactions

Three principal modes of radioactive decay have been observed among the nuclei found in nature and those produced in nuclear reactions, and the general characteristics of these will now be discussed. In the following and throughout the text, the symbol $^A_Z N$ will be used to denote the nucleus of an atom, and the symbols $^A_Z M$ or $M(Z, A)$ will be used to denote a neutral atom. Since it is possible to form nuclei or atoms in excited states, these symbols will further be restricted to imply that the nucleus or atom is in its lowest state of excitation, the so-called *ground state*. Should it be necessary to consider an arbitrary excited state, an asterisk will be appended as a superscript.

1.3.1
Alpha Decay

Radioactive decay by the emission of α-particles, nuclei of $^4_2 He$, alpha decay is a principal decay mode among the heaviest elements found in nature. Alpha decay is generally symbolized as

$$^A_Z N \rightarrow ^{A-4}_{Z-2} N + ^4_2 N \tag{1.1}$$

or, more simply,

$$_Z^A N \to {}_{Z-2}^{A-4} N + \alpha \qquad (1.2)$$

The nucleus produced as a result of the decay contains 4 fewer nucleons, 2 protons and 2 neutrons. As in all radioactive decay processes, the total of all particles in the nucleus, the sum of the neutrons and protons, is conserved. In addition, the total electrical charge is conserved.

A particularly important example of α-decay is that of the most abundant isotope of uranium in nature

$$_{92}^{238}U \to {}_{90}^{234}Th + \alpha \qquad (1.3)$$

$_{92}^{238}U$ has a *half-life*, the time required for half of the nuclei to undergo decay, of 4.468×10^9 y and those nuclei still present on earth have existed since the time that the elements were created. $_{90}^{234}Th$, on the other hand, has a half-life of only 24.1 d and it would not exist on earth at all if it were not constantly being produced by the α decay of $_{92}^{238}U$. As will be seen in Chapter IX, the decay of $_{92}^{238}U$ gives rise to a series of radioactive nuclei that accounts for a large fraction of the total radioactivity that occurs naturally on earth.

1.3.2
Beta Decay

The term β decay actually refers to three processes that all derive from the same underlying physics. Taken together, they represent the most common radioactive decay mode among all of the known nuclei.

The first mode recognized experimentally was that of β⁻ decay. This is the process in which a neutron in the nucleus is converted into a proton in the nucleus with the simultaneous emission of an electron – the β⁻ particle. Under most circumstances, with common radiation detectors, this is exactly what is observed. However, when β⁻ decay was first studied, it produced profound consternation because it did not appear to satisfy the conservation laws of energy, momentum, and angular momentum. Today we know that an additional, very weakly interacting particle is also emitted. It is known as the *antineutrino* and is symbolized as $\bar{\nu}$. Accounting for the antineutrino, β⁻ decay is correctly symbolized as

$$_Z^A N \to {}_{Z+1}^A N + \beta^- + \bar{\nu} \qquad (1.4)$$

This expression also demonstrates another conservation law that is evident in nature, the conservation of light particles. The particles of small mass, including the electron and antineutrino, belong to a family known as the *leptons*. Observation has shown that there is a conservation law applicable to particles within families; the creation or *annihilation* (destruction) of a particle must be accompanied by the annihilation or creation of an antiparticle[2]. The creation of the electron, a particle, is accompanied by the creation of the antineutrino, an antiparticle, belonging to the same family.

One very common nuclide that decays by β^- decay is $^3_1H_2$ or *tritium*. Tritium is frequently incorporated into different biologically-active molecules and thereby serves as a radioactive "tag" to aid in unraveling a wide range of problems in molecular and cell biology. Tritium is also likely to be one of the main "ingredients" in the first nuclear fusion systems designed to produce electric power.

A second common nucleus that decays in this way is $^{14}_6C_8$ with a half-life of 5730 y. It is constantly being produced by the interaction of cosmic rays in the atmosphere. Because it behaves chemically as "normal" carbon, it becomes incorporated into all living matter and it is one of the isotopes that confers natural radioactivity to the human body. $^{14}_6C_8$ has provided the basis for one of the more accurate means of determining the age, since death, of a previously living organism, because, after death, no additional $^{14}_6C_8$ is taken up by the organism. Thus, if we know the fraction of carbon in living matter that is $^{14}_6C_8$, the time since death can be determined by measurement of the fraction of the original $^{14}_6C_8$ that remains at the time of dating.

Finally, and perhaps the most important β^- decay with respect to understanding the underlying physics of the process itself, is the decay of the free neutron. Free neutrons can be produced by a wide variety of reactions and they are produced in abundance in nuclear fission reactors. The free neutron decays with a half-life of 10.37 min.

A second mode of β decay involves the emission by nuclei of *positrons*, symbolized by β^+. The positron is the antiparticle to the electron. It has exactly the same mass and magnitude of electric charge as an electron but the charge is of opposite sign. In β^+ decay, the overall effect is the transformation of a proton in the nucleus into a neutron in the nucleus with the emission of a positron and a *neutrino*, ν. The decay can then be written as

$$^A_ZN \to ^A_{Z-1}N + \beta^+ + \nu \qquad (1.5)$$

In some respects, this is quite strange. Free protons appear "stable". They do not undergo radioactive decay by any known mechanism; the protons present in the nuclei of atoms found in nature have existed since the creation of the elements themselves. And yet protons incorporated into certain nuclides seem to be able to transform by the process of positron emission. The explanation for this seeming contradiction is that it is the *nucleus as a whole* that gives rise to the decay. This is true for all radioactive transformations.

While not generally found in nature, positron emitters are easily produced by a variety of nuclear reactions. One of the positron emitters that is finding increasing use in medical technology is the nuclide $^{18}_9F_9$. It has a half-life of 1.83 h and is very easily detected. Because of its small size and the strong bonds it makes with carbon

---

[2] The concept of the conservation of leptons is under very active study today. Although not of practical importance to the studies presented here, the violation of the law of conservation of leptons would have great implications concerning our understanding of matter itself.

atoms, it can be substituted readily for a hydrogen atom in a large biologically-active molecule without changing the chemistry of the molecule to a great extent. Currently, $^{18}_{9}F_9$ is commonly incorporated into a molecule with very similar properties to glucose. So similar that the molecule acts just like glucose in the body and thus is used as a metabolic tracer in nuclear medicine procedures. It is the most widely used isotope in the powerful diagnostic imaging procedure known as *positron emission tomography* (PET) imaging.

The third mode of β decay is called *electron capture* and it can only take place when a nucleus has at least one atomic, or orbital, electron. In this decay, an orbital electron is captured by the nucleus and a proton is converted into a neutron. Because the electron is annihilated, a neutrino is created and emitted. The decay can be written as

$$^{A}_{Z}N + e^- \rightarrow ^{A}_{Z-1}N + \nu \tag{1.6}$$

Electron capture decay can be a difficult process to observe because the only radiation that must be emitted is the neutrino, which, under most conditions, has a negligible probability of detection. However, because an orbital electron has been removed from an atom's electron distribution, the remaining electrons will quickly re-arrange with a high probability that an x-ray will be emitted in the process. This can be detected with relative ease. An example is the electron capture decay of $^{55}_{26}Fe$, which is accompanied by the emission of an x-ray that is characteristic of the element manganese (Mn). Following Eq. (1.6), the decay can be written as $^{55}_{26}Fe + e^- \rightarrow ^{55}_{25}Mn + \nu$.

1.3.3
Spontaneous Fission

Spontaneous nuclear fission is only found among the heaviest elements. It is very similar to, but not the same as, the fission process that takes place in nuclear power reactors. In spontaneous fission, a nuclide undergoes decay by and of itself. In power reactors, the fission is "induced" by the interaction of a neutron with a uranium or plutonium nucleus. Fission is an extremely complicated process that is still the subject of active research.

The most common mode of spontaneous fission can be represented symbolically as

$$^{A}_{Z}N \rightarrow ^{A_L}_{Z_L}N^* + ^{A_H}_{Z_H}N^* \tag{1.7}$$

where $A = A_L + A_H$ and $Z = Z_L + Z_H$. The nucleus undergoing spontaneous fission splits to produce two nuclei, the so-called *fission fragments*, that contain all of the neutrons and protons present initially. The subscripts L (light) and H (heavy) are meant to indicate that the mass numbers of the two fragments are generally not the same. The asterisks on $^{A_L}_{Z_L}N^*$ and $^{A_H}_{Z_H}N^*$ indicate that these nuclides are generally created in some excited state and not in their ground state. However, the excited states decay rather quickly by the emission of electromagnetic radiation – γ-rays – and neutrons, and the *net* process can then be written as

$$^A_Z N \rightarrow {}^{A_L}_{Z_L}N + {}^{A_H}_{Z_H}N + (A - A_L - A_H)^1_0 n + \gamma \text{ rays} \tag{1.8}$$

where, in this case, $A - A_L - A_H$ is the number of neutrons that are emitted in the decay of the excited fragments[3]. Again, $Z = Z_L + Z_H$. The nuclei produced *after* the emission of neutrons and γ-rays, the so-called *prompt* neutrons and γ-rays, are generally referred to as *fission products*.

The spontaneous fission of a single nuclide is found to take place in literally hundreds of ways. For example, the mass numbers of the products of spontaneous fission of $^{252}_{98}Cf$ range from as low as 66 to as high as 172 and isotopes of over one-half of the elements found in nature are produced. On the average, 2–4 neutrons are emitted following fission of the most common isotopes. The half-life for spontaneous fission, the half-life that a nuclide would possess if it decayed only by this mode, is found to vary over an enormous range. For example, the isotope $^{239}_{94}Pu$, one of the isotopes responsible for power production in nuclear fission reactors, has a spontaneous fission half-life of about 5.5×10^{13} y. But the spontaneous fission half-lives of $^{241}_{95}Am$, $^{252}_{98}Cf$, and $^{259}_{102}No$ are 1.1×10^{12} y, 85.4 y and 4.5 h, respectively. Spontaneous fission is one of the decay modes that will limit the heaviest elements we can hope to create and study in the laboratory.

1.3.4
Gamma Decay

Gamma (γ) decay, or γ-ray emission, is the process in which an excited state of a nucleus transforms into a lower-energy state with the difference in energy appearing as electromagnetic radiation, the γ-ray. This is entirely analogous to the emission of x-rays when an excited electronic state of an atom decays to a lower-energy state. γ-rays are frequently emitted following α and β decay and indicate that, just as the fission fragments are generally produced in excited states, both α and β decay often lead to excited states of their product nuclei as well. In many cases, emission of a number of γ-rays of different discrete energies is observed in radioactive decay, indicating that a number of different excited nuclear states must be produced.

While most excited states of nuclei have very short half-lives – so short that for normal applications they can be considered to decay "instantly" – some have lifetimes so long that they appear as independent radioactive species, just as is found for the decay of some excited states of atoms and molecules. We call such states *metastable states* or *nuclear isomers* and will designate them by appending the mass number with the letter m. Thus ^{99m}Tc denotes a metastable state of the nuclide ^{99}Tc. While ^{99}Tc has a half-life of 2.11×10^5 y, ^{99m}Tc has a half-life of 6.01 h.

---

3) In the literature on nuclear engineering, the symbol $\bar{\nu}$ is commonly used to denote the number of neutrons emitted in fission, averaged over all fission modes. This should not be confused with the integral number of fission neutrons emitted in a specific fission of the type indicated by Eq. (1.8).

Although α decay, β decay, spontaneous fission and γ decay are usually presented on an equal footing to the principal forms of radioactive decay, a clear distinction between γ decay and the other three modes should be made. Unlike the others, γ decay does not produce a *transformation* of the nucleus to some other nuclide.

1.3.5
Nuclear Reactions

The term "nuclear reaction" is quite generic. It is used to refer to the interaction of nuclei with nuclei, individual nucleons with nuclei, nucleons with one another and even the interactions of photons and electrons with nuclei. Nuclear reactions are the fundamental means by which we probe the nature of the nuclear force, the structure of complex nuclei, and the means by which we produce radioactive nuclei for study or applications.

We will restrict our attention to low-energy nuclear reactions and, for the moment, will consider only *binary* nuclear reactions. These can be symbolized by the expression

$$^{A_1}_{Z_1}N + ^{A_2}_{Z_2}N \rightarrow ^{A_3}_{Z_3}N + ^{A_4}_{Z_4}N \tag{1.9}$$

where, as a result of conservation, $A_1 + A_2 = A_3 + A_4$ and $Z_1 + Z_2 = Z_3 + Z_4$. The reaction products may, as in the case of the fission fragments, be produced in various states of excitation and these generally decay by emission of γ-rays. If decay takes place by emission of neutrons, protons, α particles, etc., the latter are included in the reaction itself.

As an example, the reaction that is the principal source of the $^{14}_{6}C_8$ found in the environment is

$$^{14}_{7}N + n \rightarrow ^{14}_{6}C + p \tag{1.10}$$

The neutrons are produced primarily by the interaction of cosmic ray protons with the oxygen and nitrogen of the atmosphere. Production of $^{18}_{9}F_9$ for use in positron emission tomography and other applications is readily achieved with the reaction

$$^{18}_{8}O + p \rightarrow ^{18}_{9}F_9 + n \tag{1.11}$$

Nuclear reactions such as those in Eqs (1.10) and (1.11) are frequently written in the short-hand notation $^{14}_{7}N(n, p)^{14}_{6}C$ and $^{18}_{8}O(p, n)^{18}_{9}F_9$, respectively. This notation arose in the description of a typical reaction for which the heavier of the reaction partners was usually contained in a stationary *target* in the laboratory and was bombarded by the light reaction partner produced from a particle accelerator or nuclear reactor. In this case, the light particle is generally referred to as the *projectile*. The short-hand notation then symbolizes target (projectile, light product) heavy product. In keeping with this, the principal reactions leading to energy production in nuclear fission reactors are symbolized as $^{235}_{92}U(n,f)$ and $^{239}_{94}Pu(n,f)$, where f refers to fission and the fission products go unnamed.

The nuclear atom was defined experimentally by Ernest Rutherford through the study of the scattering of α particles by a gold foil. Most of the scattering events were consistent with the interaction of an α particle of charge $Z = 2^+$ with a gold nucleus of charge $Z = 79^+$. Such interactions were simply described as the elastic scattering of two point charges in their mutual Coulomb field. These reactions would generally be symbolized as $^A_Z N(\alpha, \alpha)^A_Z N$, or simply described as the elastic scattering of α particles on $^A_Z N$.

1.4
The Model-based Character of this Text

Essentially all students who might use this text have been exposed to the beauty and precision of classical mechanics and electricity and magnetism. The study of these topics is so beautiful because they are based on very well-defined and very well-tested physical laws. Newton's Laws, the conservation of energy and momentum and the various laws met with in electricity and magnetism, allow one to specify a problem precisely, solve it with the appropriate physics and then be sure that if an experiment were performed as accurately as possible, the experimental result would agree with theory within very small uncertainties. Unfortunately, and with the exception of those problems that involve only the application of conservation laws, the same cannot be said for most problems in low-energy nuclear physics. We simply do not have an expression for the interaction of two nuclear particles that describes the physics with the same sense of exactness as the gravitational interaction, the Coulomb interaction, etc. We do not have a law for the nuclear interaction. It is simply too complex. To make matters worse, we usually deal with nuclei containing many particles. And as you know, only two-body problems and a few other special cases can be solved exactly. This means that, for most of interesting problems we will treat, some sort of approximation method must be applied.

It is for these reasons that the majority of problems in low-energy nuclear physics are attacked by use of different models, each chosen to best describe the problem under consideration in a tractable manner and which, to a reasonable degree, will reflect reality to a good approximation. Some of these models are indeed very complex and are capable of providing near-quantitative agreement with experiment. Some of them are rather crude but still permit semi-quantitative agreement with experiment and a simple means for correlating a large body of data.

In this text, we will make extensive use of the simplest models that make sense and that serve as an introduction to the main ideas on which much more accurate models are based. While certainly not providing the satisfaction of an exact description, these models do have a bit of beauty attached to them. They do permit some very nice insights into an extremely complicated and very important part of the physical world around us.

1.5
Sources of Nuclear Data

There are a number of sites on the worldwide web that are particularly good sources for nuclear data and links to other useful sites. The National Nuclear Data Center at the Brookhaven National Laboratory (http://www.nndc.bnl.gov) is an excellent source of evaluated data on nuclear structure and the properties of nuclear reactions, especially neutron reactions. One can also access authoritative files of experimental and evaluated atomic masses. Some unevaluated experimental data are also accessible as well as a computer index of experimental neutron data. In addition, evaluated cross sections for the interaction of electromagnetic radiation with matter are available. Quite recently, the website has made available means by which log ft values for β transitions and internal conversion coefficients can be calculated directly.

A second site of great utility is the Nuclear Data Dissemination Home Page, sponsored jointly by the Isotopes Project of the Ernest Orlando Lawrence Berkeley National Laboratory and the Lund Nuclear Data WWW Service of LUNDS Universitet, Sweden (http://ie.lbl.gov/toi.html). From this page one can obtain direct links to the Table of Radioactive Isotopes and on-line programs that permit the examination of evaluated data on the properties of levels in essentially every nuclide that is known. In addition, it is possible to obtain drawings of level schemes of nuclei as well as the decay schemes of radioactive nuclei.

Much of the information available on the Nuclear Data Dissemination Home page is contained in the publication *Table of Isotopes* by Richard B. Firestone, edited by Coral M. Baglin and S.Y. Frank Chu, CD-Rom editor, Wiley-Interscience, New York. The publication is available in both a two-volume set or on a CD-ROM. The appendices in this reference provide numerous useful parameters in graphical form, that are handy for quick calculations.

Finally, we should add that the National Institute of Science and Technology (http://physics.nist.gov/) provides authoritative information in many areas that are important to low-energy nuclear physics and provides up-to-date information on physical constants, units and their uncertainties.

The information available from the references above and the literature in general fall into two broad classes: *experimental* data and *evaluated* data. Experimental data represent the results of individual measurements of various properties along with estimates of the uncertainties in the measurements. Evaluated data most often represent the best estimates of individual parameters or entire data sets after very careful, and often very complex, analysis of all data that exist in the literature. Both types of compilations are very useful. But, as with all experimentally-derived information, it is incumbent upon the user to ensure that data are used correctly in the context of a specific application. For general purposes, data in evaluated files can be taken as the "best" information available in the judgement of acknowledged experts. This is especially true when a number of investigators have made measurements and there is good reason to believe in the quality of the data. However, in some demanding cases or when dealing with nuclear properties that have not been well-studied experimentally, it is necessary that the user investigate the raw data and examine the publications that describe the measurements in question and their uncertainties.

One of the very unfortunate properties of many authoritative data files is that they do not present well-defined errors on the individual parameters contained in them, primarily because they are intended to be used as complete data sets. The novice and many experienced users tend to take the parameters in such compilations as fact and do not question their reliability or real accuracy. In addition, some sets of evaluated data contain estimates of parameters that are obtained by extrapolation, interpolation or model calculations. This is a real danger and something of which the user must be wary.

References

1 [1] C. Perrier and E. Segre, J. Chem. Phys 5 (1937) 712.
2 [2] J.A. Marinski, L.E. Glendenin, and C.D. Coryell, J. Am. Chem. Soc 69 (1947) 2781.

2
Nuclear Masses and Energetics of Radioactive Decay and Nuclear Reactions

2.1
Introduction

The driving force for any spontaneous radioactive decay is the difference in the internal energy of the particles present initially and those present after the transformation has occurred. For both radioactive decay and nuclear reactions, conservation of energy requires that there is no change in the total energy of the system; only the partition of energy between the particles present initially and those present finally can differ. It should be clear that the conditions prior to and after a transformation must be completely and accurately defined. A complete description of the system prior to a transformation defines the *initial state* and a complete description of the system after the transformation defines the *final state*.

In chemical reactions at constant temperature and pressure, we usually consider the energy changes that take place in terms of the change in the Gibbs free energy. A reaction will occur spontaneously if the total free energy of the products is less than that of the reactants. Although we normally have no need to be concerned with it, the energy changes are directly associated with very small differences in the masses of the products and reactants. In most nuclear transformations, however, the energy changes are generally so large that mass differences are readily measured. As a result, we normally use measured masses to determine the energy changes expected in nuclear transformations. In this chapter we will develop the quantitative relations between the energy changes in radioactive decay and nuclear reactions and the masses of the particles involved. This entails the mass–energy equivalence obtained from the Special Theory of Relativity. Because of this and the fact that we often deal with particles moving at an appreciable fraction of the speed of light, we will first review some of the principal conclusions from this theory.

2.2
Review of the Special Theory of Relativity

We consider the properties of a particle as measured in a coordinate system in which the particle is at rest, the *rest frame*, and those measured in a reference frame

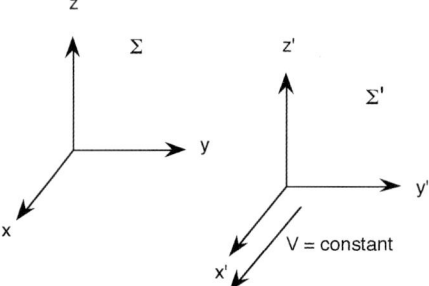

Fig. 2.1 The reference frames of interest to a simple examination of the results from the Special Theory of Relativity. The measuring apparatus is at rest at the origin of the reference frame labeled Σ, the laboratory frame. The rest frame, labeled Σ′, is the frame in which the particle of interest is at rest. Σ′ undergoes uniform translation with a velocity v parallel to the x-axis of the laboratory frame.

in which the measuring apparatus is at rest, generally referred to as the *laboratory reference frame* (Fig. 2.1).

For simplicity, we suppose that a particle is moving in vacuum with a constant velocity v parallel to the x-axis of the coordinate system Σ that is at rest in the laboratory. The measuring apparatus is located at the origin of this system. We seek to determine the relation between the properties of the particle measured by this apparatus with the corresponding properties that would be measured by an exactly similar apparatus that is moving along with the particle. For this purpose we create the reference frame Σ′, the rest frame, whose origin is fixed at the center of mass of the particle and whose axes (x′, y′, z′) are parallel to the corresponding axes of the laboratory frame. Because the Σ′ frame is fixed to the particle, it undergoes uniform translation with velocity v in the direction of the positive x-axis of Σ. We also assume that at t = 0, the origin of Σ′ coincides with the origin of Σ. Our choices for the geometry and initial conditions are not at all necessary. They merely allow the analysis to be as simple as it can be.

The special theory of relativity assumes the independence of the speed of light from the motion of the light source itself. This leads to the following relations between the coordinates (x′, y′, z′) at time t′ as measured in the rest frame Σ′, and the coordinates (x, y, z) at the corresponding time t as measured in the laboratory frame Σ;

$$\begin{aligned} x' &= \gamma(x - vt) = \gamma(x - \beta ct) \\ y' &= y \\ z' &= z \\ t' &= \gamma\left(t - \frac{xv}{c^2}\right) = \gamma\left(t - \frac{\beta x}{c}\right) \end{aligned} \quad (2.1)$$

where, $\beta = v/c$, $\gamma = 1/(\sqrt{(1-\beta^2)})$ and c is the speed of light in vacuum. These relations are known as the Lorentz transformations. They can be manipulated to give the so-called inverse Lorentz transformations that give x in terms of x' and t', etc.

If we know the position of a particle as a function of time, we can easily derive the velocity components for the particle and then combine these to get the particle's speed. In the general case, a particle may have some arbitrary velocity u in the laboratory frame and some corresponding velocity u' in the moving frame. If we let $\frac{dx}{dt} = u_x$ be the x-component of the particle velocity as measured in the laboratory frame, then the relationship for the corresponding component in the rest frame will be $\frac{d}{dt'}x' = u'_{x'}$, etc. In each case we differentiate with respect to the time appropriate to the coordinate system in which we are working.

Simple differentiation of each of the four components of the Lorentz transformations in Eq. (2.1) gives

$$dx' = \gamma(dx - vdt) = \gamma(dx - \beta c dt)$$
$$dy' = dy$$
$$dz' = dz \qquad (2.2)$$
$$dt' = \gamma\left(dt - \frac{vdx}{c^2}\right) = \gamma\left(dt - \frac{\beta dx}{c}\right)$$

In each case, we recognize that the speed v of Σ' is constant and therefore both β and γ are also constant. Clearly, the x' component of the velocity in the Σ' frame is given by the ratio of the first and last of the four relations in Eq. (2.2), i.e.,

$$u'_{x'} = \frac{d}{dt'}x' = \frac{\gamma(dx-vdt)}{\gamma\left(dt - \frac{\beta dx}{c}\right)} = \frac{dx-vdt}{dt - \frac{\beta dx}{c}} = \frac{\frac{dx}{dt} - \beta c}{1 - \frac{\beta dx}{c\,dt}} = \frac{u_x - \beta c}{1 - \frac{\beta}{c}u_x} \qquad (2.3)$$

Similarly,

$$u'_{y'} = \frac{d}{dt'}y' = \frac{dy}{\gamma\left(dt - \frac{\beta dx}{c}\right)} = \frac{\frac{dy}{dt}}{\gamma\left(1 - \frac{\beta dx}{c\,dt}\right)} = \frac{u_y}{\gamma\left(1 - \frac{\beta}{c}u_x\right)} \qquad (2.4)$$

and an exactly similar result will be obtained for $u'_{z'}$. The latter two results are important. Even though the y' and z' coordinates of the particle are identical in magnitude to the y and z coordinates if measured at the corresponding times t' and t, respectively, this is not true for the components of u' along these coordinates! The components $u'_{y'}$ and $u'_{z'}$ obviously depend upon dt' which depends upon dt. We can now find the speed u' by combining the three Cartesian components as $(u')^2 = (u'_{x'})^2 + (u'_{y'})^2 + (u'_{z'})^2$.

Given the speed, we can develop an expression for the corresponding linear momentum by means of the definitions $p = mu$ and $p' = m'u'$, respectively. To accomplish this we need the relation between the masses of a particle as measured in the two coordinate systems Σ and Σ'. This is the famous relation

$$m = \frac{m_o}{\sqrt{1-\beta^2}} \quad (2.5)$$

where m_o is the mass of the particle as measured in the rest frame[1], the *rest mass*. It must be an *invariant* quantity in the sense that, no matter what the possible reference frames are that we wish to use, the mass as measured in a frame in which a particle is at rest can never change. Eq. (2.5) shows that, in the nonrelativistic limit ($\beta \to 0$), $m \to m_o$. But as the particle speed approaches the speed of light, the mass measured in the laboratory appears to increase without limit. The effects from the apparent increase in mass of a high-speed particle are routinely seen in particle accelerators, whether for electrons, protons or other heavy ions. There is very ample experimental proof that the predictions of special relativity are accurately obeyed over particle speeds that range very nearly to the speed of light.

If the particle is at rest in Σ', we can combine the relationship in Eq. (2.5) with the speed of the particle in the laboratory frame, v, and we have the particle's momentum in that frame. Given the momentum, we can get an expression for the kinetic energy by remembering that

$$F = \frac{dp}{dt} \quad (2.6)$$

and that the kinetic energy imparted to a particle moving in a constant force field is given by

$$T = \int \mathbf{F} \bullet \mathbf{ds} \quad (2.7)$$

where $\mathbf{F}$ is the vector force and $\mathbf{ds}$ is the differential of the path length along which the force acts. For simplicity, we assume that the path is linear and that the force is directed along it. Then, if the particle was initially at rest and its final velocity is v, we have

$$T = \int_{v'=0}^{v'=v} F ds = \int_{v'=0}^{v'=v} \frac{dp}{dt} ds = \int_{v'=0}^{v'=v} \frac{dp}{dt}\frac{ds}{dt} dt = \int_{v'=0}^{v'=v} v' dp \quad (2.8)$$

where $v' = \frac{ds}{dt}$. We now use the definition of momentum to obtain the differential of the momentum in terms of the rest mass as

$$dp = d(mv') = d(\gamma m_o v') = m_o d(\gamma v') \quad (2.9)$$

---

[1] The proof of this is most easily obtained through use of a so-called invariance relation. Some of the ideas involved go beyond the level of this text and will not be presented here. Interested readers should refer to such references as P. Marmier and E. Sheldon, "Physics of Nuclei and Particles", Volume 1, Academic Press, New York (1969) or H. Goldstein, C.P. Poole, Jr., and J.L. Safko, "Classical Mechanics", Benjamin Cummings (2002).

2.2 Review of the Special Theory of Relativity

Now

$$d(\gamma v') = d\left(\frac{v'}{\sqrt{1-\frac{v'^2}{c^2}}}\right) = \frac{dv'}{\sqrt{1-\frac{v'^2}{c^2}}}\left(1 + \frac{v'^2/c^2}{1-v'^2/c^2}\right) \qquad (2.10)$$

and substitution of Eq. (2.9) into Eq. (2.8) with the use of Eq. (2.10) gives, upon integration,

$$T = m_o c^2 \left(\frac{1}{\sqrt{1-\frac{v^2}{c^2}}} - 1\right) = m_o c^2 (\gamma - 1) \qquad (2.11)$$

Eq. (2.11) directly relates the kinetic energy of a moving particle to its rest mass, and we can now write

$$\gamma m_o c^2 = mc^2 = T + m_o c^2 \qquad (2.12)$$

In the absence of an external field, the total energy of a particle is the sum of its kinetic energy and the energy equivalent of its rest mass. If E is the total energy of the particle, we have the familiar Einstein result

$$E = mc^2 = T + m_o c^2 \qquad (2.13)$$

Now there is one more important relation that we can easily obtain from Eq. (2.13) that relates the total energy to the particle's linear momentum. Starting with the definition of the momentum and using the first part of Eq. (2.13), we have

$$p = mv = \frac{E}{c^2} v \qquad (2.14)$$

Multiplying both sides by c^2 and squaring, we have

$$p^2 c^4 = E^2 v^2 \qquad (2.15)$$

But we can also use Eq. (2.13) to write

$$E^2 = (mc^2)^2 = \gamma^2 (m_o c^2)^2 = \frac{(m_o c^2)^2}{\left(1 - \frac{v^2}{c^2}\right)} \qquad (2.16)$$

Combining Eqs (2.15) and (2.16) then gives

$$E^2\left(1 - \frac{v^2}{c^2}\right) = E^2\left(1 - \frac{p^2 c^2}{E^2}\right) = E^2 - p^2 c^2 = (m_o c^2)^2, \text{ or} \qquad (2.17)$$

$$E^2 = p^2c^2 + (m_oc^2)^2 \tag{2.18}$$

The last equation completes all of the basic relativistic relations that will be needed for the material discussed in this text.

2.3
Masses of Atoms and Particles

There is no general theory that provides a means of calculating the masses of fundamental particles such as the quarks of which the neutron and proton are composed or the mass of the electron. Even if these masses could be calculated from theory, the masses of nuclei composed of many neutrons and protons cannot be calculated accurately because we cannot yet write down an analytic expression for the nuclear interaction between them. As a result, the masses of the electron, proton, neutron and all other particles must be obtained by experimental measurements. Mass measurements can now be accomplished with very high accuracy and are now so common that mass spectrometers are normally employed to follow the course of gas-phase chemical reactions, to study the composition of environmental samples, to analyze the chemical composition of solid samples after ionization of the material, and even for the very mundane (but very important) purpose of leak detection in high-vacuum systems.

The fundamental basis of accurate mass measurement is really quite simple. It relies on our ability to accelerate charged particles to well-defined energies and to create regions of space in which a well-defined magnetic field exists. To illustrate this, a sketch of the simplest scheme for measurement of the charge-to-mass ratio of a particle is shown in Fig. 2.2. We imagine that we have some means of producing particles of known mass and charge and of accelerating them to a well-defined velocity v. Over the years, various ion sources have been developed to remove one or more electrons from atoms, and to extract them into a region of an electric field which accelerates all of the ions equally. The accelerating potential is chosen to be large compared to the thermal energies of the ions at the time of their creation, but sufficiently small that v << c and relativistic effects can be neglected. The ions are then directed into a region in which a constant magnetic field **B** has been imposed.

From elementary electricity and magnetism, the force experienced by a particle of mass m, charge q and velocity vector **v** moving in a magnetic field **B** is

$$\mathbf{F} = q(\mathbf{v} \times \mathbf{B}) \tag{2.19}$$

In the present case, the particle is assumed to be positively charged, **v** is normal to **B**, and **B** is directed into the plane of Fig. 2.2. Thus F = qvB, and the particle is subject to a constant angular acceleration that forces it to traverse a circular path such that it arrives at the boundary to the magnetic field with its velocity vector normal to that boundary. We can then write

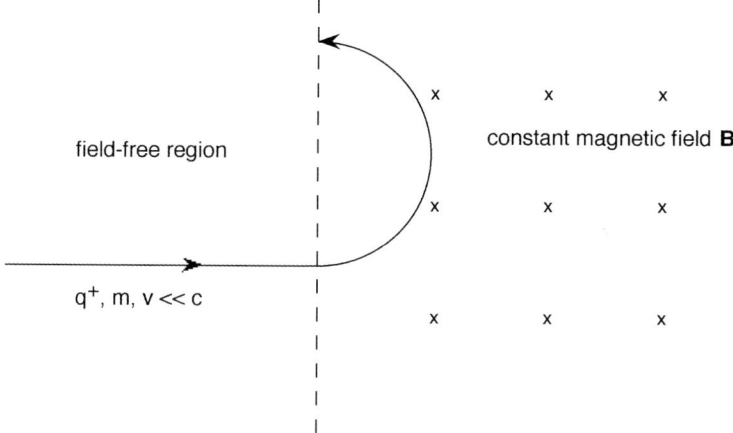

Fig. 2.2 A particle of mass m and charge q^+ is accelerated to a velocity v in a region far to the left of the vertical dashed line. To the right of this line is a uniform magnetic field of intensity B that is normal to the particle's trajectory and which points into the plane of the figure.

$$F = \frac{mv^2}{r} = qvB \qquad (2.20)$$

and

$$m = \frac{qBr}{v} \qquad (2.21)$$

If q, the electric field used to accelerate the ions to velocity v, and B are accurately known, an accurate measurement of the radius of the circular path allows a determination of the mass of the ion.

Real, high-resolution mass spectrometers are highly sophisticated devices that permit very accurate relative mass measurements. These can then be related to an absolute mass measurement of a reference ion, such as a proton, and a complete table of masses can be constructed. Given the ion masses, it is a simple matter to generate a table of atomic masses by adding the mass of the appropriate number of electrons to produce a neutral atom and correcting for the very small binding energy of these. This, of course, requires an independent measurement of the electron mass but it too can be obtained using the same principles outlined above. For our purposes, we will also need the mass of a neutron. This must be measured by different means involving nuclear reactions or radioactive decay. Nevertheless the neutron mass is also now known with high accuracy.

An extensive compilation of atomic rest masses is given in Appendix 1 and a few examples are given in Table 2.1 below.

Table 2.1 Masses of the electron, proton, neutron and several atoms.

particle	mass (kg)	mass (u)	mass (MeV c^{-2})
e	9.109537×10^{-31}	5.485803×10^{-4}	0.511004
p	1.672649×10^{-27}	1.00727645	938.28
n	1.674954×10^{-27}	1.00866501	939.5734
$^{1}_{1}$H	1.673559×10^{-27}	1.00782503	938.7910
$^{4}_{2}$He	6.645676×10^{-27}	4.00260325	3.72792×10^{3}
$^{12}_{6}$C	1.992679×10^{-26}	12.00000	1.117802×10^{4}

The second column of table gives the masses of particles and atoms in kilograms, the fundamental (SI) unit of mass. Errors in these masses are found *only* in the last significant figure shown. While represented on the fundamental mass scale, such quantities are not very convenient to use. A relative mass scale in which the mass of the atom of $^{12}_{6}$C is exactly 12 *atomic mass units*, symbolized as 12 u, has been standard for many years and is the familiar scale found in chemistry. On this scale, 1 u = 1.660566×10^{-27} kg and the masses are given in these units in the third column of the table.

We can use the mass–energy relation $E = mc^2$ to calculate the energy equivalent of particles as follows. The energy equivalent of 1 u is

$$E_{1u} = 1u(kg)c^2 = 1.660566 \times 10^{-27} \times (2.99792458 \times 10^{8})^2 \, kg \, m^2 s^{-2}$$
$$= 1.492442 \times 10^{10} \, J$$
$$= 9.315020 \times 10^{8} \, eV$$
$$= 931.5020 \, MeV$$

where we have used the equivalence 1 eV = 1.602177×10^{-19} J

In essentially all areas of low-energy nuclear physics, the energy unit of *electron volts* (eV), and its multiples *kiloelectronn volts* (keV), *megaelectron volts* (MeV), and *gigaelectron volts* (GeV) are used as the units of energy as well as for expressing the "masses" of particles. Often the usage is a bit sloppy. We frequently talk of the "mass" of the electron as about 0.511 MeV, while we really mean that the energy equivalent of the rest mass of the electron is about 0.511 MeV and the mass is correctly written as 0.511 MeV c^{-2}. In the last column of Table 2.1 the energy equivalents of the rest masses of the particles and atoms are shown.

The masses of the electron, proton and neutron are not only important in their own right, but knowledge of them is very useful for obtaining quick estimates of the energies involved in various decay processes and for calculating other parameters.

The student will be well rewarded by memorizing these masses and the energy equivalent of 1 u to 4 significant figures. For example, it is well known that the atomic masses of atoms in u are very nearly equal (numerically) to the mass number A. Then, because of the near-equality of the neutron and proton masses, one can get a quick, rough estimate of atomic masses by calculating 931.5 A MeV c^{-2}.

We can already use the few masses given in Table 2.1 to demonstrate some important points. First, note that while they are almost the same, the mass of the neutron, m_n, is actually greater than that of the proton, m_p, by $(m_n - m_p) = 1.2934$ MeV/c^2 and, more importantly, the neutron mass is greater than the mass of the hydrogen atom ^{1}H, m_H, by $(m_n - m_H) = 0.7824$ MeV c^{-2}. We can clearly conclude that if a decay mode existed that permitted the decay of a neutron into a proton plus an electron, thereby conserving electric charge, it would be favorable energetically to do so. In fact, such a decay mode exits and it is β^- decay. Accounting for the antineutrino that is also produced, we can represent the decay of the neutron symbolically by

$$^1_0 n \rightarrow {^1_1 p} + {^0_{-1} e} + \bar{\nu} + Q_{\beta^-}/c^2 \tag{2.22}$$

where the quantity Q_{β^-}, called the Q value for β^- decay, represents the energy difference between the rest mass of the neutron and the sum of the rest masses of a hydrogen atom and the antineutrino. While it is a very difficult experiment to perform – confining neutrons is quite difficult and the neutron half-life is only about 10.6 min – the β^- decay of the neutron has been well studied. Indeed, the experimental data are completely consistent with the Q value of 0.7824 MeV as calculated from the rest masses of the particles[2].

2.4
Comments Concerning "Nuclear Stability" and Energetics

2.4.1
Spontaneous Transformations and Nuclear Masses

Radioactive decay processes occur because the fundamental forces that exist within nuclei tend to drive them to a reduction in internal energy. The same is true for nuclear reactions that occur spontaneously. The interactions between the particles involved in the reaction tend to produce rearrangements between nucleons that lead to products with reduced total internal energy. In this sense, a decay or reaction is said to have the potential to occur spontaneously if the internal energy of the products of the transformation is less than that of the original nucleus or reaction partners. It must be remembered that the *total* energy is always conserved. A *spontaneous* transformation will result in the release of internal energy as radia-

---

2) Actually, this statement contains a bit of a sleight of hand. At this time we do not know the mass of the antineutrino and so we cannot formally calculate the Q value for β^- decay of the neutron. But we do know that its mass is so small that it cannot affect the mass difference significantly.

tion, kinetic energy of the final products, and/or in the form of "new" particles created in the decay process.

It is implicit in all of our discussions that, for any transformation, all known conservation laws are obeyed. In addition to energy, the total linear momentum and angular momentum will be conserved if no external forces are present, and total charge is conserved as well. There are a number of other conservation laws that are found to exist and we will discuss them as the need arises. Obviously, there must be a mechanism for the transformation to occur, and those mechanisms that do exist are inherent properties of the particles themselves and the forces that exist within and between nuclei and particles. Finally, the satisfaction of the requirements for *spontaneity* does not directly provide any information on the rate or kinetics of the possible decay or reaction. The rates depend sensitively on the characteristics of the forces existing in nuclei and, in the case of nuclear reactions, on the details of how the reaction is actually accomplished.

It should be clear from the foregoing that the determination of possible spontaneous transformations requires very precise definition of the initial and final states of a system. In the case of a neutron at rest in field-free space, the total energy balance in its decay by β^- emission can be represented as

$$m_n = m_p + m_e + m_{\bar{\nu}} + (T_p + T_e + T_{\bar{\nu}})/c^2 \tag{2.23}$$

where the m_i represent the rest masses of the particles and the T_i represent their kinetic energies. We also imply that the three particles in the final state are so far apart that they do not experience mutual interactions.

Similarly, we can write the total energy balance for the production of $^{14}_{6}C$ by the reaction of neutrons on $^{14}_{7}N$ (Eq. (1.10)) as

$$m_{^{14}N} + m_n + T_n = m_{^{14}C} + m_p + (T_{^{14}C} + T_p)/c^2 \tag{2.24}$$

where the initial state is assumed to include the nucleus $^{14}_{7}N$ at rest in its ground state and a neutron with kinetic energy T_n in field-free space, the two particles separated by a sufficiently large distance that they do not experience mutual interactions. The final state is comprised of the $^{14}_{6}C$ nucleus in its ground state and a proton with kinetic energies of $T_{^{14}C}$ and T_p, respectively, separated by a sufficiently large distance that they experience no mutual interaction.

With these examples, we can now generalize the means by which the mass–energy balances for radioactive decay processes and nuclear reactions are specified and the conditions for spontaneity defined. For simplicity, we will always assume that no external fields exist[3]. For radioactive decay, the initial state is normally taken as the bare nucleus, or the atom or ion containing it, at rest and in a well-defined state of internal excitation. The final state is comprised of the particles that

---

3) The restriction that the decay or reaction takes place in field-free space is not at all necessary. Energy balances can be written for any initial and final conditions. For most practical purposes, however, external fields are so small in comparison to the fields due to the mutual interactions of nuclei, that they can be neglected without any significant error.

exist after the decay, each in a well-defined state of excitation and with the kinetic energies found in the limit of large spatial separations of all particles. The mass–energy balance can then be written generally as

$$m^*_{initial} = \sum_{\text{all i products}} (m^*_i + T_i/c^2) \tag{2.25}$$

where the asterisks represent the state of excitation of the nucleus or particle and T_i is the kinetic energy of the ith decay product. With this notation, the condition for spontaneous decay of a nucleus is given by

$$m^*_{initial} - \sum_{\text{all i products}} m^*_i = \sum_{\text{all i products}} T_i/c^2 \geq 0 \tag{2.26}$$

So long as the mass of the initial nucleus is greater than the sum of the masses of the final nuclei and particles, the total kinetic energy of the products will be greater than zero and the decay can occur spontaneously.

Using the same notation, the mass–energy balance for a reaction in which a target, at rest in the laboratory, is bombarded by a projectile with kinetic energy $T_{projectile}$ in the laboratory (Eq. (1.9)) can be written

$$m^*_{projectile} + T_{projectile}/c^2 + m^*_{target} = \sum_{\text{all i products}} (m^*_i + T_i/c^2) \tag{2.27}$$

Here, the initial state has a total mass that is greater than the sum of the rest masses of the target and projectile by their excitation energies and by the mass-equivalent of the projectile's kinetic energy. The condition for spontaneity can then be written as

$$(m^*_{projectile} + m^*_{target}) - \sum_{\text{all i products}} m^*_i = \left(\sum_{\text{all i products}} T_i/c^2 \right) - T_{projectile}/c^2 \geq 0 \tag{2.28}$$

Once again, the condition for spontaneity is simply that the difference between the total mass of the initial and final states is equal to or greater than zero.

There are an unlimited number of possible reactions and radioactive decay processes that can be considered if we account for all possible states of internal excitation of the nuclei involved. For each one we can determine whether a spontaneous transformation can take place with the use of Eqs (2.26) or (2.28). All that is needed are the rest masses of all of the particles and nuclei involved and their excitation energies.

2.4.2
Nuclear Stability

It is common to refer to the nuclei of atoms that we encounter on earth and for which we do not sense any radioactive decay, as "stable" nuclei. Common as this might be, it is not a correct conclusion, at least without qualification. For example, normal rubidium is composed of the isotopes of $^{85}_{37}$Rb and $^{87}_{37}$Rb with atomic abundances of 72.17% and 27.83%, respectively. But from direct and indirect measurements it is known that $^{87}_{37}$Rb is *unstable* with respect to β^- decay to produce $^{87}_{38}$Sr. We normally consider it to be stable because its half-life is about 4.8×10^{10} y, roughly an order of magnitude greater than the age of the earth. The isotopes $^{82}_{34}$Se and $^{100}_{42}$Mo also appear to be "stable" with respect to all common decay modes but have recently been shown to decay by the very rare decay mode called "double β-decay". Their half-lives were measured to be $T_{1/2} = 1.08 \times 10^{20}$y and $T_{1/2} = 1.00 \times 10^{19}$y, respectively, or about 11 and 10 orders of magnitude longer than the age of the earth.

Physicists and engineers have been excited for many years about the possibility of harnessing the fusion of light nuclei to produce electric power. Large-scale experimentation with prototype fusion devices has taken place over the past several decades to determine the conditions required for construction of a practical fusion reactor. All of these experiments have studied the fusion of two $^2_1$H nuclei, *deuterons*, through the reactions

$$^2_1H + ^2_1H \rightarrow \begin{cases} ^3_2He + n \\ ^3_1H + p \end{cases}$$

or the fusion of one $^2_1$H nucleus with the nucleus of $^3_1$H, the *triton*, according to the reaction

$$^2_1H + ^3_1H \rightarrow ^4_2He + n$$

Each of these reactions is calculated to occur spontaneously with very considerable energy release as the reader should confirm. But you all know that, if you go to the right chemistry or physics stores at a research university, you can get a bottle of deuterium gas ($^2_1H_2$) and it is quite "stable". You will not find any fusion reactions taking place. This particular situation is not very different from something much more familiar to most students. The chemical reaction

$$H_2 + \frac{1}{2}O_2 \rightarrow H_2O$$

is highly exothermic and thus hydrogen and oxygen can combine spontaneously to produce water. Yet you can go to the chemistry storehouse and get bottles of hydrogen and oxygen gas, combine them in the ratio of 2 to 1 in a new container and nothing will happen for as long as you care to wait and as long as nothing external disturbs the gas mixture. However, if you use some sort of a catalyst, some platinum sponge or a spark, the reaction will occur with explosive force. The apparent stability of a mixture of hydrogen and oxygen gases under normal conditions is the result of an energy barrier that prevents reaction at any appreciable rate at room temperature.

Essentially the same is true with respect to the reaction of two deuterium atoms or a deuterium atom with an atom of tritium. While their nuclei can fuse spontaneously, an energy barrier also exists that leads to their apparent stability. (Think about the simple physics that must be going on here. What is the barrier?)

The discussion above demonstrates that *stability* is a *relative* term and does not have a clear meaning until the exact statement of a decay process or reaction is given. For example, with but a few exceptions, such as those noted above, the nuclei found in nature are stable with respect to β decay. Similarly, although the majority are not observed to decay in the laboratory and can be treated as "stable" for most purposes, essentially all nuclei found in nature with mass numbers greater than about A = 160 are unstable with respect to spontaneous α decay. And, surprising as it may seem, all nuclei of the elements of high atomic number, including Au (Z = 79), Hg (Z = 80), Pb (Z = 82), etc., are actually unstable with respect to spontaneous binary fission!

Given the statement of a decay process or nuclear reaction, stability is defined in terms of the spontaneity condition. If Eq. (2.26) is satisfied, the nucleus present in the initial state is *unstable* with respect to the decay mode being considered. If Eq. (2.28) is satisfied, the nuclei and particles in the initial state are *unstable* against the reaction being considered.

2.5
Bound and Unbound States and Their Energetics: Potential Wells

Neutral atoms in their lowest possible energy states represent the normal stable configuration of a nucleus of atomic number Z interacting with Z electrons. Suppose we have the nucleus and Z electrons, each separated from one another by infinite distances, in field-free space; any decrease in these distances will, in principle, result in the attraction of the electrons by the nucleus. As the electrons accelerate toward the nucleus, energy will be released primarily in the form of continuous electromagnetic radiation and characteristic x-rays. When all such radiation has been released and all electrons are present in the lowest possible energy states of the atom, the atom is in its ground state.

In principle, a nucleus could be formed in the same way. Given Z protons and N neutrons, each separated from one another by infinite distances, the nucleons can be brought sufficiently close to one another that they interact to form the final nucleus in its ground state. In the process, energy will also be emitted, again primarily in the form of electromagnetic radiation.

In both cases, the total energy of the atom or nucleus is less than the total energy of the independent particles from which they are composed. Systems with less total energy than the energy equivalent of the rest masses of the independent particles of which they are composed are called *bound* systems, and their energetics are conveniently summarized with the use of energy diagrams. Since the only forces we consider are those between the particles themselves, which are assumed to be conservative forces, we can represent the total energy and potential energy on the same diagram. As usual, we take the potential energy to be zero when the particles in the system do not interact, i.e., when the particles are each separated from one another by infinite distances.

Perhaps the simplest case familiar to the student is the case of an electron interacting with a proton. To a good approximation, the only force acting between the two particles that is of consequence is the Coulomb force,

$$V_C = -\frac{e^2}{4\pi\varepsilon_o r} = -k_C \frac{e^2}{r}$$

where $k_C = 4\pi\varepsilon_o$ [4]. The quantization of the interaction of the electron and proton produces the discrete states that represent the energy levels of the hydrogen atom. We can represent these schematically in the energy diagram shown in Fig. 2.3. The curve shown in the figure is the potential energy of the electron interacting with the proton in the center of mass coordinate system. As $r \to 0$, the potential energy goes to $-\infty$ and as $r \to \infty$, the potential energy goes to zero.

The lowest energy level for an electron in the hydrogen atom has a *total* energy of about -13.6 eV relative to the reference state of $r = \infty$. That is, the total energy of a hydrogen atom is smaller than that of the energy-equivalent of the sum of the rest masses of the electron and proton by 13.6 eV. If we write down the energy balance equation for the reaction of an electron with a proton to form a hydrogen atom in its ground state, we have

$$m_e + m_p = m_{^1H} + \text{b.e.}_{^1H} \tag{2.29}$$

and

$$\text{b.e.}_{^1H} = (m_e + m_p - m_{^1H})c^2 = 13.6 \text{ eV}$$

is the *electron binding energy* of the hydrogen atom or the *ionization potential* of the atom. If we wished to reverse this reaction and separate the electron and proton so they no longer interact, we would have to somehow add 13.6 eV into the system.

The hydrogen atom can exist in an infinite number of states by adding excitation energy to the atom. If the total energy of the atom, E, is less than zero, the atom is said to be in a *bound* state. Atoms can also exist in *unbound* states where $E > 0$. Such states are energetically unstable with respect to spontaneous disintegration

---

4) We will use this abbreviation throughout this text.

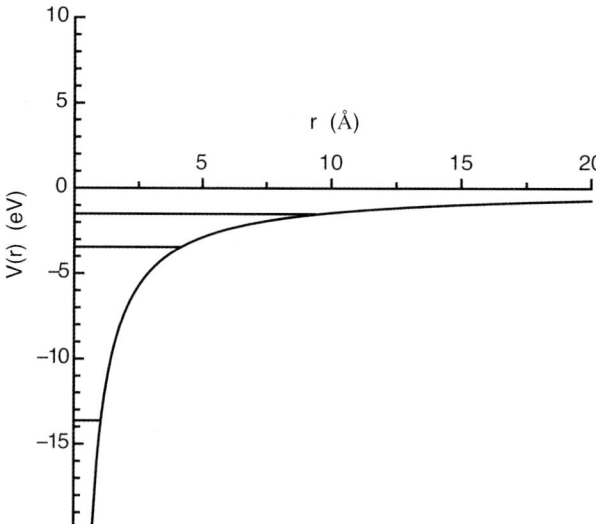

Fig. 2.3 Schematic diagram of the potential energy of the electron interacting with a proton by the Coulomb force in the center-of-mass coordinate system. The curve shown represents the potential energy as a function of the separation of the electron and proton and the horizontal lines represent the energies of the first three bound electron states of the hydrogen atom. The units of energy are eV and the radial dimension is given in Å (1 Å = 10^{-8} cm).

into a separated electron and proton. Although unstable, some of these states may live long enough that they can be studied individually.

The interactions of electrons in a neutral atom are much more simply described than the interaction of nucleons in a nucleus. Only the electromagnetic interaction is of importance. For nucleons interacting in a nucleus, however, two forces must be considered; the purely nuclear force, referred to as the *strong interaction*, which gives rise to the binding of nucleons, and the electromagnetic interaction which manifests itself by the Coulomb repulsion between protons and the emission and absorption of electromagnetic radiation. The strong interaction is much more complex than the electromagnetic interaction and, even today, no simple analytical expression has been found to describe it exactly. Regardless of these difficulties, nuclei are also quantized systems and, as such, they too possess discrete bound and unbound states.

The structure of nuclei will be discussed in some detail in later chapters of this text. For the present, it is useful to point out that the low-energy properties of certain nuclei can be described roughly as the result of the motion of a single nucleon in the average potential field[5] due to all other nucleons in the nucleus. If we plot the average potential felt by a neutron as a function of radial coordinate between the neutron and the other nucleons in the center-of-mass coordinate system, it can be represented roughly as shown in Fig. 2.4. The potential has the

---

5) A more extensive discussion of the nuclear potential is presented in Chapter IV, Section 4.2 ff.

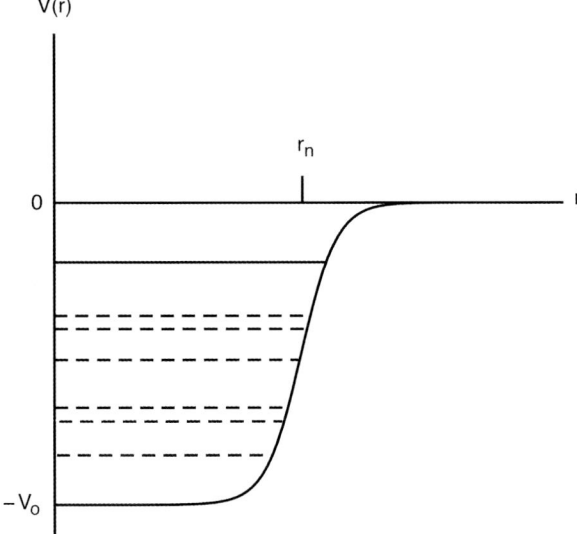

Fig. 2.4 Schematic diagram of the average potential between a neutron and all other nucleons in a heavy nucleus. The origin is taken at the center of mass of the nucleus. The dashed horizontal lines represent a few of the energy levels available for neutrons. The solid line is meant to represent the highest-lying level that is occupied by a neutron.

characteristic of a *potential well* of finite depth. At some radial dimension r_n, the potential rapidly approaches zero. That is, at dimensions somewhat larger than r_n, there is essentially no interaction between the neutron and the other nucleons. This potential is distinctly different from the Coulomb interaction that has an infinite range. To unbind an electron we must separate it from the nucleus by an infinite distance. To separate a neutron from the nucleus we do not really have to separate it by a very large radial dimension at all.

The energetics of binding of nucleons in a nucleus can be described in exactly the same manner as we have described them for the electrons in atoms. The binding energy of the most weakly bound nucleon is the energy required to separate it from the residual nucleus by an infinite distance. This energy is referred to as the *separation energy* of the nucleon.

If the potential well shown in Fig. 2.4 is approximated by the very simple function

$$V_n(r) = \begin{cases} -V_o, & (r \leq r_n) \\ 0, & (r > r_n) \end{cases} \qquad (2.30)$$

and we write the total energy as the sum of the kinetic and potential energies of the particle we are considering, i.e.,

$$E = T + V \tag{2.31}$$

then, if the particle is confined to the well, its total energy $E = T - V_o$ must be less than zero. A particle in a state near the bottom of the well will have a total energy that approaches $-V_o$ and a kinetic energy $T = E + V_o$ that approaches zero. A particle near the top of the well will have a kinetic energy that approaches the well depth V_o and a total energy that approaches zero. Clearly, particles with the smallest binding energies have the highest kinetic energy and vice versa.

The nucleus is far from as simple as the crude model discussed above might suggest. Nevertheless, it is very useful to keep such a simple model in mind and to understand its ramifications with respect to binding energies and the division between potential and kinetic energies.

2.6
Nuclear and Atomic Masses and Binding Energies

The discussion in Section 2.41 provides the basis for calculating the energetics of radioactive decay and nuclear reactions. Under normal conditions, we do not deal with "bare" nuclei that have been stripped of all their atomic electrons. Rather, we almost always deal with neutral atoms of the nuclei of interest, or at least of ions with low net ionic charge. Further, we do not normally have direct mass measurements of all known nuclei. For these reasons, the energetics of radioactive decay and nuclear reactions are usually expressed in terms of the atomic masses of nuclides. These are simply related by the masses of the nuclei and the masses and binding energies of the atomic electrons.

Let $N(Z, A)$ represent the rest mass of the (Z, A) nucleus and $M(Z, A)$ represent the rest mass of the corresponding neutral atom. If $(b.e.)_{Z,i}$ represents the binding energy of the ith atomic electron, the masses of the nucleus and atom are related by

$$N(Z, A) + Zm_e = M(Z, A) + \sum_{i=1}^{Z} (b.e.)_{Z,i}/c^2 \tag{2.32}$$

The last term in Eq. (2.32) represents the total binding energy of the atomic electrons in the neutral atom. It is the energy required to completely strip the atom of electrons. We will now use this relation to develop the standard expressions for the energetics of the common modes of radioactive decay and will illustrate its use for determining the energetics of several common nuclear reactions.

2.6.1
β⁻ Decay

For the β⁻ decay of the ground state of the nucleus $N(Z, A)$ to the ground state of $N(Z+1, A)$, the mass relationship equivalent to the general form given in Eq. (2.25) is

$$N(Z, A) = N(Z+1, A) + m_e + m_{\bar{\nu}} + (T_{N(Z+1,A)} + T_e + T_{\bar{\nu}})/c^2 \qquad (2.33)$$

The last term in Eq. (2.33) represents the mass difference between the particles present in the initial and final states and it has become standard practice over the years to call the energy equivalent of this difference the Q value for the transformation, i.e.,

$$Q_{\beta^-,\,\text{nucleus}} = T_{N(Z+1,A)} + T_e + T_{\bar{\nu}} \qquad (2.34)$$

and Eq. (2.33) is normally written in the form

$$N(Z, A) = N(Z+1, A) + m_e + m_{\bar{\nu}} + Q_{\beta^-,\,\text{nucleus}}/c^2 \qquad (2.35)$$

We can now use Eq. (2.32) to express the masses of the parent and daughter nuclei in terms of the masses of the neutral atoms and the atomic electrons. Direct substitution into Eq. (2.35) gives

$$M(Z, A) - Z m_e + \sum_{i=1}^{Z} (b.e.)_{Z,i}/c^2 =$$

$$M(Z+1, A) - (Z+1)m_e + \sum_{i=1}^{Z+1} (b.e.)_{Z+1,i}/c^2 + m_{\bar{\nu}} + m_e + Q_{\beta^-}/c^2$$

Note that the number of electron masses on both sides of the above is the same. With a little rearrangement we obtain

$$Q_{\beta^-,\,\text{nucleus}} = \qquad (2.36)$$

$$\left[M(Z, A) - M(Z+1, A) - m_{\bar{\nu}} - \left(\sum_{i=1}^{Z+1} (b.e.)_{Z+1,i}/c^2 - \sum_{i=1}^{Z} (b.e.)_{Z,i}/c^2 \right) \right] c^2$$

Eq. (2.36) is the *exact* mass–energy relation for β⁻ decay of the nucleus (Z, A). However, for both historical and practical reasons, it is not commonly used. First, the difference in the total electron binding energies of the two atoms is usually negligibly small compared to $Q_{\beta^-,\,\text{nucleus}}$. A rough estimate of this difference can be obtained from the total electron binding energy given by the Thomas–Fermi

2.6 Nuclear and Atomic Masses and Binding Energies

statistical model of the atom [1, 2]. This model, which is expected to be good in the limit of very large atoms, gives the total electronic binding energy as

$$(\text{b.e.})_Z = \sum_{i=1}^{Z} (\text{b.e.})_{Z,i} \approx 15.7 Z^{7/3} \text{ eV} \tag{2.37}$$

Using this, the difference in electron binding energies given in Eq. (2.36) can be written as

$$(\text{b.e.})_{Z+1} - (\text{b.e.})_Z \approx 15.7((Z+1)^{7/3} - Z^{7/3}) \text{ eV} \tag{2.38}$$

As an example, for the β^- decay of a heavy atom such as uranium ($Z = 92$) to produce an atom of plutonium ($Z = 93$), the Thomas–Fermi model estimates the difference in electron binding energies to be about 15.3 keV. For the decay of a medium-Z element, such as the decay of an isotope of tin ($Z = 50$) to an isotope of antimony ($Z = 51$), the difference in electron binding energies is estimated to be about 6.8 keV. Clearly, we can expect significantly smaller differences for the lighter elements. Therefore, so long as Q_{β^-} is much larger than this energy difference, electron binding energies can be neglected.

It is *standard practice* to calculate and discuss the Q value for decay of a neutral atom in the initial state to produce a neutral atom in the final state. We can then write for β^- decay

$$Q_{\beta^-} = [M(Z, A) - M(Z+1, A) - m_{\bar{\nu}}]c^2 \tag{2.39}$$

For all practical purposes, Eq. (2.39) can be simplified further. Although we do not yet know the mass of the antineutrino, all experimental data demonstrate that it is very small indeed. The current upper limit for its rest mass is about 2 eV/c² [3]. This is truly negligible for our purposes, and thus, to an excellent approximation, the Q value takes on the *standard form* found in the literature

$$Q_{\beta^-} = [M(Z, A) - M(Z+1, A)]c^2 \tag{2.40}$$

The mass–energy relations for β^- decay are usually represented in an energy level diagram in the form shown in Fig. 2.5. The ordinate is a relative energy scale and the horizontal lines represent the levels in the nuclei that are involved in the decay. It is standard practice to give the symbol for each nuclide just below the ground state and to give the level energies to the right of each level.

We emphasize that Eq. (2.40) expresses the Q value for β^- decay in terms of neutral atoms, and for almost all cases, it is an excellent approximation. But in some special cases, such as dealing very small Q_{β^-} or bare nuclei, the energy available for a decay can be quite different.

Experimental values of Q_{β^-} vary over a rather large range. One of the smallest of practical significance is that of tritium (^3_1H) for which $Q_{\beta^-} = 0.0186$ MeV. The only

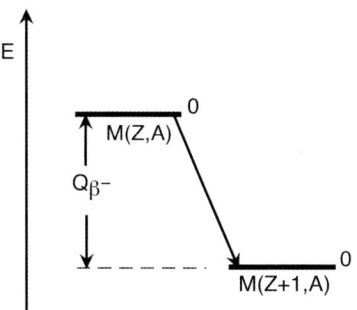

Fig. 2.5 Standard form of the energy level diagram for β^- decay of the ground state of M(Z,A) to the ground state of M(Z+1,A).

decay observed is to the ground state of $^3_2$He. The fundamentally important decay of the neutron has a more typical value of $Q_{\beta^-} = 0.782$ MeV, and Q_{β^-} for decay of the biologically important nuclide $^{14}_6$C is 0.156 MeV. In general, Q_{β^-} for nuclides with (Z,A) near those found in nature, are less than 1–2 MeV while values as large as 10–15 MeV can be found among light nuclei that are "far" from stability.

There is some standard jargon used in the discussion of radioactive decay. One speaks of the "genetic" relations between the initial and final nuclei involved. The initial nucleus is called the *parent* and the final nucleus is called the *daughter*. Thus, for example, we will refer to $^{14}_6$C as the β^- decay parent of $^{14}_7$N, etc. We will use this jargon throughout the text.

2.6.2
β^+ Decay or Positron Emission

The Q value for positron emission, Q_{β^+}, can be obtained by a procedure exactly similar to that outlined above for β^- decay. The mass–energy relation between the ground states of the parent and daughter in this case can be written

$$N(Z, A) = N(Z-1, A) + m_{e^+} + m_\nu + Q_{\beta^+}/c^2 \tag{2.41}$$

where e^+ represents the positron and ν represents a neutrino. The neutrino and antineutrino have the same mass, and we can neglect it as we have done above because it is so small. With the definition of Q_{β^+} as the energy available for decay of a neutral parent atom to the neutral daughter atom, both in their ground states, we must add Z electrons to both sides of Eq. (2.41). Then

$$M(Z, A) = M(Z-1, A) + 2m_e + Q_{\beta^+}/c^2.$$

and the Q value for positron emission becomes

$$Q_{\beta^+} = [M(Z, A) - M(Z-1, A) - 2m_e]c^2 \tag{2.42}$$

2.6 Nuclear and Atomic Masses and Binding Energies

The surprising result is that positron emission must satisfy a *threshold* condition; for a spontaneous transformation, $Q_{\beta^+} \geq 0$, and thus

$$[M(Z, A) - M(Z - 1, A)]c^2 \geq 2m_e c^2 \qquad (2.43)$$

The difference in mass between the parent and daughter atoms must be at least twice the electron rest mass for positron emission to occur spontaneously.

A physical explanation for this threshold is not difficult to discover. First, consider that one begins with the neutral atom M(Z,A). In the decay, a positron is ejected from the atom effectively reducing the total energy available for the decay by the energy equivalent of the electron rest mass. Second, an atomic electron must also be released in the process to make the neutral daughter atom of atomic number Z − 1, reducing the energy available by the energy equivalent of the electron rest mass once again. The net result is that the effective mass available in the decay is really $M(Z,A) - 2m_e$.

The energetics for β^+ decay are represented in the energy level diagram of Fig. 2.6 below.

Positron emitters are found throughout the chart of the nuclides and are especially prevalent among the lighter nuclei with mass numbers not far from those found in nature. The decay energies can vary significantly but are generally comparable to or less than those found for β^- emitters in the same region. As indicated previously, the nuclide $^{18}_{9}F$ is emerging as the "workhorse" of the powerful imaging procedure known as positron emission tomography. $^{18}_{9}F$ decays to the ground state of $^{18}_{8}O$ with a Q_{β^+} of 0.633 MeV. The nuclide $^{22}_{11}Na$ has long been a useful standard for calibration of detectors for γ radiation because of its relatively long half-life of about 2.6 years. It has a Q_{β^+} of 1.820 MeV but it decays primarily to an excited state of its daughter, $^{22}_{10}Ne$, which subsequently emits a γ-ray with energy of 1.2745 MeV. The latter is not the only intense γ-ray emitted from a source of $^{22}_{11}Na$, however. Like every positron emitter, it is a potent source of γ-rays with energies of 0.511 MeV, the energy-equivalent of the rest mass of an electron. These γ-rays arise through the process of *positron annihilation*, the interaction in which a positron combines with an electron at rest, or very nearly at rest, to produce two γ-rays, each with an energy of about 0.511 MeV. This process will be discussed further in Chapter XIV.

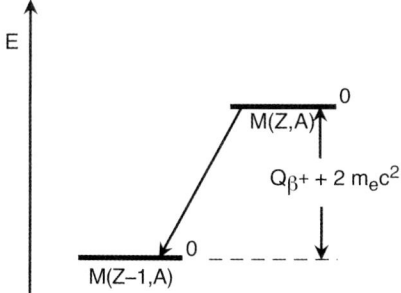

Fig. 2.6 Schematic energy level diagram for β^+.

2.6.3
Electron Capture Decay

The fact that positron emission is a threshold process means that nuclides for which

$$0 \le [M(Z, A) - M(Z - 1, A)] \le 2m_e c^2 \tag{2.44}$$

would be stable against β decay if only the mechanism of positron emission existed. However, such nuclides decay by *electron capture* (EC), where an atomic electron is captured by the nucleus with the conversion of a proton into a neutron and with the emission of a neutrino. The process cannot take place in the absence of atomic electrons and thus a nuclide satisfying Eq. (2.44) but fully stripped of its atomic electrons would indeed be stable against β decay.

Electron capture can be represented schematically as

$$\left. \begin{array}{c} \text{proton} \\ \text{in nucleus} \end{array} \right| + \text{atomic electron} \longrightarrow \left. \begin{array}{c} \text{neutron} \\ \text{in nucleus} \end{array} \right| + \nu$$

and the only radiation that must be emitted from the nucleus is the neutrino. The Q value for this process is clearly

$$Q_{EC} = [M(Z, A) - M(Z - 1, A)]c^2 \tag{2.45}$$

From Eq. (2.42) it is then seen that

$$Q_{EC} = Q_{\beta^+} + 2m_e c^2 \tag{2.46}$$

and the energy level diagram for electron capture decay is that shown in Fig. 2.6.

Nuclides that decay by electron capture are found throughout the chart of the nuclides and some of them are of technological significance. The nuclide $^{109}_{48}Cd$ decays by electron capture to an excited state of $^{109}_{47}Ag$ that sometimes decays by the emission of a γ-ray with energy of 88.0 keV. This γ-ray has great utility for adjusting and calibrating the γ cameras used in nuclear medicine imaging to insure proper operation and to minimize artifacts produced during the data processing. $^{109}_{48}Cd$ is also very valuable for energy and efficiency calibration of a wide variety of γ-ray detectors. Similarly, $^{7}_{4}Be$ decays to an excited state of $^{7}_{3}Li$ that de-excites by emission of a γ-ray with an energy of 477.0 keV, also of great use for detector calibrations.

2.6.4
Competitive Decay Modes

It should be clear from the energy level diagram in Fig. 2.6 and from Eq. (2.46) that all nuclides unstable with respect to positron emission are also unstable to decay by electron capture. When a nucleus (or any quantized system for that matter) is unstable with respect to more than one decay mode, the nucleus can and will decay by all such modes. Thus, there will be *competition* between positron emission and electron capture in the decay of all nuclides that are unstable with respect to positron emission. The relative probabilities for these vary over a wide range and depend upon such factors as the decay energy and the atomic number of the parent. Typically, positron emission tends to be much more probable among low-Z nuclides and electron capture tends to be much more probable among high-Z nuclides. In any event, a schematic of an energy level diagram illustrating the competition between positron emission and electron capture in a hypothetical case is shown in Fig. 2.7. In this example, Q_{EC} is sufficiently large that both positron emission and electron capture can take place to the ground state but only electron capture can take place to the excited state of energy E.

Competition among different decay modes is not at all rare and the case of $^{246}_{98}Cf$ shown below in Fig. 2.8 is representative of what is found among the heaviest elements. $^{246}_{98}Cf$ is unstable with respect to electron capture decay to the ground state of $^{246}_{97}Bk$, to α decay to levels in $^{242}_{96}Cm$, and to spontaneous fission. The dominant mode is α emission and it accounts for all but the 0.0002% of decay that takes place by spontaneous fission. The electron capture decay energy is very small,

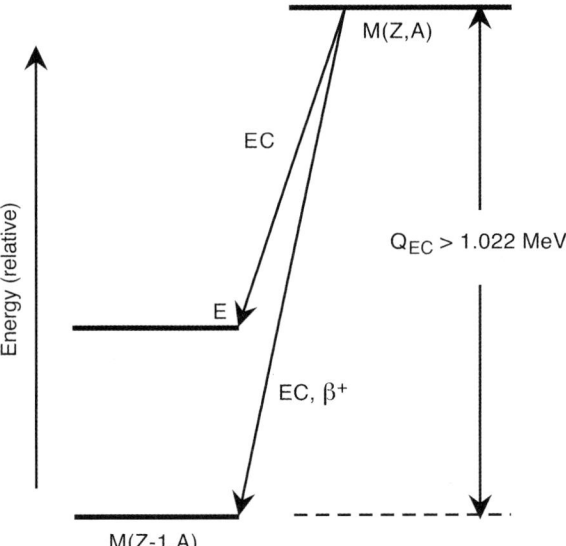

Fig. 2.7 Schematic of competition between electron capture and positron decay. Here, $Q_{EC} - E < 2m_e c^2$.

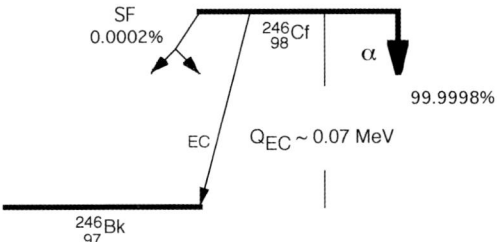

Fig. 2.8 Decay schematic for $^{246}_{98}\text{Cf}$. The absolute intensities (in percent) of decay by spontaneous fission (SF) and α-emission are given in the figure.

only about 70 keV, and it has not been observed experimentally. This does not mean that it does not occur; the decay probability is not zero. It is just so small that it is not seen in normal experiments. As illustrated here, the relative probabilities for the different modes may vary quite widely.

For the majority of radioactive nuclei, only a single decay mode is of practical significance. When more than one mode must be considered, the prediction of their relative probabilities can be fairly complicated. As we discuss radioactive decay in detail later in this text we will try to provide at least a semiquantitative understanding of the factors that affect the decay probabilities and provide qualitative discussions of the factors affecting competition.

2.6.5
α Decay

The mass–energy balance for α decay is easily written down following the prescriptions given above. In terms of the nuclei involved, this balance is

$$N(Z, A) = N(Z-2, A-4) + N(2, 4) + Q_\alpha/c^2 \qquad (2.47)$$

where $N(2, 4)$ is the α particle, the nucleus of the $^{4}_{2}\text{He}$ atom in the ground state. By addition of Z electrons to each side of the equation and neglecting the differences in the electron binding energies of the atoms, the Q value is readily found to be

$$\begin{aligned} Q_\alpha &= [M(Z, A) - M(Z-2, A-4) - M(2, 4)]c^2 \\ &= [M(Z, A) - M(Z-2, A-4) - m_\alpha]c^2 \end{aligned} \qquad (2.48)$$

Contrary to what is observed with Q values for β decay, the Q_α are found to vary over a relatively small range[6]. Among the heaviest elements, where α decay is most prevalent, Q_α is found to vary in the range of about 4–12 MeV. A few of the isotopes

---

6) A detailed discussion of the systematics of α decay energies and half-lives is presented in Chapter IX, Section 9.2.

of rare earth elements are found to undergo α decay with Q values as low as about 2.8 MeV, but nothing smaller is ever seen and α decay is not observed at all in the decay of isotopes of low- and medium-Z elements, not too far from the nuclides found in nature.

α emitters are of great importance technologically and ecologically. The majority of the naturally occurring radioactivity in the environment is the result of the α decay of the heavy elements and the other radioactive nuclides that are produced subsequent to these decays. The tailings from the mining, milling and processing of uranium ores are especially concentrated in these radioactivities and they pose special problems with respect to environmental protection. The nuclide $^{241}_{95}$Am is commonly found in smoke detectors that operate by measuring the attenuation of the α particles emitted in its decay by the particulate matter that we call smoke. α emitters are actively being examined for incorporation into biomolecules for selective delivery to and killing of cancer cells without significant damage to healthy tissues in the human body.

2.6.6
Spontaneous Fission

The mass–energy balance for binary fission can be written as

$$M(Z, A) = M(Z_L, A_L) + M(Z_H, A_H) + (Q_f')/c^2 \qquad (2.49)$$

where $A = A_L + A_H$ and $Z = Z_L + Z_H$. We have written the Q value as Q_f' to indicate that it is the Q value for a specific division of the parent nuclide. Because of the large number of ways in which fission of any nuclide actually takes place, these values can vary widely. The literature does not generally give values for the individual Q_f' but rather the energy released averaged over all of the different divisions found in the fission. We will denote this energy by Q_f. The value of Q_f for spontaneous fission generally lies in the energy range 160–180 MeV, far larger than the Q values for any other decay mode. One will note that this corresponds to almost 20% of the energy-equivalent of the rest mass of a nucleon.

While not usually considered in discussion of radioactive decay, spontaneous fission is not at all rare. It is restricted to isotopes of the heaviest elements found on earth and those of even higher Z that are produced in nuclear reactions. A large fraction of such isotopes have significant spontaneous fission probabilities and both theory and experiment suggest that this decay mode may set an upper limit to the atomic number of nuclides which we will be able to create and study.

Spontaneous fission of some isotopes of thorium and uranium are known but the probabilities are very, very small. If $^{235}_{92}$U decayed solely by spontaneous fission, it would have a half-life of about 3.7×10^{17} y, rather than the experimental half-life of about 7.1×10^8 y. On the other hand, $^{252}_{98}$Cf, with a half-life of 2.64 y, decays about 3.1% of the time by spontaneous fission, and $^{256}_{100}$Fm with a half-life of about 2.6 hours, decays by spontaneous fission in over 90% of all decays.

Spontaneous fission has found significant commercial applications because of the fact that fission leads also to the emission of neutrons. A source of $^{252}_{98}\text{Cf}$ represents a potent, portable neutron source that has been used successfully in oil well logging. In this application, the scattering of neutrons provides information that can be used to assess underground formations for the likelihood that they represent locations of significant pools of oil. $^{252}_{98}\text{Cf}$ has also found widespread use as a neutron source for the production of short-lived radionuclides used for elemental analysis of samples in a nondestructive manner.

2.7
Nuclear Reactions

The Q values for nuclear reactions can be obtained by a straightforward extension of formulations we have used above for the common modes of radioactive decay. For simplicity, we will restrict our attention to the most common reactions in which a target at rest in the laboratory is bombarded by a beam of particles or other nuclei, usually at a fixed kinetic energy. The mathematical descriptions for other arrangements are somewhat more complicated but they do not add anything essential to the physics involved. We will further restrict attention to low energies where the creation of subnuclear particles cannot occur.

With these restrictions, the mass–energy balance of Eq. (2.28),

$$(m^*_{\text{projectile}} + m^*_{\text{target}}) - \sum_{\text{all i products}} m^*_i = \left(\sum_{\text{all i products}} T_i/c^2 \right) - T_{\text{projectile}}/c^2 \geq 0$$

indicates a few significant complications. First, the general case can involve one or more excited states of the target and projectile nuclides and those of the products. Secondly, the total energy balance must include the kinetic energy of the projectile. In order to express the general mass–energy relation for nuclear reactions, it has become standard practice to write the Q value as the mass difference between the neutral atoms and particles in the initial and final states when all nuclides are in their ground states. In general, then, the reaction Q value is given as

$$Q_{\text{react}} = \left[(M_{\text{projectile}} + M_{\text{target}}) - \sum_{\text{all i products}} M_i \right] c^2 \qquad (2.50)$$

Thus, for example, the Q value for the production of $^{14}_{6}\text{C}$ by neutron bombardment of $^{14}_{7}\text{N}$,

$$^{1}_{0}\text{n} + ^{14}_{7}\text{N} \rightarrow ^{1}_{1}\text{H} + ^{14}_{6}\text{C}$$

is given by

$$Q_{n,p} = [M(0,1) + M(7,14) - M(1,1) - M(6,14)]c^2 \qquad (2.51)$$

The subscript n,p on Q is the common short-hand notation for this type of reaction. It is important to remember that one must conserve neutrons, protons and electrons. Thus, the exact inverse of the reaction $^{1}_{0}n + ^{14}_{7}N \rightarrow ^{1}_{1}H + ^{14}_{6}C$, $^{1}_{1}H + ^{14}_{6}C \rightarrow ^{1}_{0}n + ^{14}_{7}N$, would be carried out by bombarding a target of $^{14}_{6}C$ with a beam of *hydrogen atoms*, and overall neutrality would be maintained. In actual fact, this is not usually done and the $^{14}_{6}C$ target is usually bombarded with a beam of *protons*. In this case both the initial and final states have a net ionic charge of +1. While the binding energy of the hydrogen atom is entirely negligible for practical purposes, the mass of the missing electron is not.

Low-energy nuclear reactions abound in applications in science, engineering and in nuclear medicine technology. The capture of neutrons in isotopes of uranium and plutonium is the fundamental reaction that produces energy in nuclear power reactors. Neutron capture is also used to produce isotopes for medical and industrial applications and serves as the basis for one of the most sensitive elemental analysis methods developed to date – neutron activation analysis. The most elementary of reactions, elastic scattering, is used routinely for elemental analysis of the composition of thin films in the semiconductor industry. Charged particle reactions are responsible for the production of the radioactive isotopes $^{201}_{81}Tl$ and $^{18}_{9}F$, two of the most widely used isotopes in imaging of tissue function in the human body. In many hospitals and research laboratories "mini" cyclotrons make available very short-lived isotopes of carbon, oxygen and nitrogen for functional imaging of tissues. A rather detailed discussion of nuclear reactions will be presented in Chapter XIII.

References

1. H.A. Bethe and R. Jackiw, Intedrmediate Quantum Mechanics", 3rd edition, Perseus Books (1986).
2. L.I. Schiff, "Quantum Mechanics", 3rd edition, McGraw-Hill Companies (June 1968).
3. See, for example, J. Bonn, B. Bornschein, L. Bornschein, B. Flatt, Ch. Kraus, E.W. Otten, J.P. Schall, Th. Thummler, Ch. Weinheimer, Limits on neutrino masses from tritium β decay, Nuclear Physics B 110 (2002) 395.

Problems

1. We tacitly assume that the forces we consider in nuclear physics are conservative. Such a force field has the characteristic that the force is given by $\mathbf{F} = -\nabla U(x, y, z)$ as an example in Cartesian coordinates. By considering the quantity $\mathbf{F} \bullet \mathbf{v}$ and Newton's second law, show that the sum of the kinetic plus potential energy is a constant in a conservative force field.

2. Calculate the ratio of the particle speed to the speed of light in vacuum, v/c, when the particle mass is $m = (1+f)m_o$ for f = 0.01, 0.1, 1.0 and 10.0.

3. Carry out the integration of Eq. (2.8) in detail to obtain the result shown in Eq. (2.11).

4. For (a) a 1 MeV β particle, (b) a 5 MeV α particle and (c) a 30 MeV proton, where the energies are the true kinetic energies of the particles, calculate the ratio of the true kinetic energy to the kinetic energy given in the nonrelativistic limit ($T_{\text{class}} = 1/2\, m_o v^2$). To do this, find the true velocity of the particles using the relativistic expression.
(d) When is the classical approximation adequate?

5. (a) A 5 MeV proton collides with an electron at rest in the laboratory coordinate system. The collision is a "head-on" collision. Calculate the kinetic energy of the electron after the collision in the nonrelativistic approximation.
(b) A 5 MeV electron collides with a proton at rest in the laboratory coordinate system. How do the kinematics in this collision differ from those in part (a), and how does this difference influence the final momenta after the collision?

6. Singly-charged ions of ^{234}U, ^{235}U and ^{238}U that first have been accelerated through a potential V are incident upon a region of space permeated by a constant magnetic field **B**. The velocity vectors of the ions are normal to **B**. Derive an expression for the ratios of the radii $r_{234}/r_{235}/r_{238}$ of the circular orbits describing the motion of the particles in the region of the magnetic field.

7. (a) One gram of ^{238}U undergoes fission to produce ^{96}Sr and ^{142}Xe. Calculate the total energy release in MeV.
(b) The complete combustion of 1 mole of C with 1 mole of O_2 to produce CO_2 produces 393.5 kcal mole^{-1}. Calculate the ratio of the energy produced in the fission of part (a) to the energy produced in the combustion of 1 g of C to produce CO_2.
Note: the fission process is much more complicated than implied by part (a). Nevertheless, the estimate of the energy release is quite reasonable for orientation purposes.

8. ^{234}Th is known to undergo β^- decay to ^{234}Pa.
(a) Calculate the Q value for the decay of an atom of ^{234}Th.
(b) Estimate the Q value for decay of the bare ^{234}Th nucleus assuming that the total electron binding energy of an atom is given by the Thomas–Fermi approximation of Eq. (2.37).
(c) How significant is the difference in the decay of atoms and bare nuclei in this case?

9. Calculate the energy release in the fusion reactions:
(a) ^{2}H + ^{2}H = ^{4}He; (b) ^{2}H + ^{2}H = ^{3}He + n and (c) ^{2}H + ^{2}H = ^{3}H + p.

10. Calculate the energy that would be released in the decay of ^{7}Li according to $^7_3$Li = $^4_2$He + $^3_1$H.

3
Phenomenology of Radioactive Decay and Nuclear Reactions

3.1
Introduction

Many of the computations needed for applications of low-energy nuclear physics can be carried out without recourse to an understanding of the underlying physical theory beyond general principles. Further, facility with the phenomenology often eases the understanding of the more abstract, theoretical aspects of the topics covered. With this in view, the present chapter will examine some of the more important and practical aspects of the phenomenology of radioactivity and nuclear reactions. These include the decay constant for radioactive decay, the rates of change in the number of radioactive atoms involved in so-called radioactive decay chains, an introduction to statistical considerations, cross sections for nuclear reactions and the concepts of particle currents and fluxes. Additional terminology and physical units common to low-energy nuclear physics also will be introduced.

3.1.1
The Phenomenology of Radioactive Decay

Essentially everyone has been introduced to the radioactive decay law in the form

$$\frac{dn}{dt} = -n\lambda \tag{3.1}$$

that is said to represent the rate of decay of n identical nuclei of a specific radionuclide, each with the decay constant λ. Eq. (3.1) is easily integrated to give

$$n(t) = n_o e^{-\lambda t} \tag{3.2}$$

where n_o is the number of atoms present initially. The time required for half of the original atoms to have decayed, the half-life, is also simply obtained by setting $n(t_{1/2}) = 0.5 n_o$ in Eq. (3.2) and the result is

$$t_{1/2} = \frac{\ln 2}{\lambda} \tag{3.3}$$

Nuclear Physics for Applications. Stanley G. Prussin
Copyright © 2007 WILEY-VCH Verlag GmbH & Co., Weinheim
ISBN: 978-3-527-40700-2

Thus, the fraction of original atoms remaining after a decay period of two half-lives is said to be $n_o/4$, the fraction remaining after three half-lives is $n_o/8$, etc.

In a very real sense, the discussion above is not correct and it obscures one of the most fundamental characteristics of radioactivity – its inherent statistical nature. It is simply not true that the fraction of original atoms remaining at $t = t_{1/2}$ is always exactly 0.5, or that the fraction remaining after $2t_{1/2}$ is always exactly 0.25, etc. The *likelihood* that these fractions will be found in any given experiment is usually very small. As hard as it may be to grasp, radioactive decay does *not* follow the simple deterministic relations that Eqs (3.1) and (3.2) are commonly taken to imply. The fact is that the time at which any single nucleus will decay cannot be known. All that can be known is the *probability* that decay will occur during some time interval.

A simple experiment suffices to bring this point home. Suppose you obtained a small source of radioactive $^{18}_{9}F$ and fixed it in front of a detector that had a large efficiency for detecting the positrons emitted in its decay. $^{18}_{9}F$ has a half-life of about 1.830 h and one can arrange to count the source for constant, consecutive periods of, say, 10 min each. If the counts registered in each interval were plotted at the mean times of the intervals on a semilogarithmic plot of detection rate vs time, the results would look very much like those shown in Fig. 3.1.

In the figure, the points represent the experimental data in the *complete absence* of experimental error. Instead of all points coinciding with a perfectly exponential decay as predicted by Eq. (3.2), the points show inherent scatter. If you were to repeat the measurement with a new source containing exactly the same initial number of ^{18}F nuclei, you would again see very similar scatter, but the likelihood

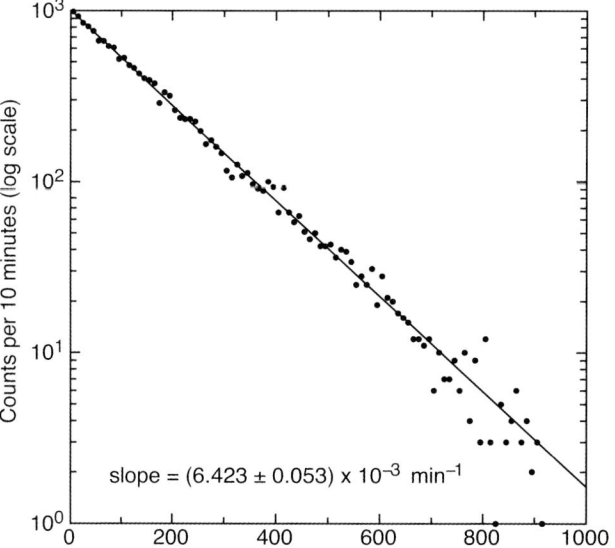

Fig. 3.1 Hypothetical experimental data from counting of a source of $^{18}_{9}F$. The literature value for the half-life is 1.8295 ± 0.0008 hr. The line shown is a least-squares fit to the "data" and gives a half-life of 1.799 ± 0.015 hr.

that the individual data points would be identical to those in Fig. 3.1 is very small. The scatter is described very well by a Poisson distribution, which can be derived from the Binomial distribution with the assumption that the probability for any nucleus to decay during the measurement interval is very small. (These two distributions and an introduction to statistical effects in experiments with radioactivity are presented in Section 3.2.) Just to be completely candid, the "data" shown in Fig. 3.1 are not real counting data at all. They are numbers obtained by statistical sampling of Poisson distributions, with means given by Eq. (3.2).

The fundamental fact is that the decay constant λ represents the *probability* per unit time that a nucleus will decay. A similar statement would apply for any time-dependent property of quantized systems. Given a nucleus with decay constant λ, the probability that it will decay in the time dt is λdt. Given n such identical nuclei, the *expected* or *mean* number that will decay during dt is $n\lambda$dt. Thus, Eq. (3.1) should really be written

$$-\left\langle \frac{dn}{dt} \right\rangle = n\lambda \tag{3.4}$$

where the symbol $\left\langle \frac{dn}{dt} \right\rangle$ stands for the mean or expected rate of change in the number of nuclei. If we have exactly n_o identical nuclei at t = 0, the number that are *expected* to remain undecayed at any time t is really

$$\langle n \rangle = n_o e^{-\lambda t} \tag{3.5}$$

or

$$\frac{\langle n \rangle}{n_o} = e^{-\lambda t} \tag{3.6}$$

It is crucial to understand that the deterministic relations of Eqs (3.1) and (3.2) represent only the time dependence of the mean or expected value. In common with most texts, we will use the deterministic forms throughout this text. However, we will try to emphasize the inherent statistical nature of decay or reaction probabilities by including statistical fluctuations in graphical representations where appropriate.

The probabilistic nature of radioactive decay is also apparent if one considers the ratio $\langle n \rangle / n_o$ in Eq. (3.6). This clearly lies in the range $1 \geq \langle n \rangle / n_o \geq 0$ as time varies in the range $0 \leq t \leq \infty$. Eq. (3.6) can then be taken to represent the probability that the *initial* nuclei have survived without decay at time t. If we take $n_o = 1$, Eq. (3.6) indicates that the probability that any single nucleus will have survived without decay for a time t is $e^{-\lambda t}$ and thus the probability p that decay has occurred is just $p = (1 - e^{-\lambda t})$.

While the half-life is the most common measure of decay probability in most applications, the mean or average is the parameter normally quoted when dealing with statistically distributed variables. This can be obtained with the aid of the schematic diagram given in Fig. 3.2. At some time t', the average number of nuclei that have not decayed is $n(t') = n_o e^{-\lambda t'}$. The number of nuclei that decay during the

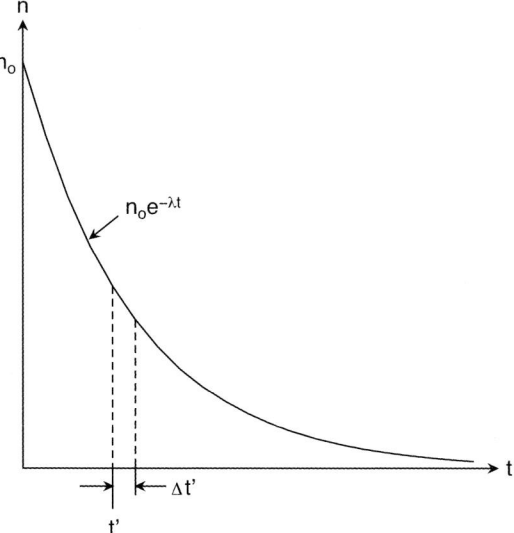

Fig. 3.2 Decay of a nuclide with decay constant λ when the initial number of atoms is n_0.

time interval t' to $t' + \Delta t'$ is, on the average, $n(t')\lambda\Delta t'$, and the time they have existed until decay is t'. The average value of the lifetime of all nuclei is then simply found by adding the products of the number of nuclides that decay during a time interval, times the length of time they have lived until their decay, and dividing the total by the total number of nuclei that have decayed. Thus

$$\langle t \rangle = \tau \approx \frac{\sum_{\text{all } i} n(t_i')\lambda t_i'\Delta t'}{\sum_{\text{all } i} n(t_i')\lambda\Delta t'}$$

$$= \frac{\int_0^\infty n\lambda t'\,dt'}{\int_0^\infty n\lambda\,dt'} = \frac{1}{\lambda} \qquad (3.7)$$

The mean lifetime $<t>$ is usually written as τ and is seen to be larger than the half-life by the factor $1/\ln 2 = 1.44$.

3.1.2
Units for Describing Radioactive Decay

The SI unit for radioactive decay rate is the Becquerel (Bq), defined as 1 disintegration per second, and abbreviated as d s^{-1} or simply s^{-1}. In usage, we say that the strength of a source is x Bq. The number of atoms of a nuclide giving a source strength of 1 Bq is just $1/\lambda$ if λ is expressed in s^{-1}. The common multiples of 10^6 and 10^9 Bq are written as MBq and GBq, respectively.

The conventional unit of radioactivity is the Curie (Ci) and is defined as exactly 3.7×10^{10} d s^{-1} or 37 GBq[1]. Because of its size, the common multiples of the Curie are 10^{-6} Ci and 10^{-3} Ci, the microcurie (μCi) and the millicurie (mCi), respectively. The reader is urged to learn the SI and conventional units as both are still in use throughout the world.

A Curie of radioactivity, while representing a very large decay rate, does not necessarily represent a very large number of atoms. A 1.0 Ci source of a radionuclide with a half-life of 1.00 y contains only about 1.68×10^{18} atoms, or about 2.80×10^{-6} moles. A source with a half-life of 1.00 d contains only 7.66×10^{-9} moles. To obtain some physical feeling for source strengths, it may be useful to consider that smoke detectors can contain on the order of 1 μCi of ^{241}Am; typical sources used to calibrate many radiation detectors may have strengths of 0.1–10 μCi; and it is not too difficult to detect and accurately quantify sources that contain as little as 10^{-12} Ci of many radionuclides.

3.1.3
Radioactive Growth and Decay

The decay of ^{18}F can be symbolized as

$$^{18}_{9}F \xrightarrow{\beta^+} {}^{18}_{8}O \text{ (stable)}$$

If we have a sample that is initially pure $^{18}_{9}F$, we say that $^{18}_{8}O$ "grows" into the sample as a result of the decay. The time dependence of both nuclides in a sample can be written

$$\frac{dn_F}{dt} = -n_F \lambda_F \; ; \; \frac{dn_O}{dt} = n_F \lambda_F \tag{3.8}$$

---

1) This magnitude for a unit of measurement seems very strange indeed. It derives from early studies of radioactive decay, when some widely available comparison standard was sought. The nuclide ^{226}Ra served as such a source. The Ci was then defined as the disintegration rate of exactly 1 g of ^{226}Ra. This was about 3.7×10^{10} s^{-1}. Because it was an experimentally determined quantity, its value varied as new and more accurate experiments were performed. The international community, needing a stable value for a standard, agreed to its present definition, which is no longer tied to the decay rate of ^{226}Ra except through history.

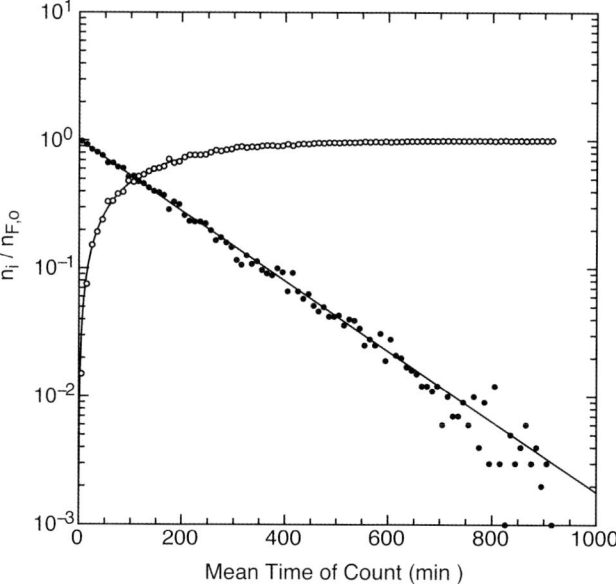

Fig. 3.3 Decay of ^{18}F and growth of ^{18}O in a sample of initially pure ^{18}F. The closed circles represent the fraction of initial ^{18}F that decay during a measurement interval and the open circles represent the fraction of the total ^{18}O atoms present at the end of each time interval.

These equations represent the *parent–daughter* relation for a radioactive parent decaying to a stable daughter nuclide. If the initial number of ^{18}F atoms is n_F^o, Eqs (3.8) are easily integrated to give

$$n_F = n_F^o e^{-\lambda_F t}$$
$$n_O = n_F^o (1 - e^{-\lambda_F t}) \tag{3.9}$$

If we plot the logarithms of the ratios n_F/n_F^o and n_O/n_F^o as a function of time using the "data" from Fig. 3.1, we obtain the distributions shown in Fig. 3.3. The statistical fluctuations in the number of ^{18}O atoms at any time t reflect the fluctuations in the number of ^{18}F atoms that have decayed up to that point. That is,

$$n_O(t) = n_F^o - n_F(t) \tag{3.10}$$

The time dependence of the *mean* number of parent and stable daughter atoms in a sample of initially pure parent is shown in Fig. 3.4 as a function of decay time in units of the parent's half-life. The fraction of initial radioactive atoms that become stable daughter atoms rises to 1/2 at a time corresponding to the parent's half-life, to 3/4 at a time corresponding to two parent half-lives, etc.

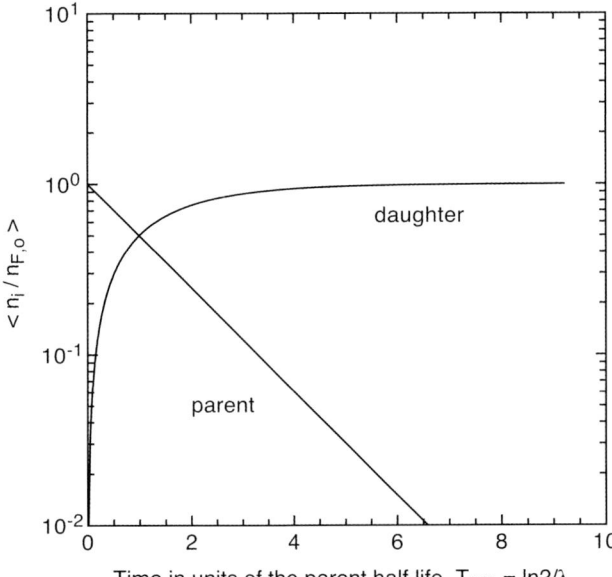

Fig. 3.4 Decay of parent and growth of stable daughter into a sample of initially pure parent as a function of time in units of the parent half-life.

A radionuclide may be so far removed from β stability that it will give rise to a radioactive daughter, granddaughter, etc. This is the general case for the radioactive products of nuclear fission. Here, the average fission product will give rise to a radioactive *decay chain* that contains 3–4 radionuclides before β stability is reached. It is therefore important to be able to determine the decay and growth of radioactivities in complex mixtures.

The simplest case to consider is the decay of a parent into a radioactive daughter that decays to a stable granddaughter;

$$n_1 \xrightarrow{\lambda_1} n_2 \xrightarrow{\lambda_2} n_3 \text{ (stable)} \tag{3.11}$$

The differential equations for the time rate of change of atoms of the three nuclides in the mixture are

$$\frac{dn_1}{dt} = -n_1 \lambda_1$$

$$\frac{dn_2}{dt} = n_1 \lambda_1 - n_2 \lambda_2 \tag{3.12}$$

$$\frac{dn_3}{dt} = n_2 \lambda_2$$

If we assume that the initial sample contains n^o atoms of the parent, integration of the first of Eqs (3.12) will yield $n_1(t) = n_1^o e^{-\lambda_1 t}$. We can now substitute this into the second of Eqs (3.12) to account explicitly for the time dependence of n_1 and write

$$\frac{dn_2}{dt} + n_2 \lambda_2 = n_1^o \lambda_1 e^{-\lambda_1 t} \tag{3.13}$$

Multiplication of this equation by $e^{\lambda_2 t}$ then gives

$$e^{\lambda_2 t}\left(\frac{dn_2}{dt} + n_2 \lambda_2\right) = \frac{d}{dt}(n_2 e^{\lambda_2 t}) = n_1^o \lambda_1 e^{(\lambda_2 - \lambda_1)t} \tag{3.14}$$

and, upon integration and solving for n_2, we have

$$n_2 = \frac{n_1^o \lambda_1}{(\lambda_2 - \lambda_1)} e^{-\lambda_1 t} + k e^{-\lambda_2 t} \tag{3.15}$$

where k is the constant of integration. If we assume that the sample contained only parent atoms initially, the condition that $n_2(t=0) = 0$ gives

$$k = -\frac{n_1^o \lambda_1}{(\lambda_2 - \lambda_1)} \tag{3.16}$$

and substitution of this into Eq. (3.15) then gives the final result,

$$n_2(t) = \frac{n_1^o \lambda_1}{(\lambda_2 - \lambda_1)}(e^{-\lambda_1 t} - e^{-\lambda_2 t}) \tag{3.17}$$

As expected, the number of daughter atoms depends on the decay properties of both it and its parent. The result in Eq. (3.17) can now be used in the third of Eqs (3.12) to obtain an expression for the growth of the stable granddaughter in the sample, an exercise that is left to the reader.

While we can substitute different choices for λ_1 and λ_2 and calculate $n_2(t)$ for any particular case of interest, it is worthwhile first to consider some limiting cases that not only provide physical insight, but also prove of great practical value.

In many cases, one deals with a very short-lived daughter of a relatively long-lived parent. For example, one of the principal products of the neutron-induced nuclear fission of $^{235}_{92}U$ is the isotope $^{144}_{58}Ce$, which has a half-life of about 284.9 d. It decays by β^- emission to $^{144}_{59}Pr$ which also undergoes β^- decay with a half-life of 17.28 min to produce $^{144}_{60}Nd$, which is β-stable. The ratio of the half-life of the parent to that of the daughter in this case is about 2.37×10^4. In the time required for a source of $^{144}_{59}Pr$ to decay by a factor of 2, only about $2.92 \times 10^{-3}\%$ of the $^{144}_{58}Ce$ atoms would have decayed. Clearly, $\lambda_{Pr} \gg \lambda_{Ce}$, and for all practical purposes, a source of $^{144}_{58}Ce$ will have a decay rate that is essentially constant over the time required for the decay rate of $^{144}_{59}Pr$ to change a great deal.

Now consider Eq. (3.17) with $\lambda_2 \gg \lambda_1$. For times $t \ll 1/\lambda_1$, $e^{-\lambda_1 t} \approx 1$, and the equation becomes

$$n_2(t) \approx \frac{n_1^o \lambda_1}{(\lambda_2 - \lambda_1)}(1 - e^{-\lambda_2 t}) \approx \frac{n_1^o \lambda_1}{\lambda_2}(1 - e^{-\lambda_2 t}) \tag{3.18}$$

Let D represent the disintegration rate of a nuclide. Because $D_2(t) = n_2(t)\lambda_2$, the disintegration rate of the daughter is given approximately by

$$D_2(t) \approx D_1^o (1 - e^{-\lambda_2 t}) \tag{3.19}$$

The decay rate of the daughter grows as a function of time with the same time dependence as the growth of the *atoms* of a stable daughter in a sample of an initially pure radioactive parent (see Eq. (3.9)). After decay times long compared to $1/\lambda_2$, but still very small compared to $1/\lambda_1$, the decay rate of the daughter approaches that of the parent. This is referred to as the case of *secular* equilibrium. In a time corresponding to the half-life of the daughter, the daughter's decay rate will be 1/2 that of the parent. After a time period of two half-lives, the daughter's decay rate will be 3/4 of that of the parent, etc. Note also that because $\lambda_2 \gg \lambda_1$, the number of daughter atoms at equilibrium is very small compared to that of the parent. In the present case,

$$\left. \frac{n_{Pr}}{n_{Ce}^o} \right|_{\frac{1}{\lambda_2} \ll t \ll \frac{1}{\lambda_1}} \approx \frac{\lambda_{Ce}}{\lambda_{Pr}} \approx 4.22 \times 10^{-5}$$

A second common occurrence is found when the half-life of the daughter is smaller than that of the parent, but not so much smaller that the parent decay rate can be neglected. If λ_2 is substantially larger than λ_1, the quantity $e^{-\lambda_2 t}$ will decrease at a greater rate than that the quantity $e^{-\lambda_1 t}$. Eventually, $e^{-\lambda_2 t} \approx 0$ and we can approximate (eq). 3.17 as

$$n_2(t) \approx \frac{n_1^o \lambda_1}{(\lambda_2 - \lambda_1)} e^{-\lambda_1 t} \tag{3.20}$$

or,

$$D_2(t) = n_2(t)\lambda_2 \approx \frac{D_1^o \lambda_2}{(\lambda_2 - \lambda_1)} e^{-\lambda_1 t} \tag{3.21}$$

After sufficiently long times, the daughter will decay with the half-life of the parent. Also, because $\frac{\lambda_2}{(\lambda_2 - \lambda_1)} > 1$, the decay rate of the daughter will be somewhat greater than that of the parent when Eq. (3.21) applies. This case is referred to as the case of *transient* equilibrium. A practical example is met in the decay of 2.748-d $^{99}_{42}\text{Mo}$ to the 6.006-h $^{99m}_{43}\text{Tc}$, and the decay of an initially pure 1 Ci source of $^{99}_{42}\text{Mo}$ and the growth and decay of its $^{99m}_{43}\text{Tc}$ daughter are shown in Fig. 3.5.

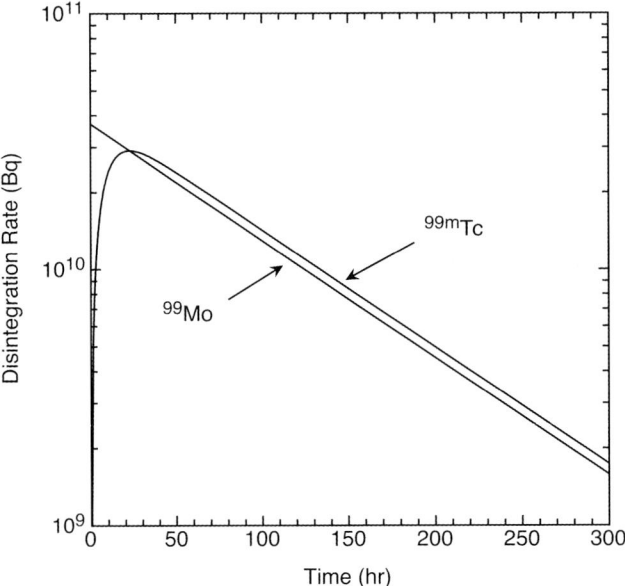

Fig. 3.5 Growth and decay of $^{99m}_{43}Tc$ and decay of $^{99}_{42}Mo$ in a sample that was initially pure $^{99}_{42}Mo$. The initial activity was assumed to be 1.00 Ci.

$^{99}_{42}Mo$ is a product of nuclear fission. It can be separated from the other fission products with very high efficiency and quite free of other radioactivities. The element molybdenum can be bound quite firmly to a column of alumina (Al_2O_3) but technetium is not bound to an appreciable degree. Thus, the $^{99m}_{43}Tc$ that grows into the sample can be washed off easily with a simple saline solution, and a source of $^{99}_{42}Mo$ fixed on an alumina column allows the removal, or "milking", of $^{99m}_{43}Tc$ over and over for as long as the $^{99}_{42}Mo$ is present.

Over the past several decades, radiochemists have learned how to incorporate $^{99m}_{43}Tc$ into molecules that serve as tracers in the human body for assessment of the function of different biological processes and tissues. Currently, it is the most widely used radionuclide in nuclear medicine imaging procedures. It emits a 140-keV γ-ray that is easily and efficiently detected. With appropriate collimation of detectors, images of tissues in a human body in which the radioactive tracer is concentrated can be obtained, and with acquisition of a number of different views, three-dimensional images of the tissues can be produced. This technique is called **Single Photon Emission Computed Tomography (SPECT)** and it is generally available in nuclear medicine departments of hospitals and clinics.

The time dependence of members of a radioactive decay chain does not depend upon the radioactive decay mode involved. A large number of decay chains with different characteristics are met with in practice, and the differential equations for all can be set up and solved by the methods outlined above. General forms for the solutions of simple decay chains of any length have been given by Bateman [1].

Regardless of the complexity of the decay chains, they can always be reduced to the Bateman forms for ease in their solutions.

3.1.4
Simple Decay Schemes and Decay Chains

As we proceed through a discussion of radioactive decay and nuclear reactions, we will find a surprising large number of processes that can occur in a single decay or reaction and a surprisingly large number of radiations that can be and are emitted. These will be dealt with in some depth in succeeding chapters but it is appropriate here to provide an example that introduces some useful concepts and introduces some of the means of representing the complexities of such processes. Because we've already introduced the simple kinetics of growth and decay of ^{144}Pr in an initially pure sample of ^{144}Ce, we choose this pair of nuclides as reasonably typical of what one may find in a simple β-decay chain.

As discussed above, ^{144}Ce decays to ^{144}Pr with a half-life of about 284.9 d. Q_{β^-} for this decay is readily calculated from the atomic masses in Appendix 1 to be 0.3186 ± 0.0020 MeV. In addition to β particles, it is found that γ-rays of at least seven different energies are emitted in the decay and this means that the β decay must populate a number of excited states in ^{144}Pr. From studies of all of the radiations emitted in the decay, the *decay scheme* shown in Fig. 3.6 has been constructed. The decay scheme is designed to show, in condensed form, the principal radiations emitted in the decay, the nuclear levels involved, and the principal characteristics of these levels. Reference to the figure shows that β$^-$ decay occurs with measurable intensity *directly* to three levels in ^{144}Pr. The two excited states populated by β$^-$ decay de-excite by γ-ray emission[2] (the vertical lines in the decay scheme) and in so doing populate two additional levels in ^{144}Pr that also decays by γ-ray emission. The intensity of β decay to the different levels is indicated along the arrows to each of the levels directly populated. Level energies are given in MeV and are shown on the top right of each level.

Consider first the β decay. Each and every nucleus of ^{144}Ce can undergo but one β decay, and each such decay can proceed to one and only one level in ^{144}Pr. Therefore, the total probability for β decay, λ_T, must be the sum of the probabilities for β decay to all energetically accessible states in ^{144}Pr. In the present case we then have

$$\lambda_T = \lambda_0 + \lambda_{0.08011} + \lambda_{0.13354} \quad (3.22)$$

β decay to the levels at 0.05903 and 0.09995 MeV have not been observed experimentally. Therefore, we can conclude that $\lambda_{0.05903}$ and $\lambda_{0.09995}$ must be significantly smaller than the decay constants to the other levels. In the general case, we can write

---

[2] In addition to γ-ray emission, these levels decay by the process of internal conversion that is discussed in Chapter XI.

$$\lambda_T = \sum_{\text{all levels i}} \lambda_i \qquad (3.23)$$

where the sum is taken over *all* daughter levels that are energetically accessible. The decay constant to each level, λ_i, is called a *partial decay constant*. The fraction of total decays that occur to any single level, the *absolute intensity* of decay to that level, is then

$$I_i = \frac{n\lambda_i}{n\lambda_T} = \frac{\lambda_i}{\lambda} \qquad (3.24)$$

where $\lambda = \lambda_T$ is the (total) decay constant reported in the literature. The relative intensities of decay to any pair of levels is just given by the ratio of their partial decay constants. In the present case, the experimental intensities, given as the percentage of all β decays, represent the experimental measurements of the relative decay constants for β decay to each level. As in most cases, the decay probabilities vary over a very large range. In practice, the range is often limited by the sensitivity of the experimental arrangement used for the measurements.

The partial decay constants can be given an additional interpretation. Because $t_{1/2} = \ln2/\lambda$, the quantity $\ln2/\lambda_i$ is the half-life that the nuclide would have if it decayed *solely* to the ith level. The quantity $\ln2/\lambda_i$ is referred to as the *partial half-life* for decay to the ith level. In the case of ^{144}Ce, the partial half-life for decay to the excited state of ^{144}Pr at 0.133 MeV is 1454 d or slightly less than four years.

Now consider the γ decay of the levels in ^{144}Pr. As excited states, they too must have their own decay constants and half-lives. Under normal circumstances, we do not measure the half-lives of the excited states because they tend to be very short, half-lives in the range of 10^{-10}–10^{-12} s are not uncommon. For practical purposes, we usually assume that once populated, an excited state will decay "instantaneously". But there are both direct and indirect methods for measuring the half-lives of excited states and all of the levels in ^{144}Pr shown in Fig. 3.6 have measured half-

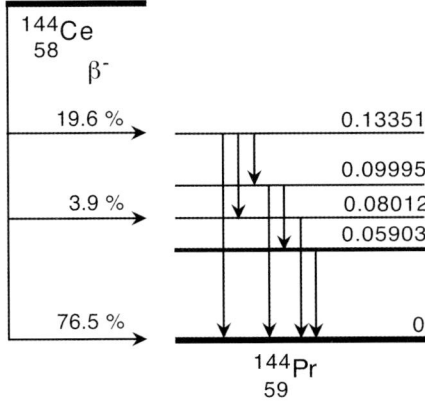

Fig. 3.6 The decay scheme for ^{144}Ce.

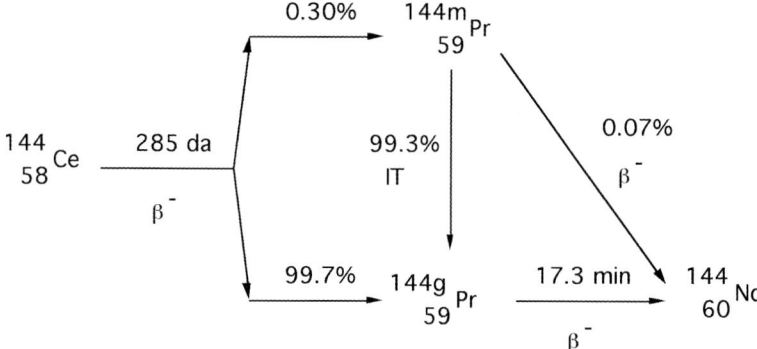

Fig. 3.7 Decay chain for decay of ^{144}Ce.

lives. For the levels at 0.08011, 0.09996, and 0.13354 MeV, the half-lives are found to be about 1.4×10^{-10}, 7×10^{-10}, and 7×10^{-12} s, respectively. However, the level at 0.05903 MeV has a half-life of 7.2 min and is an *isomeric* or *metastable state* that is relatively easily studied. As an excited state, it will have nonzero decay probabilities to any and all energetically accessible states available. In addition to decay to the ground state of ^{144}Pr, the isomeric state is known to decay to a small extent by β^- emission to ^{144}Nd. While this is not the norm, it is also far from rare. Excited states are known that decay by β^- and α decay, emission of protons or neutrons, and even by nuclear fission.

It is often useful to summarize the main decay characteristics of a parent, daughter, granddaughter, etc., in a simplified *decay chain* such as the one shown In Fig. 3.7 that begins with ^{144}Ce.

The decay chain can be used to aid in the mathematical description of the growth and decay of the different nuclides in a sample. The procedures are exactly the same as illustrated above in Section 3.13. One of the most important applications of decay chain analysis is associated with the calculation of the radioactivities produced during and after the operation of nuclear fission reactors. Nuclear reactors are prolific producers of radioactivity, and well over 300 different radionuclides can be found during their operation. They contribute to significant energy production during reactor operation as well as after a reactor has been shut down. Many of the radionuclides are so long-lived that they will persist for hundreds of thousands of years. Thus, the study of complex decay chains is not only necessary for the design of nuclear reactors but is central to the design of repositories for storage of reactor wastes that must contain them for very long periods of time. Large codes have been developed to handle the many and complex decay chains that arise from the β^- decay of the fission products and the heavier actinides that are produced through neutron reactions in nuclear power reactors. One of the more commonly used codes is ORIGIN-ARP and it is available for running on modern personal computers[3].

---

3) See, for example, http://www.ornl.gov/sci/origen-arp/

3.2
Statistical Considerations in Radioactive Decay

The statistical fluctuations inherent in radioactive decay are accurately described under almost all circumstances by Poisson statistics. The latter is a special case of the general Binomial distribution that describes the probabilities for occurrence of all possible outcomes when the occurrence of individual events is constant. In the following, the Binomial distribution is first discussed. The Poisson distribution is then derived from it in the limit that the probability of an individual event is very small. The Normal, or Gaussian distribution that is found in the limit when the mean of the Poisson distribution is large, is presented. Finally, the application of Poisson statistics to the correct interpretation of counting experiments is then presented in a simple, common experimental situation.

3.2.1
The Binomial Distribution

The Binomial distribution is the fundamental distribution that governs the probabilities for the occurrence of random events when the probabilities for different types of events are constant. That is, all individual events of a given type will each have a probability p of occurring and a probability (1–p) of not occurring. The most familiar example that is found in texts is the problem of picking a given colored ball from a box containing balls of a variety of colors. Suppose, for example, that you have a large box containing balls of different colors. The probability that a ball picked randomly will have a specific color, say blue, is just equal to the fraction of all balls in the box that are blue. If you replace this ball and randomly select another, it has exactly the same probability of being blue as the first one you picked.

Suppose you start with removing just one ball from the box. If p is the probability that a ball in the box is blue, the probability that the one you picked is blue is p and the probability that it is not blue is (1–p). Now suppose that you pick two balls. You pick the first randomly, note its color, return it to the box and pick another randomly. The probability that the two you picked were both blue is p^2 and the probability that neither was blue is $(1-p)^2$. The probability that only one was blue could have come about in two ways. If the first was blue and the second was not, the probability that this occurred is p(1–p). If the first was some other color and the second was blue, the probability that this would occur is (1–p)p. Now if all that you care about is getting just one blue ball and not the order in which it was picked, the total probability for picking just one blue ball is p(1–p) + (1–p)p, or 2p(1–p).

Now suppose you pick three balls. After each random pick, you note the color, return the ball to the box and pick another ball randomly. The probability that all three balls were blue is p^3 and the probability that none were blue is $(1-p)^3$. The probability that only one ball was blue could have been achieved in three ways, corresponding to the probabilities p(1–p)(1–p), (1–p)p(1–p), and (1–p)(1–p)p. Thus, the total probability that only one of the balls was blue, regardless of the order in which it was picked, is the sum of these, or $3p(1-p)^2$. In a similar way, you can show

Table 3.1 Probabilities for selecting r blue balls in n trials.

Number of trials	Probability that none was blue	Probability that one was blue	Probability that 2 were blue	Probability that 3 were blue	Probability that 4 were blue
1	$(1-p)$	p	–	–	–
2	$(1-p)^2$	$2p(1-p)$	p^2	–	–
3	$(1-p)^3$	$3p(1-p)^2$	$3p^2(1-p)$	p^3	–
4	$(1-p)^4$	$4p(1-p)^3$	$6p^2(1-p)^2$	$4p^3(1-p)$	p^4

that the probability that two of the balls were blue is $p^2(1-p) + (1-p)p^2 + p(1-p)p$, or $3p^2(1-p)$. Now we can put these probabilities into a table and establish a general form for the different possible results. This has been done in Table 3.1, which also includes all possible outcomes when picking four balls. It is left to the reader to show that the entries for the latter are indeed correct.

If we let P_r represent the entries in Table 3.1, a little trial and error shows that P_r is given by

$$P_r = \frac{n!}{r!(n-r)!} p^r (1-p)^{n-r} \tag{3.25}$$

where n is the number of balls picked, referred to in statistics as the number of *trials*, and r is the number of balls that were blue, the number of *successes*. Eq. (3.25) is the general form of the Binomial distribution. It is important to remember that, in addition to the assumption of the randomness of the trials, the probability of a particular event is constant, independent of the number of trials, and that we are interested only in the net outcome and not the order in which the outcome was achieved. Within these restrictions, the Binomial distribution is very general. By direct expansion of the terms in Table 3.1 or by use of Eq. (3.25), you can show that the sum of the terms for each n, the total probability that outcomes of all types will be found, is unity, as it must be. In fact, the various terms in r for a given n represent the terms in the binomial expansion of $(p + q)^n$, where $q = (1-p)$.

Now how does all of this apply to radioactive decay? First, for almost all practical purposes, nuclei in normal states of matter do not interact with one another. The shielding provided by the atomic electrons is so effective and the distances between adjacent nuclei so large that their properties are almost independent of the presence of all other nuclei. Further, the interactions that they do experience with other nuclei are so small that they cannot affect the normal decay properties of one another. Therefore, the decay probabilities of all nuclei of a given nuclide that are in the same state of excitation and in the same chemical form can be said to be identical. (For most decay modes it is not even required that the atomic forms be the same.) If we define an event as the decay of a particular nucleus, then the independence and constancy of the decay probability required for application of the Binomial distribution, are fulfilled. Now if we start with n identical nuclei and some then decay, we cannot quite fulfill the requirement that the probability of seeing a particular number of atoms decaying in a given time interval will be the same. We cannot put the nuclei back into the box in their initial states after they have decayed! However, if we consider times on the order of the differential dt, the fraction of the atoms that have decayed is negligibly small and the number of nuclei remaining undecayed is essentially constant. Therefore, if we start with a sufficiently large number of nuclei and consider times very short compared to the half-life we can fulfill the requirements for applicability of the Binomial distribution to a very high degree of approximation. We will assume that this is true in all of the cases we consider.

3.2.2
The Poisson Distribution

To apply the Binomial distribution directly to radioactive decay we start with the radioactive decay law $n = n_o e^{-\lambda t}$. As discussed in Section 3.11, we can interpret n/n_o as the probability that a nucleus will have survived decay from birth to the time t. If p is the probability that decay has occurred, $(1-p) = e^{-\lambda t}$ and $p = (1 - e^{-\lambda t})$. Substitution of these two relations into Eq. (3.25) gives

$$P_r = \frac{n!}{r!(n-r)!}(1 - e^{-\lambda t})^r (e^{-\lambda t})^{n-r} \tag{3.26}$$

In this equation, n represents the number of nuclei we have initially, say n_o. The variable r would represent the number of nuclei that decayed during the time interval t. We will call this m. The equation then says that the probability that we will see m decays during the time t is

$$P_m = \frac{n_o!}{m!(n_o - m)!}(1 - e^{-\lambda t})^m (e^{-\lambda t})^{n_o - m} \tag{3.27}$$

Now from the meaning of the decay law, Eq. (3.2), the average or expected number of nuclei that decay during the time t is $\overline{m} = n_o(1 - e^{-\lambda t})$. If we use this relation to substitute for $(1 - e^{-\lambda t})$ and $e^{-\lambda t}$ in Eq. (3.27) we obtain

$$P_m = \frac{n_o!}{m!(n_o - m)!}\left(\frac{\overline{m}}{n_o}\right)^m \left(1 - \frac{\overline{m}}{n_o}\right)^{n_o - m} \tag{3.28}$$

While this is essentially exact, we can greatly simplify the expression by use of the following conditions that almost always describe radioactive counting experiments. First, so long as the counting time is very short compared to the half-life, we generally have that the number of initial atoms will be very much larger than either

3.2 Statistical Considerations in Radioactive Decay

the mean or actual number of atoms that decay during the time interval considered, i.e., $n_o \gg \overline{m}$ and $n_o \gg m$. As a result, the ratio of the factorials $n_o!/(n_o - m)!$ can be very well approximated by

$$\begin{aligned}\frac{n_o!}{(n_o - m)!} &= \frac{n_o(n_o - 1)(n_o - 2)\ldots(n_o - m + 1)(n_o - m)!}{(n_o - m)!} \\ &= n_o(n_o - 1)(n_o - 2)\ldots(n_o - m + 1) \\ &\cong n_o^m \end{aligned} \quad (3.29)$$

We can also carry out an approximation for $(1 - \overline{m}/n_o)^{n_o - m}$, again recognizing that the quantity $(1 - \overline{m}/n_o)$ is very nearly unity. You will remember that the series expansion for e^{-x} is just

$$e^{-x} = 1 - x + \frac{x^2}{2!} - \frac{x^3}{3!} + \ldots \quad (3.30)$$

If we take $x = \overline{m}/n_o \ll 1$, only the first two terms of the expansion are needed and we then have that

$$\left(1 - \frac{\overline{m}}{n_o}\right) \cong e^{\frac{-\overline{m}}{n_o}} \quad (3.31)$$

Hence,

$$\left(1 - \frac{\overline{m}}{n_o}\right)^{n_o - m} \cong e^{\frac{-\overline{m}(n_o - m)}{n_o}} \cong e^{-\overline{m}} \quad (3.32)$$

Direct substitution of the approximations in Eqs (3.29) and (3.32) into Eq. (3.28) then yields the final result that

$$P_m \cong \frac{\overline{m}^m}{m!} e^{-\overline{m}}. \quad (3.33)$$

Written with an equality sign, the probability distribution given in Eq. (3.33) is the *Poisson distribution*. It is remarkable because it is a one-parameter distribution, the parameter being the mean $\overline{m}$. Given the mean of a Poisson distribution, we can calculate the probability for any and all possible outcomes.

We have tried to be accurate by indicating the approximate equality of the relations we have used to simplify Eq. (3.28). But the fact is, the approximations with respect to most applications to radioactive decay are really excellent. We then take the Poisson distribution as describing the statistical fluctuations in radioactive decay "exactly".

As with any probability distribution, the Poisson distribution is normalized and the proof is quite straightforward;

$$P_{\text{all m}} = \sum_{m=0}^{m=\infty} P_m = \sum_{m=0}^{m=\infty} \frac{\overline{m}^m}{m!} e^{-\overline{m}} \qquad (3.34)$$

$$= e^{-\overline{m}}\left(1 + \frac{\overline{m}}{1!} + \frac{\overline{m}^2}{2!} + \frac{\overline{m}^3}{3!} + \ldots\right) = e^{-\overline{m}} e^{\overline{m}} = 1$$

The proof that $\overline{m}$ is indeed the mean is also straightforward. The mean $\overline{m}$ of any discrete distribution is given by

$$\overline{m} = \sum_{\text{all m}} m P_m \qquad (3.35)$$

where m represents the possible values of the discrete variable and P_m are the probabilities for finding these values. Using the Poisson distribution,

$$\overline{m} = \sum_{0}^{\infty} m \frac{\overline{m}^m}{m!} e^{-\overline{m}} = e^{-\overline{m}}\left(0 + \frac{\overline{m}}{1!} + \frac{2\overline{m}^2}{2!} + \frac{3\overline{m}^3}{3!} + \ldots\right)$$

$$= \overline{m} e^{-\overline{m}}\left(1 + \overline{m} + \frac{\overline{m}^2}{2!} + \ldots\right) = \overline{m}$$

The same type of analysis can be used to determine the variance and standard deviation of a Poisson distribution. The variance of any discrete distribution is given by

$$\sigma_m^2 = \sum_{\text{all m}} (m - \overline{m})^2 P_m \qquad (3.36)$$

where σ_m is the standard deviation of the distribution. It is left as an exercise for the reader to follow an exactly similar procedure to those given above to show that the variance of a Poisson distribution is simply

$$\sigma_m^2 = \overline{m} \qquad (3.37)$$

This is an enormously useful result. The one parameter that defines the probability of the occurrence of any particular value of the variable m, $\overline{m}$, is also the variance of the distribution.

The Poisson distributions for means of 5, 10 and 20 are shown in Fig. 3.8. Note first that the Poisson distribution is not symmetrical about its mean, and the asymmetry is most pronounced when the mean is small. There is a higher probability for observing values of m much larger than the mean than would be found if the distribution was symmetrical. However, for $\overline{m} \geq 20$, the asymmetry becomes quite small. Second, the Poisson distribution has the peculiar feature that the probability of observing $\overline{m} - 1$ is identical to the probability of observing the mean of the distribution itself.

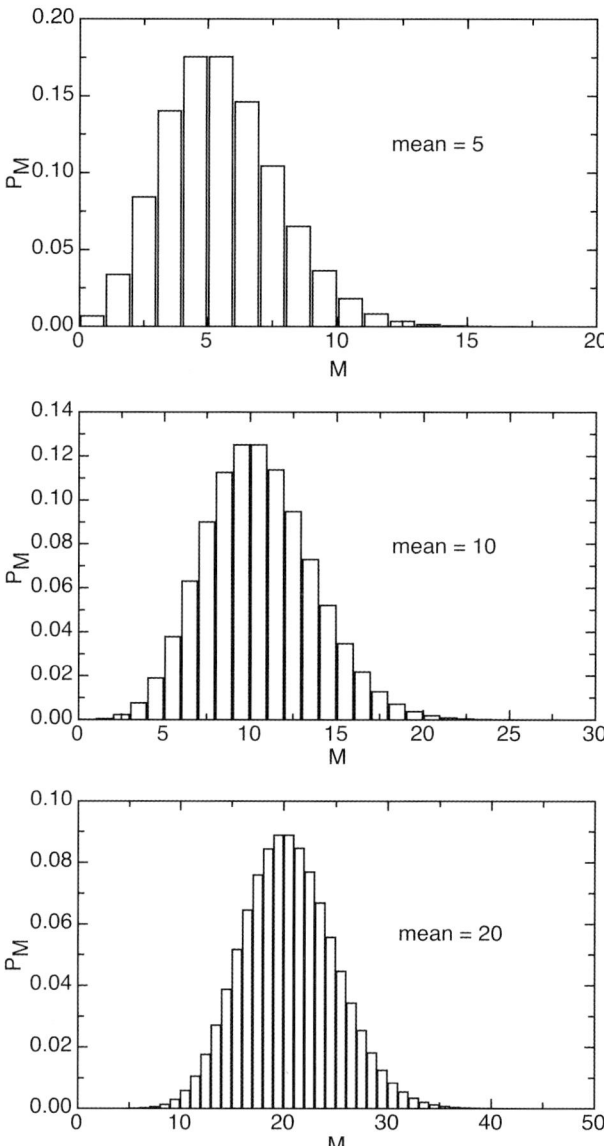

Fig. 3.8 Poisson distributions with means of 5, 10 and 20.

The characteristics of the Poisson distribution are evident in each and every measurement of radioactive decay or the outcome of a nuclear reaction. No matter what the mean value, each and every experiment that involves counting the number of decays or transformations represents a *sampling* from a Poisson distribution. That is, the Poisson distribution represents all possible outcomes for repeating the exact same experiment. The probability that a single experiment will lead to m is given by the Poisson probability of the distribution with mean $\bar{m}$. A repeat of the experiment starting with exactly the same number of atoms of the nuclide will represent a second sampling from the Poisson distribution, and repeated experiments will produce observations that, in the limit of a very large number of experiments, will reproduce the Poisson distribution itself. From the distributions in Fig. 3.8 it is seen that there can be a large probability that a single observation will not exactly produce the expected mean.

By extrapolation of the distributions in Fig. 3.8 to even larger means, it should be clear that the Poisson distribution becomes quite symmetric. In fact, it can be shown to approach a Gaussian or normal distribution of the form

$$G(m) = \frac{1}{\sigma\sqrt{2\pi}} e^{-\frac{(m-\bar{m})^2}{2\sigma^2}} \qquad (3.38)$$

where, again, $\sigma^2 = \bar{m}$. Under normal conditions where the number of disintegrations considered is larger than 25–30, the simple Gaussian distribution is quite adequate for most purposes. However, for small means and whenever one is looking far into the tail of the probability distribution, it can be very important to account for the asymmetry of the Poisson distribution.

The importance of the statistical fluctuations in radioactive decay may be appreciated a bit more by re-examination of the "data" in Figs 3.1 and 3.3. As indicated above, the "data" are really fictitious. They were generated by calculating expected means from the appropriate mathematical representations of the radioactive growth or decay, followed by random sampling from Poisson distributions with these means. It is seen quite clearly that experimental measurements can only give an *estimate* of the mean value for the decay being studied and there will always be some inherent uncertainty associated with any given measurement or set of measurements.

The fundamental issue in any experiment in which we measure radioactive decay or a nuclear reaction is how well the measurement actually represents the mean we are trying to estimate. We need a simple way to judge the uncertainty in a measurement, even after all external sources of error have been accounted for. Suppose, for example, we have one single measurement of m decays in some time interval t. In the absence of any other information, and assuming no sources of uncertainty beyond that inherent in the statistical nature of radioactive decay, we have no choice but to use the value m as our best estimate of $\bar{m}$. To express the uncertainty in our estimate, it is common practice to quote the standard deviation of the distribution that $\bar{m}$ represents, and for a Poisson distribution this is simply $\sqrt{m}$. We usually write our best estimate of the mean as $m \pm \sqrt{m}$. If m is large enough that the Poisson distribution is well-approximated by a Gaussian distribution, about 68% of all possible measurements will fall in the range $\pm \sigma$ of the mean. We usually provide an estimate of the fractional error in our estimate of the mean by calculating the ratio $\sigma_m/m = 1/(\sqrt{m})$. It is clear from this that the fractional error can be made as small as possible by arranging an experiment to give as large a value of m as possible. For example, we can make $\sigma_m/m = 0.01$ by "counting" long enough to observe 10^4 events.

3.2.3
Application of Statistical Analysis to Common Experimental Conditions

Proper attention to the statistical analysis of data is required for all experimental measurements. To begin with, we must recognize that it is very seldom that the disintegration rate of a given radioactive source is measured directly. In a typical experiment, a source will be placed at a fixed location with respect to a radiation detector that intercepts some fraction of the radiations emitted and which has a known probability for detecting the radiations incident upon it. A schematic of a simple experimental arrangement is shown in Fig. 3.9.

Radioactive sources emit radiation isotropically. Those radiations that reach the detector are emitted within the solid angle subtended by the source and the active volume of the detector. If the disintegration rate of the source is D and the fraction of decays that result in the emission of the radiation being measured is f, the rate at which radiations will be detected is given by

$$A(t) = D(t) f \frac{\Omega}{4\pi} \varepsilon \tag{3.39}$$

where $\Omega/4\pi$ is the fraction of the isotropic emissions incident upon the detector and ε is the probability that the incident radiation will actually be detected. The detection rate A(t) is commonly called the *activity* of the source and we will use this terminology to distinguish between the disintegration rate and a measurement of some fraction of it. Since the activity is directly proportional to the disintegration rate it too will be subject to the statistical fluctuations governed by the Poisson distribution.

Eq. (3.39) emphasizes the fact that the estimation of the disintegration rate is subject to uncertainties in both the solid angle Ω and the detection efficiency ε in

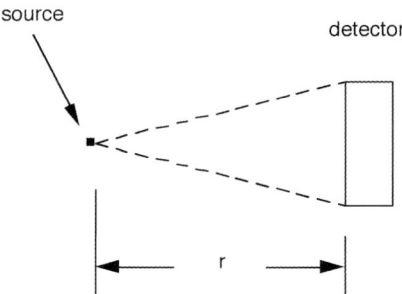

Fig. 3.9 Schematic of arrangement for measuring the decay of a radioactive source. For a point source and a detector with a circular cross section, the dashed lines represent the limits of the solid angle for which radiations emitted by the source will be intercepted by the detector.

addition to the fluctuations inherent in the Poisson distribution. Further, and in most cases, the detector will be sensitive to radiations from extraneous sources such as the natural radiation of the environment, cosmic rays, and naturally occurring radioactivity in the materials of construction of the detector itself. All of these extraneous sources produce what is referred to generically as *background radiation*. Thus, under most conditions, we do not measure the source radiations alone, but only the sum of the source radiations plus background. It is obvious that one needs some means of properly handling the uncertainties from all factors that enter into the calculation of a disintegration rate and for that purpose we introduce a brief discussion of the propagation of errors that suffices for most practical cases.

3.2.4
Propagation of Errors

Suppose we measure a quantity f(x,y) that is a function of the two independent variables x and y, and suppose that x and y are subject to statistical fluctuations of well-known forms. In the present context, x might be the detected events from decay of a radioactive source and y might be the detected events from background radiation. If the probability of finding x between x and x + dx is P(x) dx and the probability of finding y between y and y + dy is P(y) dy, the probability of finding both in these ranges is just

$$P(x, y)dxdy = P(x)P(y)dxdy \tag{3.40}$$

This represents the probability that a measurement of f will give f(x,y). We can use P(x,y) to actually obtain the probability distribution for f(x,y). More often than not, this is rather difficult and, for most applications, unnecessary. If our main goal is to get an estimate of the uncertainty in f, we can use P(x,y) directly to calculate the expected average value for f and, within a reasonable approximation, the expected standard deviation.

The average value of f, $\bar{f}(x, y)$, can be written formally as

$$\bar{f}(x, y) = \int\int_{\text{all x,y}} f(x, y)P(x, y)dxdy = \int\int_{\text{all x,y}} f(x, y)P(x)P(y)dxdy \tag{3.41}$$

If we know both P(x) and P(y), we could substitute these into Eq. (3.41) and obtain $\bar{f}(x, y)$ directly. It will prove useful, however, to take a less direct route to investigate the relation between the mean we are seeking, $\bar{f}(x, y)$, and the means $\bar{x}$ and $\bar{y}$ of the distributions P(x) and P(y). This can be accomplished by performing a Taylor series expansion of f(x,y) about the value $f(\bar{x}, \bar{y})$, substituting of the expansion into Eq. (3.41), and integrating term by term. If the expansion is taken to second order in the variables x and y, the integral appears as

$$\tilde{f}(x, y) = \int\int_{\text{all x,y}} P(x)P(y)[f(\bar{x}, \bar{y}) + (x - \bar{x})f_x(\bar{x}, \bar{y}) + \qquad (3.42)$$

$$(y - \bar{y})f_y(\bar{x}, \bar{y}) + \frac{(x - \bar{x})^2}{2}f_{xx}(\bar{x}, \bar{y}) + \frac{(y - \bar{y})^2}{2}f_{yy}(\bar{x}, \bar{y}) +$$

$$(x - \bar{x})(y - \bar{y})f_{xy}(\bar{x}, \bar{y}) + \ldots]dxdy$$

In Eq. (3.42), the quantities f_x, f_y, f_{xx}, etc., stand for the partial derivatives

$$f_x = \frac{\partial f}{\partial x}, \quad f_y = \frac{\partial f}{\partial y}, \quad f_{xx} = \frac{\partial^2 f}{\partial x^2}, \text{ etc.}$$

While it looks fairly complicated, the equation simplifies greatly with a little analysis. For example, look at the second term of the integral of Eq. (3.42).

$$\int\int_{\text{all x,y}} P(x)P(y)(x - \bar{x})f_x(\bar{x}, \bar{y})dxdy = f_x(\bar{x}, \bar{y}) \int\int_{\text{all x,y}} P(x)P(y)(x - \bar{x})dxdy$$

$$= f_x(\bar{x}, \bar{y}) \int_{\text{all x}} P(x)(x - \bar{x})dx \qquad (3.43)$$

$$= f_x(\bar{x}, \bar{y}) \left(\int_{\text{all x}} P(x)xdx - \bar{x} \int_{\text{all x}} P(x)dx \right)$$

Now the first integral on the right-hand side of the last of Eqs (3.43) is just the definition of $\bar{x}$ while the second integral is just unity, i.e.,

$$\int_{\text{all x}} P(x)dx = 1$$

Therefore the entire integral is equal to zero. Exactly the same result will be obtained by evaluating the integral of the third term in the integrand of Eq. (3.42). The only term that remains to first order in the variables is the integral of the first term on the right-hand side of Eq. (3.42).

Already we have a very important result. If f(x,y) depends *only* to first order on the variables x and y, the contributions from all higher-order derivatives vanish, and Eq. (3.42) reduces to

$$\tilde{f}(x, y) = \int\int_{\text{all x,y}} P(x)P(y)f(\bar{x}, \bar{y})dxdy = f(\bar{x}, \bar{y}) \qquad (3.44)$$

The mean value of the function is simply the value of the function calculated from the means of the independent variables themselves. For example, the addition or subtraction of two Poisson-distributed variables will have a mean equal to the result of addition or subtraction of the appropriate means. The result can be extended to a function of any number of independent variables so long as all appear only to first order.

But what if an independent variable exists to higher than first order? So long as the higher-order derivatives are sufficiently small, we can *approximate* the mean of f with

$$\bar{f}(x, y) \approx f(\bar{x}, \bar{y}) \tag{3.45}$$

Now that we can calculate our best estimate of the mean of a function of a number of statistically-distributed independent variables, the next problem is to estimate its standard deviation. We can approach this in exactly the same way as we approached the mean value for f(x,y). By definition, the variance of f(x,y) is given by

$$\sigma_f^2 = \overline{[f(x, y) - f(\bar{x}, \bar{y})]^2} \tag{3.46}$$

$$= \int\int_{\text{all x,y}} P(x)P(y)\overline{[f(x, y) - f(\bar{x}, \bar{y})]^2} dxdy$$

One can now follow a procedure exactly similar to that above to show

$$\sigma_f^2 = \overline{f^2(x, y)} - (\overline{f(x, y)})^2 \tag{3.47}$$

After some long and tedious expansions and evaluations, it can be shown that

$$\sigma_f^2 = f_x^{\,2}(\bar{x}, \bar{y})\sigma_x^{\,2} + f_y^{\,2}(\bar{x}, \bar{y})\sigma_y^{\,2} + \ldots \tag{3.48}$$

and one can derive the *exact* relations

$$\sigma_{\overline{x+y}}^{\,2} = \sigma_{\bar{x}}^2 + \sigma_{\bar{y}}^2$$
$$\sigma_{\overline{x-y}}^{\,2} = \sigma_{\bar{x}}^2 + \sigma_{\bar{y}}^2 \tag{3.49}$$
$$\sigma_{\overline{xy}}^{\,2} = \bar{y}^2\sigma_{\bar{x}}^2 + \bar{x}^2\sigma_{\bar{y}}^2$$

and the approximate relation

$$\sigma_{\overline{x/y}}^{\,2} \cong \bar{y}^2\sigma_{\bar{x}}^2 + \bar{x}^2\sigma_{\bar{y}}^2 \tag{3.50}$$

The four relations given in Eqs (3.49) and (3.50) for the elementary operations of addition, subtraction, multiplication and division, usually suffice for estimating the uncertainty in statistically-distributed quantities. The beauty of these relations is that one does not need to derive the complete probability distribution for the

3.2 Statistical Considerations in Radioactive Decay

function f whose uncertainty is to be estimated. Further, if the probability distribution for each of the independent random variables is as simple as a Poisson distribution, we can easily obtain simple relations that permit very good estimates of the standard deviation of f.

The use of the uncertainty relations given in Eqs (3.49) and (3.50) to estimate errors in radioactive decay and simple counting experiments, is straightforward. As an example, consider the simple case of the measurement of the count rate from a radioactive source as shown in Fig. 3.9. We assume that a small source is held at a fixed position adjacent to a radiation detector and the number of counts registered in the detector over a fixed period of time is measured. For simplicity, we will assume that there are no uncertainties in the measurement of time and that the efficiency of the detector is constant and has negligible error. The only variables that need be considered are the source strength itself and the ever-present background radiation. Suppose that a source is counted for time t_s and the total number of counts registered is n_{tot}. The source is removed, the background radiation is measured for a time t_b, and the count registered is n_b. We want to obtain the best estimate of the source count rate and its statistical uncertainty.

We know that the total counts n_{tot} must be the sum of the counts registered from decay of the source and the counts from background radiation but we obviously do not know either. We do have an independent measurement of the count rate from the background immediately after the measurement in the absence of the source. The best estimate we can make for the background count rate, r_b, is to take the ratio n_b/t_b. Taking $n_{tot} = n_s + n_b$, where n_s is the number of counts registered from the source during the count of length t_s, we then have

$$n_{tot} = n_s + \frac{n_b}{t_b} t_s$$

The best estimate of the average count rate from the source, its activity, is then given by

$$A_s = \frac{n_s}{t_s} = \frac{n_{tot} - \frac{n_b}{t_b} t_s}{t_s} = \frac{n_{tot}}{t_s} - \frac{n_b}{t_b}$$

This result is just the result expected intuitively. The best estimate of the source activity is just the total count rate minus our best estimate of the background count rate.

We now need to estimate the uncertainty in A_s. To obtain this, it must be remembered that the Poisson statistics apply directly to the *number of counts measured* and not to quantities derived from them. The results of the direct measurements, n_{tot} and n_b, each represent a sampling from Poisson-distributed probability distributions and it is the estimate of the means of these distributions that are required to obtain the estimated uncertainty in A_s. Using the second of Eqs (3.49) we can write

$$\sigma_{n_s}^2 = \sigma_{n_{tot}}^2 + \sigma_{n_b}^2 = n_{tot} + \frac{n_b}{t_b}t_s \qquad (3.51)$$

As a specific example, suppose that a source of hydrogen gas containing tritium was counted in an internal gas counter for 1 h and the total count registered was 343. A subsequent 2 h background measurement under identical conditions except for the absence of tritium gave 532 counts. We have

$$n_{tot} \pm \sigma_{n_{tot}} = 343 \pm \sqrt{343} = 343 \pm 18.5$$
$$n_b \pm \sigma_b = 532 \pm \sqrt{532} = 532 \pm 23.1$$

and, using Eq. (3.51) we can calculate $n_s \pm \sigma_{n_s} = 77.0 \pm 21.8$, which also represents the best estimate of $A_s \pm \sigma_{n_s}$. In this example, the background rate with mean of 266 h^{-1} is comparable to the source rate and it leads to an uncertainty for the source counts that is significantly larger than would have been found in the absence of background radiation. Because the total counts considered here are fairly large, we can take the Gaussian approximation to the Poisson distribution and then estimate that there is roughly an 18% chance that remeasurement of the source count for one hour could lead to a value of n_s greater than 101! Counting statistics are very important, and it is not a trivial matter to perform experiments properly, in order to determine source strengths with small overall uncertainties, especially when a source is counted in the presence of significant background radiation.

3.3
Reaction Cross Sections

The most common types of nuclear reactions involve the bombardment of a target containing atoms of a particular element or nuclide with a beam of particles, or projectiles, from an accelerator. The target is usually at rest in the laboratory and the projectiles have some specific energy. Unlike the case of a simple classical experiment, as in the scattering of a cue ball on one of the other balls in a game of pool, we do not know the exact trajectory of a projectile with respect to any particular nucleus in the target. Further, the target is usually a gas, liquid or polycrystalline material and therefore can be approximated as having a random spatial distribution of its atoms or molecules. As a result, we must treat the case of the incidence of projectiles with randomly spaced trajectories on target nuclei that are themselves randomly distributed. This is clearly a situation where we can only describe the reactions statistically.

The means by which we describe reaction probabilities is perhaps best approached through the simple classical example illustrated in Fig. 3.10. We suppose that we have a right circular cylinder of cross sectional area A and length l. Within the cylinder are a large number of identical, randomly distributed spheres of cross

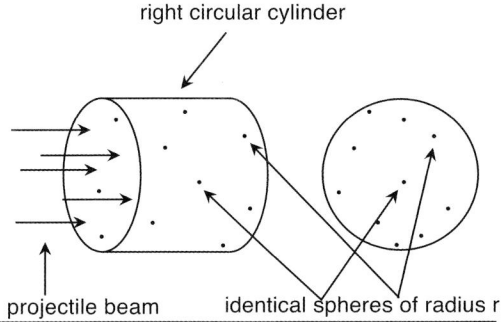

Fig. 3.10 Schematic diagram of the bombardment of randomly spaced targets by monodirectional projectiles with randomly spaced trajectories.

sectional area a and with an average number density of n cm^{-3}. The radius of each sphere is so small that if you looked down the axis of the cylinder (right part of Fig. 3.10), there would be no overlap between any of the spheres.

Now suppose that we launch a single point particle into the cylinder with a trajectory parallel to the axis of the cylinder, but located randomly within the cross sectional area shown. What is the probability that the projectile will strike one of the spheres? So long as there is no overlap of the spheres, the probability will simply be the ratio of the total projected area of the spheres – the total cross sectional area of the spheres – divided by the cross sectional area of the cylinder itself. The total number of spheres in the cylinder is just nlA and thus the probability for striking a sphere, is then

$$\frac{nlAa}{A} = nla$$

Now suppose we launch I point particles cm^{-2} s^{-1} into the cylinder. All particles have trajectories parallel to the axis of the cylinder but they are randomly distributed within the cylinder's cross section. If the probability for interaction of a single point particle is nla, the *expected* number of interactions per unit time will be

$$R = IAnla \ s^{-1} \qquad (3.52)$$

The quantity I is called the beam intensity. It is the magnitude of the *particle current density*. The number of interactions per unit time, R, is simply called the reaction or interaction rate.

But suppose we did not know the cross sectional area of the spheres within the cylinder, but did know their number density. We could use the experiment where we bombard the cylinder with a particle current I to get the cross sectional area by rearranging Eq. (3.52) to read

$$a = \frac{R}{IAnl} \text{ cm}^2 \qquad (3.53)$$

Eq. (3.53) can be taken as an expression that *defines* the cross sectional area of the spheres when they cannot be known directly. In fact, replacing the point particles with projectiles of protons, neutrons, nuclei, photons, etc., and replacing the spheres with target nuclei, Eq. (3.53) serves as the *definition* of an effective *cross section* for the interaction of projectiles with nuclei in a target. This effective cross section is simply referred to as the reaction or interaction cross section, σ, and is defined precisely as

$$\sigma = \frac{\text{reaction rate per unit area per target nucleus}}{\text{unit incident projectile current density}} \qquad (3.54)$$

The reaction cross section should not be taken literally as a measure of the physical cross sectional area of a nucleus as if it were a simple classical spherical object. It is much more accurate to describe it as an effective cross sectional area of the projectile–target combination. Nuclear reaction cross sections vary over a very wide range. They can represent an area as much as a factor of 10^6 times larger or a factor of 10^{-12} times smaller than that of a typical nucleus. Cross sections can be strongly dependent on the projectile energy and on the details of the structure of both the projectile and the target. In most cases, a particular target–projectile combination can lead to a number of different reactions, each with its own probability and hence its own cross section. As we will show in Chapter XIII, a cross section is a measure of the *probability* of a reaction and thus the total cross section for a particular target-projectile combination is the sum of the *partial* cross sections for all energetically allowed reactions. In the general case, all of the partial cross sections are energy dependent and thus the total cross section at any energy E, $\sigma_{tot}(E)$, is given by

$$\sigma_{tot}(E) = \sum_i \sigma_i(E) \qquad (3.55)$$

where the summation is taken over all reactions/interactions that are energetically possible.

The unit of measure of a nuclear reaction cross section is taken as 10^{-24} cm^2, a very small area that reflects the small dimensions of the nucleus. For historical reasons, it is called a *barn* (b). Cross sections for reactions of neutrons and protons with nuclei are typically on the order of 1b, but some neutron cross sections are known to be as large as 10^6 b. The cross sections for the interaction of high-energy photons with nuclei are typically as small as 10^{-6} b. About the smallest cross sections that are dealt with in low-energy nuclear physics involve the interaction of neutrinos with nuclei and these are as small as 10^{-20} b! It is truly very difficult to detect neutrinos.

There are a very large number of different reactions that can take place with particle beams interacting with different nuclei. Some of these are very useful for the production of desirable radioactive products and some, such as fission, are prolific sources of energy. Regardless of the projectile and its energy, there is one reaction that always takes place and that is elastic scattering. A somewhat detailed discussion of elastic scattering is provided in Chapter XIII. For the present, it will be worthwhile to consider a number of general applications of the use of cross sections for practical purposes.

One of the important issues that is always faced when considering the interaction of radiation with matter, is how the radiation is diminished or *attenuated* as it penetrates into the interior of a target. In the case that a target is very thin and the probability of interaction by the beam particles is very small, the beam intensity everywhere within the target is, for all intents and purposes, constant. This is completely analogous to the classical problem discussed above when the total cross sectional area of the spheres within the cylindrical volume represented a negligibly small fraction of the cross sectional area of the cylinder itself. In this case, known in the trade as the thin-target approximation, we can write the total reaction rate R_{tot} of a projectile beam of intensity I $cm^{-2}s^{-1}$ with a target of thickness x and nuclide density n as

$$R_{tot} = In\sigma_{tot}x \quad cm^{-2}s^{-1} \tag{3.56}$$

Because of the constancy of the reaction rate within the target, we can directly calculate the rates of any specific reaction by use of the partial cross section for that reaction. Eq. (3.56) is most commonly applicable to the interaction of neutrons with small targets that might be placed into a neutron beam for the production of some desired radionuclide.

If, however, the target is sufficiently thick, the projectile intensity incident upon atoms in the interior of the target will be less than the intensity incident at the target surface and we must account for *attenuation* of the incident beam. This is readily accomplished by reference to Fig. 3.11 where we first consider the attenuation of a neutron beam as a specific example. We assume that a beam of intensity I_o is incident on a target of thickness x. Due to attenuation, the beam intensity incident on a thin slab located at x´ and with thickness dx is reduced to I, and that exiting the slab is I – dI.

From Eq. (3.56), the intensity of the beam that interacts while traversing dx is just $In\sigma_{tot}dx$ and therefore

$$-dI = In\sigma_{tot}dx. \tag{3.57}$$

Direct integration of this equation gives

$$I(x) = I_o e^{-n\sigma_{tot}x} \tag{3.58}$$

As a direct result of the interaction of the neutrons with the target, the beam will suffer exponential attenuation.

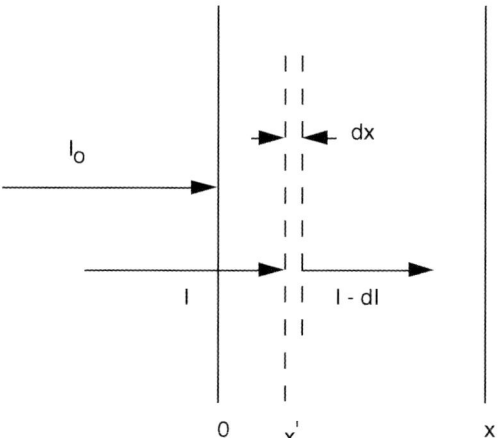

Fig. 3.11 Schematic diagram for calculating the attenuation of a neutron beam in a thick target.

It is very important to have a clear understanding of just what Eq. (3.58) says, namely, that *attenuation* will take place exponentially with the distance penetrated in a uniform, homogeneous material. The beam is attenuated by any and all energetically allowed reactions; simple elastic scattering, where the only observable changes may be the energy and direction of the neutron trajectory, *inelastic* scattering, where the interaction causes transfer of some of the kinetic energy of an incident neutron into internal excitation energy of the target nucleus but a lower-energy neutron still remains, as well as reactions in which the neutron may completely disappear as a free particle. As a result, the *attenuation* described in Eq. (3.58) is that of particles with the *exact* characteristics of the incident beam particles because any interaction will lead to a change in at least one such characteristic, even if it is just the change in direction that takes place in elastic scattering. In the parlance of nuclear physics and application areas, those beam particles that penetrate without attenuation are *uncollided* particles. Attenuation refers to the uncollided beam.

The relationship given in Eq. (3.58) also describes the attenuation of a beam of x or γ-rays. However, the interaction mechanisms are quite different. While neutrons experience significant interactions only with nuclei, high-energy photons interact almost entirely with atomic electrons. Nevertheless, photons over a wide energy range also interact by elastic and inelastic processes. For charged particles, the situation is somewhat more complex. They too interact predominantly with atomic electrons, but for particles with masses equal to or greater than that of the proton, or for high-energy electrons, their trajectories are roughly linear and the main effect of their interaction is a change in energy. Thus directional changes are of much-reduced importance in regard to attenuation.

A second general problem is the production of a radionuclide as a result of a nuclear reaction in a target sufficiently thin that attenuation can be neglected.

3.3 Reaction Cross Sections

Suppose, as a result of an interaction with a cross section σ_{reac}, a nuclide with decay constant λ is produced. In the absence of attenuation, we can write the rate of change in the number of atoms of the nuclide in the target per cm² of target area as

$$\frac{dN}{dt} = I_o n \sigma_{reac} x - \lambda t \ \text{cm}^{-2} \text{s}^{-1} \tag{3.59}$$

where n is the atom density of the target nuclide with which the projectiles interact, σ_{reac} is the cross section and x is the target thickness. In the limit that I_o is constant and the reaction rate is very small, the first term on the right-hand side of Eq. (3.59) is constant and thus the production rate of the radionuclide is constant. In form, the equation is identical to the equation describing secular equilibrium in the production of a short-lived nuclide from a very long-lived parent. If the number of radionuclide atoms present initially is zero, integration of Eq. (3.59) gives

$$N = \frac{I_o n \sigma_{reac} x}{\lambda}(1 - e^{-\lambda t}) = \frac{R}{\lambda}(1 - e^{-\lambda t}) \tag{3.60}$$

where R is the production rate of the radionuclide per cm² of target area. To obtain the total production rate, the cross sectional area of the beam must be specified.

The nuclear fission process in reactors is neutron-induced fission where the absorption of a low-energy neutron by a nucleus of ^{235}U can produce fission with a very high probability and, at the same time, produce some 2.5 neutrons in the process. As a result, there are more than enough neutrons in a reactor to supply the requirements of fission and the remainder can be used for other purposes. Very commonly, targets of stable nuclides are inserted into the reactor to produce useful radioactive nuclides, primarily by the process of neutron absorption. Whether we consider interactions with the nuclear fuel itself or a target placed into the reactor, the incident neutrons are not contained in a beam. Rather, they arrive at a target from all directions. The distribution of their velocity vectors is approximately isotropic.

Regardless of the directions from which they come, we can define the rate at which neutrons cross a hypothetical element of unit area at the position of our target. This quantity is called the *neutron flux* and it is usually given the symbol ϕ. As opposed to a beam intensity or current density that is a *vector* quantity, the neutron flux is a scalar but with the same dimensions of particles per unit area and unit time. The neutron flux can be used in exactly the same manner as the beam intensity in Eqs (3.52)–(3.60). If, for example, we were to place a mono-isotopic target of mass m into a reactor where the neutron flux was ϕ, the production rate of a radionuclide that results from a neutron reaction with cross section σ_{react} would be

$$R = \frac{m}{At.Wt.} N_A \sigma_{reac} \phi \tag{3.61}$$

where At. Wt. is the atomic weight of the target nuclide and N_A is Avagadro's number. The reader should verify that this is the same result as would be obtained with a neutron beam, a target of thickness x and atom density such that the target contains the same total number atoms as assumed in Eq. (3.61).

References

1 The original reference is H. Bateman, Proc. Cambridge Phil. Soc., 15 (1910) 423. A more useful reference is R.D. Evans, "The Atomic Nucleus", Chapter 15, McGraw-Hill, Inc., New York (1955). It is also available in reprint edition from Krieger Publishing Company (1982).

General References

Although it is quite dated, there are excellent discussions of both the decay properties of radioactive series and statistical fluctuations in nuclear processes in R.D. Evans, "The Atomic Nucleus", Krieger Publishing Company; Reprint edition (June 1982).

An excellent reference for simple error considerations in radioactive decay and other nuclear processes can be found in Glenn F. Knoll, "Radiation Detection and Measurement", John Wiley & Sons, Inc., New York (2000).

Problems

1. Calculate the energy release in (a) the α decay and (b) the β decay of ^{238}U.
(c) Accounting for the fact that ^{238}U indeed undergoes fission, what are the possible decay modes of ^{238}U?
(d) What decay modes are listed for ^{238}U in Appendix 2 and what are their absolute intensities per decay? What decay mode must control the half-life of ^{238}U?
(e) Using the experimental half-life from Appendix 2, estimate the rate of spontaneous fission of 1.0 g of ^{238}U. Express your answer in both Ci and Bq.

2. If a nuclide has a decay constant λ, what is the probability that it will *not* have decayed within a time period of $3t_{1/2}$ of its birth?

3. Given the decay chain $N_1 \rightarrow N_2 \rightarrow N_3$, where the decay constants $\lambda_1 = \lambda_2 = \lambda$, derive an expression for the number of atoms $N_2(t)$ when $N_1(t=0) = N_1^0$ and $N_2(t=0) = 0$.

4. The nucleus ^{106}Ru ($t_{1/2}$ = 386 d) undergoes β decay to produce 29.9 s ^{106}Rh. An initially pure sample of ^{106}Ru is allowed to stand for a period of three ^{106}Ru half-lives,

and at that time the ^{106}Rh is separated from it. If a source made from the separated ^{106}Rh is counted at a time corresponding to two ^{106}Rh half-lives after the separation, what is the ratio of its disintegration rate to the disintegration rate of ^{106}Ru? You can assume that the decay rate of ^{106}Ru is constant.

5. The nuclide ^{64}Cu is known to decay to both ^{64}Ni and ^{64}Zn, but to no other nuclides.
(a) Joe Bleau measured the half-life of ^{64}Cu by counting only β^- particles and found $(T_{1/2})_{Joe}$; his brother Frank measured the half-life by counting only β^+ particles and found $(T_{1/2})_{Frank}$. Theoretically, what is the relation expected between the two half-lives?
(b) Starting with n_o atoms of ^{64}Cu, Joe and Frank find that the total number of emitted β^- and β^+ particles account for only $0.57n_o$ after complete decay of the sample. What must have happened to the remaining $0.43n_o$ atoms?

6. Photons are attenuated according to the equation $I = I_o e^{-\mu_o x}$. Given this, we can say that the probability for a photon interacting between x and x + dx is $P(x)dx = ke^{-\mu_o x}dx$, where k is a normalization constant.
(a) Obtain an explicit expression for P(x).
(b) Use P(x) to determine the mean distance λ_γ traversed by photons until an interaction occurs. This is the so-called mean free path for the interaction.

7. You have 1mCi each of radionuclides with half-lives of 10^x s, where x varies by 2 in the range from −2 to 8. Calculate the number of atoms in each source.

8. Using the information given in Appendix 2, calculate the disintegration rate of 1.00 g of ^{235}U.

9. A hypothetical nuclide decays by both electron capture and positron emission. Electron capture occurs three times as frequently as positron emission when the nuclei are in neutral atoms. A source of this nuclide is counted on a detector that can only measure positrons. The initial rate of positron detection is 78.90 s^{-1} and the rate exactly 3 hr later is 17.61 s^{-1}.
(a) What is the half-life of this nuclide?
(b) A sample of the nuclide is completely ionized and placed into a cyclotron where the nuclei can rotate in the absence of free electrons. What is the half-life of the nuclide under these conditions?
(c) Use Poisson statistics to determine the uncertainties in the half-lives you computed in parts (a) and (b).

10. Using the definition of the variance given in Eq. (3.36), prove that the standard deviation of a Poisson distribution is given by Eq. (3.37).

11. During a recent measurement of the γ decay of the short-lived fission products, the number of counts registered in a 3.5-h interval with energies greater than 4.0

MeV was 87. During the same time interval the number of counts with energies greater than 3.0 MeV was 206. The counts due to background during this time in the energy interval 3.0–4.0 MeV was 3. What is the best estimate of the count rate with energies in the range 3.0–4.0 MeV in units of s^{-1} and what is its estimated standard deviation?

12. A 1.0 mg sample of natural uranium is irradiated with thermal neutrons for a period of 10 s at a thermal neutron flux of 10^{13} $cm^{-2}s^{-1}$. Calculate the number of fissions that occur, the number of ^{236}U atoms produced and the number of ^{239}U atoms produced. Assume that the fission and neutron capture cross sections of ^{235}U are 580 b and 92 b, respectively, and that the capture cross section of ^{238}U is 2.72 b. Note: neutron capture reactions are generally of the form $^{A}Z + {}^{1}n \to {}^{A+1}Z$.

13. A 10 mg sample of Al is irradiated in a thermal neutron flux of $10^{12} cm^{-2} s^{-1}$ for a period of 10.0 min. Calculate the disintegration rate of ^{28}Al 10 min after the end of the irradiation. Assume that the capture cross section on ^{27}Al is 0.231 b.

4
Nuclear Binding Energies: Empirical Data and the Forces in Nuclei

4.1
Empirical Masses and Average Binding Energies of Nucleons

The general discussion of nuclear masses and binding energies in Chapter II permits us to use empirical mass measurements to calculate the energetics of all radioactive transformations and nuclear reactions. In this chapter we will develop some general ideas concerning the important forces within nuclei that not only give some physical insight into nuclear structure, but also lead to the development of a simple model of masses with which the systematics of nuclear energetics can be understood.

To begin, the mass of any nuclide is related to the masses of the protons and neutrons from which it is composed, by the expression

$$Zm_p + (A-Z)m_n = N(Z, A) + (B.E.)_{Z,A}/c^2 \qquad (4.1)$$

where $(B.E.)_{Z,A}$ is the total binding energy of the nucleons in the nucleus and we have used the shorthand notations m_p and m_n for the masses of the proton and neutron, respectively. With the approximation that we can neglect atomic binding energies, we add the masses of Z electrons to both sides and rewrite Eq. (4.1) in terms of atomic masses as

$$ZM(1, 1) + (A-Z)m_n = M(Z, A) + (B.E.)_{Z,A}/c^2 \qquad (4.2)$$

We can use Eq. (4.2) to calculate the total nuclear binding energies directly. But as we might expect from simple classical considerations of a droplet of water, for example, the binding energies ought to increase as the mass number of the nucleus increases; the more particles bound, the greater the total binding energy. What is much more interesting and revealing is the average binding energy per nucleon, $(B.E.)_{Z,A}/A$. If the nucleus were similar to a macroscopic droplet of water, we would expect the average binding energy per nucleon to be about constant and independent of mass number. A quick look at the few average binding energies given in Table 4.1 shows, however, that this is not the case. Although the total binding energy increases with A as expected, the average binding energy varies significant-

Nuclear Physics for Applications. Stanley G. Prussin
Copyright © 2007 WILEY-VCH Verlag GmbH & Co., Weinheim
ISBN: 978-3-527-40700-2

Table 4.1 The total and average nuclear binding energies for a few nuclides.

Nuclide	(B.E.)$_{Z,A}$ (MeVc^{-2})	(B.E.)$_{Z,A}$/A (MeVc^{-2})
$^{1}_{1}$H	0	–
$^{2}_{1}$H	2.224	1.112
$^{4}_{2}$He	28.296	7.074
$^{56}_{26}$Fe	492.26	8.790
$^{238}_{92}$U	1801.73	7.57

ly. To get a more complete picture of this variation, the average binding energies of all nuclides found in nature with appreciable abundance are shown in Fig. 4.1. The average is seen to rise fairly regularly from a low of 1.112 MeV for deuterium, $^{2}_{1}$H, to a high of 8.795 MeV for $^{62}_{28}$Ni and then to decrease smoothly to a low of 7.570 MeV for $^{238}_{92}$U. To be sure there are some noticeable variations about a smooth curve, especially at $^{4}_{2}$He (B.E./A = 7.074 MeV), $^{12}_{6}$C (B.E./A = 7.680 MeV), $^{16}_{8}$O

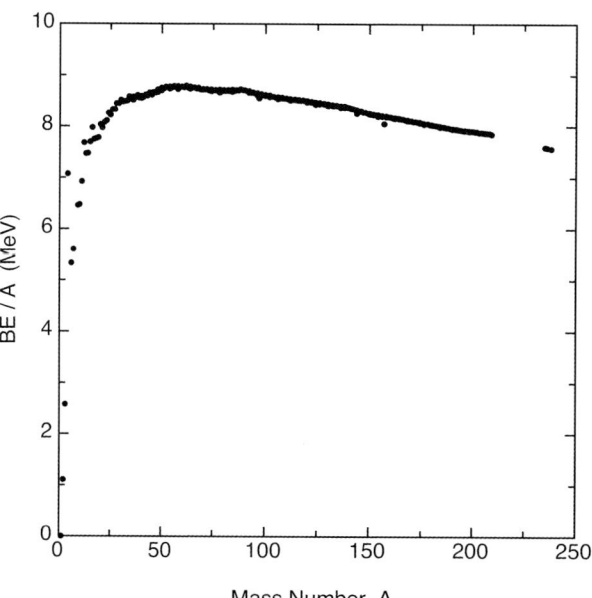

Fig. 4.1 Average binding energies of nuclides found in nature with significant atomic abundance. The few nuclides with mass numbers greater than 230 are the long-lived isotopes of thorium (Z = 90) and uranium (Z = 92).

(B.E./A = 7.96 MeV) among the lightest nuclei, in the regions centered near A = 30, 50, 88, and perhaps in the region near A = 140. In the main, however, the average binding energy varies with mass number in a fairly regular manner and, with the exception of the lightest nuclides, lies in the range of about 7.5–8.8 MeV with an average of about 8.4 MeV.

The large variation in average binding energy among the lighter nuclei has some very practical implications. For example, consider the large difference in the average binding energies between $^{2}_{1}H$ and $^{4}_{2}He$. Now we can conceive of making the $^{4}_{2}He$ nucleus by the reaction that combines two nuclei of deuterium, the fusion reaction;

$$2 \, ^{2}_{1}H = \, ^{4}_{2}He + Q/c^2 \tag{4.3}$$

The energy release from this reaction would be

$$\begin{aligned} Q &= [2M(1,2) - M(2,4)]c^2 \\ &= 2\{(M_H + m_n)c^2 - B.E._{1,2}\} - \{(2M_H + 2m_n)c^2 - B.E._{2,4}\} \\ &= B.E._{2,4} - 2B.E._{1,2} \\ &= 4(7.074) - 4(1.112) \\ &= 23.848 \quad MeV \end{aligned}$$

This is very large. It is representative of the fact that the fusion of essentially all light nuclei would result in significant energy release because the average binding energy of the nucleons in the product is significantly greater than the average binding energy in the initial nuclei. Similarly, because the average binding energies in the heaviest elements are smaller than those of medium-massed nuclei, we immediately see that *fission* of heavy nuclei will also release energy. If, for example, we were able to split a nucleus of $^{238}_{92}U$ into two nuclei of $^{119}_{46}Pd$, we can estimate that the energy release would be about 221 MeV!

The variations in average binding energies displayed in Fig. 4.1 must reflect the fundamental interactions of the nucleons and the quantized structure of the nucleus just as the electromagnetic interaction and quantization define the binding energy of electrons in atoms.

4.2
The Forces Acting Between Nucleons

We know that there are actually four fundamental forces[1], or interactions, to which nucleons and nuclei must be subject. First, nuclei have mass and therefore they must be subject to the gravitational interaction. Second, nuclei are charged and

---
1) Modern unified field theory has successfully demonstrated that the weak and electromagnetic interactions derive from the same interaction. From a practical point of view, however, these interactions can be treated quite separately.

therefore must be subject to the electromagnetic interaction. If these were the only forces that existed we could show immediately that most nuclei could not exist. The Coulomb force is so much stronger than the gravitational force that the repulsion of the protons would immediately cause most nuclei to disassemble. We can conclude that there must be at least one additional force that is *attractive* and sufficiently strong to overcome the repulsion between the protons. This force is referred to simply as the *strong* interaction or the nuclear force. Finally, there is an additional interaction that we must consider in order to account for β decay. It is so weak that its affect on nuclear properties is entirely negligible and can be forgotten except when we need to account for β decay explicitly.

How do we go about determining the relative strengths of the forces that exist in nuclei? The proper way is to consider the relative strengths of different interactions when acting on nucleons under the same dimensions and conditions. For example, the gravitational interaction between two bodies of masses m_1 and m_2 can be written

$$F_{grav} = -g_{grav} \frac{m_1 m_2}{r^2} \tag{4.4}$$

where $g_{grav} = 6.67 \times 10^{-11}$ N m^2 kg^{-2} is the so-called *gravitational coupling constant*. The Coulomb force between two charged particles q_1 and q_2 can be written as

$$F_{Coul} = k_C \frac{q_1 q_2}{r^2} \tag{4.5}$$

Now if we have two charged particles with masses m_1 and m_2 and charges q_1 and q_2, respectively, the ratio of the forces acting upon them is simply

$$\frac{F_{grav}}{F_{Coul}} = \frac{-g_{grav}}{k_C} \frac{m_1 m_2}{q_1 q_2} \tag{4.6}$$

For two protons in a nucleus, we can rewrite Eq. (4.6) in the form

$$\frac{F_{grav}}{F_{Coul}} = \frac{-g_{grav}}{k_C} \frac{m_p^2}{e^2} = \frac{-g_{grav}}{k_C \hbar c} \frac{m_p^2}{e^2/\hbar c} \tag{4.7}$$

where e is the electronic charge and $\hbar = h/2\pi$ where h is Planck's constant. Using the numerical value for g_{grav} given above, the proton mass of about 1.67×10^{-27} kg, and the value of about 1/137 for the ratio $e^2/\hbar c$, the so-called fine structure constant, we then have

$$\left| \frac{F_{grav}}{F_{Coul}} \right| \approx 8.1 \times 10^{-37}$$

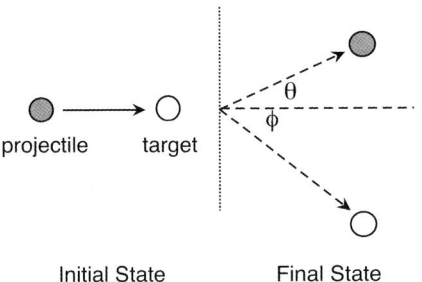

Fig. 4.2 Schematic diagram of the elastic scattering between a target and projectile in the laboratory reference frame.

Clearly, the gravitational force is negligibly small compared to the Coulomb force and we can disregard it completely when considering the interaction of nucleons in a nucleus.

What do we know about the strong interaction and how is it studied? As for all fundamental interactions, we need a probe or probes that interact by the force of interest, and in the present case, the simplest probes available are neutrons and protons. These particles can be used in scattering experiments of the type you are used to from classical mechanics, and a schematic of a simple elastic scattering experiment is shown in Fig. 4.2. For simplicity, we assume that a neutron of known kinetic energy is incident upon a target of protons that are at rest in the laboratory. After the interaction, the neutron is scattered through some angle θ and the proton recoils at some angle ϕ.

If the force acting in the interaction is conservative, as we know is true for the gravitational and electromagnetic interactions, and we carry out the experiments under conditions where no internal excitation of the scattering partners takes place, the kinematics of the scattering, the energies of the particles after scattering as a function of scattering angle, are completely defined by the conservation of kinetic energy and linear momentum. From a study of such experiments over a wide range of energies, it is found that indeed the strong interaction between a neutron and a proton is conservative. Further, if one carries out elastic scattering of protons on protons and corrects for the very well-known effects from the electromagnetic interaction, one finds again that the strong interaction between two protons is conservative. Although we cannot make a simple target of neutrons alone, all indirect lines of evidence indicate that the strong interaction between two neutrons is also conservative.

Kinematics does not account for the *probability* that a projectile will scatter through some angle θ to $\theta + d\theta$. This probability is, however, completely defined by the detailed characteristics of the force or forces acting on the particles [1], and that includes the nature of the particles themselves. For the gravitational interaction or for the Coulomb interaction between two simple point charges, this probability is described by the *differential scattering cross section* defined as

$$\sigma(\theta) \equiv \frac{\partial \sigma}{\partial \Omega_\theta}$$

where $\sigma(\theta)$ is the scattering cross section that results in the projectile appearing at an angle θ to $\theta + d\theta$, σ is the scattering cross section and Ω_θ is the solid angle between θ to $\theta + d\theta$ and we have assumed, as is often the case, that there is cylindrical symmetry in the scattering distribution about the direction of incidence of the projectile, i.e., the scattering is independent of the azimuthal angle. If the force is known, $\sigma(\theta)$ can be calculated and used to predict the intensities of projectiles that will appear at all angles θ to $\theta + d\theta$. Conversely, an experiment that measures the angular distribution of the scattered particles can be used, at least in principle, to uncover the form of the force that produced it.

When we deal with neutrons, protons and even electrons, however, we must be aware of the fact that these particles are fermions and they possess an *intrinsic angular momentum*. As strange as it may seem from our experience with classical mechanics, these particles possess angular momentum at rest. This *intrinsic* angular momentum, just as mass and charge, is an inherent property of the particles. But there is more. The intrinsic angular momentum vector cannot, in the presence of a magnetic field, align itself along the field direction. It can align itself with but two projections with respect to the field direction and we often refer to these alignments as "spin up" and "spin down" for simplicity. And that means that the force between fermions may be dependent upon both the angles of Fig. 4.2 as well as an azimuthal angle that measures the relative angles between the spin directions of the interacting pair and would be measured relative to an axis normal to the plane shown in the figure.

Two types of scattering experiments can be performed. In the simplest, the incident projectiles, neutrons or protons, will have completely random orientations of their spin directions and so will the target protons. This is the normal case that takes place in the laboratory and both the projectiles and the target protons are said to be *unpolarized*. This experiment will provide information on the average probability for scattering into the angle θ to $\theta + d\theta$. It actually provides the type of information that is needed for most experiments and applications.

The second type of experiment is where we actually produce both *polarized* projectiles and *polarized* target particles. By polarized, we mean that the projectiles have either "spin up" or "spin down" and so do the target particles (they do not both have to have the same relative orientations). This type of experiment will provide the detailed information we need to determine if the strong interaction is dependent upon the orientation of the spins of interacting fermions. If a beam of polarized neutrons with, say, "spin up" is scattered from a target of polarized protons with "spin up", one finds that the neutrons are preferentially scattered in one direction relative to the trajectories of the incident neutrons. And if the neutrons are polarized with "spin down" and the target protons still have "spin up", the neutrons are scattered preferentially in the opposite direction. Further, the *magnitudes* of the differential scattering cross sections are not the same! This indicates that the strong interaction is indeed *spin-dependent*.

One of the simplest experimental observation that points to the spin-dependence of the strong interaction is the nature of the deuteron that contains one proton and one neutron. As discussed above, this nucleus has the lowest binding energy of all nuclei found in nature. As will be discussed later in this chapter and in Chapter VII, there is ample evidence that the nuclear levels in which particles reside have an *orbital* angular momentum due to the particle motion, in addition to the intrinsic angular momentum of the particles. Further, protons and neutrons, being different particle types, have their own independent level sequences. All evidence, both experimental and theoretical, shows that the lowest energy levels in which nucleons can reside have zero orbital angular momentum. That means that when a proton and neutron reside in the lowest energy states possible in the deuteron, the condition that defines the ground state found in nature, the total angular momentum of the deuteron must be due solely to the intrinsic angular momentum of the two particles. Because the magnitude of the intrinsic angular momenta of all fermions is identical and because angular momentum is a vector quantity with only two possible states for the intrinsic angular momenta, "spin up" or "spin down", the total angular momentum of the deuteron will be zero if one of the particles has "spin up" and the other "spin down" or it will be twice the intrinsic angular momentum of a fermion. There are no other possibilities.

Experiment shows that it is the latter arrangement of the intrinsic spins of the proton and neutron, the "parallel spin orientation", that exists in the ground state of the deuteron. The deuteron has an angular momentum twice that of a fermion. Experiment also shows that the deuteron has no other bound states. If you try to make the deuteron with the zero angular momentum combination, the proton and neutron will simply not bind! In fact, the zero angular momentum configuration of the two particles can be seen in the cross section for scattering of a neutron on a proton target and it is indeed unbound.

The nature of the deuteron and what it tells us about the strong interaction will be discussed further in later chapters. This nuclide is certainly more complex than we have considered so far. Nevertheless, there is nothing that negates the discussion given here. We can say, with certainty, that the total angular momentum of the deuteron in its ground state cannot be explained unless the strong interaction is spin-dependent.

Now there is no *a priori* reason why the strong interaction between polarized proton projectiles and polarized target protons should be identical to the interaction between polarized neutrons and a polarized proton target. The analysis of the data from such an experiment is a bit more complex than that for the neutron–proton scattering because of the electromagnetic interaction between the protons and because the target and projectile are the same type of particle – we really cannot distinguish after the experiment which was the target and which was the projectile. Nevertheless, as we have noted above, the effects from the electromagnetic interaction are very well known and methods are available to handle the fact that the colliding particles are of the same type. Thus, essentially the same information can be extracted with respect to the strong interaction as in the case of neutron–proton scattering. If we perform such experiments, we find that the spin-dependence is the same with respect to the preferred direction of the scattered particles. Also, the magnitudes of cross section with both "spin up" and "spin down" orientations for the beam and with a constant orientation of the spins of the polarized protons are very nearly, but not exactly, the same as those found in the corresponding scattering experiments with neutrons and protons. We cannot perform scattering experiments in the same way with neutrons on neutron targets directly, but one can extract similar information indirectly from more complex reactions. And the results are quite similar to those described above. We can conclude from all of the experiments performed to date, that the strong interaction between a neutron and a proton or two protons or two neutrons is *almost the same*.

The nuclear force – the purely nuclear force between two nucleons – is indeed very complicated. We now understand that the complicated nature of the force derives from the fact that the neutron and proton are not really fundamental particles as appears to be true of electrons. They are each composed of three quarks, and it is the net interaction of the quarks that leads to what we measure as the nuclear force.

With all of this as preamble, what do we know about the nuclear force or its potential? Elastic scattering of neutrons on protons and protons on protons, and indirect information that bears on the interaction between two neutrons provide the following conclusions;

1. The nuclear force is short ranged. Nucleons do not experience a significant nuclear interaction when they are separated by more than about $(2-4) \times 10^{-13}$ cm. The quantity 10^{-13} is a unit referred to as a *fermi* and symbolized by fm.
2. The nuclear force is strongly attractive except at very small radial separations between nucleons where it becomes strongly repulsive.
3. The nuclear force is spin-dependent and the magnitude of the force depends upon the relative orientation of the intrinsic spin vectors of the interacting nucleons.
4. The purely nuclear interaction is very much stronger than the electromagnetic interaction. On a relative scale where $F_{Coul} = 1$, $F_{nucl} \approx 100$.
5. While it is not exactly true, the purely nuclear force between two neutrons, two protons, and between a neutron and a proton are very nearly the same. The nuclear force appears to be independent of charge, to a good approximation.

The very short range of the nuclear force is wholly different from the Coulomb or gravitational interactions, both of which have infinite range. The fact that the nuclear force becomes strongly repulsive at small separations of nucleons and tends to prevent the "overlap" of nucleons in space is not really so strange. The same behavior is found in the interaction of atoms. For example, the hydrogen atoms in a H_2 molecule have a rather well-defined equilibrium distance of separation, and if we try to reduce this, we must add energy to the system because the atoms now experience a net repulsive interaction. We can conclude that the short range of the nuclear force coupled with its repulsive nature at small distances will tend to produce some rather well-defined equilibrium distances between nucleons in a nucleus.

Apart from the complications due to the spin-dependence of the nuclear force, we have the great simplification that the force appears to be about the same regardless of the nature of the nucleon pair we are considering: two neutrons, two protons or a neutron and a proton. Differences do indeed exist, but for our purposes they are sufficiently small that we will neglect them. The errors that will be incurred by this approximation are sufficiently small that they will not affect the general conclusions we draw in our discussions.

4.3
The Average Nuclear Interaction Between Nucleons in the Nucleus and Nuclear Radii

It is possible to write down an analytical expression that accounts reasonably well for the experimental data gleaned from nucleon–nucleon scattering and use it to calculate the total interaction between the A nucleons in a nucleus, and hence calculate nuclear structure directly. This is not only difficult but cannot even be done exactly. Apart from one- and two-body problems, we cannot solve the equations of motion exactly for a general many-bodied problem. Some sort of approximation scheme must be used. In such complicated cases we usually resort to a model and give up hope of a completely accurate description of the problem. This is true whether we are considering the structure of many-electron atoms, simple molecules or complex molecular structures. The art is to choose a model that contains enough of reality that it is a good approximation. If it is, the results obtained will represent a first approximation to a complete description. Once this is found, we can use additional approximation schemes as *perturbations* to the model to include factors that were initially neglected. In this way a rather good comparison between model calculations and experimental fact can usually be obtained. In our approach, we will use the simplest possible models that make sense and which can give us the essential physics of nuclear properties and reactions. While simple, some of the models permit calculation of certain properties with surprising accuracy. Some may provide only semiquantitative agreement with experiment but can account for, and allow us to understand, the essential characteristics of radioactive decay and the systematics found in experiment.

When dealing with complex nuclei, one of the simplest ways to obtain a reasonable model approximation is to consider the average characteristics of the interaction of the nucleons. We can, for example, get such information by looking at the interaction of probes with complex nuclei. We can use neutrons, protons and electrons of various energies to probe the interaction that each experiences as it moves through the nucleus. These three particles, taken together, can give a very interesting set of data. Neutrons will experience the interaction with the nucleons solely as a result of the nuclear force. As such, they can provide information on the average strong interaction potential throughout a nucleus. Since the nuclear force is so short-ranged, neutrons will experience interactions only with the nearest neighbors they encounter. Neutrons should thus be a probe of the *density* of nucleons. If we use protons, we will probe the interaction potential due both to the nuclear and Coulomb forces, and if we use electrons, we will probe only the long-range Coulomb potential.

Distilling the data from many studies on the distribution of matter in nuclei, we will consider directly three bits of evidence that lead to the conclusions that, with the exception of the very lightest nuclei;

- nuclear matter has essentially constant density independent of the mass or atomic number of the nucleus
- the distribution of neutrons and protons within the nucleus is roughly uniform.

The first such information comes from the scattering of electrons on nuclei. This type of experiment probes only the Coulomb field from the protons to a good approximation, and directly gives the root-mean-square radius (rms) of the charge distribution $<r^2>^{1/2}$. In Fig. 4.3 are shown the root-mean-square radii obtained from a variety of experiments as a function of the cube-root of the mass number of the nucleus [2]. A good fit to the data is found with a linear function of the form

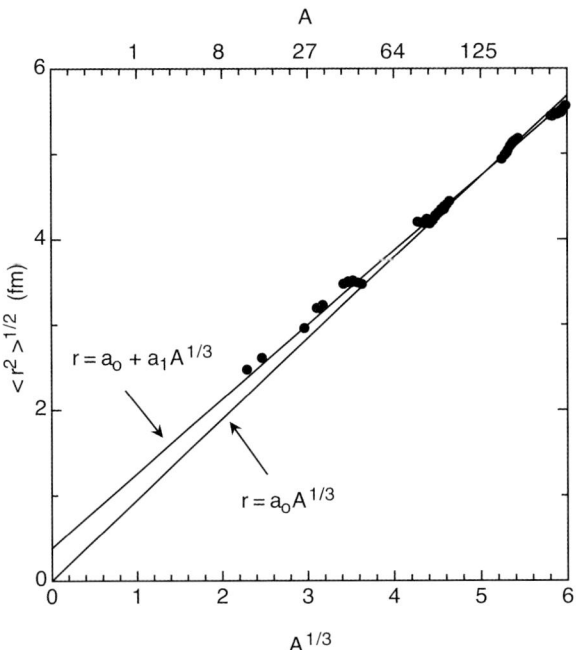

Fig. 4.3 The root-mean-square (rms) radii of some nuclides as a function of $A^{1/3}$. The data were derived from electron scattering and other experiments.

$\langle r^2\rangle^{1/2} = a_o + a_1 A^{1/3}$ that intercepts the abscissa at a root-mean-squared (rms) radius of about 0.5 fm. However, the simple linear form $\langle r^2\rangle^{1/2} = a_o A^{1/3}$ that forces the intercept to pass through the origin fits the data reasonably well for $A \geq 40$ and yields the value of $a_o = 0.9475 \pm 0.0024$ fm. The data demonstrate that nuclear radii for all but the lightest nuclei vary linearly with $A^{1/3}$ to an excellent approximation.

The approximate dependence of the rms radius of the charge distribution on $A^{1/3}$ is quite remarkable, as a very simple calculation shows. Suppose that the nucleus was a sphere of constant density. If it were a classical, continuous body it would have a volume

$$V_{nucl} = \frac{4}{3}\pi R^3 = A v_{nucleon} \tag{4.8}$$

where $v_{nucleon}$ is the average volume occupied by a single nucleon. The radius of the sphere would be given by

$$R = \left(\frac{3 A v_{nucleon}}{4\pi}\right)^{1/3} = r_o A^{1/3} \tag{4.9}$$

We are thus led to the conclusion that, to a good approximation, heavier nuclei can be described as spherical objects with *constant charge density*. This is quite different from what one might expect classically where, for example, the charge on a spherical conductor resides on the surface of the sphere. Now we can go one step further to conclude that the data imply a *constant mass density* for nuclei as well, because if the proton density and the neutron density did not vary in the same way, we could not expect to find $r \sim A^{1/3}$. As a result, we are led to conclude that nuclei can be approximated as spherical objects with both constant charge and mass densities.

If we take the simple classical approximation of a uniformly charged sphere, it would have a charge density of $\rho(r) = (Ze)/\left(\frac{4}{3}\pi R^3\right)$, and the mean-square radius of this body is given by

$$\langle r^2 \rangle = \frac{\int_0^R r^2 \rho(r) dV}{\int_0^R \rho(r) dV} = \frac{3}{5} R^2 \tag{4.10}$$

Assuming $\langle r^2\rangle^{1/2} = a_o A^{1/3}$, we then have $R = \sqrt{\frac{5}{3}} a_o A^{1/3}$, and Eq. (4.10) then gives the result

$$r_o = \sqrt{\frac{5}{3}} a_o \cong 1.22 \text{ fm} \tag{4.11}$$

The quantity r_o is simply called the nuclear *radius parameter*, and the fact that it has a magnitude of about 1.2 fm has some significance. It represents the effective radius of a single nucleon in the constant-density approximation. This corresponds to an rms radius of about 0.94 fm. Now it is known from independent measurements that the actual rms radius of the proton is 0.813 ± 0.008 fm. Thus, the very simple constant density model provides an estimate of the nucleon radius that is comparable to the charge radius of the proton.

Regardless of the accuracy with which we can fit the experimental data on nuclear charge distributions with the relation $<r^2>^{1/2} = a_o A^{1/3}$, one must be careful in interpreting the significance of the magnitude derived for r_o. It is a model-dependent quantity derived with an expression that does not express nuclear radii exactly. In fact, the values for r_o derived from different types of experiments tend to differ because they tend to measure somewhat different quantities or different weighting of quantities. Nevertheless, all of the measures of nuclear dimensions are in agreement with the assumption of the approximate constancy in nuclear density of heavier nuclei and all provide *roughly* the same magnitude for r_o. For simplicity, we will take $r_o = 1.25$ fm for the calculations performed in this text.

Now let us consider the breakdown of the constant density model for the lighter nuclei as indicated from the data in Fig. 4.3. The breakdown is actually to be expected and can be understood by reference to classical systems. Constant density can be expected for a classical body composed of incompressible components, or compressible components that are each subject to the same net force. If a body is composed of compressible components and if the forces between components are so short-ranged that each component experiences an interaction only with its nearest neighbors, we can expect that those components surrounded by the same number of nearest neighbors will experience the same net force and thus will have the same average volumes. Of course, those that reside at the surface of the body will not have the same number of nearest neighbors and therefore will not have the same effective volumes. The requirement for a constant-density approximation is then that the fraction of compressible components residing at the surface of the body must be very small. This model actually applies to ordinary condensed matter composed of atoms or molecules. Normally we deal with macroscopic quantities composed of very large numbers of atoms or molecules and constant density is taken for granted. The fraction of atoms or molecules at the surface is really negligibly small. However if we were to deal with only a few atoms or molecules of the same materials, the constant-density approximation would be found to fail.

A rough estimate of the minimum size of a nucleus for which we expect to see approximately constant density can be made as follows. Suppose we have a hypothetical body composed of elementary cubes where the range of the force between each cube is about the dimension of the side of the cube itself. In order that any single cube experiences a total net force independent of the size of the body, it must be surrounded by the full complement of nearest neighbors. In this simple example, the cube at the center of a $3 \times 3 \times 3$ array will have this full complement of nearest neighbors, but none of the others will. Thus, we must have a body containing at least 27 elementary cubes before any one of them can be said to experience a net force of interaction that will be unaffected by the addition of one or more cubes to the system. We can now infer that nuclei with mass numbers less than about 27 will not have a "core" of nucleons that experience roughly a net force of interaction, or total potential, that is unaffected by the addition of one or more nucleons. Reference to the data in Fig. 4.3 shows that this is just about where the approximation $r = r_o A^{1/3}$ begins to apply reasonably well.

The direct measurement of the mass distribution in the nucleus requires probing by the strong interaction. Many experiments that rely on different approaches have been made. The results from all of these point to the fact that the neutron density in nuclei closely follows that of the proton density. The effective radii of the neutron distributions in heavier nuclei differ from those of the proton distributions by no more than 0.1–0.2 fm.

Electron scattering experiments provide much more detail on the charge distribution in nuclei than just the mean squared radius. They give us further information on the extent to which the constant density approximation is valid and they permit us to infer a reasonable approximation for the average radial dependence of the potential that must be experienced by a nucleon. In Fig. 4.4 are shown the radial dependence of the charge distributions for a number of nuclei that sample the mass range found in nature [3]. The ordinate in the figure is the proton density in units of fm^{-3} and the abscissa is the radius (fm) relative to the center of charge of the target nucleus. Because $\int \rho(r) dV_{nucleus} = Z$, $\rho(r)$ directly gives the density of protons as a function of radial dimension. For each of the three sets of experimental data shown, there is a central region in which the proton density is roughly

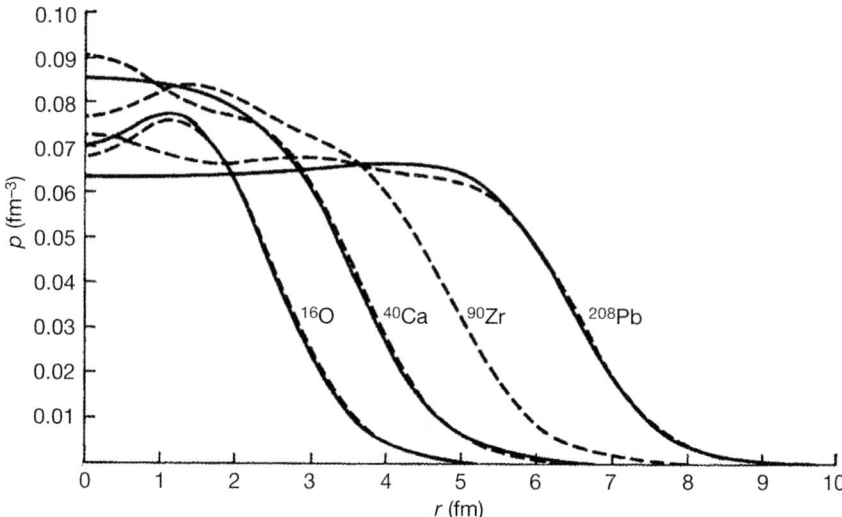

Fig. 4.4 The radial distributions of charge density in ^{16}O, ^{40}Ca, ^{90}Zr and ^{208}Pb from electron scattering (solid curves) and from model calculations (dashed curves). Experimental data for ^{90}Zr were not available at the time of this publication.

constant. In each case this region is followed by a smooth decrease to zero. The data clearly show that the constant density approximation for heavy nuclei is quite reasonable. Variations about a constant density in the nuclear interior, that must reflect the quantized structure of the nucleus, amount to less than about 10% of the average over the central region.

The smooth decrease in density at large radial dimensions also shows that the nucleus is not a sharply-bounded body. If we take the radial dimension t over which the proton density decreases from 0.9 to 0.1 of its average central value, we find $t \cong 2.3 \, \text{fm}$. This has crucial significance. As we indicated above, the nuclear force is short ranged and nucleons do not experience a significant nuclear interaction when they are separated by more than about 2–4 fm. This is just about the magnitude of t. We may infer that the probability for finding nucleons outside of the central core of constant density decreases roughly as the range of the nuclear force itself. Further, we can also conclude that a nucleon approaching the surface of a nucleus will not experience a significant interaction via the strong interaction until it has a separation from the surface that is about the same as the sum of the radii of two nucleons in contact in the hard-sphere approximation.

With all of this as background, we can take the experimental measurements of the charge and mass distributions in the nucleus to arrive at the simple approximation for the nucleon density shown in Fig. 4.5. This density distribution represents

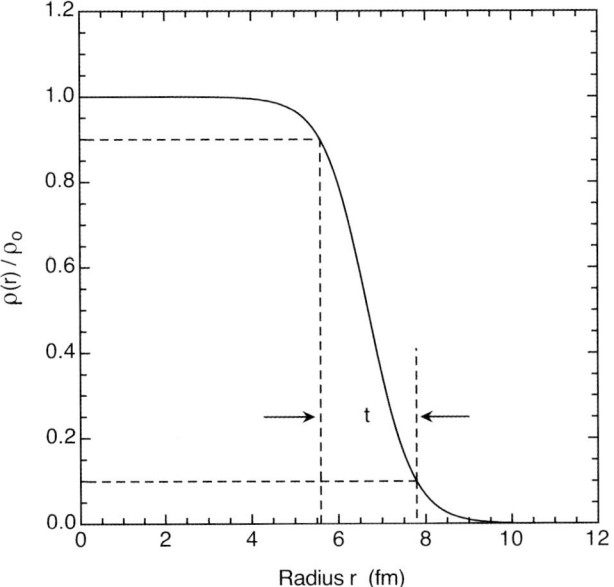

Fig. 4.5 Idealized model of the radial dependence of the nucleon density relative to the density at r = 0. The curve was calculated for A = 208 using the Saxon–Woods distribution. The dimension t is the distance required for the density to fall from 0.9 ρ_o to 0.1 ρ_o.

one of the common forms used in idealizations, and is known as the Saxon–Woods or Fermi distribution. It has the form

$$\frac{\rho(r)}{\rho_o} = \frac{1}{1 + e^{\frac{(r-R)}{a}}} \quad (4.12)$$

where R is termed the nuclear radius and is usually taken as $R = r_o A^{1/3}$ with $r_o \approx$ 1.25 fm, The parameter a is a measure of the rate at which the nuclear density approaches zero at its surface and is about 0.52–0.58 fm. The radial dimension over which the nuclear density falls from 0.9 to 0.1 of its central value is usually referred to as the *skin thickness*. In the Saxon–Woods model it has the value $t = 4a \ln 3 \approx 2.28 - 2.55 \text{fm}$. The radius parameter R in this model has a very precise meaning. At $r = R$, the nuclear density is exactly 1/2 that of the central density.

All of the foregoing gives us a surprising simple picture of the nucleus. Regardless of the complications of their quantized nature, nuclei appear to be roughly spherical bodies with uniform density, a density that is truly enormous. The relation $R = r_o A^{1/3}$ implies that the volume occupied by a single nucleon is $\frac{4}{3}\pi r_o^3$, or about 8.18 fm³. Since the mass of a nucleon is about 1.66×10^{-27} kg, the mass corresponding to 1 cm³ of nuclear matter is about 2.0×10^{11} kg, compared to the densities of ordinary matter which are 0.001–0.020 kg cm⁻³! Also, because nuclei are essentially uniformly charged, there must be a rather significant stored energy due to the Coulomb repulsion of the protons. Classical electrostatics shows that the total stored Coulomb energy in a uniformly charged sphere is given by

$$E_C = \frac{3}{5} k_C \frac{q^2}{r} \quad (4.13)$$

where q is the total charge and r is the radius of the sphere. In a nucleus containing Z protons, the total charge is Ze and the total stored energy in this approximation is simply

$$E_C = \frac{3}{5} k_C \frac{Z^2 e^2}{R} \quad (4.14)$$

Two points should be made before we use this relation. The first has to do with physics. The classical expression for the stored energy is derived with the assumption that a differential of charge dq is brought up from infinity and spread uniformly through the volume of radius r. This is followed by a second differential of charge and so on. Such a procedure treats charge as a continuous quantity. We can get away with this in the normal macroscopic world because of the very small charge of an electron and because we usually deal with enormous numbers of them over macroscopic dimensions. But this is not the case when we deal with just a few protons in the nucleus. The fact is, we should account for the discrete nature of the proton charge.

The correction that we should provide to Eq. (4.14) is easily seen if we consider the following picture. A proton is brought up from infinity and placed in a spherical container where it is free to move about. On the average it has the same probability of being found in any differential volume element throughout the sphere. The average charge density is just q/V_{sphere} but it did not cost us any energy (work) to bring it up from infinity and place it in the sphere. A second proton is brought up from infinity and it, too, is allowed to move freely about so that it also leads to an average charge density of q/V_{sphere}. If we carry this out for Z protons, Eq. (4.14) becomes

$$E_C = \frac{3}{5} k_C \frac{Z(Z-1)e^2}{R} \tag{4.15}$$

The Coulomb energy for discrete charges uniformly distributed throughout a sphere is just the stored energy for the continuous-charge model minus the energy that would have been required classically to bring up Z units of electron charge and spread each of them, individually, over the volume. The latter is called the Coulomb self-energy of the Z charge units and we can consider it as a component of the rest mass of each of the Z protons.

The second point has to do with calculations and how to do them efficiently and rapidly. This is in reference to the fact that much of the older literature is based on so-called Gaussian units where the expression in Eq. (4.15) would be written

$$E_C = \frac{3}{5} \frac{Z(Z-1)e^2}{R} \tag{4.16}$$

In the Gaussian system of units, the quantity $4\pi\varepsilon_0$ is actually contained in the unit of charge.

Now the quantity $e^2/(m_e c^2)$ in Gaussian units, or the quantity $e^2/(4\pi\varepsilon_0 m_e c^2)$ in S.I. units arises in classical electrodynamics and is called the *classical electron radius*, r_e. It has the magnitude $r_e = 2.818$ fm. Because we have already memorized the energy equivalence of the electron rest mass, the Coulomb energy can be calculated easily from

$$E_C = \frac{3}{5} k_C \frac{Z(Z-1)e^2}{R} = \frac{3}{5} \frac{Z(Z-1)}{R} \left(\frac{e^2}{m_e c^2}\right) m_e c^2 \tag{4.17}$$

Whether you use Eq. (4.17), which we recommend, or the equivalent expression in Eq. (4.15), we can use the constant-density model to easily estimate the stored Coulomb energy of a nucleus, and the results for some representative nuclides are listed in Table 4.2. Several important points should be gleaned from the table. The mean values of nuclear radii vary over the relatively small range of (1–2) fm $\leq r \leq 8$ fm. The stored Coulomb energies, on the other hand, are quite large and vary over a much wider range because of the approximately squared dependence on atomic number. For $^{238}_{92}U$, the stored energy is about 933 MeV, roughly the energy-equivalent of one neutron mass! This clearly indicates that the stored Coulomb energy must be a significant factor in the determination of the properties of nuclei.

Table 4.2 The mean radii and stored Coulomb energy of several nuclides. The radius parameter was taken as $r_o = 1.25$ fm.

Nuclide	Radius (fm)	E_C (MeV)	E_C/A (MeV)
$^{20}_{10}$Ne	3.39	22.9	1.15
$^{40}_{20}$Ca	4.27	76.9	1.92
$^{136}_{54}$Xe	6.43	384	2.83
$^{238}_{92}$U	7.75	933	3.92

This is further emphasized by the values of the stored Coulomb energy per nucleon shown in the last column of the table. They range from about 1–4 MeV over the mass range of the nuclides considered. Now we know, from Fig. 4.1, that the average total binding energy per nucleon for nuclides found in nature with $A \geq 20$ is about 7.5–8.8 MeV. This means that somehow the attractive interaction of the nucleons has to increase in such a way that it overcomes the increasingly disruptive Coulomb potential and maintains rough constancy in the total average binding energy per nucleon. And the means by which this is accomplished is clearly indicated by the composition of the nuclides found in nature as shown in Fig. 4.6.

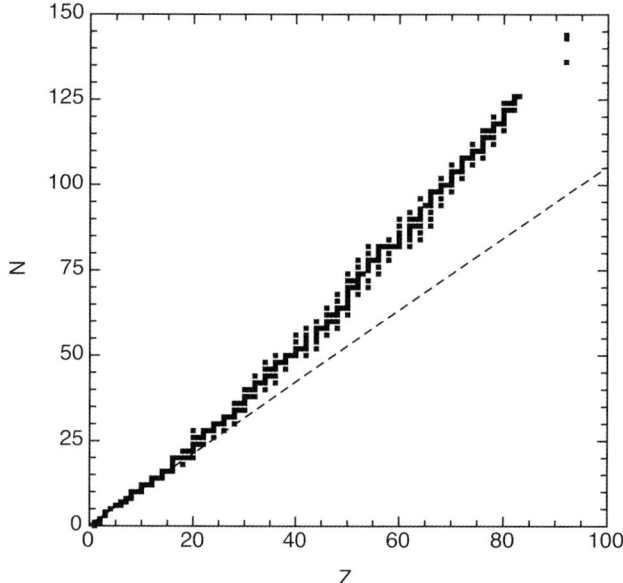

Fig. 4.6 The neutron and atomic numbers for the nuclides found in nature in significant abundance. The line shown in the graph represents N = Z.

For nuclides with Z less than about 20, N = Z. Equal numbers of neutrons and protons lead to the most stable nuclides found in nature. But with increasing Z, there is a continuous and smooth increase in the ratio of N/Z, reaching a value of 1.59 at $^{238}_{92}U$. Now for any mass number A, nuclides with smaller N/Z than found in nature are known and they are all unstable with respect to decay by electron capture and/or β^+ emission. Such decays entail the conversion of a proton into a neutron with a subsequent increase in N/Z, and therefore result in a decrease in the stored Coulomb energy. We can conclude that the N/Z ratio for nuclides found in nature are defined in large measure by the need to overcome the disruptive effect of the Coulomb force and this is accomplished by the addition of neutrons that add only a net attractive potential to the nucleus. The stability of nuclides is strongly affected by the stored Coulomb energy and the effects of this will be seen over and over as we proceed to discuss radioactive decay in more detail in later chapters.

After considering the foregoing, the reader can reasonably ask why it is that we do not see nuclei composed only of neutrons. Researchers have, in fact, sought evidence for the existence of the di-neutron and other combinations of a few neutrons. All evidence indicates that the spin dependence of the nuclear force causes the binding of neutrons, in the absence of protons, to be negative.

4.4
Quantization of the Nucleus: Pairing of Identical Nucleons

Because the nuclides found in nature must be especially stable with respect to radioactive decay, it is worthwhile to look further for clues concerning their stability. One of the more remarkable findings is that the nuclides in nature tend to have a strong preference for having even numbers of both protons and neutrons. This is seen in Table 4.3. Of the 281 nuclides found in nature in appreciable abundance and which are stable against β decay[2], only 4, or about 1.4%, have *neither* an even atomic or neutron number. Over 60% have both atomic and neutron numbers that are even and there are about the same number that have Z = even, N = odd as there are with Z = odd, N = even. The strong preference for even numbers of *identical* nucleons in the most stable nuclides must indicate that there is some *extra stability* associated with pairs of identical particles. The fact that there are about the same number of nuclei with Z or N even suggests that this extra stability is roughly the same for proton and neutron pairs. Some part of the average binding energy of nucleons must be associated with this pairing, and it referred to as the *pairing energy*.

The pairing energy has no classical analog. It arises from the quantized nature of the nucleus and the short-range character of the strong interaction. Pairing is also known to take place with electrons in atoms. For example, the lowest-energy state of the hydrogen atom is the 1s state that can be occupied by two electrons, one

---

2) A number of nuclides that exist in nature are known to be unstable with respect to β decay. They have half-lives that are comparable or longer than the age of the earth. These have been excluded from Table 4.3.

Table 4.3 Characteristics of nuclides found in nature

	Z = even	Z = odd
N = even	171	50
N = odd	56	4

with intrinsic spin up and the other down. In the helium atom, the lowest-energy state actually has the two electrons *paired* in its 1s orbital. But it costs some energy to do this because of the Coulomb repulsion between the electrons. The binding energy of helium is somewhat smaller than it would be if you could somehow turn off the repulsion between the two electrons. In heavier atoms, where orbits can be occupied by more than two electrons, one finds that the lowest-energy states for two electrons in such orbits do not have the electrons paired. Clearly, the pairing energy associated with the nuclear force is quite different. We can speculate that the nucleus must also possess discrete quantized states of which some can be occupied by more than two nucleons. Since additional stability is seen only for identical pairs, we are led to conclude that the "force" between two identical nucleons in the same state is attractive and that there must be a sequence of states for the protons and a separate sequence of states for the neutrons. All of this is consistent with the fact that neutrons and protons are fermions that obey the Pauli Exclusion Principle. As for electrons, each nucleon has an intrinsic angular momentum and no two identical particles in the nucleus can have the same set of quantum numbers.

The demonstration of the existence of a pairing energy and an estimate of its magnitude can be obtained by considering the binding energy of adjacent (even, even) and (even,odd) or (odd,even) nuclei. For this purpose, consider the very simple schematic diagram in Fig. 4.7. The figure is meant to represent an idealization of the potential wells seen by neutrons and protons in the nucleus. We show them separately because of the clear indication that neutron and protons fill levels separately. The shaded areas represent completely filled levels, which, as we will see

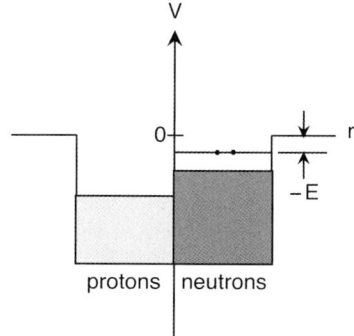

Fig. 4.7 Schematic diagram of the potential wells seen by neutrons and protons in the nucleus. The shaded areas represent levels that are completely filled with nucleons. Two neutrons occupy the highest-lying neutron level shown.

later, will always contain an even number of identical particles. The nucleus in the schematic is an (even, even) nucleus with two neutrons occupying the level with the most weakly bound neutrons, which are also the most weakly-bound nucleons in the system. The total energy of a neutron in this state, in the *absence* of the pairing energy, is –E, and this is the energy that would have to be added to remove one of these neutrons from the nucleus. However, in the presence of a pairing energy, we will have to provide the energy E plus the energy to break the binding of the neutron pair to remove a single neutron. We will symbolize this pairing energy by Δ.

The schematic suggests a simple means to demonstrate the existence of a pairing energy and to estimate its magnitude. Suppose we choose an (even, even) nucleus of mass number A. Its atomic mass is M(Z,A) and that of the isotope with one less neutron is M(Z,A – 1). The binding energy of the most weakly bound neutron in M(Z,A) is then given by

$$(B.E.)_n = [M(Z, A-1) + m_n - M(Z, A)]c^2 \qquad (4.18)$$

If we compare this to the binding energy of the last neutron in the nucleus M(Z,A–1) that now contains only a single unpaired neutron in the same level from which the first neutron was removed, we should find that the latter is smaller and the difference in binding energies would be an estimate of the pairing energy.

In Fig. 4.8 we show the binding energies of the most weakly bound neutrons in isotopes of Sn (Z = 50) and the binding energies of the most weakly bound protons in the isotones of N = 50. The data clearly exhibit an odd/even effect in the binding energies for both particles. The binding energy for a neutron (proton) in an even N (even Z) nuclide is always greater than the binding energy of a neutron (proton) in the adjacent isotone (isotope) containing one less neutron (proton). If we take the difference between the two we get an estimate of the pairing energy. These differences are shown as closed squares in the diagrams. For both nucleon types, the data show that the pairing energy is roughly constant and independent of mass number.

The mean value of the difference for the neutrons is 2.41 ± 0.08 MeV, and for the protons it is 1.94 ± 0.14 MeV.

One should not take the data in Fig. 4.8 as representative of pairing energies throughout the chart of the nuclides. But there can be no doubt that pairing energy exists and is typically on the order of 2 MeV. This is quite large and amounts to roughly 25% of the average binding energy of a nucleon in all but the lightest nuclides. The pairing effect thus also plays a large role in defining the masses of nuclei, and it plays a very large role in defining the low-energy nuclear structure of nuclei as well. It will therefore have a significant effect on radioactive decay and nuclear reactions.

The pairing effect represents the first clear quantum effect in nuclear binding that modifies the simple classical picture we have developed so far. There are many others. While we will discuss some of these later, it will prove useful for the present purposes to get a rough idea of the nature of the quantum states of nuclei with a very crude but useful simple model, the model of a particle in a quantized box. This

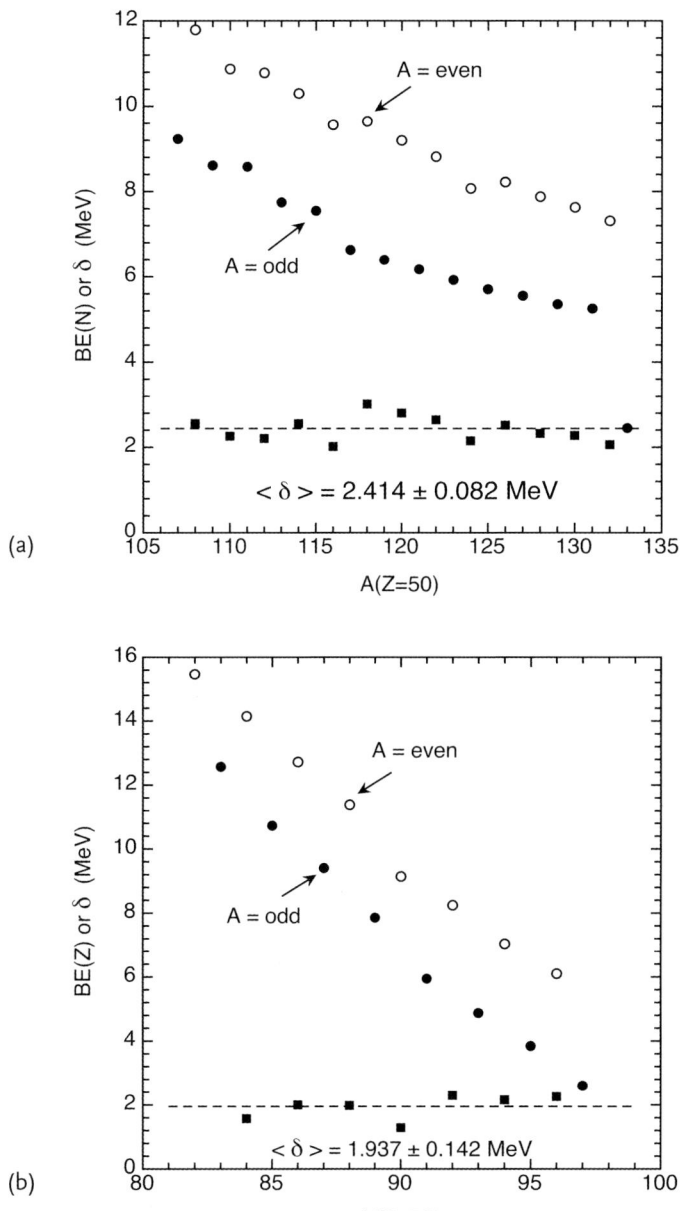

Fig. 4.8 The binding energy of the most weakly bound neutron in isotopes of Sn (a) and the binding energy of the most weakly bound proton in the isotones of N = 50 (b). The data points shown as closed squares represent the estimate of the pairing energy.

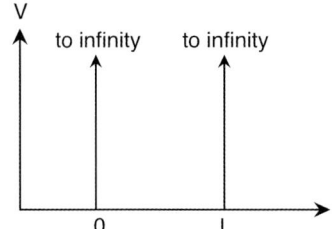

Fig. 4.9 Schematic diagram of a one-dimensional "box" with walls that are defined by infinitely large potentials.

is probably the simplest problem that can be studied in quantum mechanics but it points out some of the most essential physics that one must keep in mind when dealing with real quantized systems.

The problem we want to consider is shown schematically in Fig. 4.9. The potential diagram in the figure can be described as

$$V(x) = \begin{cases} 0, & x > 0, x < L \\ \infty, & x \leq 0, x \geq L \end{cases} \tag{4.19}$$

If a particle is placed in the region $0 < x < L$, it will be free to move in field-free space but it will not be able to move into any other region of space. It is completely confined in the "well". Thus, its wave function must vanish everywhere outside of the well. In the limit of nonrelativistic quantum mechanics, the states available to the particle are described by the wave functions ψ that are solutions to the one-dimensional Schrödinger equation

$$-\frac{\hbar^2}{2m}\nabla^2\psi + V\psi = E\psi \tag{4.20}$$

where $\hbar$ is Planck's constant divided by 2π, m is the mass of the particle, E is the total energy of the particle taken as the sum of its kinetic plus potential energies, and in one dimension, $\nabla^2 = \frac{d^2}{dx^2}$. Substituting the latter into Eq. (4.20) and rearranging gives

$$\frac{d^2\psi}{dx^2} + \frac{2m}{\hbar^2}(E-V)\psi = 0 \tag{4.21}$$

If the particle is in the box, $V = 0$ and its total energy is just the kinetic energy $E = p^2/(2m)$. With these substitutions, Eq. (4.21) becomes

$$\frac{d^2\psi}{dx^2} + \frac{p^2}{\hbar^2}\psi = 0 \tag{4.22}$$

The general solution to this equation can be written as the sum of sine and cosine terms in the general form

$$\psi = A \sin kx + B \cos kx \qquad (4.23)$$

Because the potential walls are infinitely high, the particle cannot have any probability of escaping the well and thus the boundary conditions that we can specify are that $\psi = 0$ at $x = 0$ and at $x = L$. Using the first of these, we must have $B = 0$ and thus

$$\psi = A \sin kx \qquad (4.24)$$

Using the second of these we must have $\sin kL = 0$ for any A. This condition is satisfied if

$$kL = n\pi, \quad n = 0, 1, 2, \ldots \qquad (4.25)$$

which then gives $k = n\pi/L$, $n = 0, 1, 2, \ldots$. Now we can substitute Eq. (4.24) back into Eq. (4.22) to obtain the relation between k and the linear momentum p as

$$p = \hbar k = \frac{n\pi\hbar}{L} \qquad (4.26)$$

The total energy E is therefore found to be

$$E = \frac{p^2}{2m} = \frac{n^2 \pi^2 \hbar^2}{2mL^2} \qquad (4.27)$$

Now a word of caution is in order. From the mathematics of the boundary condition at $x = L$, we found that $n = 0, 1, 2, \ldots$. But suppose that $n = 0$. Then from Eq. (4.25), $k = 0$ and from Eq. (4.24), $\psi = 0$. The wave function vanishes! And if the wave function vanishes, there can be no particle state. Therefore, the physical requirement that the particle exists leads to the final result

$$E = \frac{p^2}{2m} = \frac{n^2 \pi^2 \hbar^2}{2mL^2}, \quad n = 1, 2, 3, \ldots \qquad (4.28)$$

The conclusion reached from this physical requirement is that a particle of finite mass confined to a finite region of space cannot have zero kinetic energy. The allowed energy states for a particle in a one-dimensional box are quantized in multiples of n^2 where n is an integer. The energies of states are all multiples of the energy of the lowest-energy state $E = \pi^2 \hbar^2 / 2mL^2$. The energies decrease as the length of the box increases and as the mass of particle increases. These observations have very important qualitative consequences that apply regardless of the nature of the real potential to which a particle finds itself subject, and regardless of the fact that real potential wells have finite depths. Namely, the separation in energies of adjacent states will be small if the dimension of the well is large and the particle mass is large, but the energy separations will be large if the well dimension is small and the particle mass is small.

The relevance of all of this to the problem of nucleons in the nucleus and the pairing energy will be seen once we extend the model to three dimensions. This is actually straightforward and we will now outline how it is accomplished. A three-dimensional box that is the analog of the problem discussed above will have a potential of the form

$$V(x) = \begin{cases} 0, & x > 0, x < L \\ \infty, & x \leq 0, x \geq L \end{cases}$$

$$V(y) = \begin{cases} 0, & y > 0, y < L \\ \infty, & y \leq 0, y \geq L \end{cases} \quad (4.29)$$

$$V(z) = \begin{cases} 0, & z > 0, z < L \\ \infty, & z \leq 0, z \geq L \end{cases}$$

This is truly a cubical box where the potential everywhere outside of and on the faces of the box is infinite. Once in the box, a particle will be trapped there forever.

For a three-dimensional problem in Cartesian coordinates, ∇^2 becomes

$$\nabla^2 = \frac{\partial^2}{\partial x^2} + \frac{\partial^2}{\partial y^2} + \frac{\partial^2}{\partial z^2}$$

If we search for solutions of the Schrödinger equation of the form $\psi(x, y, z) = X(x)Y(y)Z(z)$, we can substitute this and the expression for ∇^2 into Eq. (4.20) and with a little rearrangement get

$$YZ\frac{d^2X}{dx^2} + XZ\frac{d^2Y}{dy^2} + XY\frac{d^2Z}{dz^2} + \frac{p_x^2 + p_y^2 + p_z^2}{\hbar^2}XYZ = 0 \quad (4.30)$$

In this equation, we have expanded the momentum explicitly in terms of its three Cartesian components. If we now multiply the equation by ψ^{-1}, we can collect terms and write the result as

$$\left(\frac{1}{X}\frac{d^2X}{dx^2} + \frac{p_x^2}{\hbar^2}\right) + \left(\frac{1}{Y}\frac{d^2Y}{dy^2} + \frac{p_y^2}{\hbar^2}\right) + \left(\frac{1}{Z}\frac{d^2Z}{dz^2} + \frac{p_z^2}{\hbar^2}\right) = 0 \quad (4.31)$$

Eq. (4.31) contains three parts, each of which depends only on one independent variable and each of which is identical in form. In order for the equality to be satisfied for *any* choice of (x,y,z), each of the three parts must vanish individually. Therefore,

4.4 Quantization of the Nucleus: Pairing of Identical Nucleons

$$\left(\frac{1}{X}\frac{d^2X}{dx^2} + \frac{p_x^2}{\hbar^2}\right) = 0$$

$$\left(\frac{1}{Y}\frac{d^2Y}{dy^2} + \frac{p_y^2}{\hbar^2}\right) = 0 \qquad (4.32)$$

$$\left(\frac{1}{Z}\frac{d^2Z}{dz^2} + \frac{p_z^2}{\hbar^2}\right) = 0$$

Each of the three equations above is identical in form to Eq. (4.22). The process of *separation of variables* has reduced the three-dimensional problem to three one-dimensional problems all of the same form as that which we have already solved above. Each of the solutions will then be of the form $\psi = A\sin kq$ where $q = x, y$ or z, and each will therefore have its momentum component given by $p_q = \hbar k_q = n_q \pi \hbar / L$ subject to the same integer quantization of $n_q = 0, 1, 2....$. The total linear momentum will then be given by

$$p^2 = p_x^2 + p_y^2 + p_z^2 = \frac{\pi^2\hbar^2}{L^2}(n_x^2 + n_y^2 + n_z^2) = \frac{\pi^2\hbar^2}{L^2}n^2 \qquad (4.33)$$

and the total energy of the allowed states is given by

$$E = \frac{p^2}{2m} = \frac{n^2\pi^2\hbar^2}{2mL^2}, \quad n^2 = (n_x^2 + n_y^2 + n_z^2), \quad n_i = 1, 2, 3, ... \qquad (4.34)$$

In this case, each of the quantum numbers must satisfy $n_i \geq 1$ if the wave function is not to vanish. Because of the independence of the three components, the three-dimensional particle in the box will exhibit a greater density of states (states per unit energy) as compared to the one-dimensional problem. Each set of (n_x, n_y, n_z) corresponds to a specific state of the system. Further, some states will exhibit *degeneracy in energy* in the sense that each state with the same n but different (n_x, n_y, n_z) will have identical energies. And, as we pointed out before, each state of (n_x, n_y, n_z) is also two-fold degenerate because it can be occupied by two fermions with different orientation of their intrinsic spins.

Now the three-dimensional particle in the box can serve as a simple crude model for a nucleus. We can, for example, use it to get an idea of the magnitudes of the excitation energies of states and the difference in energy between adjacent states. We can reasonably take the nuclear diameter as an estimate of L and we can use the mass of a neutron as representative of the mass of a nucleon. With these, we can write the energy of states as

$$E = \frac{n^2\pi^2(\hbar c)^2}{2m_n c^2(2r_n)^2} \qquad (4.35)$$

where r_n is the nuclear radius. If we take A = 100, evaluation of the parameters in Eq. (4.35) gives E = 1.52 n^2 MeV. The lowest-energy state of a particle has n^2 = 3 and one of the next levels will have, for example, (n_x, n_y, n_z) = (2,1,1) with n^2 = 6. Thus, the difference in energy between the lowest-lying states in the well will be about ΔE = 4.5 MeV. While we cannot take this number too seriously, we can conclude that the energy separation of states in a medium-massed nucleus should be on the order of MeV.

As a check on the quality of the order of magnitude of the energies given by this simple model, we can calculate the energy separation of energy states in the atom. If we take r_{atom} = 2 × 10^{-8} cm and use the rest mass of the electron, we find an energy separation of the two lowest-energy states in an atom of ΔE = 7.0 eV. While again one cannot take the absolute value too seriously, a few eV is indeed the right order of energy separation.

To be sure, nucleons in a nucleus cannot really be described as particles in a box and the potential well in which the nucleons exist is not infinitely deep. Nonetheless, there are sufficient similarities between the nucleus and this simple model that we can use it to get rough estimates of the energies of particle states. While we will not prove it, it can be shown that, so long as the potential well is deep enough and we limit attention to states that are not near the top of the well, the energies of states are about the same as those found in an infinitely deep well.

4.5
Quantization of the Nucleus: Asymmetry Energy

We know that the average binding energy of a nucleon in all but the lightest nuclei is about 7–8 MeV. Because the energy states must be separated by the order of MeV, the nuclear potential well would likely have a depth of some tens of MeV. This turns out to be about right as we will see later. For the present, we can use the results of the particle in a box model to understand some other general properties of the binding energies of neutrons and protons and how these affect the overall mass of nuclei. For this purpose, the schematic diagram shown in Fig. 4.7 is reproduced in Fig. 4.10 with the addition of discrete levels for both neutrons and protons. The figure represents a schematic of proton and neutron levels in a finite potential well of constant depth $-V_o$ for a neutron-rich nucleus. For simplicity, we have neglected the Coulomb potential and therefore show energy levels that are about the same for both particles. In the simple particle in a box problem in Cartesian coordinates, each level can be occupied by only two fermions. In real nuclei, however, the majority of energy levels can accommodate more than one pair of identical nucleons. This fact and the neglect of the Coulomb potential will not affect the principal argument on which we want to focus. The shaded areas in the figure represent levels that are completely filled. If the nucleus has N > Z the most weakly bound neutrons will reside in levels closer to the top of the well than the levels in which the most weakly bound protons reside. This means that the average binding energy of the neutrons will be *less* than the average binding energy of the protons, a fact

4.5 Quantization of the Nucleus: Asymmetry Energy

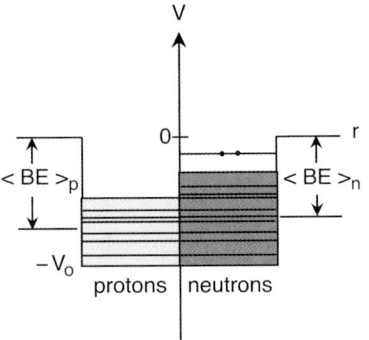

Fig. 4.10 Schematic diagram of proton and neutron levels in a potential well of constant depth $-V_0$. The average binding energies of the protons and neutrons are $<BE>_p$ and $<BE>_n$, respectively.

that we have not yet considered. If the nucleus had $Z > N$, the average binding energy of the protons will be less than for the neutrons. Now the total binding energy of the nucleus will be given by

$$(B.E.)_{Z,A} = \sum_{\text{all protons}} (B.E.)_p + \sum_{\text{all neutrons}} (B.E.)_n$$

and will clearly be dependent upon the number of nucleons of each type present as well as the actual potential in which the particles reside.

We can use the particle in a box approximation to understand, in a general way, how the total binding energy will vary with different ratios of N/Z. We will assume that the particles in our box are *non-interacting* particles. This means that the energies of the particle *states* are the same regardless of the number of particles present. Further, we will assume that all particles reside in the lowest possible energy states consistent with the Pauli Exclusion Principle. Such a system is referred to as a *degenerate Fermi gas*. The total energy of a particle confined to the box is just its kinetic energy as given by Eq. (4.35). Then, because of the finite number of particles that can reside in the quantized level, the average kinetic energy of the neutrons in a nucleus with $N > Z$ must be greater than the average kinetic energy of the protons and *vice versa*. If we can determine how the total kinetic energies of particles in the box vary with the ratio of N/Z, we will have a rough idea of how nuclear binding energies depend on this ratio.

To perform this calculation, we have to count the number of states populated by each nucleon type and sum the kinetic energies of the particles in them. An easy way to do this is by reference to the diagram in Fig. 4.11. Each of the allowed energy states for a particle in a box has a specific set of quantum numbers (n_x, n_y, n_z) that define the magnitudes of the components of linear momentum. Each such state can be occupied by two identical fermions. The actual states can be displayed in the three-dimensional space with orthogonal components (n_x, n_y, n_z) as shown in the figure. Because $n_i \geq 1$, all states must be located in the one octant of the (n_x, n_y, n_z) space shown in the figure. Each state will be represented by a point at the corner of a cube and each cube will have sides of unit dimension.

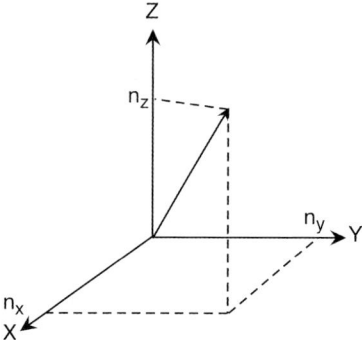

Fig. 4.11 The states for the particle in a three-dimensional box in the space of (n_x, n_y, n_z). Each state can be occupied by two identical fermions.

Equally, each state can be located by a vector of length $n = (n_x^2 + n_y^2 + n_z^2)^{1/2}$ from the origin to the point (n_x, n_y, n_z). For simplicity, we are going to treat n as a continuous variable, an approximation that can be considered valid in the limit of very large nuclei.

If the nucleus has all of the nucleons of a given type in the lowest energy states available, there will be some maximum value of n for the occupied states which we denote by n_{max} and which is given by $n_{max} = (n_x^2 + n_y^2 + n_z^2)^{1/2}_{max}$ for one particle type. In the continuous approximation, the volume of the occupied octant of the sphere in (n_x, n_y, n_z) space is just $\frac{1}{8}\left(\frac{4}{3}\pi n_{max}^3\right)$. Because each state is occupied by two particles, and the volume per state is constant, the total number of particles with $n \leq n_{max}$ is just

$$N = \frac{1}{4}\left(\frac{4}{3}\pi n_{max}^3\right) = \frac{\pi}{3}\left(\frac{p_{max}L}{\pi\hbar}\right)^3 \tag{4.36}$$

where p_{max} is the momentum corresponding to n_{max} and we have used the italic N to prevent confusion with the neutron number. In the model, L^3 is just the volume of the box. For a nucleus we can replace L^3 by the constant density approximation $V_{nucl} = \frac{4}{3}\pi R^3 = \frac{4}{3}\pi r_0^3 A$ and obtain the result

$$N = \frac{4 p_{max}^3}{9 \pi\hbar^3} r_0^3 A \tag{4.37}$$

Now this relation should be applied separately to the neutrons and protons that are assumed to fill separate level schemes. Therefore we can write

$$Z = \frac{4 p_{p,max}^3}{9 \pi\hbar^3} r_0^3 A$$
$$N = \frac{4 p_{n,max}^3}{9 \pi\hbar^3} r_0^3 A \tag{4.38}$$

4.5 Quantization of the Nucleus: Asymmetry Energy

where the subscripts "p" and "n" on the momenta indicate proton and neutron, respectively. Rearranging the equations gives the maximum momenta of the protons and neutrons as

$$p_{p,\max} = \left(\frac{9\pi\hbar^3 Z}{4r_o^3 A}\right)^{1/3}$$
$$p_{n,\max} = \left(\frac{9\pi\hbar^3 N}{4r_o^3 A}\right)^{1/3} \qquad (4.39)$$

The expressions in Eq. (4.39) show immediately that the maximum momenta of protons and neutrons will vary as $(Z/A)^{1/3}$ and $(N/A)^{1/3}$, respectively. Hence, the maximum kinetic energies will vary as $(Z/A)^{2/3}$ and $(N/A)^{2/3}$, respectively.

The total kinetic energies of the particles can be obtained by simple summing over the energies of all occupied states. Within our continuous approximation, we can take Eq. (4.37) to specify the general relation between the momentum of particles and the number of particles in the system. Differentiating Eq. (4.37) then gives

$$dN = \frac{4}{3}\left(\frac{r_o^3 A}{\pi\hbar^3}\right) p^2 dp$$

as the number of particles with momentum between p and $p + dp$. Using this we can write the summation for the kinetic energy of protons as

$$T_p = \sum_{n=1}^{n_{\max}} \frac{p_{p,n}^2}{2m_p} \approx \frac{1}{2m_p}\int_0^{n_{\max}} p_{p,n}^2 \, dN = \frac{1}{2m_p}\int_0^{p_{p,\max}} \frac{4r_o^3 A}{3\pi\hbar^3} p^4 dp \qquad (4.40)$$
$$= \frac{3}{10}\left(\frac{9\pi}{4}\right)^{2/3}\left(\frac{\hbar^2}{m_p r_o^2}\right) Z\left(\frac{Z}{A}\right)^{2/3}$$

Similarly, the total kinetic energy of the neutrons is given by

$$T_n = \frac{3}{10}\left(\frac{9\pi}{4}\right)^{2/3}\left(\frac{\hbar^2}{m_n r_o^2}\right) N\left(\frac{N}{A}\right)^{2/3} \qquad (4.41)$$
$$= \frac{3}{10}\left(\frac{9\pi}{4}\right)^{2/3}\left(\frac{\hbar^2}{m_n r_o^2}\right)(A-Z)\left(\frac{A-Z}{A}\right)^{2/3}$$

Note that the two expressions for total kinetic energy contain the same ratios of constants except for the different masses of the two particles. Numerically the latter are so close to one another that we will not incur significant error if we simply replace them by, say, the average of the two masses $m = (m_p + m_n)/2$. With this the summation of the total kinetic energy easily proceeds to give

$$T = T_p + T_n = C\left[Z\left(\frac{Z}{A}\right)^{2/3} + (A-Z)\left(\frac{A-Z}{A}\right)^{2/3}\right] \tag{4.42}$$

where $C = \frac{3}{10}\left(\frac{9\pi}{4}\right)^{2/3}\left(\frac{\hbar^2}{mr_0^2}\right) \approx 32.6\,\text{MeV}$.

Now we are almost home because Eq. (4.42) gives us nearly the expression in which we are interested. It expresses the total kinetic energy of the nucleus in the limit of a very simple quantized model. It demonstrates *approximately* how the total kinetic energy and hence the total nuclear binding energy depend upon N and Z. Although it can be used directly, it will prove useful to extend the analysis a bit farther.

We know that the most stable light nuclei have N = Z, and that as A increases, the neutron-to-proton ratios also increase to overcome the disruptive Coulomb potential. It then should prove useful to use the case of N = Z as a reference and to consider how the kinetic energy depends upon departures from this case. For this purpose, we will define the *neutron excess*, Δ, as $\Delta = (N - Z) = (A - 2Z)$. We now use this in Eq. (4.42) to obtain

$$\begin{aligned}T &= C\left[\left(\frac{A-\Delta}{2}\right)\left(\frac{A-\Delta}{2A}\right)^{2/3} + \left(\frac{A+\Delta}{2}\right)\left(\frac{A+\Delta}{2A}\right)^{2/3}\right] \\ &= \frac{C}{2^{5/3}}\left[(A-\Delta)\left(\frac{A-\Delta}{A}\right)^{2/3} + (A+\Delta)\left(\frac{A+\Delta}{A}\right)^{2/3}\right]\end{aligned} \tag{4.43}$$

We also know that for the nuclides found in nature, $N/Z \leq 1.6$, and therefore $\Delta/A \leq 0.23$. Because it is relatively small, we can get a reasonable approximation for the terms $\left(\frac{A-\Delta}{A}\right)^{2/3}$ and $\left(\frac{A+\Delta}{A}\right)^{2/3}$ by expanding them in a binomial expansion and retaining just the first few terms of each. The binomial expansion for $\left(1+\frac{\Delta}{A}\right)^{2/3}$ is, for example,

$$\left(1+\frac{\Delta}{A}\right)^{2/3} = 1 + \frac{2}{3}\left(\frac{\Delta}{A}\right) - \frac{1}{9}\left(\frac{\Delta}{A}\right)^2 + \ldots \tag{4.44}$$

to second order in (Δ/A). If the same type of expansion is accomplished for $\left(1-\frac{\Delta}{A}\right)^{2/3}$ and both are substituted into Eq. (4.43) we then obtain

$$\begin{aligned}T &= \frac{C}{2^{5/3}}\left[A - \Delta - \frac{2}{3}\left(\Delta - \frac{\Delta^2}{A}\right) - \frac{1}{9}\left(\frac{\Delta^2}{A} - \frac{\Delta^3}{A^2}\right) + \ldots \right. \\ &\quad \left. + A + \Delta + \frac{2}{3}\left(\Delta + \frac{\Delta^2}{A}\right) - \frac{1}{9}\left(\frac{\Delta^2}{A} + \frac{\Delta^3}{A^2}\right) + \ldots\right] \\ &= \frac{C}{2^{5/3}}\left[2A + \frac{10}{9}\frac{\Delta^2}{A} + \ldots\right] \\ &= \frac{C}{2^{2/3}}\left[A + \frac{5}{9}\frac{(A-2Z)^2}{A} + \ldots\right]\end{aligned} \tag{4.45}$$

The last of Eqs (4.45) is the expression we want. The total kinetic energy of nucleons in a nucleus in the degenerate Fermi gas approximation is given by the product of a constant times the mass number A, plus additional correction terms, the first of which adds an energy proportional to the square of the neutron excess divided by the mass number. This clearly shows how level quantization affects the total kinetic energy of the particles. If N = Z, the average kinetic energy of the neutrons and protons are the same. However, if N > Z the excess neutrons contribute additional kinetic energy which then reduces the average binding energy relative to the case that N = Z. Note that the result in Eq. (4.45) is *symmetric* with respect to N and Z. The total kinetic energy of the nucleus with $\Delta > 0$ (N > Z) is identical to the result obtained with $\Delta < 0$ (N < Z) for the same absolute value of Δ. Of course this applies only in the absence of the Coulomb interaction.

The implications of this analysis to the binding energy of real nuclei should be clear. In the absence of the Coulomb potential, neutron and proton levels will have very nearly the same energies. The quantization of the levels then says that the binding energy will be *lower* than expected for N = Z by an amount roughly proportional to $(N - Z)^2$. This energy has traditionally been referred to as the *asymmetry energy*. It has a marked effect on nuclear binding energies and hence on radioactive decay. It represents a very significant factor affecting nuclear stability.

We can use the expression for the kinetic energy in Eq. (4.45) for one other very good purpose. If N = Z, the total kinetic energy in our simple model is just a constant times the mass number. The magnitude of the constant is about 20.5 MeV and must represent the kinetic energy of a single nucleon in the context of the model. Now we already know, from the data shown in Fig. 4.1, that the average binding energy of a nucleon in some medium-massed nuclides not far from the nuclides found in nature, is on the order of 7–8 MeV. Because E = T + V, we can infer that the nuclear potential well in real nuclei will have a depth on the order of at least 30 MeV or so. Many lines of evidence suggest that the average nuclear potential well depth experienced by a nucleon is actually 40–50 MeV.

References

1 See, for example, H. Goldstein, C.P. Poole, Jr., and J.L. Safko, "Classical Mechanics", Benjamin Cummings (2002).
2 G. Fricke, C. Bernhardt, K. Heilig, L.A. Schaller, L. Schellenberg, E.B. Shera and C.W. De Jager, Nuclear ground state charge radii from electromagnetic interactions, Atomic Data and Nuclear Data Tables 60 (1995) 177.
3 J.W. Negele, Structure of finite nuclei in the local-density approximation, Phys. Rev. C1 (1970) 1260.

General References

The strong interaction, being the result of the net interactions between the quarks that make up the nucleons, is very complicated indeed. There are no simple discussions of the strong interaction because it is so complicated. But a nice and fairly approachable discussion of the salient properties can be found in Kenneth S. Krane "Introductory Nuclear Physics", John Wiley & Sons, New York (1987).

A simple and readable discussion of some properties of the nuclear potential can also be found in R.D. Evans. "The Atomic Nucleus", Krieger Publishing Company; Reprint edition (June 1982).

A more advanced and detailed discussion of the Fermi gas model can be found in Amos. deShalit and Herman Feshbach "Theoretical Nuclear Physics", Volume I: Nuclear Structure, John Wiley & Sons, New York (1974).

Problems

1. Prove that the Coulomb energy of a uniformly-charged sphere is given by

$$E_c = \frac{3}{5} \frac{q^2}{4\pi\varepsilon_0 r}$$

2. Taking the radius parameter as $r_o = 1.25$ fm, calculate the Coulomb energy released in the symmetric fission of ^{238}U.

3. If a wave function is indeed the description of a particular physical system, then the quantity $|\psi|^2$ measures the probability per unit dimension for finding the system within a dimension range, say q to q + dq. If the wave function is correct, and the system exists, the total probability of finding the system anywhere must be unity. That is,

$$\int_{\text{all } q} |\psi|^2 dq = 1$$

If this is satisfied, we say that the wave function is normalized.
(a) The wave function for the particle in a one-dimensional box is given by Eq. (4.24) with the restriction for k given by Eq. 4.(25) and the text following it. Check to see if Eq. (4.24) is normalized. If not, normalize it.
(b) Calculate and plot the probability density $|\psi|^2$ as a function of x in the box when n = 1 and n = 5. Make sure that you calculate and plot sufficient points to see the oscillations that are implied.
(c) Compare, qualitatively, the probability distributions from part (b) with what you expect from simple, classical mechanics.

4. Assume that you have non-interacting particles of two *different* types confined to the same one-dimensional well with infinite potential walls. There are eight particles of type 1 and four particles of type 2. Both types are fermions and both

therefore obey the Pauli Exclusion Principle independently. If the dimension of the well is 10.0 fm and if the masses of the particles are the same and equal to 930 MeV c^{-2}:

(a) Calculate the ratio of the average kinetic energy of the type-1 particles to the average for the type-2 particles. Remember that the momentum of the particles is quantized and for each momentum state there can be two identical fermions (one with "spin up" and one with "spin down").

(b) While the wave functions for a particle in a finite potential well are not the same as those in a well with infinite walls, the differences in the energies of the allowed states are fairly small if the finite well is fairly deep and we consider only the lowest-energy states in the well, those corresponding to the states with the most tightly bound particles.

With this in mind, use the results from part (a) to estimate the average binding energy of particles of type 1 and type 2 and all particles in a finite well of the form

$$V_o(r) = \begin{cases} -55 \text{ MeV}, 0 \leq x \leq L \\ 0 \text{ MeV}, x < 0, x > L \end{cases}$$

5. The figure below is a schematic representing particles of one type in a finite potential well with a depth of 18 MeV, but the levels have energies that are essentially those of the particle in a box problem with infinite walls. The level energies relative to the bottom of the well are given by $E = n^2$ MeV.

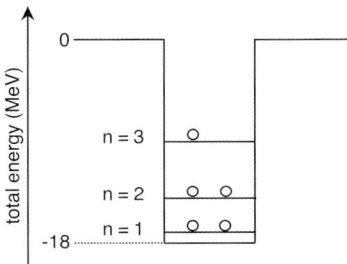

The system is bombarded with monoenergetic photons with an energy of 15 MeV. Under these conditions, it is possible for a photon to be completely absorbed by a particle and, if the energy is sufficient, it can be ejected from the well.

(a) Assuming the particles do not interact with one another, calculate the expected energies of ejected particles and sketch the expected spectrum, i.e., the relative number of emitted particles versus their energy. You may neglect the recoil energy of the system and thus the full energy of the absorbed photon is given by the sum of the particle binding energy and its final kinetic energy.

(b) Now assume that the particles interact as nucleons and thus have a pairing energy of 1.5 MeV. Repeat part (a) using this assumption.

5
The Semi-Empirical Mass Formula and Applications to Radioactive Decay

5.1
Introduction

The discussion of the properties of nuclei found in nature and the results of experiments probing the charge and mass distributions of heavy nuclei have provided a relatively simple picture of the average, gross properties of nuclei. Notwithstanding the complexity of the strong interaction, nuclei appear to be uniformly charged spheres of roughly constant density, with average nucleon binding energies that are remarkably constant for all but the lightest nuclides. The quantized structure of the nucleus is reflected in two gross effects on binding energies and hence nuclear masses. The discrete energy states of a system of identical fermions can be occupied by only a finite number of identical particles and there are independent systems of states for the neutrons and protons. Neglecting the Coulomb potential, the states for neutrons and protons have very nearly the same energies. Nuclei with $N = Z$ will then have about the same average binding energy for both types of particles, but if $N \neq Z$, the binding energies will depend upon $(A-2Z)^2/A$ to a first approximation. A pairing effect is observed between identical particles that leads to an increase of 1–2 MeV in the binding energy of a pair as compared to the binding energy of two unpaired nucleons in the same state.

These observations, and the more detailed discussions of Chapter IV, suggest a rather simple approach that can be used to understand the behavior of nuclear binding energies and masses as a function of A and Z, and thus an understanding of the gross systematics of the energetics of radioactive decay processes. We will develop a semiclassical model modified by the most general aspects of the quantized structure of nuclei. We do not expect the model to describe nuclear masses exactly, but, as we will soon see, it is remarkable for its near-quantitative descriptions of radioactive decay energetics. We will also use the formula to uncover one of the most profound aspects of nuclear structure – the existence of a shell structure in spherical nuclei – that exerts great influence on nuclear stability and decay properties.

Nuclear Physics for Applications. Stanley G. Prussin
Copyright © 2007 WILEY-VCH Verlag GmbH & Co., Weinheim
ISBN: 978-3-527-40700-2

5.2
The Semi-Empirical Mass Formula

In a number of respects, the nucleus appears to be analogous to a classical liquid drop of macroscopic size. A droplet of real water is a good example. In field-free space a droplet of water will assume a spherical shape that minimizes its total energy, and, to an excellent approximation, water has a constant density independent of the size of the drop. This arises because intermolecular forces, in rough analogy to the strong interaction, are themselves short-ranged and molecules tend to interact strongly only with their nearest neighbors. The kinetic energies of the molecules are distributed according to the Maxwell–Boltzmann distribution function, dependent on temperature but not the size of the drop. Therefore, so long as a molecule is a few molecular diameters removed from the surface of a droplet, it will, on the average, have the same average kinetic energy and binding energy as any other molecule. The molecules at the surface are certainly less tightly bound because they do not have as many nearest neighbors with which to interact; it is easier to remove a molecule from the surface than it is from the interior. However, the fraction of molecules in a macroscopic droplet that reside near the surface is negligibly small, and the average binding energy is then almost exactly the average binding energy of a molecule in the interior. The total binding energy of the droplet is then just the product of this average times the total number of molecules present.

The similarity between the gross properties of nuclei and the properties of a real liquid drop suggest that we might be able to use a "liquid drop" model as the basis of a description of nuclear binding energies. If we liken the nucleus to a drop of a classical liquid, we expect that if it were large enough, if the Coulomb force did not exist, if the pairing energy was negligible and if the masses of the neutron and proton were identical, the binding energy per nucleon would be constant, and the total binding energy would be directly proportional to the number of nucleons present. We will take this approximation to represent the leading term in an expression for the nuclear binding energy and write it as

$$a_V A \qquad (5.1)$$

Because the volume of the nucleus is given by $V_{nucl} = \frac{4}{3}\pi r^3 = \frac{4}{3}\pi r_o^3 A$, this term is referred to as the *volume term* and the coefficient a_V represents the average binding energy per nucleon in infinite, uncharged nuclear matter.

Real nuclei are relatively small. In contrast to a macroscopic droplet, a significant fraction of the nucleons present will reside at the nuclear surface and will possess reduced binding energy compared to those in the interior. The number of nucleons at the surface will be proportional to the surface area, which, for a sphere of radius R, is given by $S_{Nucl} = 4\pi R^2 = 4\pi r_o^2 A^{2/3}$ in the constant-density approximation. Thus, the total binding energy will be reduced by an amount proportional to $A^{2/3}$. Taking the proportionality constant as a_s, our model estimate for the total binding energy of finite, uncharged nuclear matter now becomes

5.2 The Semi-Empirical Mass Formula

$$a_V A - a_S A^{2/3} \tag{5.2}$$

We can easily account for the reduction in binding energy due to the stored Coulomb energy from repulsion of protons by use of Eq. (4.15), which gave

$$E_C = k_C \frac{3}{5} \frac{Z(Z-1)e^2}{R} = k_C \frac{3}{5} \frac{Z(Z-1)e^2}{r_o A^{1/3}} = a_C \frac{Z(Z-1)}{A^{1/3}}$$

The total binding energy of our uniformly charged drop of finite nuclear matter is now

$$a_V A - a_S A^{2/3} - a_C \frac{Z(Z-1)}{A^{1/3}} \tag{5.3}$$

The model does not yet account for any effects of quantization and the presence of two types of fermions that fill levels independently. We know that the total kinetic energy of a nucleus modeled as two degenerate Fermi gases is given by Eq. (4.45), i.e.,

$$T = \frac{C}{2^{2/3}} \left[A + \frac{5}{9} \frac{(A-2Z)^2}{A} + \ldots \right]$$

and thus the average binding energy of A nucleons in the absence of the Coulomb potential will be smaller if N ≠ Z than if N = Z. If we take the second term in Eq. (4.45) to approximate the increase in kinetic energy that results from this *asymmetry*, our binding energy will be reduced, in first order, by an amount proportional to $(A-2Z)^2/A$. Taking the proportionality constant as a_{asym}, the binding energy expression now becomes

$$a_V A - a_S A^{2/3} - a_C \frac{Z(Z-1)}{A^{1/3}} - a_{asym} \frac{(A-2Z)^2}{A} \tag{5.4}$$

Finally, we must account for pairing. We could do this simply by adding up the number of pairs of identical nucleons that are expected when a nucleus is in its ground state and then multiply this by a constant to get the average total pairing energy. But the fact is, *all* nucleons are paired in the ground state, except for odd particles. If N and Z are both even, we have no unpaired nucleons, if A = odd, we have one unpaired nucleon and if N and Z are both odd, we have two unpaired nucleons. This suggests that we only need to consider the number of unpaired nucleons. For example, suppose we take the expression in Eq. (5.4) to represent the binding energy of an odd-A nuclide. The adjacent (even, even) nucleus formed by adding one nucleon of the type that is odd will have one more pair and the adjacent (odd, odd) nucleus formed by adding one nucleon of the type that is even will have one less pair than the (even, even) nucleus of the same mass number. If the pairing energy is assumed to be the same for neutrons and protons, the binding energy of the adjacent (even, even) nucleus will be greater by the pairing energy δ and the binding energy of the adjacent (odd, odd) nucleus will be smaller by δ. Adding such a term to the expression in Eq. (5.4) gives an expression for the nuclear binding energy of

$$BE(Z, A) = a_V A - a_S A^{2/3} \tag{5.5}$$

$$- a_C \frac{Z(Z-1)}{A^{1/3}} - a_{asym} \frac{A - 2Z^2}{A} + \begin{cases} +\delta, \text{ if } (e,e) \\ 0, \text{ if } A = \text{odd} \\ -\delta, \text{ if } (o,o) \end{cases}$$

Eq. (5.5) is referred to as the semi-empirical expression for the nuclear binding energy. It is a general description of the binding energy of a uniformly charged drop of classical fluid corrected in the simplest way possible for two major effects of the quantized structure of the nucleus; the discrete energy spectrum of fermions and the pairing energy that is a reflection of a part of the strong interaction that is not accounted for by the volume term alone. Although the Coulomb term, as written above, is the correct expression when the discrete charge of protons is considered, the formula is most often written with the assumption that the charge can be treated continuously. In this case the binding energy formula is usually written as

$$BE(Z, A) = a_V A - a_S A^{2/3} \tag{5.6}$$

$$- a_C \frac{Z^2}{A^{1/3}} - a_{asym} \frac{(A-2Z)^2}{A} + \begin{cases} +\delta, \text{ if } (e,e) \\ 0, \text{ if } A = \text{odd} \\ -\delta, \text{ if } (o,o) \end{cases}$$

This form really represents the model binding energy in the limit of large atomic number. Because it is a *model*, it will not make much difference which form we use, and we will assume the expression in Eq. (5.6) in all our further discussions.

Using Eq. (5.6), the energy-equivalent of the rest mass of a nuclide can now be written as

$$M(Z, A)c^2 = Z M_H c^2 + (A - Z) m_n c^2 - a_V A + a_S A^{2/3} \tag{5.7}$$

$$+ a_C \frac{Z^2}{A^{1/3}} + a_{asym} \frac{(A-2Z)^2}{A} - \begin{cases} +\delta, (e,e) \\ 0, A = \text{odd} \\ -\delta, (o,o) \end{cases}$$

Eq. (5.7) is referred to as the *Weizsäcker semi-empirical mass formula*. It contains the five adjustable parameters a_V, a_S, a_C, a_{asym}, and δ, representing the volume, surface, Coulomb, asymmetry and pairing terms, respectively. These must be estimated somehow before we can proceed further. Because the mass formula neglects all but the most rudimentary representations of nuclear structure, we should expect, and indeed find, that the five parameters cannot be determined uniquely. They are found to vary somewhat with the means used to estimate them. For our present

Table 5.1 Semi-empirical parameter sets. The set shown in the second column is found commonly in the literature and other texts. The set shown in the third column was obtained from a least-squares fit of over 1000 atomic masses.

Term	Parameter	Set 1 (MeV)	Set 2 (MeV)[1]
volume	a_v	15.56	15.68
surface	a_s	17.23	18.56
Coulomb	a_c	0.697	0.717
asymmetry	a_{asym}	23.285	28.1
pairing	δ	12.0	$34A^{-3/4}$

1) W.D. Myers and W.J. Swiatecki, Nuclear masses and deformations, Nucl. Phys. 81 (1966) 1.

purposes, we are interested in understanding how well the model represents nuclear masses, on the average, throughout the entire chart of the nuclides, i.e., we are interested in a *global* fit, and how well it represents nuclides in any particular region of the chart, i.e., a *local* fit.

A global fit can be obtained by taking a large number of measured masses and obtaining the parameter set that gives the best overall fit in the least-squared sense. A set of parameters that provides an excellent fit to the some 270 nuclides found in nature in appreciable abundance is given in the second column of Table 5.1, and a set of parameters from a fit to over 1000 atomic masses throughout the chart of the nuclides is given in the third column of the table. First note that the parameters in the two sets are comparable but differ significantly from one another. The Coulomb parameters are very nearly the same, but the pairing parameters are quite different. When fitting over 1000 atomic masses, it was possible to extract a mass dependence on the pairing energy that was not searched for when parameter Set 1 was determined. Regardless of their differences, both sets of parameters are "valid".

Just how well the parameters represent empirical data is shown in Fig. 5.1, where the average binding energies shown in Fig. 4.1 for those nuclides found in nature with significant abundances, are shown along with those calculated with parameter Set 1. For simplicity, we have neglected the pairing energy, which, for this purpose, is negligible. As seen in the figure, the semi-empirical mass formula provides an excellent representation of the empirical binding energies. For the 248 nuclides with $A \geq 20$ included in the figure, the absolute difference between the empirical and calculated average binding energies is less than 0.6%. The clear ability of the model to represent the experimental data with high accuracy lends support to the assumption that the semi-empirical mass formula has a good deal of physics built into it and therefore should allow us to understand the general systematics of

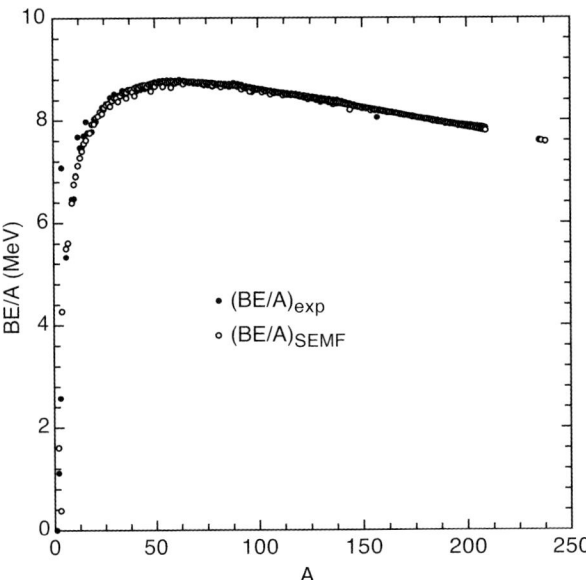

Fig. 5.1 Empirical and calculated average binding energy per nucleon for nuclides found in nature.

nuclear energetics and radioactive decay quite well, notwithstanding the fact that the parameter set is not unique.

The effects of the main terms in the semi-empirical mass formula on the binding energy per nucleon are shown graphically in Fig. 5.2 for the same nuclides included in Fig. 5.1. It is quite clear from the figure that the surface and Coulomb terms are the principal factors that cause the average binding energy per nucleon to be roughly half that expected for a nucleon in infinite uncharged nuclear matter. For example, Fig. 5.1 shows that the average binding energy of a nucleon in the (even, even) ^{40}Ca nucleus, is only about 8.3 MeV. Because the asymmetry term is zero, all of the difference between the volume term of 15.56 MeV and 8.3 MeV must be due to the Coulomb and surface terms. The reader should compute the magnitudes of these terms per nucleon and repeat the calculations for some heavy nuclide such as ^{208}Pb. While not producing as large an effect as these, the asymmetry term is clearly a major factor in the continuous and substantial decrease in the average binding energy at the higher mass numbers. We can expect, and indeed find, that this effect continues to increase throughout the region of heaviest elements that we have been able to make in the laboratory.

Just how well the semi-empirical mass formula can reproduce nuclear binding energies globally can be inferred from the average difference between the empirical total binding energies and those calculated with the parameters of Set 2. This is about ±1% of the empirical binding energy for the roughly 1000 nuclides that were fitted. It is indeed remarkable that the simple model of a uniformly charged,

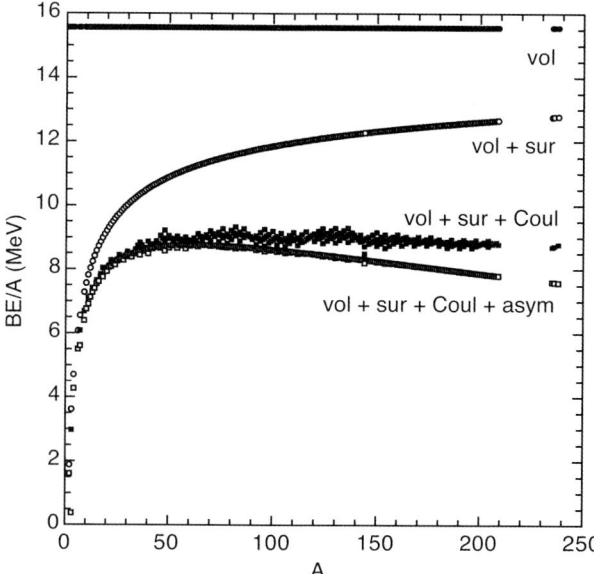

Fig. 5.2 The contributions of the first four terms in the semi-empirical mass formula to the average binding energy per nucleon for the 248 nuclides found in nature.

constant density liquid drop can yield such a good representation of nuclear binding with but two simple corrections for the details of nuclear structure. To demonstrate further what we can and cannot learn from this model, we now consider some of its implications with respect to nuclear stability and radioactive decay.

5.3
The Nuclear Mass Surface

If we take the semi-empirical mass formula to represent reality, it should be able to tell us all of the possible combinations of A and Z that lead to nuclides with binding energies greater than zero. While other combinations might exist transiently during the course of a nuclear reaction, they cannot be termed "stable" in any sense of the word.

It turns out that one of the more interesting things we can do with the mass formula is to examine what it predicts concerning the masses of isobars. We start by writing the mass formula in the slightly more compact form

$$M(Z, A)c^2 = ZM_H c^2 + (A-Z)m_n c^2 - a_V A + a_S A^{2/3} \qquad (5.8)$$
$$+ a_C \frac{Z^2}{A^{1/3}} + a_{asym} \frac{(A-2Z)^2}{A} - \delta(A)$$

where $\delta(A) = \begin{cases} +\delta, & \text{if (e,e)} \\ 0, & \text{if A = odd} \\ -\delta, & \text{if (o,o)} \end{cases}$.

If we expand the terms in Eq. (5.8) and collect those in the same power of Z we obtain

$$M(Z, A)c^2 = (Am_n c^2 - a_V A + a_S A^{2/3} + a_{asym} A) \tag{5.9}$$
$$+ Z(M_H c^2 - m_n c^2 - 4a_{asym}) + Z^2\left(\frac{a_C}{A^{1/3}} + \frac{4a_{asym}}{A}\right) - \delta(A)$$

or

$$M(Z, A)c^2 = \alpha + \beta Z + \gamma Z^2 - \delta(A) \tag{5.10}$$

where

$$\begin{aligned} \alpha &= Am_n c^2 - a_V A + a_S A^{2/3} + a_{asym} A \\ \beta &= (M_H - m_n)c^2 - 4a_{asym} \\ \gamma &= \frac{a_C}{A^{1/3}} + \frac{4a_{asym}}{A} \end{aligned} \tag{5.11}$$

The coefficients α, β, and γ can be viewed as constants for a given mass number A. As a result, Eq. (5.10) predicts that the masses of isobars vary parabolically with Z for all mass numbers. If A is odd, $\delta(A) = 0$, and all masses are predicted to reside on a single parabola. However, if A is even, the masses will reside on two parabolas. The (even, even) nuclides will lie on one parabola that lies below the parabola on which all of the (odd, odd) isobars reside. Further, because γ is always positive, the parabolas for different A will each have a minimum mass corresponding to a different Z. If we were to plot the predicted masses in three dimensions with, for example, (x,y,z) = (Z,N,M(Z,A)), we ought to see a *mass surface* that is parabolic in nature.

To demonstrate this, the masses calculated for $A \geq 20$, odd and with atomic numbers $Z = Z_A \pm 10$, where Z_A is the atomic number of the isobar with minimum mass, are shown in Fig. 5.3. The parabolic form of the mass surface is quite evident. The curvature of the surface decreases continuously with increasing mass number so that the surface becomes "less steep" or "flatter" with increasing A. This behavior is easily seen from the form of the two terms in the coefficient γ that vary as $A^{-1/3}$ and A^{-1}, respectively (Eq. (5.11)). Note also that the "valley" in the mass surface represents the minima of the parabolas for isobars and thus traces the curve of N vs. Z through the isotopes found in nature shown in Fig. 4.6.

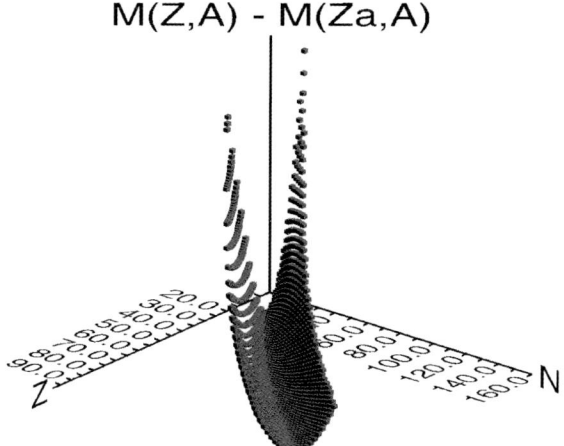

Fig. 5.3 The nuclear mass surface calculated with the semi-empirical mass formula for $A \geq 20$. The surface is shown for odd-A isobars in the range $Z_A \pm 10$ where Z_A is the atomic number of the isobar with minimum mass. The vertical scale is the calculated mass, minus the mass corresponding to the (hypothetical) nucleus that would reside on the minimum of each parabola.

5.4
The Semi-Empirical Mass Formula and β Decay

Eq. (5.10) predicts that the masses of isobars will reside on a single parabola if A = odd, and on two parabolas if A = even. Rather than deal directly with the masses themselves, it is common practice to deal with the difference between the experimental masses and the mass of A atomic mass units by defining the *mass excess*, Δ, as

$$\Delta = [M(Z, A) - A]c^2 \tag{5.12}$$

where M(Z,A) is the atomic mass on the unified scale (units of u) and A is the mass number in the same units. Δ is normally given in energy units of either keV or MeV. That is, given the atomic mass of M(Z,A) in atomic mass units and given the mass number A, multiplication of each by the energy equivalent of 1u, 931.494 MeV, and subtracting, gives the mass excess. In Fig. 5.4 are shown the experimental mass excesses for all known isobars of A = 103 and A = 104 along with parabolas calculated according to Eq. (5.10) by least-squares fits to the data (local fits).

For both mass numbers, the predictions of the semi-empirical mass formula are in very good agreement with the experimental data. The mass excesses for the A = 103 isobars are quite well-fitted with a single parabola, and those for A = 104 are well-described by two parabolas, the (odd, odd) nuclides residing on one that is

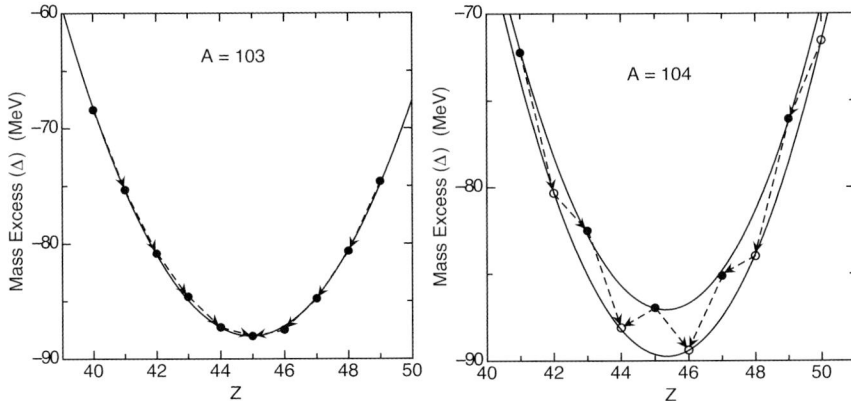

Fig. 5.4 Experimental mass excesses (MeV) for the isobars of A = 103 and A = 104. The curves shown are fits with the semi-empirical mass formula.

displaced vertically from a parabola of the exact same form on which the (even, even) isobars reside. We stress that the fits represent *local* fits, and throughout the entire range of known nuclides, with but few exceptions, local fits of similar quality are found. The predictions of the semi-empirical mass formula are generally in quite close agreement with most experimental measurements.

The connection of the foregoing to the energetics of β decay can now be shown directly. Consider first odd-A nuclides. From Eqs (2.40) and (2.45),

$$Q_{\beta^-} = [M(Z, A) - M(Z + 1, A)]c^2 = [\Delta(Z, A) - \Delta(Z + 1, A)]c^2$$
$$Q_{EC} = [M(Z, A) - M(Z - 1, A)]c^2 = [\Delta(Z, A) - \Delta(Z - 1, A)]c^2 \quad (5.13)$$

Theoretically, all nuclides with Z less than the atomic number corresponding to the minimum in the parabola have $Q_{\beta^-} > 0$ and thus are unstable with respect to β^- decay, and all nuclides with Z greater than the atomic number corresponding to the minimum in the parabola have $Q_{EC} > 0$ and thus are unstable with respect to electron capture decay. (If $Q_{EC} > 2m_ec^2$, they are also unstable with respect to positron emission.) The arrows shown in Fig. 5.4 indicate the β decay of isobars.

The atomic number corresponding to the minimum in a parabola for a given mass number, Z_A, is easily obtained in terms of the parameters of the mass formula by the differentiation

$$\frac{\partial}{\partial Z}M(Z, A)\Big|_{Z = Z_A} = 0 \quad (5.14)$$

Applying this to the form $M(Z, A)c^2 = \alpha + \beta Z + \gamma Z^2 - \delta(A)$ we have the result

$$Z_A = -\frac{\beta}{2\gamma} \quad (5.15)$$

which, for the fit to the mass excesses of the A = 103 isobars in Fig. 5.4, gives Z_A = 44.96. Note that Z_A does not correspond to an integer, and this is a general result. It represents the atomic number of the *hypothetical* most stable isobar. If A = odd, the mass formula indicates that the real nuclide with Z closest to Z_A will be the only β-stable isobar. A search through the chart of the nuclides shows that, with but one or two exceptions, there is only one isobar for each A = odd that is found in nature. Where exceptions exist, one of the two isobars is unstable with respect to β decay but has a half-life that is comparable to or greater than the age of the earth. When A = even, the decay energies are modulated by the requirement that decay take place between isobars on the two different parabolas. As seen in Fig. 5.4 for A = 104, this produces large decay energies for the (odd, odd) nuclides that are far removed from Z_A. It further leads to the possibility that there will be *more than one* β-stable isobar if A = even. For A = 104, the isobars of Z = 44 (Ru) and Z = 46 (Pd) are both stable with respect to ordinary β decay, notwithstanding the fact that the mass of ^{104}Ru is considerably greater than that of ^{104}Pd[1]. Throughout the chart of the nuclides there are usually two and sometimes three β-stable isobars for A = even.

The data for A = 104 provide further insight into the pairing energy and its effect on nuclear masses and the energetics of β decay. First, the mass parabolas for (odd,odd) and (even,even) isobars are displaced from one another by twice the pairing energy δ(A) (see Eq. (5.6) or (5.7)), i.e.,

$$M(Z_{odd}, A_{odd})c^2 - M(Z_{even}, A_{even})c^2 \quad (5.16)$$
$$= [\alpha + \beta Z + \gamma Z^2 + \delta] - [\alpha + \beta Z + \gamma Z^2 - \delta] = 2\delta$$

As a result, the masses of even-A nuclides provide local estimates of pairing energies throughout the region of known nuclides. For the A = 104 isobars, the fits to the mass excesses give 2δ = 2.365 MeV, or a pairing energy of about 1.18 MeV. This is consistent with the pairing energies found from the neutron and proton binding energies of the Z = 50 isotopes and N = 50 isotones discussed in Chapter IV. Second, we can see how the pairing energy leads to the large number of (even, even) nuclides in nature and the scarcity of (odd, odd) nuclides. Fig. 5.4 shows that it is very unlikely that an (odd, odd) nuclide will be stable. In fact, with the exception of the very light nuclides ^{10}B and ^{14}N, the (odd, odd) nuclides found with reasonable abundance in nature are all β-unstable.

Finally, the data for A = 104 show that the pairing energy can lead to nuclides that are unstable *both* to β$^-$ and EC/β$^+$ decay and $^{104}_{45}$Rh is such an example. One of the more common examples is that of ^{64}Cu whose decay is shown in Fig. 5.5.

The nuclides that we find in nature can now be understood in terms of the parabolic nature of the nuclear mass surface and its affect on β-decay energies. Anticipating the detailed discussion in Chapter X, the half-lives of β emitters generally decrease rapidly with increasing decay energy. Thus nuclides with Z far

---

1) While stable against ordinary β decay, nuclides such as $^{104}_{44}$Ru are actually unstable against the very rare decay mode called double β decay.

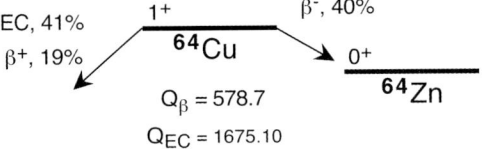

Fig. 5.5 Decay schematic for ^{64}Cu.

removed from Z_A for any mass number will tend to have very short half-lives. The conclusion is that, with the exception of a very few very long-lived nuclides, the nuclides found in nature are *stable* against β decay. The parabolic mass surface shown in Figs 5.3 is somewhat fanciful, but aptly, referred to as describing a valley whose floor winds along the line of β stability.

There is yet more that we can learn from the application of the semi-empirical mass formula to β decay of odd-A nuclides. Because of the parabolic nature of the mass surface, the mass difference between adjacent isobars is predicted to vary *linearly* with the atomic number, and thus will the decay energies (β⁻ or EC). Taking β⁻ decay of the nuclide M(Z,A) as the example, we have

$$Q_{\beta^-} = [M(Z, A) - M(Z+1, A)]c^2 \tag{5.17}$$
$$= \alpha + \beta Z + \gamma Z^2 - [\alpha + \beta(Z+1) + \gamma(Z+1)^2]$$
$$= -\beta - \gamma - 2\gamma Z$$
$$= 2\gamma\left(Z_A - Z - \frac{1}{2}\right)$$

The equation predicts that the β-stable nuclide found in nature has an atomic number $Z(Q_{\beta^-} = 0) = Z_A - \frac{1}{2}$.

The decay energies of the A = 103 isobars are shown in Fig. 5.6. They clearly follow the linear prediction quite well. Because $Q_{\beta^-} = [M(Z, A) - M(Z+1, A)]c^2 = Q_{EC}$, all isobars with negative β⁻-decay energies have positive Q_{EC}.

The linear variation of decay energy with atomic number can be used to advantage as another means of estimating one of the fundamental constants on which the semi-empirical mass formula is based, the radius parameter r_o. This comes about because of the existence of so-called *mirror nuclides*. By definition, one nucleus is the mirror of the other if the neutron number of one is the atomic number of the other, and vice versa. For the present purpose, we are interested in the mirror nuclei where the pair of nuclides (Z,N) and (Z + 1, N − 1) are related by Z = N − 1 and A = 2Z + 1. Two such pairs are $^3_1H_2$, $^3_2He_1$ and $^7_3Li_4$, $^7_4Be_3$. Using Eq. (5.8), the β⁻ decay energy for the lower-Z member of a pair can be written as

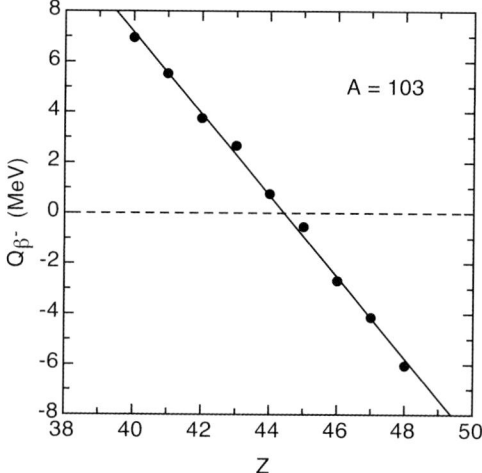

Fig. 5.6 Q_{β^-} decay energies for the A = 103 isobars.

$$Q_\beta = [M(Z, A) - M(Z + 1, A)]c^2 \tag{5.18}$$

$$= ZM_H c^2 + (A - Z)m_n c^2 - a_V A + a_S A^{2/3} + a_C \frac{Z^2}{A^{1/3}} + a_{asym}\frac{(A - 2Z)^2}{A}$$

$$- (Z + 1)M_H c^2 - (A - Z - 1)m_n c^2 + a_V A - a_S A^{2/3}$$

$$- a_C \frac{(Z + 1)^2}{A^{1/3}} - a_{asym}\frac{(A - 2Z - 2)^2}{A}$$

$$= (-M_H + m_n)c^2 - \frac{a_C}{A^{1/3}}(2Z + 1)$$

The mass difference between a mirror pair is predicted to be due solely to the mass difference between the neutron and the hydrogen atom and the difference in the Coulomb energy of the nuclei. With A = 2Z + 1, and writing $\Delta_{n,H} = (m_n - M_H)c^2 = 0.782$ MeV, we can rewrite Eq. (5.18) simply as

$$Q_{\beta^-} = \Delta_{n,H} - a_C A^{2/3} \tag{5.19}$$

This expression indicates that the β^--decay energies of the low-Z member of a mirror pair should vary linearly with $A^{2/3}$ with a slope of $-a_C$. As it turns out, except for the very lightest mirror pairs, the β^--decay energies of the low-Z members are negative because of the increase in Coulomb energy from the conversion of a neutron into a proton. As a result, most pairs will be connected by electron capture / positron emission.

The empirical β^--decay energies for mirror nuclei are shown in Fig. 5.7 as a function of $A^{2/3}$ and they follow the linear prediction of Eq. (5.19) quite well. The slope of the least-squares fit shown in the figure gives the value $a_C = 0.708 \pm$

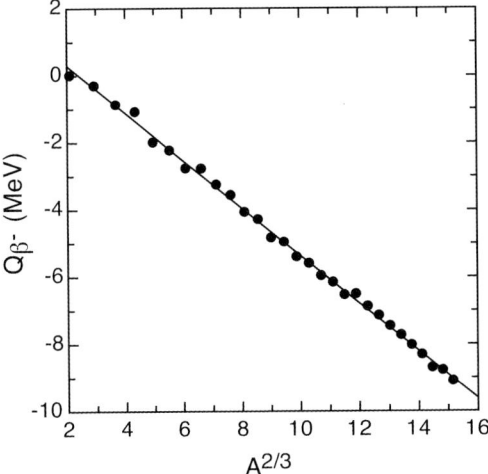

Fig. 5.7 Q_{β^-} energies for the low-Z members of known mirror pairs as a function of $A^{2/3}$.

0.006 MeV, in very good agreement with the values given in Table 5.1. Now the stored Coulomb energy in the continuous-charge approximation was given by Eq. (4.14) as

$$E_C = \frac{3}{5} k_C \frac{Z^2 e^2}{r}$$

If we use the constant-density model $r = r_o A^{1/3}$, the equation becomes

$$E_C = \frac{3}{5} k_C \frac{Z^2 e^2}{r_o A^{1/3}} = a_C \frac{Z^2}{A^{1/3}} \tag{5.20}$$

and therefore,

$$a_C = \frac{3}{5} k_C \frac{e^2}{r_o}$$

The fit to the decay energies of the mirror pairs then yields the value $r_o = 1.22 \pm 0.01$ fm, in excellent agreement with values obtained from many other types of measurements.

The semi-empirical mass formula is very successful in accounting for the energetics of β decay in most cases, but it must be remembered that it is based on a simple model that does not account for the details of nuclear structure. Although the majority of masses are quite well-described by the parabolic mass surface, there are some specific regions where fits to mass data are very much poorer than those shown in Fig. 5.4. For example, consider the data and fits for mass numbers A = 137 and A = 207 as shown in Fig. 5.8. The deviations of the experimental mass

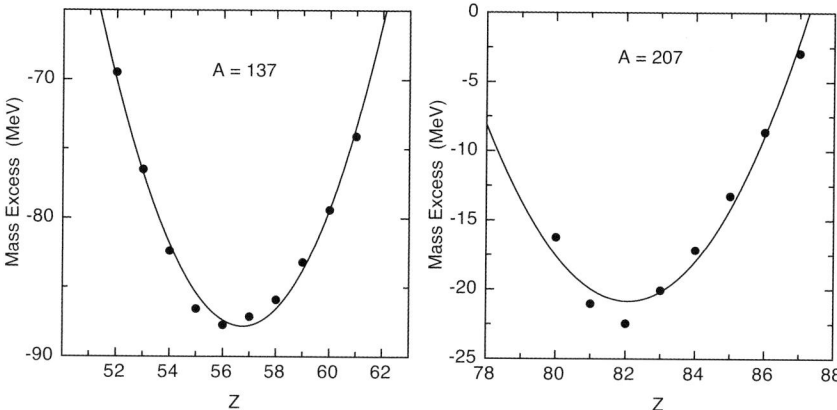

Fig. 5.8 Mass excesses and least-squares fits with the semi-empirical mass formula for isobars of A = 137 and A = 207.

excesses from the fits are very much larger than seen in Fig. 5.4. The masses of the nuclides included in the figure are very well known and, for the most part, errors in the mass excesses are comparable to, or smaller than, the sizes of the data points shown. Rather, the deviations reflect effects from the structure of nuclei that are not included in the simple liquid-drop model. In this case, the deviations are signatures of nuclear shell structure that produces especially strong binding for certain numbers of neutrons and protons. We will, in fact, use the departures from the semi-empirical mass formula to help understand nuclear structure effects as we proceed further.

5.5
The Semi-Empirical Mass Formula and α Decay

The systematics of α decay provide an excellent demonstration of both the strengths and limitations of the simple liquid-drop model of the nucleus. The Q value for α decay is given by

$$Q_\alpha = [M(Z, A) - M(Z-2, A-4) - M(2, 4)]c^2 \quad (5.21)$$

where Z and A are the atomic and mass numbers of the α-decay parent and M(2,4) is the mass of the $^4_2$He atom. Neglecting the differences in atomic electron binding energies, Eq. (5.21) can be written directly in terms of the nuclear binding energies as

$$Q_\alpha = [BE(2, 4) + BE(Z-2, A-4) - BE(Z, A)] \quad (5.22)$$

If we use the empirical value of 28.296 MeV for BE(2,4), and the semi-empirical mass formula expressions for the parent and daughter nuclides, Eq. (5.22) becomes

$$Q_\alpha = 28.296 - 4a_V + a_S[A^{2/3} - (A-4)^{2/3}] +$$
$$a_C\left[\frac{Z^2}{A^{1/3}} - \frac{(Z-2)^2}{(A-4)^{1/3}}\right] + a_{asym}(A-2Z)^2\left[\frac{1}{A} - \frac{1}{(A-4)}\right] \quad (5.23)$$

As unappetizing as the equation appears, it is actually not difficult to determine the atomic number or numbers for any mass number A for which $Q_\alpha \geq 0$. By subtracting Q_α from both sides of the equation, one obtains a simple quadratic equation in Z, and it is easy to generate the estimates shown in Fig. 5.9. The figure shows the, hypothetical, most stable isobar, Z_A, calculated from the global-fit parameters of parameter Set 2 in Table 5.1, along with the smallest integer atomic number, $Z_{min}(A)$, for which $Q_\alpha > 0$. For a given A, all nuclides with higher Z will also have $Q_\alpha > 0$. The mass formula predicts that near the valley of β stability, all nuclides with $A \geq 150$ should be unstable with respect to α emission. This is in marked contrast to β decay which is found at all mass numbers.

The region of all known nuclides is shown in Fig. 5.10 along with all β-stable nuclides and those that are known α emitters. With the exception of a few very light nuclides, which the semi-empirical mass formula cannot describe, no α emitters are found near the valley of stability below a mass number of about 150. A cluster of α emitters is found in the vicinity of N = 86, Z = 64 but, with the exception of these and a few others, none are encountered until the region of the heaviest nuclides found in nature, beginning at about N = 126, Z = 86. This is not the result of errors in the prediction of α decay energies. Indeed, in agreement with the mass

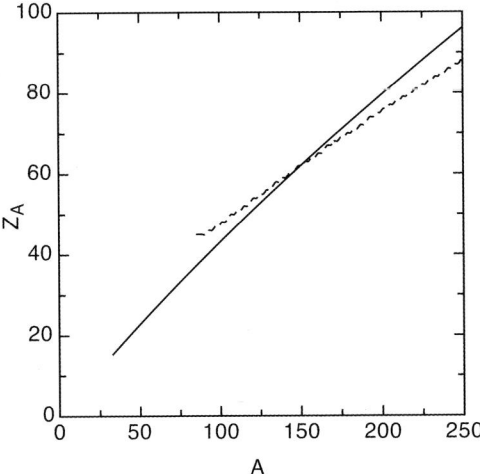

Fig. 5.9 The charge of the, hypothetical, most stable isobar, Z_A (solid line) and the smallest atomic number for which $Q_\alpha \geq 0$ as a function of the mass number A (dashed line).

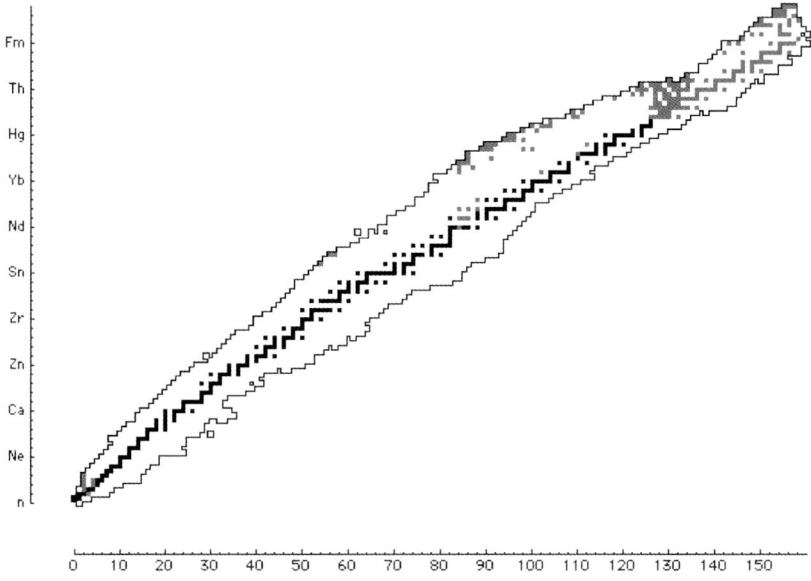

Fig. 5.10 The region of known nuclides. β-stable nuclides are shown in black and α emitters are shown in gray.

formula predictions, nearly all nuclides near the valley of stability and with mass numbers greater than about 150, do possess $Q_\alpha > 0$, but α emission is simply not *observed*. In contrast to β emitters, where decay energies range from as low as 0.0186 MeV to over 10 MeV, no α emission is observed from any nuclides if $Q_\alpha < 3.5$ MeV near A = 150 and $Q_\alpha < 4.5$ MeV in the region of the heaviest nuclides. Clearly, the α decay probabilities at low energies must be very small. The elementary physics of α decay will be discussed in some detail in Chapter IX. Suffice it to say here that the probability for α emission is extremely sensitive to both the decay energy and the Coulomb potential which it experiences as it is emitted. The latter acts as a barrier through which the α particle must penetrate.

5.6
The Semi-Empirical Mass Formula and Nuclear Fission

Application of the semi-empirical mass formula to spontaneous fission is complicated because of the very many products that are formed and the varying states of excitation that they may have. However, the most important qualitative factors underlying the energetics of fission, and a determination of which nuclides are unstable to spontaneous fission, can be obtained through analysis of the simple case of symmetric binary fission. Such a division is actually quite rare but it will make very little difference with respect to the main conclusions we draw with respect to energetics.

We begin quite generally by rewriting the mass balance in fission of Eq. (2.49) as

$$M(Z, A) = M_L(Z_L, A_L) + M_H(Z_H, A_H) + Q_f/c^2 \quad (5.24)$$

where we have assumed, for simplicity, that the two nuclides produced are in their ground states and together contain all of the neutrons and protons present in the nuclide M(Z,A). We now solve this equation for Q_f with the semi-empirical mass formula expressions for the atomic masses of the products. To simplify even further, we will neglect the pairing energy terms in these expressions because they are really relatively small and their neglect will not affect our main conclusions. With this approximation, Eq. (5.24) becomes

$$Q_f = M(Z, A)c^2 - [M_L(Z_L, A_L) + M_H(Z_H, A_H)]c^2 \quad (5.25)$$

$$= ZM_Hc^2 - (A-Z)m_nc^2 - a_V A + a_S A^{2/3} + a_C \frac{Z^2}{A^{1/3}} + a_{asym}\frac{(A-2Z)^2}{A}$$

$$- \left[Z_L M_L c^2 - (A_L - Z_L)m_n c^2 - a_V A_L + a_S A_L^{2/3} + a_C \frac{Z_L^2}{A_L^{1/3}} + a_{asym}\frac{(A_L - 2Z_L)^2}{A_L}\right]$$

$$- \left[Z_H M_H c^2 - (A_H - Z_H)m_n c^2 - a_V A_H + a_S A_H^{2/3} + a_C \frac{Z_H^2}{A_H^{1/3}} + a_{asym}\frac{(A_H - 2Z_H)^2}{A_H}\right]$$

where $A = A_H + A_L$ and $Z = Z_H + Z_L$. For symmetric binary fission, the parent must be an (even, even) nuclide and we then have $A_H = A_L = A/2$, and $Z_H = Z_L = Z/2$. With these substitutions and a little algebra, Eq. (5.25) becomes

$$Q_f/c^2 = a_S\left[A^{2/3} - 2\left(\frac{A}{2}\right)^{2/3}\right] + a_C\left[\frac{Z^2}{A^{1/3}} - \frac{2(Z/2)^2}{(A/2)^{1/3}}\right] \quad (5.26)$$

Eq. (5.26) indicates that the only terms that contribute significantly to energy release in symmetric fission of an (even, even) nucleus are the surface and Coulomb terms. The volume terms clearly cancel and it is not difficult to see that symmetric fission cannot result in nuclides with different N/Z than the parent and thus the asymmetry term cannot contribute to the mass difference. Because the surface area of the two products must be greater than the surface area of the parent, an increase in surface energy takes place and thus if energy is released it must be associated with the change in the stored Coulomb energy. Even though we have made significant simplifications, this same result applies for asymmetric fission of (even, even) nuclides and for fission of odd-A nuclides as well.

If we now set $Q_f/c^2 \geq 0$, Eq. (5.26) can be solved to give the ratio Z^2/A as

$$\frac{Z^2}{A} \geq \frac{a_S(2^{1/3}-1)}{a_C(1-2^{-2/3})} \cong 18.2 \quad (5.27)$$

Table 5.2 Z^2/A for a few nuclides found in nature.

Nuclide	Z^2/A
$^{56}_{26}$Fe	12.2
$^{84}_{36}$Kr	15.4
$^{90}_{40}$Zr	17.8
$^{102}_{44}$Ru	19.0
$^{238}_{92}$U	35.6

This result says that all nuclides for which Eq. (5.27) is fulfilled are unstable with respect to spontaneous symmetric fission. Now we can easily go through the list of nuclides that exist in nature and find out where we expect to find such fission. The ratios of Z^2/A for a few of these are given in Table 5.2. The amazing result is that nuclides in nature – those that lie on the valley of β-stability – with mass numbers greater than about 100–110, are *all* unstable with respect to spontaneous fission. This is no mistake and it is supported by direct calculation with empirical atomic masses.

Now the simple fact is that we just do not see spontaneous fission for any of the nuclides in nature, with the exception of the three isotopes of uranium ^{234,235,238}U for which the percentages of decay by spontaneous fission are 1.7×10^{-9}, 7.0×10^{-9} and 0.5×10^{-4}, respectively! If you proceed through isotopes of the elements beyond uranium you do, however, find spontaneous fission frequently and with high probability. For example, 3.09% of the decays of $^{252}_{95}$Cf occur by spontaneous fission and almost 92% of the decays of $^{256}_{100}$Fm occur by spontaneous fission. There is very strong evidence that the limit on the most massive nuclei that we can ever expect to make may be defined by the half-life for spontaneous fission.

The lack of observation of spontaneous fission where it is energetically allowed, with the exception of the very heaviest of elements, is even more dramatic than for α decay. In this case we know that there is also a barrier that acts to keep nuclei from undergoing fission.

We could go on and apply the liquid drop model as represented by the semi-empirical mass formula to look at nuclear reactions or other, more esoteric decay modes. But the discussion presented above is sufficient to point out the major aspects of the physics of nuclei that we can learn from it. In later sections we will return to these ideas where they are helpful. Nevertheless, the nucleus is really a much more complicated object than we have considered up to now, and the semi-empirical mass formula is probably the most rudimentary nuclear model we might propose. It has been surpassed by considerably more complex models that attempt to either add in factors associated with the details of nuclear structure that we have omitted, or which attempt to compute masses from reasonable representations of

5.7
Discrepancies Between Experimental Masses and those Predicted by the Semi-Empirical Mass Formula

We have pointed out in Section 4.1 that a close look at the experimental nuclear binding energies of the nuclides found in nature reveals a few rather distinct departures from the smooth variation predicted by the semi-empirical mass formula. While we have glossed over these to the present, it will be well worthwhile to look at the discrepancies between experimental and predicted masses in some detail. To begin, it is a simple matter to take the difference between the masses shown in Fig. 5.1 and the predictions of the mass formula using global parameters. We have gone a step further by calculating the differences for the roughly 1840 nuclides for which such information was available in the 1995 mass compilations. These are shown in Fig. 5.11 as a function of the atomic number Z and neutron number N. In place of the smoothly varying binding energy that is characteristic of our model, the experimental binding energies show regular and quite marked deviations at neutron or proton numbers of 28, 50 and 82, and at the neutron number 126.

It is remarkable that the peaks in the distributions are located at identical numbers of protons or neutrons. For the most part, the empirical binding energies are *larger* than the calculated values at the peaks by up to 10–15 MeV. Now the mass formula is far from perfect and thus the absolute values of the differences must be viewed with caution. Nevertheless, it is quite clear that a large number of nuclei have significantly greater binding energy and stability than predicted by our simple model.

The binding energy differences in the regions below N(Z) ≤ 30 seem less systematic than at higher numbers. This is due, in part, to the fact that the semi-empirical mass formula cannot be expected to describe the light nuclei very well. But there is ample evidence that especially large binding energy is associated with neutron or proton numbers of 2, 8 and 20 in addition to those noted above. Taken together, the numbers 2, 8, 20, 28, 50, 82, and 126 have become known as the *magic numbers*.

The type of behavior displayed in Fig. 5.11 is quite reminiscent of the first ionization energies of the elements; the energies required to overcome the binding of the most weakly bound electrons in the atoms, as shown in Fig. 5.12. The shell structure of the electrons in atoms results in increasing ionization energy as electrons are added to each shell. The highest ionization potentials are found each time a shell is completely filled, and this occurs at He (Z = 2), Ne (Z = 10), Ar (Z = 18), Kr (Z = 36), Xe (Z = 54) and Rn (Z = 86). These are the so-called rare gases

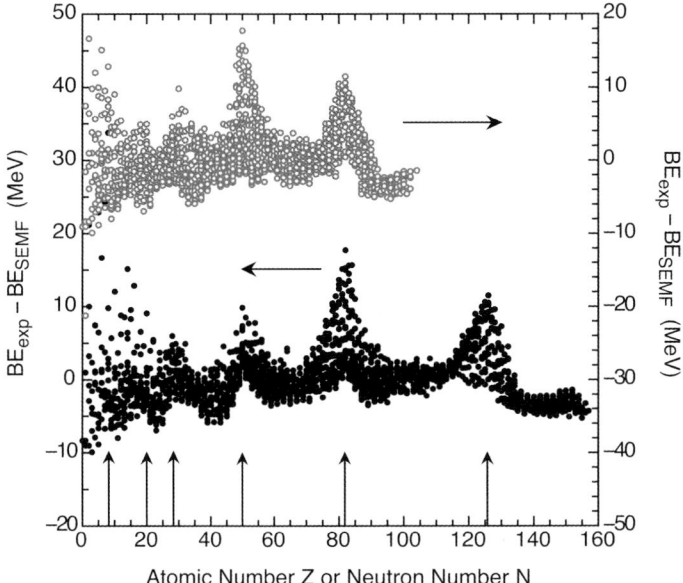

Fig. 5.11 Differences between the experimental nuclear binding energies (MeV) and those calculated with the semi-empirical mass formula. The upper set of data represent the differences as a function of atomic number (the right-hand scale) and the lower set are the differences as a function of neutron number (left-hand scale). Note the shift in the energy scales. The vertical arrows at the bottom mark the magic numbers greater than 2.

and, with the exception of the heaviest members of the group, are found to be chemically inert under almost all conditions. The next electron added after a shell is filled is very weakly bound and thus atoms of the elements corresponding to atomic numbers one greater than that of a rare gas, the alkaline earth elements Li, Na, K, Rb, Cs, and Fr, are easily ionized. In almost all chemical combinations the alkaline earths are found in the (+1) oxidation state. A comparison between Figs 5.11 and 5.12 shows some marked similarities that suggest that the variation in binding energy with neutron and proton number reflects *nuclear shell structure* in analogy with the electronic shell structure. Nuclear binding energies tend to increase sharply as the magic numbers are approached and decrease rather sharply as an additional neutron or proton is added beyond these numbers. In between the magic numbers, the binding energies tend to vary in a fairly smooth manner, just as found in the case of atomic electrons. The fact that the magic numbers are the same for neutrons and protons is consistent with the assumption that neutrons and protons fill levels separately, and with the fact that the nuclear force is much stronger than the Coulomb force. The Coulomb potential must produce some differences between the energy levels for the two particle types, but the level structure cannot be so different as to affect the magic numbers themselves.

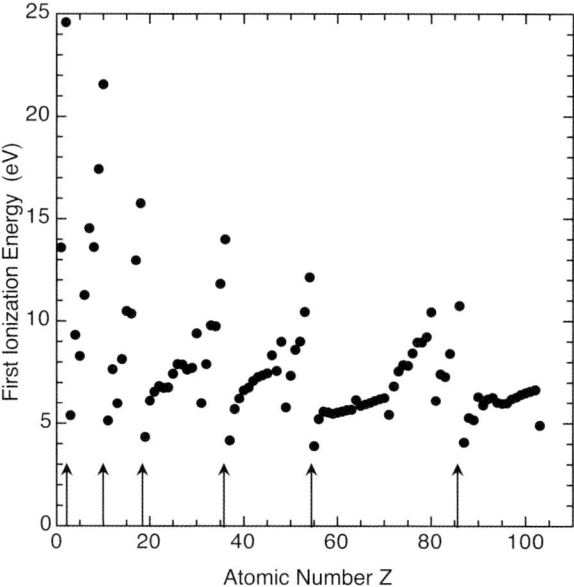

Fig. 5.12 The first ionization potentials of the chemical elements. The arrows mark the closed electron shells at the rare gases.

It is somewhat more difficult to obtain a clear picture of the binding energies of the most weakly bound proton or neutron that would correspond to the ionization energy of the neutral atom. Because of the strong pairing effect in the nucleus, the energy required for removal of a neutron or proton from an (even, even) nucleus will be significantly different from the energy required for removal of a neutron or proton from an odd-A nucleus. Such energies are referred to as the neutron (S_n) or proton (S_p) *separation energies*. We can get away from this problem simply by focusing on the energy required to remove an odd neutron from an (even, odd) nucleus, or an odd proton from an (odd, even) nucleus. Again, using the semi-empirical mass formula estimate as a reference, the differences between the experimental and calculated separation energies for odd-A nuclides are shown in Fig. 5.13.

The similarity between the data in Figs 5.12 and 5.13 is quite striking, especially for proton or neutron numbers of 50 and 82 and for neutron number 126. One sees a continuous increase in separation energy as the magic numbers are approached, followed by a sudden and sharp drop to a local minimum. The distributions near the lower mass numbers do not show such well-defined forms although one can see local minima immediately after the magic numbers of 2, 8 and 20. The difference in the binding energies of odd nucleons as one crosses a magic number lies in the range of 2 –3.5 MeV. With such energies, it should be clear that shell structure will have a marked effect on the structure of nuclei as well as the energetics and kinetics of radioactive decay and nuclear reactions. Because of the

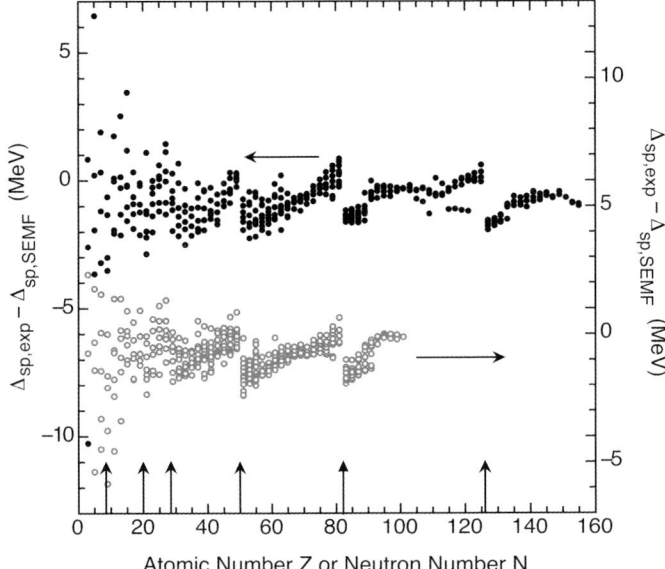

Fig. 5.13 The differences between experimental neutron (upper data set, left hand energy scale) and proton (lower data set, right hand energy scale) separation energies and those calculated with the semi-empirical mass formula. Only odd-A nuclides are considered. The arrows at the bottom of the figure mark the magic numbers greater than 2.

important role played in these processes, we will now turn to an examination of some simple models of nuclear structure and proceed to use them in later discussions to understand the details of radioactive decay and nuclear reactions.

General References

A really beautiful treatment of the liquid drop model of the nucleus can be found in R.D. Evans. "The Atomic Nucleus", Krieger Publishing Company; Reprint edition (June 1982).

Problems

1. Use the atomic masses given in Appendix 1 for $A = 131$ to perform the following.
(a) Calculate local values of the parameters α, β and γ with the masses of Te, Xe and Ba.
(b) Generate a mass parabola for $A = 131$ using the parameters from part (a). Include experimental masses for $A = 131$ on the same figure.

(c) Use the parameters α, β and γ to prepare a graph of the Q_β for β^- decay versus Z. Include in the graph the experimental Q_β obtained from the mass measurements.

(d) Use the local parameters to calculate Z_A, the charge corresponding to the, hypothetical, most stable isobar of A = 131.

(e) Calculate a least-squares fit to the experimental Q_β data and calculate Z_A corresponding to this fit. Compare this value of Z_A to that found in part (d).

(f) Use the local parameters α, β and γ to predict Q_{β^-} for ^{131}Sb and ^{131}La. Compare these with the experimental values of 3.190 ± 0.070 MeV and –4.0 ± 0.4 MeV, respectively.

(g) What is (are) the likely decay mode(s) of ^{131}La?

2. Use the global parameters of Myers and Swiatecki (Set 2 of Table 5.1) to calculate an estimate of Z_A for A = 131. Compare this with the local value from part (d) of problem 1 above.

3. (a) Calculate and plot the *two-neutron* separation energies for the barium isotopes listed in Appendix 1 as a function of neutron number. For what N is the onset of a sharp discontinuity observed? Interpret this observation physically.

(b) Why does one choose to study two-neutron separation energies rather than one-neutron separation energies?

4. Use the positron decay of ^{13}N to the ground state of ^{13}C to calculate the radius parameter r_o of the expression $r = r_o A^{1/3}$.

6
Elements of Quantum Mechanics

6.1
Introduction

The nucleus is a quantized object and we cannot really delve into its structure and decay properties without resort to quantum mechanics, even with the simplest models. Most students have been introduced to some of the fundamental aspects of the subject in introductory physics or chemistry courses, but we will need to consider some topics that are either not covered there, or that are dealt with only qualitatively. Our purpose here is not to provide a discussion of quantum theory, but to present some of the facts that are crucial to any sort of examination of nuclear structure, radioactive decay and nuclear reactions. While this will involve a significant amount of mathematics, most of it is covered in the first two years of mathematics taken by students in physical science and engineering. Where new mathematics is introduced, we will try to provide sufficient background so that it can be used at an introductory level and the physical implications inherent in it are made clear. For those readers who have the opportunity to take a course in advanced calculus for applications and a course in quantum theory at the undergraduate level, we urge them to do so. The insight gained into the fundamental characteristics of matter will be more than worth the effort.

Primarily because we can experience their ramifications in the everyday world, we usually do not question the fundamental postulates of classical physics or chemistry. While more abstract for most of us, we also accept the fundamental postulates of electricity and magnetism. Quantum theory is quite different. Except for those who have experience with the discrete radiation emitted from excited atoms or molecules, quantization and a probabilistic representation of physical laws is quite foreign to us. Nevertheless, it is not only fundamental to an understanding of matter, but it is affecting our everyday lives. The lasers that we all use as pointers and that we depend upon for playing music with compact discs or for loading information into and recording data from our computers, rely on the discrete character of the excited states of atomic systems and the probabilities for their excitation and decay. The reactions that produce power in nuclear fission reactors and the decay characteristics of the fission products that permit easy control of such devices have characteristics that are also determined by the quan-

Nuclear Physics for Applications. Stanley G. Prussin
Copyright © 2007 WILEY-VCH Verlag GmbH & Co., Weinheim
ISBN: 978-3-527-40700-2

tized nature of the reactions and decays that are involved. Further, the problems imposed by the safe disposal of nuclear wastes are fundamentally determined by the characteristics of quantized systems. On a grander scale, the power source of our sun and the very creation of the elements cannot be understood without knowledge of quantum mechanics.

The fundamental postulates of quantum mechanics are no different in principle than those of classical mechanics or electricity and magnetism. They are postulates with truth and reality associated with them in the same manner. They represent physically-tested principles that describe physical processes and observables. They represent, as far as we can tell, an accurate description of the physical universe.

6.2
Elements of Quantum Mechanics

6.2.1
The Schrödinger Equation and Conservation Laws

The fundamental postulate of quantum mechanics is that the state of a nonrelativistic system is described by a function that is a solution to the Schrödinger equation,

$$H\Psi = E\Psi \tag{6.1}$$

The functions Ψ are referred to as wave functions. H is the Hamiltonian of the system and is the sum of the kinetic and potential energies of all particles present and $E = T + V$ is the total energy of the system. The Schrödinger equation is then seen to be a statement of the conservation of energy, which may or may not be time-dependent depending on the nature of the forces that are present. We will use the upper case Ψ to represent wave functions in time-dependent problems and the lower case ψ to represent wave functions of time-independent problems. If we deal with time-independent problems, we have the analog of a static problem in classical mechanics or an equilibrium problem in chemistry. The properties of the system do not and cannot vary in time. A "stable" nucleus will be described by the time-independent Schrödinger equation. Its states are calculated in the limit that they cannot undergo decay of any form. While this might be true of the ground states of some nuclides, we know it is patently untrue of the ground states of most nuclides and all of the excited states of all nuclides. Hence, the description of most nuclear states in the time-independent limit is clearly an approximation and all problems associated with radioactive decay and nuclear reactions are fundamentally time-dependent problems. Nevertheless, under most conditions, the neglect of time dependence in calculating the properties of most nuclear states with energies less than the binding energy of a neutron or proton produces errors so small as to be negligible.

The wave functions represent as complete a description of the state of a system as is possible. All of the variables that describe a system, such as position, momentum, energy, etc., are represented in quantum mechanics as *operators*. The operators, acting upon the wave function, provide the magnitudes of the dynamical variables they represent. Some of these variables are *conserved* quantities and thus will be *constants* of the motion. It is these that we will use to characterize a system. Some of the variables, such as position and momentum, cannot both be simultaneously determined with precision. There is an inherent *uncertainty* associated with such simultaneous measurements. We will restrict our attention to those variables and their operators that are important for the problems we address and will leave others to the formal study of quantum theory.

The classical conservation laws are upheld for quantized systems over time intervals that are "long enough". The familiar conservation laws of energy, momentum and angular momentum, for an isolated system subject only to internal conservative forces, are upheld under normal conditions. However, quantum mechanics shows that these laws can be violated over times that are "short". In particular, the uncertainty in the energy of a system is related to the uncertainty in time by the Heisenberg Uncertainty Principle

$$\Delta E \Delta t \geq \hbar \qquad (6.2)$$

where E is the total energy and t is time. Clearly, for times that are short compared to $\hbar/\Delta E$, the energy of the system can violate conservation of energy by an amount ΔE. Because $\hbar$ is such a small quantity, significant differences between the energy and that expected from conservation of energy can exist for only extremely small times. For the most part, we will not have to deal with this uncertainty directly. However, we will find that levels that can decay do not have an exact "sharp" energy and this is a direct reflection of the Uncertainty Principle. The energy distributions of states are revealed with great clarity in the probabilities for certain nuclear reactions and, indeed, are fundamental to the operation of lasers.

If a particle is moving in field-free space, we are able to vary its energy and momentum essentially continuously as in the classical case. But if a particle is constrained to some region of space, such as in our particle in a box model, or in an atom, molecule or nucleus, only a restricted set of energies and momenta are possible. That is, constrained systems will have certain of their dynamical variables *quantized*. In the following, we will assume, unless otherwise specified, that we are dealing with constrained systems.

6.2.2
Elementary Properties of Operators

All measurable dynamical variables are represented by operators that can be symbolized generally as O_{op}. An operator acting on the wave function of the system, ψ, yields the magnitude of the dynamical variable it represents. For an isolated system in field-free space, acted on only by internal conservative forces, the magnitudes of certain variables will be constants of the motion. If O_{op} represents one of these variables, then

$$O_{op}\psi = O\psi \tag{6.3}$$

where O is the magnitude of the variable in the state ψ that is represented by O_{op} and it has a discrete value. In such a case, we say that ψ is an *eigenfunction* of the operator O_{op} and the quantity O is the *eigenvalue* of the operator in the state ψ. In most of the cases we consider, O_{op} will usually refer to the total energy, the angular momentum, etc.

Because the variables we use to describe real systems are themselves *real* quantities, the only acceptable operators are those whose eigenvalues are real. This may sound like a "truism" but, as we shall see in a moment, this requirement results in some very important ramifications and restrictions.

Some common operators with which we will deal are given in Table 6.1, referred to either the Cartesian coordinates in x, y, z and t, or the spherical polar coordinates r, θ, φ and t. The operators representing the spatial coordinates, time, mass and charge are simply the variables themselves. However, the operators for a component of linear momentum, the total energy, E, in the general case of a time-dependent problem, and the operators for a component of angular momentum and the square of the total angular momentum are all expressed as partial differential operators. Although it is not given explicitly in the table, it should be clear that the operator for the total linear momentum will also be in the form of a partial differential operator.

Table 6.1 The quantum mechanical operators for some common physical quantities in Cartesian coordinates (x, y, z, t) or spherical polar coordinates (r, θ, φ, t).

Physical quantity	Operator (O_{op})
x, y, z, t	x, y, z, t
r, θ, φ, t	r, θ, φ, t
charge e, mass m	e, m
component of linear momentum, p_q, q = (x, y, z)	$p_{op} = \dfrac{\hbar}{i}\dfrac{\partial}{\partial q}$
Hamiltonian H = T + V	$H_{op} = T_{op} + V_{op} = -\dfrac{\hbar^2}{2m}\nabla^2 + V_{op}$
Total Energy E (time dependent)	$E_{op} = -\dfrac{\hbar}{i}\dfrac{\partial}{\partial t}$
z-component of angular momentum, l_z	$(l_z)_{op} = \dfrac{\hbar}{i}(r \times \nabla)_z = \dfrac{\hbar}{i}\dfrac{\partial}{\partial \phi}$
square of total angular momentum, l^2	$l^2_{op} = -\hbar^2\left[\dfrac{1}{\sin\theta}\dfrac{\partial}{\partial\theta}\left(\sin\theta\dfrac{\partial}{\partial\theta}\right) + \dfrac{1}{\sin^2\theta}\dfrac{\partial^2}{\partial\phi^2}\right]$

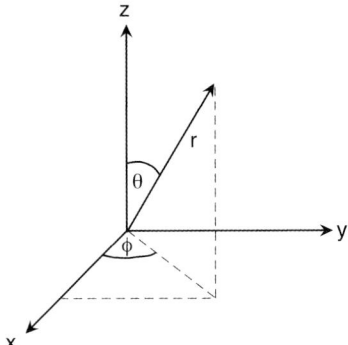

Fig. 6.1 Standard notation for the angles in a polar coordinate representation.

The form shown for the Hamiltonian operator is representative of how one can derive operators for other quantities that we may need. Writing the Hamiltonian as the sum of the kinetic plus potential energies, we use the fact that, for nonrelativistic systems, $T = p^2/2m$, and simply substitute the operator forms for both m and p. We have written the potential energy in general form because it remains unspecified here. For any given explicit form, we follow the same prescription of substituting the operators for the variables in which it is expressed, to get its operator.

We will make extensive use of angular momentum in the description of nuclear structure and for describing radioactive decay and nuclear reactions. While we will consider angular momentum further below, the important forms for the z-component and the square of angular momentum in spherical polar coordinates have been included in the table. It is not really difficult to derive these but it is a tedious bit of algebra. Further, the principal reasons for emphasis on the z-component of angular momentum will be clarified below. The expressions in terms of spherical polar coordinates make use of the standard labeling of angles as shown in Fig. 6.1.

We can let each of the operators given in Table 6.1 act, in turn, on the wave function of a particle or system of particles. If the wave function is an eigenfunction of the operator, the result of the operation will be the product of a constant times the wave function itself. The constant is the magnitude of the variable represented by the operator, exactly as indicated in Eq. (6.3).

6.2.3
Elementary Properties of Wave Functions

The wave function Ψ is usually a function of spatial coordinates (x,y,z) or (r,θ,φ), the intrinsic spins of the particles it describes, and time. The time-independent wave function ψ will be a function of the same variables except, of course, for time.

The wave function ψ of a system is given a probability interpretation. For a one-dimensional problem in the spatial coordinate q, the probability of finding the system between q and q + dq, i.e., q = (x, y, z), (r θ, φ), etc., is given by $|\psi|^2 dq$, the absolute value sign representing the fact that wave functions are generally complex quantities. If the system exists, it must be true that it can be found somewhere and thus

$$\int_{\text{all } q} |\psi|^2 dq = 1 \tag{6.4}$$

If we are dealing with a three-dimensional system acted on by internal forces that are spherically symmetric and therefore dependent only upon the radial coordinate, r, the probability of finding the system anywhere in space would be given by

$$\int_{\text{all space}} |\psi|^2 dV = \int_0^\infty 4\pi |\psi|^2 r^2 dr = 1 \tag{6.5}$$

It should now be clear that $|\psi|^2$ is a probability *density* so long as equations of the form of (6.4) and (6.5) are obeyed. To insure this, we say that all acceptable wave functions must be *normalizable*. They must satisfy Eqs (6.4) or (6.5). The wave function itself is referred to as the probability *amplitude*, in agreement with the normal use of this term in the description of classical electromagnetic waves.

There are additional restrictions that must be placed on wave functions if we are to maintain a probability interpretation for $|\psi|^2$. First, acceptable wave functions must everywhere be *single-valued* because there can be one and only one probability for finding a particle at any point in the space of the problem. Second, the wave function and its first derivative must everywhere be *finite*. Unless this restriction is met, there will be one or more values for the spatial coordinates where the probability is undefined. If a system is bound, the particles that it contains are localized in space and the probability of finding the system as $r \to \infty$ must be zero. That is, we must have that $\psi \to 0$ as $r \to \infty$. If the system is unbound, it can extend over all space and its wave function need not vanish as $r \to \infty$. In this case, the boundary condition turns out to require that the wave function be periodic as $r \to \infty$. This follows by simple consideration of the particle in the box problem that was discussed in detail in Chapter IV. There we found that the momentum eigenfunctions in each coordinate were proportional to sin kq and were subject to the quantization condition $k = n\pi/L$ where L was the length of the side of the box. An unbound particle would clearly be one in which $L \to \infty$. Finally, Eqs (6.4) or (6.5) show that it must be possible to integrate the quantity $|\psi|^2$ and thus acceptable wave functions must be *square integrable* functions.

6.2.4
Operators, Eigenfunctions and Conservation Laws

Operators and Expectation Values

With the above as preamble, it is now necessary to address some of the issues and consequences of the conservation laws. We know that the conservation laws are linear. As examples, the total energy of a system is the sum of the energies of the particles in a system and the total angular momentum is the vector sum of the angular momenta of the individual particles. Because of this, the operators representing dynamical variables must themselves be linear, i.e.,

$$A_{op}(c\psi_1 + d\psi_2) = cA_{op}\psi_1 + dA_{op}\psi_2 \tag{6.6}$$

where c and d are constants.

We have noted above that some dynamical variables will be conserved and thus are constants of the motion, and the operation of the operator representing such variables on the eigenfunction of the system will give the magnitude of these constants. Whether or not the variable a is a constant of the motion, its average or *expectation* value can be calculated from the definition

$$\langle a \rangle = \int \psi^* A_{op} \psi \, d\tau \tag{6.7}$$

Here the symbol $\langle \, \rangle$ stands for average or expectation value. The asterisk stands for the complex conjugate of ψ. The symbol $d\tau$ is a general notation for the differentials of all variables involved in the "space" of integration. For example, if ψ depends only on r, and we are integrating only over the spatial coordinates, then $d\tau = dV = 4\pi r^2 dr$. Thus, the expectation value of the radial coordinate would be calculated as

$$\langle r \rangle = \int \psi^* r_{op} \psi \, d\tau = \int_{\text{all } r} \psi^* r \psi 4\pi r^2 dr \tag{6.8}$$

Hermitian Operators

We can use the definition of the expectation value (Eq. (6.7)) to calculate the expectation value of the complex conjugate of A_{op}, A_{op}^*. If one simply takes the complex conjugate of Eq. (6.7), recognizing that $d\tau$ is real, we have

$$\langle a \rangle^* = \int (\psi^* A_{op} \psi)^* d\tau = \int \psi (A_{op}\psi)^* d\tau \tag{6.9}$$

But we know that the expectation values of all acceptable operators are real. That is, we must have

$$\langle a \rangle^* = \langle a \rangle \tag{6.10}$$

and Eqs (6.7) and (6.9) together give us

$$\int \psi^* A_{op} \psi \, d\tau = \int \psi (A_{op}\psi)^* d\tau \tag{6.11}$$

This seemingly arcane result has some very practical significance. Operators representing real quantities that obey Eq. (6.11) when the functions ψ are square integrable are known as *Hermitian* operators. We thus conclude that all acceptable operators representing real quantities in quantum mechanics are Hermitian operators.

Now we can easily show that the eigenfunctions of an Hermitian operator are *orthogonal*. Suppose that we know two nonzero square integrable functions ψ_k and ψ_n that are both eigenfunctions of the Hermitian operator A_{op}. Then

$$A_{op}\psi_k = a_k\psi_k \tag{6.12}$$

where a_k is the eigenvalue of A_{op} when the system is in the eigenstate ψ_k. Now from the discussion immediately above we can also write

$$(A_{op}\psi_n)^* = (a_n\psi_n)^* = a_n^*\psi_n^* = a_n\psi_n^* \tag{6.13}$$

If we multiply Eq. (6.12) from the left by ψ_n^* and Eq. (6.13) from the left by ψ_k and take the difference between the left-hand sides of these equations, we obtain

$$\psi_n^* A_{op}\psi_k - \psi_k(A_{op}\psi_n)^*,$$

and if we integrate this expression over all "space", we have

$$\int[\psi_n^* A_{op}\psi_k - \psi_k(A_{op}\psi_n)^*]d\tau = \int[\psi_n^* a_k\psi_k - \psi_k a_n\psi_n^*]d\tau$$
$$= (a_k - a_n)\int\psi_n^*\psi_k d\tau \tag{6.14}$$

Now the operator A_{op} is an Hermitian operator and Eq. (6.11) says that the left-hand side of Eq. (6.14) must be zero. In the general case, $(a_k - a_n) \neq 0$, and we then arrive at the important result that

$$\int\psi_n^*\psi_k d\tau = 0 \tag{6.15}$$

We will not prove it but it can be shown that the result above also applies when $(a_k - a_n) = 0$. In cases where two different eigenfunctions happen to have the same values we say the states are *degenerate*. Nevertheless ψ_n and ψ_k are distinct quantum states.

Eq. (6.15) is the definition of orthogonal functions and one usually sees the orthogonality of the eigenfunctions of operators in quantum mechanics written as

$$\int\psi_n^*\psi_k d\tau = \begin{cases} 1, \text{ if } n = k \\ 0, \text{ if } n \neq k \end{cases} \tag{6.16}$$

It also can be shown that the eigenfunctions of an Hermitian operator form a *complete* set. There are no possible eigenfunctions that are absent from this set.

Implications

The importance of the discussion above cannot be overemphasized. As you know, any reasonably smooth function can be expanded in a complete set of orthonormal functions in a unique manner. By unique we mean that the coefficients a_i of each term in the expansion can be assigned a well-defined meaning relative to the characteristics of the term it represents. Such expansions are commonly introduced in calculus courses with Fourier series. It is shown there that an arbitrary, reasonably smooth waveform can be decomposed uniquely in a set of orthogonal oscillatory functions, and that the coefficient of each term in the expansion represents the contribution of a specific frequency to the amplitude of the wave. In the present case we can now assert that, because all operators in quantum mechanics are Hermitian and the eigenfunctions of each form a complete orthonormal set, the *eigenfunctions of any Hermitian operator can be expanded uniquely in the eigenfunctions of any other operator.*[1]

This is very powerful. Suppose we have some eigenfunction ψ' that is expanded in the eigenfunctions ψ of some other Hermitian operator. We can write this as

$$\psi' = \sum_{\text{all } k} a_k \psi_k \tag{6.17}$$

Now we can write the normalization of ψ' as

$$\int |\psi'|^2 d\tau = \int \left| \sum_{\text{all } k} a_k \psi_k \right|^2 d\tau = 1 \tag{6.18}$$

To evaluate the absolute square in the middle term of Eq. (6.18), we proceed explicitly as follows;

$$\begin{aligned}(a_0\psi_0 + a_1\psi_1 + a_2\psi_2 + a_3\psi_3 + \ldots)(a_0\psi_0 + a_1\psi_1 + a_2\psi_2 + a_3\psi_3 + \ldots)^* = \\ (a_0 a_0^* \psi_0 \psi_0^* + a_0 a_1^* \psi_0 \psi_1^* + a_0 a_2^* \psi_0 \psi_2^* + a_0 a_3^* \psi_0 \psi_3^* + \ldots \\ a_1 a_0^* \psi_1 \psi_0^* + a_1 a_1^* \psi_1 \psi_1^* + a_1 a_2^* \psi_1 \psi_2^* + a_1 a_3^* \psi_1 \psi_3^* + \ldots \\ a_2 a_0^* \psi_2 \psi_0^* + a_2 a_1^* \psi_2 \psi_1^* + a_2 a_2^* \psi_2 \psi_2^* + a_2 a_3^* \psi_2 \psi_3^* + \ldots)\end{aligned} \tag{6.19}$$

The last three rows in Eq. (6.19) represent a portion of the individual terms that will be subject to integration over the space represented by $d\tau$. Remembering that the

---

[1] In this case we must be careful with the meaning of the word "unique". The expansion of an eigenfunction in the eigenfunctions of an Hermitian operator will lead to definite magnitudes of the coefficients in each term in the expansion, and definite signs for the coefficients relative to one another. However, the sign of any single term is undefined. This ambiguity does not affect the magnitudes of dynamical variables.

eigenfunctions are orthogonal, the integral of each term of the form $a_k a_n^* \psi_k \psi_n^*$ with $k \neq n$ will vanish. Only those for which $k = n$ will remain, and because $a_k a_k^* = a_k^2$, we can now rewrite Eq. (6.18) directly as

$$\int |\psi'|^2 d\tau = \int \left| \sum_{\text{all } k} a_k \psi_k \right|^2 d\tau = \sum_{\text{all } k} \int a_k^2 |\psi_k|^2 d\tau = \sum_{\text{all } k} a_k^2 = 1 \qquad (6.20)$$

Because the coefficients of an expansion in orthonormal functions are unique, and because of the normalization of the wave function, the quantities a_k^2 can now be given probability interpretations. Namely, a_k^2 represents the probability that the system described by the eigenfunction ψ' can be found in the state ψ_k that is an eigenfunction of some other Hermitian operator. This has powerful implications for our work with simple models to represent the physics of nuclei and their transformations.

Complex Nuclei and Models

When dealing with complex nuclei, we are immediately faced with a many-body problem that cannot be solved analytically. Further, we do not know an exact analytical expression for the interaction between nucleons in the nucleus. We will have to resort to some model approximations. This is where the expansion in orthonormal functions of the right choice will make a great deal of sense. Let us assume that we are interested in determining the states that are available to some nuclide of mass number A. We will look for time-independent solutions of the Schrödinger equation

$$H\psi = E\psi \qquad (6.21)$$

where E is the total energy of the system. The Hamiltonian operator represents the sum of the kinetic and potential energies of all of the nucleons. What do we assume for the potential energies? Without being explicit as to form, we can assume that the potential experienced by a single nucleon is that due to its interaction with all other nucleons individually. That is, there are no special interactions between a group of, say, four nucleons that are not included in the sum of the interactions of each nucleon with the others individually. One says that the total force experienced by a nucleon in the nucleus is the sum of *two-body* interactions only. We can easily write down the sum of the kinetic energies of all particles but summing of the potential energies must be taken with a little care to prevent double counting. This can be ensured by reference to the simple matrix shown below.

```
    1    2    3    4   ...
1       V₁₂  V₁₃  V₁₄  ...
2  V₂₁       V₂₃  V₂₄  ...
3  V₃₁  V₃₂       V₃₄  ...
4  V₄₁  V₄₂  V₄₃       ...
   ⋮    ⋮    ⋮    ⋮    ⋮
```

Each entry $V_{i,j}$ in the matrix represents the potential energy of interaction between the particle in the ith row with the particle in the jth column. Now $V_{i,j} = V_{j,i}$. We can then be sure to avoid double counting by adding the potential energies only for $j > i$. With this, we can now rewrite Eq. (6.21) for a nucleus containing A nucleons as

$$H_{op}\psi = \left[\sum_{i=1}^{A} T_{op}(i) + \sum_{i=1, j>i}^{A} V(i,j)\right]\psi = E\psi \tag{6.22}$$

This is obviously a formidable problem to solve. The question is, how can we approximate a solution to Eq. (6.22) that permits physical insight and also has a well-defined relation to the real problem? The idea is to choose some simple model that we can easily solve and understand, and see how it relates to the problem posed in Eq. (6.22).

Suppose we make the assumption that the force experienced by any one nucleon in the nucleus is just the *average* force due to all other nucleons present. Neglecting the Coulomb potential, this means that each nucleon would see the same potential and would behave as though it were an independent particle in the nucleus. This is indeed the basic assumption of the single-particle model (SPM) or independent-particle model (IPM) of the nucleus. If the average potential is written as V_{SP}, the Schrödinger equation for the ith particle would be of the form

$$H_{i,SP}\psi = E_{i,SP}\psi \tag{6.23}$$

and the total Hamiltonian for the nucleus would simply be the sum of the Hamiltonians for all A nucleons. If we sum the potentials to get the total average V_{SP}, we count the potential of particle 1 interacting with particle 2 twice. Therefore, in the extreme independent particle model approximation, the total Hamiltonian would be given by

$$H_{IPM} = \sum_{i=1}^{A}\left[T_{op}(i) + \frac{1}{2}V_{SP}(i)\right] \tag{6.24}$$

Because each of the particles behaves independently, we actually have A independent equations of the form of (6.23), and the total energy will simply be the sum of the energies of all of the independent particles.

Now, how does this model relate to the real problem? This is easily seen by adding and subtracting the total potential from Eq. (6.24) to Eq. (6.22). That is,

$$H_{op}\psi = \left[\sum_{i=1}^{A} T_{op}(i) + \sum_{i=1}^{A} \frac{1}{2} V_{SP}(i) + \sum_{i=1, j>i}^{A} V(i,j) - \sum_{i=1}^{A} \frac{1}{2} V_{SP}(i) \right] \psi = E\psi$$

$$= \left\{ \left[\sum_{i=1}^{A} T_{op}(i) + \sum_{i=1}^{A} \frac{1}{2} V_{SP}(i) \right] + \left[\sum_{i=1, j>i}^{A} V(i,j) - \sum_{i=1}^{A} \frac{1}{2} V_{SP}(i) \right] \right\} \psi = E\psi$$

$$= \left\{ H_{IPM} + \left[\sum_{i=1, j>i}^{A} V(i,j) - \sum_{i=1}^{A} \frac{1}{2} V_{SP}(i) \right] \right\} \psi = E\psi \qquad (6.25)$$

The last of Eqs (6.25) shows explicitly and formally what might be expected; the difference between the two models resides solely in the difference between their total potentials. If this difference is small, the model will closely approximate reality and the model eigenfunctions ψ_{sp} will closely approximate the eigenfunctions ψ of the real nucleus. The eigenfunctions will indeed not be exactly the same. But, because the eigenfunctions of the independent particle model form a complete orthonormal set, we can be sure that the eigenfunctions of the real nucleus can be expanded in a unique fashion in terms of the eigenfunctions of this model. Therefore, if we can understand the physics of a relatively simple model that is readily solved and is a reasonable approximation to reality, we will have a reasonable and unique description of the physics of the real nucleus.

The simple independent particle model has enough reality in it that it can be taken as a starting point. Knowledge of those aspects of nucleon–nucleon interactions that are neglected in the model, can themselves be modeled and treated as *perturbations*, or corrections, that permit closer agreement between model calculations and reality.

Conserved Quantities and Commutators

As in classical physics, the most useful means of describing a state of a system is in terms of conserved quantities, quantities that are constants of the motion, and one of the issues that we must address is: what are the conserved quantities and how do we know that they are conserved? In quantum mechanics it can be shown that variables representing conserved quantities that can be measured simultaneously with precision have the property that their operators *commute*.

The commutator of two operators A_{op} and B_{op} is symbolized by $[A_{op}, B_{op}]$ and is defined as

$$[A_{op}, B_{op}] = A_{op}B_{op} - B_{op}A_{op} \qquad (6.26)$$

Although not written explicitly, the sense of Eq. (6.26) is that the operators act on an eigenfunction and it is the result of this operation that defines the commutation properties. For example, the operation of the commutator in Eq. (6.26) on an arbitrary eigenfunction ψ can be written

$$[A_{op}, B_{op}]\psi = A_{op}B_{op}\psi - B_{op}A_{op}\psi \qquad (6.27)$$

If ψ is *simultaneously* an eigenfunction of both operators, and remembering that the operators are linear, we can write

$$\begin{aligned} A_{op}\psi &= a\psi \\ B_{op}\psi &= b\psi \\ [A_{op}, B_{op}]\psi &= A_{op}(b\psi) - B_{op}(a\psi) \\ &= ab\psi - ba\psi = 0 \end{aligned} \qquad (6.28)$$

That is, $[A_{op}, B_{op}] = 0$ and the operators are said to *commute*. This relation can obviously be repeated with all of the operators that represent variables which are constants of the motion. The overall result is that the commutators of all variable pairs that have the same eigenfunctions will be zero. A very handy result indeed.

What kind of operators can commute? Clearly all operators that are the variables themselves will commute. But what about operators, such as those for the components of linear momentum, that involve partial differential operators? Consider, for example, the commutator $[x_{op}, p_{x,op}]$.

$$\begin{aligned} [x_{op}, p_{x,op}]\psi &= x_{op}p_{x,op}\psi - p_{x,op}x_{op}\psi \\ &= x\frac{\hbar}{i}\frac{\partial \psi}{\partial x} - \frac{\hbar}{i}\frac{\partial}{\partial x}(x\psi) \\ &= \frac{\hbar}{i}\left(x\frac{\partial \psi}{\partial x} - \psi - x\frac{\partial \psi}{\partial x}\right) \\ &= -\frac{\hbar}{i}\psi \end{aligned}$$

and thus

$$[x_{op}, p_{x,op}] = -\frac{\hbar}{i} \qquad (6.29)$$

Unless the wave function was a constant, and this just does not occur, the commutator does not vanish. This means that we cannot find a state (a wave function) that is simultaneously an eigenfunction of both x_{op} and $p_{x,op}$. While we will not prove it, this means that one cannot *simultaneously* measure both x and p and expect both to have well-defined values. These variables will be connected by the Heisenberg Uncertainty Principle, such that

$$\Delta x \Delta p_x \geq \hbar \qquad (6.30)$$

This type of uncertainty is always associated with *conjugate pairs*, such as x and p_x, and t and E.

We can use these examples to state some general rules. Operator pairs that depend upon the same independent variable and one of which is in the form of a partial differential operator, will not commute. On the other hand, operator pairs that depend upon different independent variables, whether or not one or both are in the form of partial differential operators, will commute. The commutation relations have a profound effect on what we can expect to be precise quantities that describe quantum states. Important consequences of this property are discussed in the next two sections.

6.2.5
Parity

For the most part, the conserved quantities that are constants of the motion and that we will use to describe nuclear states, are those that are found as well in classical mechanics. To be sure, we will find that there are very substantial differences in how such conservation is expressed, but the general ideas remain intact. However, there are other properties that are not usually considered classically and yet express physical meaning that can be observed experimentally. The symmetry property of an eigenfunction with respect to the spatial operation of reflection through the origin of the coordinate system is probably the most important of these. It is referred to as the *parity* of the wave function and it is found to be *conserved* by both the Coulomb and strong interactions. It has very important implications with respect to nuclear structure and nuclear transformations. While at first sight it appears as no more than an abstract mathematical property, its conservation results, for example, in certain decay modes being absolutely forbidden under some conditions, and it has a marked effect on the decay constants of many others.

Everyone is familiar with the concept of even and odd functions. An even function in the variable x is symmetric in ± x, i.e., f(x) = f(–x), whereas the function is odd if f(x) = –f(–x). Similarly, a function of r can be even or odd in the same way and this is related to the parity. If $\psi(\mathbf{r}) = \psi(-\mathbf{r})$, a wave function is said to possess *even* parity. If $\psi(\mathbf{r}) = -\psi(-\mathbf{r})$, the wave function is said to possess *odd* parity. A schematic of the process of reflection through the origin is shown in Fig. 6.2.

The importance of this property is that, for the most part, the potential of a conservative force is symmetric about the origin of the coordinate system to which it is referenced. That is, V(x,y,z) = V(–x,–y,–z), or V = V(r). The potential is said to be spherically symmetric and it is a function only of the scalar magnitude of the radius parameter. With such a potential, the Hamiltonian of a particle is given by H = p^2/2m + V(r) and it is clear that H(x,y,z) = H(-x,-y,-z).

We can express the parity *formally* by defining a *parity operator*, P, that carries out the operation

$$P\psi(x, y, z) = c\psi(-x, -y, -z) \tag{6.31}$$

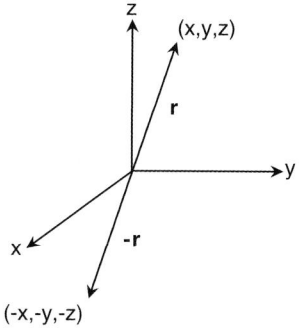

Fig. 6.2 Schematic diagram representing reflection through the origin. The vector **r** to a point in space becomes the vector −**r**.

where we assume that ψ is an eigenfunction of the operator and therefore c is the eigenvalue. If we apply the parity operator once again to the expression in Eq. (6.31), we get

$$P[P\psi(x, y, z)] = cP\psi(-x, -y, -z) = c^2\psi(x, y, z) \tag{6.32}$$

The application of the parity operator successively on any of its eigenfunctions must physically reproduce the wave function with which we started. That means that the eigenvalues of the parity operator must be defined by $c^2 = 1$, and thus c = ± 1. Eigenfunctions for which c = 1 are said to have *even* parity and those for which c = −1 are said to have *odd* parity.

Now it is not difficult to prove that the parity operator *commutes* with the Hamiltonian operator when the potential is a *scalar* function of the radial coordinate only. For an isolated system, acted on only by internal conservative forces, the total energy is conserved. Because the parity operator commutes with the Hamiltonian, we conclude that *parity is conserved* and therefore will be a constant of the motion. It is possible to extend this type of analysis to discover other dynamical variables that are conserved. It turns out that, as you would expect, angular momentum is conserved as well, but with a few twists compared to what we find in classical physics. Because of the central importance of angular momentum to the understanding of nuclear structure and nuclear transformations, we will now examine this variable in some detail.

6.3
Angular Momentum in Quantum Mechanics

Angular momentum in systems composed of fermions is complicated by the presence of the intrinsic angular momentum of the particles, in addition to the normal angular momentum that is familiar from classical physics. As one of the conserved quantities, angular momentum is one of the parameters that characterizes all states in a nucleus. Indeed, the angular momentum of states and the angular-momentum dependence of the strong interaction, give rise to the very shell structure that was introduced at the end of the previous chapter. Angular momentum conservation therefore plays a major role in all decay and nuclear reactions and in defining the lifetimes and cross sections of such processes. The intrinsic angular momentum and charge of the nucleons give rise to magnetic moments. Although small, these moments can be detected with very high sensitivity. On the microscopic scale, these moments are used to provide highly detailed information on the chemical environment of an atom and thus on the structure of complex molecules. On a macroscopic level, they are used to provide high-resolution images of human tissues in an essentially noninvasive manner. For all of these reasons, we will spend a fair bit of energy in developing the fundamentals of angular momentum in quantum mechanics and will continuously refer to the results we obtain throughout the remainder of this text.

6.3.1
Operators for Orbital Angular Momentum

The operators for the Cartesian components of angular momentum are easily obtained from the classical definitions and by the standard prescription of substituting the operator forms for the variables entering into their formulation. From the classical definition of angular momentum, $\mathbf{L} = \mathbf{r} \times \mathbf{p}$, where $\mathbf{r}$ is the radius vector of a particle and $\mathbf{p}$ is the particle's linear momentum, we can immediately use the operators in Table 6.1 to write

$$\mathbf{L}_{op} = \frac{\hbar}{i} (\mathbf{r} \times \nabla) \tag{6.33}$$

We can obtain the operator expressions for the individual Cartesian components by the simple expansion of the determinant from which the components are obtained classically. Thus,

$$\mathbf{L} = \mathbf{r} \times \mathbf{p} = \begin{vmatrix} \hat{i} & \hat{j} & \hat{k} \\ x & y & z \\ p_x & p_y & p_z \end{vmatrix} = (yp_z - zp_y)\hat{i} + (zp_x - xp_z)\hat{j} + (xp_y - yp_x)\hat{k} \tag{6.34}$$

Again using the operators from Table 6.1 we have immediately

$$L_x = \frac{\hbar}{i}\left(y\frac{\partial}{\partial z} - z\frac{\partial}{\partial y}\right)$$

$$L_y = \frac{\hbar}{i}\left(z\frac{\partial}{\partial x} - x\frac{\partial}{\partial z}\right) \tag{6.35}$$

$$L_z = \frac{\hbar}{i}\left(x\frac{\partial}{\partial y} - y\frac{\partial}{\partial x}\right)$$

In these equations, we have left out the subscript "op" and will do so in much of the remainder of the text unless it might lead to confusion.

Although it will turn out that we can consider angular momentum quite generally, we will refer to L as the *orbital* angular momentum. It is the quantum analog to the classical angular momentum with which we are all familiar.

In classical physics where particles move along well-defined trajectories, the total angular momentum and each of its components are constants of the motion for an isolated system acted on only by internal, conservative forces. But this is decidedly not the case in quantum mechanics as we will now demonstrate by use of the commutator properties of L_x and L_y. In the following, we will leave out the explicit reference to the wave function that we used in Eqs (6.27) and (6.28). From the expressions in Eqs (6.35) we can write

$$[L_x, L_y] = -\hbar^2\left(y\frac{\partial}{\partial z} - z\frac{\partial}{\partial y}\right)\left(z\frac{\partial}{\partial x} - x\frac{\partial}{\partial z}\right) + \hbar^2\left(z\frac{\partial}{\partial x} - x\frac{\partial}{\partial z}\right)\left(y\frac{\partial}{\partial z} - z\frac{\partial}{\partial y}\right)$$

$$= -\hbar^2\left[y\frac{\partial}{\partial z}\left(z\frac{\partial}{\partial x}\right) - y\frac{\partial}{\partial z}\left(x\frac{\partial}{\partial z}\right) - z\frac{\partial}{\partial y}\left(z\frac{\partial}{\partial x}\right) + z\frac{\partial}{\partial y}\left(x\frac{\partial}{\partial z}\right)\right] \quad (6.36)$$

$$+ \hbar^2\left[z\frac{\partial}{\partial x}\left(y\frac{\partial}{\partial z}\right) - z\frac{\partial}{\partial x}\left(z\frac{\partial}{\partial y}\right) - x\frac{\partial}{\partial z}\left(y\frac{\partial}{\partial z}\right) + x\frac{\partial}{\partial z}\left(z\frac{\partial}{\partial y}\right)\right]$$

This terrible looking mess actually simplifies quite nicely. By expanding the terms after the second equality, one finds that almost all of them cancel and one is left with

$$[L_x, L_y] = -\hbar^2\left(y\frac{\partial}{\partial x} - x\frac{\partial}{\partial y}\right) = i\hbar\left[\frac{\hbar}{i}\left(x\frac{\partial}{\partial y} - y\frac{\partial}{\partial x}\right)\right] \quad (6.37)$$

$$= i\hbar L_z$$

The conclusion is that the x and y components of orbital angular momentum do not commute. This means that we will *not* find a wave function that is simultaneously an eigenfunction of the two operators. Further, if you go through similar procedures, you can also prove that

$$[L_z, L_x] = i\hbar L_y$$
$$[L_y, L_z] = i\hbar L_x \quad (6.38)$$

None of the operators of the Cartesian components of orbital angular momentum commutes with one another. If we find an eigenfunction of one of the components, that component will be determined precisely for a state but the other two will not be precisely defined. However, by writing $L^2 = L_x^2 + L_y^2 + L_z^2$, you can show directly that L^2 commutes with *each* of the components of orbital angular momentum.

We thus arrive at the very important result that, because the total angular momentum is conserved for an isolated system acted on only by conservative internal forces, the square of the total, which measures the scalar length of the angular momentum vector, and one of the Cartesian components, can have the same eigenfunctions. Therefore, each state of a system will be characterized by discrete values of both the square of the angular momentum and one of the Cartesian components.

A few words are in order concerning the sources of angular momentum with which we must deal. To begin with, we have the intrinsic angular momentum of the fermions. Each fermion can occupy a state that has orbital angular momentum, and thus the total angular momentum of a nucleus must be the vector sum of the intrinsic spin and orbital angular momentum of all of the nucleons. For some nuclei, it is found that rotational motion of the nucleus as a whole is possible and in such cases the total angular momentum must account for this as well. In all cases, it is the total angular momentum that is conserved for an isolated system. Although we will not prove it, angular momentum is angular momentum, regardless of its source. Therefore, the mathematical development and the conclusions drawn above apply to all sources of angular momentum, regardless of the names we attach to them.

We now have a list of some of the properties of states of an isolated system that are conserved and can be defined precisely; the total energy, the square of the angular momentum, one of the components of angular momentum and the parity. These are the fundamental properties of states that we will refer to over and over as we proceed further.

6.3.2
Angular Momentum and Magnetic Moments

The mathematics required for the detailed development of angular momentum in quantum mechanics goes well beyond the scope of the present text. To meet our needs for manipulation of angular momenta, we will develop a simple vector model that can be made plausible through a short discussion of magnetic moments.

In classical mechanics, the motion of a charged particle in a current loop gives rise to a magnetic moment that is closely related to the angular momentum of the particle motion. Suppose we have a particle of mass m and charge q moving in a circular orbit as shown in Fig. 6.3. The angular momentum of the particle about the axis of the loop is $\mathbf{L} = \mathbf{r} \times \mathbf{p} = \mathbf{r} \times m\mathbf{v}$, where $\mathbf{v}$ is the linear velocity tangent to the loop, and the direction of the angular momentum vector is given by the right-hand rule. The magnetic moment generated by the motion is given by

$$\mu = iA \frac{\mathbf{r} \times \mathbf{v}}{|\mathbf{r} \times \mathbf{v}|} \tag{6.39}$$

where $\mu = iA$ is the magnitude of the moment, i is the current generated by the motion of the particle and A is the cross sectional area of the loop. The direction of μ is also given by the right-hand rule and is seen to be parallel to $\mathbf{L}$, if the current is positive, and antiparallel to $\mathbf{L}$ if the current is negative.

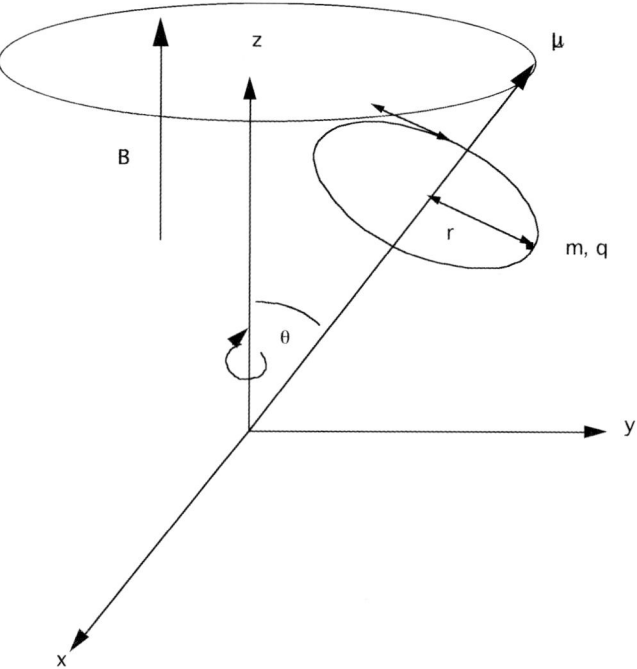

Fig. 6.3 Schematic diagram of a particle of mass m and charge q moving in a circular loop about an axis at the angle θ relative to the z-axis. The motion of the particle produces a magnetic moment μ parallel to the angular momentum vector **L**. In the presence of the magnetic field **B**, the moment μ precesses about the z-axis.

Now suppose that a uniform magnetic field **B** is superimposed parallel to and pointing in the direction of the positive z-direction. The magnetic field interacting with the magnetic moment produces a torque

$$\tau = \mu \times \mathbf{B} \tag{6.40}$$

that will produce a constant acceleration about the field direction as shown in Fig. 6.3. If the moment vector is fixed at an angle θ with respect to the z-axis, μ will rotate about the z-axis at a constant frequency. The kinetic energy of the rotation is given by

$$\mathbf{T} = \mu \cdot \mathbf{B} = |\mu||\mathbf{B}|\cos\theta \tag{6.41}$$

and the angular frequency of rotation can be obtained from this by noting that $v = r\omega$.

Fermions, such as neutrons, protons and electrons, have intrinsic magnetic moments and behave similarly to the simple classical picture discussed above. In particular, if one of these particles is placed in a magnetic field, it is found that its magnetic moment vector *never* aligns with the field direction. The same is true for the moments of complex nuclei and atoms. The vectors thus rotate or precess about the field direction at a constant frequency. Depending upon the magnitude of the moment, one finds a definite number of possible projections of the vector onto the field direction and each of these represents a distinct quantum state characterized by the magnitude of the projection of the magnetic moment along the field direction. Because the angular momentum vector is parallel to the magnetic moment, the above discussion can be taken to represent the behavior of the angular momentum vectors themselves. The number of projections of the total angular momentum vector onto the field direction is identical to that of the magnetic moment. Each projection represents a discrete quantum state of the particle or system.

Now suppose that we always viewed the angular momentum vectors in a coordinate system that rotated at the same frequency as the vectors themselves. The vectors would always be at rest and all we would see are different possible projections of the vector on the field direction. This is the simple construct we will use to visualize the allowed states of angular momentum.

6.4
The Vector Model for Angular Momentum

The intrinsic angular momentum of a fermion is referred to as the *intrinsic spin* and is usually given the symbol **s**. The magnitude of the vector is given by

$$|\mathbf{s}| = \hbar\sqrt{s(s+1)} \qquad (6.42)$$

where s is called the intrinsic spin quantum number. Theory and experiment show that s = 1/2 for all fermions and the intrinsic spin has two and only two projections with respect to a given direction in space. We will, for reasons that will become clear later, take the z-axis as our reference (Fig. 6.4). The magnitude of the projection of s on the z-axis, the z-component of intrinsic spin, is $s_z = m_s \hbar$, where $m_s = \pm 1/2$. The quantity m_s is referred to as the quantum number for the z-component of intrinsic spin. Each of these projections represents a discrete quantum state.

The fundamental physics represented by this picture is that there are two possible quantum states for a particle with angular momentum 1/2; one that gives a projection of $1/2\,\hbar$ on the z-axis in the presence of an external magnetic field and one that gives a projection of $-1/2\,\hbar$. Now, in the absence of an external field, there clearly is no real orientation of the spin vector; it can point anywhere. Nevertheless, the two states exist and are fundamental quantum states of the particle. They are degenerate in energy in the absence of an external applied field.

In addition to the intrinsic spin, a single particle may have various quantities of orbital angular momentum due to its motion in space, usually given the symbol **l**. As we noted above, the behavior of the angular momentum is the same regardless of its source and it should then come as no surprise to find that the magnitude of the orbital angular momentum is given by

6.4 The Vector Model for Angular Momentum | 153

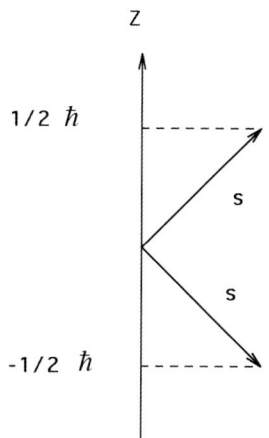

Fig. 6.4 Vector model for the intrinsic spin of a fermion. It is assumed that there is a magnetic field parallel to and directed along the positive z-direction.

$$|\mathbf{l}| = \hbar\sqrt{l(l+1)} \tag{6.43}$$

Experiment shows that the number of projections of **l** along the direction of an external magnetic field is always 2l + 1 where l is a positive integer 0, 1, 2, 3, ... In complete analogy with what we have shown for intrinsic spin, the maximum projection of |**l**| onto the z-axis is l $\hbar$, the minimum is –l $\hbar$, and all possible projections in between that differ from one another by $\hbar$ are also found. In Fig. 6.5 we show the vector model for the case that l = 2. The magnitude of the z-component of orbital angular momentum in this case is given by m_l $\hbar$, where m_l = 2, 1, 0, –1, –2, and m_l is called the quantum number for the z-component of orbital angular momentum. All of these projections represent degenerate states in the absence of an external field. Note that the number of m_l states is given by the same general formula as the number of m_s states of intrinsic spin, namely 2l + 1. It turns out that

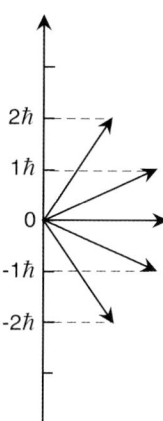

Fig. 6.5 Vector diagram for orbital angular momentum with l = 2.

the two cases above represent the only quantization of angular momentum that is allowed theoretically. As a result, the angular momentum quantum number of any system can only be integral or half-integral.

Now suppose that we have a single fermion in a state with $l > 0$. It must have a total angular momentum vector that is the sum of the vectors representing the intrinsic spin and the orbital motion. We will use the symbol j to represent the total angular momentum quantum number of a particle and thus $\mathbf{j} = \mathbf{s} + \mathbf{l}$. But how do we go about the angular momentum addition? The simplest way to see this is to examine the maximum possible values for the length of the resultant vector. This is shown in Fig. 6.6. The maximum possible projection of $\mathbf{l}$ on the z-axis is just $l\hbar$. Now the intrinsic spin vector can have only the two possible projections corresponding to the magnitudes $\pm \hbar/2$. If the projection is $\hbar/2$, the total projection of the two vectors on the z-axis is $(l + 1/2)\hbar$ (Fig. 6.6 (a)). The length of the vector corresponding to this coupling is $|\mathbf{j}| = \hbar\sqrt{j(j+1)}$ with $j = (l + 1/2)$. If the projection is $-\hbar/2$, the length of the corresponding vector is $|\mathbf{j}'| = \hbar\sqrt{j'(j'+1)}$ with $j' = (l - 1/2)$ (Fig. 6.6 (b)). Both of these couplings are found in nature and correspond to the only possible couplings allowed in quantum theory. Each of the two vectors $\mathbf{j}$ and $\mathbf{j}'$ follows the same rules as any other angular momentum vector. A vector $\mathbf{j}$ will have $(2j + 1)$ possible projections on the z-axis and each represents a single allowed state of *total* angular momentum for a single particle. A vector diagram similar to those in Figs 6.4 and 6.5 can be drawn. The projections of j on the z-axis will have magnitudes of $m_j \hbar$ where $m_j = j, (j-1), (j-2), \ldots, -(j-1), -j$.

As a specific example, suppose that a single fermion is in a state with $l = 2$. It will give rise to the two states of $j = 2 + 1/2 = 5/2$ and $j = 2 - 1/2 = 3/2$. The first will give rise to 6 states that can be occupied by the particle with z-components described by $m_j = 5/2, 3/2, 1/2, -1/2, -3/2$, and $-5/2$. The second will give rise to 4 states with $m_j = 3/2, 1/2, -1/2$, and $-3/2$. In total there are 10 states for a fermion with $l = 2$. But all of these states are *not* degenerate in energy in the absence of an external magnetic field. As you can imagine, the two different couplings can and do correspond to different energies. Just think of the interaction of two magnetic moments. Their energy of interaction will certainly depend upon their vector orientation. While all of the states corresponding to $j = l + 1/2$ are degenerate and

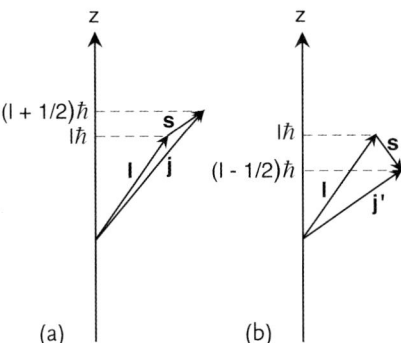

Fig. 6.6 Vector diagram for the addition of $\mathbf{l}$ and $\mathbf{s}$ for a single fermion.

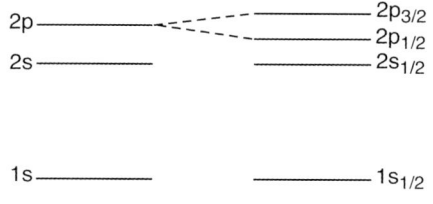

Fig. 6.7 The lowest energy levels in the hydrogen atom with and without spin orbit interaction. The levels are not drawn to scale.

all of the states of $j = l - 1/2$ are degenerate, the two sets of states will have different energies. This result may be familiar from the electron levels of the hydrogen atom (Fig. 6.7). The first state is the 1s state and corresponds to $l = 0$. The total angular momentum quantum number for an electron in this state is $j = 1/2$. The second state is the 2s state and it has the same characteristics. The third state is the 2p state corresponding to $l = 1$. It is *split* in energy due to the coupling of the orbital and intrinsic spin vectors of the electron. The state of $j = l - 1/2 = 1/2$, the $2p_{1/2}$ state, lies at a slightly lower energy than the state of $j = l + 1/2$, the $2p_{3/2}$ state. Similar splittings are found with all other states when $l > 0$. The splitting is referred to as *spin-orbit* splitting.

Spin-orbit splitting is also found for the states of neutrons and protons in nuclei. But here we see a very different picture. First, the splitting for nucleons is such that the states of $j = l + 1/2$ always lie *lower* in energy than the states of $j = l - 1/2$. Second, the energy difference between the two sets of states is *very large*. We will find that it is primarily responsible for defining the magic numbers that represent closed shells in the nucleus.

Before we leave this section we must address the problem of the coupling of the total angular momentum of a number of fermions, and here we will specifically consider the coupling of particles in the nucleus. It turns out that, to an excellent approximation, the coupling of the angular momenta of nucleons can be treated as though the total angular momentum vector of each nucleon remains intact. This type of coupling is referred to as jj coupling. We will not have to deal with the question of interactions of the individual intrinsic spins and orbital angular momenta with one another. But if this is the case, we already know how to approach the problem.

We assume first that we have two particles in states characterized by the quantum numbers j and j´. They will give rise to states with total angular momenta **J** = **j** + **j**´. We can get the number of ways in which these can combine by use of a vector diagram similar to that shown in Fig. 6.6. As shown in Fig. 6.8, the maximum projection of the total angular momentum vectors leads to a vector **J** with total quantum number $J = j + j´$. The smallest vector sum gives rise to a total angular momentum vector with quantum number $J = |j - j´|$. But there is, of course, no reason why all other possible couplings cannot be found and they all are.

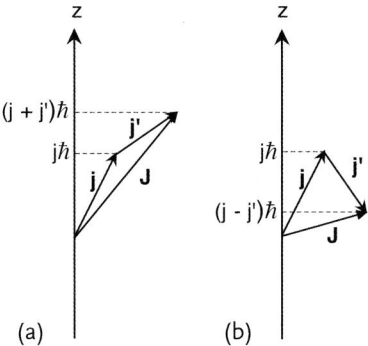

Fig. 6.8 Coupling of the angular momenta of two particles with different magnitudes of their angular momentum vectors. (a) The coupling that gives rise to the largest vector. (b) The coupling that gives rise to the smallest resultant vector.

In this case, each of the different couplings has a different energy and thus we will have a *spectrum* of states due to the presence of the two particles in these states of different j. How do we find the quantum numbers of the other states? We simply use the fact that the possible projections will differ from one another by integral multiples of $\hbar$. If the maximum sum is $j + j'$ and the minimum is $|j - j'|$, then all integral values in between will represent discrete quantum states of the two-particle system. We can summarize this by writing

$$|j - j'| \leq J \leq |j + j'| \tag{6.44}$$

As an example, the low-lying levels of some (odd, odd) nuclei can be described as due to the coupling of the odd proton and odd neutron. Suppose that one of these was in a state with $l = 4$ and $j = l + 1/2 = 9/2$, and the other was in a state with $l' = 3$ and $j' = l' - 1/2 = 5/2$. This coupling would give rise to states with angular momentum quantum numbers

$$|9/2 - 5/2| \leq J \leq |9/2 + 5/2|$$

There will then be states with quantum numbers 2, 3, 4, 5, 6, and 7. They will all have different energies.

The discussion of the coupling of the angular momenta of two fermions given above applies to the case of the coupling of a proton with a neutron in any states, or to the coupling of two *identical* particles so long as the latter are in states with *different* j. However, if we deal with the coupling of two identical particles in states belonging to the same j, a significant difference is found. As a direct consequence of the Pauli Exclusion Principle, only *even* values of the total angular momentum represent states that can exist. For two protons in a state with $l = 4$ and $j = l + 1/2 = 9/2$, for example, the only states that are allowed are states with J = 0, 2, 4, 6, and 8. It can be shown quite generally that to form states with J = odd, the two particles must have identical wave functions, and this is not possible. While there are much more sophisticated means of establishing this result, we can demonstrate it simply by considering the consequences of forming all possible couplings $m_{j1} + m_{j2} = m_J$.

Consider first the case of two identical particles in an $s_{1/2}$ or in a $p_{1/2}$ orbital. Two particles with $j = 1/2$ and $m_{j1} = 1/2$ and $m_{j2} = -1/2$ give rise to the allowed coupling of $J = 0$. To get a total angular momentum of 1 would require that both particles have the same value of m_j and this is forbidden by the Pauli Principle.

Now consider the case of two identical particles in a $p_{3/2}$ or $d_{3/2}$ orbital. In this case, the allowed values of m_j for a single particle are $3/2, 1/2, -1/2$ and $-3/2$. To determine all of the allowed couplings of the total z-component of the angular momentum for the pair, we simply write them down in a systematic way, remembering that each coupling can be found only once. Thus, for example, the coupling $m_{j1} = 1/2$, $m_{j2} = -1/2$ represents the same wave function as the coupling $m_{j1} = -1/2$, $m_{j2} = 1/2$ when we deal with *identical* particles. The allowed couplings are shown in Table 6.2. The table was begun by first considering all of the allowed couplings (total angular momentum) of m_{j2} with $m_{j1} = 3/2, 1/2, -1/2$ and $-3/2$. These are the entries in the second column. The same procedure is followed for the remaining columns. The entry x is made whenever $m_{j1} = m_{j2}$ or the coupling has already been accounted for. The table clearly shows that the only possible values of m_J are $2,1,0,-1,-2$, corresponding to the case that $J = 2$ and one additional $m_J = 0$ corresponding to the case that $J = 0$. We conclude that only states with J = even can be found for two identical particles in an orbit with $j = 3/2$. One can follow a similar procedure for all other values of j and find the same result. Although we will not be concerned with it, you can extend this method to determine the allowed couplings for more than two identical particles in the same orbit.

Finally, a practical note. Practitioners in the field quite commonly refer to an angular momentum vector by quoting the magnitude of the angular momentum quantum number that describes it, regardless of whether one is referring to the intrinsic spin of a particle, the total angular momentum of a particle or the total angular momentum of a nucleus. One will usually say that the magnitude of the intrinsic spin of a fermion is 1/2, etc. Further, although it is sloppy, the word "spin" is often used to refer not to just the intrinsic spin but to the total angular momentum as well. Most of the time, the meaning of "spin" can easily be inferred from the context in which it is used.

Table 6.2 The allowed couplings $m_{j1} + m_{j2} = m_J$ for two identical particles in an orbit with $j = 3/2$.

m_{j1} \ m_{j2}	3/2	1/2	-1/2
3/2	x	x	x
1/2	2	x	x
-1/2	1	0	x
-3/2	0	-1	-2

6.5
The Wave Functions of Many-Particle Systems

Before we end this review of some of the facts of quantum mechanics, it is important to consider a few properties of the wave functions of many-particle systems. In general, we deal with nuclei that have more than one particle that reside in different states of the quantized system. The wave function of the system must reflect this fact and must account, not only for the total angular momentum of all of the particles, but also the total parity as well. One way to see the general form that the wave function must have is to consider the case of a two-particle system such as the deuteron, ^2_1H, in its ground state. As will be shown in the next chapter, the lowest-lying state available to a fermion in any reasonable potential has an orbital angular momentum of zero, a so-called s state. Thus, for the deuteron in its ground state, both the neutron and the proton will have only the angular momentum due to their intrinsic spins, 1/2. If we ask for the probability of finding the deuteron anywhere in space, it must be given by the probability of finding both particles in their respective states. If the probability of finding particle "1" in the lowest-lying s state, the $1s_{1/2}$ state, is P_1 and the probability of finding particle "2" in the $1s_{1/2}$ state is P_2, the probability for finding the two in these states simultaneously is $P_1 P_2$. If the wave function for a two-particle system is written in the form $\psi_{1,2} \sim \psi_1 \psi_2$. The probability for finding the deuteron in space will be given by

$$\int_{\text{space}} |\psi_{1,2}|^2 dV \sim \int_{\text{space}} |\psi_1|^2 |\psi_2|^2 dV = \int_{\text{space}} |\psi_1|^2 dV \int_{\text{space}} |\psi_1|^2 dV \qquad (6.45)$$

exactly what we want. There are, of course, much more fundamental ways of arriving at this and the general result, but the above shows that the wave function of a many-particle system is expressed in terms of the *products* of the wave functions of the individual particles, i.e.,

$$\psi_N \sim \psi_1 \psi_2 \psi_3 \ldots \ldots \psi_n \qquad (6.46)$$

where N is the number of particles in the system and 1,2,3,.....,n label the individual particle states.

In general, the wave functions are quite complex because of the need to insure that they satisfy all conservation laws. We will not need to delve into these fundamentals at our level of study. But there is one important result for the form of the wave function that we should point out and will have need to use. The parity of a nucleus is a constant of the motion and the parity of each of the particles in the nucleus is also conserved. The form of Eq. (6.46) indicates that the total parity of an N-particle system is just the *product* of the parities of all of the individual particles.

General References

There are an abundance of introductory and advanced texts on quantum mechanics. The flavors of the texts vary because of the complexity of the topic and the proclivities of the authors. The interested reader is urged to try a few and find out what suits her/his taste. Rather than suggest one or more texts, I suggest that the reader consider what is offered for courses on quantum mechanics at various institutions and try them out.

Problems

1. The wave function of a particle in spherical polar coordinates (r, θ, ϕ) is $\psi = A \cos(kr)$. Does the function have even or odd parity? Prove your answer.

2. Determine the commutator for E_{op} and t_{op}, $[E_{op}, t_{op}]$.

3. Prove explicitly that $[L_z, L_x] = i\hbar L_y$.

4. A nuclear state ψ is expanded in terms of the eigenfunctions of the operator Φ_{op} and has the form

$$\psi = \frac{\sqrt{15}}{5}\varphi_1 + \frac{3}{5}\varphi_2 + \frac{1}{5}\varphi_3$$

The operator Φ_{op} represents a dynamic variable f and has all of the required properties of a proper operator in quantum mechanics. The three functions $\varphi_1, \varphi_2, \varphi_3$ are normalized eigenfunctions of Φ_{op} such that

$$\Phi_{op}\varphi_1 = 2\varphi_1$$
$$\Phi_{op}\varphi_2 = 4\varphi_2$$
$$\Phi_{op}\varphi_3 = 8\varphi_3$$

(a) Show that ψ is properly normalized.
(b) What is the expectation value for f in the state ψ?

5. Determine the total number of states available for a fermion in a state with $l = 3$ in the following ways.
(a) Determine the total number of m_l states and then the number of m_s states available to each.
(b) Consider separately the two couplings $j = l + 1/2$ and $j = l - 1/2$. For each, determine the number of m_j states available and add them. Also, make tables for each of the couplings that list the values of m_j and the possible values of m_l and m_s that can give rise to them.

6. Follow the outline of the m scheme given in Table 6.2 and determine the values of J for two identical fermions in a state of $j = 5/2$.

7. A hypothetical nuclide can be described as an inert (even, even) core with two additional identical particles in an orbit with $l = 4$ and $j = 9/2$. This orbit can contain up to 10 particles. In the lowest energy configuration, the two particles will be paired and give rise to a total angular momentum of 0, i.e., one will have the quantum number m_j and the other will have the quantum number $-m_j$. But it is possible that the two particles can be unpaired and still reside in the same orbit. The angular momenta of the two particles would thus not be zero.

(a) What are the total angular momentum quantum numbers of all configurations expected by exciting the nucleus such that the two particles remain in this orbit?

(b) The next highest single particle state has $l = 6$ and $j = 13/2$. Assuming that one of the particles is excited into this orbital, what are the total angular momentum quantum numbers of all of the possible configurations for one particle in the orbital with $l = 4$ and $j = 9/2$ and one particle in the orbital with $l = 6$ and $j = 13/2$?

7
Nuclear Structure: The Spherical Shell Model

7.1
Introduction

The structure of nuclei is clearly a very complicated matter. Modern theoretical approaches are quite complex and can, in many instances, provide very good comparisons with experimental data. In this chapter we will introduce one of the simplest models of nuclear structure that makes some sense; the simple independent particle model. The model provides insight into some of the fundamental physics involved in nuclear structure without undue complications, and it also serves as a means of understanding nuclear shell structure and empirical data on the low-lying level structure of some nuclides. We do not expect to be able to obtain quantitative accuracy. But we can get semiquantitative results that allow both an understanding of underlying principles and an understanding of the limitations that can be removed only by considerably more complex approaches.

7.2
The Independent Particle Model

We begin with the assumption that the nucleus can be treated as a spherical object in which nucleons move under the influence of an average spherically-symmetric potential due to the presence of all other nucleons. For the present, we neglect the Coulomb potential among the protons and thus explicitly treat neutrons only. The neglect of the Coulomb interaction will be rectified, in part, when we introduce the empirical single particle level diagrams for neutrons and protons.

In the center of mass coordinate system, the Schrödinger equation for a single particle can be written as

$$-\frac{\hbar^2}{2\mu}\nabla^2\Psi + V(r,s)\Psi = i\hbar\frac{\partial \Psi}{\partial t} \quad (7.1)$$

where μ is the reduced mass of the neutron and is given by

Nuclear Physics for Applications. Stanley G. Prussin
Copyright © 2007 WILEY-VCH Verlag GmbH & Co., Weinheim
ISBN: 978-3-527-40700-2

$$\mu = \frac{m_n M}{m_n + M} \cong \left(\frac{A-1}{A}\right) m_n \tag{7.2}$$

The potential is written as a function of both the radial coordinate and spin. It is actually more complicated than this but consideration of these two parameters will provide the basis for a reasonable comparison with experiment. We treat the problem in spherical polar coordinates and will consider only time-independent, or stationary solutions that can be written in the separated form

$$\Psi = \psi(r)\chi(s)T(t) \tag{7.3}$$

that is the product of functions that depend only on spatial coordinates, the spin of the particle and time, respectively. Direct substitution of this into Eq. (7.1), followed by multiplication by Ψ^{-1}, leads to

$$-\frac{\hbar^2}{2\mu}\frac{1}{\psi\chi}\nabla^2\psi\chi + V(r,s) = \frac{i\hbar}{t}\frac{\partial T}{\partial t} = \text{constant} = E \tag{7.4}$$

From the form of the left-hand side of Eq. (7.4), it is seen that the constant must be the total energy of the system, E. Thus, the technique of *separation of variables* results in the two equations

$$\frac{\hbar^2}{2\mu}\nabla^2\psi\chi + [E - V(r,s)]\psi\chi = 0$$

$$i\hbar\frac{\partial T}{\partial t} = ET \tag{7.5}$$

The second of Eqs (7.5) is easily integrated to give

$$T = Ce^{-\frac{iEt}{\hbar}} \tag{7.6}$$

where C is the constant of integration. We will need to consider this result later when we consider the problem of radioactive decay and nuclear reactions. But for now we concentrate on the first of Eqs (7.5) that provides the allowed energy states of the model in the absence of decay.

The first issue we face is the specification of the potential function. To make the problem as simple as possible, we will assume that the main part of the potential can be taken as an average, spherically-symmetric central potential that we write as V(r). The spin-dependent part is assumed to be sufficiently small in magnitude that its effects on the energies of states can be added as a "perturbation" to the results obtained using the central potential only. With this assumption, we can neglect the spin-dependent part of the wave function for the moment and rewrite the first of Eqs (7.5) in the form

$$\frac{\hbar^2}{2\mu}\nabla^2\psi + [E - V(r)]\psi = 0 \tag{7.7}$$

The assumption that the potential is spherically symmetric suggests that we consider a further separation of variables. We first search for wave functions ψ of the form

$$\psi(\mathbf{r}) = \psi(r, \theta, \phi) = R(r)Y(\theta, \phi) \tag{7.8}$$

In spherical polar coordinates,

$$\nabla^2 = \frac{1}{r^2}\frac{\partial}{\partial r}\left(r^2\frac{\partial}{\partial r}\right) + \frac{1}{r^2}\left[\frac{1}{\sin\theta}\frac{\partial}{\partial \theta}\left(\sin\theta\frac{\partial}{\partial \theta}\right) + \frac{1}{\sin^2\theta}\frac{\partial^2}{\partial \phi^2}\right] \tag{7.9}$$

Although it appears a bit complicated, the expression in Eq. (7.9) follows directly from the expression for ∇^2 in Cartesian coordinates and the definitions of (r,θ,ϕ) in terms of (x,y,z). The derivation is straightforward but tedious.

By direct substitution of Eqs (7.8) and (7.9) into Eq. (7.7) and following the routine used above, we obtain the equality

$$\tag{7.10}$$

$$\frac{1}{R}\frac{\partial}{\partial r}\left(r^2\frac{\partial R}{\partial r}\right) + \frac{2\mu r^2}{\hbar^2}[E - V(r)] = -\frac{1}{Y}\left[\frac{1}{\sin\theta}\frac{\partial}{\partial \theta}\left(\sin\theta\frac{\partial Y}{\partial \theta}\right) + \frac{1}{\sin^2\theta}\frac{\partial^2 Y}{\partial \phi^2}\right] = \lambda$$

where λ is a constant whose form and meaning is yet to be determined. We can now write the separated equations as

$$\frac{1}{R}\frac{\partial}{\partial r}\left(r^2\frac{\partial R}{\partial r}\right) + \frac{2\mu r^2}{\hbar^2}[E - V(r)] = \lambda$$

$$\left[\frac{1}{\sin\theta}\frac{\partial}{\partial \theta}\left(\sin\theta\frac{\partial Y}{\partial \theta}\right) + \frac{1}{\sin^2\theta}\frac{\partial^2 Y}{\partial \phi^2}\right] = -\lambda Y \tag{7.11}$$

Only the first of Eqs (7.11) contains the potential. Therefore the assumption that the space-dependent potential is a function solely of the scalar radial coordinate implies that all effects of the potential will be contained in the radial part of the wave function, $R(r)$. The angular part of the wave function, $Y(\theta,\phi)$, being completely independent of V, *will be universal for all $V(r)$*. The importance of this result cannot be overstated. We may not know the interaction potential between two nucleons exactly, and model approximations for $V(r)$ may differ considerably. Nevertheless, so long as we can be sure that the interaction potential does not depend on the angular coordinates θ and ϕ, the angular wave functions $Y(\theta,\phi)$ will be the same for all such potentials. As we will see shortly below, the $Y(\theta,\phi)$ define both the angular momentum and the parity of the wave functions. We may not know the energies of interactions exactly, but we can know both the angular momentum and parity exactly.

We can now go one step further in separating the variables. If we assume that $Y(\theta, \phi) = \Theta(\theta)\Phi(\phi)$, the second of Eqs (7.11) can also be separated by the same approach we have applied above. If we let the separation constant be written as v^2, we have

$$\frac{\sin\theta}{\Theta}\frac{\partial}{\partial\theta}\left(\sin\theta\frac{\partial\Theta}{\partial\theta}\right) + \lambda\sin^2\theta = -\frac{1}{\Phi}\frac{\partial^2\Phi}{\partial\phi^2} = v^2 \qquad (7.12)$$

Combining this result with the radial equation from Eq. (7.11), we have the set of equations

$$\frac{1}{R}\frac{d}{dr}\left(r^2\frac{dR}{dr}\right) + \frac{2\mu r^2}{\hbar^2}[E - V(r)] = \lambda$$

$$\sin\theta\frac{d}{d\theta}\left(\sin\theta\frac{d\Theta}{d\theta}\right) + (\lambda\sin^2\theta - v^2)\Theta = 0 \qquad (7.13)$$

$$\frac{d^2\Phi}{d\phi^2} + v^2\Phi = 0$$

Note that we have replaced the partial derivatives with total derivatives now that we have completely separated the variables.

7.2.1
The Angular Equations: Angular Momentum and Parity

Because the equations in the angular variables lead to universal functions in the spherical potential approximation, we will proceed to study these first. The last equation in the variable ϕ is not only the easiest to handle but we must solve it first to find the proper forms of the first two. It is similar to the equations which most students have met with in their calculus courses and, either by direct solution or by substituting back into the last of Eqs 7.13, it is easy to find the general solution

$$\Phi = Ae^{iv\phi} + Be^{-iv\phi} \qquad (7.14)$$

To obtain particular solutions that meet the requirements of eigenfunctions, we note that the separation of the variables implies that the Φ must be good eigenfunctions in their own right. In particular, they must be single-valued everywhere if we are to provide a probability interpretation to them. Because the range of ϕ is $0 \leq \phi \leq 2\pi$, this means that $\Phi(\phi) = \Phi(\phi + 2\pi)$. Either by direct inspection or by trial and error, it is not difficult to conclude that this requirement can be met only if v is an integer. Therefore, we replace v by the integer m and have the acceptable form of

$$\Phi = Ae^{im\phi} + Be^{-im\phi}, \; m = 0, \pm 1, \pm 2, \ldots \qquad (7.15)$$

We can now require that the Φ eigenfunctions be normalized. Thus,

$$\int_0^{2\pi} |\Phi|^2 d\phi = \int_0^{2\pi} |Ae^{im\phi} + Be^{-im\phi}|^2 d\phi = \int_0^{2\pi} (A^2 + B^2) d\phi = 1 \quad (7.16)$$

and the normalization requirement leads to

$$(A^2 + B^2) = \frac{1}{2\pi} \quad (7.17)$$

The normalization defines only the sum of the squares of the constants A and B and there is no other physical requirement that can be used to define the constants individually. This means that any set of A and B that satisfies Eq. (7.17) represents a physically-acceptable solution. We are completely free to use whatever set we choose. As a result, we might as well choose the simplest possible set that we can. Therefore, we arbitrarily take B = 0 and our eigenfunctions Φ can be written

$$\Phi = \frac{1}{\sqrt{2\pi}} e^{im\phi}, \; m = 0, \pm 1, \pm 2, \ldots \quad (7.18)$$

The Φ eigenfunctions are simple oscillatory functions whose real parts are of the form $\cos(m\phi)$. The integer m is one of the quantum numbers that describe a particular eigenfunction, and we can express its meaning immediately by recollection of the operator for the z-component of orbital angular momentum in terms of spherical polar coordinates as given in Table 6.1. Namely,

$$(L_z)_{op} = \frac{\hbar}{i} (\mathbf{r}_{op} \times \nabla_{op})_z = \frac{\hbar}{i} \frac{\partial}{\partial \phi}$$

Because the operator depends only on the variable ϕ and because the eigenfunctions Φ are universal so long as the potential is spherically symmetric, we can define the z-component of angular momentum without regard to the remainder of the wave function of a state. By direct substitution of Eq. (7.18), we find

$$(L_z)_{op} \Phi = \frac{\hbar}{i} \frac{\partial}{\partial \phi} \left(\frac{1}{\sqrt{2\pi}} e^{im\phi} \right) = \frac{m\hbar}{\sqrt{2\pi}} e^{im\phi} = m\hbar \Phi \quad (7.19)$$

As we pointed out in Chapter VI, the z-component of the orbital angular momentum can be taken as the one Cartesian component that is a constant of the motion, and it will have the magnitude $m_z \hbar$. We thus conclude that the Φ functions are the eigenfunctions of the z-component of the orbital angular momentum and that m is the z-component of the orbital angular momentum quantum number, m_z. In fact, it is because of the simple relation between the Φ and $(L_z)_{op}$ that the z-component of the angular momentum is chosen as the constant of the motion rather than either of the other two Cartesian components.

We now consider the equation in θ, the second of Eqs (7.13), which we rewrite in the form

$$\frac{1}{\sin\theta}\frac{d}{d\theta}\left(\sin\theta\frac{d\Theta}{d\theta}\right) + \left(\lambda - \frac{m^2}{\sin^2\theta}\right)\Theta = 0 \qquad (7.20)$$

We have replaced v^2 by the square of the integer m in accord with the solution of the equation in ϕ. The solution of Eq. (7.20) is considerably more involved than the solution of the ϕ equation and we will not go into all of the details of the process. The equation is a form of Legendre's equation that is dealt with in many courses on mathematical physics or engineering mathematics [1, 2]. In the present case we note that solutions that are nontrivial and finite everywhere *require* that the separation constant λ be given as the product l(l + 1) where l is a *positive* integer, and that $|m| \leq l$. With these restrictions, the solutions of Eq. (7.20) are the so-called associated Legendre polynomials symbolized as $P_l^m(\cos\theta)$. These polynomials are functions only of sin θ and cos θ, and a number of them are listed in Table 7.1 for reference. Depending upon the reference used, the Legendre functions that correspond to m = 0 may be listed separately from the associated Legendre functions that correspond to m ≠ 0. In any event, the Θ functions are simple oscillatory functions.

Because the functions with which we deal are the products $Y(\theta,\phi) = \Theta(\theta)\Phi(\phi)$, it is the normalization of the product which is the most important issue. The properly normalized functions $Y(\theta,\phi)$ are found frequently and are known as the *spherical harmonics*. They are

Table 7.1 The lowest-order associated Legendre polynomials.

l, m	$P_l^m(\cos\theta)$
0, 0	1
1, 0	$\cos\theta$
1, ± 1	$\sin\theta$
2, 0	$\frac{1}{2}(3\cos^2\theta - 1)$
2, ± 1	$3\cos\theta\sin\theta$
2, ± 2	$3\sin^2\theta$
3, 0	$\frac{1}{2}(5\cos^3\theta - 3\cos\theta)$
3, ± 1	$\frac{3}{2}(5\cos^2\theta - 1)\sin\theta$
3, ± 2	$15\cos\theta\sin^2\theta$
3 ± 3	$15\sin^3\theta$

$$Y_{l,m}(\theta, \phi) = (-1)^m \left[\frac{(2l+1)(l-m)!}{4\pi(l+m)!} \right]^{1/2} e^{im\phi} P_l^m(\cos\theta) \qquad (7.21)$$

The normalization constant shown in the square brackets can be obtained in a straightforward way and is derived in many texts on quantum mechanics and applied mathematics. What concerns us here is that these functions are the universal angular wave functions for spherically symmetric potentials. The first few of the spherical harmonics are given in Table 7.2. The fact that we have already been able to identify the quantum number m with the z-component of orbital angular momentum and that fact that $|m| \le l$, strongly suggests that l must be the orbital angular momentum quantum number. This can be shown to be the case by reference to the operator given in Table 6.1 for the square of the orbital angular momentum,

$$(L^2)_{op} = -\hbar^2 \left[\frac{1}{\sin\theta} \frac{\partial}{\partial\theta} \left(\sin\theta \frac{\partial}{\partial\theta} \right) + \frac{1}{\sin^2\theta} \frac{\partial^2}{\partial\phi^2} \right]$$

Table 7.2 The spherical harmonics of lowest order in l.

l, m	Y(θ, φ)
0, 0	$\frac{1}{\sqrt{4\pi}}$
1, 0	$\sqrt{\frac{3}{4\pi}} \cos\theta$
1, ±1	$\mp \sqrt{\frac{3}{8\pi}} \sin\theta e^{\pm i\phi}$
2, 0	$\sqrt{\frac{5}{16\pi}} (3\cos^2\theta - 1)$
2, ±1	$\mp \sqrt{\frac{15}{8\pi}} \cos\theta \sin\theta e^{\pm i\phi}$
2, ±2	$\sqrt{\frac{15}{32\pi}} \sin^2\theta e^{\pm 2i\phi}$
3, 0	$\sqrt{\frac{7}{16\pi}} (5\cos^3\theta - 3\cos\theta)$
3, ±1	$\mp \sqrt{\frac{21}{64\pi}} (5\cos^2\theta - 1) \sin\theta e^{\pm i\phi}$
3, ±2	$\sqrt{\frac{105}{32\pi}} \cos\theta \sin^2\theta e^{\pm 2i\phi}$
3, ±3	$\mp \sqrt{\frac{105}{192}} \sin^3\theta e^{\pm 3i\phi}$

If this is compared to the second of Eqs (7.11), it is immediately seen that we can rewrite that equation as

$$(L^2)_{op} Y(\theta, \phi) = \lambda \hbar^2 Y(\theta, \phi) = l(l+1)\hbar^2 Y(\theta, \Phi) \tag{7.22}$$

where we have used the fact that $\lambda = l(l+1)$ for solutions to the Legendre equation that are acceptable wave functions. Eq. (7.22) shows that the length of the orbital angular momentum vector for a particle in a state l is indeed $\hbar\sqrt{l(l+1)}$ as we had indicated in Chapter VI. The spherical harmonics are thus shown to be the eigenfunctions of the operator for the square of the orbital angular momentum and for the z-component of the orbital angular momentum, the two quantities that can be measured as constants of the motion for a particle that does not possess intrinsic spin.

It is not difficult to demonstrate that the spherical harmonics also define the parities of states in a spherically symmetric potential. We do this by direct examination of the inversion or reflection through the origin of the coordinate system. As a result of the parity operation, the only possible change in the wave function is its sign; if the sign remains the same, the wave function has *even* parity and if it changes it has *odd* parity.

Consider the diagrams in Fig. 7.1. In the top part of the figure, the radius vector to a point in space for the wave function $\psi(\mathbf{r})$ is shown. As a result of the parity operation, the wave function is reflected through the origin and the corresponding radius vector in the function $\psi(-\mathbf{r})$ is shown as a dashed line. If we examine the coordinates corresponding to $\mathbf{r}$ and $-\mathbf{r}$, it is immediately seen that any change in sign must come from the change in sign of the angular parts of the wave function because the sign of the coordinate $\mathbf{r}$ does not change. In the Fig. 7.1 (b) is the view of the upper diagram looking down along the z-axis from above, and thus the vectors shown represent projections of $\mathbf{r}$ and $-\mathbf{r}$ in the x–y plane. The angle ϕ of the vector $\mathbf{r}$ is clearly carried into the angle $\pi + \phi$ in this inversion process. Figure 7.1 (c) shows the view of the upper diagram looking toward the origin along the positive x-axis. Remembering that θ is the angle between r and the positive z-axis that remains constant for rotations about that axis, it is seen that θ is carried into the angle $\pi - \theta$ by the parity operation. Thus the coordinate transformation that is produced by the parity operation can be summarized as

$$(r, \theta, \phi) \rightarrow (r, \pi - \theta, \pi + \phi) \tag{7.23}$$

Now it is a straightforward matter to apply the changes in angles to the spherical harmonics given in Table 7.2 to determine their parities. In the case that $l = 0$, only one spherical harmonic exists, corresponding to the only allowed value of $m_l = 0$. Since Y(0,0) is just a constant, and it must have even parity.

The spherical harmonic Y(1,0) is proportional to $\cos \theta$. Because $\cos \theta = -\cos(\pi - \theta)$, Y(1,0) must have odd parity. The functions Y(1,± 1) are proportional to $\sin\theta e^{\pm im\phi}$. Now $\sin \theta = \sin(\pi - \theta)$ and

7.2 The Independent Particle Model

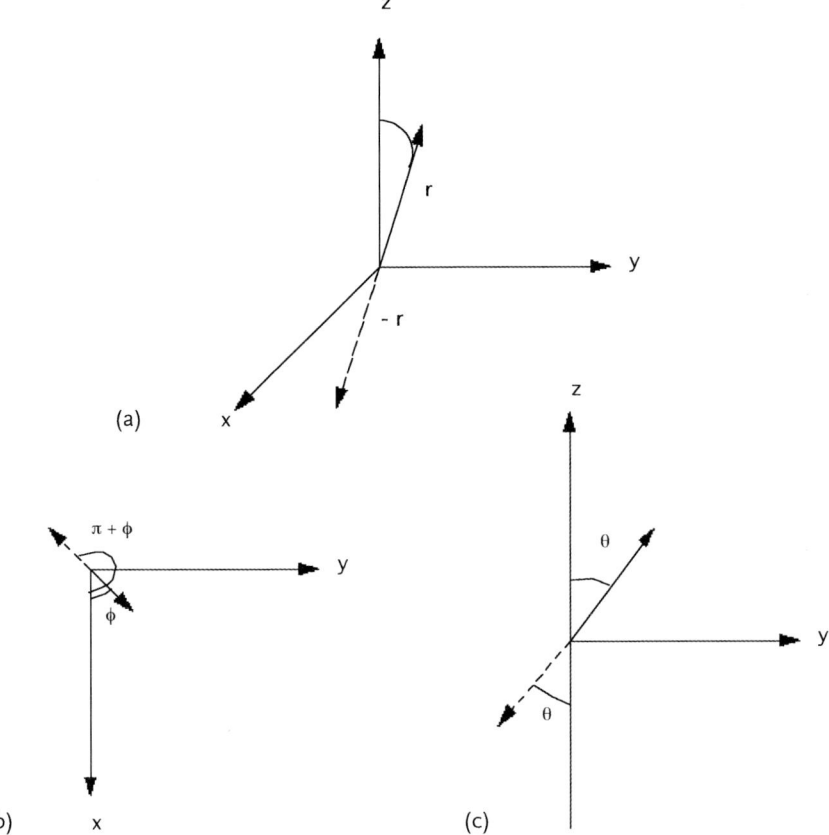

Fig. 7.1 The angular relations between the radius vector to a point of the wave function ψ(**r**) and to the same point in the function ψ(−**r**) produced by the parity operation (a). The vector **r** is shown as a solid line and the vector −**r** is shown as the dashed line. The relations between the angles θ and φ of **r** and the angles for −**r** are shown in (b) and (c).

$$e^{i(\pi + \phi)} = (\cos\pi + i\sin\pi)e^{i\phi} = -e^{i\phi} \tag{7.24}$$

Therefore $\sin\theta e^{\pm im\phi}$ must change sign as a result of the parity operation and the functions Y(1,± 1) also have odd parity. All three eigenfunctions of l = 1 have odd parity.

If one extends the same analysis to the spherical harmonics corresponding to l = 2 one finds that they all have even parity and if the analysis extended to higher values of l, the same pattern of results is found. The parities of all eigenfunctions of a given l are identical and are *even if l = even* and *odd if l = odd*.

We have, in summary, a very powerful result. For spherically symmetric potentials that depend only on the scalar radial coordinate, the spherical harmonics represent the angular parts of the wave functions of stationary states. They determine both the orbital angular momentum, the z-component of angular momentum and the parity of the states. These are three of the conserved quantities that can be measured precisely to characterize a state.

7.2.2
Some Properties of the Wave Functions

Now that we have developed some of the formal mathematics, it is perhaps worthwhile examining the wave functions of states in a spherically symmetric potential as they now stand. Formally, we have $\psi = R(r)Y(\theta, \phi)\chi(s)$. Even though we do not yet know the forms of $R(r)$ and $\chi(s)$, they can only act to define the magnitude of ψ. At any r the shape of the wave function in space will be determined only by $Y(\theta,\phi)$. Therefore, the *relative* probability of finding a particle in space can be determined without knowledge of $R(r)$ and $\chi(s)$.

The probability of finding a particle with coordinates r to $(r + dr)$, θ to $(\theta + d\theta)$ and ϕ to $(\phi + d\phi)$ is given by

$$P(r)dr d\theta d\phi = |\psi(r, \theta, \phi, s)|^2 dV(r, \theta, \phi) \qquad (7.25)$$

where dV is a differential volume element. In spherical polar coordinates, the volume element can easily be defined by reference to the diagram in Fig. 7.2. The surface area shown in the figure is given by the product $(r\, d\theta)\,(r \sin\theta\, d\phi)$. Thus

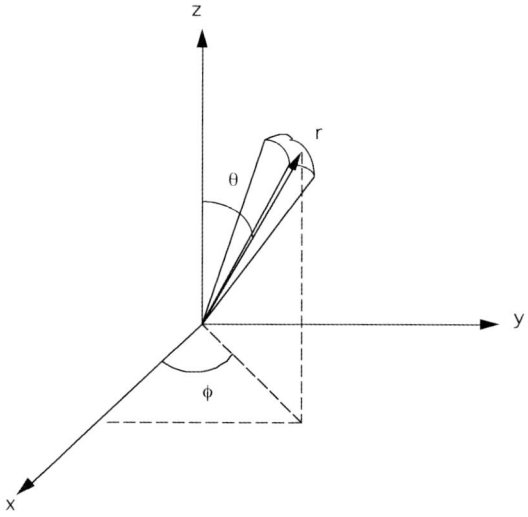

Fig. 7.2 The relations between spherical polar and Cartesian coordinates.

$$dV(r, \theta, \phi) = r^2 \sin\theta \, dr \, d\theta \, d\phi \tag{7.26}$$

and the probability of finding a particle between r and (r + dr), θ and (θ + dθ) and φ and (φ + dφ) is

$$P(r) dr \, d\theta \, d\phi = |\psi(r, \theta, \phi, s)|^2 r^2 \sin\theta \, dr \, d\theta \, d\phi \tag{7.27}$$

We can write the expression for Y(θ,φ) given in Eq. (7.21) in simplified form as

$$Y(\theta, \phi) = c(l, m) P_l^m(\cos\theta) e^{im\phi} \tag{7.28}$$

and if we use this to evaluate $|\psi|^2$ we obtain

$$\begin{aligned}|\psi|^2 &= |R(r)\chi(s)|^2 c^2(l, m) [P_l^m(\cos\theta) e^{im\phi}][P_l^m(\cos\theta) e^{-im\phi}] \\ &= |R(r)\chi(s)|^2 c^2(l, m) [P_l^m(\cos\theta)]^2 \end{aligned} \tag{7.29}$$

The result is that the probability of finding a particle in an orbit of specific l and m is independent of the azimuthal angle φ. It depends only on r and θ. Now we also know that each state of a fermion can be occupied by two particles with antiparallel intrinsic spins. Each of these will have *identical* spatial probability distributions. It should be clear that because of the short range of the strong interaction, paired nucleons will interact more strongly on the average than unpaired nucleons. Indeed, this is the origin of the pairing effect.

We can also demonstrate one other important fact. The probability for finding particles in a completely filled orbit of a given l is independent of the angular coordinates. For example, consider the total probability for finding a particle in a state of l = 1. From Table 7.2 this can be written symbolically as

$$\begin{aligned} P(r) dr &= \iint_{\theta \, \phi} \sum_{m=-1}^{1} |\psi_{l=1, m}(r, \theta, \phi, s)|^2 dV(r, \theta, \phi) \\ &\sim \iint_{\theta \, \phi} \left(\frac{3}{4\pi} \cos^2\theta + \frac{3}{8\pi} \sin^2\theta + \frac{3}{8\pi} \sin^2\theta\right) dV(r, \theta, \phi) \\ &\sim \iint_{\theta \, \phi} \frac{3}{4\pi} dV(r, \theta, \phi) \end{aligned} \tag{7.30}$$

Without solving the problem completely, we can already see from the last of Eqs (7.30) that the probability of finding a particle when the three states of m_l are filled is independent of the angular coordinates. This is, by definition, a *spherically symmetric* probability distribution. While it is a bit more complicated, you can repeat this calculation with higher l and find the same result. We can then assert that completely filled orbits have spatial distributions that are spherically symmetric. This has very important implications for our assumption that nuclei are spherical. As we will soon see, closed shells of nucleons correspond to situations in which all of the states below the shell are completely filled. That means that nuclei with completely filled shells are indeed expected to be spherical.

Now the fact is, we deal with fermions and it is the *total* angular momentum of a state, j, that is conserved. Because the spin-dependent part of the wave function does not depend upon the spatial coordinates, the conclusion we have drawn above will not change. But we must be careful here because of the spin–orbit coupling that causes a large energy difference between the couplings of $j = l + 1/2$ and $j = l - 1/2$. The energy difference is so large that the state $j = l + 1/2$ may actually lie above a magic number and thus in a higher-energy shell. While it is beyond the level of this text, if one goes through the appropriate mathematics you can demonstrate that the spin–orbit coupling does not change the situation. A completely filled j state is also spherically symmetric.

To provide a feeling for the probability distributions of different particle states, the quantities $|Y_{l,m}(\theta, \phi)|^2$ are shown in Fig. 7.3 for the spherical harmonics given in Table 7.2. These figures give relative probabilities of finding a particle per unit volume when in the different $Y_{l,m}(\theta, \phi)$ states. Note the systematics in the states of different l but constant m.

7.2.3
The Radial Equation and the Centrifugal Potential

With the angular parts of the stationary eigenfunctions in hand, we now turn our attention to the radial eigenfunctions, and thus to the form or forms we should consider for the interaction potential. But before we begin this, it is important to examine the radial equation that was obtained through the process of separation of variables. For this purpose we rewrite the first of Eqs (7.13) as

$$\frac{d}{dr}\left(r^2 \frac{dR}{dr}\right) + \frac{2\mu r^2}{\hbar^2}\left[E - V(r) - \frac{l(l+1)\hbar^2}{2\mu r^2}\right]R = 0 \tag{7.31}$$

where we have replaced λ by $l(l + 1)$ as was found in the solution to the Θ equation, and where l is identified as the orbital angular momentum quantum number. In addition to the potential V(r) that we started with, we now have the additional term $l(l + 1)\hbar^2/2\mu r^2$. This represents the quantum mechanical equivalent of the centrifugal potential found classically when a particle is subject to angular acceleration about an axis. To see this, consider the case of a particle moving in a circular orbit due to an attractive force (Fig. 7.4). In this case there is a force directed outward in the radial direction, the centrifugal force, that must be balanced exactly by an equal and oppositely-directed force, the centripetal force, which maintains the radial coordinate of the particle. If the particle moves at the constant angular frequency ω about the z-axis, it is subject to a constant centrifugal force that can be written as

$$F_{cent} = ma = \frac{mv^2}{r} \tag{7.32}$$

7.2 The Independent Particle Model

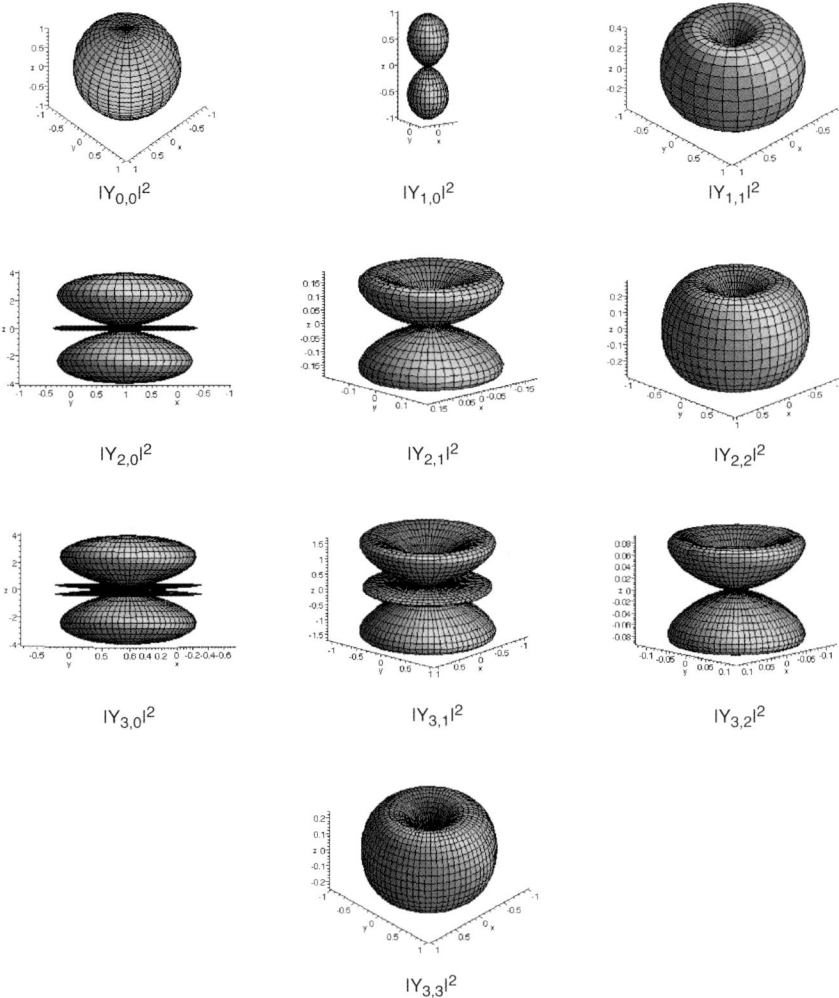

Fig. 7.3 Three-dimensional renderings of the squares of the spherical harmonics, $|Y_{l,m}(\theta, \phi)|^2$, for $l \leq 3$ and $|m| \leq l$. The surfaces represent relative probabilities per unit volume.

where v is the particle velocity tangential to the particle's orbit. The rotational motion produces a constant angular momentum $L = mvr$, and we can therefore write the force as

$$F_{cent} = \frac{L^2}{mr^3} \qquad (7.33)$$

Now the centrifugal force is conservative and therefore it can be related to a centrifugal potential by

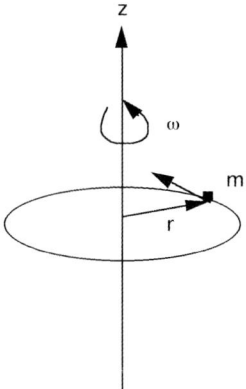

Fig. 7.4 Schematic diagram of a particle of mass m moving in a circular loop about the z-axis at a constant frequency ω.

$$F_{cent} = -\frac{d}{dr}(V_{cent}) = \frac{L^2}{mr^3} \tag{7.34}$$

Integration of this equation gives the centrifugal potential as

$$V_{cent} = \int \frac{L^2}{mr^3} dr = \frac{1}{2}\frac{L^2}{mr^2} + C \tag{7.35}$$

where C is the constant of integration. We can evaluate this by noting that, as $r \to \infty$, the potential must go to zero, and thus

$$V_{cent} = \frac{1}{2}\frac{L^2}{mr^2} \tag{7.36}$$

If we now replace m by the reduced mass μ and L^2 by $\hbar^2 l(l+1)$, we indeed have the additional term in the potential that appeared as a result of the separation of variables.

The presence of the centrifugal potential has a marked effect on the probabilities of finding a particle at or near the origin of the coordinate system. It should be quite clear that, except in the case l = 0, the total effective potential experienced by a particle as the origin is approached is infinite and the probability of finding a particle with $l \neq 0$ at the origin is zero. Further, the probability of finding a particle in the vicinity of the origin will decrease with increasing angular momentum as l(l + 1). This effect has a very strong influence on the probabilities for radioactive decay processes and on the cross sections for nuclear reactions, as we will see in subsequent chapters.

7.2.4
Models for the Average Central Potential in the Independent Particle Approximation

Now let us return to the problem of the form we should take for V(r). The place to begin is a reconsideration of the information in Chapter VI on the mass and charge distribution in complex nuclei and the range of the nuclear interaction. To a reasonable approximation, the mass and charge densities in heavy nuclei are found to be approximately constant in the nuclear interior and fall smoothly to zero over a dimension of a few fermi (see Fig. 4.4). The idealized variation of mass density with radius was given in Eq. (4.12) as

$$\rho(r) = \frac{\rho_o}{1 + e^{(r-R)/a}}$$

where $R = r_o A^{1/3}$ and a is about 0.55 fm. We also know that the range of the nuclear interaction is 1–2 fm, of the same order as the dimension found for the nuclear density to fall from its central value to zero. Because of the short range of the interaction, it is reasonable to expect that the average strength of the nuclear interaction experienced by a particle will be directly proportional the density of nuclear matter. This implies that a realistic form for the average central potential can be written as

$$V(r) = \text{constant} \cdot \rho(r) = \frac{-V_o}{1 + e^{\frac{(r-R)}{a}}} \tag{7.37}$$

This form is known as the Saxon–Woods potential (Fig. 7.5). In order to match the known spacings of levels found experimentally, it is found that V_o must lie in the range 40–50 MeV. Unfortunately, even this simple form cannot be dealt with analytically. But it does suggest two very simple approximations that are relatively easily handled and provide results that are at least in semiquantitative agreement with experiment. First, for a nucleus of A = 100, the nuclear radius is about 5.8 fm when this potential has dropped to 1/2 of its central value. If we assumed that the potential was constant at its central value up to the nuclear radius where it then vanished, we would have the approximation of a simple spherical potential well that can be handled analytically without too much difficulty. This approximation is also shown in Fig. 7.5. While crude, this model is at least reasonable, especially for the heaviest nuclides that have the largest nuclear volumes. It is, however, a rather poor approximation for the lighter nuclides. For example, the Saxon–Woods potential is shown in Fig. 7.6 for a nucleus of A = 10. Because of its small size, a light nucleus has virtually no interior region where the potential is constant. On the other hand, the harmonic oscillator potential of the form

$$V(r) = -V_o + kr^2 \tag{7.38}$$

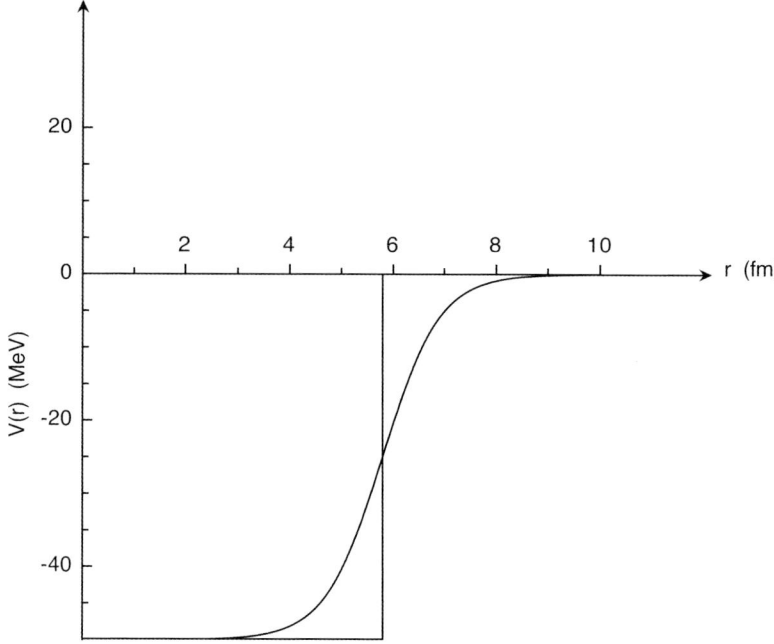

Fig. 7.5 Schematic diagram of the Saxon–Woods potential (solid line) calculated for a nucleus of A = 100. Also shown is the spherical potential well approximation.

shown as the dashed line in Fig. 7.6, can represent the potential fairly well. It has approximately the shape of the Saxon–Woods potential up to the simple hard-sphere radius (about 2.69 fm) and well beyond. It is obviously a poor approximation at large radii where it increases without limit. But it does have the nice property that a particle subject to such a potential is completely confined to the well, a property that simplifies the mathematical solution greatly. This potential is know as the *isotropic* harmonic oscillator potential (IHO).

To illustrate this last point, let us examine the solution to the finite spherical potential well shown in Fig. 7.5 in a schematic manner. Such a potential can be written generally as

$$V(r) = \begin{cases} -V_o, & r < r_n \\ 0, & r \geq r_n \end{cases},$$

where r_n is the nuclear radius. Because the potential is finite, even a particle in a bound state will have a wave function that extends over the entire range $0 \leq r \leq \infty$. We clearly have two distinct regions of constant potential and we therefore have to solve Eq. (7.31) in each region. If the wave functions in the two regions are written $R_{in}(r)$ and $R_{out}(r)$, respectively, we need to assure that the entire solution and its derivatives are continuous. This can be accomplished by the requirements that at $r = r_n$,

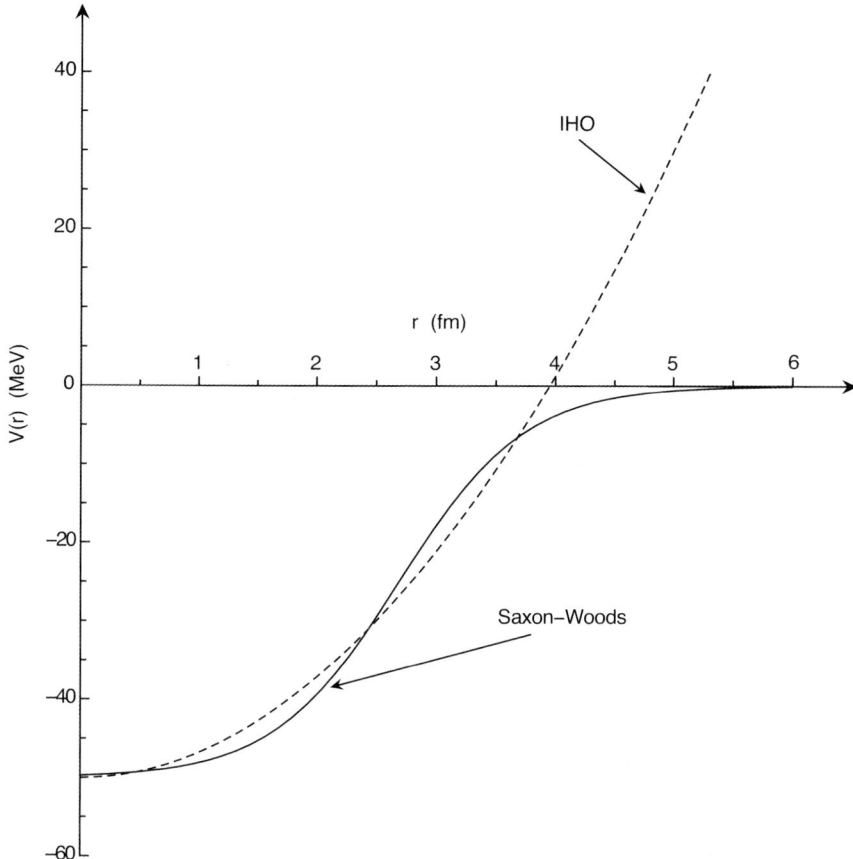

Fig. 7.6 Saxon–Woods potential and harmonic oscillator potential for A = 10.

$$R_{in}(r_n) = R_{out}(r_n)$$
$$\left.\frac{dR_{in}}{dr}\right|_{r_n} = \left.\frac{dR_{out}}{dr}\right|_{r_n} \tag{7.39}$$

All of this can and has been done. However, it is clear that the mathematics is much more complex than in the case where a particle is confined in a well of infinite depth. Only one solution of the radial equation is necessary to get the wave functions in this case.

Fortunately, there is a simple fact that allows us to use both a spherical potential well of infinite depth and the isotropic harmonic oscillator potential to obtain solutions that are at least reasonable as a first approximation. So long as we deal with levels that are not near the top of a finite potential well, the probability for finding the particle outside of the well is very, very small. And that means that the wave functions "almost" vanish at the edge of the well. And if this is the case, the energies of the levels in the finite well will not differ significantly from those in the infinitely deep well.

We should be careful here to understand what can and cannot be learned with the simple potentials suggested above. We cannot expect to calculate the energies of nuclear states quantitatively but can expect to gain an understanding of the general properties of the levels and something of their energy and spin distributions, at least in the vicinity of ground states of nuclei. Our models do not permit particles to be removed from a nucleus and thus they cannot account properly for the binding energies of nucleons. These limitations, while significant, do not pose barriers that will cloud the main results we obtain with the simple models.

7.2.5
The Infinite Spherical Potential Well

We begin the examination of the independent particle model by determining the states of a particle in an infinite spherical potential well of the form

$$V(r) = \begin{cases} 0, & r < r_n \\ \infty, & r = r_n \end{cases} \tag{7.40}$$

It is useful to manipulate the radial wave function R(r) in Eq. (7.31) a bit because it permits one to recast the equation into a form that is more readily identified with equations that are well-known in engineering physics. In place of the function R(r), we define a new function

$$v(r) = r^{1/2} R(r) \tag{7.41}$$

By direct substitution into Eq. (7.31) along with $V(r) = 0$, a little algebra then gives

$$\frac{d^2 v}{dr^2} + \frac{1}{r}\frac{dv}{dr} + \left(k^2 - \frac{(l + 1/2)^2}{r^2}\right)v = 0 \tag{7.42}$$

where $k^2 = (2\mu/\hbar^2)E$. The quantity $(\hbar k)^2$ is seen to be just the square of the linear momentum of the particle in the center-of-mass coordinate system.

Eq. (7.42) is a form of Bessel's equation that is well known. It has solutions that lead to radial wave functions of the form

$$R_l(r) = A\sqrt{k} j_l(kr) = A\sqrt{\frac{\pi}{2r}} J_{l+1/2}(kr) \tag{7.43}$$

The functions $j_l(kr)$ are known as spherical Bessel functions and the functions $J_{l+1/2}(kr)$ are known as the Bessel functions of half-integral order. While the latter are probably more familiar (if, in fact, you have ever heard of Bessel's equation) it

Table 7.3 Spherical Bessel Function of order 0–3.

l	$j_l(kr)$
0	$\dfrac{1}{kr}\sin kr$
1	$\dfrac{1}{(kr)^2}\sin kr - \dfrac{1}{kr}\cos kr$
2	$\left(\dfrac{3}{(kr)^3} - \dfrac{1}{kr}\right)\sin kr - \dfrac{3}{(kr)^2}\cos kr$
3	$\left(\dfrac{15}{(kr)^4} - \dfrac{6}{(kr)^2}\right)\sin kr - \left(\dfrac{15}{(kr)^3} - \dfrac{1}{kr}\right)\cos kr$

is the spherical Bessel functions that we will examine directly. First and foremost, the $j_l(kr)$ are not very complicated functions. The first few of these are listed in Table 7.3. They are simple oscillatory functions in sin kr and cos kr modulated by polynomial functions in kr. Note that the highest order in kr that appears in a function is l + 1.

There are a number of properties of these functions that are of direct interest. First, they can all be derived from the function of l = 0 by use of the *recurrence relation*

$$j_l(kr) = (-1)^l (kr)^l \left(\frac{1}{kr}\frac{d}{d(kr)}\right)^l j_0(kr) \qquad (7.44)$$

Second, all of the functions go to zero as $kr \to \infty$ and therefore satisfy one of the requirements for being acceptable eigenfunctions. If you take the limit of each function as $kr \to 0$, you find

$$\lim_{kr \to 0} j_l(kr) = \begin{cases} 1, l = 0 \\ 0, l \neq 0 \end{cases} \qquad (7.45)$$

All of the functions go to zero as $kr \to 0$ with the *exception* of $j_0(kr)$. Third, there is a very simple *asymptotic* form as kr gets small, namely,

$$j_l(kr) \to \frac{(kr)^l}{1 \cdot 3 \cdot 5 \cdot 7 \cdot \ldots (2l+1)} = \frac{(kr)^l}{(2l+1)!!}, \quad kr \to 0 \qquad (7.46)$$

The quantity (2l + 1)!! in the denominator is called a *double factorial*. It is given by the general form n(n-2)(n-4)... 1. The result in Eq. (7.46) indicates that the magnitude of the spherical Bessel functions of very small kr will be in the order $j_0(kr) \gg j_1(kr) \gg j_2(kr)\ldots$.

7 Nuclear Structure: The Spherical Shell Model

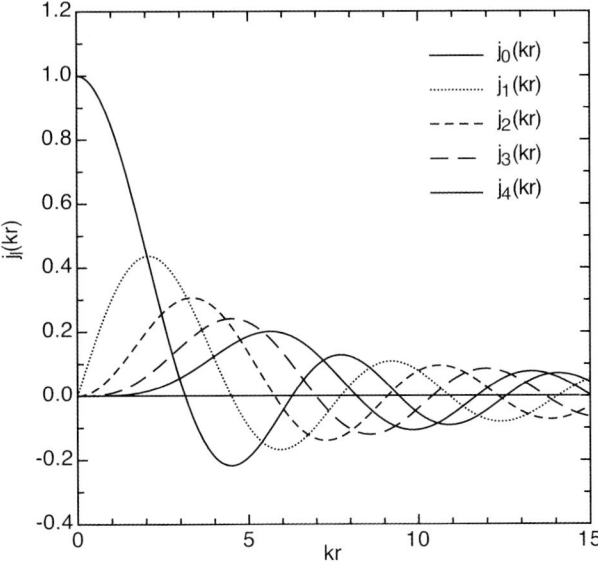

Fig. 7.7 The spherical Bessel functions $j_l(kr)$ for $l = 0$–4.

The spherical Bessel functions of lowest order are shown graphically in Fig. 7.7. It is plainly evident that the only function that does not go to zero at the origin is $j_0(kr)$, just as expected from the presence of a centrifugal potential. The magnitudes of the functions near the origin for $l > 0$ also decrease rapidly as l increases, again just as expected from the centrifugal potential. Also, the oscillatory nature of the functions produces an infinite number of crossings of the kr-axis for each of the functions as $kr \to \infty$. The value of the kr-coordinate when $j_l(kr) = 0$ is called a *zero* of the Bessel function.

The relationship between the characteristics of the spherical Bessel functions and the energies of states of different orbital angular momentum is directly tied to the boundary condition that we place on the radial wave functions. Namely, because the potential wall is infinite at $r = r_n$, the wave functions must go to zero at that point. That is, from Eq. (7.43),

$$R_l(r_n) = A\sqrt{k}j_l(kr_n) = 0 \tag{7.47}$$

Because the momentum of a particle can never be zero if it is confined to a finite region of space, $k > 0$ always and therefore, to satisfy the equation, we must have

$$j_l(kr_n) = 0 \tag{7.48}$$

But the values of the argument kr_n at which this occurs are just the zeros of the spherical Bessel functions.

Suppose we let $\chi_{i,l}$ represent a zero of a spherical Bessel function of order l where i is an index that simply counts the zeros beginning at the smallest value of the argument kr_n. The boundary condition of Eq. (7.48) then implies

$$kr_n = \chi_{i,l} \tag{7.49}$$

Because the potential is zero in the well, the total energy of the particle is just its kinetic energy. With the relation $p = \hbar k$, we can now write

$$E = \frac{p^2}{2\mu} = \frac{\hbar^2 k^2}{2\mu}$$
$$= \frac{\hbar^2 \chi_{i,l}^2}{2\mu r_n^2} \tag{7.50}$$

The energies of all states in a spherical potential well with infinite walls are simply defined by the squares of the zeros of the spherical Bessel functions. Each zero represents a discrete particle state in this system. For each value of l there will be a *spectrum* of allowed states, each with the same orbital angular momentum, and the total of all such states for all l represents the entire spectrum of the allowed states for a single particle. To get a feeling for what this spectrum looks like, a few of the zeros of the spherical Bessel functions of lowest order are given in Table 7.4. The zeros of $j_0(kr)$ occur at $n\pi$ and thus are equally spaced. But this is not true for the spacing of the zeros for any other l. Note that the energies of states of the same i increase with increasing l. The lowest-lying level in the well, the state with the lowest total energy, is predicted to be the first $l = 0$ state. Not only is this in agreement with experiment, but it is a characteristic of *all* reasonable potential functions. The lowest-energy bound state corresponds to zero orbital angular momentum.

We can use the information in Table 7.4 to get our zeroth-order estimate of the single-particle levels for a nucleon in the nucleus, in the limit that its intrinsic spin can be neglected, as shown in Fig. 7.8. While we must be very careful not to confuse this with the levels of a real nucleus, there are some generalities that we can point

Table 7.4 The first four zeros of the Spherical Bessel Functions of order l = 0–3.

i	$\chi_{i,0}$	$\chi_{i,1}$	$\chi_{i,2}$	$\chi_{i,3}$
1	π	4.493	5.763	6.987
2	2π	7.725	9.095	10.41
3	3π	10.90	12.32	13.69
4	4π	14.06	15.51	18.68

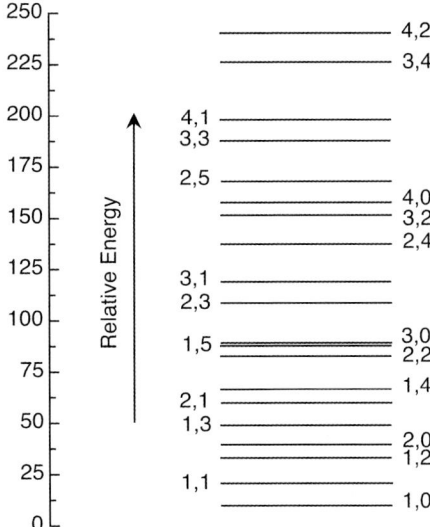

Fig. 7.8 Energy levels for states of l = 0–5 in an infinite spherical potential well in units of $(2\mu r_n^2/\hbar^2)E = \chi_{i,l}^2$. The levels are labeled with i, l on the left for l = odd and on the right for l = even.

out with some assurance. First, the single particle level sequence will contain the assorted angular momenta in a very characteristic pattern, and the energy spacing between levels can vary significantly. The energies of levels will be dependent upon the mass number. We can estimate the scale for the energies by calculating the quantity $(2\mu r_n^2)/\hbar^2$. For a medium-massed nucleus of A = 100, $(2\mu r_n^2)/\hbar^2$ is about 1.62 MeV^{-1}. The energy of the lowest level in the well (1,0) is then about 6.08 MeV, that of the second level is about 12.4 MeV, etc. We can conclude that in a simple spin-independent central potential, single-particle levels in a medium-massed nucleus would be separated by several MeV.

From the early days of the study of atomic spectra, physicists developed a system for designation of the orbital angular momentum of quantized states that is

Table 7.5 Spectroscopic designations for levels of different l.

l	Spectroscopic designation
0	s
1	p
2	d
3	f
4	g
5	h
6	i

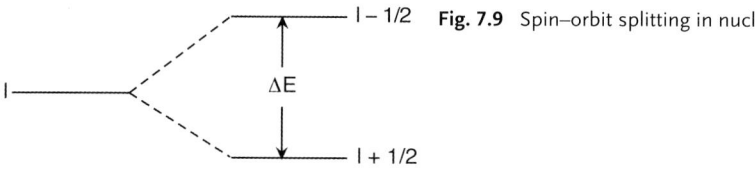

Fig. 7.9 Spin–orbit splitting in nuclei.

universally used both there and in nuclear physics. Instead of specifying l directly, a letter designation is used as shown in Table 7.5. For example, we will speak of an s state to designate a state of l = 0, etc.

Now we must consider the spin-dependence of the nuclear interaction. One of the many surprising findings during the early days of the study of nuclear structure was the very strong spin–orbit coupling in nuclei that was of the opposite sense to that found in atoms. For both neutrons and protons, the spin–orbit coupling produces a marked lowering of energy of the l + 1/2 coupling and a marked raising of the energy of the l − 1/2 coupling relative to energy in the absence of a spin-dependent interaction (Fig. 7.9). Although the spin–orbit interaction between electrons in the atom is well understood and can be calculated quite accurately with relativistic quantum mechanics, there is still no completely satisfactory theory that permits a similar calculation for the nuclear interaction. Nevertheless, it is quite clear that the interaction is proportional to **l** · **s** and for rough estimates, it may be taken to be

$$V_{so} \cong \kappa \mathbf{l} \cdot \mathbf{s} \tag{7.51}$$

where κ is a constant. We will adopt this form for our work.

Now we must evaluate the quantity **l** · **s** and get a numerical value for κ. The evaluation of the dot product is quite simple. First, it must be remembered that both **l** and **s** are operators in quantum mechanics and so is their sum **j** = **l** + **s**. Further, the quantity j^2 is a constant of the motion. If ψ is the wave function of the state we consider, then

$$j_{op}^2 \psi = (\mathbf{j} \cdot \mathbf{j})\psi = \hbar^2 j(j+1) \tag{7.52}$$

We can expand **j** · **j** explicitly as

$$\begin{aligned} \mathbf{j} \cdot \mathbf{j} &= (\mathbf{l} + \mathbf{s}) \cdot (\mathbf{l} + \mathbf{s}) = \mathbf{l} \cdot \mathbf{l} + \mathbf{s} \cdot \mathbf{s} + \mathbf{l} \cdot \mathbf{s} + \mathbf{s} \cdot \mathbf{l} \\ &= \mathbf{l} \cdot \mathbf{l} + \mathbf{s} \cdot \mathbf{s} + 2\mathbf{l} \cdot \mathbf{s} \end{aligned} \tag{7.53}$$

In this equation we have made use of the fact that l and s must commute because they do not depend on the same variables; l is a function of the spatial coordinates but s is a function only of the intrinsic spin of a nucleon. If we now let the expanded operator expression in Eq. (7.53) act on ψ we obtain

$$(\mathbf{j} \cdot \mathbf{j})\psi = \hbar^2 j(j+1)\psi$$
$$= (\mathbf{l} \cdot \mathbf{l} + \mathbf{s} \cdot \mathbf{s} + 2\mathbf{l} \cdot \mathbf{s})\psi \quad (7.54)$$
$$= [\hbar^2 l(l+1) + \hbar^2 s(s+1) + 2\mathbf{s} \cdot \mathbf{l}]\psi$$

and therefore

$$\mathbf{l} \cdot \mathbf{s} = \frac{\hbar^2}{2}[j(j+1) - l(l+1) - s(s+1)] \quad (7.55)$$

It is easy to show that if $j = l + 1/2$, the quantity in brackets in Eq. (7.55) is just l and if $j = l - 1/2$ it is $-(l + 1)$. Therefore our approximation for the spin–orbit potential becomes

$$V_{so} \cong \frac{\kappa \hbar^2}{2} \begin{cases} l, & j = l + 1/2 \\ -(l+1), & j = l - 1/2 \end{cases} \quad (7.56)$$

The magnitude of κ is most easily found by direct reference to experiment. Anticipating the level diagrams that we will discuss shortly, we will use a parameterization that accounts roughly for the experimental data on nuclides with $N(Z) \geq 50$. This gives

$$V_{so} \cong -1.9 A^{-2/3} \begin{cases} l, & j = l + 1/2 \\ -(l+1), & j = l - 1/2 \end{cases} \text{MeV} \quad (7.57)$$

for the spin–orbit interaction for illustrative purposes.

The effect of the spin–orbit interaction on the level scheme in the infinite spherical potential well is startling. In Fig. 7.10 we show the levels calculated for A = 100 without spin–orbit coupling (Fig. 7.8) along with the levels including spin–orbit coupling as calculated with the approximation of Eq. (7.57). The level density with spin–orbit coupling has essentially doubled and in place of the roughly 4–5 MeV separations we found neglecting the intrinsic spin, we now have something more like 2 MeV or so. Further, some of the levels cluster rather closely with relatively large energy gaps between the clusters. A sort of shell structure is beginning to emerge that is indeed very different from the shell structure of atoms. Further, the spin–orbit coupling sometimes causes states of a given l to be split so greatly in energy that one or more states of different l reside between them. To the extent that level structure and decay properties depend on the states in which the most weakly bound nucleons reside, this result would say that nuclides differing by just a few nucleons can exhibit very different characteristics. In actual fact, they often do. Although the nucleus is not a collection of nucleons moving independently in an infinitely deep spherical potential well, the general points we have made stand and we will see that these results bear a reasonable relation to the level diagrams found experimentally.

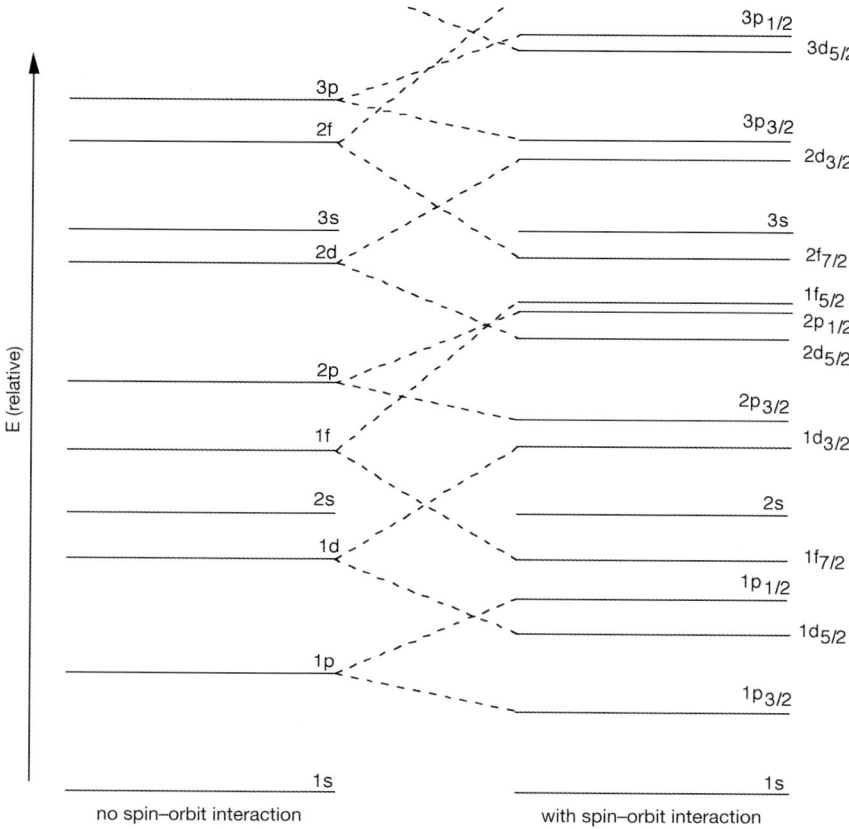

Fig. 7.10 Energy levels for nucleons in an infinite spherical potential well with and without a simple approximation for spin–orbit coupling. The levels that arise as a result of addition of the spin–orbit interaction are connected by dashed lines to the level in the absence of the spin–orbit interaction.

7.2.6
The Isotropic Harmonic Oscillator

Although it leads to some peculiarities, the simple harmonic oscillator potential is a reasonable representation of the Saxon–Woods potential for light nuclides (see Fig. 7.6) if we only consider levels not too close to the top of the well. As you will recall from classical mechanics, a mass m subject to a force of the form $F = -kx$ will experience harmonic oscillations about $x = 0$ with an angular frequency $\omega = \sqrt{k/m}$. The potential energy for this oscillator is simply $V_x = kx^2$. The equivalent oscillator in quantum mechanics has equally spaced energy levels that are given by $E_n = \hbar\omega(n + 1/2)$ where $n = 0, 1, 2, \ldots$. The quantity $\hbar\omega$ is usually referred to as a *phonon* and the energy when $n = 0$ is called the zero-point energy.

This simple expression describes the low-energy vibrations of diatomic molecules and is at least a first-order description of the vibration of atoms that are bound by covalent bonds in molecules.

In the case of the nucleus, we deal with a three-dimensional problem and the potential can be written in the form

$$V_{IHO} = -V_o + \frac{1}{2}\mu\omega^2 r^2 \tag{7.58}$$

where V_o represents the depth of the potential well and we have substituted the reduced mass of a nucleon for m. This is completely analogous to the form of the one-dimensional potential. Now if this was substituted into the Schödinger equation and r^2 replaced by $x^2 + y^2 + z^2$, we could go through the process of separation of variables and reduce the problem to three one-dimensional harmonic oscillators, each of which has the same angular frequency ω. Such a three-dimensional oscillator is referred to as an *isotropic* harmonic oscillator and that is why the potential in Eq. (7.58) is written with the subscript "IHO". Each of the equations will lead to quantized energies of the same form and it is not difficult to see that the total energy will be given by

$$E_n = -V_o + \hbar\omega(n_x + n_y + n_z + 3/2) = -V_o + \hbar\omega(n + 3/2) \tag{7.59}$$

where

$$n = n_x + n_y + n_z \tag{7.60}$$

Again, the states will be equally spaced in energy but they will possess degeneracy because states of the same n but different (n_x, n_y, n_z) will have the same energy.

If, in place of the Cartesian coordinates, the problem was solved using spherical polar or cylindrical coordinates, the quantization of orbital angular momentum arises directly and the energy is given by

$$E_{i,l} = -V_o + \hbar\omega(2i + l + 3/2) = -V_o + \hbar\omega(n + 3/2) \tag{7.61}$$

where l is the orbital angular momentum quantum number that we have met with above, n = 2i + l, and the new quantum number i is also a positive integer 0, 1, 2,....The energies of states in Eqs (7.59) and (7.61) are identical. We have the same degeneracies but we can now associate an explicit orbital angular momentum with each state. It is an easy matter to obtain the relative energies and angular momentum characteristics of the level spectrum of the IHO and this is shown in Table 7.6. In the first column of the table are the values for n, i, l that are allowed by the relation n = 2i + l and the quantization of both i and l.

For n = 0 or 1, only one level exists but for higher l, degeneracy always occurs and it increases as l increases. The remarkable finding is that where degeneracies occur all of the levels have the *same parity* because all of the levels are characterized by either even l or odd l. The second and third columns give the energies of the levels

Table 7.6 The levels of the isotropic harmonic oscillator for $n \leq 5$.

N i l	$E_{i,1} = E_n$	$E_n^\dagger$	Degeneracy (2l +1)	Total number of fermions
0 0 0	$-V_o + \frac{3}{2}\hbar\omega$	0	1	2
1 0 1	$-V_o + \frac{5}{2}\hbar\omega$	$\hbar\omega$	3	6
2 0 2 2 1 0	$-V_o + \frac{7}{2}\hbar\omega$	$2\hbar\omega$	5 1 } 6	12
3 0 3 3 1 1	$-V_o + \frac{9}{2}\hbar\omega$	$3\hbar\omega$	7 3 } 10	20
4 0 4 4 1 2 4 2 0	$-V_o + \frac{11}{2}\hbar\omega$	$4\hbar\omega$	9 5 1 } 15	30
5 0 5 5 1 3 5 2 1	$-V_o + \frac{13}{2}\hbar\omega$	$5\hbar\omega$	11 7 3 } 21	42

and the energies relative to the energy of the lowest lying level, respectively. In the fourth column is the *degeneracy* of each level, that is, the number of m_l states for each l. We do this because the total number of fermions that can occupy a level is just twice this number and it is given in the last column of the table.

Now you ought to see something rather striking. The first level is an s-state as is the first level in any reasonable potential well. It can be occupied by two particles. It is separated by one phonon from the next level that is a p state that can contain six particles. If the two levels are completely filled we will have eight nucleons. The p state is separated by one phonon from the degenerate s and d states that, when filled, will contain a total of 12 fermions, and a nucleus that has all levels filled through the 2s and 1d states by identical nucleons will have a total of 20 nucleons. The numbers of 2, 8 and 20 are the first three magic numbers known empirically to represent closed shells for both neutrons and protons! The isotropic harmonic oscillator naturally has a shell structure that reproduces the lowest magic numbers found for nuclei. Now we have not yet considered the spin–orbit coupling that we know can have a marked effect on the clustering of levels. Therefore one must be careful not to draw too strong a conclusion from this correspondence.

A comparison of the level diagrams for the IHO and the spherical potential well is shown in Fig. 7.11 in the absence of spin–orbit coupling for simplicity. On the left of this figure are the energy levels of the IHO, each labeled by the designations for the degenerate states that each contains. For reference, the cumulative number

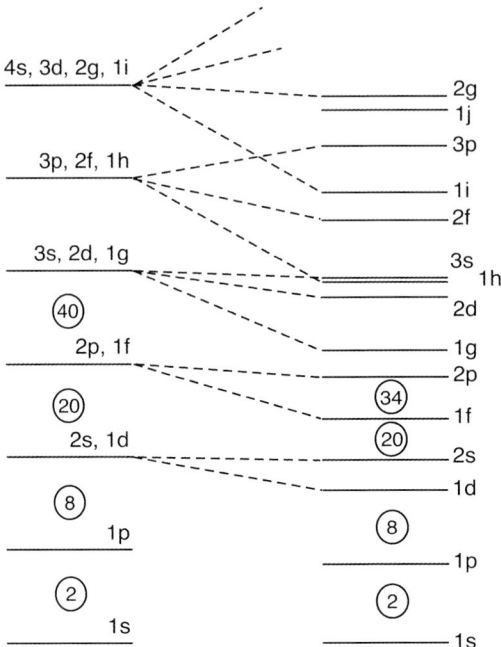

Fig. 7.11 Comparison of the levels in the IHO, and infinite spherical potential well in the absence of spin-orbit interaction. The correspondence between levels of the two potentials is indicated by the dashed lines.

of identical nucleons that fill all states are shown for the first four shells. On the right of the figure are the levels of the infinite spherical potential well with similar notations. Two very important points are immediately evident. First, we can make a one-to-one correspondence between the levels in the two potentials. This is a very general result. While the energies of the levels will certainly vary, any reasonable spherically-symmetric potential well will yield the same levels as any other. No new levels are introduced. Second, there is a very definite effect on the energies of degenerate states in the IHO as one goes to the spherical potential well. Namely, the degeneracy in l is removed such that the state of highest l in a shell of the IHO has the lowest energy in the spherical potential well, the next highest l will have the next lowest energy, etc. This result is again found by all reasonable central potentials including the Saxon–Woods form. We can then expect that these general characteristics will be reflected in the single-particle level diagram for nucleons in real nuclei.

The simple potentials we have treated above were chosen to investigate the general properties of levels and some of their systematics with a minimum of mathematical complications. The Saxon–Woods potential is much more realistic and is one of the means by which single-particle level schemes are obtained in practice. In addition to this, however, a modified isotropic harmonic oscillator potential is also used and because of its importance in understanding the levels in nonspherical or deformed nuclei, it is worthwhile to discuss this briefly.

A major drawback of the isotropic harmonic oscillator potential is the degeneracies of all levels in a shell in the absence of spin–orbit interaction. This does not exist in real nuclei. The comparison between the levels in the IHO, and those in the spherical potential shows that the degeneracies are removed in a very definite order; the energies of the higher-angular momentum states are lowered and there is a monotonic increase in the energies as l increases. If one could artificially add a term to the IHO potential that produced this effect, one could then maintain the general form of the potential. This approach was taken by Nilsson early in the 1960s. Because the energies are dependent on l and because the square of the orbital angular momentum is conserved, Nilsson chose to add a term proportional to l^2 to the IHO to obtain a radial potential of the form

$$V_{Nil} = -V_o + \frac{1}{2}\mu\omega^2 r^2 + Cl^2 \qquad (7.62)$$

where the constant C is chosen empirically to reproduce the level spacing found experimentally for states belonging to a single shell in the IHO. This form has been remarkably successful in reproducing the main features of the levels in both spherical and deformed nuclei and will be discussed further in Sections 8.6 and 8.7.

7.3
The Single-Particle Levels of Spherical Nuclei

The single-particle levels of real nuclei can be determined by a study of the properties of nuclides in their ground states. Additional information on these and low-lying excited states can also be obtained from the study of radioactive decay and certain nuclear reactions. The results from such work have provided empirical information on the sequence of the energy levels and their spins and parities.

In order to understand the ideas involved, we once again consider the degenerate Fermi gas model of non-interacting particles in a box. We assume that an even number of identical fermions are contained in a three-dimensional box and that they occupy the lowest possible energy states available. Each energy state will then be filled by two particles. If we define the *occupancy* of a state as unity if filled by two particles and zero if completely empty, and if we plot the occupancy as a function of the energy of the states for a single nucleon type, we will find a distribution such as that given schematically in Fig. 7.12. All levels below some maximum energy, referred to as the *Fermi Energy*, E_f, are completely filled and all higher-lying levels are completely empty. If we try to excite this system with the *minimum* possible energy, the only excitation that can take place is the excitation of one of the particles with $E = E_f$ to the next higher-lying level. If we try to excite one of the fermions in a lower-lying level with such an energy, it will not be possible because the next level is completely filled and the Pauli Exclusion principle guarantees that we cannot have more that two particles in any state. Thus we can be sure that the lowest

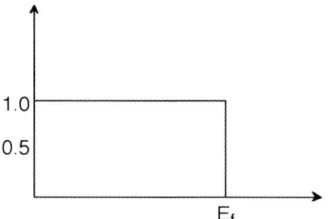

Fig. 7.12 Schematic diagram of the occupancy of levels for a degenerate Fermi gas in the particle in the box model. It is assumed that the gas contains an even number of identical fermions.

possible excitation above the ground state for this system will involve the excitation of a particle from a state with $E = E_f$. We can then determine the energy difference between the state with the Fermi energy and the next highest-lying state. If, however, we were to excite the system with an energy sufficient to excite more than one particle, it may be very difficult to interpret the result.

If we now consider real nuclei with interacting nucleons, the character of any general excited state can be quite complex. Nevertheless, if we are dealing with an odd-A nucleus in its ground state, all particles will reside in the lowest energy states available, and all but the odd nucleon will be paired. If we excite the nucleus with the minimum possible energy, we can generally assume that it is the odd particle that is being excited and the excitation energy will represent the difference between the energies of two single particle states. This is especially true for odd-A nuclides where the neutron and proton numbers are very near to the magic numbers. In such cases the energy required to excite all but the odd nucleon is especially large.

As a specific example, consider the nucleus $^{17}_{8}O$. Experimental measurements show that the spin and parity of its ground state is $5/2^+$, where the total angular momentum quantum number j is given with the superscript + indicating that the parity is even. Now how do we interpret this? We assume that all nucleons are in the lowest possible energy states and, because of the strong pairing energy, we can assume each proton is paired with a second proton that has an equal and opposite angular momentum vector. Thus the total angular momentum of the protons is zero. We also can say the same for 8 of the 9 neutrons. Further, because each pair must have even parity, the total parity of all of the paired nucleons must be even. That means that the total angular momentum of $^{17}_{8}O$ in its ground state must be due solely to the angular momentum of the 9th unpaired neutron and the parity of this state must also be even. We therefore assign the 9th neutron to a $5/2^+$ single-particle state. Because the parity is even, the orbital angular momentum quantum number l in this state must be even and therefore l = 2. The state of the 9th neutron is taken as a $d_{5/2}$ state.

We can next perform various nuclear reactions to produce $^{17}_{8}O$ in a state with the lowest allowed excitation energy, the first excited state, and we can, in various ways, determine its total angular momentum and parity. It turns out to be a state at 0.8708 MeV above the ground state and it is a $1/2^+$ state. Now we again need to make some assumption about how to interpret this information. It is possible that by adding energy to the nucleus we have simply raised the 9th neutron to the next

7.3 The Single-Particle Levels of Spherical Nuclei

single-particle neutron state. This is equivalent to saying that the eight protons and first eight neutrons are so tightly bound that it is not possible to excite them with so little energy. On the other hand, it might be true that the added energy went into unpairing a proton or neutron pair so that they no longer have equal and opposite angular momenta. Which is the correct interpretation is dependent upon the binding energies of the pairs and the energy levels that are available to them. In the present case, there are two facts that suggest that it is the first possibility that makes sense. First, we know from our examination of the separation energies of neutrons and protons that the pairing energy is likely to be 1–2 MeV, roughly twice the energy that has been put into the $^{17}_{8}$O nucleus. Second, and more importantly, we know empirically that eight is a magic number that implies especially strong binding. Taken together it is quite reasonable to assume that we have excited the ninth neutron to the next single-particle state. Because the parity is even we conclude that the state must be an $s_{1/2}$ state. Now we can try to continue playing this game. However, the next level we find in $^{17}_{8}$O lies at an energy of 3.055 MeV above the ground state. This energy is so large that it is possible that more complex excitations are involved than just simply the further excitation of the 9th neutron.

For every complex nucleus, no matter how "simple" we think it is, there will be at most a few levels above the ground state that we can attribute to simple single-particle states, or states which approximate this character. For light nuclei, we can be sure from our study of the particle in a box problem (Chapter IV), that the level spacing will be significantly larger than the spacing in the heaviest nuclei. Therefore, we can expect to find more low-lying levels with single-particle character in heavier nuclei before we will be troubled by much more complex excitations. But regardless of mass, we will not be able to determine the sequence of but a few single-particle states for any individual nucleus.

What we can do is determine the relative spacings of one or a few single-particle states for a series of nuclei spanning the range of masses that are found in nature. By combining these, we can arrive at an empirical determination of the sequence of single-particle states in all nuclei that refer to the spacings expected near the ground states. With this in view, the *empirical* single-particle level diagram for neutrons is shown in Fig. 7.13, along with the level diagram for the isotropic harmonic oscillator.

Although the lines connecting them have not been drawn, the reader can easily show that there is a one-to-one correspondence between the levels of the isotropic harmonic oscillator and those of the empirical single-particle level diagram. The closed shells at the magic numbers are shown in the latter and they are produced as a result of the natural grouping of levels and the strong spin–orbit coupling. The first shells at 2, 8 and 20 arise naturally even in the absence of spin–orbit coupling. The closed shell at 28, however, results from the strong spin–orbit coupling of the 1f orbital. The energy of the $f_{7/2}$ level falls quite low while the increase in energy of the $f_{5/2}$ level causes it to fall in a cluster of other levels that arise from the n = 3 harmonic oscillator shell. Similarly, the shells at 50, 82 and 126 arise because of the strong spin–orbit coupling of the 1g, 1h and 1i orbitals, respectively. The levels within each of these three shells include all of the levels of a single harmonic

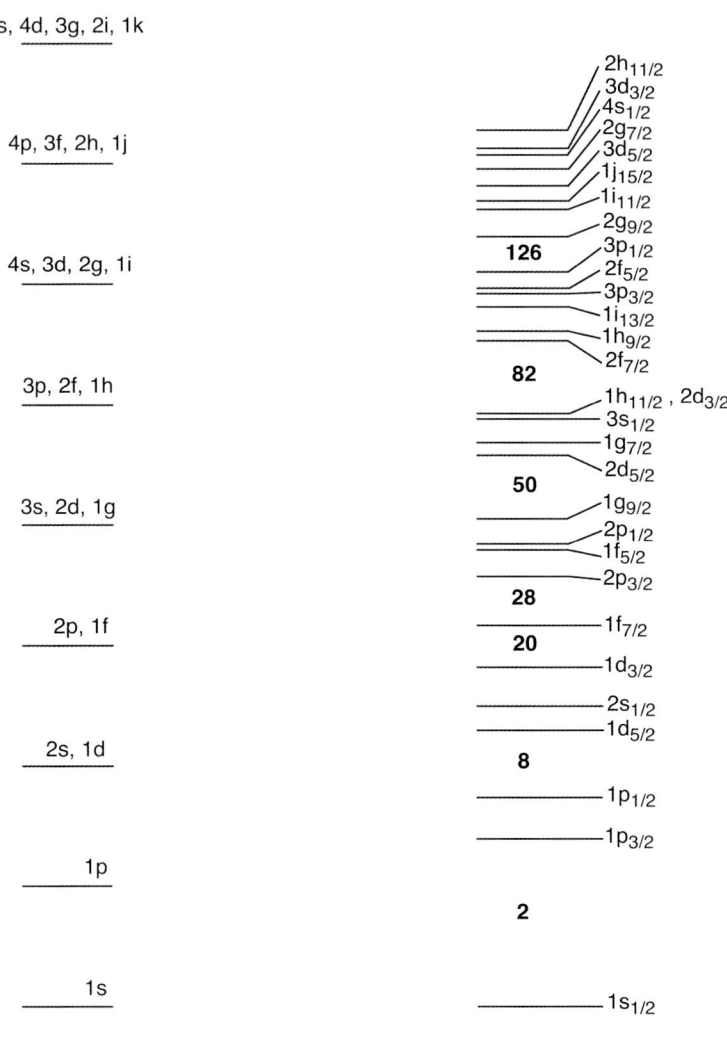

Fig. 7.13 Empirical single-particle level diagram for neutrons (right) along with the levels of the isotropic harmonic oscillator. To a good approximation the single-particle level diagrams for both neutrons and protons are the same for $N(Z) \leq 50$.

oscillator shell except for the level of the highest l with coupling $j = l + 1/2$, as well as one level from the next harmonic oscillator shell that has the highest l and the coupling $j = l+1/2$. This simple structure has a marked effect on the properties of nuclei, because it means that all of the levels in a shell have the same parity except for that of the level of highest j. The empirical level diagram for protons with $Z \leq 50$ is essentially the same as that for neutrons, as we might expect, because of the

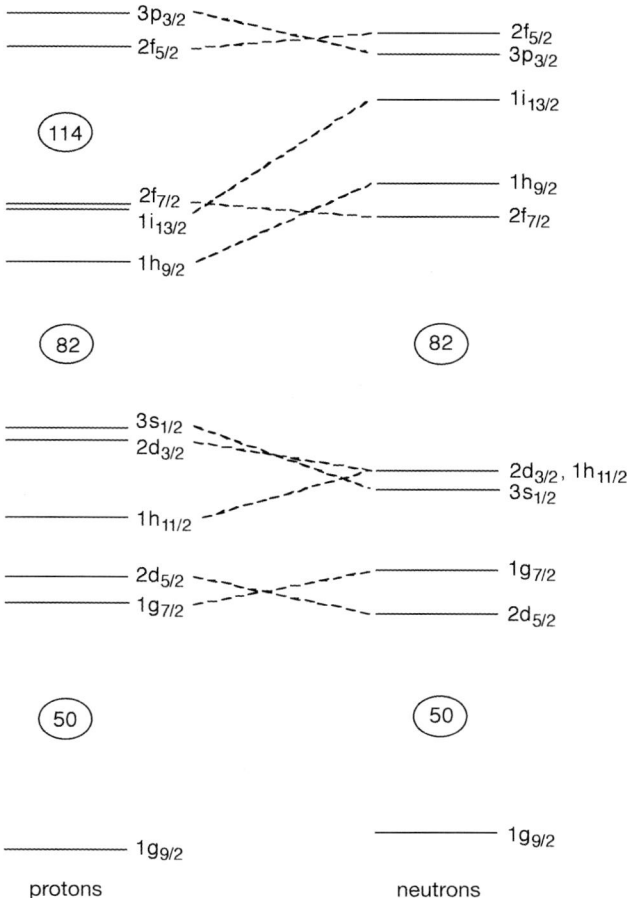

Fig. 7.14 The approximate single-particle level diagrams for protons and neutrons when Z(N) > 50.

relative weakness of the Coulomb interaction in comparison with the strong interaction.

With increasing Z, Coulomb repulsion begins to have a marked effect as seen in Fig. 7.14. Consider first the levels below the closed shell of 82. Beginning with the neutron levels and tracing through to the proton levels, a clear pattern is established. Levels with higher-l have reduced energies and levels with lower-l have increased energies in the proton scheme relative to their energies in the neutron scheme. For example, the $2d_{5/2}$ and $1g_{7/2}$ levels have inverted order in the proton level diagram, relative to those in the neutron diagram. Similarly, while the $2d_{3/2}$ and $1h_{11/2}$ levels are nearly degenerate in energy for neutrons, the $1h_{11/2}$ level lies much lower in energy for the protons while the $2d_{3/2}$ level is increased in energy.

To understand this behavior, consider the degenerate levels that belong to a single shell in the isotropic harmonic oscillator potential in the absence of spin–orbit coupling but including the Coulomb potential of the protons, which we have neglected up till now. In the absence of the Coulomb interaction, the probability of finding a particle in these levels will vary significantly with radial displacement from the center of mass of the system. In all cases, the higher the value of l the larger is the mean radial location of the probability distribution. Now consider the effects of an added Coulomb potential. Although the correct quantum mechanical calculation is beyond what we want to explore, we can get a good qualitative understanding of what the effect ought to be by considering the radial dependence of the Coulomb potential of a classical uniformly charged sphere. This is just the problem we considered when we developed the semi-empirical mass formula. If the radius of the sphere is R and the total charge within the sphere is Q, it is not difficult to show that within the sphere,

$$V_{Coul}(r) = \frac{Q}{2R}\left(3 - \frac{r^2}{R^2}\right), \qquad r \leq R \tag{7.63}$$

The potential at the surface is only 2/3 as large as it is at the center of the sphere. If, in the absence of charge, a particle experienced a constant attractive interaction independent of radius and if a uniform positive charge distribution is now added, it is easy to see that a positively charged particle located at the surface of the sphere would experience a reduced Coulomb energy and thus a higher net binding than could be found at any other location.

The comparison with the nucleus should be fairly obvious. The roughly constant density of both mass and charge within the nucleus suggests that proton levels of different l that are degenerate in energy in the absence of charge will have different energies in the presence of charge. Those with the highest probability for finding a proton at larger mean radial locations, the orbits of higher angular momentum, will have lower total energies.

The effects of the Coulomb energy on the ordering of proton levels can be so large that shell structure may be modified. While the single-particle neutron level diagram is known quite well at large N, we really do not have very good *direct* information on the proton level diagram above about Z = 90. However, indirect information allows the extrapolation to the higher atomic numbers shown in Fig. 7.14. And if the extrapolated level structure turns out to be correct, it points to the fact that the next closed shell for protons might be at Z = 114 and not at Z = 126, as for the neutrons. You can see that the lowering of the energies of the $1h_{9/2}$ and $1i_{13/2}$ orbitals relative to their locations in the neutron level diagram cause the change in the location of the large energy gap that represents the location of a shell. Because closed shells confer increased stability, the nuclei of some elements near Z = 114 might be especially stable in a relative sense. As of this writing, a few atoms of the elements Z = 114 and Z = 115 have been reported.

Before we proceed to consider the agreement of the simple independent particle model with experiment, it is worthwhile to examine the general energy differences between levels within a shell and the magnitudes of the shell gaps that define the magic numbers. We can start by considering the magnitude of $\hbar\omega$ in the IHO that reproduces roughly the level spacing found in experiment. It can be shown that the oscillator frequency ω is related to the size of nuclei and a not-too-difficult calculation leads to the estimate [3]

$$\hbar\omega \cong 41 A^{-1/3} \text{ MeV} \tag{7.64}$$

Thus, for a typical nucleus of mass number A = 100, the magnitude of $\hbar\omega$ is about 8.83 MeV. This would be the shell gap if the nuclear potential were really that of an isotropic harmonic oscillator. Now a nucleus of A = 100 near stability will have a neutron number of about 55. By comparing the relative spacings of the harmonic oscillator levels shown to the left in Fig. 7.13 with those of the empirical level sequence shown to the right, it is seen that the shell gap at N = 50 is roughly 1/2 $\hbar\omega$, or about 4 MeV or so. If we consider nuclei near the doubly magic $^{208}_{82}\text{Pb}$, the shell gap at N = 82 is about the same, and the shell gap at N = 126 is perhaps about 3 MeV. Similarly, a comparison of the level diagrams in Fig. 7.14 shows that the shell gap at Z = 82 is significantly smaller than at N = 82. For light nuclei, the shell gap at N(Z) = 28 is about 4.5 MeV.

Examination of the level diagrams shows that the total energy spanned by the levels within a shell is roughly the same as the energy of an adjacent shell gap. Therefore, because the shells near N = 50 and 82 contain 4 and 6 levels, respectively, the average energy difference between single-particle states is expected to be about 1 MeV and 0.5 MeV, respectively. As we compare the predictions of the independent particle model with experiment, it will be very useful to keep the magnitudes of these quantities in mind.

7.4
Comparison of the Predictions of the Single-Particle Model with Experiment

In a real sense, we are really cheating a bit when we take the level diagrams in Figs 7.13 and 7.14 and apply them to determine the properties of low-lying levels of nuclei because they represent the result of decades of experimental studies that have been used to define the sequences themselves, and thus must have built-in predictive value. Nevertheless, the schemes do permit very good correlations of some properties of nuclear states and they also can be used to understand the limitations of this very simple approach to nuclear structure.

The simplest comparison of experiment with the model can be made by examining the ground-state spins and parities of nuclei for which the model is expected to be a "good" one. As we indicated above, we can assume that in the ground states, identical nucleons are paired as far as possible. Because the angular momentum and parity of a pair must be 0^+, we are immediately led to the prediction that the ground state spin and parity of *all* (even, even) nuclei must be 0^+. We actually do not need the shell model to tell us this: all we need is a strong pairing energy. This expectation is found without exception for the ground-state spin and parity of *all known* (even, even) nuclei. We can then predict that the ground-state spin and parity of all odd-A nuclides is that of the state of the odd nucleon. It is an experimental fact that for most odd-A nuclides not too far from closed shells, the single-particle diagrams do indeed predict the ground-state spins and parities. A few words of caution are important here. First, the level diagrams in Figs 7.13 and 7.14 are not exact. Especially when the energy spacing between two single-particle states is small, it is sometimes found that the simple model does not correctly predict the spin and parity but in such cases it is often found that one of two closely-lying states is indeed the correct one. If one is far removed from the magic numbers, the simple spherical shell model breaks down dramatically. And if the model does not apply we cannot expect it to give meaningful predictions. These regions will be discussed in some detail in the next chapter.

We have yet to discuss nuclei of the (odd, odd) type. In such cases we must deal with the addition of the angular momentum of the two odd nucleons. We know from the discussion in Chapter VI that the possible angular momentum quantum numbers from coupling of the two vectors will be given by $|j_n - j_p| \leq j \leq |j_n + j_p|$, where j_n and j_p are the total angular momentum quantum numbers of the neutron and proton states, respectively. We also know that the parity of these states must be given by the product of the parities of the neutron and proton states. Now, without a more complete treatment of the nucleon–nucleon interaction we cannot make a clear-cut prediction of the ground-state spin, but we can at least give the range of possible angular momenta and we can make a clear prediction of its parity. In almost all cases for (odd, odd) nuclei not too far from the magic numbers, the parity is correctly predicted and the ground-state spin is one of the values expected.

Now we should consider how well this simple model predicts the properties of low-lying excited states of nuclei. Here we must be careful to remember the discussion above concerning the excited states of $^{17}_{8}O$. We can only expect the simple model to apply when no other excitations can occur. If this is the case, we imply that the remaining nucleons are so tightly bound in pairs that they cannot be excited. In effect, the remaining nucleons behave as if they were coupled together in a spherical inert core. Pictorially we have something like that shown in Fig. 7.15.

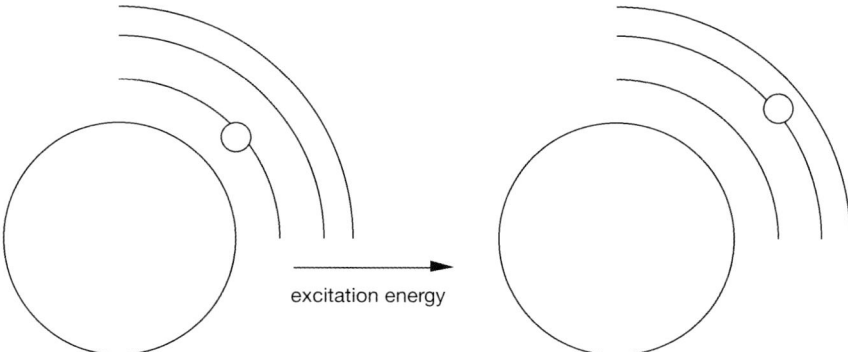

Fig. 7.15 Schematic diagram of the excitation of a single particle in the "inert core" approximation.

7.4 Comparison of the Predictions of the Single-Particle Model with Experiment

If we expect this model to be at all applicable, we ought to try and test it on odd-A nuclei that are immediately adjacent to (even, even) doubly magic nuclei, for the latter have the largest excitation energies of any nuclei.

We begin by considering the two nuclei $^{41}_{20}$Ca and $^{41}_{21}$Sc. This is a mirror pair and we should expect the levels of the two nuclei to be very similar. Further, each has one particle outside of the doubly magic N(Z) = 20 core. Our simple model above would then suggest that we treat these nuclides as an (even, even) $^{40}_{20}$Ca core outside of which is a single neutron or proton, respectively. Because the Coulomb energy is relatively small, the level diagrams for neutrons and protons will be about the same. The level diagram in Fig. 7.13 indicates that the 21st particle should reside in the $1f_{7/2}$ level leading to the prediction of a ground-state spin and parity of $7/2^-$ for both nuclides. Further, if the $^{40}_{20}$Ca is sufficiently tightly bound, the first excited state should be the $2p_{3/2}$ state and we should see a spin and parity of $3/2^-$. However, note that the first excited state requires that we promote the 21st particle across the shell gap of N(Z) = 28, that is about as large as the gap at N(Z) = 20. This means that we are just at the verge of losing our ability to consider the core as inert and we may begin to see much more complicated excitations than just the single-particle excitations we have been discussing.

In Fig. 7.16 are shown the experimental level diagrams for both $^{41}_{20}$Ca and $^{41}_{21}$Sc. These have been produced through evaluation of experimental data and they are presented in the, more or less, standard form found in the literature. Each horizontal line represents a known level, with the experimental spin and parity written to the left and the energy relative to the ground state to the right, in keV. The spins and parities and energies of levels are displaced for easy reading. The diagrams for the two nuclides display a number of very similar features. The ground state of each has the spin and parity of $7/2^-$, in agreement with the single-particle level diagram. In both cases, the first excited state has spin and parity of $3/2^-$, also in agreement with our expectations. There is ample experimental evidence from nuclear reaction studies to indicate that the ground and first excited states have "single-particle" character as a whole.

In both nuclei one also sees that the second excited state has even parity. This is not in agreement with the next single-particle state expected, the $1f_{5/2}$ state. In fact, a state with even parity cannot be expected without a more complex type of excitation, and we should not be surprised by finding this. Note that the energies of the first excited states are already at 1.7–2.0 MeV above the ground state and this is about the pairing energy. That means that at higher energies we can certainly expect more complex excitations. Indeed, starting at about 2 MeV one observes a very large number of excited states separated by much smaller energies and such a high level density cannot be expected in spherical nuclei without rather complex excitations. We can conclude that the experimental data at low energies, where complex excitations cannot take place, indicate level schemes of these mirror nuclei that are very nearly the same and in agreement with the single-particle model.

If the single-particle model is a reasonable approximation, we should find that the addition of pairs of protons (neutrons) to a nucleus will not change the state of the odd neutron (odd proton) and hence the ground state spin and parity should

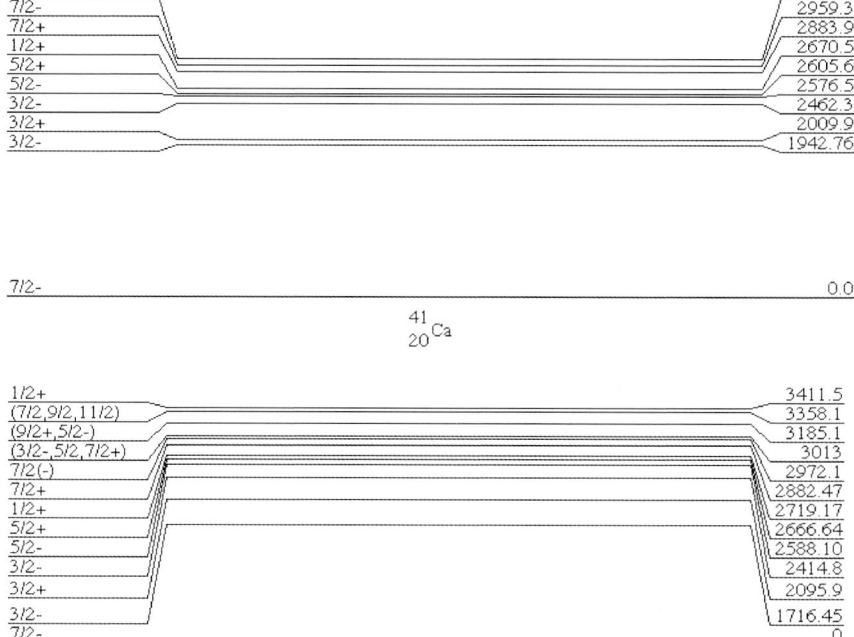

Fig. 7.16 The level diagrams of $^{41}_{20}$Ca and $^{41}_{21}$Sc. Experimental spins and parities of levels are written to the left, and level energies above the ground state are given to the right.

not change. This is generally found to be true. An especially simple case is that of $^{49}_{21}$Sc where four neutron pairs have been added to form a closed shell of neutrons and its level scheme is shown in Fig. 7.17. $^{49}_{21}$Sc can be considered as the doubly magic core $^{48}_{20}$Ca$_{28}$ plus the 21st proton. If the core was truly inert, the low-lying levels of $^{49}_{21}$Sc should have properties similar to those of $^{41}_{21}$Sc. The ground-state spin and parity of $^{49}_{21}$Sc is indeed identical to that of $^{41}_{21}$Sc but the first and second excited states have positive parity and cannot be explained by the simple excitation of the odd proton. However, the third excited state has the spin and parity of 3/2⁻ expected for the single-particle 2p$_{3/2}$ state, and experimental data indicate that the level has, for the most part, the character expected for a single-particle excitation. Again, because of the large single-particle excitation energies, the 2p$_{3/2}$ level lies some 3.1 MeV above the ground state, not only much larger than the pairing energy, but comparable to the energy gap between major shells. With such large energies, it is not surprising that one sees the very high density of excited states shown in the figure.

The character of the closed shell at N = 28 can also be examined with the level structure of $^{49}_{20}$Ca, which can be considered as the $^{48}_{20}$Ca$_{28}$ plus the 29th neutron.

7.4 Comparison of the Predictions of the Single-Particle Model with Experiment

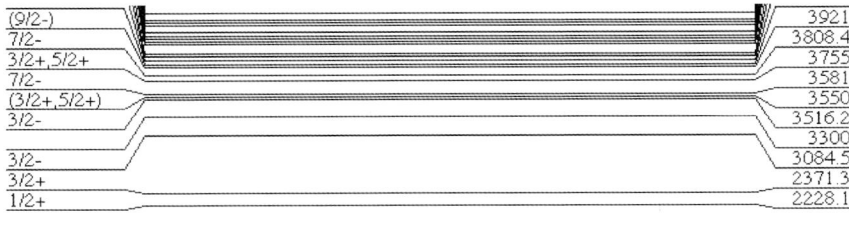

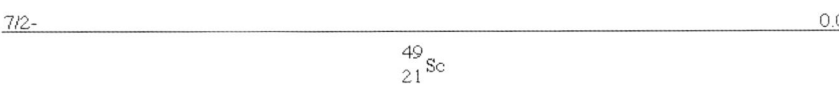

Fig. 7.17 The level scheme of $^{49}_{21}\text{Sc}$.

The level scheme is shown in Fig. 7.18. The ground state with spin and parity of $3/2^-$ is just what is expected from the single-particle level diagram of Fig. 7.13. The first excited state of $1/2^-$ is known to have the character of a $2p_{1/2}$ state. The single-particle level diagram shows that the $2p_{3/2}$ and $1f_{5/2}$ states are expected to lie very close in energy and the real effects of nuclear structure can cause energy shifts that lead to differences in the observed level sequence in such cases.

The examples considered above illustrate rather nicely, both the usefulness and limitations of the simple shell model. In general, the lowest energy levels can be rationalized rather well, but once there is sufficient energy available to produce more complex excitations, the model is too simple to account for the spectrum of states. Our study of the particle in the box problem leads to the expectation that the

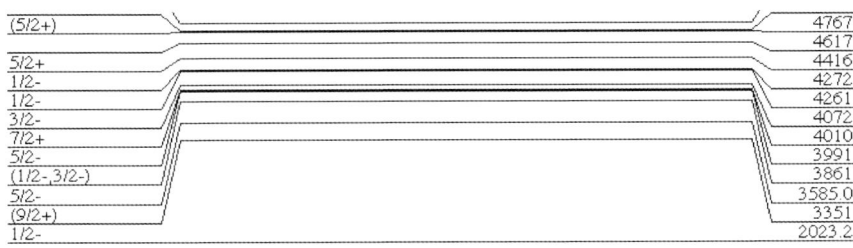

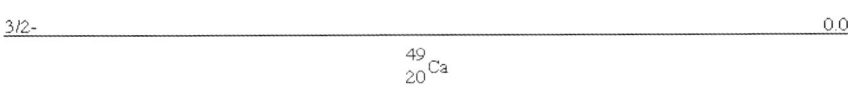

Fig. 7.18 The level scheme of $^{49}_{20}\text{Ca}$.

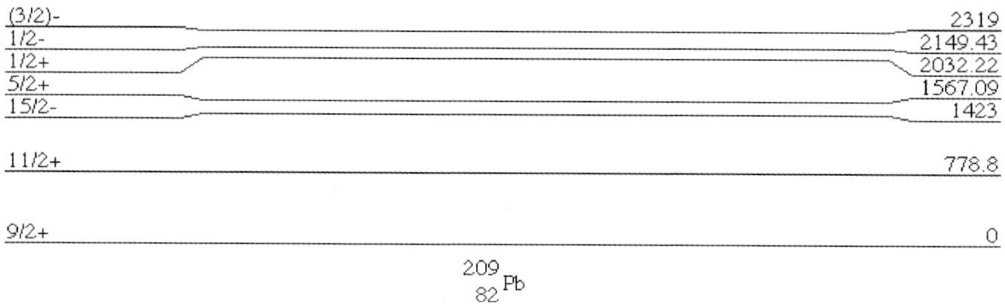

```
(3/2)-                                              2319
1/2-                                                2149.43
1/2+                                                2032.22
5/2+                                                1567.09
15/2-                                               1423

11/2+                                               778.8

9/2+                                                0
                        209
                        ₈₂Pb
```

Fig. 7.19 The level diagram of $^{209}_{82}$Pb.

single-particle level spacing will decrease significantly with increasing mass number and hence the size of the nucleus. This suggests that the level structures of nuclei in the vicinity of the heaviest doubly magic nucleus $^{208}_{82}$Pb should provide an even better test of the simple shell model ideas. In Fig. 7.19 is shown the level diagram of the nucleus $^{209}_{82}$Pb. Reference to the single-particle level diagram in Fig. 7.14 suggests that the state occupied by the 127th neutron in the ground state should be the $2g_{9/2}$ state and that the next higher single neutron states should be the $1i_{11/2}$, $1j_{15/2}$, and $3d_{5/2}$ states. The spins and parities of the first four levels in $^{209}_{82}$Pb are exactly in accord with this level sequence and other experimental evidence indeed supports the single-particle character of the levels. Note that the first three excited states lie below 1.6 MeV of excitation energy. The energy separation between these states is on the order of 0.5 MeV as opposed to the roughly 2 MeV seen in the calcium and scandium nuclei we have discussed above.

The type of agreement between the single-particle model and the level structure of $^{209}_{82}$Pb is found for essentially all nuclei in the vicinity of $^{209}_{82}$Pb, and the low-lying levels of $^{209}_{83}$Bi that correspond to the excitations of the 83rd proton are shown as

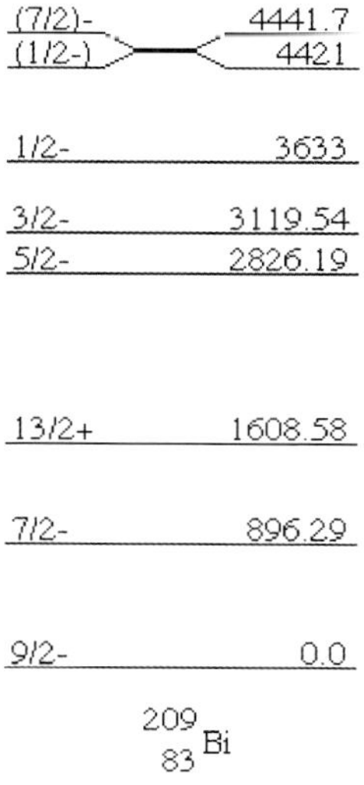

```
(7/2)-          4441.7
(1/2-)          4421

1/2-            3633

3/2-            3119.54
5/2-            2826.19



13/2+           1608.58

7/2-            896.29


9/2-            0.0
         209
         ₈₃Bi
```

Fig. 7.20 Levels in $^{209}_{83}$Bi that have been identified as single-proton excitations.

our final example in Fig. 7.20. The spectrum is quite remarkable in two respects. First the levels shown below 1.61 MeV are all of the known levels in this energy range. They have been shown to of single-particle character and they have the spins and parities expected from the level scheme shown in Fig. 7.14 but with a slightly different sequence. In the energy range 1.61–3.633 MeV, there are some 12 levels known. Those whose energies and spins and parities are given are also known to be of single-particle character and they correspond to the levels that would reside in the next higher shell. Indeed, the shell gap in this case is no more than about 1.2 MeV or so and is not very well defined.

References

1. See, for example, G. Arfken, "Mathematical Methods for Physicists", Academic Press, Inc. (1985).
2. An extremely useful reference is the Handbook of Mathematical Functions with Formulas, Graphs, and Mathematical Tables, edited by Milton Abramowitz and Irene Stegun of the U.S. National Bureau of Standards. It is available over the worldwide web.
3. See, for example, Samuel S.M. Wong,"Introductory Nuclear Physics", 2nd edition, John Wiley and Sons, Inc., New York (1998).

General References

The classic reference for the nuclear shell model is M.G. Mayer and J. H. D. Jensen, "Elementary Theory of Nuclear Shell Structure", John Wiley & Sons., New York (1955). Very readable papers on the substance that led to development and acceptance of the shell model are M.G. Mayer,"On Closed Shells in Nuclei. II", Phys. Rev. 75 (1949) 1969 and M.G. Mayer, "Nuclear Configurations in the Spin–Orbit Coupling Model. I. Empirical Evidence", Phys. Rev. 78 (1950) 16.

A very nice concise description of the shell model as we know it today is given in Kenneth S. Krane, "Introductory Nuclear Physics", John Wiley and Sons, New York (1987).

Problems

1. Refer to Table 7.2 and show that the spherical harmonic $Y_{2,1}$ is normalized.

2. Recalling the De Broglie relation between the wavelength of a particle and its momentum, and the relation between the momentum of a particle and the propagation number k, show that the function $\Psi = \Psi_o e^{i(kx-\omega t)}$, representing a free particle traveling in the positive x direction is an acceptable solution of the time-dependent Schrödinger equation.

3. Explicitly carry out the separation of the variables in the second of Eqs (7.11) to obtain the second and third equations of Eq. (7.13).

4. Use the shell model level diagrams in Figs 7.13 and 7.14 to determine the expected total angular momentum quantum numbers and parities of ^{7}Li, ^{15}O, ^{31}P, ^{89}Y, ^{107}Ag, ^{131}Sb, ^{143}Ce and ^{205}Pb.

5. Go to the Table of Isotopes and look up the level diagrams of the adjacent nuclei (^{88}Y, ^{89}Y, ^{90}Y), (^{128}I, ^{129}I, ^{130}I) and (^{208}Bi, ^{209}Bi, ^{210}Bi). Compare quantitatively the level densities within 1 MeV of the ground states of the members of the triads. What do you conclude concerning the level densities (level spacings) of odd-A nuclei relative to those in adjacent odd, odd nuclei?

6. ^{34}Cl has a ground state with a half-life of 1.526 s and spin and parity of 0^+, and an isomeric first excited state with a half-life of 32.00 min and spin and parity of 3^+. Using the shell model level diagrams, demonstrate that the spins and parities of these two states are consistent with the assumption that they arise from the same particle configuration with different angular momentum couplings.

7. Use the single-particle shell model to predict the total angular momentum quantum numbers and parities of the ground and first excited states of $^{99}_{42}$Mo.

8. We have pointed out that, so long as one considers levels that are well below the top of a potential well of finite depth, the energies of eigenstates will not differ very much from those of the infinite potential well. You can demonstrate this with the following exercise.

(a) For a finite spherical potential well of depth $-V_o$, the solutions of the Schrödinger equation for the regions $r < r_n$ and $r \geq r_n$ with use of the continuity conditions given in Eq. (7.40), gives the transcendental equation

$$\cot\left[\frac{r_n}{\hbar}\sqrt{2\mu(V_o - |E|)}\right] = -\sqrt{\frac{|E|}{V_o - |E|}}$$

that defines the energies of bound s states. Here $|E|$ is the absolute energy of an eigenstate, r_n is the nuclear radius within the model approximation and μ is the reduced mass of a single particle.

Assuming $V_o = 50$ MeV, $r_n = 5.8$ fm and replacing the reduced mass by the mass of a neutron, determine the energies of the bound s states for a neutron in the potential. You can solve the equation by either an iterative method, a root-finding method or a graphical method. Compare these results with those found from the equivalent well of infinite depth, both in terms of their energies relative to the bottom of the well and the differences in the energies of the states within the same well.

(b) Repeat the analysis for p states in the same wells. In this case the transcendental equation defining the eigenenergies in the finite well is

$$\frac{r_n\sqrt{2m_nc^2(E+V)}}{\hbar c} \cot\left(\frac{r_n\sqrt{2m_nc^2(E+V)}}{\hbar c}\right) = 1 - \left(\frac{E+V}{E}\right)\left(\frac{r_n\sqrt{2m_nc^2|E|}}{\hbar c}\right)$$

9. The nucleus $^{18}_{9}\text{F}$ has a ground-state spin and parity of 1^+. It is assumed that the ground state can be described as an odd neutron and an odd proton outside of the (even, even) doubly-magic core of $^{16}_{8}\text{O}$. From the single-particle level diagram for spherical nuclei, the next levels available to an odd particle are $d_{9/2}$, $s_{1/2}$ and $d_{3/2}$. Model calculations indicate that the ground-state wave function is approximately

$$\psi_{gs} = 0{,}732\,\psi(d^2_{5/2}) + 0{,}477\,\psi(d_{5,2}d_{3/2}) + 0{,}464\,\psi(s^2_{1,2})$$
$$-0{,}131\,\psi(d^2_{3,2}) - 0{,}009\,\psi(d_{3,2}s_{1,2})$$

In this representation, Coulomb repulsion has been neglected and either odd particle can exist in any of the possible levels indicated. The wave function $\psi(d^2_{5/2})$ means that there are two particles in the single-particle $d_{5/2}$ orbital, and similar interpretations apply to the other components.

(a) Determine whether the wave function is normalized.
(b) What is the probability that both odd particles will be found in $d_{3/2}$ single-particle states?
(c) Which of the five components could *not* contribute to the wave function of a 1^+ state if both particles were identical?

8
Nuclear Shapes, Deformed Nuclei and Collective Effects

8.1
Introduction

The simple independent particle model discussed in Chapter VII has provided a rather nice correlation of the properties of low-lying levels of many nuclei not too far removed from closed shells. But the model fails badly in accounting for the low-energy properties of nuclides far removed from closed shells, not even reproducing ground state spins and parities in many cases. When the model fails badly, it is often found that level densities near the ground state are considerably greater than can be accounted for by any reasonable spherically-symmetric potential.

Fundamental to everything we have done so far is the assumption that nuclei are spherical, or approximately so. While reasonable, we have not considered any experimental evidence that provides information to support this assumption. A way to accomplish this is by consideration of the electric field of nuclei and a simple classical example demonstrates how we should proceed.

Suppose we have a spherical body that is uniformly charged. If the body is fixed in a laboratory reference frame, the electric potential can be measured as a function of the spatial coordinates r, θ, and ϕ. If the total charge of the body is Z and the charge on the test body is q, the Coulomb force at any point will be given by $F_{Coul} = -k_C q Z e / r^2$ and the electric potential will be given by $V_{Coul} = k_C Z / r$, where r is the center-to-center distance between the body and the test charge. If all measurements are made at a fixed r, we will find a constant force or potential of interaction. If, however, the charge distribution was not spherically symmetric, the potential will depend upon at least one of the variables θ and ϕ as well. This is clearly a means by which we can determine the charge distribution in an object.

For simplicity, we will assume a charge distribution that possesses symmetry about the z-axis so that the electric field at a point P outside of the body will depend only on r and θ as shown in Fig. 8.1. We will also assume that the charge density ρ of the body is constant. If we have a differential volume element dv, it will contain a total charge ρdv and the electric potential at P due to this charge will be $dV_{Coul}(P) = k_C \rho dv / d$.

Nuclear Physics for Applications. Stanley G. Prussin
Copyright © 2007 WILEY-VCH Verlag GmbH & Co., Weinheim
ISBN: 978-3-527-40700-2

8 Nuclear Shapes, Deformed Nuclei and Collective Effects

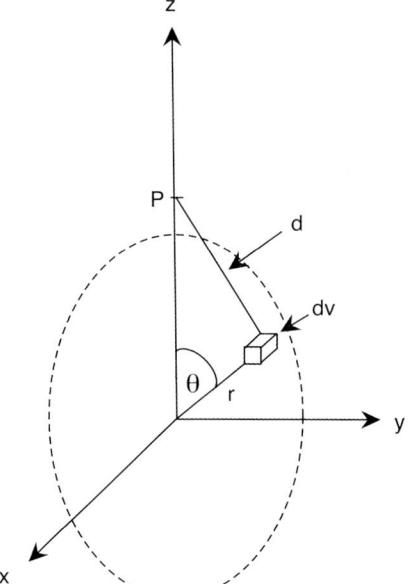

Fig. 8.1 Schematic diagram of the electric potential at a field point P due to a nonspherical distribution of charge (dashed line) centered at the origin of the coordinate system.

The electric potential from the entire body at the point P can then be written as

$$V_{Coul}(P) = k_C \int_{object} \frac{\rho dv}{d} \qquad (8.1)$$

We can express the distance d in terms of r and the coordinate of P on the z-axis, z, by use of the Law of Cosines, $d^2 = z^2 + r^2 - 2rz\cos\theta$, and Eq. (8.1) can be rewritten as

$$V_{Coul}(P) = k_C \int_{object} \frac{\rho dv}{(z^2 + r^2 - 2rz\cos\theta)^{1/2}} \qquad (8.2)$$

While we could express dv and z in terms of the spherical polar coordinates and proceed to integrate, it will be much more useful to first carry out an expansion of the denominator in terms of the binomial series for the general quantity $(a + x)^n$,

$$(a + x)^n = a^n + na^{n-1}x + \frac{n(n-1)}{2!}a^{n-2}x^2 + \ldots \quad , (x^2 < a^2)$$

Writing the denominator in the integrand of Eq. (8.2) as

$$(z^2 + r^2 - 2rz\cos\theta)^{1/2} = z\left[1 + \left(\frac{r}{z}\right)^2 - \frac{2r}{z}\cos\theta\right]^{1/2}$$

$$= z\left[1 + \frac{r}{z}\left(\frac{r}{z} - 2\cos\theta\right)\right]^{1/2}$$

and letting $n = -1/2$, $a = 1$, and

$$x = \frac{r}{z}\left(\frac{r}{z} - 2\cos\theta\right)$$

the denominator can be written in the form

$$d^{-1} = z^{-1}\left[1 + \frac{r}{z}\cos\theta + \frac{r^2}{2z^2}(3\cos^2\theta - 1) + \ldots\right] \tag{8.3}$$

Now each of the terms in this expansion can be identified with a Legendre polynomial $P_n^0(\cos\theta)$ times the quantity $(r/z)^n$ (see Table 7.1). Therefore, the integral in Eq. (8.2) can now be written as

$$V_{Coul}(P) = k_C \int_{object} \frac{\rho dv}{(z^2 + r^2 - 2rz\cos\theta)^{1/2}}$$

$$= k_C \int_{object} \frac{\rho}{z}\left[1 + \frac{r}{z}\cos\theta + \frac{r^2}{2z^2}(3\cos^2\theta - 1) + \ldots\right]dv \tag{8.4}$$

$$= k_C \int_{object} \frac{\rho}{z}\left(\sum_{n=0}^{\infty}\left(\frac{r}{z}\right)^n P_n^0(\cos\theta)\right)dv$$

The terms with $n = 0, 1, 2, 3, \ldots$ in the summation are referred to, respectively, as the monopole, dipole, quadrupole, octupole, etc., terms. The first term, the monopole term, is of the form

$$k_C \int_{object} \frac{\rho}{z}dv = k_C \frac{1}{z}\int_{object}\rho dv = k_C \frac{q}{z} \tag{8.5}$$

where q is just the total charge within the body. The electric potential represented by q/z is just that from a point charge or spherically-symmetric charge located at the origin. The second term, the dipole term, is of the form

$$k_C \int_{object} \frac{\rho}{z}\left(\frac{r}{z}\right)\cos\theta\, dv = \frac{k_C}{z^2}\int_{object}\rho r\cos\theta\, dv = \frac{k_C}{z^2}\int_{object}\rho z'dv \tag{8.6}$$

Because we are carrying out the integration within the nucleus we have replaced the quantity $r\cos\theta$ by z' where the prime signifies "within the object". Now ρdv is just the charge within the differential volume element dv and the integral therefore represents the summation of the product of charge times its z-coordinate. This is just the *electric dipole moment* of a charge distribution.

The third term in the expansion can be written as

$$k_C \int_{\text{object}} \frac{\rho}{z}\left(\frac{r}{z}\right)^2 \frac{1}{2}(3\cos^2\theta - 1)dv = \frac{k_C}{z^3} \int_{\text{object}} \frac{r^2}{2}(3\cos^2\theta - 1)\rho dv \tag{8.7}$$

In this case, the charge in each volume element is multiplied by the factor $(3z'^2 - r^2)$ and the integral clearly depends upon the *volumetric distribution* of charge within the object. It should therefore provide direct information on whether the distribution is spherical or not.

Each term in the expansion of the electric potential (Eq. (8.4)) is dependent upon the distance between the point of measurement and the object itself. However, the factor multiplying $\frac{1}{z^n}$ is dependent only upon the charge distribution within the object and thus represents an *intrinsic* property of the object. These factors are referred to as the *multipole moments* of the charge distribution and thus the quadrupole moment is

$$Q = k_C \int_{\text{object}} \frac{r^2}{2}(3\cos^2\theta - 1)\rho dv \tag{8.8}$$

The characteristics of the quadrupole moments corresponding to different shapes with axial symmetry can be understood with a bit of analysis. We begin by rewriting Eq. (8.8) as

$$Q = k_C \int_{\text{object}} \frac{1}{2}(3r^2\cos^2\theta - r^2)\rho dv \tag{8.9}$$

$$= k_C \int_{\text{object}} \frac{1}{2}[3z'^2 - (x'^2 + y'^2 + z'^2)]\rho dv$$

where primes again indicate that we are integrating only over the volume of the object. Suppose first that we have spherical symmetry. The integral in the equation will vanish. This is most easily seen by considering that for a sphere, $<x> = <y> = <z>$ and thus, on the average, $\langle 3z'^2 - (x'^2 + y'^2 + z'^2)\rangle = 3\langle z'^2\rangle - 3\langle z'^2\rangle = 0$. However, if we have an object that is distorted from spherical symmetry such that the range of x' and y' is either larger or smaller than the range of z', the object will have a *nonzero* quadrupole moment.

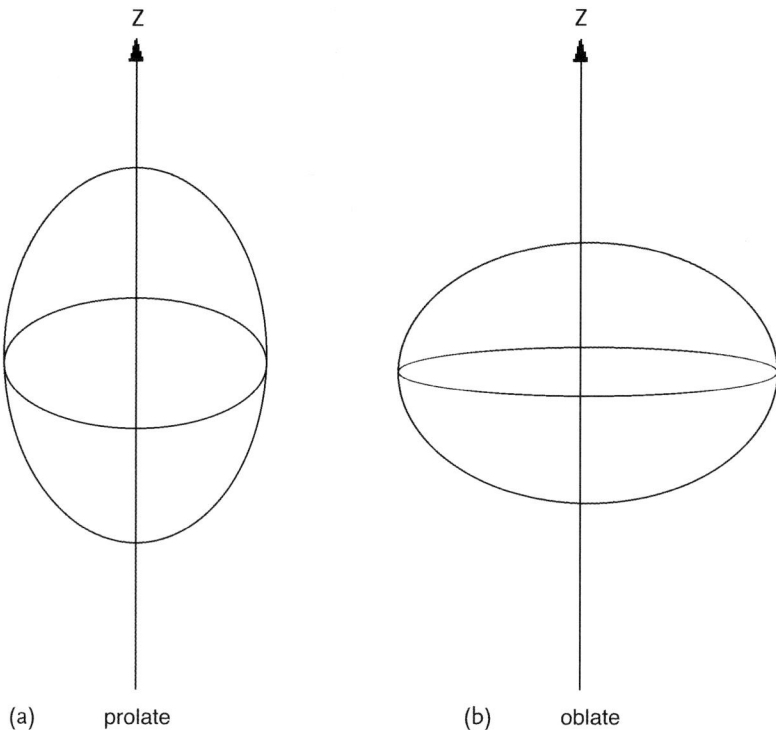

(a)   prolate   (b)   oblate

Fig. 8.2 Prolate and oblate ellipsoids of revolution.

Suppose that the object has a shape produced by the rotation of an ellipsoid about the axis of symmetry. It will have a circular cross section in planes normal to that axis. Depending on the orientation of the semi-major and semi-minor axes, two possible shapes can result as shown in Fig. 8.2, where (a) has its semi-major axis parallel to the z-axis and is referred to as a *prolate* ellipsoid of revolution and (b) has its semi-minor axis parallel to the z-axis and is referred to as an *oblate* ellipsoid of revolution.

These two shapes will give rise to very different quadrupole moments. If the object had a prolate shape, the first term in the parentheses of Eq. (8.9) will always be greater than the second because, on the average, z'^2 is larger than both x'^2 and y'^2. Therefore the quadrupole moment of a prolate ellipsoid of revolution will always be *positive*. For an oblate shape, the average value of z'^2 will be less than the averages of both x'^2 and y'^2 and the quadrupole moments for oblate ellipsoids of revolution will be *negative*.

All of the foregoing can be taken over to describe the electric potential of nuclei and gain an understanding of their shapes. We need to convert the quadrupole moment into an operator and evaluate its expectation value. Formally, if we had Z protons in the nucleus, each with a wave function ψ_i, the quadrupole moment relative to the axis of symmetry could be calculated from the expression,

$$\langle Q \rangle = \sum_{i=1}^{Z} \int_{\text{nucleus}} e\psi_i^* \frac{r^2}{2}(3\cos^2\theta - 1)\psi_i \, dv \qquad (8.10)$$

$$= e\sum_{i=1}^{Z} \int_{\text{nucleus}} \psi_i^* \frac{r^2}{2}(3\cos^2\theta - 1)\psi_i \, dv$$

For historical reasons, however, nuclear quadrupole moments relative to the axis of symmetry are defined as

$$\langle Q_o \rangle = \sum_{i=1}^{Z} \int_{\text{nucleus}} \psi_i^* r^2 (3\cos^2\theta - 1)\psi_i \, dv \qquad (8.11)$$

equivalent to multiplying the right-hand side of Eq. (8.10) by the factor $2/e$. $\langle Q_o \rangle$ is referred to as the *intrinsic quadrupole moment*. Because the integrands in Eq. (8.11) have the dimensions of areas, nuclear quadrupole moments are reported in units of *barns*.

Unfortunately, the intrinsic quadrupole moment cannot be measured directly. We can only align a nucleus to produce the $2j + 1$ possible projections of the total angular momentum vector on a space-fixed axis and the angular momentum vector itself is not aligned to the axis of symmetry. While we will not go through the details here, it is possible to relate a measured quadrupole moment for a specific projection m_j to the moment relative to the symmetry axis of the nucleus. The experimental moments reported in the literature are those measured when the projection of the angular momentum onto a space fixed axis has $m_j = j$. In addition, the characteristic symmetries of the nuclear wave function do not permit observation of the electric quadrupole moments of nuclei with angular momentum quantum numbers of 0 or 1/2. Only those with total angular momenta $J \geq 1$ have observable quadrupole moments.

With these comments in view, the experimental electric quadrupole moments of the ground states of nuclei $\langle Q_{\text{nucl}} \rangle$ are shown in Fig. 8.3. One is immediately struck by the fact that the majority of these are *not* zero, in marked contrast to the simple spherical assumption. Further, there is a fairly regular and very similar pattern in the variation of $\langle Q_{\text{nucl}} \rangle$ with N or Z. In the immediate vicinity of a magic number, the quadrupole moments are small and tend to be negative just after a magic number. Far from magic numbers, however, the quadrupole moments are relatively large and positive. The majority of the experimental data are consistent with the assumption that most nuclides have *prolate* shapes. The fact that odd-N nuclides display very similar systematics indicates that the quadrupole moments cannot be due solely to a single nucleon.

We expect closed shells of identical nucleons to be spherically symmetric. But, with the exception of a nucleus with a single nucleon in an s-orbit outside of an (even, even) "core", all other configurations will be nonspherical because the orbits are nonspherical. Although we will not prove it, it can be shown that a single

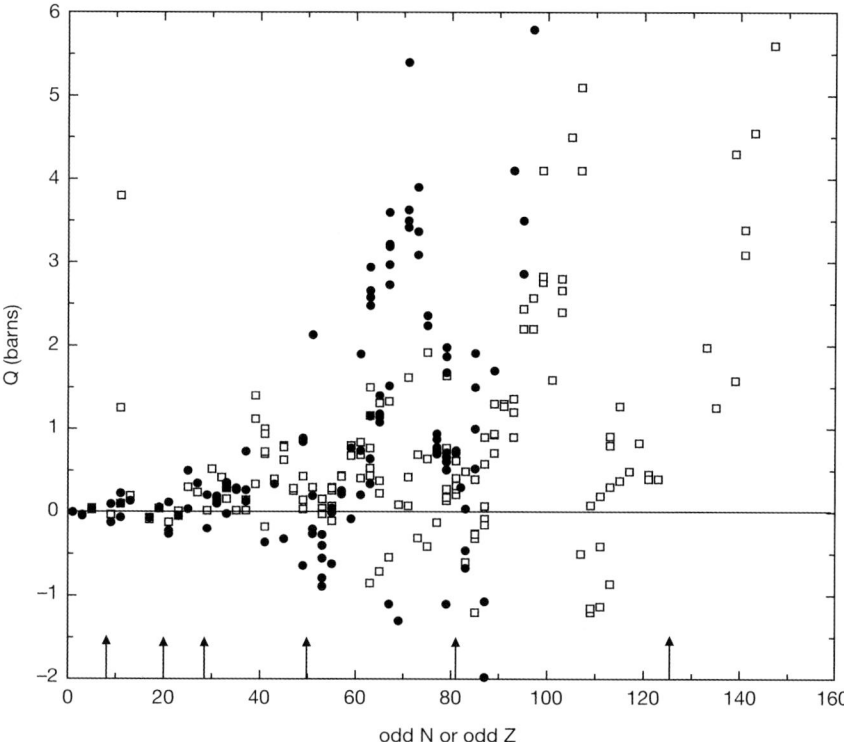

Fig. 8.3 Experimental electric quadrupole moments of odd-A nuclides in their grounds states. Nuclides of odd-Z are shown as filled circles and those of odd-N are shown as open squares. The arrows mark the magic numbers that are greater than 2.

particle in an orbit of total angular momentum j will have an electric quadrupole moment of

$$\langle Q_{SP} \rangle = -\langle r^2 \rangle \frac{2j-1}{2(j+1)} \tag{8.12}$$

where $\langle r^2 \rangle$ is just the average of the square of the radial displacement of the particle in its orbit [1]. This indicates that the electric quadrupole moment for a nucleus that has a single particle outside closed shells will be negative. Qualitatively, the dependence of the moment on $\langle r^2 \rangle$ makes sense. If we take the simple constant density estimate that $r_n = r_o A^{1/3}$, we can easily estimate the magnitude of the moment from a single particle. For a nucleus of mass number A = 100, $\langle r^2 \rangle \cong 34$ fm^2 = 0.34b. Because

$$\frac{1}{4} \leq \frac{2j-1}{2(j+1)} \leq 1$$

for all states with measurable electric quadrupole moments we can be sure that this is within a factor of 4 of the magnitude that any single particle will have. As particles are added to the orbit, they will each contribute to the total quadrupole moment.

Theory shows that, as an orbit continues to get filled, the quadrupole moment changes sign once it is more than half-filled. Further, the quadrupole moment of an orbit that contains one less than the total number that can occupy that orbit, an orbit that contains "one hole" in it, is predicted to have a quadrupole moment that is identical in magnitude but opposite in sign to the moment for an orbit containing just one nucleon.

Now consider, again, the data in Fig. 8.3. Whether positive or negative, the quadrupole moments of spherical nuclei with a single unpaired nucleon outside of an (even, even) core cannot be larger than a few tenths of a barn. Thus the majority of nuclei that are removed from closed shells of neutrons or protons must have a number of nucleons that are moving in nonspherically symmetric orbits, and those that are far from a magic number appear strongly *deformed*. There can be no doubt that the majority of deformed nuclides are prolate objects.

If a nucleus is deformed, the potential seen by a single nucleon cannot be spherically symmetric. The potential must depend upon at least one angular coordinate in addition to the radial coordinate. All available evidence indicates that the nuclear density is the same whether the nucleus is spherical or deformed and thus the simple idea of a potential that is roughly constant throughout the central part of deformed nuclei still seems reasonable. While an attempt to devise a simple potential to describe single-particle motion in deformed nuclei should be attacked, we will defer this for the moment because it is rather complicated and because there is a very simple idea that leads directly to observables that are clearly characteristic of nuclear deformations. This comes about by an extension of the liquid drop model that we discussed in detail in Chapters IV and V.

8.2
Collective Excitations

The liquid drop model views the nucleus as a uniformly charged spherical droplet of constant density. A classical liquid drop in field-free space can be made to undergo vibrations about its normally spherical shape by subjecting it to a small impulse of energy. The vibrations will be harmonic and would continue indefinitely if it were not for the viscous or frictional forces between the molecules that eventually convert the energy into heat. The vibrations of the droplet are harmonic because all of the molecules in the liquid undergo motions that are *correlated* with one another as compared to their normal random motions due to thermal energy. In addition to vibrations, the drop can also be made to rotate about its center of mass. Here again the motions of the molecules are correlated. Such correlated motions as vibrations and rotations are commonly referred to as *collective* motions.

Vibrations and rotations of quantized systems are very well known. One of the simplest cases is that of a diatomic molecule. Level diagrams of diatomic molecules display features that can be nicely separated into three different categories. First, there is a series of levels due to *electronic excitations* where one or more electrons are excited to higher energy levels. Typically, these excitations require energies on the order of a few electron volts. Second, for each of the electronic states, including the ground state, there is a series of approximately equally spaced levels and these are *vibrational states* built upon the electronic state. This is just what is expected for a simple quantized one-dimensional harmonic oscillator. One can picture the vibration of the two atoms about their center of mass as a variation of the bond length. A quantum of vibrational energy is typically on the order of 0.1 eV or so. Finally, for each vibrational state there is a series of states that can be associated with the *rotation* of the molecule in space. The lowest-energy rotational states typically have energies on the order of 0.01 eV.

One of the principal reasons why the spectra of diatomic molecules are so nicely separated into electronic, vibrational and rotational states, is that the energies required for the excitations vary from one another by roughly an order of magnitude. That is, the total energy of the system can be written approximately as

$$H_{total} \cong H_{rotation} + H_{vibration} + H_{electronic} \qquad (8.13)$$

If the molecule is in the ground state and you add only about 0.01 eV, it is not possible to carry out a vibrational or electronic excitation and thus all that can take place is a rotational excitation. If only enough energy is added to excite a vibration, only rotations with the atom in its ground electronic state and a single vibration can be produced. In essence, and to a good approximation, the lowest-energy rotations and vibrations are independent of one another. To a good approximation, you can excite the lowest-energy rotations without affecting the equilibrium distance between the nuclei of the two atoms. It is not difficult to see that, if a large amount of energy is added, however, this may not be true. Under such conditions many types of excitation can be produced and the spectrum can get very complicated indeed.

8.3
Rotational Excitations in (Even,Even) Nuclei

Now we try to apply the same notions to the nucleus. If collective motions of the nucleons exist, the discussion of the preceding paragraph suggests that the lowest-energy excitations should be rotations of the nucleus as a whole, and if the energies of these are sufficiently small, no other types of excitation should take place. It is not difficult to understand what the energy spectrum of rotational states ought to be in the absence of any other excitations. We simply use the same approach we have used before. Write down the classical Hamiltonian for the system, convert it into the appropriate operator form and, if possible, solve it analytically.

The total energy of a nonrelativistic body undergoing classical rotations about an axis is given by

$$H = T + V = \frac{1}{2}\Im\omega^2 \tag{8.14}$$

where $\Im$ is the moment of inertia of the body about the axis of rotation and ω is the angular frequency. The total energy is just the kinetic energy of rotation. If the axis of rotation is taken as the z-axis, the moment of inertia is given by

$$\Im = \Im_z = \int_{\text{volume}} \rho r_\perp^2 dv \tag{8.15}$$

where ρ is the mass density within the body and $r_\perp$ is the magnitude of the radius vector normal to the axis of rotation. We can express H in terms of the angular momentum of rotation by remembering that $L = \Im_z \omega$. With this, Eq. (8.14) becomes

$$H = \frac{1}{2}\frac{L^2}{\Im_z} \tag{8.16}$$

Now let us convert the Hamiltonian into operator form. L^2 is just the square of an angular momentum and as we have pointed out repeatedly, it makes no difference what the source of the angular momentum is; the same properties apply. However, to make sure that there is no confusion, we will use R^2 for the square of the angular momentum due to rotations. Now the operator for the moment of inertia is just the moment of inertia itself as can be seen from the definition in Eq. (8.15). The operators for each of the variables in the definition are just the variables themselves.

Before we can proceed, we need to consider the wave function of the system. We picture the nucleus as a deformed body undergoing rotations. If we had a classical, deformed liquid drop and started it rotating with the addition of a very small amount of energy, its shape would remain the same. However, the greater the energy of the rotation the greater the centrifugal force and the drop would tend to elongate in the direction normal to the axis of rotation because it is deformable. But if this takes place, the moment of inertia will change and will be dependent upon the energy of rotation. If this were to take place in a nucleus, the potential seen by a nucleon would change because the shape of the nucleus has changed. Thus, if the nucleus is considered to be deformable as a result of rotation we will have a very complicated problem on our hands. If, however, we make the assumption that the nucleus can be approximated as a *rigid body*, its shape will be independent of rotation, the moment of inertia will be constant and we can treat the rotations independently of all other interactions. In this case the wave function of the nucleus can be written as the product of the wave function in the absence of rotation times, a wave function that represents the rotation of a rigid body. We will make this assumption for simplicity and use the comparison with experimental data to tell us how well this assumption is found in reality.

8.3 Rotational Excitations in (Even, Even) Nuclei

The simplest case to consider is the case of an (even, even) nucleus in its ground state. The total angular momentum must be due solely to the angular momentum of rotation. If we represent the rotational eigenfunction by $\mathcal{D}$, the rotational part of the Schrödinger equation becomes

$$H\mathcal{D} = \frac{1}{2}\frac{R^2}{\Im}\mathcal{D} \tag{8.17}$$

where the subscript on the moment of inertia has been dropped for generality. Now the Hamiltonian, or the total energy E_{rot}, is just the kinetic energy. Further, if $\mathcal{D}$ is an eigenfunction of R^2, then

$$R^2\mathcal{D} = I(I+1)\hbar^2\mathcal{D} \tag{8.18}$$

where I is the quantum number for the rotational angular momentum. Therefore, the rotational energy eigenvalues must be

$$E_{rot} = \frac{\hbar^2}{2\Im}I(I+1) \tag{8.19}$$

where $I = 0, 1, 2,\ldots$. Now there are two more issues that we must consider. First, and contrary to the classical case, a spherical nucleus cannot possess a rotational spectrum. This arises because a spherical nucleus must have a wave function that is independent of angular coordinates and this symmetry makes rotations unobservable. Symmetry arguments also show that if the nucleus has cylindrical symmetry, as we have assumed, only rotations about axes normal to the symmetry axis lead to observables. Second, symmetry shows that if we deal with a nucleus with zero angular momentum in its ground state, the only allowed rotational states are those for which I is *even*. These results then indicate that a rigid rotator with zero angular momentum in its ground state should possess a rotational spectrum with the spin sequence 0 (ground state), 2, 4, 6,.... Because of the symmetry of the rotational eigenfunctions, the parity of the rotations is even and therefore all of the states will have even parity. The prediction is then that an (even, even) deformed nucleus, considered as a rigid body, will have a rotational spectrum built upon the ground state with the characteristics shown in Fig. 8.4. In the figure the energies are given to the right of each level in units of $(2\Im/\hbar^2)E_{rot}$.

The predicted rotational spectrum can be compared with the low-energy spectra of nuclei that are strongly deformed in their ground states. As seen from Fig. 8.5, where the rotational spectra built upon the ground states are shown for $^{8}_{4}Be$, $^{24}_{12}Mg$, $^{154}_{64}Gd$ and $^{238}_{92}U$, such deformed nuclei are found throughout the chart of known nuclides when N or Z are "far" from magic numbers. Qualitatively, the spectra are just what the model predicts, and the rotational spectra in the heavier nuclei can be followed to quite high angular momenta. Strikingly, the energy separation of the states decreases dramatically as the mass number increases. While the energy differences between the lowest members of the rotational band

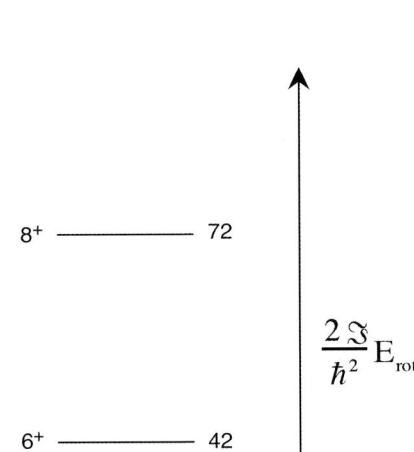

Fig. 8.4 Rotational band built upon the ground state of a deformed, (even,even) nucleus in the rigid rotor approximation.

in ^{8}Be are on the order of several MeV, the energy differences in ^{238}U are on the order of 0.01–0.1 MeV. This can be understood qualitatively from the increasing size of the nuclei that leads to an increasing moment of inertia. A rough estimate of the relative spacings of levels can be obtained from the classical moment of inertia of an ellipsoid of revolution of mass M about its axis of symmetry

$$\mathfrak{J}_o = \frac{M}{5}(a^2 + b^2) \tag{8.20}$$

Here a and b are the dimensions of the semi-major and semi-minor axes. We make the additional crude assumption that the four nuclides addressed in Fig. 8.5 have exactly the same deformation and hence the same ratios of a/b. Let a = R + ΔR and b = R. The geometric mean radius of the body is $\bar{R} = (a^2 b)^{1/3}$. Using this expression, $(a^2 + b^2)$ becomes

$$(a^2 + b^2) = \bar{R}^2 \left[\frac{2}{(1+k)^{2/3}} + \frac{2k}{(1+k)^{1/3}} + k^2 \right] \tag{8.21}$$

8.3 Rotational Excitations in (Even, Even) Nuclei

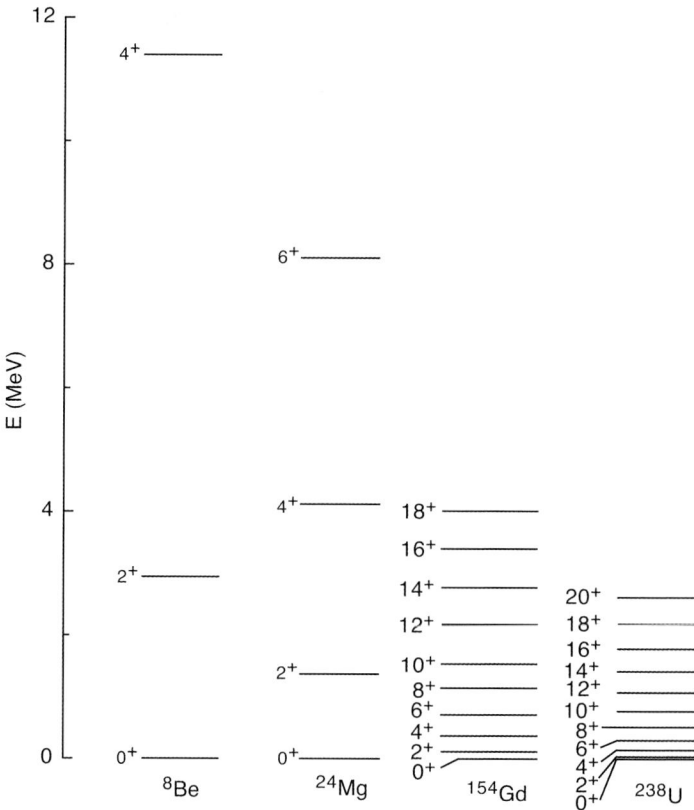

Fig. 8.5 Rotational spectra of $^{8}_{4}\text{Be}$, $^{24}_{12}\text{Mg}$, $^{154}_{64}\text{Gd}$ and $^{238}_{92}\text{U}$.

where $k = \Delta R/R$. The expression in brackets will be the same for all four nuclides if they have the same ratio of a/b. If we equate the geometric mean radius to the constant density expression and express the mass of a nucleus approximately as $M \cong Am_u$, Eq. (8.20) can be written as

$$\Im_o \cong \frac{Am_u}{5} r_o^2 A^{2/3} \left[\frac{2}{(1+k)^{2/3}} + \frac{2k}{(1+k)^{1/3}} + k^2 \right] \quad (8.22)$$

$$= \text{constant} \cdot A^{5/3}$$

We then have, for example, $\Im_o(^{238}_{92}\text{U})/\Im_o(^{8}_{4}\text{Be}) \cong (238/8)^{5/3} = 28.6$ and we then expect that the level spacings in ^{238}U will be about 1/28.6 times the level spacing in ^{8}Be. Experimentally, the ratio of the excitation energies of the 2^+ states in the rotational bands is found to be 0.021. With all of the simplifying assumptions we have made, not the least of which are the assumptions that the nuclei can be treated as rigid rotors and both have the same extent of deformation, this must be taken as rather good agreement. The data in Fig. 8.5 also indicate that the rotational bands

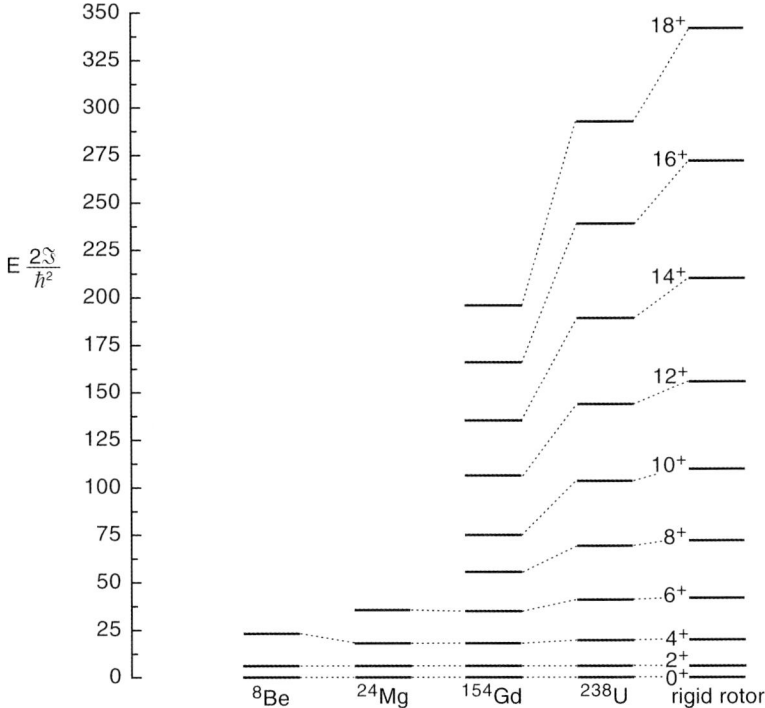

Fig. 8.6 Comparison of the energy spacings of the ground state rotational bands in $^{8}_{4}$Be, $^{24}_{12}$Mg, $^{154}_{64}$Gd and $^{238}_{92}$U with the spacings of a rigid rotor.

in heavy nuclei persist to very high angular momenta. For the light nuclei the moment of inertia is so low that even the first few members of the rotational band have energies comparable to single-particle energies and pairing energies. For the heavier nuclides, the large moments of inertia permit many rotational states to exist with energies below a few MeV.

The quality with which the simple rigid rotor approximation is able to describe rotational states in these nuclei is shown in Fig. 8.6. To compare the experimental spectra with the model, the energy difference between the ground and 2^+ level for each nuclide was used to calculate the quantity $\hbar^2/2\Im$ and the experimental energies of all states were then divided by this quantity. If the spectra followed the rigid rotor model, the result should be $I(I + 1)$ for each state. In the figure, the levels determined experimentally are shown as solid horizontal lines. Levels of the same spin and parity for different nuclides are connected by dotted lines and compared to the levels of the rigid rotor shown to the right in the figure. Clearly, nuclei *do not* behave as perfectly rigid rotors, but the approximation is remarkably good, especially for the lower spin states. With the exception of ^{8}Be, the experimental energies are seen to lie systematically lower than those expected for a rigid rotor and the discrepancy increases as the angular momentum increases. This is just what is expected for a deformable nucleus for which the moment of inertia will increase with increasing angular momentum. In fact, more sophisticated models that account for the changes in the moment of inertia are able to provide excellent agreement with rotational bands in essentially all well-deformed nuclides. For the heavier nuclides such as ^{238}U, where moments of inertia are very large and the energy differences between the levels in a rotational band are small, the rigid rotor approximation is fairly good even at rather high spins. The energy of the 18$^+$ level in ^{238}U is, for example, within 15% of that expected for a rigid rotor, and the discrepancy between the model predictions and experiment decreases rapidly for lower members of the rotational band. In addition to rotational bands built upon the ground states, bands are seen that are built on excited states as well.

8.4
Rotational Excitations in Odd-A Nuclei

The description of rotational excitations in odd-A nuclides is somewhat more complicated because of the existence of the angular momentum of the unpaired nucleon. Nevertheless, if we assume the same limiting approximations that we have taken for (even, even) nuclei, we can obtain a rather simple expression that includes the principal parts of the physics involved and provides about the same quality of agreement with experiment. To do this, we must recognize that there are three angular momentum vectors to be considered; the angular momentum of the nucleus in the absence of rotations, **J**, the angular momentum of the rotational motion, **R**, and the total angular momentum of the nucleus, **I** = **J** + **R**. These are shown relative to the axis of symmetry for a prolate ellipsoid of revolution in Fig. 8.7 along with the z-axis of the laboratory coordinate system. The Cartesian coordinates of the body-fixed axes are labeled "1", "2", and "3", with the "3" axis

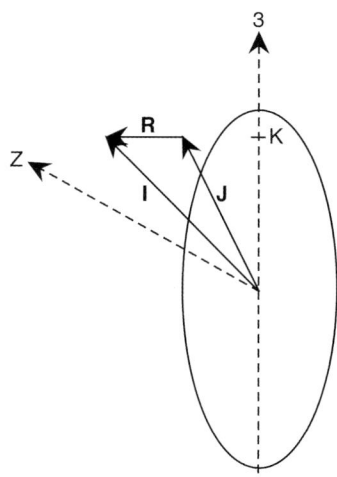

Fig. 8.7 Angular momentum vectors and axes relevant for describing rotational motion in an axially-symmetric body. K is the quantum number representing the projection of both the vectors **I** and **J** on the body-fixed symmetry axis "3".

representing the axis of symmetry. The laboratory spatial coordinates are labeled in the usual manner as x, y, and z. In this restricted case, the rotational angular momentum vector is normal to the "3" axis and the quantum number representing the projections of both **I** and **J** on this axis is K.

Following the ideas presented in Section 8.3, we can write the total Hamiltonian for such an object as

$$H = H'_{int} + H_{rot} \tag{8.23}$$

where H'_{int} is meant to represent the Hamiltonian for the motion of nucleons in the absence of rotations. Now H_{rot} can be written quite generally as

$$H_{rot} = \frac{\hbar^2}{2\mathfrak{J}_1} R_1^2 + \frac{\hbar^2}{2\mathfrak{J}_2} R_2^2 + \frac{\hbar^2}{2\mathfrak{J}_3} R_3^2 = \frac{\hbar^2}{2\mathfrak{J}}(\mathbf{R}^2 - R_3^2) + \frac{\hbar^2}{2\mathfrak{J}_3} R_3^2 \tag{8.24}$$

where the subscripts 1, 2 and 3 refer to the three-body fixed axes and we recognize that symmetry requires that the moment of inertial about the 1 and 2 axes are the same. Because we can only observe the total angular momentum, we make the substitutions $\mathbf{R}^2 = (\mathbf{I} - \mathbf{J})^2$ and $R_3 = (I_3 - J_3)^2$ and, after expanding $\mathbf{R}^2$ in terms of its components, and a bit of rearranging, we obtain

$$H_{rot} = \frac{\hbar^2}{2\mathfrak{J}}(\mathbf{I}^2 + \mathbf{J}^2 - 2\mathbf{I}\cdot\mathbf{J}) + \left(\frac{\hbar^2}{2\mathfrak{J}_3} - \frac{\hbar^2}{2\mathfrak{J}}\right)(I_3 - J_3)^2 \tag{8.25}$$

Because the projections of **I** and **J** on the "3" axis are identical, we can immediately set the last term to zero. Now, if we expand the dot product in the first term we get

$$H_{rot} = \frac{\hbar^2}{2\mathfrak{J}}(\mathbf{I}^2 + \mathbf{J}^2 - 2I_3 J_3) + \frac{\hbar^2}{\mathfrak{J}}(I_1 J_1 + I_2 J_2) \tag{8.26}$$

Within the limits of the model we have used, this is the correct expression with which we have to deal. The presence of the second term in Eq. (8.26) presents some complications. It actually represents the quantum-mechanical equivalent of the Coriolis force that is met with in rotating coordinate systems in classical mechanics. Fortunately, it is found that this term can be neglected in first order for all rotational states, except for those where J = 1/2. In essence, we will then make the approximation that **I** and **J** are sufficiently large that we can neglect their projections onto axes normal to the "3" axis and be willing to give up a reasonable description of the rotational states for which J = 1/2.

We can now write the total Hamiltonian in the form

$$H = H'_{int} + \frac{\hbar^2}{2\mathfrak{J}} \mathbf{J}^2 + \frac{\hbar^2}{2\mathfrak{J}}(\mathbf{I}^2 - 2I_3 J_3) \tag{8.27}$$

The second term on the right-hand side of Eq. (8.27) really has nothing to do with observables that depend on rotation and we can just as well combine it with H'_{int} to write

$$H_{int} = H'_{int} + \frac{\hbar^2}{2\mathfrak{J}}J^2$$

Because the projections of **I** and **J** on the "3" axis are identical for the symmetry we have assumed, we can then write the total Hamiltonian, with the neglect of the Coriolis term, as

$$H = H_{int} + \frac{\hbar^2}{2\mathfrak{J}}(I^2 - 2J_3^2) \tag{8.28}$$

We now need to recognize that H_{int} only operates on the internal motion of the nucleons and thus provides the energy E_o in the absence of rotations. I^2 will have the eigenvalues $I(I+1)$ representing the general eigenfunctions of an angular momentum operator, and J_3^2 will give the square of the quantum number K. Thus, operating on the wave function of a state with completely separable internal and rotational modes, Eq. (8.28) yields

$$E(I, K) = \frac{\hbar^2}{2\mathfrak{J}}I(I+1) - \frac{\hbar^2}{2\mathfrak{J}}2K^2 + E_o \tag{8.29}$$

Eq. (8.29) represents our approximation for the energies of rotational levels, E(I,K), when the nucleus, in the absence of rotations, has the angular momentum quantum number J ≠ 0. Because of this, there is no restriction on the evenness or oddness of K or I and thus K can have the quantum numbers K = J, J + 1, J + 2, etc. I, representing the total quantum number of a state can then have the values I = K, K +1, K +2, etc.

Each rotational band in an odd-A nucleus will be built upon a state in the deformed nucleus that has the angular momentum quantum number J = K. Such states are called the *band heads*. The band heads represent the ground state and all excited states of the nucleus in the absence of rotations, the *intrinsic* states of the system. For each band, the rotational levels built on it will have total angular momentum quantum numbers of I = K, representing the intrinsic state, and I = K + 1, K +2, etc., representing the rotational excitations built on the intrinsic state. Our result is expected to be reasonable so long as K > 1/2.

As an example of how well our model accounts for reality, the levels in the deformed nucleus $^{175}_{72}\text{Hf}$ are shown in Fig. 8.8 along with the level energies calculated with Eq. (8.29) for the rotational states built upon the ground state of spin $5/2^-$. To perform the calculations, the energy difference between the ground and first-excited states was used to estimate the value of $\hbar^2/2\mathfrak{J}$ and the energies of all other rotational states were calculated using this value. For spins up to as high as $25/2^-$, our very simple model reproduces experimental level energies with errors of

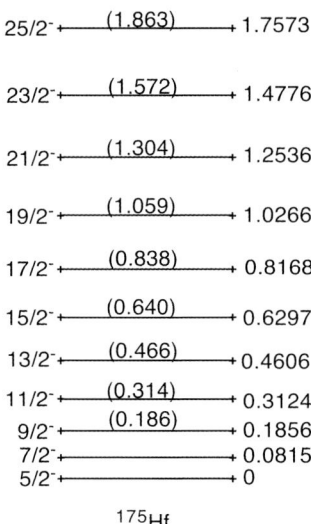

Fig. 8.8 Rotational band built upon the 5/2⁻ ground state of ^{175}Hf. Spins and parities are given to the left and experimental level energies to the right of each level. Energies calculated with Eq. (8.29) are shown in parentheses on the top of each level.

no more than about 6%. We see all of the predicted levels and the discrepancies between experimental and calculated energies follow in the same manner as we found for (even, even) nuclei. Rotational bands built on other excited states of ^{175}Hf and throughout the well-deformed rare earths and actinides are also reproduced with similar quality. Collective rotations in all well-deformed nuclei can be considered quite well understood.

8.5
Vibrational Excitations in Nuclei

Having shown that deformed nuclei possess rotational excitations, it is natural to look for vibrational excitations. The problem here is a bit more complicated because the energies involved are sufficiently large that they approach the pairing energy and the energies of single-particle excitations. If vibrations occur, and if, as we know to be true, the nuclear density remains constant, it must be true that the nucleus will undergo *changes in shape* as a result of the vibrations. Thus, vibrations must produce a change in the potential seen by the individual nucleons and we can expect it to be much more difficult to apply a very simple model and obtain good quantitative agreement with experiment. Nevertheless, there are a number of nuclei with spectra that are in reasonable agreement with a simple vibrational model and we will use these to illustrate such collective effects.

If we consider the simple model of harmonic vibrations such as are seen in the spectra of diatomic molecules, we expect that the lowest vibrational states will be separated equally in energy and we can define the *quantum* of vibrational energy that separates adjacent levels. Now quanta may or may not carry angular momentum. If the angular momentum is zero, then each of the vibrational excitations will produce a single discrete level and the angular momentum of each will be the angular momentum of the ground state upon which the vibrations are built. However, if the quantum has nonzero angular momentum, then the simple harmonic model predicts that the addition of a quantum of vibrational energy can lead to a number of degenerate states with angular momenta that satisfy the conservation laws.

In order to determine what angular momenta vibrational quanta can have, we can make use of the fact that a vibration must lead to a shape change in the nucleus. If we can describe the shape in some series expansion that automatically contains angular momentum, we can seek the effect of each term on the possible observables. To do this we can make use of the spherical harmonics. Because they form a complete orthonormal set, the radial coordinate of any shape can be described by an expansion to give $r = r(\theta,\phi,t)$. Thus, for example, the radius of the deformed objects shown schematically in Fig. 8.2 can be written at any time t as

$$R(\theta, \phi, t) = R_o \left[1 + \sum_{\lambda=0}^{\infty} \sum_{\mu=-\lambda}^{\lambda} \alpha_{\lambda,\mu}(t) Y_{\lambda,\mu}(\theta, \phi) \right] \tag{8.30}$$

In place of the quantum numbers l and m that we have used previously to represent the orbital angular momentum for a single particle and its projection on the z-axis, we use the symbols λ and μ to avoid any confusion. But the restrictions on the quantum numbers are the same. The time variation of the radius to any point on the surface of the body is reflected in the time dependence of the expansion coefficients $\alpha_{\lambda,\mu}$. Because the nucleus does not have the sharply defined surface of a classical body, $R(\theta,\phi,t)$ is considered to apply to a surface that encloses some definite fraction of the total probability of finding all nucleons. While Eq. (8.30) looks a bit formidable, it is really rather simple, especially if the object has the cylindrical symmetry we have been considering here. In this case, the radial coordinate will be independent of the angular coordinate ϕ and the only terms that enter into the summation are those with $\mu = 0$. The spherical harmonics then reduce to the Legendre polynomials and the general expression of Eq. (8.30) can be simplified as

$$R(\theta, t) = R_o \left[1 + \sum_{\lambda=0}^{\infty} \alpha_{\lambda,0}(t) Y_{\lambda,0}(\cos\theta) \right] \tag{8.31}$$

We can now proceed to examine the physical picture represented by each of the terms in the expansion. For simplicity we assume that the original nucleus is spherical with the radial dimension R_o. This is just the result of Eq. (8.31) when all of the expansion coefficients $\alpha_{\lambda,0}(t)$ vanish. Now suppose all coefficients are zero except for the first, $\alpha_{0,0}$, that carries no angular momentum. From Table 7.2, $Y_{0,0} = 1/(\sqrt{4\pi})$ and the radius is then given by $R(\theta, t) = R_o[1 + \alpha_{0,0}(t)/\sqrt{4\pi}]$.

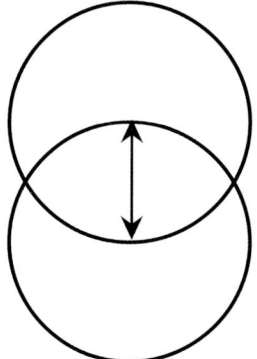

Fig. 8.9 The displacement of a sphere along the z-axis as a result of an oscillation of the form

$$R(\theta, t) = R_o\left[1 + \sqrt{\frac{3}{4\pi}}\alpha_{1,0}(t)\cos\theta\right].$$

The nuclear shape will still be spherical but it will oscillate in size as the expansion coefficient oscillates. This implies that the nuclear volume changes and hence that the nuclear density must change. This would be in disagreement with experimental fact and we conclude that a vibrational quantum of $\lambda = 0$ will not give rise to physically meaningful observables.

Now consider the case where all of the expansion coefficients are zero except for that corresponding to the term $\lambda = 1$. From Table 7.2,

$$Y_{1,0} = \sqrt{\frac{3}{4\pi}}\cos\theta$$

This would produce a radial coordinate of

$$R(\theta, t) = R_o\left[1 + \sqrt{\frac{3}{4\pi}}\alpha_{1,0}(t)\cos\theta\right]$$

If the original nucleus was spherical with radius R_o, a vibrational quantum of $\lambda = 1$ would again produce a spherical shape but one that is displaced from the origin as shown in Fig. 8.9. The center of mass of the nucleus is shifted along the vibrational axis, but the shape is still the same and so is the radius. This describes the oscillation of the nucleus as a whole and the only way in which this can occur is by the action of some external force. Again, this type of vibration does not give rise to observables in which we have interest.[1]

Now consider an oscillation of the $\lambda = 2$ type, a so-called quadrupole vibration. For our cylindrically symmetric nucleus this will lead to a radial coordinate of the form

---

1) We should point out that there is another type of $\lambda = 1$ vibration that does result from internal forces and therefore does produce observables that are well known. So far, we have tacitly assumed that a vibration would act on all of the nucleons equally, and this is what appears to take place at low energies. But one could consider the possibility that a vibration of the neutrons *relative* to the protons could take place and such vibrations have indeed been found. They give rise to the so-called giant dipole resonances found at some tens of MeV in excitation energy.

$$R(\theta, t) = R_o\left[1 + \alpha_{2,0}(t)\left(\frac{3}{2}\cos^2\theta - \frac{1}{2}\right)\right] \qquad (8.32)$$

For relatively small positive values of $\alpha_{2,0}(t)$, the quadrupole term produces a prolate shape as a shown in Fig. 8.10. If $\alpha_{2,0}(t) < 0$, an oblate shape is produced. If the nucleus was originally spherical, we can picture a vibration as transforming the nucleus into either shape. If the nucleus was originally prolate in shape, we can picture the vibration as producing a more prolate or more spherical shape, depending on the energies required to produce such changes. We therefore conclude that quadrupole vibrations are the lowest-order vibrations that could lead to observables in the low-energy spectra of nuclear states and so long as we consider only small vibrations, we expect them to be described, at least approximately, as simple harmonic vibrations. Because $\lambda = 2$, a quantum of vibrational energy will have an angular momentum quantum number of 2 and a relative parity that is even. If we consider an (even,even) nucleus in its 0^+ ground state, the addition of a quadrupole phonon will then produce an excited state with angular momentum and parity of 2^+. If we add a second quadrupole phonon of the same energy, we can now produce states that are the result of the coupling of the angular momenta of the two phonons and we predict degenerate states within the range 0^+–4^+. If one goes through the appropriate analysis of the wave functions that are represented by the different total angular momentum projections, one finds that only the states with angular momenta of 0, 2 and 4 can exist. One can go on to consider the addition of a third phonon, etc., but from a practical viewpoint we can stop here without much loss. In any event, the idealized spectrum of states we can expect to see from small quadrupole vibrations is shown in Fig. 8.11.

The model we have used to arrive at the spectrum of Fig. 8.11 really assumes that vibrations do not affect the internal structure of the nucleus. But, as pointed out above, if the shape changes, the potential experienced by the nucleons must also change and thus the vibration must affect the internal structure of the nucleus. This dependence can manifest itself in the inequality of the energy separation of

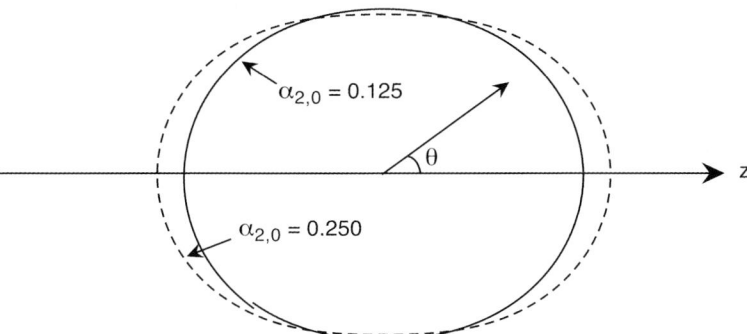

Fig. 8.10 Prolate deformations of the form of Eq. (8.32) with $\alpha_{2,0}(t) = 0.125$ (solid line) and 0.250 (dashed line), respectively.

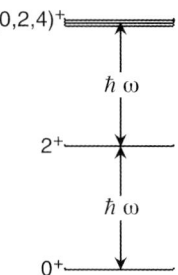

Fig. 8.11 Idealized harmonic spectrum from a quadrupole vibration with one and two phonons.

the one- and two-phonon states as well as in the removal of the degeneracy of the states from the two-phonon excitation. Just as in the case of rotational spectra, we can expect that only certain regions of N,Z will display this simple signature of vibrational motion.

A survey of the lowest-energy excitations of all known (even, even) nuclei points out a remarkable fact. With but few exceptions, the first-excited state has spin and parity of 2^+, and in most cases where the first-excited state has a different spin and parity, there is a 2^+ level at a slightly higher energy. The actual excitation energies of the first 2^+ levels in (even, even) nuclei are shown in Fig. 8.12. The energies generally trend from several tens of MeV for the lightest nuclei down to a few tens of keV for the heaviest nuclei, and very clear patterns are observed in different mass regions. The most striking features are the peaks in the vicinity of A = 90, 140 and 210 where the excitation energies are 1–2 MeV. These are the regions of nuclides near the isotones of N = 50, the doubly magic Z = 50, N = 82, and the doubly magic Z = 82, N = 126, respectively. The next striking features are the very low excitation energies in the regions A = 150–180 and A > 220. It is in these regions, far from the magic numbers, where one finds strongly deformed nuclei with well-defined rotational bands. Throughout the data one sees some other structures that can be associated with isotopes of a fixed mass number.

For most of these nuclides, a 4^+ level is found that lies at higher energy. To search for signatures of possible vibrations, the ratio of the energies of the 4^+ levels to the energies of the corresponding 2^+ levels shown in Fig. 8.12 are shown in Fig. 8.13. These data show, with remarkable clarity, the regions of strongly deformed nuclei that exhibit rotation spectra in the mass ranges A = 150–180 and A > 220. In both cases, the ratios $E(4^+)/E(2^+)$ are quite close to the value of 3.33 expected for the ideal rigid rotor. One is also struck by the fact that in the vicinity of nuclei with N = 50 (A ~ 90), N = 82 and Z = 50 (A ~ 140), and N = 126, Z = 50 (A ~ 210), the ratios are just slightly greater than one, indicating that the energies of the two levels are nearly degenerate. This is in marked disagreement with the expectations of both the rigid rotor or the harmonic oscillator model. In fact, vibrations in these tightly-bound spherical nuclei are unknown. We will not prove it, but it can be shown with a relatively simple model, that the low-lying spectra of such nuclei are well-described by the angular momentum couplings of two identical particles in the same orbit, and the energies of allowed 2^+ and 4^+ states are about the same. Finally,

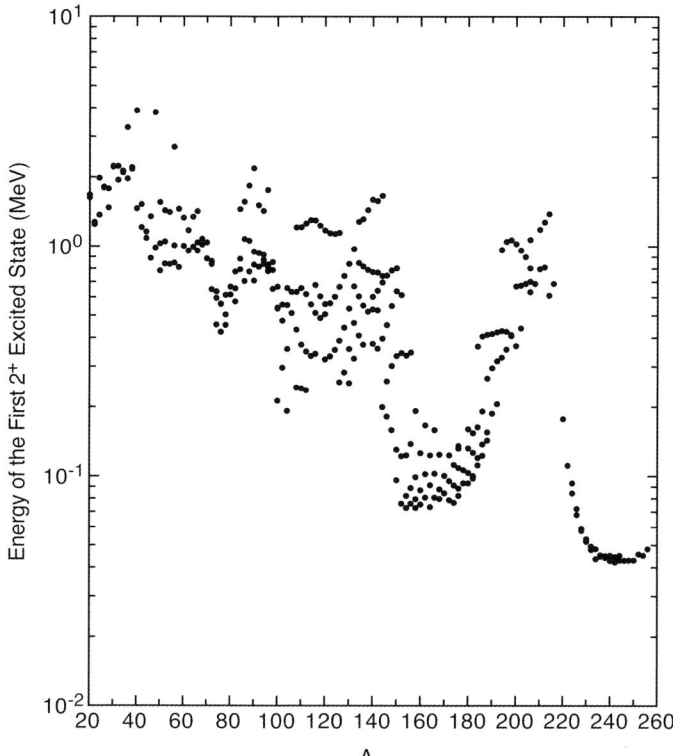

Fig. 8.12 The excitation energies of the first 2^+ states in (even, even) nuclei (log scale) as a function of mass number.

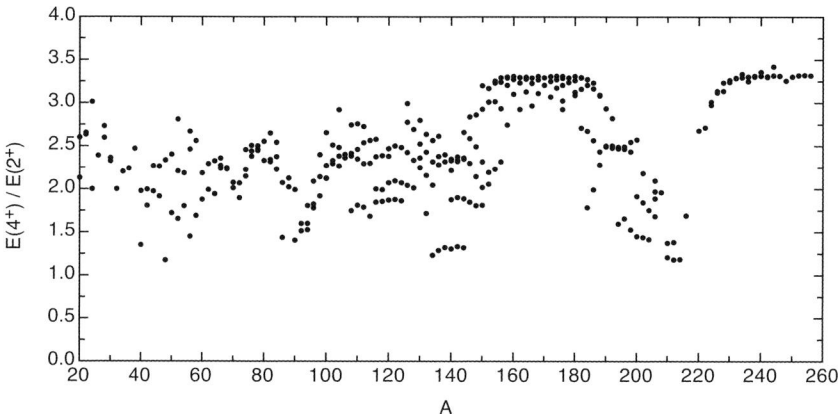

Fig. 8.13 The ratio of the energy E(4+) of the first 4^+ excited state to the energy E(2+) of the first 2^+ excited states in even, even nuclei.

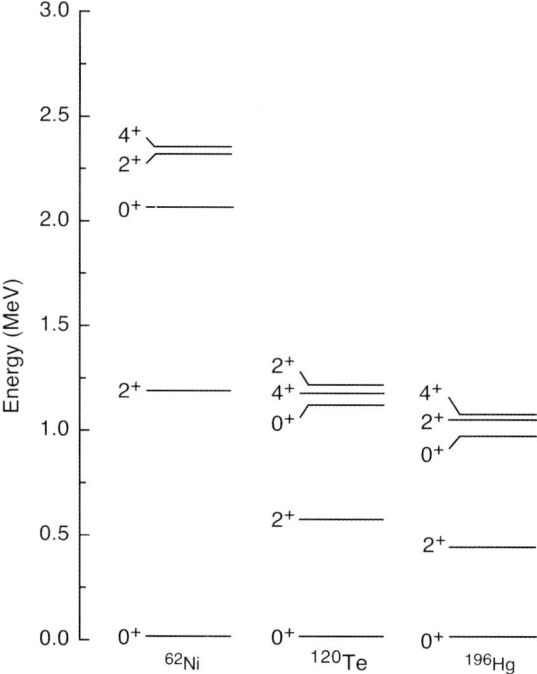

Fig. 8.14 Apparent vibrational structures in the low-energy spectra of $^{62}_{28}\text{Ni}$, $^{120}_{54}\text{Te}$ and $^{196}_{80}\text{Hg}$ from one and two phonon excitations.

while there is no region where the ratio 2 expected for the simple harmonic oscillator is seen cleanly, there are a number of nuclides with mass numbers in the range A < 40, A = 60–70, A = 110–130, and A = 200–220 that have ratios within about 10–20% of the simple oscillator model. An examination of the spectra of some of these does indeed indicate rough agreement with the expectations of vibrational spectra. The levels shown for ^{62}Ni, ^{120}Te and ^{196}Hg in Fig. 8.14 are the lowest energy levels known in these nuclides. They represent all of the levels expected from one- and two-phonon vibrations and they have the approximate energy separations predicted by an ideal harmonic vibrator. The energy range of the 0^+, 2^+, and 4^+ levels assigned to the states of a two-phonon vibration is small compared to the energy of the first 2^+ level, they are not degenerate but almost so, and their mean energy is roughly twice that of the first 2^+ level. There are many other examples of apparent vibrational structure that are quite similar to those shown in Fig. 8.14 but there are also many others in the vicinity of these nuclides for which the simple vibrational structure is not quite so clear or for which it is simply not at all apparent. In many nuclei, there is evidence for a three-phonon vibration, especially for nuclides that are spherical or nearly so. However, all of the states that are expected to arise from such a vibration are not clearly seen. Generally, at the excitation energies at which these levels should occur, level spectra are very complex. In some respects it is indeed surprising that one is even able to see the simple structure expected from a two-phonon vibration because the energies at which they lie, in the vicinity of 1–2 MeV, is comparable to the pairing energy of neutrons and protons as well as the energy difference between single-particle states in these regions. Nevertheless, there is ample evidence to suggest that the collective mode of vibration is a significant factor governing the spectra of many nuclei. More sophisticated models that account for both vibrational and other excitations have been quite successful in describing the level structures, even in cases where the spectrum from the simple harmonic oscillation is not clearly evident.

Having developed the evidence for existence of nuclear deformations and the effects of collective motion, we must acknowledge that the potential experienced by individual nucleons in such cases cannot be spherically symmetric and we now turn to an examination of the nature of the potential and its effect on the level structure of deformed nuclei.

8.6
Nuclear Structure in a Deformed Potential

The motion of a particle in a potential that is not spherically symmetric is, in general, quite complicated. However, if we start with the basic assumptions that have been used to describe particle motion in a spherically symmetric potential and consider potentials that have cylindrical symmetry, the fundamental differences can be seen qualitatively without much difficulty. All experimental evidence suggests that the nuclear density remains constant at the same value in deformed nuclei and thus the assumption of roughly constant average potential throughout the nucleus still applies. If the deformed nucleus has the shape of a prolate or oblate object, the principal difference will be the effective extent of the potential in the axial direction and directions normal to that axis as shown in Fig. 8.15. Considering only the central potential, the Hamiltonian for this system can be written schematically in Cartesian coordinates as

$$H = H_x + H_y + H_z = 2H_\perp + H_\parallel \tag{8.33}$$

where the subscripts $\perp$ and $\parallel$ refer to the directions normal to and parallel to the axis of symmetry, the z-axis. If we take as our potential model the equivalent of a finite potential well with constant potential within the well, we would then have

$$V_\parallel = \begin{cases} -V_o, & z \leq a \\ 0, & z > a \end{cases}$$

$$V_\perp = \begin{cases} -V_o, & (x^2 + y^2)^{1/2} \leq b \\ 0, & (x^2 + y^2)^{1/2} > b \end{cases} \tag{8.34}$$

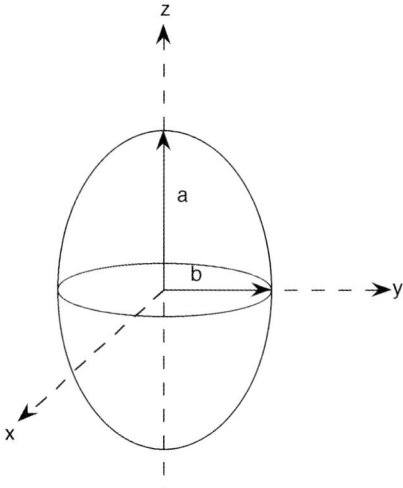

Fig. 8.15 Schematic drawing of a prolate ellipsoid of revolution with semi-major and semi-minor axes of a and b, respectively.

While the potential is everywhere constant, the linear dimensions in which a particle is constrained to move is different in the two directions. Because of the form of the potential, the Schrödinger equation will be separable along the three coordinate axes and while the energy eigenfunctions might have the same form, the energy eigenvalues in the directions parallel to and normal to the symmetry axis can be expected to be different. This can be inferred from our discussion of the energy states of a particle in a cubical box (Chapter IV), where, when the length of a side of the box was L, the energy eigenvalues were of the form $E_n = n^2\pi^2\hbar^2/2mL^2$. Thus, for a prolate shape, the states of particles moving, on the average, parallel to the symmetry axis, should have energies smaller than those corresponding to motion normal to the symmetry axis. As usual, the less constrained a particle is in space, the lower will be its energy states.

To see qualitatively how the energies of states might vary with "deformation", we can compare the energies of states in a cubical box with those in a rectangular parallelepiped of length a and square cross section of b^2. To maintain the analogy with the constant-density assumption, the volume of the box must be held constant. If a side of the cubical box has length L, the constant-density requirement means that $L^3 = ab^2$. To maintain direct reference to the cubical case, we can write the energy of a state in the "deformed" box as

$$\begin{aligned} E_n &= \frac{n_x^2\pi^2\hbar^2}{2mb^2} + \frac{n_y^2\pi^2\hbar^2}{2mb^2} + \frac{n_z^2\pi^2\hbar^2}{2ma^2} \\ &= \frac{\pi^2\hbar^2}{2mL^2}L^2\left[\frac{n_x^2}{b^2} + \frac{n_y^2}{b^2} + \frac{n_z^2}{a^2}\right] \\ &= \frac{\pi^2\hbar^2}{2mL^2}\left(\frac{a}{b}\right)^{-1/3}\left[(n_x^2 + n_y^2)\left(\frac{a}{b}\right) + n_z^2\left(\frac{a}{b}\right)^{-1}\right] \end{aligned} \qquad (8.35)$$

where the n_i each vary over the range 1, 2, 3,...... This relation shows schematically how the energies of states will vary as a function of "deformation". If a = b, the problem reduces to the simple particle in a cubical box. For a "prolate" box, a > b and for an "oblate" box, a < b.

It is a simple matter to calculate the energies of states as a function of a/b in units of $\pi^2\hbar^2/2mL^2$ and these are shown in Fig. 8.16. In the figure, the energies of states with a/b = 1 are just those of the cubical box, which, in this crude approximation, is analogous to the case of a spherically symmetric potential. The dependence of the energy of a level on a/b varies considerably for different levels. Some are seen to increase or decrease monotonically with increasing a/b. Some have energies that vary roughly symmetrically about a/b = 1, and some show minima in the region of a/b < 1. If one chooses a fixed value of a/b ≠ 1, it is seen that the *sequence* of the levels can differ appreciably from that found in the case of a = b. The qualitative

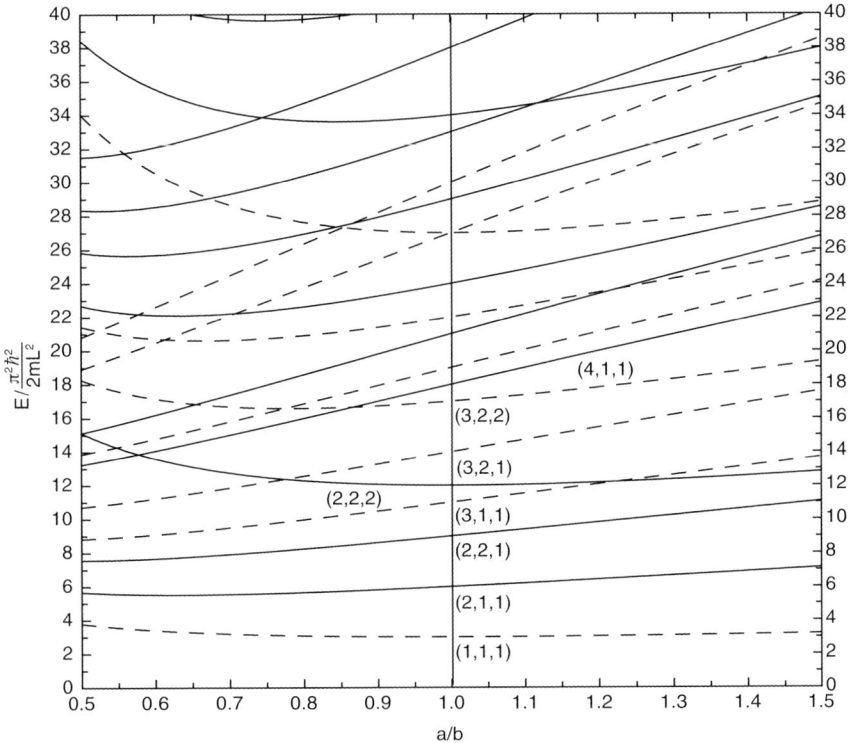

Fig. 8.16 The energy levels of a particle in a box of constant volume with square cross section as a function of the ratio of the length of the box in the z-direction to the length in the x- and y-directions, a/b. The energies are given in units of $\pi^2\hbar^2/2mL^2$. For convenience in reading, levels corresponding to n_x = odd are shown as dashed lines and those for which n_x = even are shown as solid lines. The energy levels of the cubical box correspond to a/b = 1.0. The quantum numbers (n_x, n_y, n_z) for the lowest-lying levels are shown just below the lines representing these states.

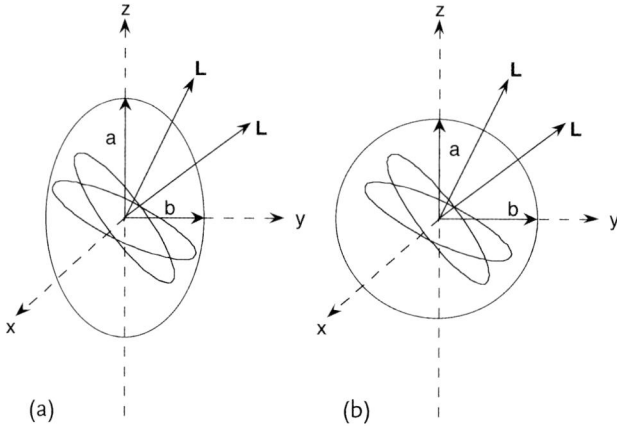

Fig. 8.17 Schematic diagrams of classical particles with fixed total angular momentum but different orientations with respect to the z-axis in an prolate (a) and spherically symmetric (b) potential. The ovals represent projections of the circular paths of the particles that lead to the angular momentum vector L normal to the planes of the paths.

conclusion is that deformation should substantially affect the energy differences between adjacent states and, depending upon the extent of deformation, the level ordering itself.

The simple particle in a box model does not allow a direct understanding of the effects of deformation on states of different angular momentum. Nevertheless, it is not difficult to see qualitatively what the effects will be by reference to the simple vector model of angular momentum presented in Chapter VI. In Fig. 8.17 is a schematic diagram of two different orientations of an orbital angular momentum vector **L** for a classical particle moving in a circular orbit so that **L** is directed normal to the plane of the orbit.

Consider first Fig. 8.17 (b) that represents motion in a spherically symmetric potential. Regardless of the orientation of the particle's path, that is, regardless of the projection of the angular momentum vector on the z-axis, the potential experienced by the particle in the two orbits illustrated will be identical. The projection of **L** can be changed arbitrarily without affecting the energy of the particle. Now consider Fig. 8.17 (a) that represents particle motion in a deformed potential with axial symmetry. The situation here is quite different. As the angle between **L** and the z-axis changes, the potential experienced by the particle will vary and thus will the total energy of the particle.

We can now carry over these considerations to the case of a spinless particle in a quantized system with orbital angular momentum **l** and z-component $l_z = m_l \hbar$ by recognizing the fact that, although the particle's orbit is not restricted to a plane, the motion will be concentrated in planes normal to the angular momentum vector.

8.6 Nuclear Structure in a Deformed Potential

Regardless of the orientation of **l**, and hence m_l, the potential experienced by the particle in a spherically symmetric potential will be the same. This is the fundamental reason why the 2l + 1 states of different m_l for a given l are degenerate in a spherically symmetric potential. However, in the deformed potential, states with different $|m_l|$ will experience different potentials and thus their energies will differ. *The deformed potential removes the energy degeneracy of the substates for a given l.* This has a profound effect on the level diagrams for motion of a fermion in a deformed potential, because it means that the level density must be much larger than the level density in spherical nuclei. In place of a level diagram defined by states of the coupling j = l ± 1/2, the states in a deformed potential will be defined by $|m_j|$. And for each of these, the energies will vary as a function of deformation as illustrated schematically in Fig. 8.16. We can even use the particle in a deformed box problem and the orientation picture in Fig. 8.17 to understand qualitatively how the energies of these substates will differ. Consider the "prolate" box. Because the energy of states with the angular momentum vector more closely aligned to the z-axis would be higher than those with the vector more closely normal to this axis (because of the different dimensions of the sides of the box), it is easy to see that levels with larger projections of **j** on the z-axis must have *higher* energies than those with *smaller* projections. Therefore for the states of a given j that are degenerate in energy in a spherically symmetric potential, states of higher $|m_j|$ must have *higher* energies than those of lower $|m_j|$. The conclusion is that the degeneracy in energy will be removed to give the sequence shown schematically in Fig. 8.18. We expect the energies of states to decrease in the order $E(|m_j|=j) > E(|m_j|=j-1) > E(|m_j|=j-2) > ...$ The opposite would be expected if the deformation were of the oblate form.

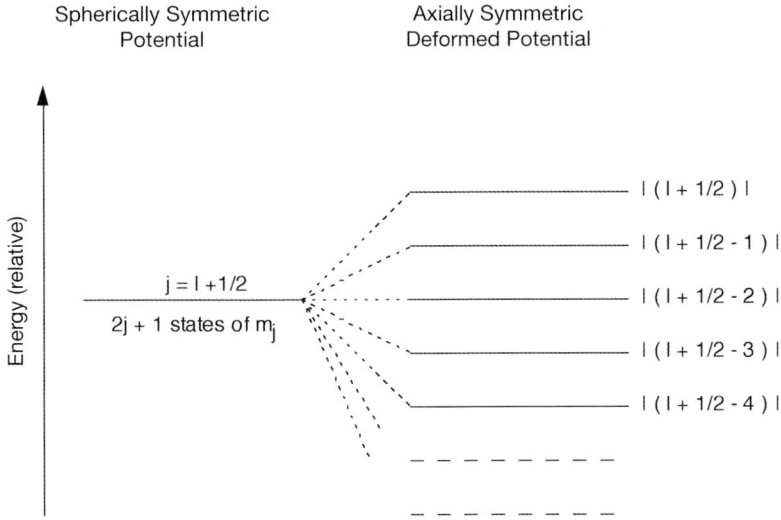

Fig. 8.18 Schematic diagram of the removal of degeneracy of states of the same l and different $|m_j|$ in a deformed potential. It is assumed that the deformation leads to a prolate shape.

With this background, we are now ready to present the highly successful Nilsson model for deformed nuclei, that is the analog of the single-particle shell model for spherical nuclei.

8.7
The Nilsson Model

The model developed by Nilsson [2] to describe single-particle states of deformed nuclei, while complex in detail, is based on many of the simple ideas that we have discussed previously. To begin with, one must choose some model for the average central potential experienced by a nucleon and arrange for it to account explicitly for deformation. Nilsson assumed that deformed nuclei were axially symmetric objects and therefore that the single particle central potential was axially symmetric. The analytical form of the potential was taken as that of an *anisotropic* harmonic oscillator to reflect deformation. The oscillator frequency, proportional to the classical spring constant, along the z-axis differs from that in the x,y plane. This is analogous to our "deformed" particle in a box model. As we indicated above, one must be careful to insure that deformation maintains constant nuclear density. In the Nilsson model, this means that the frequencies parallel to and normal to the symmetry axis must be related because, taken together, they define the effective size of the nucleus. (To see this, consider a diatomic molecule. The larger the frequency of vibration, the larger will be the mean separation between the two nuclei.) In Cartesian coordinates, constancy in volume is then obtained by requiring that $\omega_x \omega_y \omega_z$ = constant. For an axially symmetric potential $\omega_x = \omega_y$.

The oscillator frequencies not only define the effective volume of the nucleus but they also define the extent of the deformation; the greater the difference between the two frequencies, the greater the deformation. Nilsson chose to use the oscillator frequency that would correspond to a spherically symmetric potential as a reference, and then related this directly to the oscillator frequencies of the deformed potential while requiring that the nuclear volume remain constant. To describe the shape of the deformation, Nilsson assumed the simple quadrupole form discussed in some detail in Section 8.5. With this, the Hamiltonian for a particle moving in the simple central potential can be written

$$H_{central} = -\frac{\hbar^2}{2\mu}\nabla^2 + \frac{1}{2}\mu\omega_o^2 r^2 - \mu\omega_o^2 r^2 \frac{4}{3}\sqrt{\frac{\pi}{5}}\delta Y_{2,0}(\theta,\phi) \quad (8.36)$$

This expression has a rather simple structure. The first and second terms on the right-hand side represent nothing more that the kinetic and potential energies of a single particle of reduced mass μ in an isotropic harmonic oscillator potential with angular frequency ω_o, both of which we have dealt with in Chapter VII. The solutions to this problem provided the level sequence shown in Fig. 7.13.

The third term in Eq. (8.36) represents all of the effects of deformation. The simple quadrupole shape is represented by the spherical harmonic $Y_{2,0}(\theta,\phi)$ (see

Eq. (8.32)). The quantity δ is directly related to the extent of deformation and hence the relation between ω_z and ω_x ($=\omega_y$). It is given by $\delta = \Delta R/R_o$, where R_o is the radius of the undeformed nucleus (Eq. (8.31)) and ΔR is just the difference between the semi-major and semi-minor axes of the deformed object. Eq. (8.36) is then seen to be the Hamiltonian for the motion of a spinless particle in a deformed potential. It reduces nicely to the Hamiltonian for the motion of a spinless particle in an isotropic harmonic oscillator potential as δ goes to zero.

If we wish to provide a model that contains enough of the physics of the nuclear interaction to compare favorably with experiment, there are two more issues that must be dealt with. First, the model must account for the strong spin–orbit coupling found experimentally. This is easily handled by addition of a term proportional to $\mathbf{l} \cdot \mathbf{s}$ as was used in the spherical shell model. Second, we know that the isotropic harmonic oscillator has the special property that all of the states corresponding to a single shell but with different l are degenerate in energy. This is in conflict with experiment and is not found with other, reasonable potential models. We also know that the degeneracy is removed in the latter such that levels with higher l lie lower in energy than those with lower orbital angular momentum quantum numbers. Nilsson solved this problem by adding an additional term, proportional to $\mathbf{l}^2$, to the potential. This not only removes the degeneracy in l, but it also has the additional advantage that the $\mathbf{l}^2$ operator commutes with both the Hamiltonian for the isotropic harmonic oscillator and the spin–orbit term. With these additional terms the Nilsson Hamiltonian becomes

$$H_{central} = -\frac{\hbar^2}{2\mu}\nabla^2 + \frac{1}{2}\mu\omega_o^2 r^2 - \mu\omega_o^2 r^2 \frac{4}{3}\sqrt{\frac{\pi}{5}}\delta Y_{2,0}(\theta,\phi) + C\mathbf{l}\cdot\mathbf{s} + D\mathbf{l}^2 \quad (8.37)$$

Unfortunately, the presence of the third term prevents an analytical solution and one must rely on machine computations to determine the eigenfunctions and energy eigenvalues of the various allowed levels. Regardless, the fundamental issue is how to ensure that the parameters ω_o, C and D are chosen properly, to reflect reality as much as possible. Nilsson's approach was quite simple. From our knowledge of the single-particle levels in spherical nuclei and how they vary with mass number, one can derive the effective magnitude of the angular frequency ω_o and its mass dependence. The most common correlation found from such a study gives the oscillator quantum $\hbar\omega_o$ as

$$\hbar\omega_o = 41 A^{-1/3} \text{ MeV} \quad (8.38)$$

Given this, a consistent set of ω_o, C and D can be obtained by requiring that the Nilsson model reproduce the spherical shell model in the limit that $\delta \to 0$. Then, to the extent that the deformation included in the model reflects real deformed nuclei, the Nilsson model should express the single-particle level sequence for deformed nuclei of arbitrary deformation in exact analogy to the single-particle level sequence in spherical nuclei.

Level sequences obtained with the Nilsson model are summarized in what are commonly referred to as *Nilsson diagrams*, several of which will now be discussed. In Fig. 8.19 is the Nilsson diagram for nuclei with N or Z less than 50. The ordinate of the diagram is given in units of the oscillator quantum $\hbar\omega_o$. The abscissa, ε_2, is a deformation parameter that is related to δ by

$$\varepsilon_2 = \delta + \frac{1}{6}\delta^2 + \frac{5}{18}\delta^3 + \ldots \tag{8.39}$$

Typically, $\delta \leq 0.3$ for nuclei not too far from the valley of β stability, and thus ε_2 is essentially identical to δ for our purposes. When $\varepsilon_2 = 0$, we have a spherically symmetric potential and the level spacings reflect the best estimate of the level spacing of spherical nuclei not far from magic numbers. For $\varepsilon_2 < 0$, nuclei are deformed as oblate objects and for $\varepsilon_2 > 0$ nuclei are deformed as prolate objects. Each of the levels is labeled with four numbers of the form $\Omega[N\ n_z\ \Lambda]$. Ω is the quantum number for the z-component of the total momentum of a state, and the numbers in brackets represent the quantum numbers of the isotropic harmonic oscillator shell from which the level arises in the absence of deformation. N represents the quantum number of the isotropic harmonic oscillator shell from which the state arises, n_z represents the quantum number along the z-axis of the anisotropic harmonic oscillator and Λ represents the z-component of orbital angular momentum in the absence of deformation. These numbers are commonly referred to as "asymptotic" in the sense that they would have precise interpretations in the limit that the third term in Eq. (8.37) can be neglected. However, the presence of this term causes the wave functions of the deformed states to be rather complex. The total angular momentum quantum number j that we have in the spherical limit is no longer a "good" quantum number for states in a deformed potential and neither is the orbital angular momentum quantum number l.

For zero deformation, all of the levels corresponding to an orbit of fixed j are degenerate. As deformation carries the nucleus into a prolate shape, this degeneracy is removed and, in all cases, the energy ordering for small deformations is exactly that found from our simple physical argument above; lowest energy for the smallest Ω, next lowest for the next smallest, etc. For small oblate deformations the ordering is exactly the opposite. Both of these variations are easily seen by tracing the two deformed levels that arise from the $1p_{1/2}$ state and the three levels that arise from the $1p_{3/2}$ state starting at $\varepsilon_2 = 0$. At larger deformations, the level ordering can vary significantly from this simple picture, especially for oblate shapes. For prolate nuclei with ε on the order of 0.1 or greater, it is clearly seen that the high level density due to the removal of degeneracy leads to a complete loss of the shell structure found in spherical nuclei and further, there may be "crossing" of levels that lead to a different sequence of spins and parities from those found with small deformations.

With larger neutron or proton numbers, the deformation can lead to very complicated level diagrams. For example, in Fig. 8.20 is the Nilsson diagram for neutron numbers $50 \leq N \leq 82$. Prolate deformations greater than about 0.1 are seen

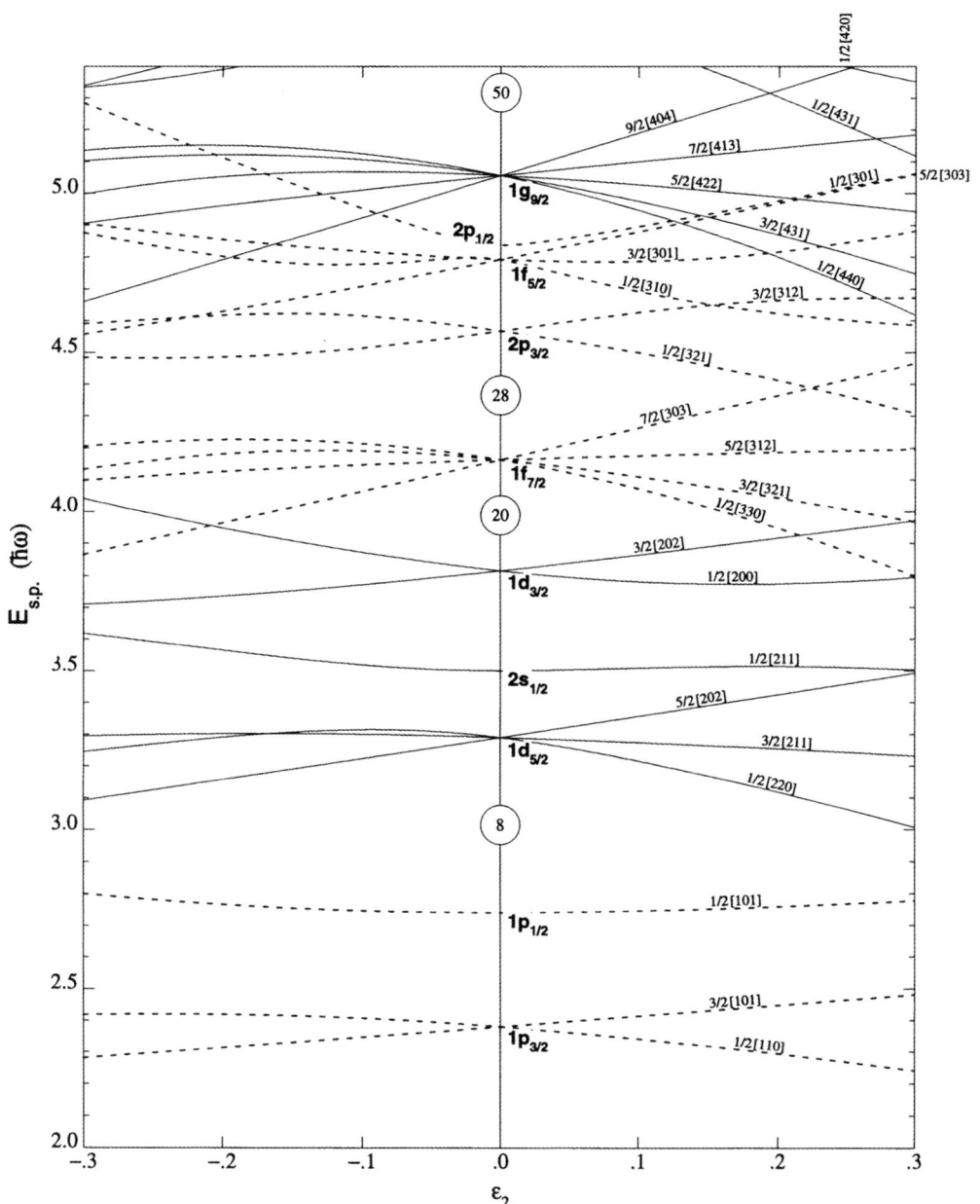

Fig. 8.19 The Nilsson diagram for neutrons and protons with N(Z) ≤ 50. The ordinate is $E_{s.p.}$ ($\hbar\omega$).

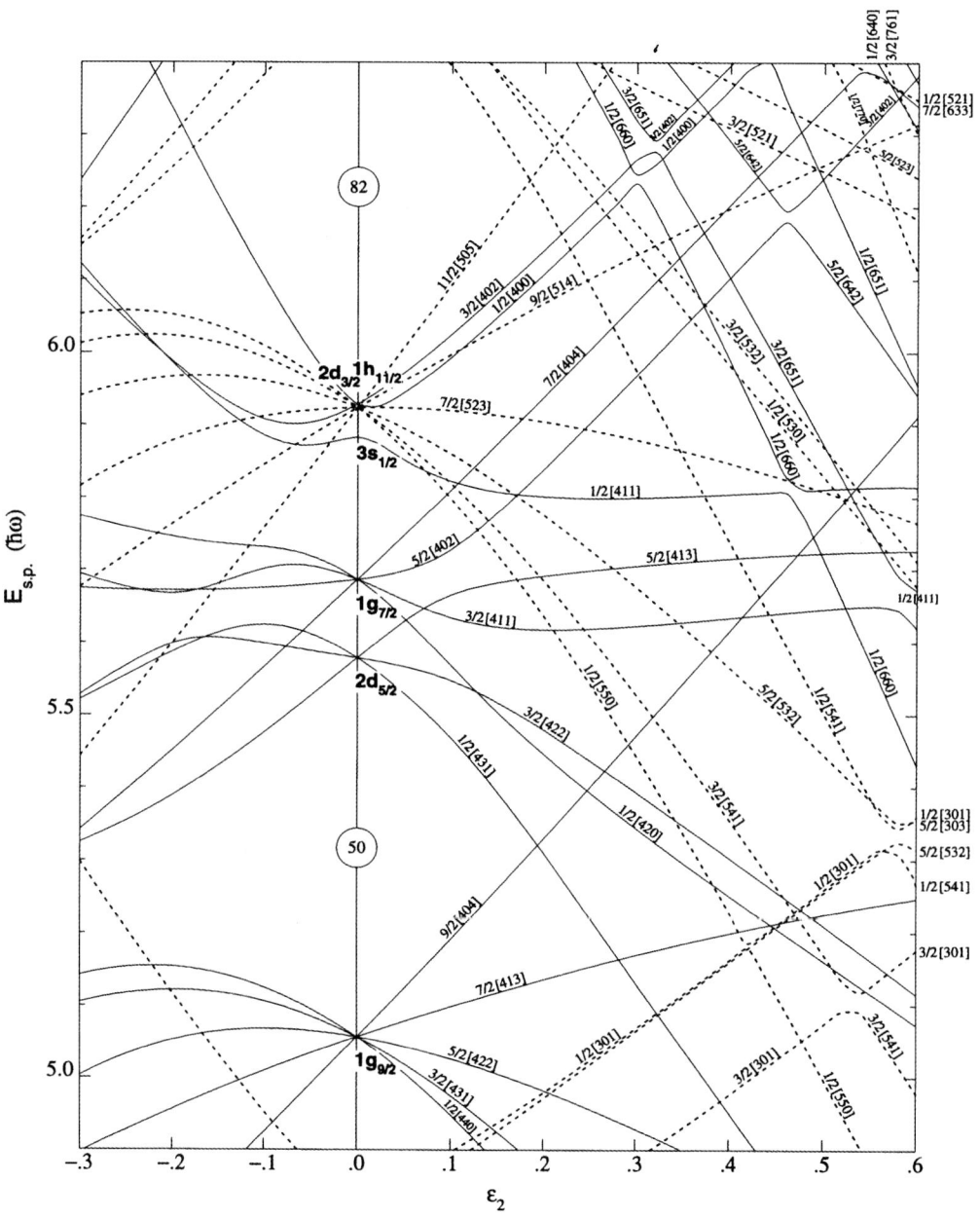

Fig. 8.20 The Nilsson diagram for neutron numbers $50 \leq N \leq 82$.

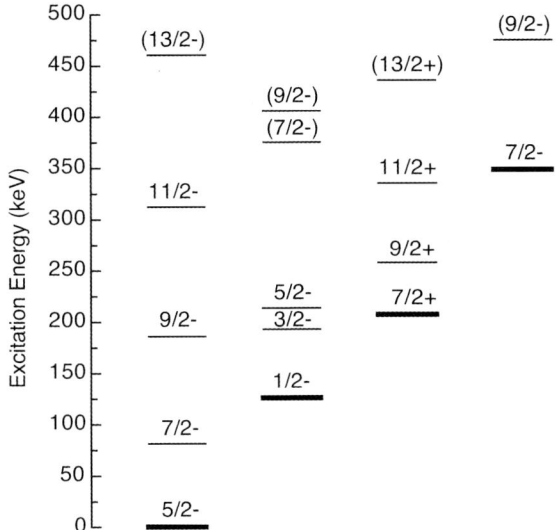

Fig. 8.21 Known levels in ^{175}Hf below 500 keV. For ease in reading, the levels have been segregated horizontally according to rotational bands built upon intrinsic states.

to lead to very complex level diagrams, and the variation of the energy of a state with increasing deformation can be quite complicated.

The real question to pose is how well the Nilsson diagram accounts for the level structure of deformed nuclei. The fact is that it does remarkably well. It is able not only to provide rather accurate level sequences when compared to experiment, but it also points out that, although the spherical shell structure is lost, significant gaps do appear between adjacent levels in certain cases and thus leads to the creation of a *shell structure in deformed nuclei*.

As an example of intrinsic excitations and the rotational bands built upon them in a deformed odd-A nucleus, Fig. 8.21 represents the level diagram for all known levels in ^{175}Hf below 500 keV. For simplicity, the intrinsic states, shown as bold lines, and the rotational levels built on them have been displaced horizontally from one another. We will leave it up to the reader, by reference to the available data sets, to demonstrate that the simple rotational model discussed above accounts quite well for the rotational states built upon the $5/2^-$ and $7/2^+$ intrinsic states. Our purpose here is to determine the extent to which the Nilsson model can account for the intrinsic states. Noting that the deformation parameter δ is roughly 0.25 for nuclides in this region, corresponding to a value of $\varepsilon_2 \sim 0.26$, we can search the Nilsson diagram for odd N = 103 as shown in Fig. 8.22. By carefully tracing the region near $\varepsilon_2 \sim 0.26$ and $E_{s.p.} \sim 6.1$, the reader will indeed find that the level predictions are in reasonable accord with experiment.

The increased level densities and the presence of rotational bands that can be built upon the ground and each excited state lead to very dense and quite compli-

240 | 8 Nuclear Shapes, Deformed Nuclei and Collective Effects

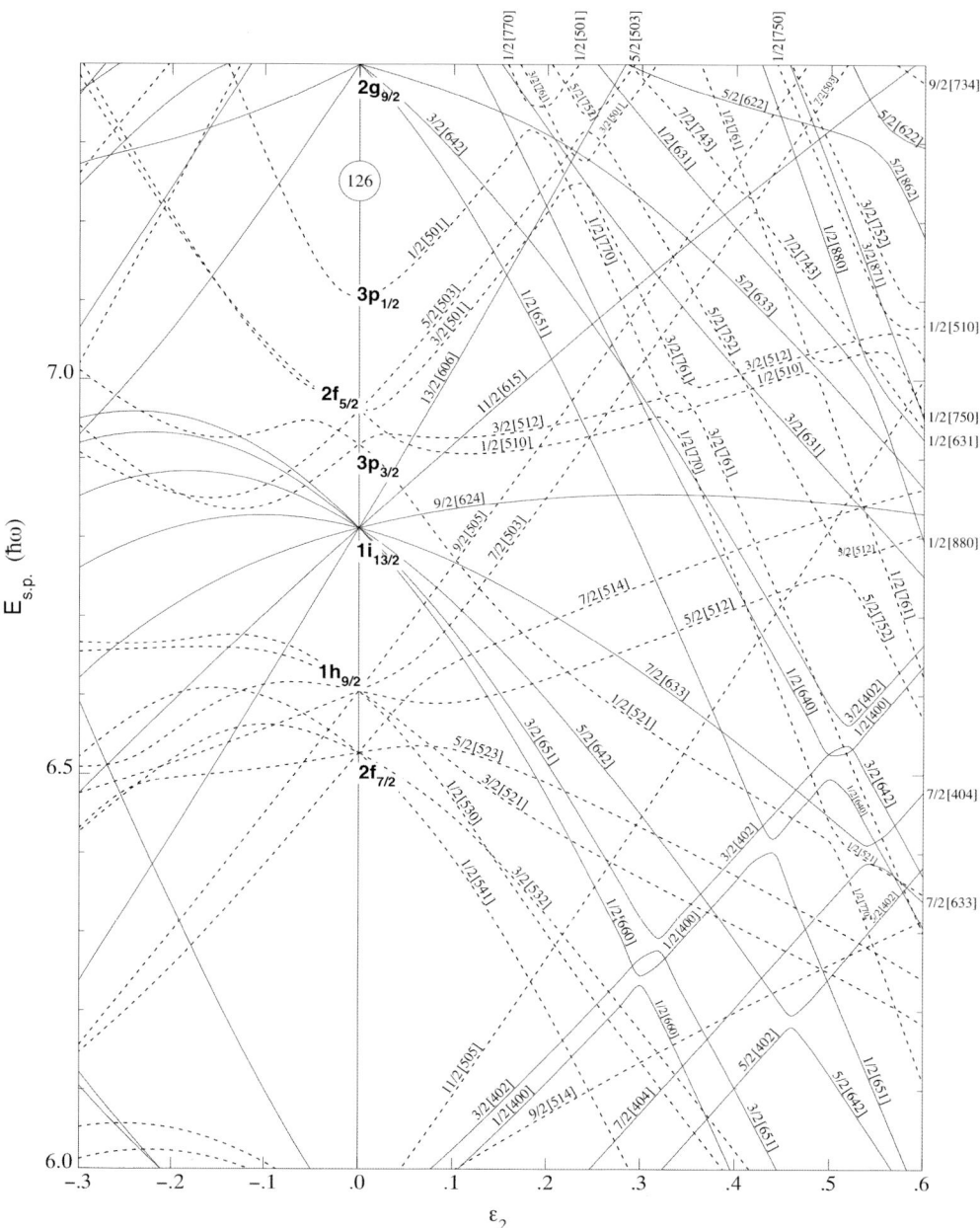

Fig. 8.22 Nilsson diagram for $82 \leq N \leq 126$.

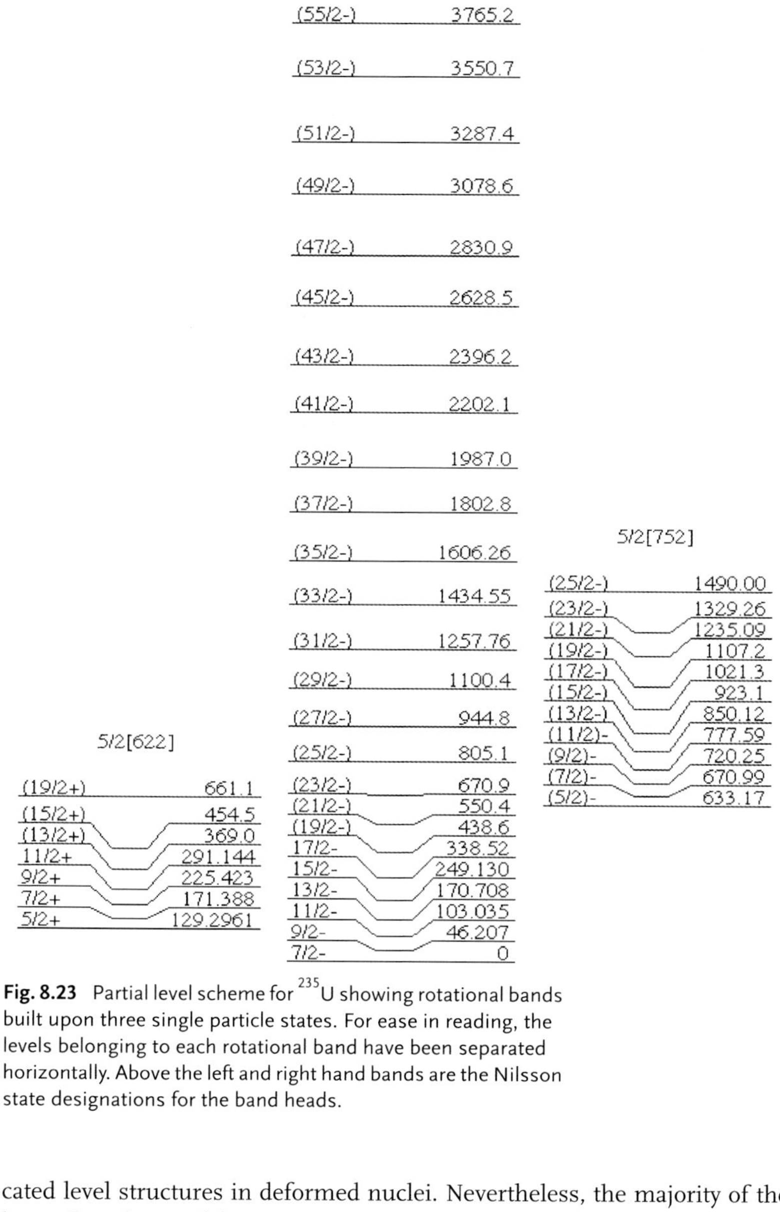

Band 5/2[622]

J^π	Energy
(19/2+)	661.1
(15/2+)	454.5
(13/2+)	369.0
11/2+	291.144
9/2+	225.423
7/2+	171.388
5/2+	129.2961

Middle band

J^π	Energy
(55/2−)	3765.2
(53/2−)	3550.7
(51/2−)	3287.4
(49/2−)	3078.6
(47/2−)	2830.9
(45/2−)	2628.5
(43/2−)	2396.2
(41/2−)	2202.1
(39/2−)	1987.0
(37/2−)	1802.8
(35/2−)	1606.26
(33/2−)	1434.55
(31/2−)	1257.76
(29/2−)	1100.4
(27/2−)	944.8
(25/2−)	805.1
(23/2−)	670.9
(21/2−)	550.4
(19/2−)	438.6
17/2−	338.52
15/2−	249.130
13/2−	170.708
11/2−	103.035
9/2−	46.207
7/2−	0

Band 5/2[752]

J^π	Energy
(25/2−)	1490.00
(23/2−)	1329.26
(21/2−)	1235.09
(19/2−)	1107.2
(17/2−)	1021.3
(15/2−)	923.1
(13/2−)	850.12
(11/2)−	777.59
(9/2)−	720.25
(7/2)−	670.99
(5/2)−	633.17

Fig. 8.23 Partial level scheme for ^{235}U showing rotational bands built upon three single particle states. For ease in reading, the levels belonging to each rotational band have been separated horizontally. Above the left and right hand bands are the Nilsson state designations for the band heads.

cated level structures in deformed nuclei. Nevertheless, the majority of these can be well understood by comparison to the appropriate Nilsson diagram and by considering the rotational bands that are present. As another example, a partial level diagram for ^{235}U showing three rotational bands built upon single-particle states is shown in Fig. 8.23. The level scheme is quite complex and to provide a simple means of examining the level systematics, levels belonging to a single rotational band have been offset from each other. In the figure, only three of the large number of rotational bands known are shown. The first member of each band, the band head, is a single particle neutron state in the deformed potential. The ground state, with spin and parity of $7/2^-$ represents the state of the 143rd neutron in ^{235}U. The designation of the Nilsson state corresponding to the heads of the remaining bands are shown above the bands. The rotational bands are well defined and can be accounted for very nicely by the simple rigid rotor model discussed in Section 8.4. Remarkably, the Nilsson diagram provides an excellent summary of the single-particle neutron states found in this spectrum as well as in the spectra of the majority of the actinides.

The models of nuclear structure presented in this text represent a few of the simplest models that have been developed over the last decades. Other much more sophisticated models have been developed that permit an understanding of many additional aspects of nuclear structure, as well as of the transition rates for different modes of radioactive decay and the details of nuclear reactions. These are left for more advanced study. Now we will proceed with a study of radioactive decay and use our nuclear models as a means of correlating data and pointing out some of the effects of the underlying nuclear structure on radioactive decay properties.

References

1. See for example, R.R. Roy and B.P. Nigam, "Nuclear Physics", John Wiley and Sons, New York (1967).
2. The original reference is S.G. Nilsson, "Binding states of individual nucleons in strongly deformed nuclei", Kongelige Danske Videnskabernes Selskab, Matematisk-Fysiske Meddelelser, 29 (1955) 68.

General References

A very nice discussion of deformed nuclei and their level structure is given in H. Frauenfelder and E.M. Henley, "Subatomic Physics", Prentice Hall (1991).

Problems

1. Go to the Table of Isotopes and view the level diagram for ^{148}Sm.
(a) For the first six levels assigned to the ground-state rotational band, compare the experimental energies with those obtained from the simple rigid rotor model using the experimental energy difference between the 0^+ and 2^+ levels to determine the magnitude of $\hbar^2/2\Im$. Note that the 8^+ member of the band already has an excitation energy of about 2715 keV.
(b) Note that this band has been characterized experimentally to a spin of 26 at an excitation energy of 8659.5 keV. How does this energy compare with the neutron binding energy of ^{148}Sm?

2. Go to the Tables of Isotopes and view the level diagram for ^{238}Pu. Compare the experimental energies of all known levels in the ground-state rotational band to those obtained from the simple rigid rotor model using the experimental energy difference between the 0$^+$ and 2$^+$ levels to determine the magnitude of $\hbar^2/2\Im$. What can you conclude?

3. Go to the Table of Isotopes and examine the low-lying level diagram of ^{245}Cm. The ground state has been assigned as the 7/2[624] Nilsson state and the ground state of ^{249}Cf has been assigned as the 9/2[734] Nilsson state.
(a) Trace through the appropriate Nilsson diagram to demonstrate that these assignments are consistent with a deformation parameter of about $\delta = 0.26$. You will need the level diagram for $N \geq 126$ and it can be found at http://ie.lbl.gov/toip-df/nilsson.pdf

In order to locate the correct region, locate the region of $\delta = 0.26$ at the bottom of the figure, count up the number of states that arise from below and are not filled at that point, correct for the number of levels that arise from above and are filled, and then count up vertically until you have found the predicted levels for the neutron states in question.
(b) Predict the energy of the 17/2$^+$ member of the $K = 5/2^+$ band whose ground state lies at 0.25280 MeV above the ground state of ^{245}Cm.

4. The first-excited state of ^{238}U lies at an energy of 44.9 keV above the ground state. This nuclide also has a rotational band built on a 0$^+$ excited state at 927.2 keV above the ground state and the 2$^+$ member of this band lies at an energy of 966.1 keV. Calculate the ratio of the moment of inertia ^{238}U in the ground-state band to that in the band built on the excited 0$^+$ state.

5. The excitation energies of the first 2$^+$ levels in $^{154}_{64}$Gd and $^{238}_{92}$U are 123.07 and 44.92 keV, respectively. Assuming that the deformations of the two nuclei are the same, compare the ratio of level energies to the ratio estimated from a comparison of the moments of inertia estimated classically.

6. Systems with very high angular momentum can be expected to depart quite significantly from our simple idealized models. To illustrate this quite clearly, use the experimental level energy for the first-excited member of the ground state rotational band in ^{235}U and calculate the energies predicted by the rigid rotor model for members with spins up to 53/2. Compare your results to those shown in Fig. 8.23.

9
α Decay and Barrier Penetration

9.1
Introduction

α decay has played a major role in the development of low-energy nuclear physics and our understanding of nuclear structure. It has proven to be a crucial tool for the discovery of the heaviest elements now known, some isotopes of which have been defined conclusively by the observation of as little as a single decay. α decay is also of great practical importance. The long-lived α emitters produced during the normal operation of nuclear power reactors represent perhaps the single most important issue with respect to the safe disposal of used reactor fuel. There is great interest in the possibility that labeling of new, highly-specific radiopharmaceuticals with α-emitting nuclides might provide an important tool for selective killing of malignant cells in the human body.

The great prevalence of α decay among the heaviest elements results in probably the most important source of radioactivity in the natural environment. Nuclides such as $^{238}_{92}U$ are not only radioactive, but their daughters and other descendents are also radioactive. The decay of a long-lived parent such as $^{238}_{92}U$ gives rise to a long and complex decay chain that is referred to as a radioactive decay series. Four such series are known, but only three have parents with sufficiently long half-lives that they now contribute to naturally occurring radioactivity. All nuclides in each series decay by either α decay, β⁻ decay or both. Because α decay changes the mass number by four units and β decay does not change the mass number, all members of a particular series have mass numbers given by the general relation 4n + i where both n and i are integers. The series that begins with $^{238}_{92}U$ has n = 59 and i = 2, and is thus the "4n + 2 series". It can be considered as typical and is shown schematically in Fig. 9.1.

The half-life of $^{238}_{92}U$, 4.47×10^9 yr, is comparable to the lifetime of the earth and is a remnant of nucleogenesis. Apart from $^{238}_{92}U$, all radioactive nuclides in the series have lifetimes short compared to the age of the elements and thus they exist only because they are being regenerated by the continuous decay of $^{238}_{92}U$. One finds that all of the longer-lived members decay by α emission but not all of the α emitters have long half-lives. The stable end product is $^{206}_{82}Pb$ and all three of the natural decay series end in an isotope of this element. The 4n + 2 series contains

Nuclear Physics for Applications. Stanley G. Prussin
Copyright © 2007 WILEY-VCH Verlag GmbH & Co., Weinheim
ISBN: 978-3-527-40700-2

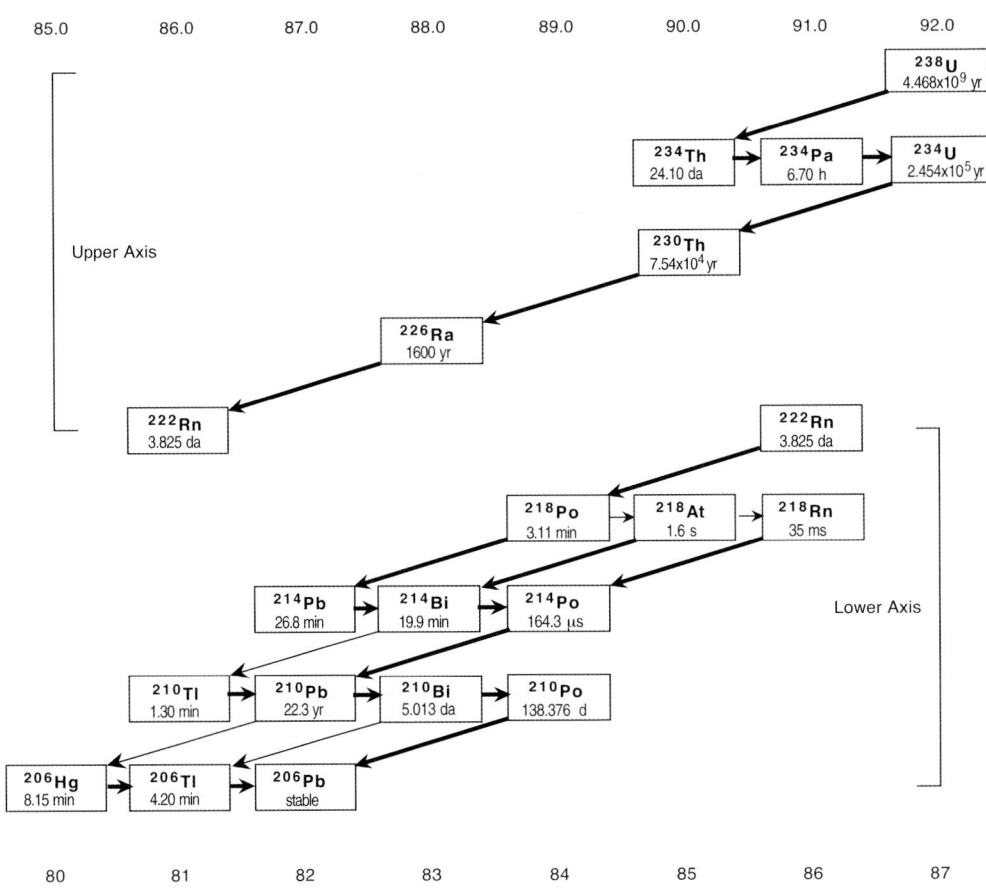

Fig. 9.1 The 4n + 2 natural decay series. The arrows sloping downward represent decay by α-emission and the horizontal arrows represent decay by β emission. When a nuclide is known to decay by both α and β emission, the heavy arrows represent the dominant decay mode and the light arrows represent the weak decay mode.

10 nuclides that decay mainly by α emission and 9 that decay mainly by β⁻ decay. In addition to producing the long-lived $^{230}_{90}\text{Th}$, the 4n + 2 series gives rise to $^{226}_{88}\text{Ra}$ with a half-life of 1600 years. Having a chemistry similar to that of calcium, radium is found in trace quantities in most calcium minerals and, of course, in rock containing trace quantities of uranium. As a result, radium is found in essentially all buildings and structures comprised of concrete in its many forms. α decay of $^{226}_{88}\text{Ra}$ directly produces the rare gas isotope $^{222}_{86}\text{Rn}$ with a half-life of about 3.8 days. Because it is a rare gas and does not form stable compounds under normal conditions, $^{222}_{86}\text{Rn}$ can diffuse out of materials into the atmosphere. It then acts as an atmospheric source for all of its radioactive descendents. Most of these are

isotopes of elements that do not form gaseous compounds, and they are found most often absorbed on small particles that act as aerosols, and deposit throughout the environment.

The decay rates of the members of 4n + 2 decay series can be determined by application of the rate equations that were introduced in Chapter III in the discussion of radioactive growth and decay. Considering the earth and its atmosphere as a closed system, secular equilibrium has been established between $^{238}_{92}U$ and each member of the decay series. Hence, most members will have a decay rate equal to that of $^{238}_{92}U$. We can get a rough estimate of the decay rate of each, therefore, by simply considering the total mass of $^{238}_{92}U$ in the earth's crust, estimated to be about 4 grams per metric ton (4 parts per million by weight). This turns out to be about 5×10^4 Bq or 1.3 µCi per metric ton. Therefore, if we neglect the transport of activity from the earth into the atmosphere, we can estimate that the total radioactivity from the 19 isotopes in the 4n + 2 series, per metric ton of the earth's crust, is about 20 µCi.

As discussed in Chapter II, the energetics of α decay can be summarized by the equation

$$M(Z, A) = M(Z - 2, A - 4) + M(2, 4) + Q_\alpha/c^2 \quad (9.1)$$

Because of the relatively large mass of the α particle, a substantial fraction of the total decay energy is imparted to the daughter nucleus as *recoil* energy. The relation between the kinetic energy of the α particle and Q_α is easily obtained through simple energy and momentum conservation. With the parent nucleus initially at rest in the laboratory, conservation of momentum requires that

$$p_p = p_d + p_\alpha = 0 \quad (9.2)$$

where the subscripts p, d, and α refer to the parent, daughter and α particle, respectively. Thus

$$p_d = -p_\alpha \quad (9.3)$$

and the kinetic energy of the daughter atom, T_d, is then given by

$$T_d = \frac{p_d^2}{2M_d} = \frac{p_\alpha^2}{2M_d} \cdot \frac{2M_\alpha}{2M_\alpha} = \frac{M_\alpha}{M_d} T_\alpha \quad (9.4)$$

This, added to the kinetic energy of the α particle, must equal the Q_α. Thus,

$$Q_\alpha = T_\alpha + T_d = T_\alpha \left(1 + \frac{M_\alpha}{M_d}\right) \quad (9.5)$$

and the kinetic energy of the α particle is then given by

$$T_\alpha = Q_\alpha\left(\frac{M_d}{M_d + M_\alpha}\right) \tag{9.6}$$

The expression in Eq. (9.6) is exact. But for many applications, one can replace the atomic mass in mass units by the mass numbers themselves, and the equation is often approximated by

$$T_\alpha \cong Q_\alpha\left(\frac{A_d}{A_d + A_\alpha}\right) = Q_\alpha\left(\frac{A-4}{A}\right) \tag{9.7}$$

where A is the mass number of the α-decay parent.

Eq. (9.6) assumes, as usual, that the α particle and daughter nuclide are both formed in their ground states. Because the first-excited state of the α particle is near 20 MeV, one can be sure that only its ground state is produced. However, it is quite common that α decay takes place to both the ground and one or more excited states of the daughter. If decay takes place to an excited state, the kinetic energy of the α particle is given by Eq. (9.6) after correcting Q_α for the excitation energy of the daughter nucleus.

Among the nuclides found in nature, α decay is observed only among the heaviest elements and a few isotopes of the rare earth elements. Typical α emitters among the heavy elements are nuclides such as $^{210}_{84}$Po, $^{235,\,238}_{92}$U, and $^{241}_{95}$Am that have Q_α in the range 4.5–8.0 MeV, and typical α emitters among the rare earth elements are $^{146,\,147}_{62}$Sm and $^{148}_{64}$Gd, which have Q_α in the range 2.5–3.5 MeV.

If we consider a typical α emitter among the heavy nuclides, say of mass number 240, we find that T_α/Q_α is about 0.98, and the kinetic energy of the daughter will be on the order of 0.1 MeV. This has practical significance with respect to the transport of an α-decay daughter in the environment, because the daughter nucleus will penetrate a significant distance in matter before it comes to rest. Thus, some fraction of the daughter atoms produced near the surface of a source will have sufficient energy to recoil into an adjacent medium, and if the daughter nuclide is itself radioactive, this is a second mechanism by which radioactivity can be transported into the external environment.

9.2
Q_α and α Decay Half-Lives

Although the Q values for α decay can be calculated directly from the measured atomic masses, it will prove useful to first examine the predictions of the semi-empirical mass formula. As shown in Chapter V, a comparison of empirical masses with the predictions of the mass formula gave a clear indication of the shell structure of nuclei and we can expect that nuclear structure effects in α decay, if any, should be reflected through a similar comparison. For the present purpose, we will restrict attention to the α decay of (even, even) isotopes of elements with atomic numbers in the range 76–100. The mass formula predictions for Q_α are shown in Fig. 9.2 for all nuclides for which α decay has been observed experimentally and masses have been measured [1]. The Q values are predicted to vary smoothly and systematically. For each element, the mass formula predicts that Q_α decreases with increasing N at a roughly constant rate, while for isotones, the Q_α are predicted to increase with increasing Z also at a roughly constant rate. And very importantly, the Q_α are predicted to vary over the relatively small range of about 3.5–9.5 MeV, a factor of less than 3.

How well these predictions compare with experiment can be seen by examination of Fig. 9.3 which shows the Q_α calculated with the empirical atomic masses. The α decay of all isotopes included in the figure have been observed. For the elements Os - Pb, the variations of Q_α with both N and Z are quite similar to those predicted by the semi-empirical mass formula. The magnitudes are somewhat smaller than predicted but this is due in large part to our use of global parameters in the mass formula calculations rather than local parameters chosen to best fit masses among the heavy elements. However, for isotopes of the elements Po - U, the experimental values vary in a markedly different fashion as the magic number N = 126 is approached. Shell structure has a profound effect on α-decay energies in this region and it is primarily due to the neutron shell closure. A smaller effect due to the proton shell at Z = 82 is seen in the somewhat larger separation between the lines connecting the Pb and Po isotopes as compared to other elements.

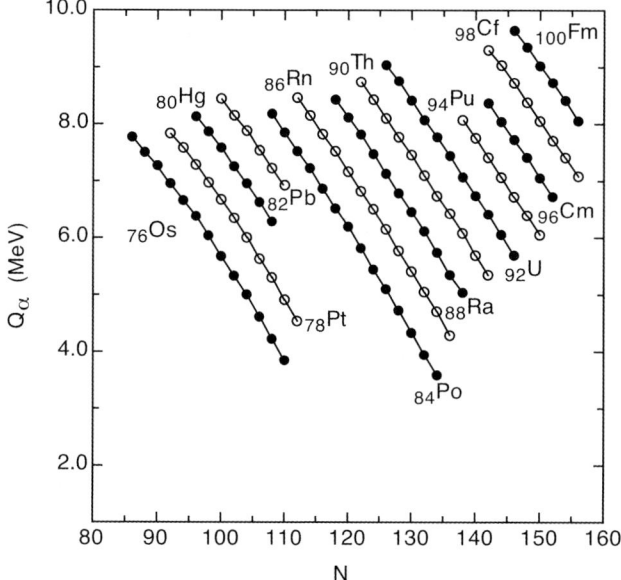

Fig. 9.2 Q_α as a function of neutron number, N, for selected (even, even) isotopes of the elements Os (Z = 76) to Fm (Z = 100). The Q_α were calculated with the semi-empirical mass formula using the global parameters of Eq. (5.10).

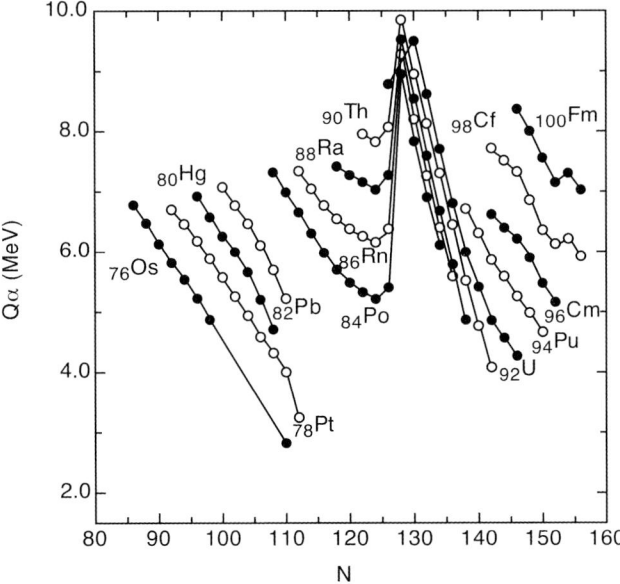

Fig. 9.3 Experimental Q_α as a function of neutron number, N, for selected (even,even) isotopes of the elements Os (Z = 76) to Fm (Z = 100).

The α-decay half-lives of the nuclides contained in Fig. 9.3 are shown in Fig. 9.4 as a function of Q_α. They are startling. For a change in decay energy by a factor of about 2.5, the half-lives change by a factor of about 10^{26}! The half-lives decrease quite systematically with increasing decay energy and increase quite systematically with increasing Z. Remarkably, the strong dependence of decay energies on shell structure near N = 126 is not reflected in the half-lives when viewed as a function of Q_α. We can take this to imply that, for these nuclides, the details of nuclear structure must exert only a small effect on decay probability in comparison to that exerted by decay energy.

The data shown in Fig. 9.4 are the experimentally measured half-lives divided by the fraction of decays that occur by α emission, i.e., they are the partial half-lives for α emission. Nuclides very near the valley of β stability are known to be pure α emitters. For most proton-rich nuclides quite far from the valley of stability, where other decay modes are likely, there are many lines of evidence that suggest that α decay is certain to be the dominant decay mode. For nuclides in between, both α emission and β decay are possible and are observed in many cases.

It is also important to note that the partial half-life for α decay is itself not necessarily a simple quantity to interpret. With Q_α large compared to the energy separation of levels of spherical nuclei (~ 0.1–0.5 MeV) and very large compared to the energy separation of rotational levels in the deformed heavy elements (~0.01–0.1 MeV), decay can take place to a large number of levels with varying Q_α- E_x where E_x is the energy of the excited state to which decay occurs. Thus, the partial half-life

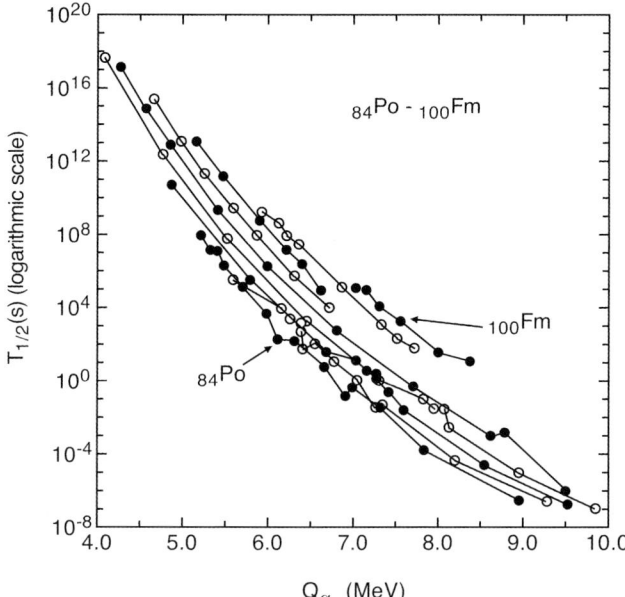

Fig. 9.4 Logarithm of the partial half-lives for α-decay of even, even isotopes of Po - Fm as a function of Q_α.

for α decay is a measure of the total probability for α decay to all levels that are energetically accessible, and not just a measure of the decay from the ground state of the parent to the ground state of the daughter. But from the very strong energy dependence of the half-life on decay energy as shown in Fig. 9.4, it is easy to see that we expect transitions with high decay energy to dominate and therefore most of the α decay will be expected to take place to, or very near, the ground states of the daughters.

As a specific and very typical example, the experimental decay scheme for α decay of ^{238}U is shown in Fig. 9.5. The daughter, ^{234}Th, is a typical actinide isotope.

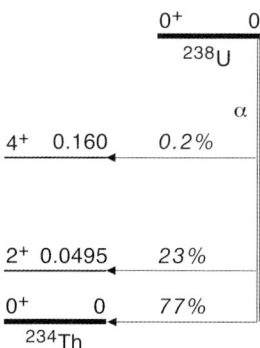

Fig. 9.5 Decay scheme of ^{238}U showing the α-branching to the ground and first two excited states of ^{234}Th. Level energies are given in MeV.

It is deformed in its ground state and displays a rotational band that is reasonably well described by the simple rigid rotor model discussed in Chapter VIII. In the figure, the spins and parities of levels are shown to the left and it is seen that those of the first and second excited states have the expected properties of the first two members of a rotational band built on the 0^+ ground state. The Q_α is 4.2703 MeV. Assuming that ^{238}U is initially at rest in the laboratory, the energies of α particles emitted in the decay to the ground, first and second excited states are easily calculated to be (Eq.(9.7)) 4.198, 4.150, and 4.041 MeV, respectively. They vary by less than 4%. On the other hand, and in keeping with the extreme sensitivity of decay probability on decay energy, the percentage of total decays to the ground state is 77%, to the first-excited state is 23% and to the second excited state is only 0.2%. We can then conclude that a decrease in decay energy from 4.198 MeV to 4.150 MeV reduces the decay probability by a factor of (77/23) or about 3.3, while a decrease from 4.198 MeV to 4.041 MeV decreases the decay probability by a factor of roughly 385! To be candid, it is not just the decay energy that is changing but the angular momentum of the α particle as well. However, as we will discuss later in this chapter, the effect of angular momentum on α decay is rather small and the dominant factor is very definitely the energy available for the decay.

This example demonstrates that the half-lives for α decay of the deformed (even, even) isotopes of the heavy elements are dominated by decay to the ground and first-excited states of a deformed daughter and, for almost all practical purposes, the decay to any other excited states can be safely neglected. Further, because the energies of the first-excited states are typically 0.03–0.05 MeV, we can expect that roughly 75% of the decay will occur to the ground state.

While we have restricted our attention to the decay of (even, even) nuclei, very similar properties are seen in the decay of odd-A and (odd, odd) nuclei as well. Here, however, nuclear structure effects are somewhat more strongly evident. For typical odd-A decays, one finds that the ground state to ground state transitions are generally not the dominant ones. The majority of the decay intensity proceeds to one or more excited states, typically within 0.1–0.2 MeV of the ground state and therefore the main conclusions from the discussion given above apply as well. In all cases the overriding factor that governs α-decay probabilities is the decay energy. To understand how this arises and to develop a model for α-decay probabilities we must consider two issues. First, if an α particle is to escape from the nucleus it somehow must already exist or it must be formed at the instant of decay. Even though we have treated nuclei in their ground states as though all of the nucleons occupied the lowest single-particle energy levels available, this is really only a very simple and idealized model. Second, if the nucleus really contained α particles to begin with, what is it that causes their emission probability to be so exquisitely sensitive to the decay energy? In an attempt to make plausible the highly successful model that was developed early in the last century by George Gamow, we will examine these two issues qualitatively in the following section.

9.3
Binding of Valence Nucleons and the Potential for Interaction Between an α Particle and a Heavy Nucleus

The α particle is by far the most tightly bound of all of the lighter nuclei. With an average binding energy of 7.074 MeV, a nucleon in the α particle is not bound by much less energy than an average nucleon in the ^{238}U nucleus, which is 7.570 MeV. Clearly, it is completely possible that the more weakly-bound nucleons in ^{238}U have binding energies that are so small that they approach or are even less than the binding energies of nucleons in an α particle. A rigorous examination of the question of the existence of α particles in heavy nuclei will obviously require a complex nuclear structure calculation that is well beyond the scope of this text. However, we can take a rather simple approach that ought to allow us to make a reasoned guess concerning the possibility that two neutrons and two protons inside a heavy nucleus might form an α particle with appreciable probability.

Using ^{238}U as an example, we begin with the assumption that, in zeroth order, its 92 protons and 146 neutrons will be bound as pairs in the lowest single particle energy states available. Removal of a single neutron or proton will require an energy that represents the sum of the binding energy of a nucleon in a state plus its pairing energy. On the other hand, if we remove a pair of nucleons, the energy required will account not only for the binding energy of the two nucleons but the pairing energy as well. We can easily calculate the separation energy of a pair of like nucleons by use of the atomic masses given in Appendix 1. The separation energy for a pair of protons from ^{238}U is given by the equation

$$S_{2p}/c^2 = [M(^{236}Th) + 2M_H - M(^{238}U)] \tag{9.8}$$

and that for the separation of two neutrons is given by

$$S_{2n}/c^2 = [M(^{236}U) + 2m_n - M(^{238}U)] \tag{9.9}$$

With the exception of ^{236}Th, the masses of the nuclides and particles in these equations are known. Although the atomic mass of ^{236}Th can be estimated with the semi-empirical mass formula, we will use the estimated value of $\Delta(^{236}Th) = 46305$ keV found in Ref. [1]. With this, the two-proton separation energy of ^{238}U is found to be 13.58 MeV and the two-neutron separation energy is found to be 11.28 MeV. Thus, the sum total of the binding and pairing energies of the last four nucleons in ^{238}U amounts to about 24.86 MeV for an average of 6.21 MeV per nucleon, well below the average of 7.074 MeV per nucleon in the free α particle.

Now assume that it was possible to create and measure (or calculate) the mass of a ^{238}U nucleus for which we were certain that all neutrons and protons were paired and resided in the lowest single-particle energy levels available. The binding energies discussed above suggest that if we now allowed the most weakly bound pairs of neutrons and protons to form an α particle with the same total binding energy of the free α particle, the mass of the ^{238}U nucleus would be smaller. It is then conceivable that, rather than reside as pairs in their respective energy levels, the most weakly bound pair of neutrons and protons in ^{238}U exist as an α particle in the nucleus with nonzero probability.

This suggestion is not peculiar to ^{238}U. A similar calculation for the nuclide ^{210}Po leads to an estimate of the average binding energy of 5.76 MeV for the four most weakly bound nucleons that could be the constituents of an α particle, again well below the average in the free α particle. Similar calculations can be done for other heavy nuclides with the same or similar results. We can then generally conclude that, all other things being equal, it would be energetically favorable for an α particle to form within these heavy nuclei rather than have all nucleons, as we have supposed to the present, exist as separate pairs of identical nucleons in their respective sequences of single-particle levels.

If we suppose for the moment that an α particle existed in a heavy nucleus as a bound entity, it is natural to ask what the potential of interaction would be between it and the remaining nucleus. In the classical limit of a uniformly charged sphere, the particle would certainly experience a Coulomb interaction similar to that experienced by a pair of protons. Taking the host nucleus, the α-decay daughter, as a uniformly charged sphere, we can easily calculate the Coulomb potential between it and the α particle as a function of radius. But what of the strong or nuclear interaction? This is not so straightforward to specify. In principle, the interaction between an α particle and other nucleons can be studied by examination of the scattering of α particles on heavy nuclei in much the same way as the average nucleon–nucleon interaction can be studied by examining the interaction of high-energy protons on heavy nuclei. Unfortunately, it is found that α particles are strongly absorbed by nuclei and it is very difficult to know what the interaction potential is except for the region very near the nuclear surface.

In the process of decay, when the α particle is sufficiently far outside the nucleus that it is beyond the range of the nuclear force, it must experience only the Coulomb potential between it and the daughter nucleus. At smaller distances of separation, it must experience an attractive nuclear interaction with the daughter nucleons. Taken together, we can represent the potential energy diagram of an α particle and a residual daughter nucleus roughly as is shown in Fig. 9.6. To provide a realistic scale for energies and dimensions, the calculations represented in the figure assume a potential that applies to a "daughter nucleus" with Z = 90 and A = 234. The Coulomb potential, V_C, was calculated for a point particle interacting with a uniformly charged sphere of radius $r = r_o A^{1/3}$. The central part of the strong interaction, $V_{central}$, was taken as a Saxon–Woods potential with depth of $V_o = -48$ MeV. For simplicity, it was also assumed that the angular momentum of the system was zero. The total potential is shown as the full line in the figure. Remarkably, at radial dimensions less than the sum of the radius of the daughter and the alpha particle, about 9.35 fm, the Coulomb potential is so strong that it almost completely compensates for the strong interaction as represented by the single-particle potential. One must be careful not to take this result too seriously because the single-particle potential is most assuredly not the exact potential that would be experienced by an α particle in a heavy nucleus. Nevertheless, it is also rather certain that

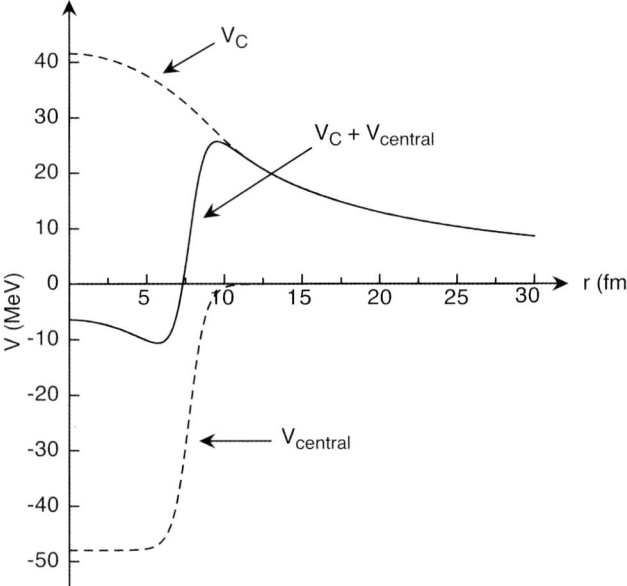

Fig. 9.6 Schematic of the potential experienced by an α particle interacting with a heavy nucleus. For this example it was assumed that the daughter nucleus was a sphere corresponding to a mass number A = 234 and atomic number 90. The central potential was approximated with a Saxon–Woods potential with a depth of 48 MeV. It was assumed that the total angular momentum of the system is zero.

the potential experienced by the α particle would be no stronger per nucleon than the single-particle potential. The conclusion is that, if an α particle traversed the interior of a heavy nucleus, it might experience a rather small net interaction potential.

With the same potential approximation, consider the case where α decay takes place with the parent nucleus at rest in the laboratory. The center of mass is also at rest. After emission, when the radial distance between the α particle and the daughter is very large, the α particle is traveling in field-free space and its total energy is its kinetic energy. Now regardless of where the particle is, its total energy, $E_\alpha = T_\alpha + V$, must remain constant. In Fig. 9.7 is a schematic of the same total potential as shown in Fig. 9.6 and a representation of the total energy of a typical α particle of 6 MeV. If we continue to view the problem classically we immediately come to the conclusion that, if an α particle were present within the nucleus, it would exist in a potential well and experience a barrier where $V > E_\alpha$. It could never be emitted; α emission corresponding to realistic conditions in the heavy elements is completely forbidden classically. Conversely, if a 6.0 MeV α particle was incident on a nucleus of a heavy element, it could never penetrate into the nucleus and indeed, it could not even get very close to it. Even at a radial separation of about

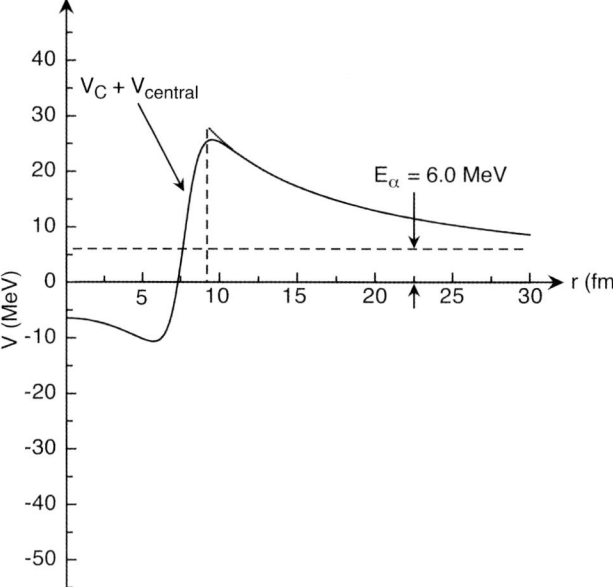

Fig. 9.7 Schematic diagram of the total potential between an α particle and daughter nucleus when the total energy of the α particle in the center of mass system is E = 6.0 MeV. The vertical dashed line marks the Coulomb potential when the α particle and the daughter are separated by the sum of their radii.

30 fm, a distance of about three times the sum of the radii of the α particle and the daughter, the Coulomb potential is still significantly greater than the total energy of the α particle. Most probably, the incident particle would simply undergo elastic scattering in the repulsive potential. It is straightforward to determine where this will happen. If the incident α particle is far from the nucleus, its total energy, E_α, is just its kinetic energy. As it approaches the nucleus, the kinetic energy is reduced and the potential energy increases until all of the energy is potential. At this point we must have

$$E_\alpha = k_C \frac{2(Z-2)e^2}{b} \tag{9.10}$$

Here, Z is the atomic number of the parent nuclide and b is the radial dimension at which the total energy is potential energy. In the present example we find b = 43 fm. The Coulomb potential acts as a *barrier* against α decay and low-energy α-induced reactions. The fact that, contrary to the classical expectation, α decay does indeed occur is purely a quantum mechanical effect that is fundamentally related to the fact that the wave function of a particle extends everywhere throughout space unless the particle is confined in a potential well with infinite walls. This was alluded to during our discussion of the finite spherical potential well. In that case, we were able to neglect the difference between finite and infinite wells because the probability of finding a nucleon outside the potential well is negligibly small for nucleons residing in the lowest energy levels available. As we will see shortly, it is also true that the probability of finding an α particle outside the Coulomb barrier of a heavy nucleus is also very small. But it is just this very small probability that leads to α decay and therefore we must consider this in some detail. Before we do so in the next section, it is worthwhile to reconsider some of the discussion presented above and re-frame the problem of α decay qualitatively in terms of quantum mechanics.

The wave function of the ground state of a heavy nucleus represents the function that describes the lowest energy state of the system. We have, up till now, assumed that all but the last odd nucleon of a given type will be paired with an identical particle and that the pairs will reside in the lowest energy states available. While reasonable, the fact that the model *confines* all nucleons to the lowest possible energy states should, with a little reflection, raise some question in the readers mind because, as we know, confining particles leads to higher energies compared to the case where the particles are unconfined. For example, in a simple semiconductor the energy levels available to the electrons are divided into well-defined valence and conduction bands that are separated by a band gap with energy large compared to the energy separation of levels in either bands that are immediately adjacent to the gap. Electrons occupy the lowest energy states available, i.e., states in the valence band, *only* in the limit of the absolute zero of temperature. At higher temperatures, some of the electrons will achieve sufficient energy that they will occupy levels above the band gap and thus give rise to electrical conductivity. The fact is that the lowest energy state of the system at temperatures above the absolute zero leads to some electrons residing in states that are not the lowest in energy.

The same general ideas can be applied qualitatively to the nucleus. Many lines of evidence, both theoretical and experimental, demonstrate that pairs of identical nucleons are not completely constrained to reside in the lowest energy states available. If we describe the wave function of the ground state of an arbitrary (even, even) nucleus by an expansion in the eigenfunctions of the independent particle or Nilsson models, one finds that there are always a number of terms required. The first and dominant term is usually the simple configuration expected when all identical pairs reside in the lowest energy states, and many of the remaining terms represent configurations where pairs of identical nucleons occupy excited states in the vicinity of the lowest energy states available. The same is true for odd-A nuclei, but in these cases even more complicated particle configurations are often required to explain experimental data. The fact is, our simple single-particle model does not reflect all of the details of the strong interaction. However, as discussed in Section 6.24, the wave function of the ground or an excited state can always be written as

$$\Psi = \sum_n a_n \psi_n \qquad (9.11)$$

where the ψ_n represent specific particle configurations of the independent particle or Nilsson models. Recalling that the squares of the expansion coefficients, $|a_n|^2$, represent the probability of finding the system in the states ψ_n, we find that the ground states of most heavy (even, even) nuclei have very significant contributions from configurations where pairs of neutrons and/or protons occupy a number of higher-lying levels. Given this, it is not difficult to see that the complete wave function could contain a term or terms that represent the correlated motion of a pair of neutrons with a pair of protons that we can picture as a "nascent" α particle. In this sense it is essentially certain that α particles "exist" in heavy nuclei.

9.4
The Wave Functions for Particles in Finite Potential Wells and Barrier Penetration

The nature of the wave functions for particles confined to finite potential wells and the probability for barrier penetration can most simply be presented by studying some simple one-dimensional problems. To begin, we consider the one-dimensional step potential shown in Fig. 9.8. We assume that a particle of mass m moves in the presence of a potential of the form

$$V(x) = \begin{cases} 0, & (x < 0) \\ V_o, & (x \geq 0) \end{cases} \tag{9.12}$$

and has a total energy $E < V_o$. The region where $V(x) = 0$ we will call region I and the region where $V(x) = V_o$ we will call region II. In both regions we must solve the Schrödinger equation

$$\frac{d^2\psi}{dx^2} + \frac{2m(E-V)}{\hbar^2}\psi = 0 \tag{9.13}$$

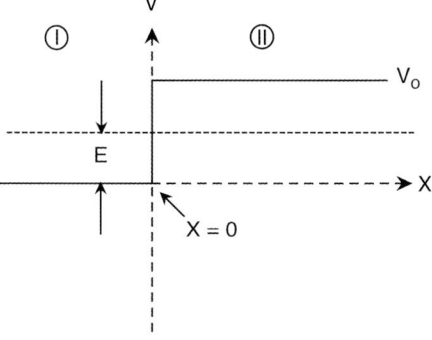

Fig. 9.8 Potential diagram for a particle moving in one dimension in the presence of a potential step of height V_o that extends from the origin to infinity.

9.4 The Wave Functions for Particles in Finite Potential Wells and Barrier Penetration

When $V = 0$ we have a so-called "free" or unconstrained particle and the Schrödinger equation is simply

$$\frac{d^2\psi}{dx^2} + \frac{2mE}{\hbar^2}\psi = 0 \tag{9.14}$$

If we define $k^2 = 2mE/\hbar^2$, Eq. (9.14) has the general solution

$$\psi = A'\sin kx + B'\cos kx \tag{9.15}$$

or, equivalently,

$$\psi = Ae^{ikx} + Be^{-ikx} \tag{9.16}$$

The first term in Eq. (9.16) represents a wave traveling in the positive x direction while the second term represents a wave traveling in the negative x direction. Keeping this in mind, we now consider a particle in region I that is initially moving from left to right. As in the classical case, it will undergo reflection at $x = 0$ and the general form of its wave function will be

$$\psi_1 = Ae^{ik_1x} + Be^{-ik_1x} \tag{9.17}$$

where $k_1^2 = 2mE/\hbar^2$. But in this case we must consider the possibility that the wave function will extend into region II where $V = V_o$. The Schrödinger equation here is

$$\frac{d^2\psi_2}{dx^2} + \frac{2m(E-V_o)}{\hbar^2}\psi_2 = 0 \tag{9.18}$$

Because $V_o > E$, the second term is less than zero. If we define $k_2^2 = 2m(V_o - E)/\hbar^2$, the equation becomes

$$\frac{d^2\psi_2}{dx^2} - k_2^2\psi_2 = 0 \tag{9.19}$$

with the general solution

$$\psi_2 = Ce^{k_2x} + De^{-k_2x} \tag{9.20}$$

Because the wave function must everywhere be finite we must have $C = 0$ and the wave function simplifies to

$$\psi_2 = De^{-k_2x} \tag{9.21}$$

Eq. (9.21) says that, in general, the wave function will not vanish at the potential step at x = 0 but will extend under the barrier as a simple decreasing exponential function.

We can determine the relation between A, B and D by the continuity requirement at x = 0. Thus,

$$\psi_1(0) = \psi_2(0)$$
$$\psi'_1(0) = \psi'_2(0) \tag{9.22}$$

Direct substitution, differentiation and a little algebra yield the following results;

$$D = A + B; \quad \frac{B}{A} = \frac{1 - ik_2/k_1}{1 + ik_2/k_1}; \quad \frac{D^2}{A^2} = \frac{4}{1 + \frac{V_o - E}{E}} \tag{9.23}$$

The assumption in this problem is that, initially, a single particle is moving in the positive x direction toward the barrier at x = 0. As a result of the interaction with the barrier, the wave function at x < 0 is the summation of incident and reflected waves. The ratio of the intensity reflected at x = 0 to the intensity incident from the left is

$$\frac{|B|^2}{|A|^2} = 1 \tag{9.24}$$

That is, the entire incident amplitude is reflected at the barrier, exactly as in the classical case. Nevertheless, the wave has penetrated below the barrier and has nonzero amplitude there. From the third of Eqs (9.23) it is seen that the amplitude of the penetrating wave will be zero only in the limit that $V_o \to \infty$ for any arbitrary $E < V_o$, or if $E \to 0$ for any arbitrary finite V_o. The strange part about the penetrating wave is that it corresponds to $V_o > E$ and therefore to a negative kinetic energy!

To provide some insight into how this relates to the problem of α emission, an entire wave disturbance is shown in Fig. 9.9 on a scale that represents an α particle with total energy of 5 MeV incident on a potential step of height 10 MeV. The simple oscillatory function to the left of the barrier joins smoothly to an exponential under the barrier at x = 0. Note that the amplitude of the wave under the barrier becomes very small indeed within 5–6 fm of x = 0. This will obviously vary with the phase of the wave but the general point is still valid. The wave will decrease rapidly with the distance penetrated under the barrier but, because it decreases exponentially, it will always have a finite value so long as x remains finite.

The total reflection that takes place at the potential step is due to the fact that the barrier extends over the range $0 < x < \infty$. There is no possibility that the wave can find a region for x > 0 where $E > V_o$. If, however, the barrier width was finite, the situation changes considerably. Consider the schematic of Fig. 9.10, for which the barrier height remains at V_o but only exists over the range $0 \leq x \leq a$. For x > a, the potential vanishes. Now, because the amplitude of the wave under the barrier will

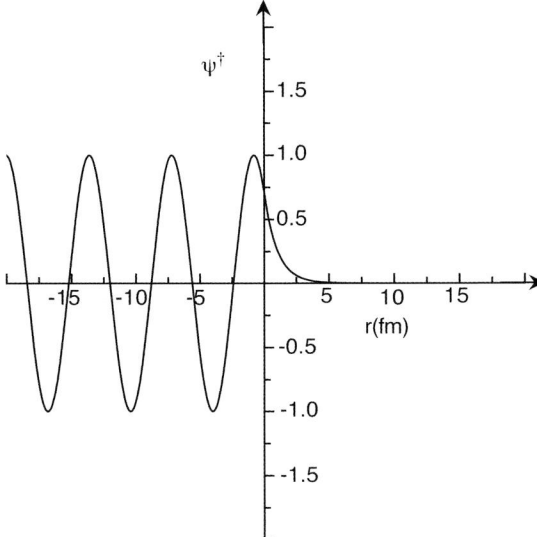

Fig. 9.9 Schematic of the wave function for a particle incident upon the step potential of Fig. 9.8. For the sake of definiteness, it was assumed that an α particle with a kinetic energy of 5 MeV was incident upon a barrier of height $V_o = 10$ MeV. The ordinate, $\psi^\dagger$, gives the amplitude of the wave in relative units.

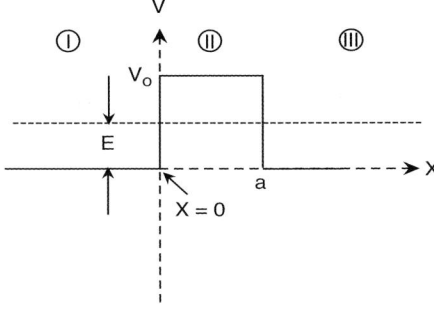

Fig. 9.10 Potential diagram for a particle moving in one dimension in the presence of a potential step of V_o that extends over the finite range $0 < x < a$.

be greater than zero at x = a, the wave can *penetrate* into region III. If this takes place we have the case of a purely outgoing wave because there is no potential there to cause reflection. The qualitative conclusion is that a particle incident from the left in region I can actually become an outgoing particle in region III. The particle, therefore, can indeed escape the barrier even though its total energy is less than the barrier height. This process is referred to as *quantum mechanical tunneling* or *barrier penetration*.

Now let us make this quantitative. Referring to Fig. 9.10, we assume, as in the previous example, that a particle of energy E < V_o is incident upon the barrier at x = 0, and the wave to the left of the barrier, region I, will be of the same form as given in Eq. (9.17). Under the barrier in region II, the wave will also have the same general form as given in Eq. (9.20). However, because the barrier is of finite thickness and reflection can occur at the potential discontinuity at x = a, the coefficient C will not vanish. Examination of Eq. (9.20) shows that the term $Ce^{k_2 x}$ represents a wave that would decrease exponentially with decreasing x, just what is expected of a reflected wave. Finally, if the wave penetrates into region III, there will be a purely outgoing wave, and because the total energy of the particle has not changed, the total energy E in region III must be the same as in region I, just the kinetic energy of the free particle. In fact, the wave function in region III can differ only from the incoming part of the wave in region I by its amplitude. We can therefore immediately write that

$$\psi_3 = Ge^{ik_3 x} = Ge^{ik_1 x} \tag{9.25}$$

So long as |G| > 0, it must be true that the particle has some probability of penetrating the barrier. From Eq. (9.17) we can see that if |G| = |A|, the outgoing wave in region III would be identical to the wave that was incident upon the barrier at x = 0. If |G| < |A|, the implication is that there is some probability less than unity that a particle presenting itself at x = 0 will actually penetrate the barrier and be found in region III.

In order to proceed further, we need to review the definition of a current density in classical mechanics and then translate it into quantum mechanics. Consider the case of a beam of particles, all of the same energy, moving with velocity v as shown in Fig. 9.11. While the trajectories are assumed to be parallel, the particles are incident randomly on a cross sectional area A normal to the beam axis. We define the magnitude of the *particle current density*, j, as the number of particles, n, crossing a unit area normal to their trajectories per unit time, e.g., particles cm^{-2} s^{-1}. At the steady state we can also describe the beam in another way. If we have a unit cross sectional area and consider the volume element swept by moving this surface a unit distance along the particle axis, there will always be a constant number of particles in the volume so long as the current is constant. That is, the particle density, n cm^{-3}, will be constant. If the particle velocity is v then nv particles will traverse the unit area normal to v per unit time and therefore the current density is j = nv cm^{-2} s^{-1}. If we are dealing with a one-dimensional problem, n will

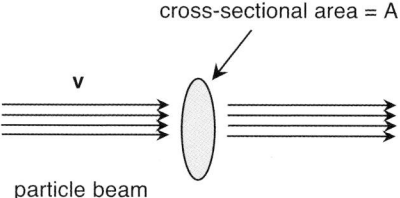

Fig. 9.11 Schematic representation of a monoenergetic, monodirectional particle beam with cross sectional area A.

9.4 The Wave Functions for Particles in Finite Potential Wells and Barrier Penetration

represent the number of particles per unit length and the current density nv will be the number of particles flowing past a point per unit time.

To translate this into the equivalent quantum mechanical expression, remember that $|\psi|^2$ represents the probability density for finding a particle or a system in the dimensions of the problem. In three dimensions, $|\psi|^2$ is the probability per unit volume. For the one-dimensional problem at hand, $|\psi|^2$ is the probability per unit length in the x direction. As a result, the current or the intensity corresponding to one particle moving in the x direction is simply $|\psi|^2 v$.

The sense of the problem we are considering is that a particle is incident upon or collides with the barrier at x = 0 and is represented by the wave function $\psi_{1,\text{incident}} = A e^{ik_1 x}$. As a result of the interaction with the barrier, a wave of the form $\psi_3 = G e^{ik_1 x}$ appears to the left of the barrier. Then, because $v_1 = v_3$, the ratio

$$T = \frac{|\psi_3|^2 v_3}{|\psi_{1,\text{incident}}|^2 v_1} = \frac{|\psi_3|^2}{|\psi_{1,\text{incident}}|^2} \tag{9.26}$$

represents the ratio of the particle intensity escaping the barrier, to the intensity incident upon the barrier and therefore the probability that the particle penetrates the barrier per collision. The quantity T is referred to as the *transmission coefficient*. The result in Eq. (9.26) is very general. It applies formally to any type of barrier and has wide application in radioactive decay and in nuclear reactions as well.

We can now use this relation to determine the probability for barrier penetration in our one-dimensional problem. All we need is an expression that relates the amplitudes A and G. This is easily accomplished by use of the continuity conditions at x = 0 and x = a. Working from right to left, we first relate G to the coefficients of the wave under the barrier (see Eq. (9.20)) using the continuity requirement at x = a. Thus

$$G e^{ik_1 a} = C e^{k_2 a} + D e^{-k_2 a}$$
$$G i k_1 e^{ik_1 a} = C k_2 e^{k_2 a} - D k_2 e^{-k_2 a} \tag{9.27}$$

By adding and subtracting Eqs (9.27), expressions for C and D in terms of G are obtained as

$$C = \frac{G}{2}\left(1 + \frac{ik_1}{k_2}\right) e^{ik_1 a - k_2 a}$$
$$D = \frac{G}{2}\left(1 - \frac{ik_1}{k_2}\right) e^{ik_1 a + k_2 a} \tag{9.28}$$

We can now relate C and D to A and B by use of the continuity conditions at x = 0,

$$A + B = C + D$$
$$A i k_1 - B i k_1 = C k_2 - D k_2 \tag{9.29}$$

These lead directly to the relations

$$A = \frac{C}{2}\left(1 - \frac{ik_2}{k_1}\right) + \frac{D}{2}\left(1 + \frac{ik_2}{k_1}\right)$$
$$B = \frac{C}{2}\left(1 + \frac{ik_2}{k_1}\right) + \frac{D}{2}\left(1 - \frac{ik_2}{k_1}\right) \tag{9.30}$$

At this point we could substitute the expressions for C and D from Eqs (9.28) into Eqs (9.30) and obtain the exact relation between A and G. However, there is a practical simplification that can be made. If we take the ratio of Eqs (9.28) we find

$$\frac{C}{D} = \left(\frac{k_2 + ik_1}{k_2 - ik_1}\right) e^{-2k_2 a} \tag{9.31}$$

Note that C/D decreases exponentially with increasing barrier thickness a. That is, the amplitude of the wave reflected at x = a in region II will decrease exponentially with increasing a. Now in practical cases, the barrier is sufficiently thick that we can, to an excellent approximation, take $C \cong 0$ and substitution of the expression for D in Eqs (9.28) into the first of Eqs (9.30) then gives

$$A \cong \left(\frac{G}{4}\right)\left(1 + \frac{ik_2}{k_1}\right)\left(1 - \frac{ik_1}{k_2}\right) e^{ik_1 a + k_2 a} \qquad a \gg 0 \tag{9.32}$$

The transmission coefficient through the barrier can now be written as

$$T = \frac{|\psi_3|^2}{|\psi_{1,\text{incident}}|^2} = \frac{|G|^2}{|A|^2} = \frac{16 e^{-2k_2 a}}{\left(1 + \frac{k_2^2}{k_1^2}\right)\left(1 + \frac{k_1^2}{k_2^2}\right)} \tag{9.33}$$

The transmission coefficient for this simple rectangular barrier is an exponential function that depends sensitively on the barrier width and on the barrier height relative to the total energy of the particle (see the definition of k_2 just after Eq. (9.18)). The first factor in the denominator becomes very large as E becomes small, whereas the second factor approaches unity in the same limit. To obtain an estimate of the magnitude that can be expected for transmission coefficients in α decay, we can calculate T for the same assumptions used for the calculations shown in Fig. 9.9, namely E = 5 MeV, M = m_α and V_o = 10 MeV. For a = 10 fm, T = 3.83×10^{-6}. For a = 50 fm, a dimension that is more nearly what we will find applicable to the α decay of the heavy elements, T = 3.20×10^{-30}! The transmission coefficient through a thick barrier is very small indeed.

To be sure, the magnitudes of transmission coefficients will depend sensitively on the form of the potential barrier, but the basic physics involved and the general dependence of transmission coefficients on the parameters of the problem are quite well presented with the simple rectangular barrier. With this in view, we will

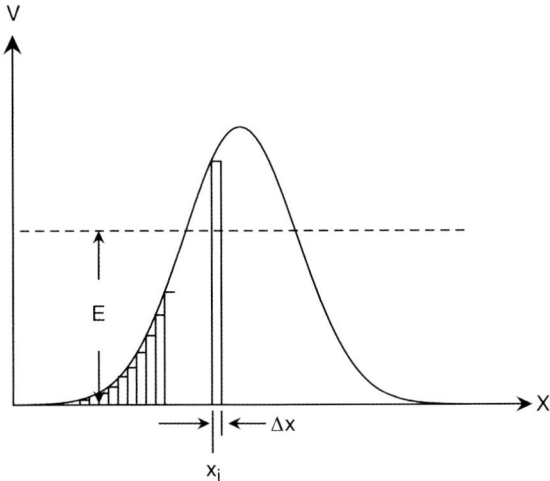

Fig. 9.12 Schematic drawing of a potential barrier of arbitrary shape considered as the summation of a series of barriers of constant width.

now present a very simple approximation to estimate the penetration through a barrier of arbitrary shape.

In principle, a barrier of arbitrary shape can be considered as composed of a series of barriers of differential width Δx and height equal to the height of the real barrier as shown in Fig. 9.12. If it is true that the probability for penetration of any one of the individual barriers is very small and if the penetration of any one does not affect the penetrability of any other, the penetration of the entire barrier will be approximated by the product of the probabilities for penetration of all of the barriers.

Now consider the transmission coefficient given in Eq. (9.33). If we substitute the explicit expressions for k_1^2 and k_2^2, we find

$$T = \frac{16E(V_0 - E)}{V_0^2} e^{-2k_2 a} \tag{9.34}$$

and taking the logarithm of both sides, we have

$$\ln T = -2k_2 a + \ln \frac{16E(V_0 - E)}{V_0^2} \tag{9.35}$$

In most practical cases, the magnitude of $16E(V_0 - E)/V_0^2$ is on the order of unity whereas the magnitude of $2k_2 a$ is on the order of 10^2. Thus the second term in Eq. (9.35) is sufficiently small that it can be neglected, and the transmission through a barrier of thickness Δx can be written as

9 α Decay and Barrier Penetration

$$T \cong e^{-2k_2 \Delta x} \tag{9.36}$$

The transmission through any barrier of arbitrary V(x) is then approximately

$$T \cong \exp\left(-2 \int k_2 dx\right) = \exp\left(-2 \int \sqrt{\frac{2m[V(x) - E]}{\hbar^2}} dx\right) \tag{9.37}$$

This simple approximation is generally quite good so long as the probability for penetration is very small. It can be shown to be in agreement with the results of much more sophisticated developments of the transmission coefficient.

We now apply the idea of barrier penetration to a very simple model that is remarkable in its ability to provide semiquantitative estimates of the decay constant for α decay.

9.5
A Simple Model for α Decay

The arguments presented in Section 9.3, along with the sketch of the potential given in Fig. 9.7, suggest a very simple model for the estimation of the decay constant for α decay. Although we are certain that this cannot be exactly true, let us assume that an α particle exists with unit probability in heavy nuclei. The fact that the potential experienced by the α particle within the daughter nucleus cannot be too large suggests that, for simplicity, we assume it is zero. The potential diagram in Fig. 9.7 shows that, just beyond the range of the nuclear potential well, the residual potential is that of the Coulomb potential alone. If we assume that both the α particle and the daughter nucleus are spherical objects, we can take this to suggest that we treat the potential for radial dimensions at and beyond the sum of the radii of the two particles as just the Coulomb potential and we are then led to the simple problem shown schematically in Fig. 9.13. The idealized potential is taken as

$$V(r) = \begin{cases} 0, & r < a \\ V_C(r), & r \geq a \end{cases} \tag{9.38}$$

where $a = r_\alpha + r_d$ and the subscript d denotes the daughter nucleus. As given, the model applies only to cases where the angular momentum l_α of the emitted α particle is zero. The total energy of the α particle is E.

We now substitute the expression for the potential in Eq. (9.38) into the transmission coefficient given in Eq. (9.37) and obtain

$$T = \exp\left[-2 \int_a^b \sqrt{\frac{2\mu}{\hbar^2}[V_C(r) - E]} dr\right] \tag{9.39}$$

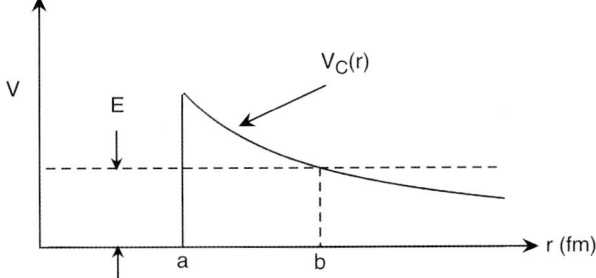

Fig. 9.13 Idealized model for α decay. The potential is taken as the Coulomb potential between the α particle and the daughter nucleus over the range r ≥ a. Both nuclei are assumed to be spherical objects.

In this equation,

$$\mu = \frac{m_\alpha m_d}{m_\alpha + m_d}$$

is the reduced mass of the α particle. The Coulomb potential between the α particle and the daughter nucleus is simply

$$V_C(r) = k_C \frac{2(Z-2)e^2}{r} \tag{9.40}$$

where Z is the atomic number of the parent nuclide. The transmission coefficient is frequently written in the form $T = e^{-2G}$ where

$$G = \int_a^b \left[\frac{2\mu}{\hbar^2}(V_C(r) - E) \right]^{1/2} dr$$

is known as the *Gamow factor*.

If T represents the probability for barrier penetration per presentation at the barrier, we need only estimate the rate at which the α particle presents itself at the barrier and we will have an estimate of the decay constant. The velocity of the α particle in the well is just $v = \sqrt{2E/\mu}$. If it is moving outward from the origin and collides with the potential wall at x = a and does not escape, it will be reflected back to the origin where it will again begin to move outward. It will then traverse a distance 2a between each unsuccessful collision with the barrier at x = a. Therefore the frequency with which it collides with the barrier is f = v/2a, and the probability per unit time for escape, the decay constant, is then

$$\lambda_\alpha \cong fT = \frac{v}{2a} e^{-2G} \tag{9.41}$$

We now have our simple model estimate of the decay constant for α decay when the angular momentum change in the decay is zero.

It is not difficult to show that the Gamow factor can be written in a simple analytical form. To do this, we can make use of the fact that at x = b the energy E must be equal to the Coulomb potential as given in Eq. (9.10). As a result we can write

$$\frac{V_C(r)}{V_C(b)} = \frac{b}{r} \tag{9.42}$$

and thus

$$V_C(r) = \frac{b}{r} V_C(b) = \frac{Eb}{r} \tag{9.43}$$

Substitution of Eq. (9.43) into the expression for G then gives

$$G = \sqrt{\frac{2\mu E}{\hbar^2}} \int_a^b \left(\frac{b}{r} - 1\right)^{1/2} dr \tag{9.44}$$

Letting $b/r = \sec^2\theta$, G can now be evaluated as

$$G = b\sqrt{\frac{2\mu E}{\hbar^2}} \left[\cos^{-1}\left(\sqrt{\frac{a}{b}}\right) - \left(\sqrt{\frac{a}{b}}\right)\left(\sqrt{1 - \frac{a}{b}}\right)\right] \tag{9.45}$$

9.6
Application of the Model to the Decay of Even-Even Nuclei

With the expression given in Eq. (9.41), we are now in a position to see just how well the simple model can account quantitatively for experimental data. We will first examine the decay of the ground state of ^{238}U to the ground state of ^{234}Th in order to outline the calculations and to provide a feeling for the magnitudes of the quantities involved. Because our model was derived with the assumption that the angular momentum change in the decay is zero, let us formally show that this is the case for the present transition. By conservation of angular momentum we can write

$$\mathbf{j}_{238} = \mathbf{j}_{234} + \mathbf{j}_\alpha \tag{9.46}$$

where the $\mathbf{j}_i$ are angular momentum vectors and the numerical subscripts are the mass numbers of the parent and daughter nuclides, respectively. The ground-state spin and parity of ^{238}U is $I^\pi = 0^+$ and we then must have $-\mathbf{j}_{234} = \mathbf{j}_\alpha$. Now the ground state of ^{234}Th also has $I^\pi = 0^+$ and thus the total angular momentum of the α particle must be zero as well. Finally, because the spin of the ground state of the α particle is also zero, its orbital angular momentum must be zero.

9.6 Application of the Model to the Decay of Even-Even Nuclei

The experimental data for α decay of ^{238}U are shown in Fig. 9.5. Neglecting experimental errors, the Q_α = 4.2703 MeV and E_α is then 4.198 MeV (Eq. (9.6)). In the hard-sphere approximation, the radius of ^{234}Th is $r_{234} = 1.25(234)^{1/3} = 7.70$ fm. The radius of the α particle cannot be estimated well from the simple constant-density model but its root-mean-square radius has been measured to be about 1.61 fm. We then calculate the equivalent hard-sphere radius of $r_\alpha = 2.08$ fm by use of Eq. (4.11). With this, we then estimate the parameter a, representing the radial onset of the Coulomb interaction, as 9.78 fm.

The radial dimension b at which the $E_\alpha = V_C$ is readily calculated as

$$b = k_C \frac{2(Z-2)e^2}{E} = \frac{2(Z-2)}{E}\left(\frac{e^2}{m_e c^2}\right) m_e c^2 \quad (9.47)$$

$$= \frac{2 \cdot 90 \cdot 2.818 \cdot 0.511}{4.198} = 61.74 \text{ fm}$$

In this case, the radial dimension at which $E = V_C$ is roughly six times the sum of the radii of the α particle and ^{234}Th, and the barrier thickness is about 52 fm. If we now substitute a and b into Eq. (9.45), the Gamow factor is calculated to be $G = 43.61$. This is huge. The transmission coefficient through the barrier is then $T = 1.32 \times 10^{-38}$! Clearly, the probability of α emission through the barrier per collision is very, very small.

The velocity of the α particle within the well is $v = 4.79 \times 10^{-2} c = 1.44 \times 10^9$ cm s^{-1} and the frequency with which it collides with the barrier at $r = a$ is then $f = 7.34 \times 10^{20}$ s^{-1}, a very large rate. The model estimate for the decay constant $\lambda_{\alpha,model}$ is then 9.70×10^{-18} s^{-1} and the corresponding half-life is $t_{1/2,\,model} = 7.15 \times 10^{16}$ s = 2.26×10^9 y. Now the experimental half-life of ^{238}U is 4.468×10^9 y and the partial half-life for decay to the ground state is 5.803×10^9 y. Our simple model calculation estimates the decay probability within a factor of 2–3.

In the general case, if you were to produce a model that provided estimates of experimental data to within a factor of three or so, the very first thing that would be suggested is that you go back and try again. But in the present case we must consider the agreement between experiment and the model calculation to be remarkably good. This is because of the extreme sensitivity of the decay constant to the form of the potential function, and the very simple assumptions we have made in developing the model. For example, consider the radius parameter a, which we have taken as the sum of the hard-sphere radii of the α particle and the daughter nucleus. We know that the range of the nuclear force is on the order of 1–2 fm and it is reasonable to consider that our estimate of a could be too small by about 1 fm. If we recalculate the decay constant for ^{238}U with the assumption that $a = 8.7$ fm rather than 7.7 fm, the decay constant becomes 9.29×10^{-20} s^{-1}, smaller by a factor of about 7. The fact is that adjustment of the radius parameters alone could easily permit us to get "agreement" with experiment.

Just how well the model can explain the decay probabilities for α emission from (even, even) nuclides on a global scale can be seen by comparing the calculated and experimental half-lives for all of the nuclides shown in Fig. 9.4. This comparison is shown in Fig. 9.14 in the form of a log-log distribution. Calculated half-lives that agreed with experiment exactly would all lie on the dashed line shown in the figure. In general, the half-lives estimated with the model are within factors of 3-10 of the experimental values over the entire range of about 32 orders in magnitude. This strongly supports the essential ideas contained in the model.

Alpha decay is certainly a much more complicated process than we have outlined here, and it has taken rather sophisticated theory to obtain truly quantitative comparisons with experiment. It is, in fact, rather remarkable that the neglect of the details of nuclear structure still permits us to model decay probabilities so well. That nuclear structure does play an important role is easily demonstrated from the data in Fig. 9.14. If we calculate the ratio of the experimental to the model half-lives for the same nuclides considered in Fig. 9.14 and plot them as a function of neutron number, we find the distribution shown in Fig. 9.15. The very sharp peak in the vicinity of the magic number $N = 126$ demonstrates that shell structure has a marked effect on decay rates over and above that which can be ascribed to the mass effects already accounted for in the experimental Q_α. There is also a suggestion of a second peak near $N = 152$ reflecting a shell structure that is predicted from the Nilsson model for deformed nuclei.

There can be no doubt that our model of α decay accounts for the principal effects of barrier penetration that contain the main dependence on decay energy, atomic number and the size of nuclei. It has, however, many deficiencies and

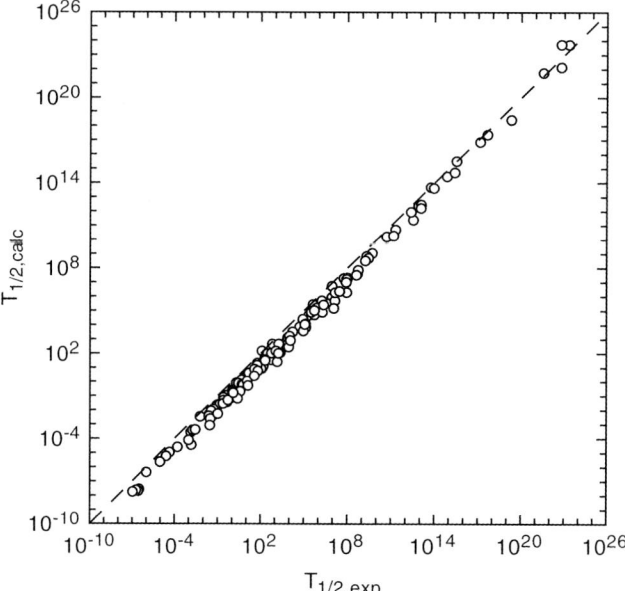

Fig. 9.14 Calculated α decay half-lives (log scale) as a function of experimental total partial α decay half-lives (log scale) for all nuclides shown in Fig. 9.4. The dashed line represents a linear function with a slope of one and an intercept of zero.

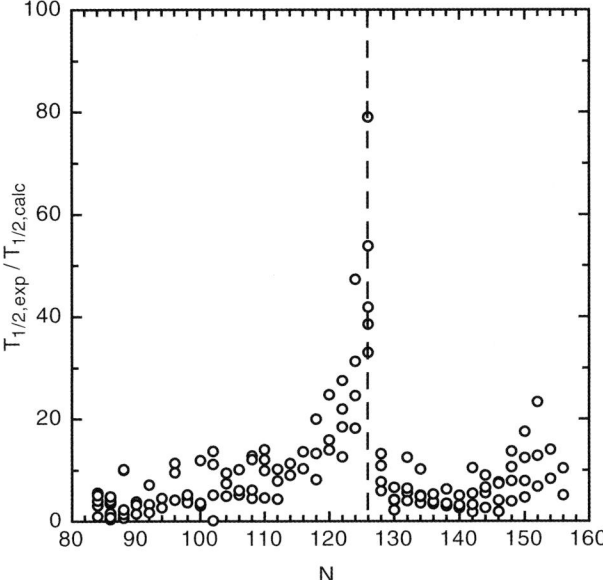

Fig. 9.15 The ratio $T_{1/2,\text{exp}}/T_{1/2,\text{calc}}$ as a function of neutron number for the nuclides shown in Fig. 9.4.

should be viewed for no more than what it is – a very simple model that permits a rather nice correlation of experimental data and demonstrates some of the gross aspects of the physics of α decay. It does not account for the probabilities that α particles exist in the nucleus, or their formation probabilities at the instant of decay, and certainly does not account for the details of nuclear structure. We have also treated all nuclei as spherical objects whereas the majority of the actinides are indeed strongly deformed. In this regard it may be somewhat surprising that the largest deviations from the model predictions are for nuclei that we know to be essentially spherical. These deviations must reflect not only the issue of nuclear shape but inherent differences in the probabilities for formation of the α particle in decay of the various nuclides. For example, in the vicinity of the magic numbers we should expect that the separation energies of pairs of neutrons and protons will reach a maximum and that level spacings are large. In such cases, the probabilities of forming an α particle in the nucleus must be considerably smaller than in nuclei where the binding energies of pairs and level spacings are significantly smaller. Qualitatively, this is just the variation that is reflected in the data shown in Fig. 9.15.

9.7
Angular Momentum Effects in α Decay

Apart from the ground-state to ground-state decay of (even, even) nuclei, α emission will entail orbital angular momenta greater than zero, and we now turn our attention to modifying the simple model developed in Section 9.5 to include angular momentum effects.

The radial equation found in Chapter VII through the separation of variables is

$$\frac{-\hbar^2}{2\mu r^2}\frac{d}{dr}\left(r^2\frac{dR}{dr}\right) + \left(V(r) + \frac{l(l+1)\hbar^2}{2\mu r^2}\right)R = ER \tag{9.48}$$

where R is the radial wave function. The centrifugal barrier, $l(l+1)\hbar^2/2\mu r^2$, arose naturally and in the present case should simply be added to our model potential to account for orbital angular momentum. To demonstrate, qualitatively, the effect of the centrifugal potential on the total potential experienced by an α particle, the centrifugal potential corresponding to $l = 3$ is shown in Fig. 9.16 along with the Coulomb and central potentials previously given in Fig. 9.6. The centrifugal potential is, as expected, very large near the center of mass of the system but reduces to a very small value in the vicinity of the edge of the potential well. It is so small, in fact, that it can represent but a very small perturbation to the Coulomb potential and therefore it will have a relatively small effect on decay probabilities. Because it

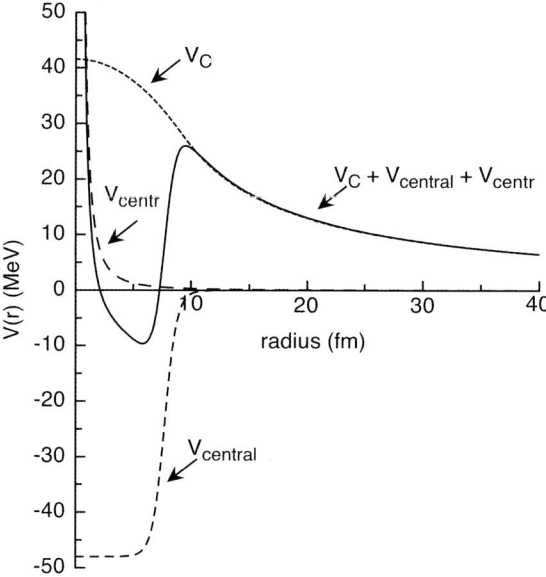

Fig. 9.16 The total potential for an α particle and its components, including a centrifugal potential (V_{centr}) corresponding to $l = 3$. See discussion of Fig. 9.6.

Table 9.1 Relative transmission coefficients for the ground-state to ground-state decay of ^{238}U for different assumed values of the relative angular momentum quantum number of the emitted α particle.

l_α	$T^\dagger$
0	1.000
1	0.846
2	0.608
3	0.373
4	0.196
5	0.0084

increases the height and thickness of the barrier, it must act to decrease barrier penetrability and therefore decrease the decay probability. Within the context of our model, we can get a rough picture of how the angular momentum barrier affects decay probabilities by calculating the relative penetrabilities for a fixed energy as a function of orbital angular momentum with all other nuclear properties held constant. As an example, we consider the ground-state to ground-state α decay of ^{238}U with an energy of 4.198 MeV and calculate the transmission coefficients for *hypothetical* transitions of l = 1–5 relative to the real transition that corresponds to l = 0. All we do is to simply add the centrifugal potential to the expression for G in Eq. (9.41) and carry out the integration numerically. The relative transmission coefficients, which, in this case, represent the relative α decay probabilities as a function of l are given in Table 9.1. While the relative transmission coefficients given in the table should only be taken as rough approximations, they show that for fixed energy and other nuclear properties, the centrifugal barrier exerts a relatively small addition to the hindrance of α emission. We expect only a factor of about two reduction in transmission probability for l_α of about 2–3 compared to the case of l_α = 0. The centrifugal barrier exerts a "fine" adjustment to the transmission coefficients calculated with neglect of angular momentum.

An estimate of how well the simple model, including angular momentum effects, can explain such experimental data is obtained by examination of Table 9.2

Table 9.2 Experimental and calculated α-intensities in the decay of ^{238}U.

Level Energy (MeV)	J^π	E_α (MeV)	I_{exp} (%)	I_{model} (%)
0	0^+	4.198	77	81
0.0495	2^+	4.150	23	19
0.160	4^+	4.041	0.2	0.6

that shows the fraction of decays of ^{238}U that take place to the ground and first two excited states – the branching ratios – along with the estimates from the model calculation. Comparison of the last two columns of the table show that the model explains, at least semiquantitatively, the combined effects of reduced decay energy and angular momentum. More sophisticated models are able to do a much better job but at the expense of a considerable increase in complexity, especially in attempts to gain an understanding of the effects of underlying nuclear structure on the decay process.

9.8
Decay of Odd-A Nuclides and Structure Effects

Before we leave our discussion of α decay, it is important to bring out some general characteristics of the decay of odd-A nuclides because of the structure effects that are immediately apparent and because these lead to some practical consequences with respect to the emission of penetrating γ radiation along with the α particles. A simple means to approach the problem is to consider the experimental data on the decay of two typical α emitters, $^{235}_{92}$U and $^{241}_{95}$Am. A partial decay scheme for $^{235}_{92}$U is shown in Fig. 9.17. Some 21 levels in $^{231}_{90}$Th are known or suspected to be populated by α decay of $^{235}_{92}$U. Because of the great sensitivity of modern detectors, it has been possible to see an α-decay branch with an intensity as low as about 10^{-3}% in this case, and even smaller intensities have been measured in the decay of other nuclei. $^{231}_{90}$Th is a deformed nucleus and that leads to the very high level density shown in the figure. Some 18 levels in $^{231}_{90}$Th are known with excitation energies less than 500 keV, leading to an average energy separation of about 28 keV, very much smaller than is found in spherical nuclei. The levels in $^{231}_{90}$Th can be identified with levels predicted by the Nilsson model and the rotational states that are built upon them. The energy dependence of the α intensities shown to the right of the levels in Fig. 9.17 are remarkably different from those found in the decay properties of (even, even) nuclides. Rather than the majority of decays leading to the ground state, the decay of $^{235}_{92}$U locates the principal part of the α intensity at two levels with excitation energies of 205.3 keV and 236.9 keV, respectively. The simple barrier penetration model predicts intensities to the ground and first-excited states that are larger by factors of about 15 and 9, respectively, than those found experimentally. Further, the experimental data show decay intensities to the levels at 278.0 keV and even 387.8 keV that are comparable to the intensities to the ground and first-excited states. There can be little doubt that the very different behavior of α decay in this case must be the result of nuclear-structure effects.

A schematic of the α decay of $^{241}_{95}$Am is shown in Fig. 9.18. Levels in $^{237}_{93}$Np with energies as high as 800 keV are populated directly by α emission but the intensity to the majority of these is very small. For simplicity we show only the intensities to the five low-lying levels that account for more than 99.6% of all decays. Once again, the intensity pattern is very different from that expected from the simple barrier-

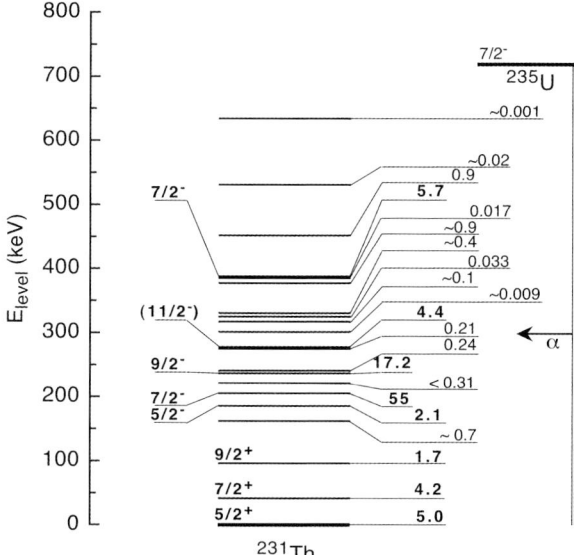

Fig. 9.17 Partial decay scheme for the α decay of $^{235}_{92}$U to levels in $^{231}_{90}$Th. Absolute α intensities (%) are given to the right on each level. Levels populated with intensities greater than 1% are shown in bold face along with their spins and parities.

penetration model. The majority of the α intensity goes to two levels at energies of 59.54 and 102.96 keV with very little intensity to the ground and first-excited states.

To understand the possible structure effects that might be at play, we must first discuss the problem of how the angular momenta and parity of α transitions affect the decay probabilities. We can write the conservation equation for angular momentum generally as

$$\mathbf{j}_p = \mathbf{j}_d + \mathbf{j}_\alpha = \mathbf{j}_d + \mathbf{l}_\alpha \tag{9.49}$$

The second of the equalities in Eq. (9.49) arises because the angular momentum of the ground state of the α particle is 0. Because $\mathbf{l}_\alpha = \mathbf{j}_p - \mathbf{j}_d$, and the angular momenta of the initial and final states can take any allowed projection on a space-fixed z-axis, the allowed values of the orbital angular momentum quantum number of the α particle are given by

$$|j_p - j_d| \leq l_\alpha \leq |j_p + j_d| \tag{9.50}$$

In addition to conservation of angular momentum, we must also conserve parity. Because the parity of a wave of definite orbital angular momentum is defined by the angular momentum quantum number, conservation of parity between the initial and final states must be given by

9 α Decay and Barrier Penetration

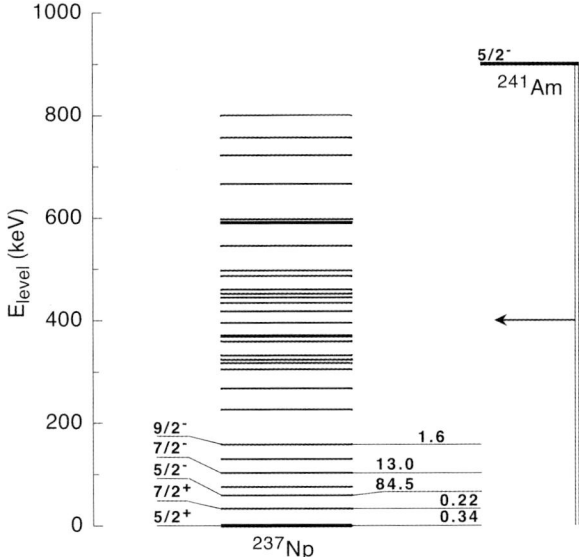

Fig. 9.18 Partial decay scheme for the α decay of $^{241}_{95}$Am to $^{237}_{93}$Np. All levels populated directly by α decay are shown. For simplicity, the absolute intensities (%) are shown only for the levels for which $I_\alpha > 1\%$. (See caption to Fig. 9.17.)

$$\pi_p = \pi_d(-1)^{l_\alpha} \tag{9.51}$$

Thus, in any given decay, the angular momentum quantum number of the emitted α particle can be anything within the range specified by Eq. (9.50) that satisfies the parity requirements of Eq. 9.(51). For the case of the decay of the ground state of $^{235}_{92}$U (7/2⁻) to ground state of $^{231}_{90}$Th (5/2⁺), for example, Eq. (9.50) permits $1 \leq l \leq 6$ while parity conservation demands that the orbital angular momentum quantum number be odd. Therefore, $l_\alpha = 1, 3$ and 5. Apart from any structure effects that may be present, the decay can occur by emission of α particles with any of these angular momenta and therefore the decay constant will be the sum of the decay constants for three transitions. The range in angular momentum that is allowed increases the number of ways in which decay can occur and therefore increases the total decay probability. A rough idea of how much the decay constant can be affected is obtained from the relative transmission coefficients given in Table 9.1. If we sum the transmission coefficients for the three angular momenta, we find that the total transmission coefficient is about 1.2 times that expected from the emission of an α particle with zero angular momentum. While this approach does not provide very accurate decay rates, it does show that the additional degrees of freedom cannot account for very large changes in decay rates. It cannot be a significant factor in the departures from the penetrability model seen in the examples of the decay of $^{235}_{92}$U and $^{241}_{95}$Am.

To get at an explanation for the experimental observations, consider once again the qualitative ideas that give rise to the likelihood that α particles "exist" in heavy nuclei. Because of the strong pairing energy between identical nucleons, all nucleons, with the exception of the odd particle, will be paired. The pairs will tend to be found in the lowest energy states available but they are not constrained to do so. The lowest total energy of the system will be found when the pairs have non zero probabilities of occupying some of the first few excited states as well, and, because of its large binding energy, it is relatively easy to envision that a pair of neutrons and a pair of protons can be combined to form an α particle. Crucial to the movement of a pair of identical nucleons between levels is the fact that a pair has a total angular momentum and parity of 0^+. Thus the movement of a pair from one level to another does not change the angular momentum and parity of the system even if the levels differ in their parity and angular momentum quantum numbers. The conservation of these two quantities is maintained. But single nucleons cannot so easily be excited without changing the spin and parity of the nucleus, only those excited states with the same spin and parity of that of the odd nucleon can be considered and it is not so common to find adjacent low-lying levels with the same spin and parity. The conclusion is that the most likely way in which the odd nucleon can be combined to form an α particle is by breaking an existing pair such that the state of the new odd particle has the same spin and parity as that of the original odd nucleon. This is clearly relatively improbable. Now it is certainly true that more complex excitations than those considered here can and do take place, but the general arguments given above suggest that they will tend also to be less probable than the excitation of pairs of identical particles.

These ideas lead to the expectation that the formation of an α particle in an (even, even) or odd-A nuclide does not involve the odd particle in first order and that the odd nucleon will remain in its original state. If we assume this, we conclude that the formation of an α particle in an odd-A nuclide will tend to follow essentially the same path and with similar probability that we would find in the (even, even) nucleus that is formed by removal of the odd nucleon. The odd nucleon is little more than a spectator. If we now remember that we are dealing with deformed nuclei that are well-described by the Nilsson model and that in the deformed potential, a single-particle state can be occupied by only two particles, we immediately see that α decay of an odd-A nuclide should indeed proceed to an *excited state* in the daughter as shown schematically in Fig. 9.19. The schematic is drawn to represent an odd-A nucleus with N = odd and Z = even in the approximation that

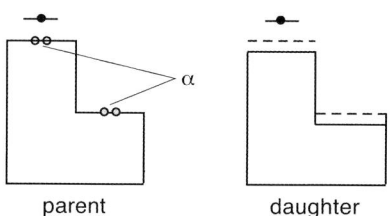

Fig. 9.19 Schematic diagram of the formation of an α particle from paired neutrons and protons in an odd-A nucleus. The left-hand side of each diagram represents completely filled neutron levels in a deformed nucleus and the right-hand side represents the filled levels in the daughter nucleus.

all nucleons occupy the lowest-lying levels available. To the left of each diagram representing the parent and the daughter are the neutron levels and to the right are the proton levels. Taking the odd neutron as a spectator, we illustrate the formation of an α particle from the most weakly bound pairs of neutrons and protons. With the emission of the α particle, the daughter nucleus is left with the neutron in the first-excited state of the daughter. Clearly, other pairs of particles can be involved in the formation of an α particle but the same conclusion is drawn – the daughter will, most probably, be left in an excited state, a state that represents the excitation of the odd nucleon.

Although we have not discussed the Nilsson model in very great detail, it can be shown that the most likely levels to which decay of an odd-A nuclide occurs are band heads and rotational states built upon them that have wave functions corresponding to or closely connected to the wave function of the Nilsson state that is occupied by the odd nucleon in the parent ground state. This implies that the parity of states populated with the highest intensity will be the same as the parity of the parent ground state and, from our discussion of the centrifugal barrier, we can expect and do indeed find that the levels populated with the highest intensity correspond to small l_α.

The structure effect discussed above has a significant practical consequence. The fact that odd-A nuclides tend to decay to levels at energies of 0.1–0.2 MeV tends to result in the emission of intense γ-rays with similar energies. In the case of $^{235}_{92}U$, for example, the intense decay to the 205.3 keV level in $^{231}_{90}Th$ is followed by emission of a two-photon cascade, the second of which has an energy of 185.7 keV and an absolute intensity of about 53%. The decay of $^{241}_{95}Am$ is accompanied by the emission of, among others, a γ-ray with energy of 59.5 keV and an absolute intensity of about 35.7%. Both of these find very practical use, the first in detection and quantification of the $^{235}_{92}U$ content of a specimen and the second both as a γ-ray standard for calibration of photon detectors and as source for exciting x-ray emission in samples that can be used for the purpose of elemental analysis. Conversely, the decay of the (even,even) actinides is usually accompanied by much weaker photon emission at lower energies.

References

1. G. Audi and A.H. Wapstra, "The 1995 update to the atomic mass evaluation", Nucl. Phys. A595, 409 (1995).

General References

The problem of barrier penetration is covered in essentially all introductory texts on quantum mechanics. Choose the text that serves you best.

Some very good references on α decay can be found on the worldwide web. A straight forward and readable discussion can also be found in P. Marmier and E. Sheldon, "Physics of Nuclei and Particles", Academic Press, Volume I, New York (1969).

Problems

1. A sample of ^{238}U was separated from all elemental impurities 30 y prior to counting. Estimate the *total* decay rate of the sample per g.

2. For each of the three nuclides ^{226}Ra, ^{240}Pu and ^{243}Am calculate: (a) the Q value for α decay; (b) the laboratory kinetic energies of the α particles emitted when decay takes place to the ground state of the α decay daughter; and (c) the Coulomb barriers between the α particle and the α-decay daughter when the two nuclei are just in contact in the hard-sphere limit.

3. Calculate the kinetic energies of the α particles emitted to the ground and first three excited states of ^{237}Np by α decay of ^{241}Am. (See Fig. 9.18 for level properties.)

4. Beginning with Eq. (9.27), carry out the derivation of Eq. (9.33) explicitly.

5. Using the same parameters that were used to prepare the wave disturbance shown in Fig. 9.9, calculate and plot the wave disturbance in regions I, II and III of Fig. 9.10. Use a = 4 fm.

6. Go to the Table of Isotopes and look up the decay properties of ^{8}Be. Use the simple barrier penetration model to estimate the half-life of this nuclide. For this calculation you should assume that the hard-sphere radius of the α particle is 2.08 fm.

7. Determine the allowed angular momenta of α particles in the decay of ^{241}Am to the ground and first three excited state of ^{237}Np. See Fig. 9.18 for details.

8. Decay by emission of nuclei such as ^{12}C, ^{14}C, etc., is energetically possible in the decay of many of the isotopes of the heaviest elements. Consider the case of the decay of ^{226}Ra. Estimate the ratio of the decay probability for ^{12}C emission to that for α emission when both decays would lead to the ground states of the respective daughters.

9. Go to the Table of Isotopes and display the α-decay schemes of ^{242,243}Cm. Compare their characteristics to those of the (even, even) and odd-A nuclides discussed in this chapter. What do you conclude?

10. A hypothetical deformed nucleus has $Q_\alpha = 6.0$ MeV and a ground-state spin and parity of 1^+. The nuclide produced in α decay also has a ground-state spin and parity of 0^+. Can a ground-state to ground-state α decay occur? Discuss the ramifications of the observation of such a transition.

10
β Decay

10.1
Introduction

β decay is one of the more complex and intriguing topics of low-energy nuclear physics. Historically it caused great consternation when physicists were confronted with a decay mode that appeared to violate several of the cherished and well-proven conservation laws. To find a way out, it took one of the great physicists of his era, Wolfgang Pauli, to propose the existence of a new particle, now known as the neutrino, that had not been observed in any experiment, when he made his insightful suggestion in 1930. At that time, the particle we now call the neutron was also not even known. It was discovered by Chadwick in 1932. In a letter dated December 4 1930, Pauli wrote (translated from the original German text)

> "... I have hit upon a desperate remedy to save the 'exchange theorem' of statistics and the law of conservation of energy. Namely, the possibility that there could exist in the nuclei electrically neutral particles, that I wish to call *neutrons*, which have spin 1/2 and obey the exclusion principle and which further differ from light quanta in that they do not travel with the velocity of light. The mass of the *neutrons* should be of the same order of magnitude as the electron mass and in any event not larger than 0.01 proton masses. The continuous beta spectrum would then become understandable by the assumption that in beta decay a *neutron* is emitted in addition to the electron such that the sum of the energies of the neutron and the electron is constant ...
>
> I agree that my remedy could seem incredible because one should have seen these *neutrons* much earlier if they really exist. But only the one who dare can win and the difficult situation, due to the continuous structure of the beta spectrum, is lighted by a remark of my honoured predecessor, Mr. Debye, who told me recently in Bruxelles: 'Oh, It's well better not to think about this at all, like new taxes'. From now on, every solution to the issue must be discussed. Thus, dear radioactive people, look and judge."

Nuclear Physics for Applications. Stanley G. Prussin
Copyright © 2007 WILEY-VCH Verlag GmbH & Co., Weinheim
ISBN: 978-3-527-40700-2

The student should read "neutrino" for "neutron" in Pauli's text and we have added italics to emphasize this. This most famous of notes should be taken as the thinking of a very bright physicist doing his best to understand the unknown.

Pauli's postulate was so compelling and allowed the conservation laws to be maintained that it was accepted and used to further our knowledge of β decay and other fundamental physics. Roughly 30 years later, the interaction of the neutrino with matter was observed and its existence conclusively proven. But the very means by which the particle was detected led to serious questions related to our understanding of the nature of our sun and the means by which fusion produces the energy that it radiates. This problem has led once again to a very serious challenge to our understanding of the fundamentals of particle physics that is under intense investigation today. Within the past several years, new experiments on the properties of the neutrino have made it clear that our understanding of the nature of the forces between fundamental particles must be modified and it is very likely that the reverberations from the discovery and study of β decay will continue to contribute greatly in the quest to understand the fundamentals of physical interactions.

The theory of β decay is rather mathematical even in its most basic formulation. It will involve the application of the results of time-dependent perturbation theory that is itself less than transparent. However, it is important to go through this carefully, not only to understand β decay, but also because it will be used to understand γ decay and nuclear reactions. We will begin by reviewing the fundamental issues of the conservation laws that required the bold hypothesis of Pauli, and then proceed to a discussion of some of the practical issues associated with the three decay modes of β^-, β^+, and electron capture (EC). This will be followed by a discussion of the simplest formulation of β decay, the theory of allowed decay by Fermi, and the correlations that can be derived from it. For simplicity, discussion of electron capture will be delayed until the discussion of β^+ and β^- decay has been completed. Finally, we will discuss β–decay schemes and some of the practical applications of the general theory that have been developed.

10.2
β Decay and Conservation Laws: The Neutrino and the Weak Interaction

The problems encountered with the fundamental conservation laws are easily demonstrated by considering the normal observables in the decay of ^{3}H to ^{3}He as shown in Eq. (10.1).

$$^3\text{H} = {}^3\text{He} + \beta^- + Q_{\beta^-} \tag{10.1}$$

The expression in Eq. (10.1) must quantitatively express the mass or energy relation in β decay of ^{3}H. The value of Q_β calculated from this expression is about 0.0186 MeV. If the decay took place as written, we can expect that the kinetic energy of the emitted β^- particles would be very nearly equal to the Q_β because of the very large mass difference between an electron and the nucleus of ^{3}He

$(m_e/M_{^3He} \approx 1/5511)$. Thus the recoil energy imparted to the ^{3}He nucleus can be safely neglected. However, careful measurements of the decay of ^{3}H show that the emitted β^- particles are *not* mono-energetic, but have kinetic energies in the range $0 \leq T_{\beta^-} \leq Q_{\beta^-}$. Because no other particles or radiations are normally observed, it is implied that the equality sign in Eq. (10.1) is not correct, i.e., energy is not conserved. And if energy is not conserved, linear momentum cannot be conserved.

There is also a serious problem with the conservation of angular momentum if Eq. (10.1) is correct. We know from experiment that the ground-state spins and parities of both ^{3}H and ^{3}He are $1/2^+$. Because the β^- particle is a fermion, it also has an intrinsic spin of $1/2$ and, regardless of its orbital angular momentum, it must have a total angular momentum quantum number that is half-integral. Now conservation of angular momentum requires that $\mathbf{I}_{^3H} = \mathbf{I}_{^3He} + \mathbf{j}_{\beta^-}$ and because the ground-state spin of ^{3}H is half-integral, the sum of the angular momentum quantum numbers of ^{3}He and the β^- particle must be half-integral. But the coupling of two vectors described by half-integral quantum numbers can only give a vector with a total angular momentum quantum number that is *integral*.

The issues of the violation of the conservation of energy and momentum led Pauli to suggest that in addition to the β^- particle, a second particle that is a fermion must be emitted, a particle that has such a small probability for interaction with matter that it goes undetected under normal conditions. Today we call this particle the antineutrino, $\bar{\nu}$, and write the normal equation for β^- decay as

$$M(Z, A) = M(Z + 1, A) + \beta^- + \bar{\nu} + Q_{\beta^-}/c^2 \qquad (10.2)$$

The emission of an antineutrino rather than a neutrino results from an additional conservation law that appears to be necessary, and that is the conservation of *leptons* (light particles). A lepton cannot be created or destroyed without its antiparticle being created or destroyed[1].

With the emission of two leptons, energy and linear momentum can now be shared such that

$$Q_{\beta^-} = T_{\beta^-} + T_{\bar{\nu}} + m_{\bar{\nu}}c^2 \qquad (10.3)$$

again neglecting the very small recoil energy of the daughter nucleus. Pauli's original postulate was that the antineutrino had a small rest mass. The best experimental measurements from which the mass of the antineutrino can be inferred directly, indicate an upper limit of perhaps 2 eV/c^2, negligibly small compared to the mass of electrons and nuclei and negligible in comparison to essentially all Q_β that are of interest. Several recent and ongoing experiments have proven that neutrinos must have *some* mass although the actual magnitude is still unknown. While this is truly an exciting result, it is clear that the existence of neutrino mass is entirely negligible with respect to ordinary β decay and we will neglect it in our

---

1) The conservation of leptons is currently being challenged by experiments on double β decay. Specifically, experiments are being conducted that are searching for a decay mode that would occur only if the neutrino and antineutrino were identical.

further discussions. We can therefore, to an excellent approximation, treat neutrinos as massless particles that move at the speed of light with total energies given by

$$E_\nu^2 = p_\nu^2 c^2 \tag{10.4}$$

The decay modes of positron emission and electron capture decay are governed by the same interaction that gives rise to β^- decay and they can be written as

$$M(Z, A) = M(Z-1, A) + \beta^+ + \nu + Q_{\beta^+}/c^2$$
$$M(Z, A) = M(Z-1, A) + \nu + Q_{EC}/c^2 \tag{10.5}$$

Because the positron is an antiparticle, *neutrino* emission must accompany positron decay. Similarly, because electron capture results in the disappearance of a particle, a neutrino must also be emitted to maintain conservation of leptons.

β decay also raises a number of other issues that we should consider. Using β^- decay as an example, we picture the decay process as

$$n|_{nucleus} \rightarrow p|_{nucleus} + \beta^- + \bar{\nu} \tag{10.6}$$

It is natural to ask where the β^- particle and antineutrino come from; are they present in the nucleus and just released with some probability as we have assumed in the case of α decay? Or are they formed at the instant of decay? It is not difficult to show that it must be the latter. If the β^- particle existed in the nucleus prior to decay, it would somehow have to be bound there. Using the simple particle in the box model as a crude approximation, we can estimate the kinetic energy that the electron would have within the nucleus. Taking the lowest state in the box with n = 1 and assuming a nucleus with mass number A = 200, the kinetic energy of the bound electron is estimated to be about 1.5 GeV! Not only is this enormous compared to the nuclear potential, but we only observe emitted electrons with energies less than about 10 MeV in the majority of cases. There is just no doubt that the leptons could not be confined in the nucleus and therefore we must conclude that they are created at the instant of decay.

β decay must occur as a result of the existence of some force, or interaction. Many lines of evidence and arguments indicate that it is not the result of the Coulomb, strong or gravitational interactions. It can be associated with a new type of interaction referred to as the *weak* interaction. The weak interaction also governs the interaction of neutrinos with matter and, because it is so weak, it is extremely difficult to detect them. Were it not for the fact that electrons possess charge and an appreciable mass, we would indeed have a difficult time detecting these as well. The weak interaction is part of the total interaction between nucleons in the nucleus but because it is so weak, it is entirely negligible except when dealing with β decay itself. The real states of a nucleus will not differ from those determined by neglect of the weak interaction by very much at all. The difference is so small that a simple first-order approximation for the perturbation caused by the weak interaction is extremely accurate in describing the process of β decay and we now turn to a description of the Fermi theory of allowed decay that is based on this approach.

10.3
The Fermi Golden Rule No. 2

First-order time-dependent perturbation theory is treated in most texts on quantum mechanics and an outline of the theory is given in Appendix 4. For the present, we will present the decay constant given by this theory, the so-called Fermi Golden Rule No. 2, and try to make its form plausible.

We consider the problem of the β^- decay of the ground state of a nuclide to the ground state of the (Z + 1, A) daughter as shown in Fig. 10.1. The wave function of the parent, in the absence of the weak interaction, is ψ_i and that of the daughter is ψ_f. These wave functions are time-independent, such as those found from our spherical potential well calculations. If decay is to occur, however, these can only be approximations. States that can decay must be time-dependent. If they are and decay occurs, the wave function must somehow progress from ψ_i to ψ_f. In the limit that the interaction that produces the decay is very weak, it must be true that the time-independent approximation is nevertheless very good. Before decay takes place, the wave function appears, for all intents and purposes, as stationary. As soon as decay occurs, the nucleus is transformed into the daughter whose state also appears to be stationary, even if it too might be unstable to β^- or some other decay mode. This suggests that we can *formally* consider the decay process to be due to some operator that converts the initial, almost stationary, nuclear state $\Psi_{o,i}$ into the final, almost stationary state $\Psi_{o,f}$ in the form

$$O_{op} \Psi_{o,i} \to (\Psi_{o,f})_{total} \tag{10.7}$$

where $(\Psi_{o,f})_{total}$ represents the complete time-dependent wave function of the final state, including the daughter nucleus, the emitted electron and the antineutrino. The approximation that the nuclear wave functions are very nearly stationary is at the heart of first-order perturbation theory.

The transformation from the initial to the final state is due to the weak interaction. We can, as we have done up till now, use the potential of this force to describe it, and we take the potential of the weak interaction, H_β, as the operator that produces the transformation. In quantum mechanics, the expression in Eq. (10.7) is written in the form

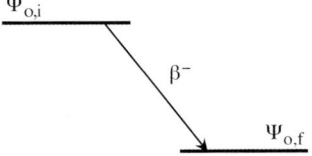

Fig. 10.1 The β^- decay of the ground state of a parent ($\Psi_{o,i}$) to the ground state of the daughter ($\Psi_{o,f}$).

$$H_\beta|\Psi_{o,i}\rangle \rightarrow |\Psi_{o,f}\rangle \tag{10.8}$$

If we view the left-hand side of Eq. (10.8) as a wave function itself, then the quantity

$$\langle\Psi_{o,f}|H_\beta|\Psi_{o,i}\rangle = \int \Psi_{o,f}^* H_\beta \Psi_{o,i} d\tau \tag{10.9}$$

will be a measure of how close the eigenfunction $H_\beta|\Psi_{o,i}\rangle$ is to the eigenfunction $|\Psi_{o,f}\rangle$. (Here, again, the variable $d\tau$ represents all of the variables contained in the wave functions over which we must integrate and $\langle\Psi_{o,f}| = \Psi_{o,f}^*$ is the complex conjugate of $|\Psi_{o,f}\rangle$.) If the two are very similar, we expect the integral in Eq. (10.9) to be relatively large. If not, the integral will be relatively small.

Now what we are after is the probability per unit time for β^- decay to take place, i.e., the decay constant. If Eq. (10.9) is a measure of how well the *amplitude* of wave functions compare, it is plausible that the *probability* for the transformation will be proportional to $|\langle\Psi_{o,f,total}|H_\beta|\Psi_{o,i}\rangle|^2$, and if one goes through the details of the theory, this is exactly what is found. This expression has units of energy squared, and because systems are quantized in units of $\hbar$ (energy–time), it is also plausible that the decay probability will be proportional to $|\langle\Psi_{o,f,total}|H_\beta|\Psi_{o,i}\rangle|^2/\hbar$.

We must be careful in describing a transformation such as occurs in β^- decay where the electron and antineutrino are "free particles". In quantum mechanics, they are still quantized and therefore we must have definite states into which they are emitted. As it turns out, for any specific set of initial and final nuclear states, there are a number of different states into which the leptons can be emitted and we must account for all of these. To do this, we introduce the so-called density of final states, $\rho(E)$, that describes the number of final states available to the emitted particles. Combining this with the expression above, the full quantum mechanical treatment shows that the first-order perturbation approximation for the decay constant is

$$w = \frac{2\pi}{\hbar}|M_{f,i}|^2 \rho(E) \tag{10.10}$$

where

$$M_{f,i} = \langle\psi_{o,f,total}|H_\beta|\psi_{o,i}\rangle = \int \psi_{o,f,total}^* H_\beta \psi_{o,i} d\tau \tag{10.11}$$

Eq. (10.10) is the Fermi Golden Rule No. 2. It expresses the decay probability from a specific initial to a specific final state in the limit that these states can be treated as stationary. The quantity $M_{f,i}$ is commonly referred to as the *matrix element* between the initial and final states. This derives from the matrix formulation of the interaction between any two states that is met with in more advanced discussions of quantum mechanics. We will treat it only in the form of the explicit integrals that must be solved.

Just how all of this applies in the case of β decay is developed in the following.

10.4
The Fermi Theory of Allowed β Decay [1]

We begin with the integral on the right-hand side of Eq. (10.11) and, for the present, we will treat the process of β^- decay explicitly. The initial state of the system, $\psi_{o,i}$, is some β^- unstable nuclide, which, for the present, we assume is in its ground state. Similarly, the final state of the system, $\psi_{o,f,total}$, is taken as the ground state of the daughter nucleus, $\psi_{o,f}$, the emitted electron, ψ_e, and the emitted antineutrino, $\psi_{\bar{\nu}}$. We will have to take the complex conjugates of these to substitute them into Eq. (10.11).

Now how do we express H_β? As the potential of the weak interaction, it has some magnitude and range. As an operator, it also must contain some part that converts a neutron in the nucleus into a proton with the creation of an electron and antineutrino. The simplest representation we can take is to assume that the strength of the weak interaction is some constant, g. As for the range, it is assumed that the interaction is *local*. This means that the decay, along with the creation of the electron and antineutrino, takes place at the site of the decaying neutron. The latter, of course, has some probability of being found everywhere within the nuclear volume, the probability being defined by its wave function. The decay can therefore take place spacially with the same probability distribution.

Suppose we let Q_n represent an operator that converts a neutron into a proton. We can then formally represent the operator H_β as

$$H_\beta = g\psi_e^* \psi_{\bar{\nu}}^* Q_n \tag{10.12}$$

With this, Eq. (10.11) becomes

$$M_{f,i} = g \int \psi_{o,f}^* \psi_e^* \psi_{\bar{\nu}}^* Q_n \psi_{o,i} d\tau \tag{10.13}$$

The integrand can now be read to represent the action of the operator Q_n on the initial nucleus to produce the final nucleus and the lepton pair.

The integral in Eq. (10.13) contains the wave functions of both the nuclei and the leptons. We can, however, obtain a significant simplification in its form by examination of the wavelengths of the leptons relative to the dimensions of the nucleus. The simplest wave function to treat is that of the antineutrino. Because of its very weak interaction with matter, its wave function is essentially that of a free particle, a plane wave. We therefore can write

$$\psi_{\bar{\nu}} = A e^{i(k_{\bar{\nu}} \cdot r)} = A e^{\frac{i}{\hbar}(p_{\bar{\nu}} \cdot r)} \tag{10.14}$$

The wave function must be normalized in some arbitrarily large volume V and this gives

$$\int \psi_{\bar{\nu}}^* \psi_{\bar{\nu}} dV = A^2 \int dV = A^2 V = 1 \tag{10.15}$$

We then have the normalized wave function

$$\psi_{\bar{\nu}} = \frac{e^{i(k_{\bar{\nu}} \cdot r)}}{V^{1/2}} \tag{10.16}$$

Now consider the wavelengths of typical antineutrinos. Because the neutrino mass is negligibly small, we can write

$$E_{\bar{\nu}} = h\nu = \frac{hc}{\lambda_{\bar{\nu}}} \tag{10.17}$$

or,

$$\lambda_{\bar{\nu}} = \frac{hc}{E_{\bar{\nu}}} = \frac{2\pi\hbar c}{E_{\bar{\nu}}} \tag{10.18}$$

Generally, $0 < Q_{\beta^-} < 10$ MeV and therefore $0 < E_{\bar{\nu}} < 10$ MeV. Using these limits, we use Eq. (10.18) to calculate that $125 \text{ fm} < \lambda_{\bar{\nu}} < \infty$. But nuclear radii lie in the range $1 < r_n < 9$ fm. Clearly, , $\lambda_{\bar{\nu}} \gg r_n$ and over the dimensions of the nucleus, the *amplitude* of the antineutrino wave function is essentially *constant*. If we make the approximation that the amplitude is constant, the so-called *long-wavelength approximation*, we can then replace the wave function by its value at any place within the nucleus, and we might just as well take the value at the origin of the coordinate system. Therefore, we make the approximation that

$$\psi_{\bar{\nu}} \approx \psi_{\bar{\nu}}(0) = \frac{1}{V^{1/2}} \tag{10.19}$$

With this approximation the antineutrino wave function can be removed from the integral in Eq. (10.13).

If the electron were not charged, the very same long-wavelength approximation could be applied to approximate its wave function by a constant. Even though the rest mass is not zero, it is so small relative to the largest Q_{β^-} we usually encounter that the range in electron wavelengths is very nearly the same as that calculated above for the antineutrinos. However, the electron must interact with the Coulomb field of the protons. Because of the attraction, the electron wave will be distorted from that of a plane wave such that its amplitude in the vicinity of the nucleus is increased. On the other hand, if we were dealing with positron emission, we can see that the repulsion between it and the protons will tend to decrease its amplitude near the nucleus. In either case, the distortion from a plane wave will be energy dependent; the wave functions of lower-energy particles will suffer greater distortion than the wave functions of higher-energy particles.

Fermi introduced a very simple idea that allows us to use the long-wavelength approximation and at the same time maintain the proper amplitude of the electron wave function as determined by the Coulomb field of the nucleus. Let $|\psi_e(0)|_Z^2$ and $|\psi_e(0)|_0^2$ represent the probabilities for finding an electron of some kinetic energy E_e at the origin of the coordinate system in the presence of Z protons and in the absence of nuclear charge, respectively. Now define the Fermi function $F(Z, E_e)$ as

$$F(Z, E_e) = \frac{|\psi_e(0)|_Z^2}{|\psi_e(0)|_0^2} \tag{10.20}$$

Because $F(Z,E_e)$ does not depend in detail on the nature of the wave function of the decaying neutron and can be calculated quite well with rather general assumptions concerning the Coulomb field of the nucleus, we can use Eq. (10.20) to replace the electron wave function with the amplitude of a plane wave at the origin, times a quantity that is essentially independent of the nuclear wave functions. As a result, we can now write

$$\psi_e^*(0) \approx \frac{F(Z, E_e)^{1/2}}{V^{1/2}} \tag{10.21}$$

where we have used the amplitude of the undistorted plane wave at the origin as in the case of the antineutrino. Substituting the expressions for the wave functions of both leptons (Eqs (10.19) and (10.21)) into Eq. (10.13) then gives

$$M_{f,i} = \frac{g}{V} F(Z, E_e)^{1/2} \int \psi_{o,f}^* Q_n \psi_{o,i} d\tau \tag{10.22}$$

and we can now use this in (10.10) to find

$$w = \frac{2\pi}{\hbar} \frac{g^2}{V^2} F(Z, E_e) \left| \int \psi_{o,f}^* Q_n \psi_{o,i} d\tau \right|^2 \rho(E) \tag{10.23}$$

The equation is getting simpler but is still rather complex. It says that the decay probability from some specific initial nuclear state to some specific final state is given by the product of some constants, the Fermi function, which, presumably, can be calculated independently and without any detailed knowledge of β decay, the square of the matrix element that is determined solely by nuclear wave functions, and the number of ways in which the decay can be accomplished as represented by the density of final states $\rho(E)$.

To proceed further, we first consider what the density of final states is in this problem. Although it seems rather vague, it is closely related to the problem of the density of states of the particle in a box that we met in Chapter IV. First, we remember that the initial and final nuclei will each be in definite states, and thus $\rho(E)$ is defined solely by the leptons. The electron and the antineutrino are emitted as "free particles" with substantial kinetic energies. If we dealt with such a problem in classical mechanics we would say that the leptons were emitted into the *continuum* where their kinetic energies could be varied continuously subject to the requirement that their sum is equal to the total decay energy. But in the present case we are dealing with a quantized system and even though the leptons are not bound, they still must be represented by properly normalized wave functions. As such, we know that their momenta must be quantized in units of $\hbar$ and thus the continuum for a quantized system does not allow the continuous variation found in classical mechanics. The leptons must be emitted into a "continuum" of quantized states. In a Cartesian representation, the particles are emitted into a "box" of volume V that we can take as cubical with sides L, i.e., $V = L^3$. From Eq. (4.26) we know that each component p_q where, for example, q = x, y, or z, is given by $p_q = n_q \pi \hbar / L$ and that the square of the total momentum is given by $p^2 = n^2 \pi^2 \hbar^2 / L^2$ where $n^2 = n_x^2 + n_y^2 + n_z^2$. Following the development given in Section 4.5, the total number of states available with momenta 0 to p_{max} is (Eq. (4.36))

$$N = \frac{\pi}{3}\left(\frac{p_{max}L}{\pi \hbar}\right)^3$$

As will be seen shortly, we are only interested in the number of states of a given spin projection, and thus the number of states we consider is 1/2 that given above. Now because of the very small size of $\hbar$ compared to the momenta we are considering, we can treat the number of states between p and p + dp as a continuous variable, to an excellent approximation, and then write the number of such states as

$$dN = \frac{4\pi V}{(2\pi \hbar)^3} p^2 dp \tag{10.24}$$

This relation holds for both of the leptons. Therefore, if the electron is emitted with momentum p_e to $p_e + dp_e$ and the antineutrino is emitted with momentum $p_{\bar{\nu}}$ to $p_{\bar{\nu}} + dp_{\bar{\nu}}$, the total number of states available will be

$$dN_{tot} = dN_e dN_{\bar{\nu}} = \frac{16\pi^2 V^2}{(2\pi \hbar)^6} p_e^2 dp_e p_{\bar{\nu}}^2 dp_{\bar{\nu}} \tag{10.25}$$

This result has a profound effect on decay probabilities. There is no reason why any of the possible momentum states of a lepton is any more probable to be filled than any other, and if all are equally probable, the fact that the number of states varies as p^2 indicates that higher momenta and higher kinetic energies will be strongly favored in the emission process from this effect alone. Of course the momenta of the leptons are restricted by the requirement that the sum of their kinetic energies must equal the total decay energy, and therefore the number of states available will be dependent upon the division of energy between the two particles.

Because we normally observe the electrons, it makes sense to express Eq. (10.25) in terms of the electron kinetic energy. To do this we write

$$E_o = E_e + E_{\bar{\nu}} = E_e + p_{\bar{\nu}} c \tag{10.26}$$

where E_o is the total decay energy and E_e is the kinetic energy of the electron. Thus

$$p_{\bar{\nu}} = (E_o - E_e)/c \tag{10.27}$$

10.4 The Fermi Theory of Allowed β Decay

For a fixed electron kinetic energy, a differential in the antineutrino momentum is related to a differential in the total decay energy by

$$dp_{\bar{\nu}} = dE_o/c \tag{10.28}$$

We can now use the expressions in Eqs (10.27) and (10.28) to write Eq. (10.25) as

$$dN_{tot} = dN_e dN_{\bar{\nu}} = \frac{16\pi^2 V^2}{(2\pi\hbar)^6 c^3}(E_o - E_e)^2 p_e^2 dp_e dE_o \tag{10.29}$$

We now get a clear definition of the density of final states for this decay mode. Within a differential of the total decay energy, the density of states available to the emitted leptons is

$$\frac{dN_{tot}}{dE_o} \equiv \rho(E_o) \tag{10.30}$$

The density of final states in the Fermi Golden Rule represents the number of ways in which the leptons can be emitted while conserving energy. We then have

$$\rho(E_o) = \frac{16\pi^2 V^2}{(2\pi\hbar)^6 c^3}(E_o - E_e)^2 p_e^2 dp_e \tag{10.31}$$

Note that this is the density of states available *only* for electrons emitted with momenta p_e to $p_e + dp_e$. If this is substituted directly into (10.23) we find

$$w = \frac{32\pi^3}{\hbar(2\pi\hbar)^6 c^3}g^2 F(Z, E_e)\left|\int \psi_{o,f}^* Q_n \psi_{o,i} d\tau\right|^2 (E_o - E_e)^2 p_e^2 dp_e \tag{10.32}$$

We can now give a precise meaning to the transition probability w. It represents the probability for β⁻ decay when the electron is emitted with a momentum p_e to $p_e + dp_e$. If λ is the total decay constant, $w = \frac{d\lambda}{dp_e}dp_e = d\lambda$. We then arrive at the final expression for the β⁻ decay probability in the limit of the Fermi theory of allowed β⁻ decay,

$$\lambda = \frac{32\pi^3}{\hbar(2\pi\hbar)^6 c^3}g^2 \int_{E_e=0}^{E_o} F(Z, E_e)\left|\int \psi_{o,f}^* Q_n \psi_{o,i} d\tau\right|^2 (E_o - E_e)^2 p_e^2 dp_e \tag{10.33}$$

This expression, while complicated, can be analyzed to give some surprisingly simple results with respect to both the spectrum of the β⁻ particles and the nuclear states that are populated with high probability.

The term *allowed* has a specific meaning that we will discuss in some detail later. For the moment it suffices to point out that our treatment of the lepton wave functions permits nonzero decay probabilities *only* if the probability of finding each of the leptons at the origin of the coordinate system is itself nonzero.

10.5
β Spectra

The expression for the decay constant in Eq. (10.33) provides a direct prediction of the spectrum of β^- particles emitted in the decay. The integral $\left|\int \psi_{o,f}^* Q_n \psi_{o,i} d\tau\right|^2$ is dependent only on nuclear wave functions whereas the integral over the electron momentum is essentially independent of the nuclear states. This means that the decay constant can be written as

$$\lambda = \left|\int \psi_{o,f}^* Q_n \psi_{o,i} d\tau\right|^2 \frac{32\pi^3}{\hbar(2\pi\hbar)^6 c^3} g^2 \int_{E_e=0}^{E_o} F(Z, E_e)(E_o - E_e)^2 p_e^2 dp_e \qquad (10.34)$$

It is thus the product of two independent factors in the allowed approximation: a part dependent solely on the nuclear states and a part dependent only on the lepton characteristics and the interaction of the emitted electron (or positron) with the Coulomb field of the nucleus. This means that the spectrum of the emitted leptons depends only on the Coulomb field and the total decay energy, i.e.,

$$\frac{d\lambda}{dp_e} \sim F(Z, E_e)(E_o - E_e)^2 p_e^2 \qquad (10.35)$$

Now the quantity $d\lambda/dp_e$ is just the probability for decay with the electron momentum in the range p_e to $p_e + dp_e$ and it therefore represents the *momentum spectrum* of electrons that will be emitted, $n(p_e) \sim d\lambda/dp_e$. All that we need to calculate β spectra are the Coulomb distortion factors $F(Z,E_o)$. But even without these, we can begin to understand what β spectra must look like by a simple analysis of the right-hand side of Eq. (10.35). Consider first the limit of $Z = 0$ so that $F(Z,E_o) = 1$. The spectral shape is now determined by the factors $(E_o - E_e)^2 p_e^2$. At electron energies small compared to E_o, the quantity $(E_o - E_e)^2$ varies very slowly but the quantity p_e^2 varies quite rapidly. On the other hand, as the electron energy approaches E_o, it is the factor $(E_o - E_e)^2$ that varies rapidly while p_e^2 varies very slowly. We are then led to expect that the β^- spectrum will vary roughly as p_e^2 at low energies and as $(E_o - E_e)^2$ at the highest energies. In Fig. 10.2 the predicted shape of the β^- spectrum is shown in the limit that $F(Z,E_o) = 1$.

The spectrum rises sharply at low energies and reaches a peak, the most probable energy, at E/E_o of about 0.35 E_o. The quadratic dependence on $(E_o - E_e)^2$ is evident as the energy approaches E_o. Because $F(Z,E_o)$ has been taken to be unity, this spectrum would apply to either β^- or positron emission.

The calculation of Fermi functions in all but the very low-energy (nonrelativistic) limit is a bit complex and we will not delve into the mathematical details. Rather,

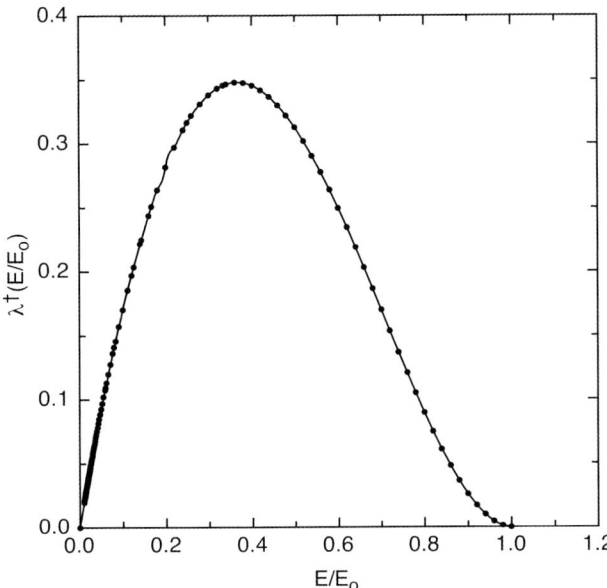

Fig. 10.2 The relative decay probability, $\lambda^\dagger$, for electrons with energies $0 \leq E \leq E_o$ as a function of E/E_o. The Fermi function has been assumed to be unity.

we will settle for examination of the results of the detailed calculations shown in Fig. 10.3, where it has been assumed that the electron or positron sees the Coulomb field of a uniformly charged sphere in the constant-density approximation. The presence of atomic electrons has been neglected. For β^- decay, the probability of finding the electron at the origin, essentially the center of the nucleus, is always greater than that expected in the absence of the Coulomb field, and it is clearly seen that the attraction between the protons and the β^- particle increases this probability to the greatest extent for smaller kinetic energies. For high Z and low energies, the probabilities for finding a β^- particles at the origin are increased by as much as two to three orders of magnitude compared to the case of Z = 0.

The effect of the Coulomb field on the wave functions of positrons is even more dramatic at the low energies where reductions in probabilities for finding the positron at the origin are reduced by over four orders of magnitude. The effect of the Coulomb field is relatively small for positron kinetic energies of 1 MeV but, curiously, the effect increases at higher energy and increasing atomic number. This can be traced to the relativistic effects of the Lorentz contraction and time dilation.

The Coulomb field of the nucleus will produce significant modifications to the spectrum of emitted particles shown in Fig. 10.2, and the modifications will be strongly dependent on both atomic number and decay energy. As examples, the spectra shown in Fig. 10.4 represent calculations of the hypothetical β^- decay of a nucleus with Z = 49 and the hypothetical β^+ decay of a nucleus with Z = 51 to the same daughter such that the Q-values for both decays are identical and equal to

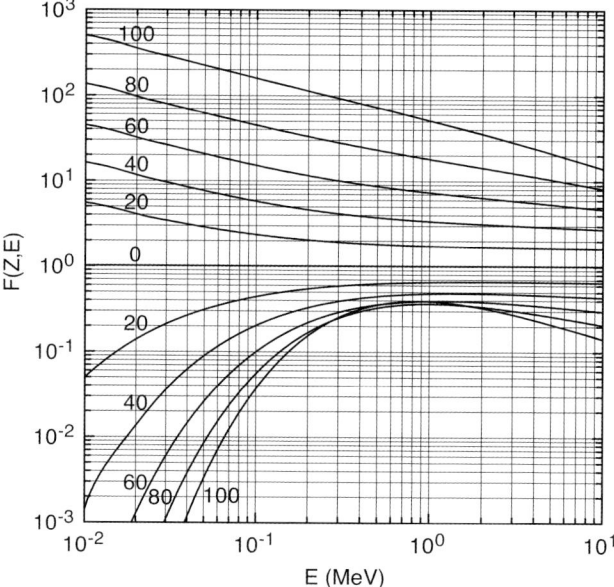

Fig. 10.3 Log–log plot of the Fermi function F(Z,E) versus the electron kinetic energy. F(Z,E) for β^- decay are represented by the curves with F(Z,E) > 1 and the F(Z,E) for β^+ decay are represented by the curves with F(Z,E) < 1. The atomic number of the decay *daughter* is indicated adjacent to each the curve.

1.00 MeV. The calculations represent the evaluation of the right-hand side of Eq. (10.35). Compared to the case of F(Z,E) = 1, the spectrum from β^- decay contains a much higher intensity of low-energy particles while the β^+ spectrum possesses considerably fewer low-energy particles and is more nearly symmetrical in shape.

The functions shown in Fig. 10.4 are the relative decay probabilities per unit energy and the area enclosed by each of the functions is unity. Reference to Eqs (10.34) and 10.35 shows that the spectral distributions in the figure would be expected to be found in counting experiments where the total decay probabilities of each of the three hypothetical nuclides were identical. But, in fact, they will not be the same because of the Fermi function (Fig. 10.3). The increased probability of finding an electron at the nucleus due to the attractive Coulomb interaction leads to a the total decay probability for β^- decay for any Z > 0 that will be larger than for Z = 0 if both decays have the same value of Q_{β^-}. Similarly, the total decay probability for β^+decay will be smaller than for any Z > 0 because of the repulsion between the positron and protons in the nucleus. Indeed, the actual integrals on the right-hand side of Eq. (10.34) for the three cases considered in Fig. 10.4 are 1472 (β^- decay), 73 (β^+ decay) and 223 (Z = 0), respectively. Due solely to the Coulomb potential of the nucleus, the decay probability for positron emission is only about 0.05 that of β^- decay in this hypothetical case. To be sure, the relative

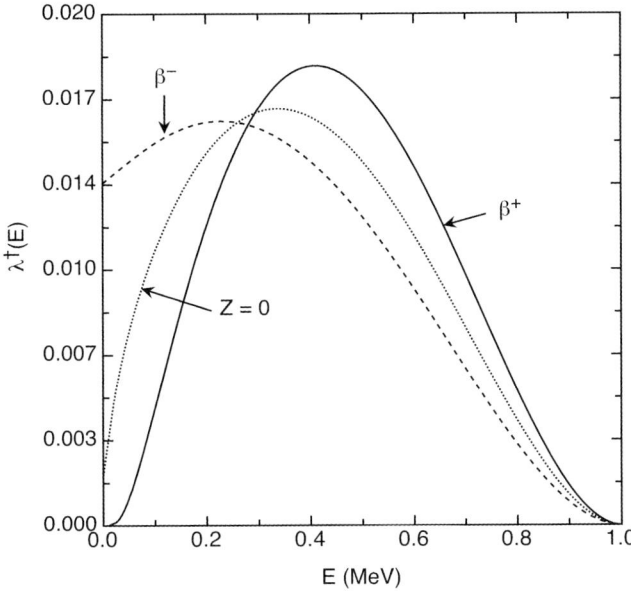

Fig. 10.4 β spectra for the hypothetical decay of nuclides with Z = 49 (β⁻) and Z = 51(β⁺) to a daughter with Z = 50. It is assumed that the Q values for β⁻ decay and positron emission produce maximum kinetic energies of both particles of 1 MeV. For comparison, the spectrum expected in the absence of the Coulomb field of the nucleus is also shown (Z = 0). The ordinate in the figure is the relative decay probability per unit energy and the area enclosed by each curve is unity.

effects are very sensitive to the atomic number of the daughter and to the Q values. But the very strong reduction in the probability for finding a positron near the origin suggests that positron emission among the heaviest of elements ought to occur with greatly reduced decay probabilities compared to the probabilities among lower-Z nuclides. Indeed, it is a fact that positron emission is very rare among the heaviest elements.

Before we go on to the actual calculation of decay probabilities for β⁻ decay and positron emission, we should comment on our neglect of the presence of atomic electrons. Under normal conditions, we deal with the decay of neutral atoms or ions with low ionic charge and the Coulomb field experienced by β⁻ particles and positrons is the total field due to the protons in the daughter nucleus and the atomic electrons. But the neglect of the atomic electrons is easily seen to be relatively small. First, the leptons are created within the nuclear volume. The Coulomb field experienced from the protons is very high because they are concentrated in the small volume of the nucleus with linear dimensions of 2–9 fm. Atomic electrons, on the other hand, are contained in a much larger volume that has linear dimensions on the order of 1–2 Å, or $(1-2) \times 10^5$ fm. If we assume that the atom can be approximated as a uniformly charged sphere, it is not difficult to get an estimate of the ratio of the potentials due to the electrons and the protons. In elementary electromagnetism it is shown that the electrostatic potential from a uniform spherical distribution of charge q is given by

$$V_C = \begin{cases} k_C\left(-\dfrac{qr^2}{2R^3} + \dfrac{3q}{2R}\right) & r \leq R \\ k_C\dfrac{q}{r} & r \geq R \end{cases} \quad (10.36)$$

where R is the radius of the charge distribution. Thus, the ratio of the magnitudes of the potentials from the atomic electrons and the protons at the common center of the two uniform spherical charge distributions is

$$\frac{V_e(0)}{V_p(0)} = \frac{R_p}{R_e} \approx (1-4) \times 10^{-5} \quad (10.37)$$

and the ratio of the potentials at the nuclear surface is about the same magnitude. Hence, because of the rather diffuse distribution of the electrons, we can safely expect that the potential from them is quite small. Qualitatively, the inclusion of the atomic electrons will somewhat reduce the probability of finding the β^- particle and increase the probability of finding a positron at the origin of the coordinate system. Although small, the most accurate calculations of the Fermi function include the presence of the atomic electrons.

10.6
Decay Probabilities for β^- and β^+ Decay

The expression for the decay constant in Eq. (10.34) contains the integral

$$f(Z, E_o) = \int_{E_e = 0}^{E_o} F(Z, E_e)(E_o - E_e)^2 p_e^2 dp_e \quad (10.38)$$

and with knowledge of $F(Z, E_e)$ these integrals can now be performed. The function $f(Z, E_o)$ is referred to as the integrated Fermi function and it contains all of the effects of the emitted leptons on the decay probability. Because there is no reference to the parent or daughter nucleus other than the specification of Z, the $f(Z, E_o)$ are essentially universal functions that can be calculated once and for all and tabulated.[2] In place of the integral in Eq. (10.38) one usually finds a "dimensionless" form that we will present because it leads to important implications for the product of

---

2) There is the slightest of error in this statement because the calculation of $F(Z, E_e)$ entails knowledge of the charge distribution of the nucleus and this depends upon mass number and the actual shape of the daughter nuclide. However, the errors incurred by neglect of this isotope-specific property are entirely negligible in most cases.

the constants in the expression for λ. For this purpose we define the quantity W as the *total* energy of a particle in units of the energy-equivalent of electron rest mass. Because E_e in Eq. (10.38) refers to the kinetic energy of the electron or positron, we then write

$$W = \frac{E_e + m_e c^2}{m_e c^2} \tag{10.39}$$

Using this to substitute for E and E_o in the integrand, and using the expression

$$p^2 c^2 = W^2 (m_e c^2)^2 - (m_e c^2)^2 \tag{10.40}$$

to substitute for the electron momentum and its differential, Eq. (10.38) becomes

$$f(Z, W_o) = \frac{(m_e c^2)^5}{c^3} \int_{W=1}^{W_o} F(Z, W)(W_o - W)^2 (W^2 - 1)^{1/2} W \, dW \tag{10.41}$$

Because F(Z, W) is itself dimensionless, the integral is now completely dimensionless.

Logarithmic plots of the integrated Fermi function are shown in Fig. 10.5, calculated with the neglect of screening from the atomic electrons. The magnitude of the function is seen to vary over a large range, and the dependence on decay energy is considerably stronger than the dependence upon the atomic number of the daughter nucleus. In rough analogy, $f(Z, E_o)$ plays a role in β decay that is quite similar to the role played by the transmission coefficient in α decay. But the effect is much weaker. For a factor of 2 change in the α-decay energy, we found that the decay probability changed by roughly 22 orders of magnitude. In β decay, a factor of 2 change in decay energy is seen to produce roughly a factor of 10 or so change in the decay probability. Nevertheless, and apart from all other factors that can affect the decay constant, $f(Z, E_o)$ is seen to account for 10–12 orders of magnitude change in decay probability as the decay energy varies over the range generally found experimentally.

If the expression in (10.41) is substituted into (10.34), we have the result

(10.42)

$$\lambda = \frac{g^2 m_e^5 c^4}{2\pi^3 \hbar^7} \left| \int \psi_{o,f}^* Q_n \psi_{o,i} d\tau \right|^2 \int_{W=1}^{W_o} F(Z, W)(W_o - W)^2 (W^2 - 1)^{1/2} W \, dW$$

and the collection of constants is usually written as

$$\Gamma = \frac{g^2 m_e^5 c^4}{2\pi^3 \hbar^7} \tag{10.43}$$

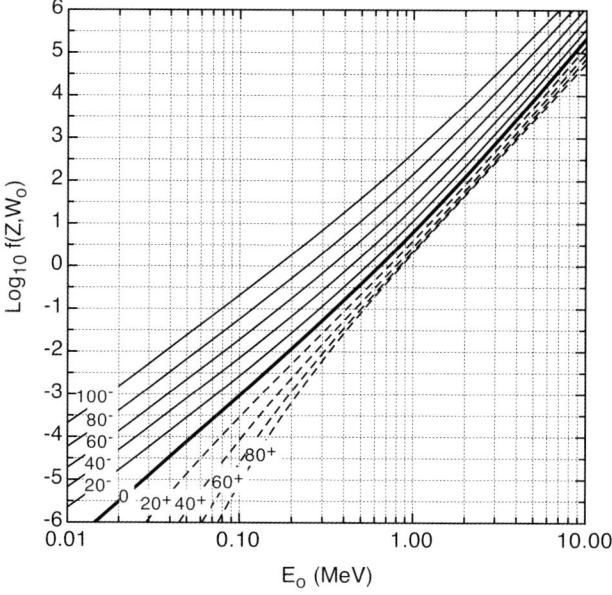

Fig. 10.5 Log-log representation of the integrated Fermi function f(Z,W₀) calculated with neglect of the screening from atomic electrons. The abscissa represents the maximum energy of the emitted particles. Each curve is labeled by the atomic number of the decay daughter with superscript of – or + for β⁻ and β⁺ decay, respectively.

Dimensional analysis shows that g has units of energy times volume. (This is perhaps most easily seen by examining Eq. (10.23) and recognizing that w must have units of inverse time.) With this, Γ must have the dimensions of inverse time and therefore represents a frequency or rate that is characteristic of β decay. In the literature, the factor g is called the vector coupling constant, G_V, when we deal with the simplest form of the operator Q_n known to exist (see Section 10.72). Its magnitude has been determined from detailed studies of the decay of mirror nuclei, and the best current value is $G_V = 1.4155 \times 10^{-62}$ J m³. The corresponding value of Γ is

$$\Gamma \cong 1.1286 \times 10^{-4} \, \text{s}^{-1}$$

and leads to the characteristic time

$$\frac{1}{\Gamma} \cong 8861 \, \text{s}$$

This is a very long time relative to the time required for nucleons to move about the nucleus and is indicative of the weakness of the force that leads to the decay itself (hence the name "weak" interaction). Apart from the magnitude of the integral involving the nuclear wave functions, which is expected to be on the order of unity or less, the product $\Gamma f(Z, W_o)$ provides an estimate of the order of magnitude of β-decay lifetimes. With $f(Z,W_o)$ in the range 10^{-5} to 10^{+5} (Fig. 10.5), we see that the decay constants will lie roughly in the range 10^{-9}–10 s^{-1} with nuclear matrix elements of order unity.

The decay constant given by (10.42) is often written in the short-hand notation

$$\lambda = |M_{f,i}|^2 \Gamma f(Z, W_o) \tag{10.44}$$

Alternatively, one can substitute the half-life for the decay constant and rearrange this equation to read

$$ft_{1/2} = f(Z, W_o) t_{1/2} = \frac{\ln 2}{\Gamma |M_{f,i}|^2} \tag{10.45}$$

The quantity $ft_{1/2}$, simply referred to as the "ft value", has become the standard means by which specific β-decay transitions are described. It is referred to as the *comparative half-life* and inspection of Eq. (10.45) shows why this makes sense. Within the limits of the allowed theory, the product $ft_{1/2}$ would be a constant for all β transitions if the matrix elements $M_{f,i}$ for all transitions were identical. That is, to the extent that nuclear structure effects are not very significant, all transitions would have about the same comparative half-life. Conversely, to the extent that nuclear-structure effects produce large variations in the magnitudes of matrix elements, the comparative half-lives would be expected to vary widely. Nuclear-structure effects that would tend to reduce the rates of transitions would be reflected in large comparative half-lives and structure effects that would tend to enhance the rates of transitions would be reflected in relatively small comparative half-lives. From the magnitude given above for $1/\Gamma$, we can expect ft values to have magnitudes measured in thousands of seconds and, as a result, it is common practice to actually quote the magnitude of $\log_{10} ft$ where the integrated Fermi function is multiplied by the experimental half-life (or partial half-life) in units of seconds.

Before we proceed to examine the $\log_{10} ft$ found experimentally, it is first reasonable to discuss some of the implications of the allowed theory and to examine, in the simplest cases, what the matrix elements for β transitions are expected to be. For both of these purposes we will make use of the simple shell-model approximation that was developed in Chapter VII.

10.7
Some Implications of the Simple Theory of Allowed β Decay

10.7.1
Angular Momentum Effects

The allowed theory presents a simple picture of the β-decay process. As a result of the weak interaction a nucleon in the nucleus undergoes decay with the creation of a lepton pair (electron and antineutrino, or positron and neutrino) at the location of the nucleon undergoing the transition. Because of their large wavelengths relative to the dimensions of the nucleus, the wave functions of the leptons can be taken as constants over the dimensions of the nucleus with appropriate corrections for interaction of the charged leptons with the Coulomb field of the protons. Because the weak interaction is so weak, the initial and final nuclear states can be taken as stationary states with no significant error. These states must have well-defined angular momentum and parity and both must be conserved.

This simple picture has profound implications on decay probabilities when the leptons are forced to be emitted with nonzero orbital angular momenta. This is most easily seen by considering the neutrinos whose wave functions are undistorted plane waves of the form $\psi_{\bar{\nu}} = Ae^{i(\mathbf{k}_{\bar{\nu}} \cdot \mathbf{r})} = Ae^{i(\mathbf{p}_{\bar{\nu}} \cdot \mathbf{r})/\hbar}$ (Eq. (10.14)). Because of the presence of the product $\mathbf{p}_{\bar{\nu}} \cdot \mathbf{r}$ in the exponent, it should be clear that the wave function can be expressed in terms of the angular momentum that it carries, which, of course, is quantized. As a result, it makes sense to find a means of describing the wave function directly in terms of the angular momentum quantum number l. This can be accomplished by use of two of our old friends, the spherical harmonics and the spherical Bessel functions. We state without proof, that the direct expansion of $e^{i(\mathbf{k}_{\bar{\nu}} \cdot \mathbf{r})}$ gives

$$e^{i(\mathbf{k}_{\bar{\nu}} \cdot \mathbf{r})} = \sum_{l=0}^{\infty} (2l+1) i^l j_l(kr) \sqrt{\frac{4\pi}{2l+1}} Y_{l,0}(\cos\theta) \tag{10.46}$$

This rather messy looking expression is actually quite simple. It says that an outgoing plane wave can generally be expressed as the superposition of waves described by spherical Bessel functions of definite orbital angular momentum, each modulated by a spherical harmonic corresponding to the same angular momentum that is independent of the azimuthal angle ϕ. The presence of the quantity i^l is really of no great concern because we will be interested in probabilities that are proportional to the squares of the wave functions. Of particular interest here is that the *radial* dependence of the antineutrino wave function is completely contained in the properties of the spherical Bessel functions that were displayed in Fig. 7.7, and, for convenience, the same figure is reproduced below as Fig. 10.6. It is immediately evident that if the magnitude of the argument kr is small, the amplitudes of wave functions corresponding to l > 0 will all be small compared to that for l = 0 and the amplitudes will decrease rapidly as l increases. Because the

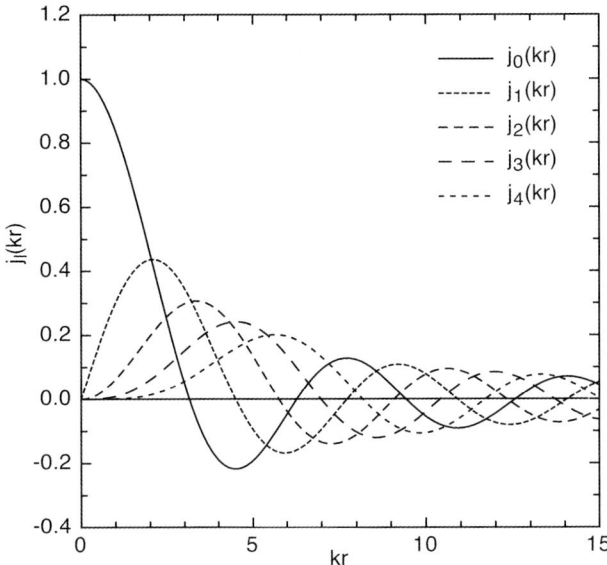

Fig. 10.6 The spherical Bessel functions $j_l(kr)$ for $l = 0$–4.

probabilities of finding antineutrinos within the nuclear volume are proportional to the squares of the amplitudes, the probabilities will decrease even more rapidly with increasing l. We can easily estimate just how large kr can be. The maximum radius we deal with is just the maximum nuclear radius and the maximum value of k will be given by the maximum possible kinetic energy of the antineutrino, the Q value for the decay. Taking a mass number of 250 and a Q value of 10 MeV, we find $kr \leq 0.40$. Thus the low-energy behavior of the spherical Bessel functions will indeed govern the amplitudes of the antineutrino wave functions. Further, everything we have concluded concerning the behavior of the neutrino wave functions can be taken over and applied to the wave functions of the charged leptons because the effect of the Coulomb field of the protons has already been included in the Fermi function $F(Z, E)$. In fact, in the limit of a point nucleus, all lepton wave functions *vanish* except for those corresponding to $l = 0$; the only decay that can occur in this case is that for which no orbital angular momentum is carried by either of the leptons. These are the so-called *allowed* transitions, and it is for this reason that the simple Fermi model is referred to as the theory of allowed β decay.

Because of the behavior of the spherical Bessel functions for small arguments, we can conclude that the fastest possible transitions, the most probable transitions, are those for which the leptons carry away no orbital angular momentum whatsoever. When orbital angular momentum must be carried away, much smaller decay constants must be found. A rough estimate of the relative reductions in decay probabilities can be obtained by comparing the squares of the amplitudes of the lepton wave functions at the nuclear radius in a typical case. This is relatively easily accomplished with the well-known approximation

$$x^{-1}j_l(x)\Big|_{x\to 0} \to \frac{1}{1\cdot 3\cdot 5\ldots(2l+1)} \tag{10.47}$$

For a typical case of A = 100, and Q_{β^-} = 4 MeV, one finds that the relative magnitudes of the squares of the numerical coefficients of each of the first three terms in the expansion of Eq. (10.46) are $1/0.017/9.5\times 10^{-5}$ and similar reductions are found for succeeding terms. Thus, all other things being equal, the transition probability will be reduced by roughly *two to three orders of magnitude* for each unit increase in the angular momentum quantum number for the transition. Angular momentum exerts a very much larger effect on β decay than it does on α decay.

10.7.2
Nuclear Matrix Elements: Fermi Transitions

The last factor that we must consider is the nuclear matrix element. To do this properly involves a detailed study of the possible forms of the operators Q_n in Eq. (10.42). This can be found in a number of advanced texts but will not be considered here. Rather, we can use the discussion above and the simple single-particle model to arrive at the two possible forms for Q_n in a very plausible way. Although we will use β^- decay as our example, everything can be carried over to positron emission as well.

We consider the decay of a single neutron in a nucleus. The discussion above strongly suggests that we consider only the case where the emitted lepton pair carries away no orbital angular momentum. If this is the case, there cannot be any difference between the orbital angular momenta of the state of the initial neutron and the state of the final proton. In the single-particle model approximation, we can write the wave function of a particle symbolically in the form ψ_{n,l,j,m_j}. The subscript n numbers which of the states of (l, j, m_j) that is being considered, starting with the first such state found in the potential well. The forgoing says that $l_n = l_p$ and, because of the large energy differences between states corresponding to different n, it is not unreasonable to assume that $n_n = n_p$ as well. Assuming this, we are left with the issue of the intrinsic spins of the emitted leptons. They can be either parallel or antiparallel to one another. If the latter were the case, the total angular momentum of the lepton pair is zero, and there cannot be any difference between the total angular momentum of the initial and final states of the nucleons involved in the decay. That is, the wave function of the state of the proton formed is identical to the state of the initial neutron. The two particles have space and spin properties that are identical to one another. The operator for such a change would have the properties that it simply converts a neutron into a proton in the same space and spin state and creates a lepton pair with antiparallel spins. We show this pictorially in Fig. 10.7. In this schematic, we show the decay of an odd-A nuclide that has one neutron outside of a doubly-magic core. The operator we are considering produces a proton with the same wave function outside the same doubly-magic core. The final nucleus is identical to the initial nucleus except for the exchange of the odd nucleon.

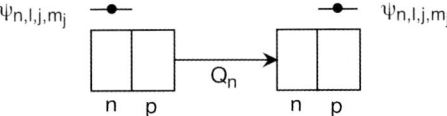

Fig. 10.7 Pictorial representation of the decay produced by the Fermi operator.

Now it is easy to see what the matrix element would be in this case. Letting ψ_{core} represent the doubly-magic core,

$$M_{f,i} = \int \psi^*_{n,l,j,m_j} \psi^*_{core} Q_n \psi_{n,l,j,m_j} \psi_{core} d\tau \qquad (10.48)$$

Because the operator is just converting a particle that resides in the state ψ_{n,l,j,m_j} into another particle with the same wave function, this integral cannot have any magnitude other than unity. That is,

$$M_{f,i} = \int \psi^*_{n,l,j,m_j} \psi^*_{core} \psi_{n,l,j,m_j} \psi_{core} d\tau = \int |\psi_{n,l,j,m_j}|^2 |\psi_{core}|^2 d\tau = 1 \qquad (10.49)$$

Now the fact is that this simplest of operator forms is one of the two that are theoretically possible and it is indeed found experimentally. It is known as the Fermi operator and is frequently written symbolically as $Q_F = \int 1$. The Fermi operator transforms the entire nucleus into a state that differs from the initial state by the replacement of a neutron with a proton and the only changes that can take place are those due to the changes in the Coulomb field.

Given the Fermi operator, we can now actually calculate what the values for $\log_{10} ft$ ought to be for such decays. If there is only one nucleon that can decay, we should find

$$\log_{10}(ft)\big|_F = \log_{10}\left(\frac{\ln 2}{\Gamma}\right) \approx 3.8 \qquad (10.50)$$

A search through the experimental data on the β decay of the lighter nuclei shows a number of clear examples where decay by the Fermi operator is the only possible choice. We can guarantee this if we look at decays involving $0^+ \to 0^+$ transitions. For such transitions there cannot be any net angular momentum carried away by the leptons. Further, if we examine nuclides where $N \approx Z$ we are likely to find cases where the final nucleon state is not occupied and thus is available for the decay. In Fig. 10.8 are shown partial decay schemes for the positron decay of five (even,even) nuclides. In each case there is an excited 0^+ level in the daughter that permits the $0^+ \to 0^+$ transition. And for all of these cases, the value of $\log_{10} ft$ derived experimentally is 3.5, very close to the value predicted by our very simple model.

It is usual to describe most transitions by a set of so-called *selection rules* that give the angular momentum and parity changes associated with them. For Fermi transitions, there is no difference between the angular momentum of the initial and final nuclear states.

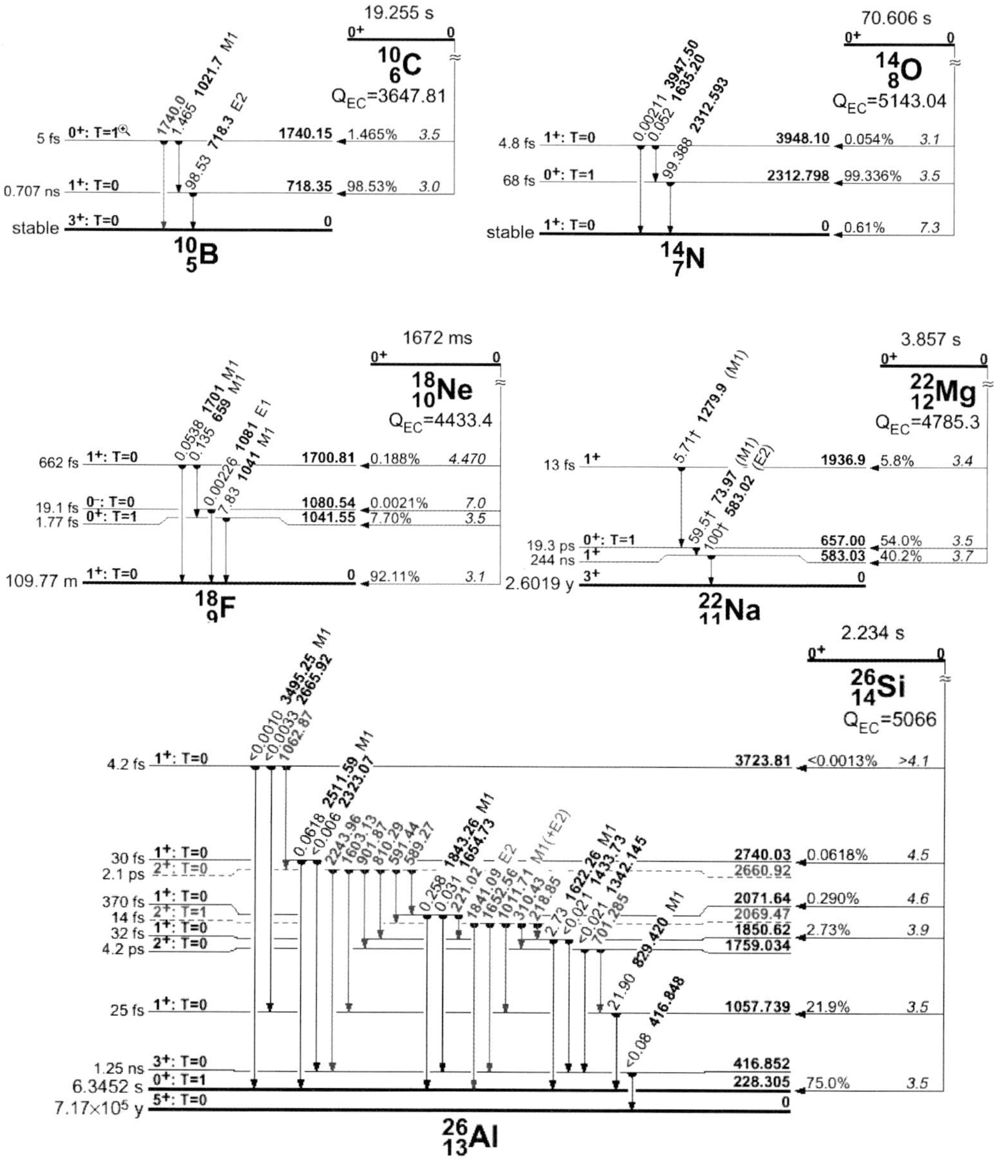

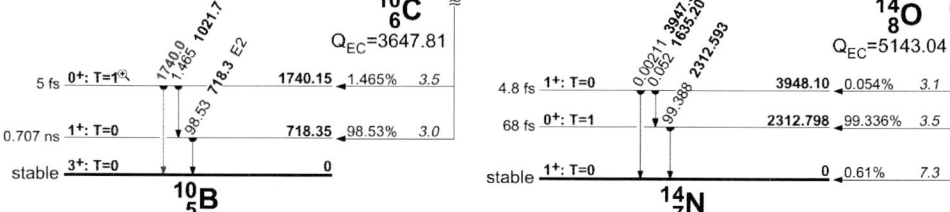

Fig. 10.8 Decay schemes of $^{10}_{6}$C, $^{14}_{8}$O, $^{18}_{10}$Ne, $^{22}_{11}$Mg and $^{26}_{14}$Si, each decaying by a fast $0^+ \to 0^+$ transition.

Because no orbital angular momentum is carried by the emitted leptons, they must be emitted as s waves with respect to the decaying nucleus. As a result the selection rules for Fermi decay are summarized by the simple statement $\Delta J = 0$ (no) where "no" stands for no change in parity between the nuclear states.

10.7.3
Nuclear Matrix Elements: Gamow–Teller Transitions

We now consider the possibility that the leptons are emitted with parallel spins. This means that there must be an angular momentum change of unity between the initial and final nuclear states. But this does not necessarily imply that the angular momentum quantum number of the two states will differ. Remember that angular momentum conservation means that the *vector sum* of the angular momenta of the particles in the final state must equal the angular momentum vector in the initial state. Therefore, in addition to the case where the angular momentum quantum number of the two states differ by unity, we can also have the case where the angular momenta of the states are the same but are connected by the momentum triangle shown in Fig. 10.9.

What is implied with respect to the state of the final nucleon if the leptons carry away one unit of angular momentum? We still maintain that no orbital angular momentum is carried by the leptons and thus we still have $l_n = l_p$. It is possible that angular momentum conservation is maintained by the simple vector addition shown in Fig. 10.9 and that the wave functions of the initial and final nucleons are identical, just as in the case of Fermi decay. But it is also possible that the angular momentum change is accomplished by a reorientation of the orbital and intrinsic spin angular momentum vectors. This means, for example, that a neutron in a state with $j = l + 1/2$ can decay to a proton in a state $j = l - 1/2$. Whereas a neutron in a $p_{3/2}$ state can decay only to a proton in a $p_{3/2}$ state by Fermi decay, an operator that permitted leptons to be emitted with parallel spins would allow decay to produce a proton in a $p_{1/2}$ state as well. This is shown pictorially in Fig. 10.10.

As it turns out such an operator does indeed exist and it is called the Gamow–Teller operator. It is proportional to the spin operator **s**. In our notation, it is written as $Q_n \boldsymbol{\sigma}$ where $\boldsymbol{\sigma} = 2\mathbf{s}$ is called the Pauli spin operator. The matrix elements of this operator are rather complex to derive and we will not go into them in detail. Suffice it to say that in the simple single-particle limit they too are on the order of unity when only a single particle can undergo the transition in question.

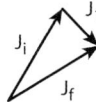

Fig. 10.9 Conservation of angular momentum with no change in the angular momentum quantum numbers of the initial and final nuclear states. Here $J_i = J_f$ and J_1 represents the total angular momentum of the emitted lepton pair corresponding to an angular momentum quantum number of $J = 1$.

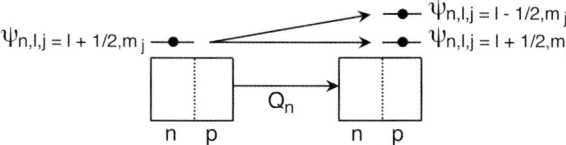

Fig. 10.10 Decay by an operator that creates leptons with parallel spins.

To demonstrate the existence of the Gamow–Teller operator we need to find examples where the only means by which β decay can occur is through its existence. We can do this by looking for cases where the initial and final states have angular momentum quantum numbers that differ by unity and that the only way this can occur is by emission of the leptons with parallel spin vectors. This is exactly what must occur in transitions of the type $0 \leftrightarrow 1$ where the parity of the two nuclear states is the same. We can find examples in each of the decay schemes shown in Fig. 10.8. For each parent decay, there is at least one 1^+ level in the daughter that is populated with a $\log_{10} ft$ between 3.0 and 3.5, somewhat faster than the Fermi transitions in the same nuclides. But note also that not all of the $1^+ \to 0^+$ transitions are so fast. For example, the decay of ^{14}O to the ground state of ^{14}N is slower by over four orders of magnitude than the decay to the 1^+ level at 3.9478 MeV. This must mean that the wave function of the final state somehow differs significantly from that of the initial state.[3] This is a very important example of one of the most common findings concerning β decay; the transition rates are very sensitive to the details of the structure of the parent and daughter nuclides. While Gamow–Teller transitions in light nuclei are among the fastest known, Gamow–Teller transitions in heavier nuclei are always slower, generally by 1–2 orders of magnitude.

The selection rules for Gamow–Teller transitions allow for a difference in the angular momentum quantum numbers of the initial and final nuclear states of 0 or 1 and of course both states must have the same parity. However, because it is impossible to obtain a total lepton angular momentum quantum number of unity when both the initial and final nuclear states have zero angular momentum, the selection rules for Gamow–Teller transitions are written as $\Delta J = 0,1$(no) except that transitions of the type $0^+ \to 0^+$ are absolutely forbidden.

10.8
Classification of β Transitions and Experimental $\log_{10} ft$

Thousands of β-decay transitions have been studied experimentally and are found to have widely varying half-lives, even for transitions with the same ΔJ, parity changes and decay energies. These observations show that the details of nuclear structure have a very strong effect on the nuclear matrix elements for β decay. Nevertheless, there are some general and useful correlations that permit at least a broad overview of the characteristics of β transitions and in Table 10.1 are shown classifications from an authoritative review [2]. The first two columns give the changes in angular momentum and parity between the initial and final nuclear states. The third column gives the name assigned to each class of transitions. The first entry corresponds to the allowed transitions that we have discussed in the previous two sections of this chapter. They are transitions for which the leptons

---

3) Although not treated here, the difference between the probabilities of decay to the two 0^+ levels in ^{14}N and certain other transitions in light nuclei can be traced to the near charge independence of the nuclear force. This gives rise to the existence of an additional, approximately conserved quantum number called the *isospin*.

Table 10.1 Classification of β transitions.

ΔJ	Δπ	Name	Range of $\log_{10} ft$	Comments
0,1	no	allowed	≥ 2.9	generally 3.0–4.5
$0^+ \to 0^+$	no	isospin forbidden	≥ 6.5	found among the light nuclei
0,1	yes	first-forbidden, non-unique	≥ 5.1	(Z ≥ 80) generally 6.0–8.0
0,1	yes	first-forbidden, non-unique	≥ 5.9	(Z ≤ 80)
2	yes	first-forbidden, unique	≥ 8.5	
2	no	second-forbidden, non-unique	≥ 11.0	
3	no	second-forbidden, unique	≥ 12.8	

carry no orbital angular momentum and they include the transitions brought about by both the Fermi and Gamow–Teller operators. The next group labeled "isospin forbidden" essentially represents $0^+ \to 0^+$ transitions where the initial and final states of the decaying nucleon do not correspond to the same space and spin-wave functions. The third and fourth groups labeled "first-forbidden non-unique" represent the great bulk of β transitions found experimentally. They correspond to transitions for which the lepton pair is forced to carry away one unit of orbital angular momentum. The succeeding groups are each characterized by increasingly smaller transition rates. The term "non-unique" is assigned to transitions where $\Delta J^{\Delta\pi} = 0^-, 1^-, 2^+, 3^-, \ldots$ whereas the term "unique" is assigned to transitions with the opposite parity change.

The fourth column gives the lower limit of the $\log_{10} ft$ found experimentally for each transition type. This is because nuclear structure, as reflected in the nuclear matrix elements, can cause significant and sometimes very large reductions in transition rates but they cannot cause an increase above the rates set by the most favorable structure configurations. The lower limits display a very definite pattern. The minimum ft increase by about 1-3 orders of magnitude with each increase in the degree of forbiddenness. This is just what is expected from the decrease in the probabilities for finding leptons with increasing l within the nuclear volume (Section 10.71).

10.9
Electron Capture Decay

Electron capture decay is complicated by the fact that we must account for the wave functions of the atomic electrons. Because the weak interaction is a local interaction, decay must take place at the site of the decaying nucleon. Thus, the probability for electron capture decay will be proportional to the probability for finding atomic electrons within the nuclear volume. This will generally be highest for the most tightly bound electrons that have the smallest mean radii, such as those in the $1s_{1/2}$, $2s_{1/2}$, $2p_{1/2}$, and $2p_{3/2}$ orbitals. Because electron binding energies increase with atomic number, we expect and indeed find that the probabilities for finding the electrons within the nuclear volume are much larger for high-Z nuclides. This is a second reason why electron capture is generally favored in high-Z nuclides when positron emission is energetically allowed.

Electron capture can be written symbolically as

$$p|_{nucleus} + \text{atomic electron} \rightarrow n|_{nucleus} + \nu \tag{10.51}$$

and in a ground-state to ground-state transition, the only particle that is emitted is the neutrino that is not detected under normal conditions. Detection of the decay is usually accomplished by measurement of the radiations associated with the rearrangement of the electrons in the final atom. If decay leaves the daughter nucleus in an excited state, detection is usually accomplished by measurement of γ-ray emission. Before we delve into the simple theory of electron capture decay, we first review the processes of x-ray emission and Auger electron ejection, and the spectrum of atomic radiations following electron capture.

10.9.1
X-ray Emission

Consider a neutral atom of a nuclide that is unstable with respect to electron capture. We assume that both the atom and the nucleus are in their respective ground states. In principle, any of the atomic electrons is available for capture so long as the decay energy is larger than the binding energy of the electron in the parent atom. If Q_{EC} is the energy-equivalent of the mass difference between the ground states of the parent and daughter atoms, the energy available for capture of an electron from the ith orbital in the parent atom is $Q_{EC} - \varepsilon_i$ where ε_i is the binding energy of the electron. The energy balance in electron capture decay is then

$$Q_{EC} - \varepsilon_i = [M(Z, A) - M(Z-1, A)]c^2 - \varepsilon_i = E_{\nu, i} \tag{10.52}$$

where we have neglected the truly negligible recoil energy imparted to the daughter nucleus.

In Table 10.2 we have listed the binding energies of the most tightly bound electrons in selected atoms. The first column in the table lists the normal spectroscopic designation of the atomic orbitals with the same interpretation as given for single-particle states in the nucleus, except for the leading digit that represents the ordering of the states in the atom. The second column lists the designation of the orbits according to symbols common to atomic spectroscopy starting with K and following in alphabetical order. The first atomic shell is the K shell and is the $1s_{1/2}$

Table 10.2 Binding energies of the most tightly bound atomic electrons in selected elements (keV).

Orbital	Spect. symbol	$_{20}$Ca	$_{26}$Fe	$_{50}$Sn	$_{82}$Pb	$_{92}$U
$1s_{1/2}$	K	4.0381	7.1120	29.2001	88.0045	115.6020
$2s_{1/2}$	L_1	0.4378	0.8461	4.4647	15.8608	21.7580
$2p_{1/2}$	L_2	0.3500	0.7211	4.1561	15.2000	20.9480
$2p_{3/2}$	L_3	0.03464	0.7081	3.9288	13.0352	17.1680
$3s_{1/2}$	M_1	0.0437	0.0929	0.8838	3.8507	5.5480
$3p_{1/2}$	M_2	0.0254	0.0540	0.7564	3.5542	5.1810
$3p_{3/2}$	M_3	0.0254	0.0540	0.7144	3.0664	4.3040
$3d_{3/2}$	M_4		0.0036	0.4933	2.5856	3.7260
$3d_{5/2}$	M_5		0.0036	0.4848	2.4840	3.5500
$4s_{1/2}$	N_1			0.1365	0.8936	1.4410

orbital. The second is the L shell and the subscripts 1, 2 and 3 refer to the $2s_{1/2}$, $2p_{1/2}$ and $2p_{3/2}$ orbitals, or subshells, respectively. Electrons in the K shell have binding energies that range from about 13.6 eV in hydrogen to well over 100 keV in the heaviest elements. Electrons in the L shell are less tightly bound by factors of 5–10 as one proceeds from calcium (Ca) to uranium (U). This results predominantly because the presence of the K-electrons "screen" the L electrons from the full Coulomb field of the nucleus. The effect of screening by both the K and L electrons is clearly evident in the binding energies of electrons in the M and higher shells as well.

The data in Table 10.2 show that the difference in the decay energy available for electron capture in different shells is not very large among the low-Z elements but is quite significant in the higher-Z elements. For elements with Z ≥ 80, for example, the decay energy must be larger than about 80 keV in order that K capture can occur. If Q_{EC} is smaller than this, capture can only occur from the L and less tightly bound shells.

The variation in electron binding energies reflects, in part, the probabilities for finding the electrons within the nuclear volume and thus the probability that electron capture from different shells can occur. The relative probabilities calculated for finding electrons within the nuclear volume for the shells given in Table 10.2 are given in Table 10.3. For each atom, the probability for finding the K electrons within the nuclear volume is seen to be a factor of 6–10 larger than for electrons in

Table 10.3 Relative probabilities for finding atomic electrons within the nuclear volume [3].

Shell	$_{20}$Ca	$_{26}$Fe	$_{50}$Sn	$_{82}$Pb	$_{92}$U
K	1.342e-2	3.2636e-2	0.403	5.511	12.234
L_1	1.229e-3	3.466e-3	4.994e-2	0.8781	2.153
L_2	3.229e-6	1.618e-5	1.112e-3	6.647e-2	0.2248
L_3	1.179e-5	5.603e-5	2.713e-3	7.268e-2	0.173
M_1	1.973e-4	5.649e-4	1.045e-2	0.2125	0.5332
M_2	3.609e-7	2.245e-6	2.4e-4	1.759e-2	6.141e-2
M_3		7.645e-6	5.964e-4	2.055e-2	5.166e-2
M_4		1.602e-9	1.347e-6	1.728e-4	5.813e-4
M_5			4.464e-6	4.346e-4	1.298e-3
N_1			2.23e-3	5.849e-2	0.1547

any other shell. For electrons within the L, M, N shells, etc., those in the "1" subshell, the electrons in that shell's s-orbit, have the largest probability of being found within the nuclear volume. Further, the probabilities increase dramatically with increasing atomic number. We can conclude from these calculations that, for the neutral atoms, K capture will be the most probable so long as it is energetically allowed, and that the relative probabilities for capture will generally follow the order K > L > M > N…

With this in view, we now consider the effect of electron capture on the daughter atom. Regardless of the state of excitation of the nucleus, the direct result of the capture is the production of a vacancy in one of the inner electron shells. Because Z has been reduced by one, the atom is electrically neutral. However, it is very highly excited. This state will de-excite by any and all decay processes that are allowed. The mode of de-excitation that is most commonly discussed is that of x-ray emission in which an electron from a less tightly bound state drops down to fill the vacancy and the difference in binding energy between the two states is emitted as a photon called a characteristic x-ray.

X-rays are named for the shell in which the initial vacancy was produced. Hence, a K x-ray can be emitted whenever a vacancy is produced in the K shell, regardless of the subshell from which the electron that fills the vacancy arises. L x-rays will be emitted when electrons from the M, N,… shells fill an initial vacancy anywhere in the L shell, etc. As you can imagine, there are a very large number of x-rays that can be emitted. It is not terribly important for our purposes that all of these names be

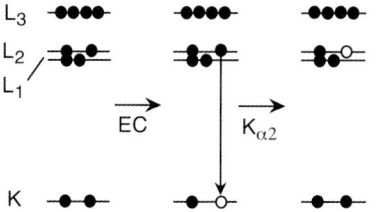

Fig. 10.11 Schematic representation of the change in the electron structure of an atom as a result of electron capture in the K shell followed by emission of a $K_{\alpha 2}$ x-ray. The separation in energy of adjacent levels is not to scale.

given [4]. Suffice it to say that the names are associated with both the energies and relative intensities of grouping of the x-rays. For example, the subscript "α" on K refers to x-rays that arise from K vacancies being filled by L electrons. The subscripts "1" and "2" stand for the orbital from which the electron that fills the vacancy arises, namely the L_2 ($2p_{1/2}$) and L_3 ($2p_{3/2}$) subshells, respectively.

As an example, the diagram in Fig. 10.11 represents $K_{\alpha 2}$ x-ray emission following K capture in which an L_2 electron fills the vacancy in the K shell. Although one might think that there are three possible x-rays that can be emitted following K capture when the vacancy is filled by L electrons, there are only two. This arises because, as will be discussed further in the next chapter, photons must carry at least one unit of orbital angular momentum relative to the emitting center. Because the L_1 shell is the 2s shell, x-ray emission in the transition of a 2s electron into the 1s orbital cannot occur. The only transitions that can give rise to x-ray emission involve L_2 ($2p_{1/2}$) and L_3 ($2p_{3/2}$) electrons.

In the absence of any energy dependence on the probability for x-ray emission, one would expect that the intensity of these two x-rays, $K_{\alpha 2}$ and $K_{\alpha 1}$, would be in the ratio of 1/2 because there are twice as many electrons in the L_3 shell as in the L_2 shell. This is essentially what is found experimentally because the energy difference between the two p subshells is so small.

10.9.2
Auger Electron Ejection

Not all vacancies in an inner electron shell give rise to the emission of x-rays. In place of photon emission, the difference in binding energies between the initial and final states of the electron can be transferred directly to a second electron resulting in its ejection from the atom. No intermediate photon is emitted. This process is commonly referred to as Auger electron ejection[4], an example of which is shown in Fig. 10.12. There a vacancy in the K shell is filled by an electron from the L_1 shell with the ejection of a L_2 electron, a so-called KL_1L_2 transition. The kinetic energy of the ejected electron is given by

---

4) In the special case where an initial vacancy in the L shell is filled by electron ejection and results in a vacancy in a less tightly bound L subshell, the process is referred to as a Coster–Kronig transition.

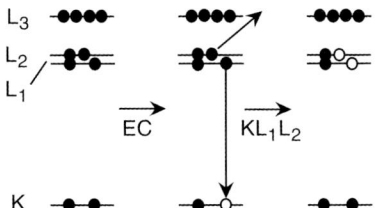

Fig. 10.12 Schematic diagram representing an Auger electron ejection process. In this case an initial vacancy in the K shell is filled by an electron from the L_1 shell with the ejection of an L_2 electron. The transition is referred to as a KL_1L_2 transition.

$$E_e = \varepsilon_K - \varepsilon_{L_1} - \varepsilon_{L_2} \tag{10.53}$$

where the ε_i are the binding energies of the electron subshells[5].

The probabilities for emission of x-rays and Auger electrons are quite complicated to calculate but they have been measured and tabulated. The parameter used to describe the relative probability for x-ray emission is the *fluorescence yield*. It is defined as the probability for emission of a particular x-ray given an initial vacancy in an electron shell, and is given the symbol ω_i, where i represents the shell or subshell in which the initial vacancy is created. Formally,

$$\omega_i = \frac{\lambda_x}{\lambda_x + \lambda_e} \tag{10.54}$$

where λ_x and λ_e represent the decay probabilities for emission of x-rays and the total probability that the initial vacancy will be filled by Auger electron ejection, respectively. The fluorescence yields for the K and L shells are shown in Fig. 10.13. The fluorescence yield for the K shell is negligible for the lightest elements but increases rapidly with Z starting around calcium (Z = 20). For the heavier elements the majority of vacancies in the K shell will be filled by x-ray emission. On the other hand, the probabilities for x-ray emission when a vacancy is created in the L shell are relatively small everywhere except for the heaviest elements.

The variation in the fluorescence yield has implications with respect to the measurement of electron capture decay, to problems where the deposition of energy in a material is an important issue, and to problems where the chemistry of the daughter atom is important. The small fluorescent yield among the light elements makes the observation of electron capture decay quite difficult because the most common detectors generally available are photon detectors and thus very weak signals can be expected. For example, the nuclide ^{55}Fe decays by electron capture with $Q_{EC} = 231.38$ keV. It decays to the ground state of ^{55}Mn in almost 100% of all decays. (Decay does occur to a ^{55}Mn level at an excitation energy of 125.85 keV but only in 1.3×10^{-7} of all decays!) The most intense x-rays emitted have energies

---

[5] Some care must be taken in performing the calculations indicated in Eq. (10.53). The electron binding energies given in Table 10.2 and normally quoted refer to the binding energies of neutral atoms. However, once a vacancy is created by Auger electron ejection, the binding energies are those of the resulting ion.

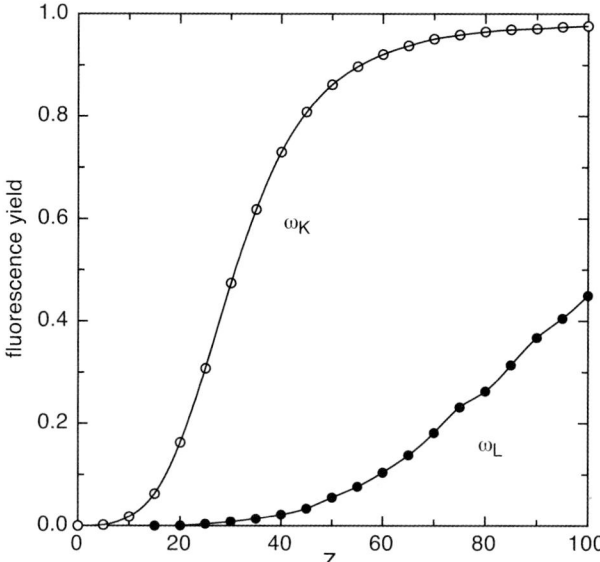

Fig. 10.13 K shell and average L shell fluorescence yields for the elements Z ≤ 100.

of only about 0.59 keV. It is usually rather difficult to measure x-rays with energies below about 1–2 keV, and when fluorescence yields are very small the measurement problem can become severe.

An introduction to the interaction of radiation with matter will be presented in Chapter XIV. For the present, we note only that the linear dimensions needed to attenuate photons is usually much larger than the dimensions required to attenuate charged particles of similar energies. When the fluorescence yield is small and most of the observable energy in an electron capture decay is represented by ejected electrons, most of the energy will be deposited much closer to the site of decay than in the case where most of the energy is carried away by x-rays. This can have important implications with respect to radiation damage in biological materials.

Extrapolation of the fluorescence yields shown in Fig. 10.13 to less tightly bound electron orbitals, implies that vacancies produced in shells beyond the L shell will generally undergo de-excitation primarily by the Auger process. This is essentially what is found in experiment. Now consider an atom of medium Z, say 50, in which a vacancy is created in the K shell. The daughter atom is neutral but highly excited. About 83% of the time, the vacancy will be filled by x-ray emission and about 17% of the time by Auger electron ejection. If x-ray emission occurs, a neutral atom is produced mostly by an L electron filling the K vacancy. The hole in the L shell will clearly be filled mostly by Auger electron ejection. But the ion will still most probably be in an excited state and the most probable additional steps in the de-excitation process will again be Auger electron ejection, leading to higher and higher electronic charges. If the hole in the K shell was filled by Auger electron ejection, we would immediately have an ion of charge +1. It, too, will decay mostly by Auger electron ejection, and so on. Indeed, the complete decay process usually leads to net electronic charges very much larger than those corresponding to any of the common oxidation states of the elements in question and that means that the ion will represent a highly reactive chemical state. The fate of the ion in a material will depend very much on its environment and the thermodynamics of the possible reactions that it can undergo. This, again, can have important implications when the decay occurs in or adjacent to molecules of biological importance.

10.10
Elementary Theory of Electron Capture

The theory presented in Sections 10.3 and 10.4 for β^- and positron decay contains most of what is needed to develop an expression for the electron capture decay probability in the long-wavelength approximation. There are two main differences that we must consider. First, the wave function of the captured atomic electron must be taken into account in addition to the wave functions of the initial proton and the neutron produced by the decay. Second, the only lepton created is the neutrino and the density of final states will therefore differ from that derived in Section 10.4. But the *nuclear* matrix element will be not be affected. The interaction that produces the decay is still the weak interaction and the only operators are the Fermi and Gamow–Teller operators. Again we use the Fermi Golden Rule No.2 as our starting point.

We begin by writing down a schematic representation of the matrix element for electron capture as

$$M'_{f,i} = \langle \Phi_{f,\text{total}} | H_\beta | \Phi_{i,\text{total}} \rangle = \int \Phi^*_{f,\text{total}} H_\beta \Phi_{i,\text{total}} d\tau \qquad (10.55)$$

where the wave functions Φ_j represent the entire system, nucleus plus atomic electrons, and we use the prime on the symbol for the matrix element to emphasize that the integrals contain both nuclear and atomic electron wave functions. For H_β we substitute its strength constant g times the operator Q_p that converts a proton into a neutron, and recognize that it produces a neutrino, $\psi_v(r)$, in the final state. We take Eq. (10.55) to represent a transition in which a specific electron is captured and a specific proton is converted into a neutron. As usual, there may be more than one proton that can undergo decay and we must evaluate the matrix element for each and sum them. In addition, there may be more than one electron that can be captured and we must sum over the matrix elements involving each of these as well.

Letting $\psi_{\text{sub},i}$ represent the wave function of the ith electron in the subshell "sub", the matrix element for the capture of this electron can then be written

$$M'_{f,i} = g \int \psi^*_v \psi^*_n Q_p \psi_p \psi_{\text{sub},i} d\tau \qquad (10.56)$$

10.10 Elementary Theory of Electron Capture

We use the long-wavelength approximation to replace the wave functions of both the initial electron and final neutrino by their wave functions at the origin and remove them from the integral. That is,

$$M'_{f,i} = g\psi_v^*(0)\psi_{sub,i}(0) \int \psi_n^* Q_p \psi_p d\tau \tag{10.57}$$

We now evaluate the density of final states for the neutrino emitted into the "continuum". From (10.24), the number of states available for a neutrino with momentum p_v to $p_v + dp_v$ is

$$dn_v = \frac{4\pi V}{(2\pi\hbar)^3} p_v^2 dp_v \tag{10.58}$$

Neglecting the very small rest mass, we can write

$$E_v = p_v c \tag{10.59}$$

Recognizing that the neutrinos are mono-energetic with energies $(E_{EC} - \varepsilon_{sub})$, where E_{EC} is the energy available for the decay and ε_{sub} is the binding energy of the electron, we can use Eq. (10.59) to rewrite (10.58) as

$$\rho(E) = \frac{4\pi}{(2\pi\hbar c)^3}(E_{EC} - \varepsilon_{sub})^2 \tag{10.60}$$

We can now use this and the matrix element from Eq. (10.57) to substitute into the Golden Rule to get directly

$$\lambda_{sub,i} = \frac{g^2}{\pi\hbar^4 c^3}|\psi_{sub,i}(0)|^2|M_{f,i}|^2(E_{EC} - \varepsilon_{sub,i})^2 \tag{10.61}$$

where the prime has been removed from the matrix element to indicate that it now represents the nuclear matrix element alone.

Before we consider the wave functions for the captured electrons it will prove useful to write Eq. (10.61) in a form similar to that of (10.44). With the definitions of Γ (Eq. (10.43)) and the dimensionless energy W (Eq. (10.39)) we find directly

$$\lambda_{sub,i} = \Gamma \frac{2\pi^2 \hbar^3}{m_e^3 c_e^3}|\psi_{sub,i}(0)|^2|M_{f,i}|^2(W_{EC} - W_{sub,i})^2 \tag{10.62}$$

By comparison of Eq. (10.62) with (10.44), it is seen that the quantity

$$\frac{2\pi^2 \hbar^3}{m_e^3 c_e^3}|\psi_{sub,i}(0)|^2(W_{EC} - W_{sub,i})^2$$

plays the same role in electron capture as the integral Fermi function plays in $\beta^\pm$ decay. It contains all of the effects of the leptons on the decay probabilities. It is then natural to write this as the equivalent of the integrated Fermi function $f(Z,W_o)$, i.e.,

$$f_{EC}(Z, W_{EC}) = \frac{2\pi^2 \hbar^3}{m_e^3 c_e^3} |\psi_{sub,i}(0)|^2 (W_{EC} - W_{sub,i})^2 \qquad (10.63)$$

This is a very simple expression. The contribution from the wave function of the captured atomic electron is just a constant, and we then expect to see a quadratic dependence of $f_{EC}(Z, W_{EC})$ on $(W_{EC} - W_{sub,i})$, the neutrino energy. Further, as the decay energy increases we expect to see $f_{EC}(Z, W_{EC})$ vary in proportion to W_{EC}^2 in the practical cases where $W_{EC} \gg W_{sub,i}$.

We are now left with the evaluation of the electron wave function and this is not simple. It must be obtained from detailed computations of the atomic structure of polyelectronic atoms. Nevertheless, this has been accomplished with high accuracy and has led to quite accurate values for $f_{EC}(Z, W_{EC})$. In reality, one does not use the simple point approximation discussed here. The calculations are performed by integrating over the volume of the nucleus to give the integral probability for finding the various atomic electrons within the nuclear volume. By summing the resulting matrix elements for each of the electrons, the total electron capture decay probability can then be calculated.

For our purposes we want to use a simple model to demonstrate the dependence of f_{EC}, and hence λ_{EC}, on both Z and W_{EC}. The simplest reasonable approach is to make use of the long-wavelength approximation, the Bohr model of the atom and restrict attention to just the K electrons. While it neglects screening of the field of the nucleus by other electrons as well as relativistic effects that are important at higher Z, it should provide the right order of magnitude for $f_K(Z, W_{EC})$ and roughly the right dependence on the atomic number of the decay parent.

In most modern physics texts it is shown that the wave function for a single electron in the field of a point nucleus of atomic number Z is

$$\psi_K(r) = \psi_{1s}(r) = \frac{1}{\sqrt{\pi}} \left(\frac{Z}{a_o}\right)^{3/2} e^{-Zr/a_o} \qquad (10.64)$$

where

$$a_o = \frac{\hbar^2}{k_C m_e e^2} \qquad (10.65)$$

is the radius of the first Bohr orbit. At the origin, the wave function for the K electron is simply

$$\psi_K(0) = \frac{1}{\sqrt{\pi}} \left(\frac{Z}{a_o}\right)^{3/2} \qquad (10.66)$$

Substituting Eq. (10.65) to express a_o, and using the relation $r_e = k_C(e^2/m_e c^2)$, a little algebra gives the result

$$|\psi_K(0)|^2 = \frac{(Zr_e)^3}{\pi}\left(\frac{m_ec^2}{\hbar c}\right)^6 \tag{10.67}$$

The dependence on the atomic number to the third power is a reflection of the strong decrease in the average radius of the K orbit as the atomic number increases and the subsequent increase in the probability for finding the electron at and near the nuclear center. The probability for electron capture should increase dramatically with increasing Z. Note that the dimensions of $|\psi_K(0)|^2$ are inverse volume as should be expected. Direct substitution of the result of Eq. (10.67) into (10.63) gives the approximation

$$f_K(Z, W_{EC}) \approx 2\pi(Zr_e)^3\left(\frac{m_ec^2}{\hbar c}\right)^3(W_{EC} - W_K)^2 \tag{10.68}$$

for a single K electron. If we substitute this into (10.63) and substitute the result into Eq. (10.61) along with multiplication by 2 to account for the two K electrons, we obtain

$$\lambda_K \approx 4\pi\Gamma(Zr_e)^3\left(\frac{m_ec^2}{\hbar c}\right)^3|M_{f,i}|^2(W_{EC} - W_K)^2. \tag{10.69}$$

The predictions of this simple approximation are shown in Fig. 10.14 along with those produced by sophisticated calculations. The general features of each curve reflect the characteristics expected from the simple model. First, the functions $f_K(Z, W_{EC})$ must vanish below the K electron binding energy. The curves are seen to rise smoothly and approach a constant slope when the decay energy becomes much larger than the K binding energy. The magnitudes of $f_K(Z, W_{EC})$ span a range that is quite similar to the range spanned by the log ft shown in Fig. 10.5 for $\beta^\pm$ decay.

The quality of the $f_K(Z, W_{EC})$ produced by our simple model varies significantly with Z but generally reflects the right order of magnitude found from the more sophisticated calculations. At low Z, the model tends to overestimate f_K by about a factor of three, primarily because of the neglect of screening of the nucleus by other electrons. For medium-Z elements both calculations give very similar results, while at high Z, the simple Bohr model estimate underpredicts f_K primarily because of the neglect of rather strong relativistic effects.

We could go on to consider the calculation of the decay probabilities for capture of L and M electrons but these should be approached only with the more sophisticated models. Suffice it to say that the capture probabilities decrease in the order K > L > M... as should be expected. Typically, K capture is an order of magnitude more probable than L capture so long as the decay energy is large compared to the K-electron binding energy and in most cases, higher-order capture can be safely neglected.

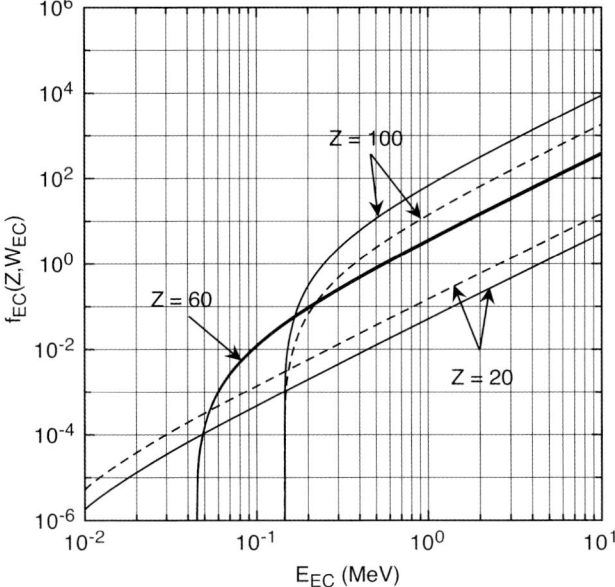

Fig. 10.14 Log–log display of the equivalent of the integrated Fermi function for electron capture of K electrons. The solid curves are based on the calculations of Bambynek, et al. [3] and the dashed curves are based on the simple Bohr model approximation.

10.11
Ratio of Electron Capture to Positron Emission

If $Q_{EC} \geq 2m_ec^2$, the same decay can take place by both positron emission and electron capture. If such competition occurs, the nuclear matrix element will be the same and hence the relative probabilities for these decay modes will be dependent solely upon decay energies, the probabilities for finding electrons within the nuclear volume and the density of final states available for the emitted leptons. If positron emission is possible, the binding energies of the atomic electrons are generally negligible in comparison to Q_{EC}, except for the highest-Z elements. Also, as we have discussed above, K capture will be the dominant mode for electron capture and we can then use our simple Bohr model to estimate the ratio of the decay probabilities. Using the ratio of Eqs (10.69) and (10.44), we immediately find

$$\frac{\lambda_K}{\lambda_{\beta^+}} \approx 4\pi(Zr_e)^3 \left(\frac{m_ec^2}{\hbar c}\right)^3 \frac{(W_{EC} - W_K)^2}{f(Z, W_{EC} - 1)} \tag{10.70}$$

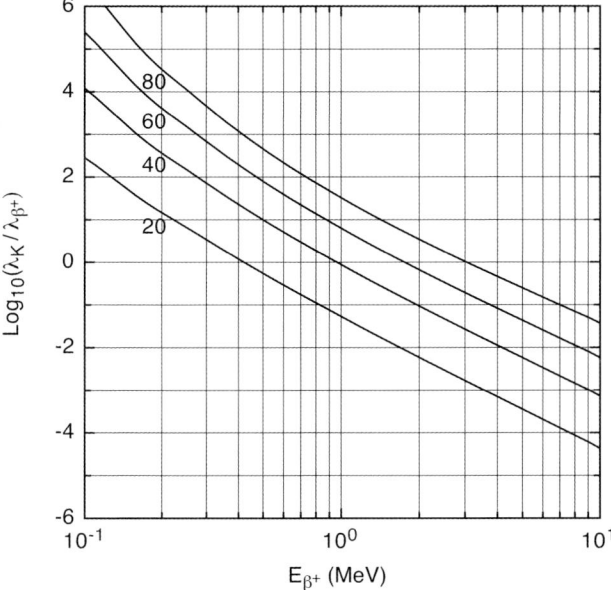

Fig. 10.15 Logarithmic display of the ratios of K-capture to positron emission from the simple model calculations. The abscissa is the decay energy for positron emission. The electron capture decay energy Q_{EC} is greater by $2m_ec^2$.

K capture will tend to dominate the decay as the atomic number becomes large, especially at lower energies. This comes about for two reasons, the most obvious being the dependence of the electron capture decay probability on the third power of Z. But as important is the strong Coulomb repulsion between a positron and the nucleus that reduces the probability for finding low-energy positrons within the nucleus. Reference to Fig. 10.5 shows that, for high Z, the strong repulsion is especially evident within several hundred keV of the emission threshold.

In Fig. 10.15 are shown approximate values for the ratios $\lambda_K/\lambda_{\beta^+}$ obtained from our model calculations of $f_{\beta^+}(Z, W_o)$. These are rough but nevertheless produce ratios that differ from the more accurate calculations by only a factor of 2–4. The ordinate in the figure is the energy available for positron emission. That means that Q for the competing K capture will be larger by $2m_ec^2$. Regardless of the atomic number, electron capture always dominates at energies near the threshold for position emission. For positron decay energies less than about 200 keV, K capture will dominate except for the very lightest nuclides. Even for positron decay energies as large as 1 MeV, K capture is observed to dominate the decay in the higher-Z elements. The overall conclusion is that electron capture will be the dominant decay mode at low energies throughout the chart of the nuclides. For nuclides near the region of β stability, positron emission will compete successfully only among the lighter nuclides.

10.12
β-Decay Schemes

The models we have developed can be used to provide a rather good qualitative and sometimes semiquantitative understanding of the details of the decay of some β-unstable nuclides. In most cases, the strong dependence of decay probability on the details of the wave functions of nuclear states introduces complications that simply have not been considered here and quantitative agreement will not be obtained. Nevertheless, our models do serve as a starting point for correlating experimental data, and in the following pages we will examine a number of decay schemes to illustrate how the simple shell model and the allowed theory of β decay can be used for this purpose.

The simplest case to consider is the decay of the neutron itself. It decays by $β^-$ emission with a half-life of 10.37 min. There are no excited states of the proton, and thus there is one and only one possible state that can be formed. The decay is summarized in Fig. 10.16 where the decay energy is given in MeV. In this particular case, we can be absolutely sure that both the neutron and proton are in $1s_{1/2}$ states. In general, one usually writes down the *configuration* of the neutrons and protons in both the parent and daughter states and the decay mode that connects them. The configuration is simply a shorthand notation for describing the states occupied by the nucleons and in the present case, we have

$$n: 1s_{1/2} \to p: 1s_{1/2}.$$

The decay takes place without any orbital angular momentum change and therefore it is an allowed decay. It can take place by both the Fermi and Gamow–Teller operators. Because we know the initial and final states exactly, we can expect the β transition to have a $\log_{10} ft$ of about 3.0–3.5. Experimentally, it is found to be $\log_{10} ft = 3.0$. It is one of the "fastest" β transitions known and is often referred to as a "super allowed" transition.

A second and very simple decay is that of tritium with the decay scheme shown in Fig. 10.17. There are no bound excited states of ^{3}He, as you can easily conclude

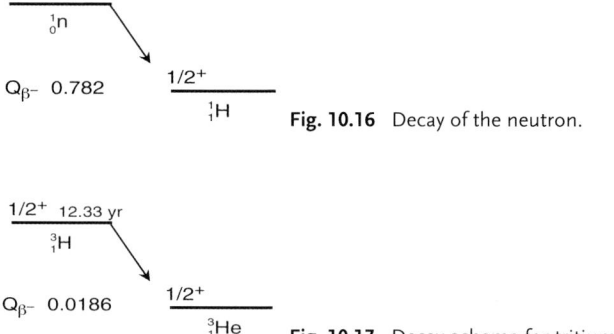

Fig. 10.16 Decay of the neutron.

Fig. 10.17 Decay scheme for tritium.

by use of the particle in the box model. We can write down the particle configurations involved by use of the simple shell model level diagram given in Fig. 7.13 and symbolize the decay as

$$\text{n: } 1s^2_{1/2} \text{ p: } 1s^1_{1/2} \rightarrow \text{n: } 1s^1_{1/2} \text{ p: } 1s^2_{1/2}$$

The two neutrons in tritium should be paired in the $1s_{1/2}$ orbital and the single proton is also in a $1s_{1/2}$ orbital. Just the reverse is true for ^{3}He; these nuclides form a simple mirror pair. It is then easy to see that, in the limit of the simple shell model, the n → p transition in this decay should be identical to that found in the decay of the free neutron. Now although we do not necessarily know the wave functions exactly, we can be quite sure that this picture must be quite close to the truth. We therefore expect that this too is a "super allowed" decay and should have a $\log_{10}$ ft of 3.0–3.5. Experimentally, $\log_{10}$ ft = 3.1 and the transition is very nearly as "fast" as that for the neutron decay. That the half-life of ^{3}H is so very much longer than that of the neutron is due almost entirely to the very low decay energy of 0.0186 MeV as compared to 0.782 MeV for the neutron. Because the Fermi function must be close to unity in both decays, it is the very much smaller density of states available to the leptons emitted in the decay of tritium that causes the half-life to be so much longer than that of the free neutron.

The electron capture decay of $^7_4$Be shown in Fig. 10.18 is a slightly more complicated case but it also shows somewhat more of how our models can be used to understand and correlate experimental data. The nuclides $^7_4$Be and $^7_3$Li are again mirror nuclides. Electron capture decay is observed to both the ground and first-excited states of $^7_3$Li in roughly the ratio of 8.6/1. Because $Q_{EC} < 2m_ec^2$, positron emission is forbidden. Both of the observed transitions satisfy the selection rules for allowed transitions.

We can again use the single-particle shell model to write down the particle configurations for both nuclides as follows;

$$\text{ground state of } ^7_4\text{Be - n: } 1s^2_{1/2}1p^1_{3/2} \text{ p: } 1s^2_{1/2}1p^2_{3/2}$$

$$\text{ground state of } ^7_3\text{Li - n: } 1s^2_{1/2}1p^2_{3/2} \text{ p: } 1s^2_{1/2}1p^1_{3/2}$$

Clearly the ground-state configuration of $^7_3$Li can be formed in electron capture by conversion of a $p_{3/2}$ proton into a $p_{3/2}$ neutron. This again is an allowed transition

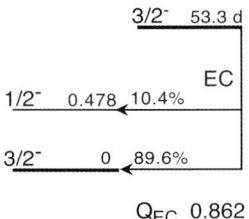

Fig. 10.18 Decay scheme of $^7_4$Be. The γ-ray transition between the level at 0.478 MeV and the ground state is not shown for simplicity.

and can occur by both the Fermi and Gamow–Teller operators. Again, because the nucleus is simple, we can expect that the shell-model representation is a reasonable approximation and we expect this decay to be "super allowed" with $\log_{10}$ ft of 3.0–3.5. It is found to be about 3.3.

Now consider the first-excited state of ^7_3Li at 0.478 MeV. It has the spin and parity of $1/2^-$. Examination of the single-particle level diagram shows that the $1p_{1/2}$ state lies just above the $1p_{3/2}$ state and one can imagine that the first-excited state could be derived from the ground-state configuration by simple excitation of the $1p_{3/2}$ proton into this level, i.e., n: $1s^2_{1/2} 1p^2_{3/2}$ p: $1s^2_{1/2} 1p^1_{3/2}$ → n: $1s^2_{1/2} 1p^2_{3/2}$ p: $1s^2_{1/2} 1p^0_{3/2} 1p^1_{1/2}$. If this were the case, it can be seen that a fairly complicated decay process would be required to populate it. There are two $p_{3/2}$ protons in the ground state of ^7_4Be. To reach the configuration guessed for the excited state, one of the protons would decay into a $p_{3/2}$ neutron while the other would simultaneously have to be excited into the $1p_{1/2}$ state. This is obviously not something that is directly possible by the Fermi or Gamow–Teller operators and, if it did occur, we can expect it to be a rather hindered process. But experiment shows that the $\log_{10}$ ft is 3.5.

A tell-tale problem that strongly suggests that our guess for the configuration of this state is wrong is that the excitation energy of the level is so low. This is a very small nucleus. Reference to the level energies expected from the spherical potential well model would suggest that the energy separation between the $1p_{1/2}$ and $1p_{3/2}$ levels ought to be about 2–3 MeV and not 0.5 MeV. Indeed the next excited state of ^7_3Li is found at an energy of about 4.63 MeV. The strong inference is that this level is not due to the simple proton excitation considered above.

Remember that the first-excited state of any nucleus is that state which requires the minimum energy to achieve it. Is there a less energetically costly way to exite ^7_3Li? The answer is yes and it is fairly easy to see what it could be. In the ground state there are two $1p_{3/2}$ neutrons in a level that can be occupied by four particles. These two particles are paired and it will cost on the order of 1 MeV or so to break the pair. Now as we discussed in Chapter VI, the possible total angular momentum quantum numbers for two identical particles in an orbit of angular momentum j are the even values between the limits of 0 and 2j. For two $p_{3/2}$ neutrons that means that the couplings could be 0, the paired coupling in the ground state, and 2, both with even parity. Suppose that we form the latter coupling. We could write the configuration of the nucleus as

$$\text{n: } 1s^2_{1/2} 1p^2_{3/2}\Big|_{2^+}, \text{ p: } 1s^2_{1/2} 1p^1_{3/2}$$

Now, rather than the total angular momentum of the nucleus being due to a single unpaired nucleon, the angular momentum will be given by coupling the angular momenta of the uncoupled neutron pair and the lone $p_{3/2}$ proton. The angular momentum states that are possible are $2 \pm 3/2$, or $1/2, 3/2, 5/2$ and $7/2$, all with *odd* parity. All of these configurations will indeed be found as excited states, not necessarily as pure "single particle excitations" but they will be represented in the excited states of the system.

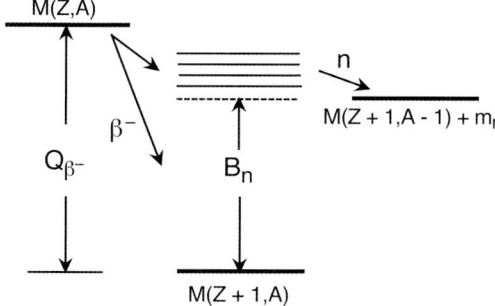

Fig. 10.21 Schematic diagram of the energetics necessary for β⁻-delayed particle emission. The dashed level at the excitation energy B_n is hypothetical.

sion. Once the neutron unbound state is populated, it decays in a time extremely short compared to the half-life of the β⁻ decay parent. Thus, neutron emission follows the half-life of the parent; neutron emission is "delayed" until β⁻ decay occurs.

There is really nothing magic about the neutron emission. In principle, one can start with the nucleus M(Z + 1,A) in its ground state and, by some means, add energy until it is excited above the neutron binding energy. Even though the nucleus is "particle unbound", it still has discrete states that will decay by any and all processes that are allowed. In general, the two most important decay modes for levels with energies not too much larger than B_n, are neutron emission and γ decay, the former being much more probable in most cases. The real difference between β⁻-delayed neutron emission and other means of exciting the daughter nucleus is that the β⁻ decay will generally populate states only by allowed transitions and thus it "selects" out only a portion of all the excited states that exist. This restricts the spins and parities of the levels that, in turn, select the spins and parities of the neutrons that are emitted.

The β⁻-decay parent is commonly referred to as the delayed neutron precursor and a very large number of them have been identified among the products of nuclear fission. The longest-lived precursor is 55.7 s $^{87}_{35}$Br with Q_{β^-} = 7.17 MeV. The β⁻-decay daughter $^{87}_{36}$Kr has 51 neutrons, just one greater than the magic number of 50 and the binding energy of the 51st neutron is about 5.15 MeV. Some 2.6% of all decays of $^{87}_{35}$Br lead to delayed neutron emission. The next longest-lived precursor is 24.5 s $^{137}_{53}$I and it leads to neutron emission from the daughter $^{137}_{54}$Xe in about 6.4% of all decays. One will note, once again, that nuclear-shell structure is at play in this decay.

β-delayed particle emission is not restricted to neutrons. A large number of β-delayed proton emitters are known among the relatively light neutron deficient nuclides. The process is very much the same as described above. The highest energy states in the daughter are populated almost solely by allowed transitions. However, the presence of the Coulomb barrier hinders proton emission significantly, and proton emission is generally not seen for levels just above the proton

What makes this excitation interesting is that it is a characteristic of all known N = 49 and Z = 49 nuclides. They all show the same level sequences, a consequence of shell structure. Now the angular momentum difference between the two states is 4, and as we will see in the next chapter, photons can be described to an excellent approximation as undistorted plane waves. As such, they can be expressed in terms of waves of definite angular momentum as was shown in Section 10.71 for the neutrino wave functions. And, just as in the case of the neutrinos, the behavior of the spherical Bessel functions at small values of the arguments kr implies that the emission of photons that carry away large quantities of angular momentum ought to be highly hindered. All of this will be shown to be the case and that is the cause of very long half-life for the first-excited state in $^{85}_{36}Kr$. The decay by photon emission to the ground state is so slow that 79% of all decays take place by β^- decay. Remember, any and all decay modes that are energetically allowed and are not forbidden by conservation laws, do indeed occur.

10.13
β-Delayed Particle Emission

The discussion of β decay cannot be considered complete without at least a brief introduction to the phenomenon of β-delayed particle emission. This is not only an interesting decay mode from the point of view of nuclear science, but it is the single most important factor that makes possible the safe control of the fission chain reaction in nuclear power reactors. It was first predicted to exist and indeed found among the fission products themselves. We will use β^- decay as our example.

A hint of what might be possible is already evident in the decay scheme for $^{85}_{35}Br$ (Fig. 10.19). We have noted that a number of levels at rather high excitation energy were populated by normal allowed transitions (see the decay to the levels at 2.03196 and 2.13734 MeV in $^{85}_{36}Kr$). Together they account for about 2.65% of all decays. Now we know from the semi-empirical mass formula that, as we move away from the valley of β stability, the decay energy is expected to increase linearly with the difference between the atomic number of the parent and the atomic number of the hypothetical most stable isobar, Z_A. Further, the more neutron-rich the isotope, the lower will be the binding energy of the most weakly bound neutron. In the event that relatively fast allowed transitions continue to be found to levels at high excitation energy in the β^- decay daughter, we can reach the situation shown schematically in Fig. 10.21 where

$$Q_\beta > B_n|_{daughter}. \tag{10.71}$$

In this event, any decay to levels with excitation energies $E_x > B_n$ leave the daughter with sufficient energy that a neutron can be emitted. We say that these levels are particle-unbound levels and in most cases they do indeed decay by neutron emission with high probability to levels near the ground state in the neutron decay daughter M(Z + 1, A – 1). This decay sequence is called β^--delayed neutron emis-

daughter are energetically accessible and direct β decay to the ground and 10–12 excited states has been observed experimentally. The absolute β intensity, the fraction of total decays to each level, is shown to the left on the arrow leading to each level. To the right of each of these is the $\log_{10} ft$ determined from the experimental data. You will note that the $\log_{10} ft$ span the range 4.8–8.0 and include a number of transitions for which the changes in spins and parities signify that the transitions are allowed. Some of these populate levels at rather high energies in $^{85}_{36}\text{Kr}$. The $\log_{10} ft$ indicate that the transitions are slower than the "super allowed" transitions by 1–3 orders of magnitude. This is typical in all but the lightest nuclides. The retardation in the decay is due to a number of nuclear-structure effects, the generalities of which are now fairly well understood.

One also sees a number of first-forbidden transitions with $\log_{10} ft$ in the range of 7–8, also quite typical. Notwithstanding the large number of levels actually populated, the principal part of the β decay (96%) goes to the $1/2^-$ level at 0.305 MeV in $^{85}_{36}\text{Kr}$. This is via an allowed transition and owes its large intensity not only to the relatively low $\log_{10} ft$ but also to the very high decay energy. You will note that no decay is reported to the ground state of $^{85}_{36}\text{Kr}$. Such a transition would correspond to $\Delta J = 3$ (yes) and we could expect a $\log_{10} ft$ of 13 or greater for it. It is simply too improbable to be observed experimentally.

The ground-state spins and parities of both the parent and daughter can be understood from the shell-model level diagrams. Of particular interest here is the spin and parity of the first-excited state of $^{85}_{36}\text{Kr}$. In the shell-model limit, the ground-state spin and parity should be that of the unpaired 49th neutron, one less than the magic number of 50. In Fig. 10.20 are shown the single particle neutron levels in the vicinity of the magic number $N = 50$. With 48 paired particles filling all lower levels, the 49th neutron is predicted to lie in the $g_{9/2}$ level and this is consistent with the experimental values of $9/2^+$ for the ground state. It is very unlikely that the first-excited state of $^{85}_{36}\text{Kr}$ would be found by raising the odd neutron to the $2d_{5/2}$ level. That would entail crossing the shell gap and require adding perhaps 3–4 MeV to the nucleus. Rather, it would be less costly energetically to simply break the pair of neutrons in the $2p_{1/2}$ orbital and excite one into the $2g_{9/2}$ orbital where it would pair with the odd neutron to completely fill that level. As it turns out, the pairing energy increases significantly as the angular momentum of an orbit increases and thus the pairing energy in the $2g_{9/2}$ orbital more than compensates for the cost of breaking the pair in the $2p_{1/2}$ orbital and makes such an excitation rather low in energy. This is essentially what occurs in exciting the $^{85}_{36}\text{Kr}$ nucleus from the ground state to its first-excited state.

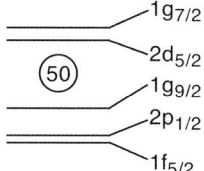

Fig. 10.20 Shell-model level diagram in the vicinity of the closed shell at $N(Z) = 50$.

Now suppose that the $1/2^-$ member of this set was the first-excited state of ^7_3Li. We could symbolize it as

$$(\text{n: } 1s^2_{1/2}1p^2_{3/2}\big|_{2^+}\ \text{p: } 1s^2_{1/2}1p^1_{3/2})\big|_{1/2^-}$$

and the electron capture decay to it would be symbolized as

$$\text{n: } 1s^2_{1/2}1p^1_{3/2}\ \text{p: } 1s^2_{1/2}1p^2_{3/2} \to (\text{n: } 1s^2_{1/2}1p^2_{3/2}\big|_{2^+}\ \text{p: } 1s^2_{1/2}1p^1_{3/2})\big|_{1/2^-}$$

The decay is easily achieved in the same way as the ground-state to ground-state transition occurs, by the simple transformation of a $p_{3/2}$ proton into a $p_{3/2}$ neutron. There is no need for any other nucleon to change its state. If this is true we ought to expect a decay to the first-excited state that is also "super allowed", just what is observed. We have not really proven that we know the structure of the $1/2^-$ level in ^7_3Li but we have used empirical and elementary knowledge of nuclear structure to make plausible explanations for the decay. By the way, the explanation, while naive, is essentially correct.

As a final example of the application of these general ideas, consider the decay of $^{85}_{35}\text{Br}$ to levels in $^{85}_{36}\text{Kr}$ as shown in Fig. 10.19. This is a fairly complex decay scheme but one that is typical of those found throughout the region of medium- and heavy-massed nuclides. With a decay energy of 2.87 MeV, many excited states in the

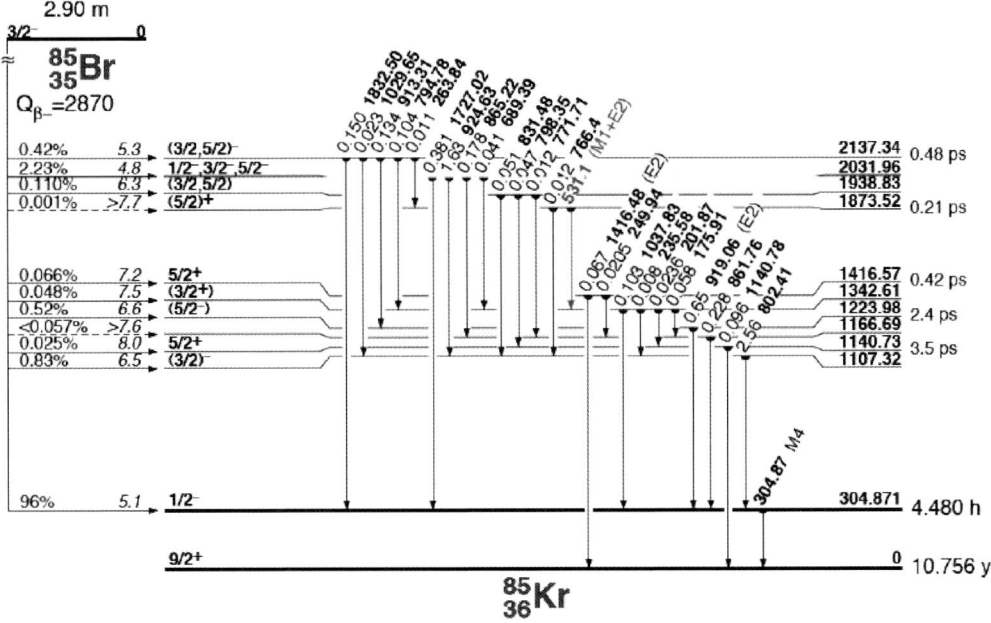

Fig. 10.19 Decay scheme for $^{85}_{35}\text{Br}$.

binding energy in the daughter. β-delayed fission is also known. A number of the neutron-deficient isotopes of the heaviest elements undergo electron-capture decay to levels that then undergo spontaneous fission very rapidly. This is somewhat akin to the process of neutron-induced fission where neutron absorption is used to provide the excitation energy required for fission to proceed with high probability.

10.14
Comments on Fermi Transitions

Decay by the Fermi operator is restricted to the lightest nuclei and heavier neutron-deficient nuclides. It can participate in electron capture and positron emission but it cannot play a role in the β^- decay of almost all nuclides. This can be understood without much difficulty. Suppose we were somehow able to turn off the Coulomb repulsion between protons so that only the strong interaction was at play in the nucleus. Because the nuclear force is charge-independent to a good approximation, all particle–particle interactions are the same and we can easily see that the level spectrum of the neutrons and the protons would be identical except for effects due to the very small mass difference between the two particle types (about 0.13%).

In Fermi decay, the daughter proton state is identical to that of the initial neutron. Therefore, in the absence of the Coulomb interaction, the only change that would take place in β^- decay would be the *reduction* in mass of the daughter compared to the parent by the neutron–proton mass difference, or 1.293 MeV. This is shown schematically in part (a) of Fig. 10.22. But the fact is, the Coulomb interaction is present and the stored Coulomb energy will be larger after decay because of the increase in atomic number by one. If ΔE_{EC} is the difference in the stored Coulomb

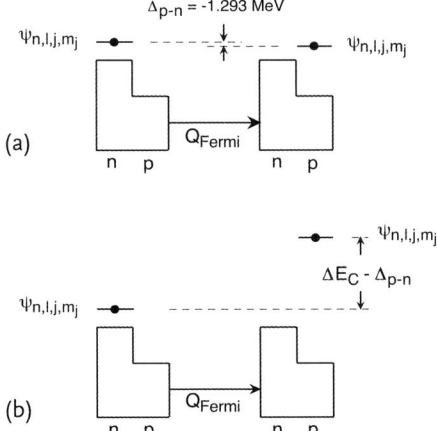

Fig. 10.22 Schematic representation of β^- decay by the Fermi operator in the absence (a) and in the presence (b) of the Coulomb interaction.

energy, the total energy of the daughter nucleus will differ from that of the parent by $\Delta E_{EC} - \Delta_{n-p}$. Clearly, if $\Delta E_{EC} > \Delta_{n-p}$, the mass of the daughter will be greater than that of the parent and Fermi decay would not be energetically allowed. This is shown schematically in part b of Fig. 10.22.

We can estimate the difference in Coulomb energy as follows. From Eq. (4.13)

$$\Delta E_C = k_C \frac{3}{5} \frac{e^2}{r}[(Z+1)^2 - Z^2] = k_C \frac{(2Z+1)e^2}{r} \tag{10.72}$$

and for the lighter nuclides where N = Z at the valley of β-stability,

$$\Delta E_C = k_C \frac{3}{5} \frac{(A+1)e^2}{r_o A^{1/3}}. \tag{10.73}$$

One finds already at A = 40, $\Delta E_{EC} - \Delta_{n-p}$ = 7.5 MeV. It is quite clear that, for all but the lightest nuclides, Fermi decay is energetically forbidden in β^- decay. The result of all of this is that neutron-rich nuclei have longer half-lives than one might expect from the previous discussions.

Although we have approached β decay within the context of the shell model, the Fermi operator actually applies to the nucleus as a whole. Namely, the wave functions of the parent and daughter nuclides as a result of decay by the Fermi operator are the same. While we will not discuss the matter further, we point out that experiments verify that a state in the daughter that is the *analog* of the parent ground state is found at just the excitation energy we have calculated here.

References

1 Every interested student should read the book by E. Fermi, with J. Orear Jay and A.H. Rosenfeld, "Nuclear Physics: A Course Given by Enrico Fermi at the University of Chicago", University of Chicago Press (1984).
2 S. Raman and N.B. Gove, Phys. Rev. C7 1973 (1995).
3 W. Bambynek, H. Behrens, M. H. Chen, B. Crasemann, M. L. Fitzpatrick, K. W. D. Ledingham, H. Genz, M. Mutterer and R. L. Intemann, "Orbital electron capture by the nucleus", Rev. Mod. Phys. 49 (1977) 77.
4 See, for example, http://www4.nau.edu/microanalysis/Microprobe/Xray-NamingTransitions.html.

General References

Discussions of β decay will be found in most texts on nuclear physics and chemistry. A large number of them are rather phenomenological in nature or sketchy on theory because the subject is simply rather complex. Others are very advanced in nature. Two references that the student might find of interest are found in K.S. Krane, "Introductory Nuclear Physics", John Wiley and Sons, New York (1987), and

Problems

1. (a) Derive a relativistic expression for the wavelength of an electron.
(b) Calculate the ratio of the wavelength of a 10 MeV electron to the radius of a nucleus with A = 100.

2. The nuclide ^{37}Ar decays by electron capture directly to the ground state of ^{37}Cl. Nevertheless, the ^{37}Cl atom may be left in an excited state. Given a detector that can "see" *all* of the radiations that can be emitted in the decay, write down all of the signals that can be expected to be observed.

3. ^{64}Cu is a well-known nuclide that is unstable with respect to both β^- and EC/β^+ decay. (a) Calculate the Q values for these three decay modes.
(b) What are the energies of the neutrino (β^+ decay) and antineutrino (β^- decay) when the charged leptons have kinetic energies of 1/2 the corresponding Q values?
(c) Neglecting the possible excited states of the daughter atom, what is the energy of the neutrino emitted in the EC decay?

4. ^{87}Br is the delayed neutron precursor that is the longest lived of the precursors found among the fission products. It has a ground-state spin and parity of 3/2$^-$. Consider only those β^- decays that leave the neutron emitter ^{87}Kr with an excitation energy $E_x \geq B_n$, and assume that all such states decay by neutron emission to leave ^{87}Kr in its ground state.
(a) Assuming that only allowed decay occurs, determine the spins and parities of the states populated in ^{87}Kr.
(b) Determine the total angular momenta and orbital angular momenta of the neutrons that can be emitted from the states given in part (a).

5. First-forbidden non-unique β transitions are characterized by the selection rules $\Delta J = 0,1(\text{yes})$. Demonstrate that the selection rules imply that the total angular momenta of the emitted leptons *cannot* be an even multiple of $\hbar$.

6. ^{89}Sr undergoes β^- decay to the ground state of ^{89}Y.
(a) Use the spherical shell model to predict the ground state spins and parities of both nuclides and use these to deduce the classification for this β transition.
(b) The first-excited state of ^{89}Y, with spin and parity of 9/2$^+$, can also be populated in the β^- decay of ^{89}Sr. Determine the classification for this β^- transition. Discuss qualitatively what you expect for the intensity of this transition relative to the transition to the ground state.

7. (a) Assuming that $F(Z,E_o) = 1$ and that $m_{\bar{\nu}} = 0$, create a Fermi–Kurie plot of the form

$$\sqrt{\frac{N(p/p_o)}{p^2}}$$

versus (E/E_o) as in Fig. 10.2. Be sure to use the appropriate relativistic form for the electron momentum.

(b) Calculate the average energy $< E/E_o >$ in this approximation.

8. ^{133}Sn is a fission product.
(a) Calculate the Q value for β^- decay of ^{133}Sn to ^{133}Sb.
(b) Calculate the binding energy of the last neutron in ^{133}Sb. Is ^{133}Sn likely to be a delayed neutron precursor?
(c) Predict the ground-state spins and parities of both ^{133}Sn and ^{133}Sb, give the classification for the ground-state to ground-state β^- transition and estimate the half-life of ^{133}Sn if it decayed solely by this transition.

11
γ Decay and Internal Conversion

11.1
Introduction

The majority of all bound excited states of nuclei decay by the emission of γ-rays, which range in energy from a low of perhaps 10 keV to a maximum of the particle binding energy, or roughly 8 MeV. Photon emission also takes place from particle-unbound states, such as are formed during the course of a nuclear reaction, but generally with much lower probability than neutron emission.

The energy of a γ-ray is, to an excellent approximation, equal to the difference in energies of the two nuclear states involved in the decay. Except for rather special applications, the recoil energy of the nucleus can be entirely neglected. The probability for photon emission is strongly dependent upon the photon energy, its angular momentum and the detailed properties of the nuclear states involved in the decay. As Auger emission is a "radiationless" competitor to x-ray emission, so is there a radiationless competitor to γ-ray emission and it is called *internal conversion*. In this process, the difference in energy between the two nuclear states is transferred directly to an atomic electron, ejecting it from the atom. The electron is referred to as a *conversion electron*. In analogy with Auger emission, the smaller the energy difference between the nuclear states, the greater is the probability that decay will occur by internal conversion. Unlike atomic states, however, the shell structure in the nucleus frequently leads to a low-lying first-excited state with an angular momentum quite different from that of the ground state as was seen in the case of $^{85}_{36}$Kr (Fig. 10.19). In such cases, the probability of decay by internal conversion is enhanced and it can become the dominant decay mode even though the transition energy is not very small. If internal conversion occurs, a vacancy will be produced in one of the more tightly bound electron shells and x-ray and Auger electron emission will follow.

In the simple case that decay of a nuclide leads only to the ground and first-excited states of a daughter, the spectrum of radiations emitted from the daughter nucleus and atom is relatively simple. If the parent decay populates a number of excited states in the daughter, however, the spectrum can become quite complex; each excited state can lead to the emission of γ-rays, conversion electrons, x-rays and Auger electrons.

Nuclear Physics for Applications. Stanley G. Prussin
Copyright © 2007 WILEY-VCH Verlag GmbH & Co., Weinheim
ISBN: 978-3-527-40700-2

The theoretical development of the emission of electromagnetic radiation from the nucleus is quite complicated and goes well beyond the level of this text. There is no simple or easy way to go about it. Introductions to the theory are generally found in texts on quantum mechanics but the full theory is found only in advanced texts. The presentation we will provide is meant to cover some of the fundamental ideas involved. We will not be very rigorous nor will we be complete. The results from simple models for photon emission probabilities will be introduced and used to understand, at least semiquantitatively, some of the more important facts concerning both γ-ray emission and internal conversion. Following this development, we will consider the qualitative application to a number of decay schemes in order to point out how they can be used to understand or correlate the spectrum of emitted radiations.

11.2
The Angular Momentum of Photons and Conservation Laws

Unlike fermions that obey Fermi–Dirac statistics, photons are *bosons* that have integral total angular momenta and have the property that any number of them can occupy the same quantum state in a system. Photons must be emitted with at least one unit of angular momentum with respect to the emitting center, and the spin of a photon is $1\hbar$. That photons cannot be emitted with zero angular momentum can be understood, qualitatively, by the following argument. Photons are electromagnetic waves that possess oscillating electric and magnetic vectors normal to one another, with both normal to the direction of propagation of the radiation. The momentum of a photon, $p = E/c$, can be considered to be carried equally, on the average, by both the electric and magnetic vectors. As a result, you cannot find a coordinate system in which both of the vectors will contribute zero angular momentum. That is, because of the *tranverse* nature of electromagnetic waves, it is impossible to find a coordinate system in which the photon will not have at least one unit of angular momentum.

If J_i and J_f are the angular momentum quantum numbers of the initial and final states connected by γ-ray emission and L_γ is the angular momentum quantum number of the emitted photon, conservation of angular momentum requires

$$|J_i - J_f| \leq L_\gamma \leq |J_i + J_f|, \quad (L_\gamma \neq 0)$$

and conservation of parity requires

$$\pi_f \pi_\gamma = \pi_i$$

An immediate consequence of the requirement that the photon have at least one unit of angular momentum is that single-photon emission cannot take place between two states, each of which has zero angular momentum. But decay can occur by internal conversion or other higher-order processes (e.g., two-photon emission, etc.).

11.2 The Angular Momentum of Photons and Conservation Laws

A second consequence of the conservation laws is that the decay of a level may occur by emission of photons carrying a range of angular momenta. For example, decay of a state with $J_i = 2$ to a state with $J_f = 2$ can occur by emission of photons with $L_\gamma = 1, 2, 3$, or 4. The relative probabilities for emission of each of these will depend upon the operators for creation of each, and the wave functions of the nuclear states.

The wavelengths of γ-rays emitted in the decay of bound states of nuclei, as well as most γ-rays emitted in nuclear reactions, are quite large compared to nuclear radii. One can perform a calculation similar to that performed for neutrinos (Section 10.4) and find that typically,

$$\frac{\lambda_\gamma}{R_n} \geq 20$$

where λ_γ and R_n are the photon wavelength ($E_\gamma \leq 10$ MeV) and nuclear radius ($R_n \leq 8$ fm), respectively. This means that the long-wavelength approximation will be applicable and we will discuss the results from theoretical development of photon emission probabilities in the limit of this approximation.

As we alluded to in the last chapter, photons can be described as plane waves that can be expanded in products of spherical harmonics and spherical Bessel functions as was illustrated in the case of the neutrino wave function (Section 10.7.1). Because of the small size of the nucleus and the characteristics of the spherical Bessel functions for small arguments, we can expect that, all other things being equal, the probability for creating photons within the nuclear volume will decrease with increasing angular momentum. This leads to the expectation that, generally speaking, only photons with the lowest possible angular momentum will be emitted with high probability and this is actually found to be the case, with one notable exception that will be discussed later.

Among the many factors that define the complexity of photon emission is the dual nature of the electromagnetic field. The nucleus has, of course, charge from the protons, and the nonspherical nature of the orbits of most of these means that the electric field will possess multipole moments that have discrete angular momenta. We have considered some of the elements of these in our discussion of the static electric field of the nucleus and how they permit the definition of the shape of the nuclear charge distribution (Section 8.1). Classically, oscillating electric fields radiate electromagnetic energy with angular momenta defined by the moments of the fields. The emission of photons by the nucleus is, in part, the result of the changes in the moments of the field as the nucleus undergoes decay from one state to another. In addition, as a result of the motion of the protons, there are currents in the nucleus that generate magnetic fields and the magnetic field will also possess multipole moments. It is also known classically that oscillating magnetic fields themselves radiate electromagnetic energy, and this radiation carries angular momentum defined by the moment of the field. Therefore, in addition to the emission of photons by the changes in electric fields due to a nuclear transition, there will also be photon emission as a result of the changes in the magnetic field. Complicating all of this is the fact that the individual nucleons possess intrinsic magnetic moments that can change during a transition and thus will participate in the emission of γ-rays.

During a transition from one state to another, it is possible that photon emission can take place by changes in either or both the electric and magnetic fields. But the radiations from these fields are not identical with respect to their relative parities. It turns out that the parity of the radiation emitted by an oscillating electric dipole is opposite to that of the radiation that is emitted from an oscillating magnetic dipole. The same is true for all higher moments as well. Thus, whereas α decay from a 1^+ to a 0^+ state cannot take place because the $l_\alpha = 1$ wave has odd parity, photon emission between such states can take place by the changes in the magnetic field induced by the transition. Magnetic dipole radiation has even parity with respect to the emitting system.

With these qualitative points in view, we will now provide a brief overview of the theory of photon emission and point out implications that are important to an understanding of γ-ray emission.

11.3
Introduction to the Theory of Photon Emission

11.3.1
The Radiation Field and Matrix Elements for Photon Emission

The very first issue that must be faced is the characterization of an electromagnetic field in terms of the photons that are its quanta. The place to begin is by writing down the Maxwell equations. They are

$$\nabla \times \mathbf{B} = \mu_o \varepsilon_o \frac{\partial \mathbf{E}}{\partial t} + \mu_o \mathbf{j}$$

$$\nabla \times \mathbf{E} = -\frac{\partial \mathbf{B}}{\partial t} \quad (11.1)$$

$$\nabla \cdot \mathbf{B} = 0$$

$$\nabla \cdot \mathbf{E} = \frac{\rho}{\varepsilon_o}$$

to which we add the continuity equation

$$\nabla \cdot \mathbf{j} = -\left(\frac{\partial \rho}{\partial t}\right) \quad (11.2)$$

In these equations, **B** and **E** are the vector magnetic and electric fields, respectively, **j** is the current density of the electric charge and ρ is the charge density. It is common to express **B** and **E** in terms of a vector potential **A** and a scalar potential ϕ defined by

$$\mathbf{E} = -\frac{\partial \mathbf{A}}{\partial t} - \nabla \phi \qquad (11.3)$$

$$\mathbf{B} = \nabla \times \mathbf{A}$$

Direct substitution of Eqs (11.3) shows that the second and third of Eqs (11.1) are satisfied directly. Substitution into the first and fourth of Eqs (11.1) with some manipulation involving vector identities gives

$$\nabla(\nabla \cdot \mathbf{A}) - \nabla^2 \mathbf{A} = -\frac{1}{c^2}\left(\frac{\partial^2 \mathbf{A}}{\partial t^2}\right) - \frac{1}{c^2}\nabla\left(\frac{\partial}{\partial t}\right)(\phi) + \mu_o \mathbf{j}$$

$$\frac{\partial}{\partial t}(\nabla \cdot \mathbf{A}) + \nabla^2 \phi = -\frac{\rho}{\varepsilon_o} \qquad (11.4)$$

These equations can be considered to define the electromagnetic field in terms of the potentials $\mathbf{A}$ and ϕ. Unfortunately they do not provide a unique definition. This can be rectified by taking $\nabla \cdot \mathbf{A} = 0$, a requirement that can be shown to leave the structure of the problem unaffected. With this addition, Eqs (11.4) become

$$-\nabla^2 \mathbf{A} + \frac{1}{c^2}\left(\frac{\partial^2 \mathbf{A}}{\partial t^2}\right) + \frac{1}{c^2}\left(\frac{\partial}{\partial t}\right)(\nabla \phi) = \mu_o \mathbf{j}$$

$$\nabla^2 \phi = -\frac{\rho}{\varepsilon_o} \qquad (11.5)$$

We can take Eqs (11.5) as the complete definition of an electromagnetic field. While the first of these looks messy, it has a rather simple underlying structure and meaning. To see this, consider what the field looks like in the absence of charges and currents, i.e., in the absence of sources to the field. With $\rho = 0$, the second of Eqs (11.5) gives $\nabla^2 \phi = 0$, Laplace's equation. In the case when the charge is everywhere zero, the solution to this equation is what you would expect intuitively; ϕ, the electrostatic potential, vanishes. That means that the field is completely described by the vector equation

$$\nabla^2 \mathbf{A} - \frac{1}{c^2}\left(\frac{\partial^2 \mathbf{A}}{\partial t^2}\right) = 0 \qquad (11.6)$$

This is a three-dimensional wave equation, and its general solution can be shown to be

$$\mathbf{A} = \mathbf{A}_o e^{i(\mathbf{k} \cdot \mathbf{r} - \omega t)} + \mathbf{A}_o^* e^{-i(\mathbf{k} \cdot \mathbf{r} - \omega t)} \qquad (11.7)$$

where $\omega = |\mathbf{k}|c = |\mathbf{k}|/(\sqrt{\mu_o \varepsilon_o})$. The solution is seen to be the superposition of outgoing and incoming plane waves. The frequency of a wave is ω and $|\mathbf{k}| = 1/\lambda$

is the magnitude of the propagation vector of the wave. The electromagnetic field in the absence of sources and currents is described by incoming and outgoing photons. This is called the radiation field. The sense of the calculations that are outlined in the following is that the presence of the nucleus does not perturb this field appreciably. The field is described by the vector potential **A** and we want to describe how the nucleus interacts with it.

We begin by writing down the Hamiltonian for a proton in the nucleus in the absence of an electromagnetic field and we will neglect its intrinsic magnetic moment for the present. To prevent confusion, we use the italic m to represent the reduced mass of the proton and use μ to represent its magnetic moment. The Hamiltonian is then written as

$$H = -\frac{\hbar^2}{2m}\nabla^2 + V(r) = \frac{p_n^2}{2m} + V(r) \tag{11.8}$$

where V(r) is the nuclear potential due to the strong interaction alone and the subscript n on p_n signifies that this is the momentum of the proton due to its interaction with the nuclear force alone.

If an electric field is present, the proton would experience an additional force

$$\mathbf{F} = e\mathbf{E} = \frac{d(m\mathbf{v})_e}{dt} = \frac{d\mathbf{p}_e}{dt} \tag{11.9}$$

where the subscript "e" refers to the interaction of the proton charge with the electric field alone. We can now relate the momentum due to this field and the vector potential **A** using (11.3). Therefore,

$$\mathbf{E} = -\left(\frac{\partial \mathbf{A}}{\partial t}\right) = \frac{1}{e}\frac{d\mathbf{p}_e}{dt} \tag{11.10}$$

and

$$\mathbf{p}_e = -e\mathbf{A} \tag{11.11}$$

In the presence of both the strong interaction and the electric field, the total proton momentum **p** is the vector sum of the two momenta from both interactions,

$$\mathbf{p} = \mathbf{p}_n + \mathbf{p}_e = \mathbf{p}_n - e\mathbf{A} \tag{11.12}$$

and the Hamiltonian is

$$H = \frac{1}{2m}(\mathbf{p}_n - e\mathbf{A})^2 + V(r) \tag{11.13}$$

To be complete, we must now consider the intrinsic magnetic moment of the proton, which is given by

11.3 Introduction to the Theory of Photon Emission

$$\mu_s = g_s\left(\frac{e\hbar}{2m}\right)s \tag{11.14}$$

Here **s** is the intrinsic spin vector and g_s is the so-called spin g-factor. This is a constant that is the ratio of the actual magnetic moment of the proton to that which would be expected classically. Now the energy of interaction of a magnetic moment with a magnetic field **B** is just $-\mu \cdot \mathbf{B}$ and therefore the total Hamiltonian for a single proton becomes

$$H = \frac{1}{2m}(\mathbf{p}_n - e\mathbf{A})^2 - g_s\left(\frac{e\hbar}{2m}\right)\mathbf{s}\bullet\mathbf{B} + V(r) \tag{11.15}$$

If we now expand the first term on the right-hand side, (11.15) can be written in the form

$$\begin{aligned} H &= \frac{\mathbf{p}_n^2}{2m} + V(r) + \frac{1}{2m}[e^2\mathbf{A}^2 - e(\mathbf{p}_n\cdot\mathbf{A} + \mathbf{A}\cdot\mathbf{p}_n)] - g_s\left(\frac{e\hbar}{2m}\right)\mathbf{s}\bullet\mathbf{B} \\ &= H_o + H_1 \end{aligned} \tag{11.16}$$

The Hamiltonian is seen to be composed of two parts. The first two terms represent H_o, the Hamiltonian for the interaction of the proton with all other nucleons due solely to the strong interaction. The remaining terms represent the interaction of the proton with the radiation field. It is this part that we can call the perturbation potential, H_1, that gives rise to photon interactions.

In general, the interaction of a proton with the radiation field must be responsible for both the emission and absorption of electromagnetic radiation by a nucleus. Both are treated in the same first-order perturbation approximation that we have used for the development of the decay constant in β decay. The Fermi Golden rule (Eq. (10.10)) says that the probability for a transition will be proportional to the square of the matrix element $|\langle\Psi_f|H_1||\Psi_i\rangle|^2$. For photon emission, we recognize that the potential must produce a photon in the final state, and for photon absorption, it must cause the disappearance of a photon that is present in the initial state. From the form of H_1, it is clear that it is the vector potential **A** that must carry out the creation or "destruction" of the photon. If we consider **A** as an operator, its matrix element between the initial and final states is of the form

$$\begin{aligned} \langle\Psi_f|\mathbf{A}|\Psi_i\rangle &= \int\Psi_f^*\mathbf{A}\Psi_i d\tau \\ &= \langle\Psi_f|(\mathbf{A}_o e^{i(\mathbf{k}\cdot\mathbf{r}-\omega t)} + \mathbf{A}_o^* e^{-i(\mathbf{k}\cdot\mathbf{r}-\omega t)})|\Psi_i\rangle \\ &= \langle\Psi_f|\mathbf{A}_o|\psi_\gamma\rangle|\Psi_i\rangle + \langle\Psi_f|\langle\psi_\gamma|\mathbf{A}_o^*|\Psi_i\rangle \end{aligned} \tag{11.17}$$

The second equality was obtained by direct substitution of Eq. (11.7) for **A**. Now consider the two terms in the third of the equalities. In photon absorption, the perturbation potential acts on an initial state that contains the nucleus and an incident photon to produce a final state that is an excited nucleus; the operator *annihilates* the photon in the initial state. This is clearly represented by the first term where $|\psi_\gamma\rangle = e^{i(\mathbf{k}\cdot\mathbf{r}-\omega t)}$. The second term contains the complex conjugate of $\mathbf{A}_o$ and the complex conjugate of an outgoing plane wave, $\langle\psi_\gamma| = e^{-i(\mathbf{k}\cdot\mathbf{r}-\omega t)}$. Thus, the operator $\mathbf{A}_o^*$ acts on the initial state to create a photon in the final state. It is the operator $\mathbf{A}_o^*$ that we are interested in here.

The potential H_1 in Eq. (11.16) is seen to contain the term $(1/2m)e^2A^2$. This must represent the process of emission or absorption of *two* photons. It is only of significance at very high field strengths and we can safely ignore it in our present discussions. We then have, for the perturbation potential,

$$H_1 = -\frac{e}{2m}(\mathbf{p}_n \cdot \mathbf{A} + \mathbf{A} \cdot \mathbf{p}_n) - g_s \frac{e\hbar}{2m}\mathbf{s} \bullet \mathbf{B} \qquad (11.18)$$

with, in the case of photon emission,

$$\mathbf{A} = \mathbf{A}_o^* e^{-i(\mathbf{k}\cdot\mathbf{r}-\omega t)} \qquad (11.19)$$

The operators $\mathbf{A}$ must satisfy Eq. (11.6), the wave equation. Substitution of Eq. (11.19) into the latter gives

$$\nabla^2\mathbf{A} + \frac{\omega^2}{c^2}\mathbf{A} = 0, \text{ or} \qquad (11.20)$$

$$\nabla^2\mathbf{A} + k^2\mathbf{A} = 0 \qquad (11.21)$$

This rather simple vector equation is known as the vector Helmholtz equation. Its solution is a bit complicated and we will not solve it in detail. But the solutions are found in terms of the solutions to the scalar equation and we can learn a good deal from examining these. Writing the scalar equivalent to Eq. (11.21) in the form

$$\nabla^2 u + k^2 u = 0 \qquad (11.22)$$

it is seen that this is the form of the wave equation for a particle moving in the absence of a potential (see, for example, Eq. (7.7) when the potential vanishes). It has solutions of the form

$$u_{LM} = j_L(kr)Y_{LM}(\theta, \phi) \qquad (11.23)$$

where, as usual, $L = 0, 1, 2,....$, and $|M| \leq L$. (If you thought you would get away from our old friends the spherical harmonics and the spherical Bessel functions, you were wrong, and will be again.) These functions look very similar to those that are the solutions to particle motion in a spherically symmetric potential. They clearly must satisfy the conditions

$$L_z u_{LM} = M u_{LM}$$
$$L^2 u_{LM} = L(L+1) u_{LM} \qquad (11.24)$$

11.3 Introduction to the Theory of Photon Emission

and, of course, they have the normal parities of even, if L = even, and odd, if L = odd. The result is that the operators **A** naturally arise as a set of orthogonal operators of well-defined angular momentum and parity and therefore represent the composition of the electromagnetic field in terms of multipole moments. Each operator will represent the emission of radiation of a definite multipole.

If one examines the solution of the vector equation, one finds that it has two sets of acceptable solutions. Both are characterized by well-defined angular momentum and parity but one set has the usual parity assignments indicated above, while the other has the opposite parity assignments. The fact that two sets of solutions arise should come as no surprise. The vector potential represents both the electric and magnetic field vectors and we know that the time dependence of both can give rise to photon emission. The two sets of solutions represent these fields and what we learn is that for the same multipole order, magnetic radiation has a parity opposite to that of electric radiation. It is standard practice to label the moments and the emitted radiation by EL or ML to distinguish between the two, and the parities of these are given by

$$\text{EL:} \quad \pi = (-1)^L$$
$$\text{ML:} \quad \pi = (-1)^{L+1} \tag{11.25}$$

Electric dipole radiation, emitted between two states with different parity, is denoted by E1 radiation. Magnetic quadrupole radiation, emitted between two states that differ in parity, is denoted by M2 radiation, etc.

The full mathematical development of the multipole operators is complicated, and we will settle for writing down the matrix elements for photon emission and examining their structure. These matrix elements can be written in general form

$$\langle \psi_f | H_1 | \psi_i \rangle = \left(\frac{\hbar c}{\varepsilon_0 R_0} \right)^{1/2} \left[\frac{(L+1)}{L} \right]^{1/2} \frac{1}{(2L+1)!!} \left(\frac{\omega}{c} \right)^{L+1/2} \langle \psi_f | O_{LM}^{EL(ML)} | \psi_i \rangle \tag{11.26}$$

where the multipole operators are

$$O_{LM}^{EL} = er^L Y_{LM}^* - \frac{i\mu e \omega}{2m(L+1)} (\boldsymbol{\sigma} \times \mathbf{r}) \cdot [\nabla(r^L Y_{LM})]^*$$
$$O_{LM}^{ML} = \frac{e\hbar}{mc(L+1)} \mathbf{L} \cdot [\nabla(r^L Y_{LM})]^* + \frac{e\hbar}{2mc} \mu \boldsymbol{\sigma} \cdot [\nabla(r^L Y_{LM})]^* \tag{11.27}$$

In these expressions, ω is the frequency and $\hbar\omega$ the energy of the emitted photon, respectively. The quantity $\boldsymbol{\sigma}$ is met with in the matrix formulation of the intrinsic spin of a fermion and it is essentially a dimensionless representation of the intrinsic spin vector **s**. The vectors **s** and $\boldsymbol{\sigma}$ are simply related by $\mathbf{s} = 1/2\hbar\boldsymbol{\sigma}$. All other parameters should be clear except for the factors R_0 and $(2L+1)!!$. Because we are considering the emission of a photon into the continuum, we have to have some volume in which its wave function is normalized. It is customary to take an arbitrarily large sphere of radius R_o for this purpose. The meaning of the term $(2L + 1)!!$ will be explained shortly.

The rather gruesome expression for the matrix elements has some parts that are simple and provide insight into their overall forms. To begin with, note that the wave function of a photon is not shown explicitly anywhere in Eq. (11.26). The wave functions in the matrix element represent only the initial and final nuclear states. That means that the factors preceeding the matrix element on the right-hand side of the equation must contain the photon characteristics.

We can demonstrate that the factor $(\omega/c)^L/(2L + 1)!!$ derives directly from the long wavelength approximation for the emitted photon. The solutions to Eq. (11.22) (Eq. (11.23)) are proportional to the spherical Bessel functions. As usual, we will need to determine the probability for finding the emitted photon within the nucleus. The argument of the Bessel function, kr, as we have shown previously in the discussion of β decay, will always be small. Therefore, we can consider the asymptotic form given in Eq. (7.46), i.e.,

$$j_L(kr)|_{kr \to 0} \to \frac{(kr)^L}{(2L + 1)!!} \tag{11.28}$$

where, again, the symbol $(2L + 1)$ is the so-called double factorial and is given by

$$(2L + 1)!! = (2L + 1) \cdot (2L - 2) \cdot (2L - 4) \cdot \ldots \cdot 1 \tag{11.29}$$

Now $k = \omega/c$, and we then see that (11.28) is just the long-wavelength approximation for that portion of the wave function of the emitted photon that is given by $j_L(kr)$ apart from the factor r^L. Because we must integrate over the radial coordinate, this factor must appear in the matrix element and if you look at the terms in the operators of Eq. (11.27) you will find it. As for the spherical harmonic part of the wave function, you will find it as its complex conjugate in each term of the operators, just what is needed for a complete description for a photon present in the final state.

The multipole operators are also complex in form but we can get some insight into their structure and action just by considering the functions we have met with before. As the simplest example, consider the operator for electric dipole radiation. Note that it, like all of the other operators, is a function of both L and M. There will then be three such operators for electric dipole emission and if you use the expression for $Y_{1,0}(\theta, \phi)$ given in Table 7.2 you will quickly arrive at the expression for the first term in the operator

$$er^L Y_{1,0}^*(\theta, \phi) = er\sqrt{\frac{3}{4\pi}}\cos(\theta) = \sqrt{\frac{3}{4\pi}}ez \tag{11.30}$$

where $z = r\cos(\theta)$ is the z-coordinate in the nucleus. The factor $(\boldsymbol{\sigma} \times \mathbf{r}) \cdot [\boldsymbol{\nabla}(r^L Y_{LM})]^*$ in the second term can be given a fairly simple meaning. With the result of (11.28), the factor in brackets is just the product of constants and the gradient of z. But the latter is just the unit vector in the z-direction. Therefore, apart from constants, the factor $(\boldsymbol{\sigma} \times \mathbf{r}) \cdot [\boldsymbol{\nabla}(r^L Y_{LM})]^*$ is just the z-component of the quantity $(\boldsymbol{\sigma} \times \mathbf{r})$, $(\boldsymbol{\sigma} \times \mathbf{r})_z$. The operator is then seen to be

$$O^E_{1,0} = \sqrt{\frac{3}{4\pi}} ez - \frac{i\mu e\hbar\omega}{4mc}\sqrt{\frac{3}{4\pi}}(\boldsymbol{\sigma} \times \mathbf{r})_z \quad (11.31)$$

What can this operator do? Consider just the first term. Because the operators for charge and position are the variables themselves, they are just multiplicative. Therefore the evaluation of the first term in the matrix element $\langle \psi_f | O^E_{1,0} | \psi_i \rangle$ must involve an integral of the form $\int \psi_f^* z \psi_i d\tau$. Because $z = r\cos(\theta)$ and $Y_{1,0} = \cos(\theta)$ has odd parity, this integral will vanish unless the parity of the initial and final nuclear states differ. The emitted photon must have odd parity. We then conclude that the first term in the electric dipole operator requires the selection rule $\pi_i \neq \pi_f$ for the nuclear states. Because the photon carries one unit of angular momentum and because angular momentum conservation requires that $|J_i - J_f| \leq l_\gamma \leq |J_i + J_f|$, the two states can have angular momenta that differ by only 0 or 1. The same type of analysis can be performed on the second factor in the operator with similar results, and the reader should try this as an exercise. You can also show that the other two electric dipole operators require the same angular momentum and parity changes, and that all of the odd-l electric multipoles will have the same parity characteristics.

Similar analyses can be performed on other operators as well but this is all we will consider. The principal point of this exercise is to demonstrate that one can look at the rather complex expressions involved in the theory of photon emission and at least qualitatively understand their basic structures and the physical effects of their actions.

11.3.2
Matrix Elements and Transition Rates

We can use the general expression for the matrix element in Eq. (11.26) and begin to write down the decay constant for photon emission using the Fermi Golden Rule. To do this we must consider the density of states for the emitted photon. The formulation of the matrix element assumes that the photon wave function is normalized in a sphere of radius R_o. If the photon is entirely contained in the sphere it must be true that its amplitude goes to zero at R_o. If you examine the characteristic of the spherical Bessel functions in the limit of very large arguments, you can easily demonstrate that this requirement leads to the relation

$$k = \frac{\pi(n + L/2)}{R_o}, \quad n = 0, 1, 2, \ldots \quad (11.32)$$

We use this as follows. The photon energy can be written as

$$E_\gamma = \hbar\omega = \hbar c k \quad (11.33)$$

Therefore

$$dk = \frac{dE}{\hbar c} = \frac{\pi}{R_o} dn \qquad (11.34)$$

and the density of final states becomes

$$\rho(E_f) = \frac{dn}{dE_\gamma} = \frac{R_o}{\pi \hbar c} \qquad (11.35)$$

Substituting this and the matrix element of (11.26) into the Fermi Golden Rule gives the probability $w_{i \to f}$ for the transition from the nuclear state i to the state f

$$w_{i \to f} = \frac{2}{\varepsilon_o \hbar} \frac{(L+1)}{L[(2L+1)!!]^2} \left(\frac{\omega}{c}\right)^{2L+1} |\langle \psi_f | O_{LM}^{E(B)} | \psi_i \rangle|^2 \qquad (11.36)$$

Now $w_{i \to f}$ applies to a decay involving definite magnetic substates in the initial and final states. Because we usually do not have "oriented" nuclei and therefore only observe the result of the decay of all possible combinations of initial and final states, we observe only the average transition rate. To obtain this, we must perform a sum over the transition rates for all possible m_i and m_f, subject to the requirement that $m_i - m_f = M$, where M is the magnetic quantum number of the emitted photon. Dividing by the total number of these, $2j_i + 1$, we obtain the average transition probability. Thus, the theoretical prediction for the observable decay constant is formally given by

$$\lambda_{i \to f}^{EL(ML)} = \frac{1}{(2j_i+1)} \sum_{m_i, m_f, M} w_{i \to f}^{EL(ML)} \qquad (11.37)$$

for all EL or ML transitions. While we will not consider the details of the computation, we can see what it involves. The matrix elements in (11.36) are just integrals over the spatial coordinates r, θ, and ϕ. But the integrals over the angular coordinates involve integrals over the spherical harmonics that are universal functions for central potentials, and they can be done, once and for all, for all possible combinations of angular momenta of the proton in the initial and final states and for each and every L of the photon. These have been calculated for practical cases, and we will write them symbolically $S(j_i, L, j_f)$. They typically have magnitudes in the range 1–10, and when $j_i = j_f$, are generally about 1–2. We can then write the decay constant for the emission of EL radiation, when a single proton makes a transition between two states in a central potential, neglecting the spin-dependent parts of the operators for simplicity, as

$$\lambda_{i \to f}(EL) = \frac{2(L+1)}{L[(2L+1)!!]^2} \left(\frac{e^2}{\varepsilon_o \hbar}\right) \left(\frac{\omega}{c}\right)^{2L+1} |\langle R_f | r^L | R_i \rangle|^2 \frac{S(j_i, L, j_f)}{4\pi} \qquad (11.38)$$

Having integrated over the angular coordinates, the only remaining integral to be performed involves the radial parts of the wave functions of the initial and final nuclear states, R_i and R_f, respectively. In principle, these are obtained by solution of the Schrödinger equation using realistic forms for the nuclear potential. But such computations go well beyond our interests, which are to gain estimates of transition rates in the limit of very simple models that can then be used to correlate data and provide a qualitative understanding of the underlying physics.

For simplicity, we will follow the simple approximation of Blatt and Weisskopf [1] that takes the wave function of a proton as a constant throughout the nuclear volume. With this simplest of functions, normalization of a radial function,

$$\langle R_f | R_f \rangle = \int_0^{R_n} R_f^* R_f 4\pi r^2 dr = 1 \tag{11.39}$$

gives the normalization constant as $\left[\frac{4}{3}\pi R_n^3\right]^{-1}$. The same must be true for the normalization of the radial function for the initial state, and the radial integral in Eq. (11.38) is now found with ease to be

$$\langle R_f | r^L | R_i \rangle = \int_0^{R_n} R_f^* r^L R_f 4\pi r^2 dr = \frac{3 R_n^L}{L+3} \tag{11.40}$$

With this result, the expression for the decay constant for electric radiation becomes

$$\lambda_{i \to f}(EL) \approx \frac{2(L+1)}{L[(2L+1)!!]^2} \left(\frac{e^2}{\varepsilon_0 \hbar}\right) \left(\frac{\omega}{c}\right)^{2L+1} \left(\frac{3 R_n^L}{3+L}\right)^2 \frac{S(j_i, L, j_f)}{4\pi} \tag{11.41}$$

and we are almost done. The last step that will prove useful is to combine all of the constants and put the variables in units that are commonly used. If one substitutes $E/\hbar$ for ω and uses units of MeV for energy and fermi for linear dimensions, the resulting constants can be combined to yield

$$\lambda_{i \to f}(EL) \approx \frac{4.4 \times 10^{21}(L+1)}{L[(2L+1)!!]^2} \left(\frac{3}{3+L}\right)^2 \left(\frac{E}{197.3}\right)^{2L+1} R_n^{2L} S(j_i, L, j_f) \; s^{-1} \tag{11.42}$$

The frequency factor of about $4.4 \times 10^{21} \; s^{-1}$ establishes a fundamental timescale for photon emission. Indeed, it is characteristic of the time required for a nucleon to move a distance comparable to the nuclear diameter. A similar treatment for magnetic multipole radiation with the same approximations leads to the result

$$\tag{11.43}$$

$$\lambda_{i \to f}(ML) \approx \frac{0.19 \times 10^{21}(L+1)}{L[(2L+1)!!]^2} \left(\frac{3}{2+L}\right)^2 \left(\mu L - \frac{L}{L+1}\right)^2 \left(\frac{E}{197.3}\right)^{2L+1} R_n^{2L-2} S(j_i, L, j_f) \; s^{-1}$$

Both expressions are quite similar and predict the same dependence of transition rates on energy and about the same dependence on angular momentum. Regardless of the crude assumptions we have made concerning the radial wave functions of the nuclear states, we can be assured that the dependencies on E and L will not be affected. The dependence on the ratio $(E/197.3)^{2L+1}$ predicts a very strong variation of transition rate on photon energy for a fixed angular momentum as well as a very strong variation on angular momentum for a fixed photon energy. The angular momentum dependence is also strongly affected by the remaining factors in L in Eqs. (11.42) and (11.43) that we write as

$$f(EL) = \frac{(L+1)}{L[(2L+1)!!]^2(3+L)^2}$$

and

$$f(ML) = \frac{(L+1)}{L[(2L+1)!!]^2(2+L)^2}$$

respectively. They are almost identical and it is easy to show that f(EL) and f(ML) each decrease by about two-orders of magnitude with each unit increase in L. This, in turn, has a profound affect on the allowed photon transitions that will actually be *observable* with appreciable intensity.

As a specific example, consider the case of the decay of a 5^+ level to a 5^- level by photon emission. Conservation of angular momentum permits all of the multipoles E1, M2, E3, M4, E5,...., E9, M10 to be emitted. Note that as a result of parity conservation, the succeeding multiples of the same type, E or M, will differ in angular momentum by two units. Now if we take the ratios of the decay constants for the allowed EL radiation, for example, Eq. (11.42) tells us they will be in the ratio $\lambda(E1) / \lambda(E3) / \lambda(E5) /..... = f(E1) / f(E3) / f(E5) /.... \sim 1 / 2.4 \times 10^{-4}/ 1.2 \times 10^{-8}/....$, because all other factors are the same for each transition except for the factor $S(j_i, L, j_f)$ whose neglect will not affect the qualitative conclusion we have drawn. Clearly, one expects that only the lowest-order electric multipole will be emitted with appreciable probability. An exactly similar result is found for the allowed magnetic multipole transitions.

To go further, we must consider estimates of the actual decay constants for photon emission and we will do so in the limit of the simple single proton model using the expressions that have been developed by Weisskopf [1] and Moskowski [2] and which have become the standards for expressing the rates of all γ-ray transitions. They are routinely used for rough estimates of transition rates and also as a means of correlating transition rates found experimentally and calculated with more sophisticated nuclear models. In Table 11.1 are the so-called Weisskopf single-particle estimates for the half-lives of γ-ray transitions. Because they are calculated on the basis of idealized single-proton initial and final states, they are expected to represent the fastest possible transition rates of each multipolarity in the limit of spherical nuclei. The estimates demonstrate the dependence of the transition probabilities on energy to the power (2L + 1). The mass dependence is

11.3 Introduction to the Theory of Photon Emission

Table 11.1 Weisskopf estimates of the half-lives, $t_{1/2}$(s), for photon emission due to single-proton transitions, when the photon energy, E_γ, is in keV.

Multipole order (L)	Electric $t_{1/2}$(s)	Magnetic $t_{1/2}$(s)
1	$\dfrac{6{,}76 \times 10^{-6}}{E_\gamma^3 A^{2/3}}$	$\dfrac{2{,}20 \times 10^{-5}}{E_\gamma^3}$
2	$\dfrac{9{,}52 \times 10^{6}}{E_\gamma^5 A^{4/3}}$	$\dfrac{3{,}10 \times 10^{7}}{E_\gamma^5 A^{2/3}}$
3	$\dfrac{2{,}04 \times 10^{19}}{E_\gamma^7 A^{2}}$	$\dfrac{6{,}66 \times 10^{19}}{E_\gamma^7 A^{4/3}}$
4	$\dfrac{6{,}50 \times 10^{31}}{E_\gamma^9 A^{8/3}}$	$\dfrac{2{,}12 \times 10^{32}}{E_\gamma^9 A^{2}}$
5	$\dfrac{2{,}89 \times 10^{44}}{E_\gamma^{11} A^{10/3}}$	$\dfrac{9{,}42 \times 10^{44}}{E_\gamma^{11} A^{8/3}}$

directly related to the appearance of the nuclear radius to the powers 2L and (2L − 2) in Eqs (11.42) and (11.43), respectively, and the assumption of constant density for all nuclei.

To provide a scale with which to compare predicted half-lives, the Weisskopf estimates have been used to calculate half-lives for the electric and magnetic multipoles of L = 1–5 and photon energies in the range 0.05–10 MeV for a nucleus of mass number A = 100. These are shown in logarithmic form in Fig. 11.1. Over this range, the half-lives, and thus the transition probabilities, are predicted to vary by over 40 orders of magnitude. For the same energy and photon angular momentum, ML radiation is predicted to take place with probabilities that are roughly two orders of magnitude smaller than EL radiation. At lower energies, the relative probabilities for emission of competing radiations (e.g., E1, M2 or M1, E2, etc.) are predicted to decrease by 6–8 orders of magnitude with a unit increase in the angular momentum and by roughly 5 orders of magnitude at higher energies. The clear inference is that when competing multipoles are possible, only the lowest-order multipole should be observed experimentally with significant intensity unless specific structure effects alter the relative transition rates dramatically.

The magnitudes of the predicted half-lives are of great practical significance. At higher energies and low multipolarites, half-lives are all very short. The estimates predict lifetimes of levels that decay by high-energy, low-multipole photons in the range 10^{-19}– 10^{-4} s. On the contrary, however, photon transitions of low energy and high multipolarity are predicted to have half-lives that can be very, very large. For example, an excited state in a nucleus of A = 100 that could decay *only* by emission of an M3 photon with an energy of 0.1 MeV is predicted to have a single-proton half-life of about 1.4×10^3 s, or about 24 min. Such long half-lives, and ones that are

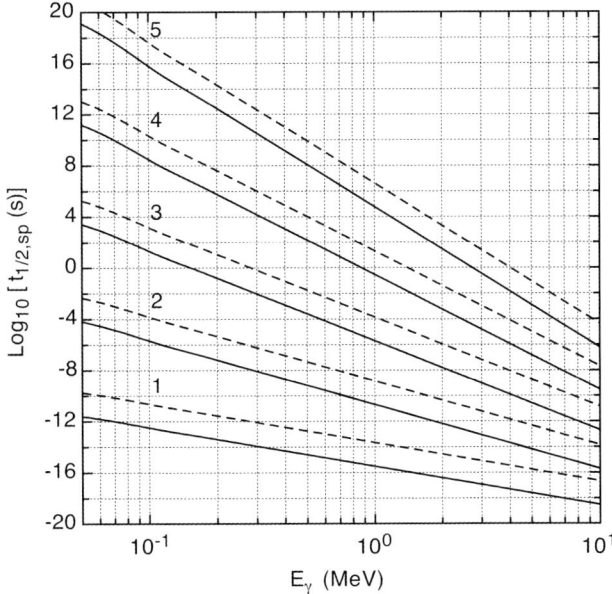

Fig. 11.1 Half-lives calculated with the Weisskopf single-particle estimates for a nucleus of A = 100. The half-lives for EL emission are shown as solid lines while those for ML emission are shown as dashed lines. The multipolarity L is shown just above each EL–ML pair.

much larger, are indeed found experimentally. Excited states of nuclei with lifetimes that are easily measured are referred to as *isomeric* or *metastable* states. They are found throughout the chart of the nuclides and tend to be concentrated in regions where nuclear-structure effects prevent decay by low-order multipoles.

Most photon transitions in spherical nuclei are found to take place with smaller probabilities than those given by the Weisskopf estimates because of the complicated nature of real nuclear states. Nevertheless, the general predictions of the Weisskopf estimates are well-supported by experiment. The fastest experimental instrumentation generally available can measure lifetimes as short as 0.1 ns, and indirect methods can sense lifetimes as short as about 10^{-21} s. In the general case, such levels can be considered to decay "instantaneously". But the longer-lived nuclear isomers are readily observed under most experimental conditions, and some of them are of great technological importance. Because of this and because they provide excellent tests of the quality of the model predictions, we will look at the characteristics of one particular class of nuclear isomers in the following section.

11.4
Examples of Nuclear Isomerism

As pointed out in the discussion of the β decay of ^{85}Br in Chapter X, odd-A nuclides of Z(N) = 49 are predicted by the spherical shell model to have ground and first-excited states with spins and parities of $9/2^+$ and $1/2^-$, respectively, corresponding to the odd particle residing in a $1g_{9/2}$ or $2p_{1/2}$ state. If such a nucleus was produced in its first-excited state, conservation of angular momentum and parity would require that the decay by photon emission is of M4 or E5 multipolarity. From the results displayed in Fig. 11.1, it is quite clear that only the M4 radiation should be seen experimentally. Indeed, in all cases where the experimental measurements have been made, it has been shown that the observed transitions are all of the M4 type.

In Figs 11.2 and 11.3 are shown the characteristics of the ground and first-excited states of a number of the indium isotopes (Z = 49) and the N = 49 isotones. In each and every case, and for all other know indium isotopes and N = 49 isotones, it is seen that the ground and first-excited states have the spins and parities predicted by the spherical shell model. In some cases, one or both of the spin and parity assignments is given in parentheses. This notation indicates that the quantity has not been proven absolutely, but is the most probable value allowed by experiment. When decay from the isomeric state to the ground state has been observed, a vertical arrow is drawn between the two states and is labeled "IT", which stands for *isomeric or internal transition*. Such decays take place by a combination of photon emission and internal conversion, with relative probabilities that depend, in most cases, upon the transition energy and the atomic and mass numbers of the parent isotope only. Concentrating on the lifetimes of the isomers, one finds that they

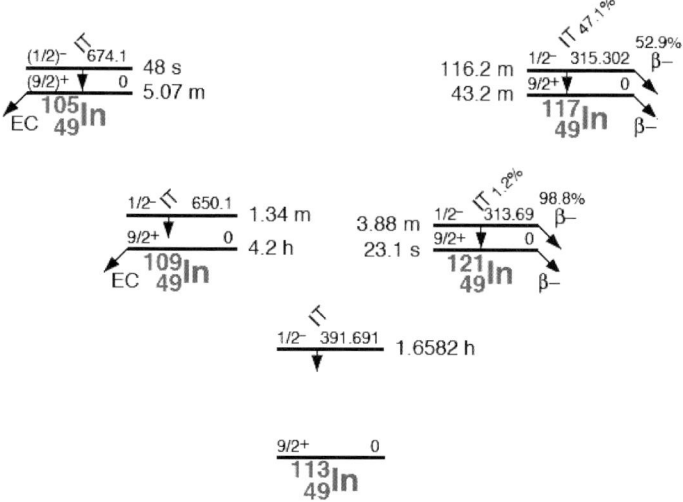

Fig. 11.2 Isomers in the Z = 49 isotopes of indium.

11 γ Decay and Internal Conversion

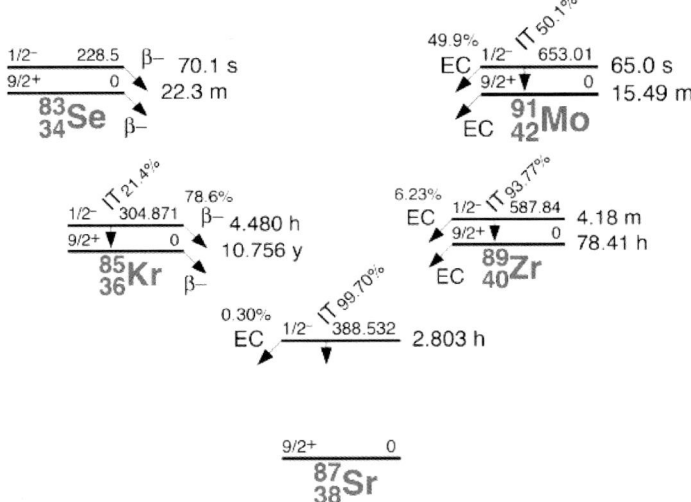

Fig. 11.3 Isomers in the N = 49 isotones.

range from about ten seconds to a number of hours. For all of the more neutron-rich In isomers, decay occurs by both isomeric transitions to the ground state and also by independent β⁻ decay to levels in the Sn (Z = 50) daughters. The isomeric states in a number of the neutron-deficient nuclides are also seen to decay by electron capture.

To examine how well the simple single-particle model accounts for the decay probabilities of these isomeric states, experimental data and theoretical calculations of internal conversion coefficients (see Section 11.6.1) have been used to derive the partial half-lives for decay by photon emission. In Fig. 11.4 are shown the logarithms of partial half-lives for the M4 transitions from isomeric states in those indium isotopes for which sufficient experimental data are available. The data are clearly in excellent agreement with the exponential variation expected from the Weisskopf model. If the half-lives for γ-ray emission are multiplied by $E_\gamma^9 A^2$, the Weisskopf estimate predicts that this should be a constant equal to 2.12×10^{32}. The average value derived for the nine data points shown in the figure is $(2.18 \pm 0.41) \times 10^{31}$.

In order to interpret this result correctly, we must remember that Weisskopf assumed that a single proton in an initial state decayed to an s state that was, initially, completely empty. If, rather than just one, two protons were present in the initial state, the probability for decay to the initially empty s state would be twice the single-particle estimate. On the other hand, if only one proton was present in the initial state, but the final s state was not empty initially but contained one proton, the decay probability would be only one-half that of the Weisskopf estimate; there is only one magnetic substate available rather than the two in the empty s state. In the context of the decay of the odd-A indium isomers, the initial $1g_{9/2}$ state contains 10 protons and the $2p_{1/2}$ final state initially contains one proton. Therefore the probability for decay will be 5 (10 × 1/2) times the probability given by the Weisskopf estimate. Thus, the result of our analysis of experimental data would imply that the constant for a single-proton transition would be $(1.09 \pm 0.20) \times 10^{31}$, within a factor of 2 of the prediction. This must be taken as remarkable agreement considering the simplicity of the model.

Fig. 11.4 also includes data on the decay of isomers in the N = 49 isotones and it is seen that they are in very good agreement with the model and almost indistinguishable from the Z = 49 data. The mean value for the constant to be compared with the Weisskopf prediction is $(2.17 \pm 0.40) \times 10^{31}$, essentially identical to the mean value for the indium isobars. Now if you have been following along carefully, this result should, at first sight, be very surprising. In the N = 49 isotones, it is a *neutron* that is undergoing the transition. It has no charge and only its magnetic moment could give a contribution to the operator for the M4 transition. Yet the transition rates in the N = 49 isotones are essentially indistinguishable from those for the indium isobars! The data we have treated here is but a small subset of all of the known data on M4 transitions but the same conclusions are reached when all experimental data are considered.

Clearly, the situation must be more complex than contained in the simple picture we have been considering up to now. It surely is, but a very simple, qualitative explanation provides one of the principal reasons for the near equivalence of the transition rates. Consider what happens in the center of mass coordinate system of

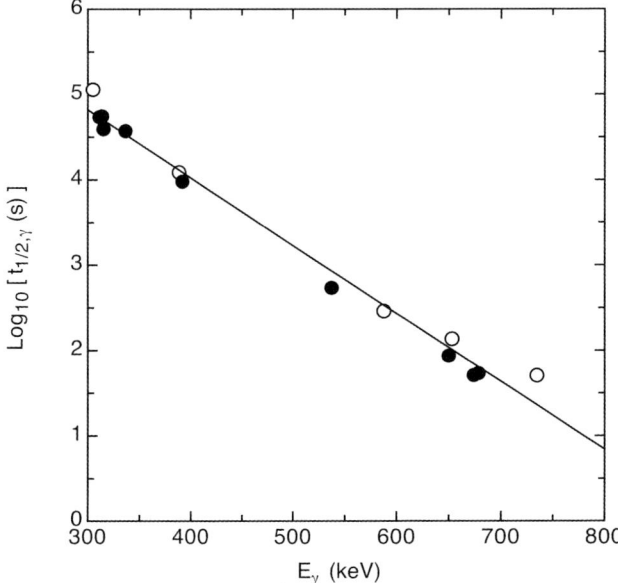

Fig. 11.4 Semi-logarithmic plot of the half-lives for γ-ray emission (s) in the decay of the odd-A indium isomers (solid circles) and N = 49 isotones (open circles) for which sufficient experimental data are available.

the nucleus when a nucleon undergoes a transition between two states. In the transition, its orbit changes and that means, for example, that its mean radial coordinate must change. But if the nucleus is originally at rest, the remaining nucleons must also undergo a change in their average coordinates if momentum is to be conserved. That is, the remaining nucleons will undergo a *recoil* during the transition, regardless of whether the particle that is changing state is a proton or neutron. The conclusion is that it is the *entire* nucleus that is undergoing the transition and all particles then give rise to the changes that are produced in the transition. Because of the near equivalence in mass of the neutron and proton, the currents and charges involved in the transition are therefore quite similar.

11.5
Some General Observations

Literally thousands of γ-ray transitions have been studied and systematics for the rates of these in comparison to the Weisskopf estimates have been developed. We will not go into all of the details but will point out some of the main conclusions that are important to keep in mind when considering the spectrum of γ-rays emitted in radioactive decay.

11.5.1
E1 Transitions

With the exception of light nuclei, E1 transition rates are generally found to be much smaller than predicted by the Weisskopf model. Typical values for the ratio of experimental to predicted transition probabilities are in the range 10^{-2}–10^{-5}. We say that they are strongly *hindered* transitions. This is not to be taken as a failure of the model, but reflects the real complexities of nuclear structure. Within the framework of the spherical shell model it is not difficult to demonstrate that E1 radiation is generally not expected as a transition between low-lying states. E1 transitions take place between levels that differ by one unit in angular momentum and have opposite parities. An examination of the empirical shell model level scheme shows that all of the levels within a major shell have the same parity except for one which has high spin. With the exception of this, the levels correspond to those from a single shell in the isotropic harmonic oscillator. It is then easy to see that in the single particle limit, E1 radiation is not expected at low energies. One would have to have levels separated by at least the energy of the shell gap to see E1 radiation in this limit. Now the facts are that E1 radiation is indeed seen as a major decay mode for very high-lying levels that lie above the shell gap and they are hindered to a much smaller degree. A practical consequence of the hindrance is that one often observes transitions of other multipolarities in competition with E1 decay between low-lying levels.

11.5.2
E2 and M1 Transitions

The Weisskopf estimates predict that when M1 and E2 radiation can compete in a decay, the M1 radiation should completely dominate except for the highest photon energies (see Fig. 11.1). But M1 transition rates are commonly hindered by factors of $10–10^3$ compared to the Weiskopf estimate, while E2 transition rates are often enhanced by factors of 10–20. As a result, it is quite common to find both M1 and E2 radiation emitted when both are allowed. The retardation of the M1 rates can be rationalized by detailed examination of the M1 transition operators and the character of single-particle states. The enhancement of E2 transition rates is due to the electric quadrupole moments of most nuclei because they are not completely spherical. In fact, E2 transition rates in the decay of the strongly deformed rare earth and actinide nuclides can be two orders of magnitude larger than the Weisskopf single-proton rate.

11.5.3
Other Transitions

Higher-order transitions are more the exception than the rule, and are usually not found except for the special cases of nuclear isomerism in spherical nuclei and in some deformed nuclei, for similar reasons. For the majority of the excited states formed in α or β decay, regardless of their spins, one usually finds one or more lower-lying levels to which decay can proceed via photon emission of low multipolarity.

In the average case, the general observations given above provide a good qualitative understanding of what radiations can be expected to be observed and reasonable estimates of the lifetimes of levels that may be isomeric. Rather detailed structure calculations are required to approach a quantitative comparison with experimental data and that lies well beyond our interest.

11.6
Internal Conversion

The process of internal conversion, where the energy difference between two levels is transferred directly to an atomic electron, always competes with photon emission. One can demonstrate that no intermediate photon is involved by the fact that conversion electrons are emitted in $0^+ \rightarrow 0^+$ transitions where photon emission is absolutely forbidden. In general, internal conversion is negligibly small when transition energies are large and multipolarities are small, but it can be the dominant decay mode when transition energies are small and multipolarities are large. Because the energy separation of low-lying levels in deformed nuclei is so small, internal conversion often is relatively intense even in competition with E2 radiation.

The spectrum of conversion electrons is comprised of monoenergetic lines that represent transfer of energy to electrons bound in the different atomic shells. If the transition energy is $E_t (= E_\gamma)$, and ε_i is the electron binding energy in the ith shell, the energy of the emitted conversion electron will be

$$E_e = E_t - \varepsilon_i \tag{11.44}$$

If sufficient energy is available to permit conversion in all electron shells, the energies of the conversion electrons will then vary in the order $E_K < E_{L_1} < E_{L_2} < E_{L_3} < \ldots$ As in electron capture, we expect that the probability for internal conversion will be dependent upon the probability for finding an atomic electron within and near the nucleus and thus expect that, given sufficient decay energy, the probability for internal conversion with electrons in the K shell will be greater than that for electrons in the L shell, etc. Further, because the probabilities for finding electrons in the nucleus increases with atomic number, we might also expect that internal conversion probabilities will increase greatly with Z. Both of these speculations are actually found experimentally, but not directly for the reasons specified. It is the strength of the interaction between the nucleus and the electrons that directly defines the probability for internal conversion as we shall see shortly.

It has become standard practice to express the probability for decay by internal conversion by its ratio to the probability for the competing photon emission. This ratio is called the *internal conversion coefficient* and is defined as

$$\alpha_i = \frac{\lambda_{e,i}}{\lambda_\gamma} \tag{11.45}$$

where the subscript i refers to the electron shell or subshell involved, and $\lambda_{e,i}$ and λ_γ are the probabilities for internal conversion from that shell or subshell, and photon emission, respectively. The total probability for internal conversion is the summation of the probabilities for conversion in each of the electron shells, and is given by

$$\alpha_T = \sum_{\text{all } i} \frac{\lambda_{e,i}}{\lambda_\gamma} = \frac{1}{\lambda_\gamma} \sum_{\text{all } i} \lambda_{e,i} = \frac{\lambda_e}{\lambda_\gamma} \tag{11.46}$$

If decay can occur only by emission of a single photon or by internal conversion, the total decay probability will then be

$$\lambda = \lambda_e + \lambda_\gamma = \lambda_\gamma (1 + \alpha_T) \tag{11.47}$$

and the half-life of the level will be

$$t_{1/2} = \frac{\ln 2}{\lambda_\gamma (1 + \alpha_T)} \tag{11.48}$$

This expression indicates that the half-life of a level will always be smaller than expected solely on the basis of photon emission. More importantly, if the total internal conversion coefficient is large, the half-life for the level can be much smaller than expected from photon emission alone.

11.6.1
Elementary Theory of Internal Conversion

The complete theory of internal conversion is well beyond the level of this text. However, a simplified treatment points out some of the essentials and we will follow this approach [3]. One of the principal points to consider is that the nuclear transition probability is the same for both photon emission and internal conversion. Internal conversion arises from the direct transfer of energy from the nucleus to an atomic electron via the electromagnetic field. The initial state of the system is normally the neutral atom with the nucleus in the excited state of interest. The final state is the +1 ion formed by ejection of the conversion electron with the nucleus in its final state, and an unbound electron with kinetic energy $E_\gamma - \varepsilon_i$. We can use the wave functions for these in the Fermi Golden Rule and calculate the decay probability for a specific pair of initial and final states with the density of final states being that for the conversion electron. We must, of course, consider all of the magnetic substate combinations that are permitted in the nuclear transition and then calculate the average probability through an appropriate summing process. Schematically, the average decay probability for internal conversion can be written as

$$\lambda_e = \frac{1}{(2j_i + 1)} \sum_{m_f, m_i} S_e \frac{2\pi}{\hbar} |\langle \Psi_f \psi_f | H' | \Psi_i \psi_i \rangle|^2 \rho(E_f) \tag{11.49}$$

In the above, the summation over all of the magnetic substates of the initial and final nuclear states, m_i and m_f, that are permitted by conservation of the z-component of angular momentum is shown explicitly. The symbol S_e is meant to symbolize a similar summation over all of the electron states involved directly in the transition. The matrix element contains both the nuclear Ψ and electron wave functions ψ in their initial and final states, respectively. The perturbation potential, H', is the electromagnetic interaction and thus it is the same as that given above in our outline of the theory of photon emission.

In our simplified model, we restrict attention to internal conversion in competition with electric multipole radiation. In this case, only the leading term in the expression for the multipole operators given in Eq. (11.27) need be considered in a first approximation. This means that we consider the interaction due solely to the Coulomb field and we will then take the potential for the interaction that gives rise to internal conversion as

$$H' = -k_C \sum_{\text{all e,p}} \frac{e^2}{|\mathbf{r}_p - \mathbf{r}_e|} \tag{11.50}$$

where r_p and r_e are the radius vectors to the proton and electron, respectively, and the summation should be taken over all of the protons and electrons in the system. Direct substitution into the matrix element of Eq. (11.49) then gives

$$\langle \Psi_f \psi_f | H' | \Psi_i \psi_i \rangle = - k_C e^2 \langle \Psi_f \psi_f | \sum_{\text{all e,p}} |\mathbf{r}_p - \mathbf{r}_e|^{-1} | \Psi_i \psi_i \rangle \tag{11.51}$$

Now the integral implied by the matrix element must be carried out over the full range of the electron coordinates and not just over the nuclear coordinates. That is, it must be taken over the range $0 \leq |\mathbf{r}_p - \mathbf{r}_e| \leq \infty$. On the other hand, we know that the principal part of the electron density will be found outside of the nucleus. Although it is a definite approximation and can lead to difficulties in some cases, the main aspects of the matrix elements can be demonstrated with the assumption that we can neglect that part of the integral within the nucleus itself. This leads to substantial simplifications because so long as we are outside of the nucleus, the potential can be expanded in terms of the electric multipole moments we considered previously in Chapter VIII, during our examination of the shapes of nuclei. In the present case, we must obtain a relation between the radius vectors to a proton and electron as shown in Fig. 11.5. The circle shown in the figure represents a cross section through the center of the nucleus. The radius vector to a proton has the polar and azimuthal angles α, β, respectively, and the radius vector to the electron has the corresponding angles γ, δ. The planar angle between the two vectors is θ_{pe} and we can first use the law of cosines to write

$$|\mathbf{r}_p - \mathbf{r}_e|^2 = |\mathbf{r}_p|^2 + |\mathbf{r}_e|^2 - 2|\mathbf{r}_p||\mathbf{r}_e|\cos\theta_{pe} \tag{11.52}$$

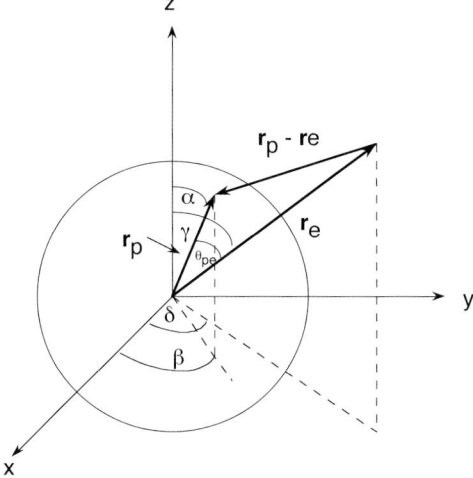

Fig. 11.5 Vector diagram relating the locations of a proton and an electron. The angles α, β and γ, δ are the polar and azimuthal angles for the proton and electron vectors, respectively.

11.6 Internal Conversion

Inverting and taking the square-root of the equation, we can follow the same prescription that was used in Chapter VIII to expand $|\mathbf{r}_p - \mathbf{r}_e|^{-1}$ in terms of Legendre polynomials and obtain

$$|\mathbf{r}_p - \mathbf{r}_e|^{-1} = \sum_{l=0}^{\infty} \frac{r_p^l}{r_e^{l+1}} P_l(\cos\theta_{pe}) \tag{11.53}$$

To perform the integration implied in (11.51), we will have to deal with integration over both the radial and angular coordinates of the wave functions. This means that we must relate the angle θ_{pe} to the angle pairs α,β and γ,δ of the two radius vectors. This is not quite so difficult as it first appears because there is a well-known relation called the addition theorem for the spherical harmonics that is derived in texts on mathematical methods for physics and engineering. We state it here, without proof, in the form

$$P_l(\cos\theta) = \frac{4\pi}{2l+1} \sum_m Y^*_{l,m}(\alpha,\beta) Y_{l,m}(\gamma,\delta) \tag{11.54}$$

In this equation, the angle sets α,β and γ,δ are the polar and azimuthal angles θ,ϕ that characterize a vector in two coordinate systems, where the second is obtained from the first by a simple rotation in space. We have used the same labels in Fig. 11.5 to point out the fact that if we consider only the *directions* of the radius vectors to the proton and electron, the direction of one can also be obtained from the other by a simple rotation of the coordinate system. Because the magnitudes of the radius vectors to the proton and the electron have been accounted for separately, we can use this relation to express θ_{pe}. We then write

$$P_l(\cos\theta_{pe}) = \frac{4\pi}{2l+1} \sum_m Y_{l,m}(\theta_e,\phi_e) Y^*_{l,m}(\theta_p,\phi_p) \tag{11.55}$$

and substitution of this into Eq. (11.53) gives

$$k_C \frac{e^2}{|\mathbf{r}_p - \mathbf{r}_e|} = \sum_l \sum_{|m|\leq l} k_C e^2 \frac{4\pi}{2l+1} \frac{r_p^l}{r_e^{l+1}} Y_{l,m}(\theta_e,\phi_e) Y^*_{l,m}(\theta_p,\phi_p) \tag{11.56}$$

Of particular interest here is the quantity $er_p^l Y^*_{l,m}(\theta_p,\phi_p)$. Reference to Eq. (11.27);

$$O^{EL}_{LM} = er^L Y^*_{LM} - \frac{i\mu e\omega}{2m(L+1)}(\boldsymbol{\sigma}\times\mathbf{r})\cdot[\nabla(r^L Y_{LM})]^*$$

shows that it is the first term in the expression for the operator of electric multipole radiation. By considering only the Coulomb field, the second term in the operator does not play a part and we can now write (11.56) as

$$k_C \frac{e^2}{|\mathbf{r}_p - \mathbf{r}_e|} \approx k_C \sum_l \sum_{|m| \le l} e O_{LM}^{EL} \frac{4\pi}{2l+1} \frac{Y_{l,m}(\theta_e, \phi_e)}{r_e^{l+1}} \tag{11.57}$$

This is a very interesting result. The operator O_{LM}^E acts only on the protons in the nucleus, whereas the ratio $Y_{l,m}(\theta_e, \phi_e)/r_e^{l+1}$ is solely associated with the electron coordinates. These two facts mean that, within the level of the approximations made, the matrix element in (11.51) is separable in terms of proton and electron coordinates and can be written as

$$\langle \Psi_f \psi_f | H' | \Psi_i \psi_i \rangle \approx k_C e \langle \Psi_f | \sum_{l,m} O_{LM}^E | \Psi_i \rangle \langle \psi_f | \sum_{l,m} \frac{4\pi}{2l+1} \frac{Y_{l,m}(\theta_e, \phi_e)}{r_e^{l+1}} | \psi_i \rangle \tag{11.58}$$

The first matrix element is exactly that which is found in the expression for the emission of the photons that compete with internal conversion. If Eq. (11.58) is substituted into Eq. (11.49), the total decay constant for internal conversion is obtained. If the ratio of this to the decay constant for photon emission of the same multipolarity is taken, the result is the total internal conversion coefficient, α_T. But because the nuclear matrix elements in each are identical, the expression for α_T will be completely independent of the nuclear states. That means that all of the internal conversion coefficients, the total as well as the coefficients for each electron shell or subshell, will be independent of the details of the nuclear transition and can be calculated and used without regard to the specific transition under study. In the majority of cases, this is true. This is quite powerful and makes internal conversion coefficients quite useful. However, while the neglect of the integration over the dimensions of the nucleus is reasonable, there are a number of cases where this neglect is a sufficiently poor approximation that it can lead to significant and substantial error, at least with respect to the use of conversion coefficients for defining the multipolarity of a radiation. From the practical viewpoint of obtaining good estimates of the total internal conversion coefficient, the approximation is not of real concern.

Detailed tables of internal conversion coefficients can be found in the literature and very useful figures can be found in such references as the Table of Isotopes. For our purposes, we want to demonstrate the results of such calculations to point out their general characteristics and to show the conditions under which internal conversion is significant. In Figs 11.7–11.10 are graphs of the E(M)1–E(M)4 K-conversion coefficients for nuclides of $Z > 10$. As the reader scans through these figures, it will be seen that there is a significant change in the ordinate of the graphs. The lower limit of each is 10^{-7} but the upper limit increases by three orders of magnitude as one proceeds through the multipoles $L = 1-4$ over the energy range of 0.01–10 MeV. The dependence of the K-conversion coefficient on the atomic number is quite large, increasing by 3–4 orders of magnitude as Z varies in the range $10 \le Z \le 100$.

Similar calculations have been performed for the three L subshells and these are also available in the literature. Typically, the total probability for conversion in the L shell is roughly 0.3 that for conversion in the K shell and conversion in the M shell is much smaller. To a good approximation, and except for the case where the decay energy is less than the binding energy of a K electron, the K conversion coefficient will permit a reasonable estimate of α_T to be made. The strong dependence of both the internal conversion coefficient and the decay constant for photon emission on decay energy leads to the fact that the half-lives of levels that can decay by only these two processes will be strongly reduced from the predictions of the Weisskopf model at lower energies and especially at higher Z. It is common, in place of Fig. 11.1, to display half-lives corrected for the presence of internal conversion. Such half-lives are displayed in Figures 11.11 and 11.12 for electric and magnetic multipoles, respectively. As can be seen, half-life estimates are strongly reduced for transition energies less than about 400 keV and multipole order 2 or greater.

11.7
Decay Schemes

We now build upon the use of decay schemes that was introduced in Chapter X to present data on radioactive decay and to provide a means of correlating data with our simple models. We begin by a reconsideration of the decay of 53.12-d $^{7}_{4}$Be that includes all of the information normally presented (Fig. 11.6). On each level to the left is the spin and parity assignment measured experimentally, and to the right is the level energy in keV relative to that of the ground state. Either above or to the left of a level is its half-life, if measured, and "stable" refers to stability against normal decay processes. The β-decay transitions, electron capture in this case, are shown as the arrows connected to the vertical line originating from the ground state of the decay parent. The atomic number of the nuclides considered increases from left to right and thus a β⁻-decay parent will be given to the left of the daughter level scheme and the arrows representing decay to individual levels will point from left

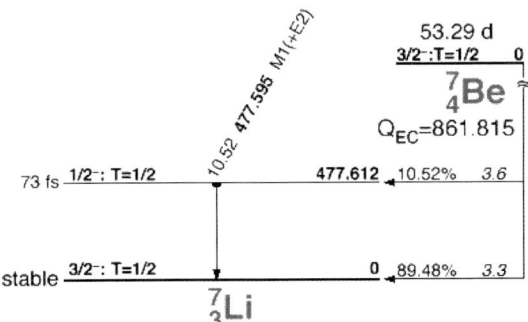

Fig. 11.6 Decay scheme for 53.12-d $^{7}_{4}$Be.

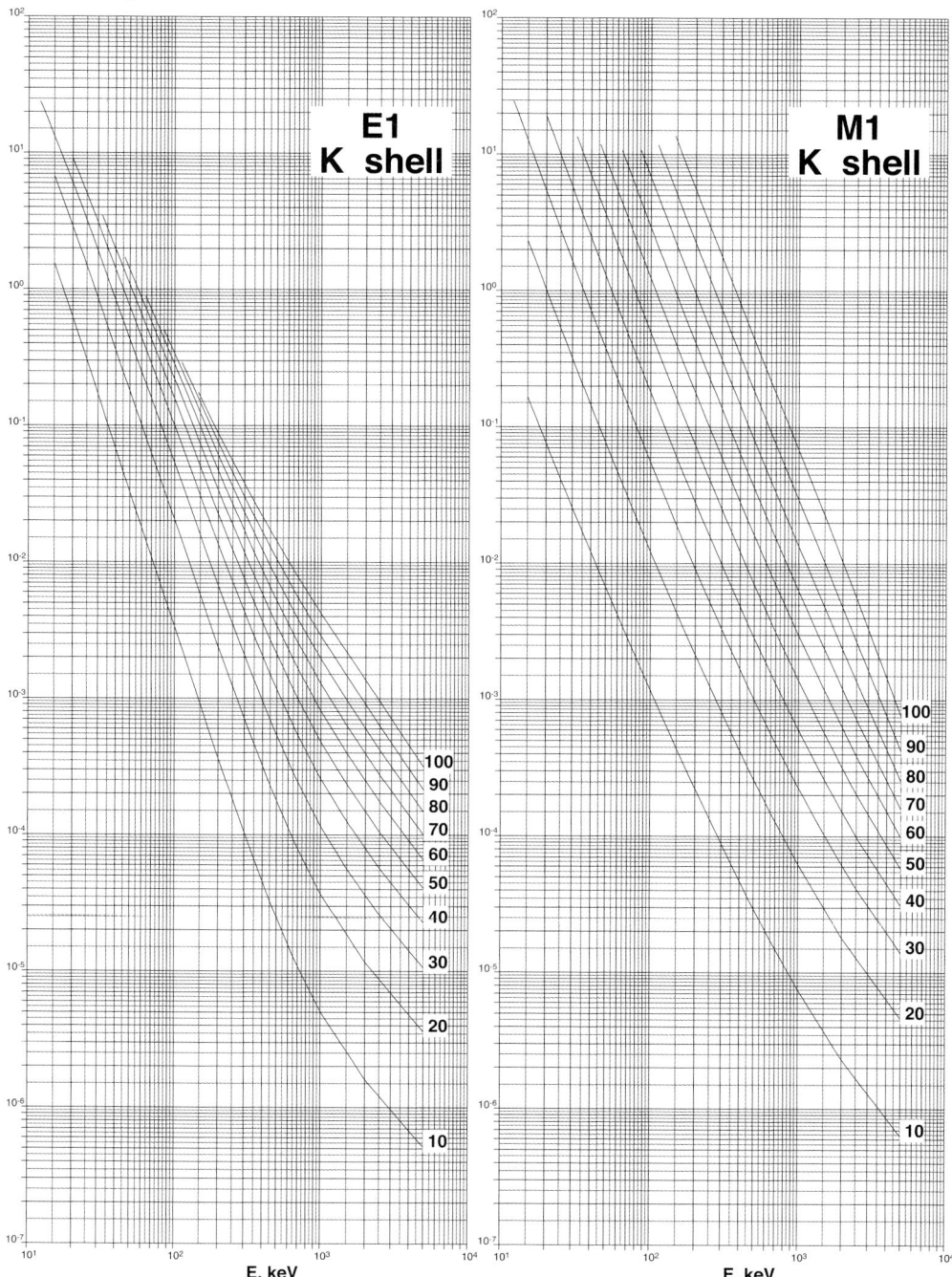

Fig. 11.7 E(M) 1 internal conversion coefficients for Z > 10.

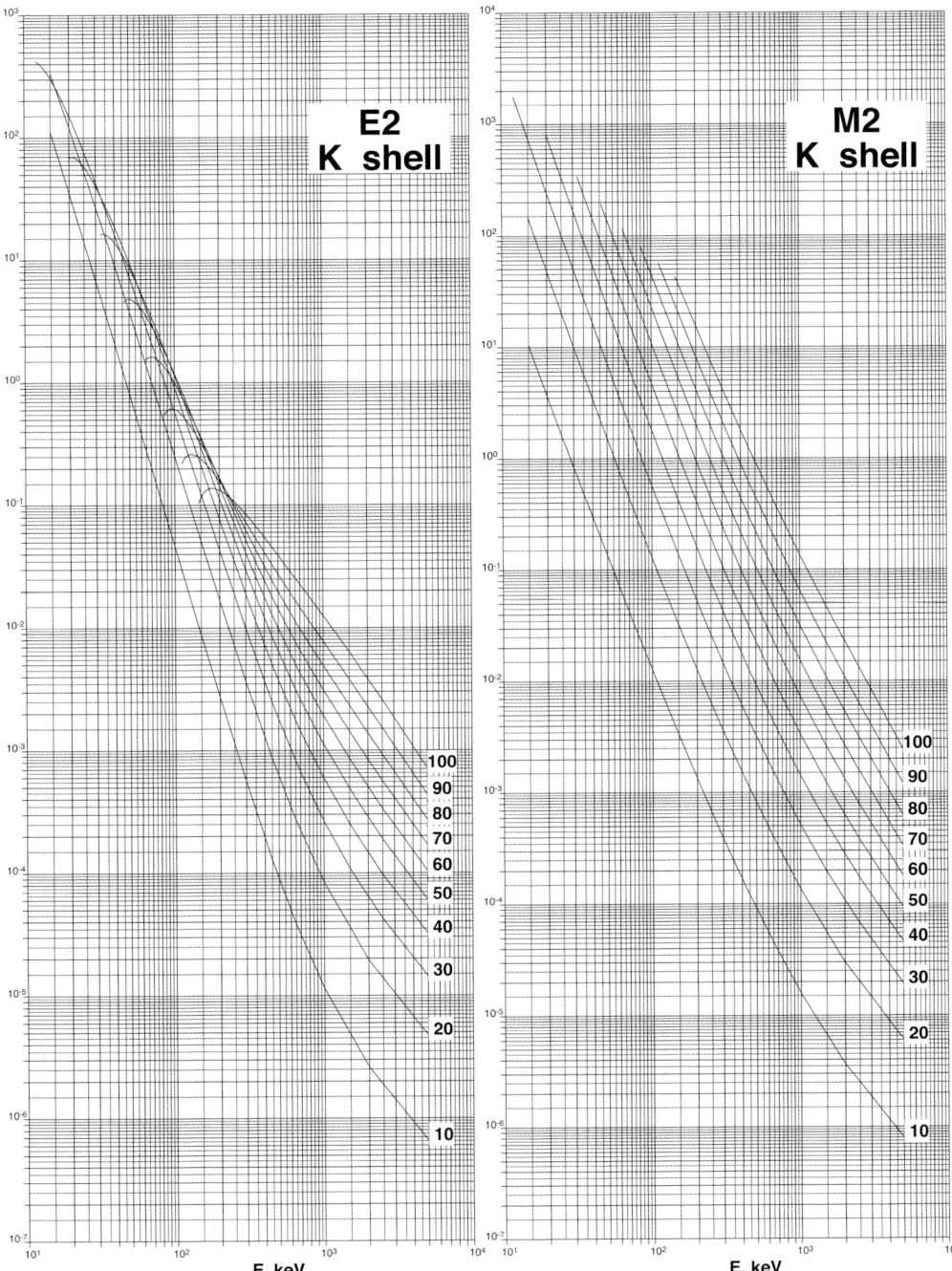

Fig. 11.8 E(M)2 internal conversion coefficients for Z > 10.

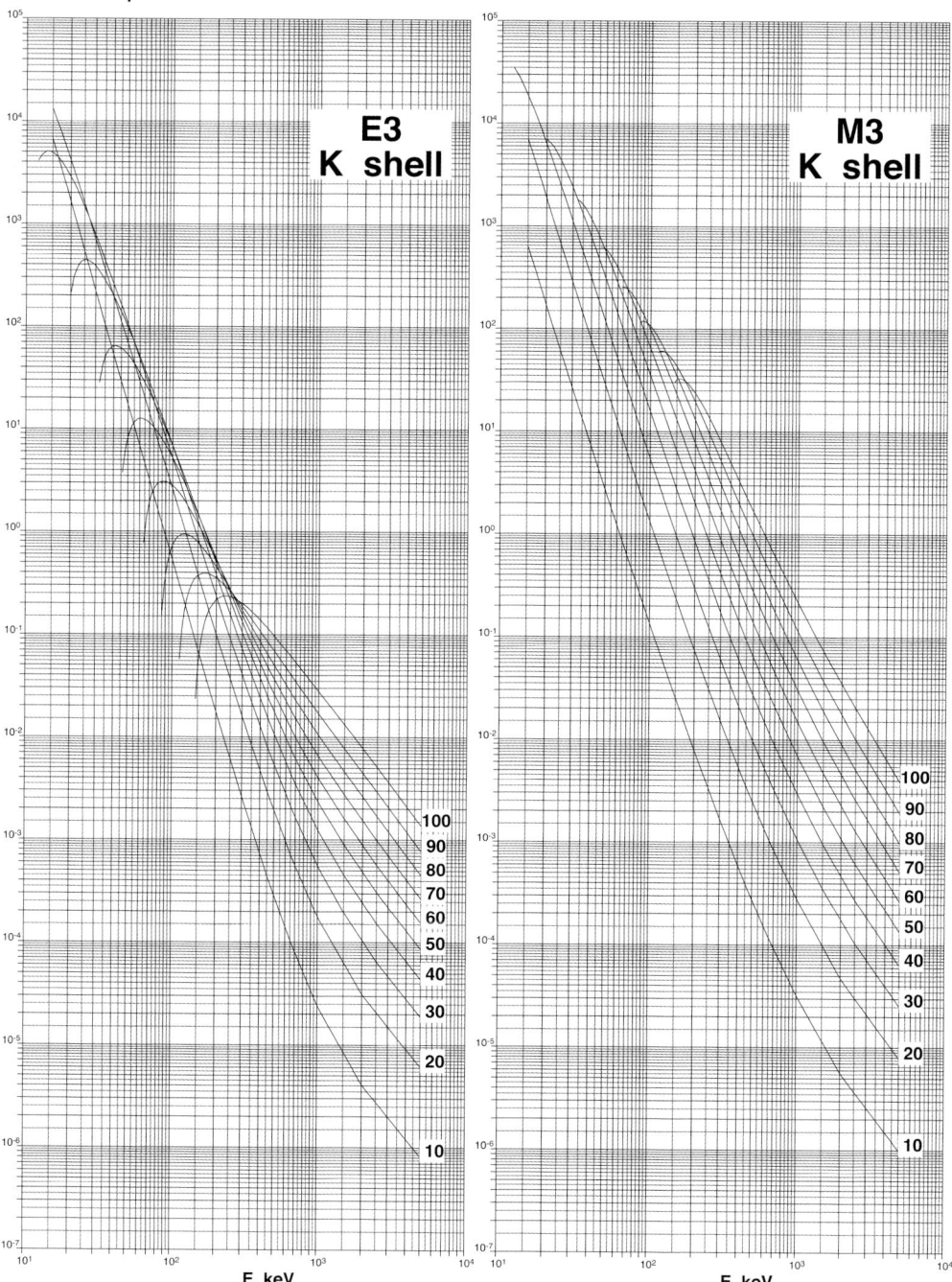

Fig. 11.9 E(M)3 internal conversion coefficients for Z > 10.

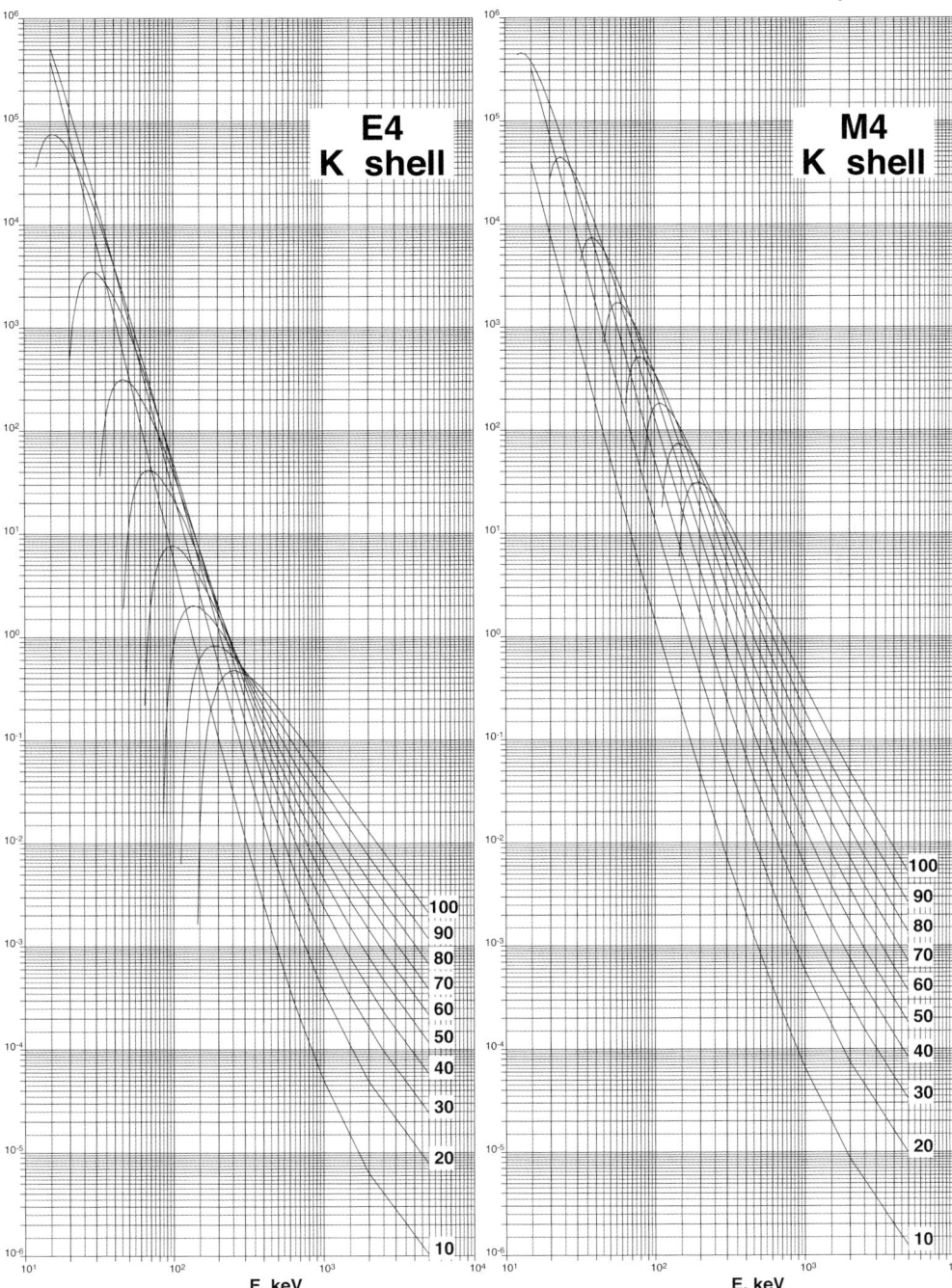

Fig. 11.10 E(M)4 internal conversion coefficients for Z > 10.

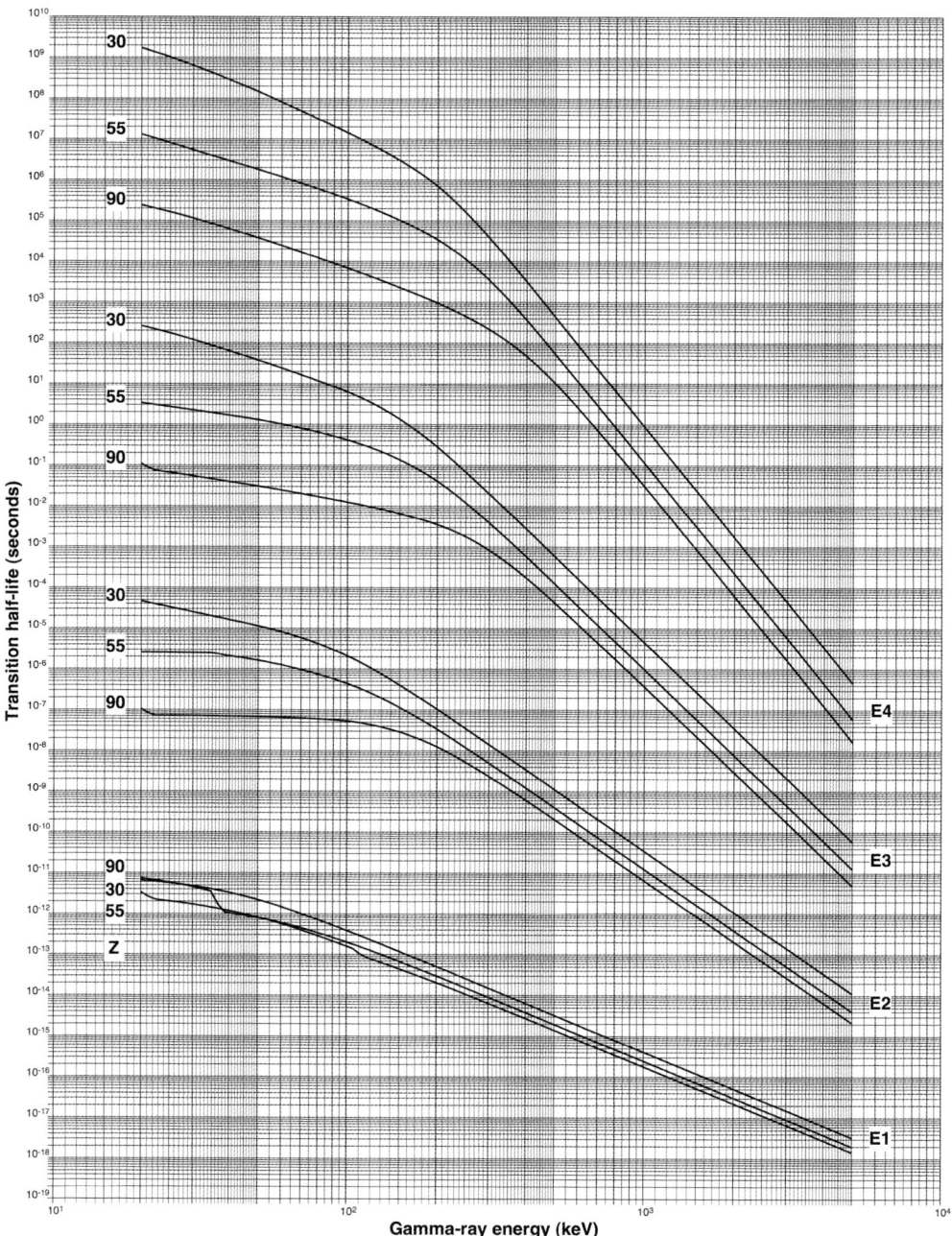

Fig. 11.11 Electric multipole transition rates (s). These represent the Weisskopf single proton transition rates corrected for internal conversion.

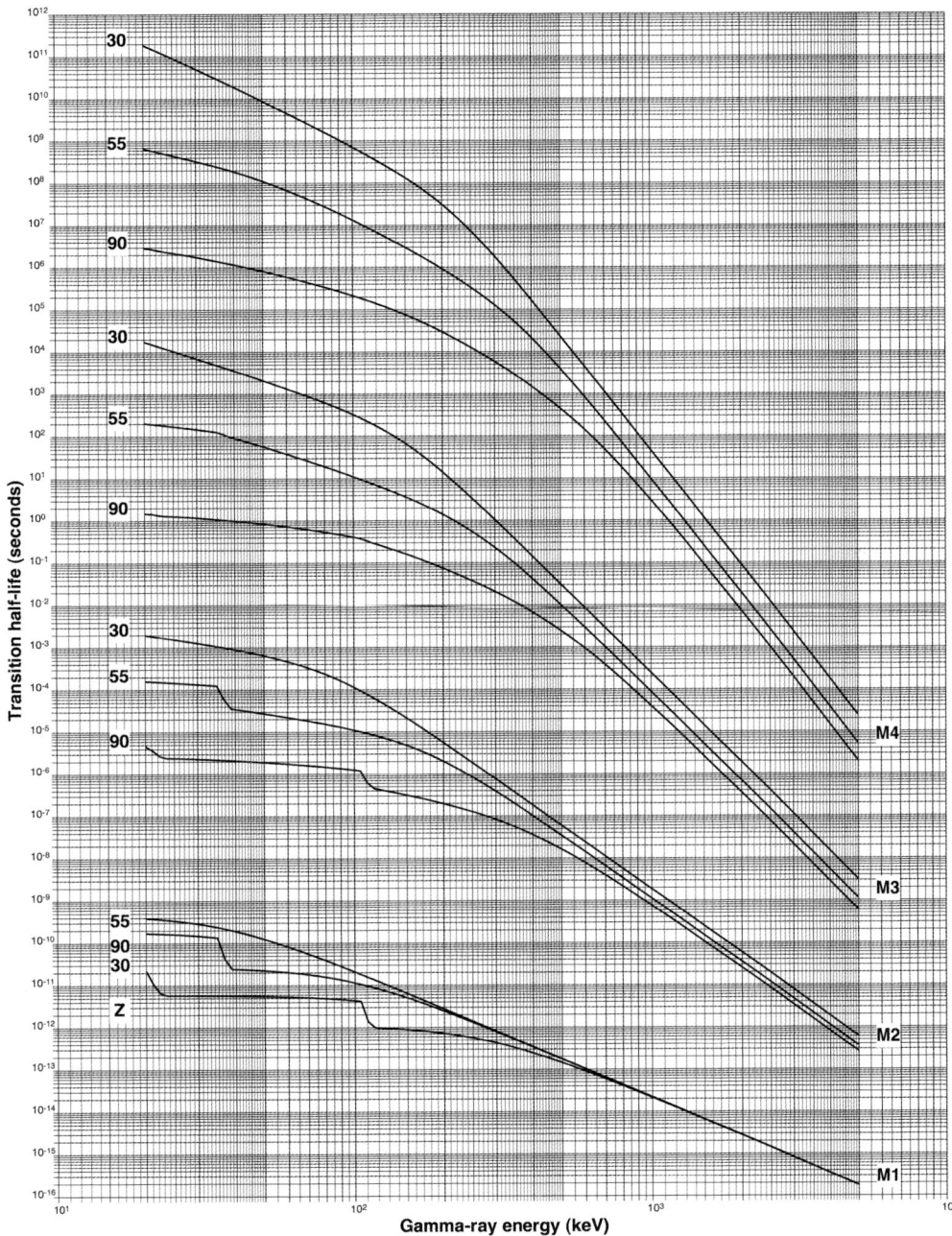

Fig. 11.12 Magnetic multipole transition rates (s). These represent the Weisskopf single proton transition rates corrected for internal conversion.

to right. The first number on each of these is the absolute decay intensity (%), i.e., the fraction of the parent decay that leads to the transition in question, and the second, shown in italics, is the experimental $\log_{10}$ ft for the decay.

In the present case, the electron capture decay energy is shown just below the ground state of the parent. The 477.6 keV photon emitted in the decay of the first-excited state is shown as the vertical arrow pointing from the latter to the ground state. Its energy, the transition energy, is given in bold face type at an angle above the base of the arrow. Preceeding the energy is the absolute intensity of the *photon* (%) found in the decay, and following it is the measured multipolarity of the transition.

The electron capture decay itself and the structure of the three nuclear levels shown were discussed previously in Chapter X and the reader should review the material presented there. Briefly, the two transitions shown are both superallowed and can be explained as due to the decay of a $p_{3/2}$ proton in $^{7}_{4}$Be to a $p_{3/2}$ neutron in $^{7}_{4}$Li. For the present, we want to concentrate on decay of the first-excited state in $^{7}_{4}$Li. The spins and parities of the ground and first-excited states are $3/2^-$ and $1/2^-$, respectively. The allowed transitions in decay of the first-excited state are then M1 and E2 only. As discussed earlier in this chapter, these frequently compete with one another and the experimental data indicate that both multipolarities contribute to the decay. From Figs 11.7 and 11.8, the K conversion coefficients for a transition of this energy in a nucleus of Z = 10 are roughly 3×10^{-5} and 8×10^{-5}, respectively. We then expect that the photon intensity should be essentially the same as the electron capture intensity that populates the excited state, and the absolute intensity of the 477.6 keV photon shown in the decay scheme is identical to that of the electron capture transition to the first-excited state. This means that, within experimental errors, the excited state does not decay by any other mode.

The Weisskopf single-particle estimates (see Table 11.1) for the half-lives of M1 and E2 transitions of about 480 keV are 2.2×10^{-13} s and 2.9×10^{-8} s, respectively. The measured half-life is 73 fs, or 7.3×10^{-14} s, reasonably close to the single-particle estimate for the M1 transition. We expect then that the contribution of E2 radiation in the decay must be very small. The experimental data indicate that the E2 contribution is only 5×10^{-3}%. Under most experimental conditions such a small contribution would not be seen.

As a second, and somewhat more complex example, we consider the β^- decay of $^{42}_{19}$K along with the electron capture/positron emission of $^{42}_{21}$Sc to levels in $^{42}_{20}$Ca shown together in Fig. 11.13. Q_{β^-} for decay of $^{42}_{19}$K is over 3.5 MeV and levels in $^{42}_{20}$Ca with energies over 3.4 MeV are populated. Notice that with the exception of the level at 3.447 MeV, to which decay occurs with a $\log_{10}$ft of 5.0, all other β^- transitions are highly hindered. However, in spite of the highly-hindered transition to the ground state ($\log_{10}$ft = 9.5), the majority of the β^- intensity leads directly to this level. This can be attributed in large measure to the overriding effect of the very large decay energy in comparison to other β^- transitions.

Experiment has uncovered two isomers of $^{42}_{21}$Sc and the decay of both is shown in the figure. The decay of the isomeric state is given in the lower part of the figure for convenience. The ground state of $^{42}_{21}$Sc has been shown to have a spin and parity

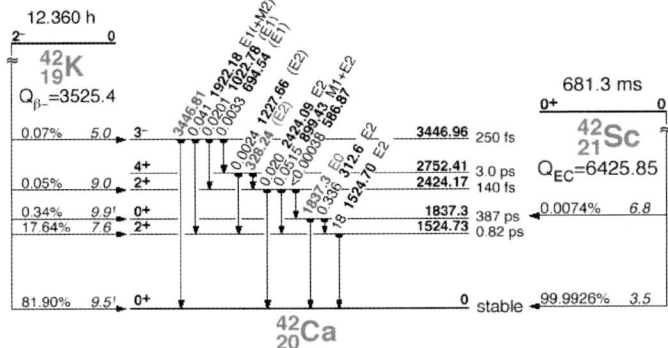

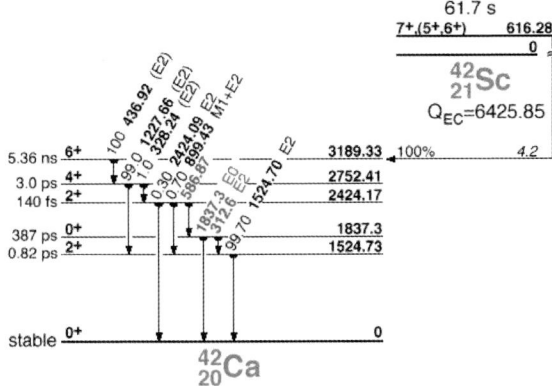

Fig. 11.13 Decay of $^{42}_{19}$K and the isomers of $^{42}_{21}$Sc to levels in $^{42}_{20}$Ca. (a) The decay of ground states of both $^{42}_{19}$K and $^{42}_{21}$Sc. (b) The decay of the isomeric state of $^{42}_{21}$Sc.

of 0^+, and almost all of its decay goes to the 0^+ ground state of $^{42}_{20}$Ca. This transition is superallowed. With a decay energy of over 6.4 MeV, the half-life of the ground state is only 0.68 s. Positron emission accounts for essentially all of the ground-state decay. Reference to the EC/β^+ ratios given in Fig. 10.15 suggests that K capture should occur in only about 10^{-4} of all decays.

The isomeric state with a half-life of 61.7 s lies above the ground state by about 616 keV, but the isomeric transition to the ground state has not been observed experimentally. Essentially all decay goes solely to the 6^+ level in $^{42}_{20}$Ca at about 3189 keV by positron emission. The lack of an observable isomeric transition follows directly from the fact that the transition is likely to correspond to a transition of ΔL = 7 (no). Even if the transition was of multipole order M5, the Weisskopf estimate of the half-life would be about 10^{12} s and clearly could not compete successfully with the observed positron transition.

The nuclei involved in these decays are not far removed from closed shells and we should be able to rationalize the decay schemes in terms of the single-particle model. To begin, we first consider the structure of $^{42}_{20}$Ca. It can be considered in zeroth order as the inert, doubly-magic $^{40}_{20}$Ca core plus two neutrons. Reference to the level diagram in Fig. 7.13 suggests that the 21st and 22nd neutrons will be paired in the $1f_{7/2}$ orbital in the ground state. The $1f_{7/2}$ orbital, when filled, completes the N = 28 shell, and the next lowest-lying orbital, the $2p_{3/2}$ orbital, lies considerably above it. Given the large energy difference, we would predict the low-lying excited states of $^{42}_{20}$Ca to be due to the various couplings of the two neutrons in the $1f_{7/2}$ orbital that give rise to levels with spins and parities of 0^+, 2^+, 4^+, and 6^+. Although we have not developed a model for the energy spectrum of states, it can be shown that the level energies should also increase in the same order with the 0^+ level lying considerably below the average energy of the other levels. Levels with these characteristics are indeed seen among the excited states. In addition, however, there are two other states of spin and parity 0^+ and 2^+, respectively, and that means that at least one other excitation must be considered. The origin of these can be understood when we consider that energies of 3–4 MeV are comparable to the energy of the shell gap. This idea is supported by the most energetic level populated in the decay of $^{42}_{19}$K, the 3^- level at about 3.447 MeV. At such large energies, it is clear that complex excitations can take place.

With this general analysis, we now turn to the ground state of $^{42}_{19}$K and its decay to $^{42}_{20}$Ca. $^{42}_{19}$K is an (odd, odd) nucleus and the single-particle level diagram would suggest that the odd particles lie in the $d_{3/2}$ proton orbital and the $1f_{7/2}$ neutron orbital. These can couple to give levels with spins and parties $(2-5)^-$. Indeed, the spin 3, 4 and 5 members of this set are found just above the ground state of $^{42}_{19}$K. It is then quite reasonable to take the ground-state configuration of $^{42}_{19}$K to be $(n: 1f^3_{7/2}\ p: 1d^3_{3/2})|_{2^-}$. To form the ground state of $^{42}_{20}$Ca with the configuration of $(n: 1f^2_{7/2}\ p: 1d^4_{3/2})|_{0^+}$, we would have to have the neutron decay n: $1f_{7/2} \to$ p: $1d_{3/2}$, corresponding to a change in orbital angular momentum of the particle orbits by two units. The change in the angular momentum and parity of the nuclear levels themselves is $\Delta J = 2$ (yes) corresponding to a first-forbidden unique transition with an expected $\log_{10} ft \geq 8.5$ (Table 10.1). This is just about what one finds experimentally. From our discussion of the structure of $^{42}_{20}$Ca we would expect that decay to all of the lower-lying excited states should be forbidden and this is exactly what is observed. The only allowed decay is to the 3^- level and it takes place with a $\log_{10} ft$ that is typical of allowed transitions among the heavier nuclides.

We now turn to the decay of the (odd, odd) $^{42}_{21}$Sc isomers. The shell model leads to the expectation that both the 21st neutron and 21st proton are in $1f_{7/2}$ orbitals in the ground state. Again, while we cannot predict the spin of the ground state, we can suggest that it has even parity and a spin in the range 0–7. Experimentally, the ground state has been shown to be 0^+. Further, we can expect that the other levels from coupling of the odd particles will lie low in energy and it is quite reasonable to associate the isomer with the spin 7 member of this group. Therefore the two would be expected to have the configuration (n: $1f^1_{7/2}$ p: $1d^4_{3/2} 1f^1_{7/2})|_{(0,7)^+}$. The 0^+ ground state of $^{42}_{20}$Ca could simply be formed by the decay p: $1f_{7/2} \to$ n: $1f_{7/2}$ from the 0^+ ground state of $^{42}_{21}$Sc. This should be allowed or superallowed and that is exactly what is found experimentally. Additionally, the same proton to neutron

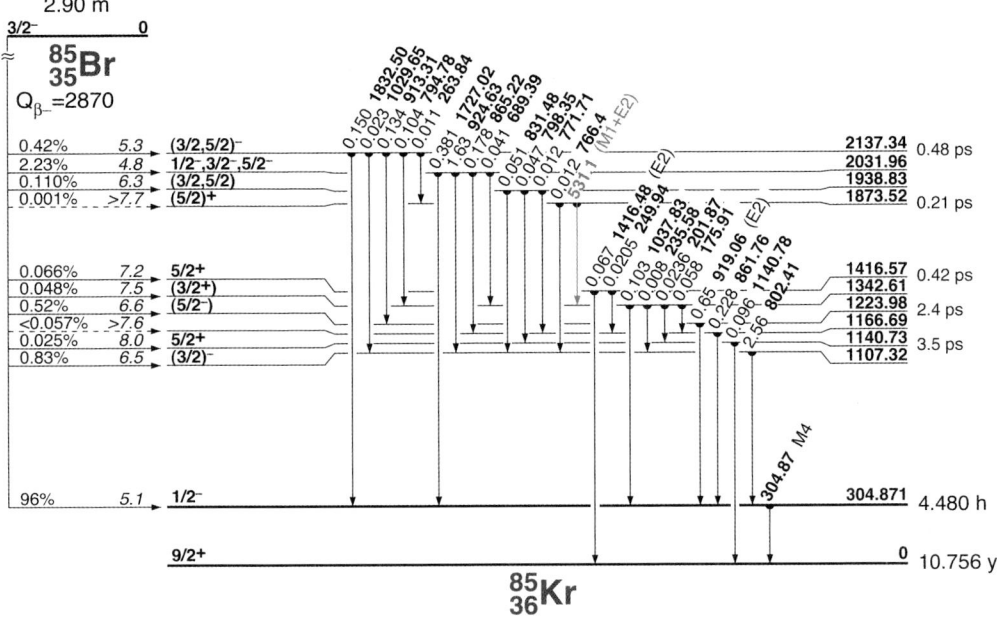

Fig. 11.14 Decay scheme for 2.90 min $^{85}_{35}$Br.

transition in decay of the isomer could populate the 6^+ member of the $f^2_{7/2}$ configuration and we indeed see an allowed decay to the 6^+ level in $^{42}_{20}$Ca at 3189 keV.

Nuclei just two nucleons removed from a doubly-magic nucleus have relatively simple structures and the example given above shows that one can generally rationalize many of the experimental observables quite well. (We leave it to the reader to examine the lifetimes of the levels in $^{42}_{20}$Ca relative to the predictions of the Weisskopf single-proton estimate.) Most nuclei are much more complex and are not nearly as amenable to such simple interpretations. Nevertheless, some of the characteristics of the decay to the ground state and at least a few of the low-lying levels can be understood in most cases in a simple way.

There are two more examples of decay schemes that are worth presenting. The first is the decay of $^{85}_{35}$Br as an example of schemes of medium complexity (Fig. 11.14). The decay populates some 12 levels in $^{85}_{36}$Kr with appreciable intensity and some 25 γ-rays are known to be emitted. Rationalizing such a complex scheme completely is just not possible with our simple models. However, as shown in Chapter X and in our discussion of nuclear isomers, some of the main features can be understood reasonably well. From a practical viewpoint such schemes present difficulties as well as opportunities. Notwithstanding the complexity of the scheme, it is relatively easy to measure the γ-ray spectrum with sufficient resolution to observe lines representing all or almost all of the individual transitions. Such a spectrum is so peculiar to the nucleus undergoing decay that it serves as a "fingerprint" that identifies the parent uniquely. On the other hand,

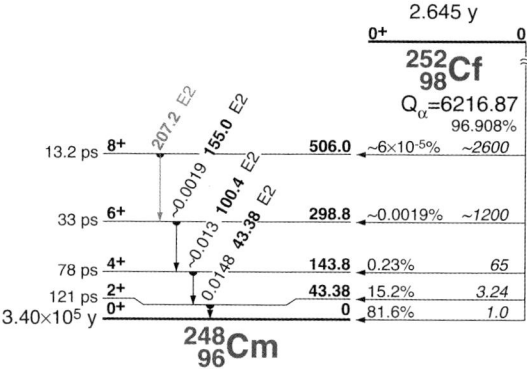

Fig. 11.15 The α-decay scheme of $^{252}_{98}\text{Cf}$ to levels in $^{248}_{96}\text{Cm}$. The numbers in italics to the far right of each level are so-called "hindrance factors". They are essentially the ratio of the α intensities expected from our simple α-decay model to those found experimentally. Note also that the remaining decay intensity is due to spontaneous fission.

the β⁻ spectrum will be continuous and quite complex and essentially useless for identification purposes.

The last example that we will present is the α-decay scheme of $^{252}_{98}\text{Cf}$. This is typical of the type of decay found among the heaviest elements (Fig. 11.15). The scheme demonstrates the population of members of a rotational band built upon the ground state of the daughter. Note that while almost all of the α intensity goes to the ground and first-excited states in $^{248}_{96}\text{Cm}$, α-decay measurements are so sensitive that they can easily see transitions with absolute intensities as small as 6×10^{-7}. Because of background radiation, it is usually very difficult to observe photons with such low intensities and thus the only γ-ray we might expect to see under most conditions is that emitted in the decay of the first-excited state. But while about 15% of the α decay populates this state, the photon intensity is only about 0.0148%. The principal decay mode of the level must be internal conversion and examination of the total internal conversion coefficient in this case indicates that it is roughly 10^3. Therefore, the most intense photons we can expect to see are x-rays following the internal conversion.

Because of the very high atomic number, the K-binding energies in the heavy elements are quite large. In the present case it is about 128 keV. This means that internal conversion can only take place in the L and higher shells. The maximum energy of the L x-rays for Z = 96 is about 23 keV. Reference to the fluorescence yields in Fig. 10.13 indicates that the total yield of L x-rays will be no more than about 0.4 per initial vacancy. Because there are many L x-rays possible, the fluorescence yield will be spread over a number of transitions and therefore none of them will be very intense. In fact the most intense x-ray has an energy of 19.2 keV and an absolute intensity of only about 2.9%.

This analysis makes a very important point. In the α decay of (even,even) isotopes of the heavy elements, photon intensities will be small and in many cases mostly associated with low-energy L x-rays. In the α decay of odd-A nuclides, as we have pointed out before, structure effects tend to produce intense decay to levels located some 100–200 keV above the ground state and some intense low-energy γ-rays are usually observed.

References

1 J.M. Blatt and V.F. Weisskopf, "Theoretical Nuclear Physics", John Wiley and Sons, New York (1952).
2 S.A. Moszkowski, in "Alpha-, Beta- and Gamma-Ray Spectroscopy", edited by K. Siegbahn, North-Holland Publishing Company, Amsterdam (1965).
3 M.A. Preston, "Physics of the Nucleus" Addison-Wesley, Inc., Reading, Massachusetts (1962).

General References

The subject of radiative transitions in nuclei is, as has been demonstrated in this chapter, very complex. There are no "simple" approaches to the general problem. Reasonable introductions can be found in most texts on quantum mechanics. Approachable discussions can also be found in P. Marmier and E. Sheldon, "Physics of Nuclei and Particles", volume I, Academic Press, New York (1969) and K.S. Krane, "Introductory Nuclear Physics", John Wiley & Sons, New York (1987).

Problems

1. (a) Derive an expression for the recoil energy of a nucleus with mass number A following emission of a γ-ray with energy E_γ. For simplicity, you can assume that the recoil energy is very small compared to E_γ.
(b) Under most situations, we neglect the recoil energy of the nucleus following the emission of a γ-ray. Calculate the recoil energy when photons of energy 0.1, 1.0 and 10.0 MeV are emitted by nuclides with mass numbers A = 10 and A = 100. Noting that it is usually very easy to measure the energy of a γ-ray to an error (± 1σ) of about 0.2 keV, estimate when the recoil energy can be safely neglected.

2. Use the Weisskopf single-particle estimates (Table 11.1) to calculate the lifetimes of states in nuclides with mass numbers A = 10, 100 and 250 that decay solely by E2 radiation with an energy of 0.01, 0.10, 1.0 and 10.0 MeV, respectively. Repeat for M1 radiation of the same energies.

3. The low-lying level scheme of a hypothetical nuclide is shown below. The level energies are given in MeV.

11 γ Decay and Internal Conversion

```
 9/2+  _____           2.622
11/2+ _____|_____    2.567
 7/2+ _____|_____    2.530


 5/2-  _____           1.744
 3/2-  _____           1.507


 9/2+  _____           0.909




 1/2-  _____           0
```

(a) For the decay of each excited state to the ground state, write down the multipolarities of all transitions that are allowed by the conservation laws.

(b) Copy the level scheme to your work page and draw in, as vertical lines with arrows at the ends, the γ-ray transitions you expect to be *observed* if each of the levels were populated in some way. Label each transition with its expected multipolarity. If significant internal conversion is expected, note this and estimate the fraction of the transition intensity that is expected to appear as conversion electrons.

(c) Assume that the levels could be populated by β^- decay of a parent whose ground-state spin and parity are $9/2^+$. Indicate the type of β transition that would be expected to take place in the direct population of each level by β^- decay.

(d) Assume that $Q_{\beta^-} = 3.00$ MeV. Estimate qualitatively which β transitions you expect to see and which are likely to be the most intense.

4. ^{144}Pr has an excited state with spin and parity of 3^-. It decays to the ground state (spin and parity of 0^-) with an experimental half-life of 7.2 ± 0.3 min. Assuming that the total conversion coefficient is three times the K-conversion coefficient, estimate the half-life of the level if it decayed solely by photon emission.

5. A hypothetical (even, even) nucleus of $Z \sim 60$ has the following low-lying level structure.

```
2+     0.401
2+     0.324
3-     0.212

0+     0.0
```

The nuclide is populated by both β^- decay and EC decay.

(a) The β⁻-decay parent has a spin and parity of 5⁻ and a $Q_{\beta-}$ of 0.530 MeV. Determine which level will be most strongly populated by β⁻ decay and estimate the half-life of the parent if it decayed solely by this transition.

(b) Excluding transitions of *very low* probability, draw in the γ-ray transition/transitions you expect to be observed as a result of decay of the β⁻-decay parent. Label the transition(s) with its (their) multipolarities.

(c) Estimate the total half-life for decay of the 3⁻ level at 0.212 MeV to the ground state.

Now consider the decay of the EC parent with a spin and parity of 1⁺ and a Q_{EC} of 0.620 MeV.

(d) Specify which EC transition(s) is (are) expected to be observed with significant intensity in the decay of the EC parent.

(e) Given the result from part d, draw in the γ-ray transitions that you expect to observe with significant intensity following decay of the EC parent.

12
Nuclear Fission

12.1
Introduction

The discovery of nuclear fission in 1939 by Otto Hahn and Fritz Strassmann [1], following years of collaborative work with Lisa Meintner, ushered in an era of hope and fear unlike anything human kind had experienced previously. Within less than five years, the discovery was applied to the production and successful deployment of a prototype of a weapon of mass destruction tens of thousands of times more powerful than anything that had ever been known. Within less than fifteen years the door was opened to the development of weaponry that could literally destroy a large fraction of the surface of the earth, and the era of the nuclear terror known as the Cold War was opened.

From the beginning, the use of fission as a power source was discussed, and by 1941 Enrico Fermi [2] and co-workers had demonstrated the first controlled chain reaction based on neutron-induced fission. Following several decades of design and construction of various types of demonstration plants of relatively small size, the nuclear power industry began to grow rapidly during the period 1960–70. In the United States today something greater than 100 commercial nuclear reactors supply about 20% of the nation's electric power. In Japan, roughly 40% of the electric power production is from nuclear plants, and in France, more than 70% of the electricity is generated by nuclear fission.

After decades of slow growth or contraction, nuclear power is again being examined for renewed investment and development throughout the world. This is being driven by population growth and the desire of most countries to achieve a living standard comparable to that enjoyed in the wealthier nations of the world; living standards are strongly tied to the abundance of electric power. In addition to a requirement for economic viability, a significant increase in the installation of nuclear power generation requires that the public and private sectors deal intelligently with the need for operational safety of the highest quality and with the issue of the long-term management of high-level radioactive waste in the form of spent reactor fuel. We urgently need a cadre of educated persons who understand the details of nuclear fission and its applications if this phenomena is to fulfill its potential for providing safe and reliable electric power in the present century.

Nuclear Physics for Applications. Stanley G. Prussin
Copyright © 2007 WILEY-VCH Verlag GmbH & Co., Weinheim
ISBN: 978-3-527-40700-2

12.2
The Discovery of Nuclear Fission

The discovery of nuclear fission is a story of great perseverance, missed opportunities and above all, the astonishing brilliance of those working with radioactivity at a time when not much was known and every day seemed to bring about a new finding or idea that pointed to added richness in the makeup of the atomic nucleus, the radiations that it can emit and the nature of the heaviest elements then known. The discovery of the neutron by Chadwick was perhaps the most important key that opened the door to fission, and it was soon found that the combination of an α emitter with light elements produced a useful source of free neutrons from the (α,n) reaction. At that time, $^{226}_{88}\text{Ra}$ was a relatively common material in most laboratories that dealt with radioactivity and, although the neutron sources it produced were relatively weak, they were sufficient to begin to examine the reactions of neutrons with the stable isotopes of elements available in chemistry and physics laboratories. It was quickly discovered that exposure of many elements to neutrons resulted in the production of β^--radioactive nuclides through the capture of a neutron. One of the most active groups engaged in such work was in Rome under the direction of Enrico Fermi. Fermi and his group irradiated almost anything they could get their hands on. The ubiquitous production of β^- radioactivity was quickly recognized as a possible means of creating elements beyond those known on earth. The argument was simple. A substance that decayed by β^- emission was know to produce a nuclide with an atomic number increased by one. Therefore, if the element with the highest known atomic number was exposed to neutrons it was possible, as a result of neutron capture, to produce a nucleus of a new element. Fermi's group was probably the first to irradiate a sample of uranium with neutrons and study the radioactivities thereby induced. They found and reported two such activities. As far as the records show, there was no attempt to perform any kind of experiment to try to identify the chemical identity of the new radioactivities.

Studies of the new radioactivities were taken up in a number of different laboratories, the most notable under the direction of Marie and Joliot Curie in Paris and under the direction of Otto Hahn in Berlin. What these researchers and their co-workers brought to bear were strong foundations in the chemical behavior of radioactive elements and the expertise to devise and carry out experiments with microscopic quantities of a material in a manner that would permit understanding the substance's chemical behavior relative to the well-known behavior of macroscopic quantities of common elements. These so-called *tracer* experiments proved crucial to the discovery of fission. They first led to the suggestion that the chemical behavior of the activities reported by the Roman group was similar to elements of lower atomic number than uranium, and they were thought to have been produced through the emission of one or more α particles following the neutron capture. Such reactions had not been observed previously and were very difficult to reconcile with what was then known concerning the nature of radioactivity and the nucleus. Although a paper by Noddack had been published, containing the suggestion that a heavy nucleus might be induced into splitting to form nuclei of much lower atomic number, this apparently was never given much attention by any of the three major groups then involved in the studies of the radioactivities induced by neutron irradiation of uranium.

Curiously, the chemical experiments also isolated radioactivities with half-lives that differed from those reported by Fermi and his co-workers. The situation was very confusing. As a result of a number of radiochemical experiments, the main activities found as a result of neutron irradiation of uranium were attributed to isotopes of radium and actinium. However, through the tedious and painstaking chemical separations performed by Fritz Strassmann, it was shown that these radioactivities could be separated from known isotopes of radium and actinium and had chemical properties essentially identical to those of barium and lanthanum. Fission of uranium had been discovered, probably by the most difficult route one could concoct for its discovery. Within a very short time, physical experiments confirmed that fragments of very high kinetic energy were produced following neutron absorption in uranium, and a fundamental theoretical understanding of the process was developed. It would take many more years before the direct observation of *spontaneous* fission was reported but, by then, the great similarity of the two fission processes was recognized. Because of their similarity, we will discuss the two processes together.

Nuclear fission is a very complex phenomenon and produces a wide range of nuclei and other radiations. In the following we will first introduce the essential ideas of why and how fission can occur following the simple arguments put forth originally by Bohr and Wheeler, and by introducing the concept of the fission barrier. The general nature of the distribution of the fission products in mass and atomic number will then be presented and the central role played by nuclear shell structure in defining these distributions will also be presented. The remaining parts of the chapter will be devoted to a discussion of the characteristics of the prompt neutrons and γ-rays emitted in fission. An examination of the behavior of the cross sections for neutron-induced fission will be presented in Chapter XIII.

12.3
The Liquid-Drop Model and Nuclear Fission: The Nuclear Potential Energy Surface

During our discussion of the implications of the semi-empirical mass formula in Chapter V, we demonstrated that essentially all nuclei near the valley of β stability with $Z^2/A \geq 18.2$ were energetically unstable with respect to symmetric fission. However, spontaneous fission is not observed experimentally until $Z^2/A \geq 36$. The liquid-drop model indicates that there are two main factors that control the energetics of fission, the surface energy of the drop and its stored Coulomb energy. If one starts with a spherical nucleus and then distorts it into a prolate or oblate shape while maintaining constant density, the surface energy must increase but the Coulomb potential energy will decrease because the mean distance between any two volume elements in the object will increase. While the total energy release in binary fission depends upon the net difference between these two terms for the parent and the two fission fragments, the probability that fission will take place must depend upon the difference in the surface and Coulomb energies at each step along the path of increasing deformation. If distortion from the spherical shape by some differential amount results in a net increase in the energy of the system, that change would not take place spontaneously in a classical system. The system would experience an *energy barrier* to deformation and hence to a path that could lead to fission. (The system can, of course, overcome this barrier by quantum mechanical barrier penetration as seen in the case of α decay, but for the moment we will restrict attention to the semiclassical liquid-drop model.) Therefore, for fission to occur in the classical sense, it must be true that the first and each successive differential in distortion must lead to a system that has a lower potential energy than the initial spherical object. These qualitative arguments suggest a simple approach to estimate the conditions where spontaneous fission will occur. Starting with a sphere, we can calculate the energy difference between it and a slightly distorted sphere in the context of the liquid-drop model. If the total energy of the distorted object is lower than that of the sphere, the latter is unstable and, classically, spontaneous fission can occur. In order to maintain some realism, the distortion must take place at the constant density found in real nuclei. We can repeat these calculations for spheres corresponding to various ratios of Z^2/A to determine when instability is reached.

A very simple way to perform the calculation is to consider distortions in the form of ellipsoids of revolution. Such shapes are representative of those that have axial symmetry and will tend to produce the minimum increase in surface area for a given distortion from sphericity. For a prolate ellipsoid of revolution with semi-major axis a and semi-minor axis b (Fig. 12.1), the volume and surface area are given by

$$V = \frac{4}{3}\pi a b^2$$
$$S = 2\pi b^2 + 2\pi \left(\frac{ab}{e}\right)\sin^{-1}e$$
(12.1)

where e is the eccentricity of the ellipse that defines the surface of the volume. In terms of the semi-empirical mass formula, the surface energy is just $E_S = (a_s/4\pi r_0^2)S$. The relation between the semi-major and semi-minor axes can be written in terms of the eccentricity as

$$b^2 = \frac{a^2}{e^2}(1-e^2)$$
(12.2)

The requirement that the density remain constant means that the volume of the initial spherical nucleus must be the same as that of the ellipsoid of revolution obtained by the deformation. If the radius of the initial nucleus is taken as R, we then have

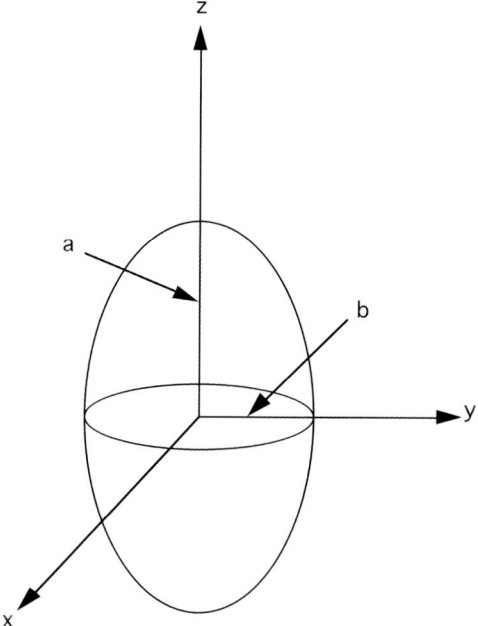

Fig. 12.1 Ellipsoid of revolution with semi-major axis a and semi-minor axis b. The cross section normal to the z-axis is circular.

$$\frac{4\pi}{3}R^3 = \frac{4\pi}{3}ab^2 \tag{12.3}$$

and the radius can be written in terms of the semi-major axis as

$$R^3 = \frac{a^3}{e^2}(1-e^2) \tag{12.4}$$

Although we will not prove it, it can be shown that the stored Coulomb energy of a prolate ellipsoid of revolution is

$$E_{\text{Coul, ellip}} = \frac{3}{10}\frac{Z^2e^2}{(a^2-b^2)^{1/2}}\log\frac{a+(a^2-b^2)^{1/2}}{a-(a^2-b^2)^{1/2}} \tag{12.5}$$

With these relations, we can now write down the difference in energy between an ellipsoid of revolution of any eccentricity and the undistorted sphere in terms of the semi-empirical mass formula. The surface energy of a sphere is simply $E_{\text{sur, s}} = a_s A^{2/3}$ and the stored Coulomb energy is $E_{\text{Coul, s}} = 3/5(Z^2e^2/R)$. As a result, the difference in energy between the two objects is

$$\Delta E = \Delta E_s + \Delta E_{Coul} = a_s \left[\frac{1}{4\pi r_0^2} \left(2\pi b^2 + 2\pi \frac{ab}{e} \sin^{-1} e \right) - A^{2/3} \right] \quad (12.6)$$

$$+ \frac{3}{10} \frac{Z^2 e^2}{(a^2 - b^2)^{1/2}} \left(\log \frac{a + (a^2 - b^2)^{1/2}}{a - (a^2 - b^2)^{1/2}} \right) - \frac{3}{5} \frac{Z^2 e^2}{r_0 A^{1/3}}$$

If $\Delta E < 0$ for all eccentricities, a spontaneous change from sphericity can occur. If $\Delta E > 0$ for even an infinitesimally small eccentricity e, then a spontaneous division of the sphere will not be possible and "fission" will not take place classically.

To apply Eq. (12.6), we will follow convention and express our results in terms of the parameter ε that is related to the semi-major and semi-minor axes of the ellipsoid by

$$a = R(1 + \varepsilon)$$
$$b = R/(1 + \varepsilon)^{1/2} \quad (12.7)$$

The second of Eqs (12.7) is simply obtained from the first by use of the constant density requirement. The eccentricity e and ε are related by

$$e = \left[1 - \frac{1}{(1 + \varepsilon)^3} \right]^{1/2}. \quad (12.8)$$

For a sphere $\varepsilon = 0$, while for $\varepsilon = 1$, the ratio of the semi-major to semi-minor axes of the ellipsoid is $a/b = 1.41$.

The obvious starting point is to ask why nuclides near $A = 100$ and the valley of stability, the first nuclides for which Q_f is positive, do not undergo spontaneous fission. In Fig. 12.2 (a) are shown the surface and Coulomb energies calculated for the most stable isobar at $A = 150$ with $Z = Z_A = 62.58$ as a function of ε. The two energy terms are seen to have about the same magnitude for a sphere. As the sphere is distorted, the Coulomb energy begins to drop and the surface energy begins to increase, but there is a significant difference in the rates of change of the two energies. In part (b) of the figure are the differences between the sum of these energies for the ellipsoid of revolution and the sum for the initial sphere. One notes immediately that the slightest deformation causes the potential energy to increase and it continues to increase for all larger deformations. The conclusion is clear. Because $\Delta E > 0$ for all deformations, there is an energy barrier that precludes a path that can lead to spontaneous fission without energy being added to the system. The classical sphere is stable against deformation. In this case the fission parameter is $Z^2/A = 26.1$.

The lightest nucleus for which spontaneous fission has been observed is $^{235}_{92}U$. The partial half-life for this decay mode is about 1.0×10^{17} s corresponding to an almost vanishing small decay probability. In Fig. 12.3 are shown the results of calculations for a classical sphere of $A = 235$ and $Z = 92$ of the same type as those shown in Fig. 12.2. Both the surface and Coulomb energies have increased significantly but, because of the squared dependence on the atomic number, the Cou-

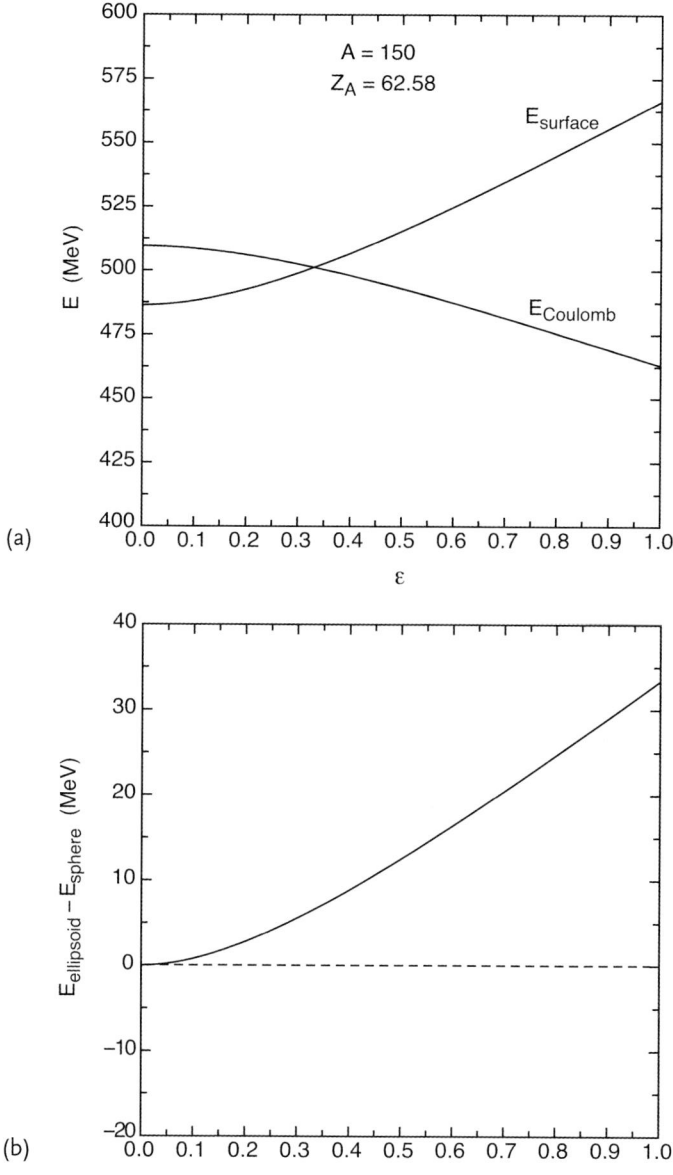

Fig. 12.2 (a) The surface and Coulomb energies for ellipsoids of revolution with $A = 150$ and $Z = Z_A = 62.58$ as a function of the distortion factor ε. (b) The difference between the sum of the surface and Coulomb energies for ellipsoids of revolution and a sphere of the same density as a function of the distortion factor ε.

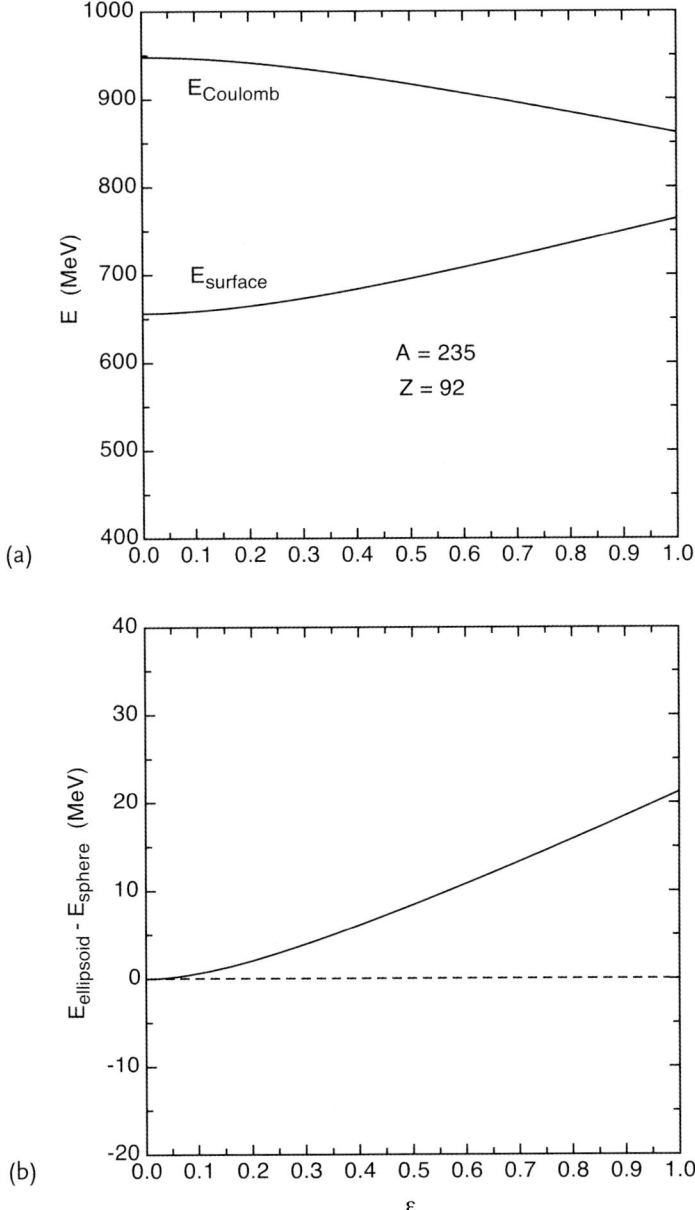

Fig. 12.3 (a) The surface and Coulomb energies for a ellipsoid of revolution with mass number A = 235 and atomic number Z = 92 as a function of the distortion factor ε. (b) The difference between the sum of the surface and Coulomb energies for a sphere and ellipsoid of revolution as a function of the distortion factor ε. In both cases the volumes of the sphere and ellipsoids are identical.

lomb energy has increased to a greater extent. The difference between the energies of the ellipsoids and sphere show the same general variations with ε as seen previously, namely ΔE > 0 for all deformations. Again we conclude that a path toward spontaneous fission is classically impossible for $^{235}_{92}$U with a fission parameter of $Z^2/A = 36.0$. But there is a significant difference between the ΔE in Figs 12.2 and 12.3. Notwithstanding the much larger Coulomb energy, the differences between the energies of the ellipsoids and the sphere are significantly smaller for the classical model of $^{235}_{92}$U at all deformations. Because spontaneous fission is indeed observed, it must be true that barrier penetration must be responsible for the decay of the real nucleus.

We could continue to increase mass and charge and, if we did, we would indeed see a continuous decrease in the energy difference between the ellipsoidal distortions and the initial spherical shapes. But it perhaps makes more sense to ask, at what value for the fission parameter is spontaneous fission allowed classically? Within the liquid-drop model, this is very easy to do. Because the only parameters that enter into the decision are Z and A, it does not make any difference what mass number we choose. All that needs to be considered is the ratio Z^2/A. For some fixed small distortion, we simply calculate the energy difference between the slightly distorted ellipsoid of revolution and the initial sphere as a function of Z^2/A. We have chosen A = 236 and have calculated the energy differences with a distortion parameter of ε = 0.01 over the range in $36 \leq Z^2/A \leq 56$, and these are shown in Fig. 12.4. For such a small distortion, the energy difference varies almost linearly with Z^2/A and becomes zero at $Z^2/A \approx 50$, very close to the oft-quoted values of

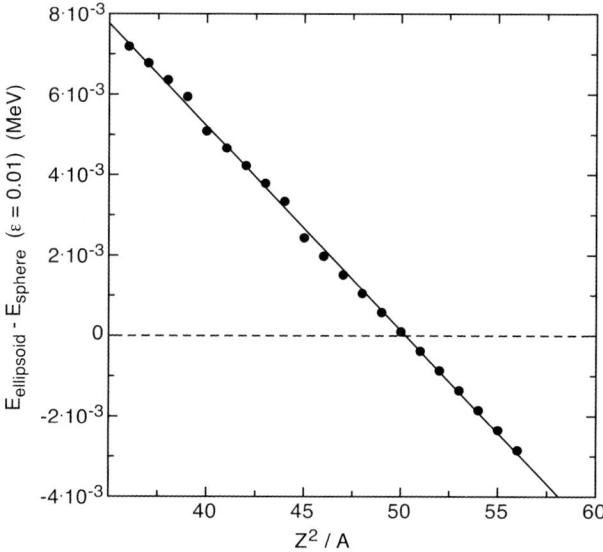

Fig. 12.4 The difference in surface and Coulomb energies between a sphere and an ellipsoid of revolution for a distortion parameter of ε = 0.01 as a function of the fission parameter Z^2/A.

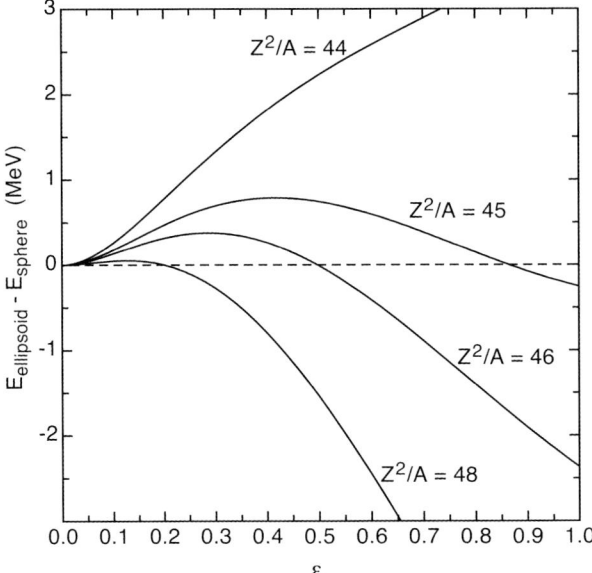

Fig. 12.5 The difference in surface and Coulomb energies between a sphere and an ellipsoid of revolution as a function of ε for fission parameters just below the critical value of $Z^2/A \approx 50$.

47–49. This means that for $Z^2/A \geq 50$, the slightest distortion produces a decrease in the potential energy of the system. The classical sphere is unstable to distortions and a path to spontaneous fission is present. The magnitude of Z^2/A at which the change in total energy is zero is sometimes referred to as the *critical* value for spontaneous fission.

The behavior of the energy differences between the ellipsoids of revolution and an undistorted sphere just below $Z^2/A = 50$ is quite interesting. In Fig. 12.5 are shown the energy differences as a function of ε for $Z^2/A = 44$, 45, 46 and 48. Over the range in ε shown, the potential energy increases continuously with increasing distortion from sphericity for $Z^2/A = 44$, but for $Z^2/A \geq 45$, a region of increased potential energy is followed by one of decreasing potential, which then becomes negative at larger distortions. These results demonstrate that, in the vicinity of the critical value, a finite potential energy barrier is predicted classically. If the barrier is surmounted, a path for fission is available. Note that the magnitude of the barrier is predicted to be on the order of 1 MeV.

How do these observations apply to the fission of real nuclei? First, the magnitude and shape of the barrier calculated with the semi-empirical mass formula cannot be expected to be quantitatively correct. Both are very sensitive to the shape assumed for the deformation and we have not considered anything but the most rudimentary effects due to nuclear structure. Further, we have assumed that the initial nuclear shape of a nucleus is spherical whereas essentially all nuclides of the heaviest elements that undergo fission are more or less strongly deformed in their ground states. What we can expect is that the classical calculations can serve as a guide for qualitatively understanding some of the gross aspects of fission. We can infer that spontaneous fission is controlled in large measure by the presence of a fission barrier, much as in the case of α decay. For $Z^2/A < (Z^2/A)_{crit}$, a barrier will exist that can be penetrated by quantum mechanical tunneling. We know that the probability for barrier penetration will be very sensitive to the details of the shape and thickness of the barrier, and we can expect fission probabilities to vary strongly with the magnitude of the barriers. In the event that $Z^2/A = (Z^2/A)_{crit}$, the fission barrier will vanish and a nucleus so formed can be expected to dissociate within times characteristic of the motions of nucleons in the nucleus. Calculations of the shapes of the barriers are not simple and rather than attempt to calculate them, we will turn to an examination of empirical data to understand how the ideas introduced above are reflected in real nuclei.

12.4
Empirical Data on Spontaneous and Neutron-Induced Fission

Because the classical model suggests that the magnitude of the fission barrier will depend upon the fission parameter, we can attempt to look for a correlation between the probabilities for spontaneous fission and Z^2/A. In Fig. 12.6 the logarithms of known partial half-lives for spontaneous fission of ground states are

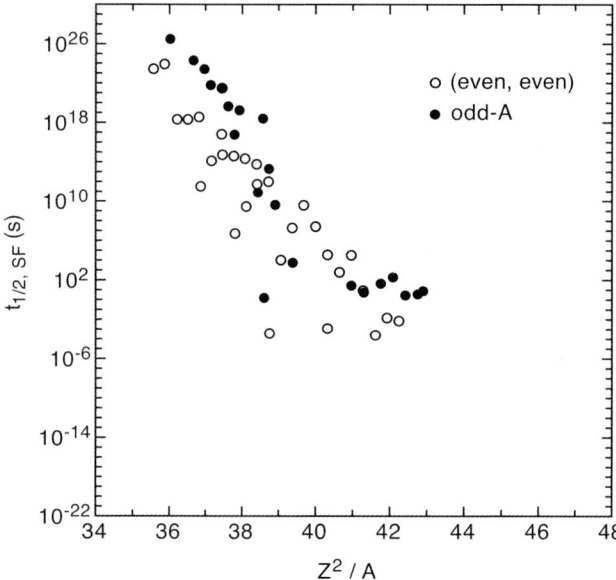

Fig. 12.6 Experimental values of the partial half-lives in seconds for spontaneous fission of ground states of (even, even) and odd-A nuclides as a function of the fissility parameter Z^2/A.

shown as a function of the fission parameter for (even, even) and odd-A nuclides. The data are really quite remarkable. For $36 \leq Z^2/A \leq 43$, the experimental half-lives decrease by about 31 orders of magnitude, an amazing sensitivity to the fission parameter. The data show considerable scatter and there are definite systematics in the half-lives of isotopes of both (even, even) and odd-A nuclides. On a gross scale, it is seen that the half-lives decrease roughly exponentially with increasing Z^2/A. If one imagines an exponential fit to the data, it is easy to see that the half-life extrapolates to times characteristic for motion of nucleons in the nucleus ($\sim 10^{-22}$ s) at $Z^2/A \sim 45$–47, very near the instability limit suggested by the simple classical calculations discussed above.

The rough quantitative agreement between the predictions of the simple model and experiment should not cloud the fact that fission is a very complicated process. The scatter in the data shown in Fig. 12.6 indicates that order of magnitude variations about a simple exponential function exist and that must indicate that nuclear-structure effects are very important in defining fission probabilities. The very fact that the heavy nuclei are deformed means that the level structure in a deformed potential is intimately involved in determining the nuclear shape and the deformation that leads to a minimum in the total energy of the nucleus. Departures from this shape, whether to a lesser or more deformed shape, must lead to a net higher total energy. This can only come from the internal energy of the system – an exchange between the kinetic and potential energies of the nucleons. Any change in shape will automatically result in a change in the level structure of the nucleus as seen by reference to the Nilsson diagrams in Chapter VIII. Such changes will actually determine the potential variation as one proceeds from the equilibrium ground state to a more deformed shape that is in a direction that could ultimately lead to fission.

Based on the results of the model calculations, we can produce a schematic diagram of the potential energy with increasing deformation as shown in Fig. 12.7 (a). The nuclei of the heavy elements are generally strongly deformed in their ground states, and the equilibrium deformation must be that which leads to a minimum total energy for the nucleus. Any change in shape, whether to a more spherical or more deformed form, must represent a state of increased potential energy. As a result, the variation in potential energy with changes in deformation in the vicinity of the equilibrium shape must describe a potential well with the equilibrium deformation at its minimum. If fission is to occur, the potential energy must eventually reach a maximum with increasing deformation and then must decrease more or less rapidly. Penetration through the barrier permits the nucleus to undergo fission.

This simple picture must be altered substantially when one takes into account the change in the level structure of a nucleus as the deformation changes. You will remember that the Nilsson model clearly indicates that both the absolute and relative energies of levels in a deformed potential depend upon the extent of deformation. Some levels increase in energy and some decrease in energy with increasing deformation. These energy changes frequently lead to a modulation of the shape of the barrier such that a second potential well appears with increasing

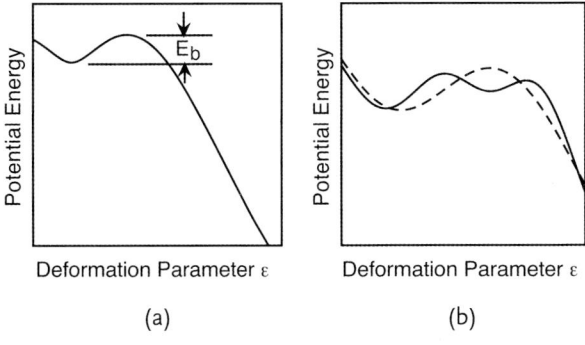

Fig. 12.7 (a) Schematic diagram of the potential energy of the nucleus as it is deformed from its equilibrium ground state in the limit of the liquid drop model. The barrier between the equilibrium deformed state and the maximum in the potential function is the fission barrier E_b. (b) Comparison of the liquid-drop potential energy (dashed line) with the potential energy that includes corrections from the variation in single-particle level structure in the deformed potential (solid line).

deformation as shown schematically in Fig. 12.7 (b). It is not difficult to guess, qualitatively, what can happen in such a case. The nucleus will exist in the first potential minimum in the ground state. If, by some means it can be excited into the second well at the larger deformation, it will actually see a smaller fission barrier with a much larger probability for spontaneous fission. Such a nucleus will possess a *fission isomer* frequently referred to as a *shape isomer*. Fission isomers are well-established in many nuclei and demonstrate the intimate relation between the details of nuclear structure in a deformed potential and fission probabilities.

It may seem very surprising, but shell structure in spherical nuclei has an even more startling effect on the way in which fission takes place than the structure effects discussed above and this is graphically demonstrated in the distribution of the products of the fission process.

Nuclear fission, whether spontaneous or induced, takes place in literally hundreds of ways. As discussed in Chapter II, the dominant fission mode is binary fission in which two fragments are produced that contain all of the nucleons of the initial nucleus. Almost all of the fragments are formed in excited states and many have excitation energies that exceed the particle binding energy. Within a very short time after fission, perhaps 10^{-19}–10^{-12} s, the fragments decay by the emission of *prompt* neutrons and γ-rays to produce nuclides in their ground or low-lying isomeric states. We will call the nuclei produced immediately after the prompt emissions and before radioactive decay of the ground and isomeric states the *primary* fission products or *fission fragments*. On the average, more than two neutrons are emitted per fission and thus the fission products will have a slightly smaller number of nucleons than the parent nucleus, but not by much.

Because the ratio of N/Z in the heavy nuclides is greater than that in medium heavy nuclides along the line of β stability, the fission products will generally decay by $β^-$ emission. Any nuclide found among the fission products can then have been produced in two ways; as a direct result of fission and as a result of β^- decay. In general, the fission products will be members of β^- *decay chains* with the last member, that with the largest atomic number, being the β-stable isobar of that mass number when A = odd. In the case of A = even where more than one stable isobar exists, the stable isobar with the smallest atomic number is for most practical purposes the last member of the chain.

With the exception of the very small fraction of decays that lead to delayed neutron emission, β^- decay preserves the mass number, and thus at any time after fission one can make measurements of the fraction of all fission products that have a given mass number immediately after prompt neutron emission. The actual experimental data are usually quoted as the fraction (percent) of all fission products that have a mass number A and, because the fission is binary, the total fraction summed over all mass numbers will be 2. These data are referred to as *mass* yields or *chain* yields. By now, very accurate sets of chain yields have been obtained for the most common neutron-induced fission reactions and a few of the nuclides that decay by spontaneous fission. Probably the most famous and extensive data set is from the fission of $^{235}_{92}U$ with thermal neutrons where the fissioning nucleus is actually $^{236}_{92}U^*$. The term thermal neutron refers to the very low-energy neutrons that are produced when higher-energy neutrons lose energy by scattering and achieve an energy distribution that roughly corresponds to a Maxwell–Boltzmann distribution at the temperature of the medium in which they exist. At room temperature, the average kinetic energy in a Maxwell–Boltzmann distribution at equilibrium is about 0.038 eV, very small indeed relative to nuclear excitation energies.

The mass-yield distribution for fission of $^{235}_{92}U$ with thermal neutrons is shown in Fig. 12.8. It is remarkable in many respects. First, fission products with masses in the range $66 \leq A \leq 172$ are found and correspond to isotopes of the elements chromium (Z = 24) through thulium (Z = 69), or almost half of the chemical elements found in nature. The total number of different nuclides found as fission products, including relatively long-lived isomers, is well in excess of 400. Second, the probability for finding a given mass number varies over 10 orders of magnitude. The major portion of the entire yield is, however, contained in the two relatively narrow *mass peaks* that range from about $80 \leq A \leq 107$ and $122 \leq A \leq 155$, respectively. Within both, the highest chain yields are roughly 6–7%. Although we have used the simple case of symmetric fission to discuss the basic energetics of the process, symmetric fission is seen to be very improbable. Mass yields near the symmetric case of A = 118 are about a factor of 600 smaller than those found in the mass peaks. This cannot be explained by the simple semiclassical liquid-drop model and must be attributed to some strong underlying nuclear structure considerations.

When binary fission occurs, there must be some probability distribution for sharing of the protons and neutrons between the two fragments. For each mass number there must then be some yield distribution among the isobars that correspond to it. In order to examine this, we define the *independent fission yield* of a nuclide, $IY_{A,Z}$, as its yield in fission after prompt neutron emission and before any

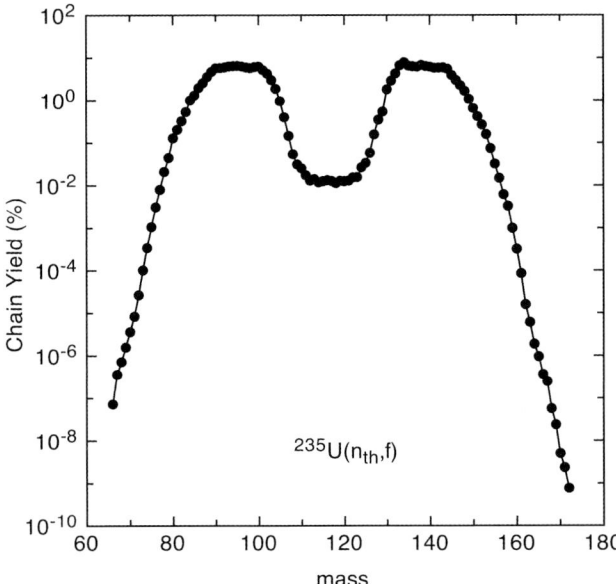

Fig. 12.8 The mass-yield distribution for the fission products of the thermal fission of $^{235}_{92}\text{U}$.

β^- decay has occurred. The chain yield, CY_A, is just the sum of the independent yields of the isobars at that mass number, or

$$\text{CY}_A = \sum_{\text{all } Z} \text{IY}_{A,Z} \tag{12.9}$$

The independent yields for the mass chains A = 95, 115, and 135 in the thermal fission of $^{235}_{92}\text{U}$ are shown as a function of atomic number in Fig. 12.9. They are quite representative of the distributions found for all mass chains, and their character is notably different from that seen in the mass-yield distribution. For each chain, the principal part of the chain yield is contained in at most 3–4 isobars.

The curves shown in the figure represent least-squares fits with Gaussian or normal distributions of the form

$$G'(Z) = A' e^{-\frac{(Z-Z_p)^2}{2\sigma^2}} \tag{12.10}$$

In each fit, the three parameters A', Z_p, and σ were all free parameters. Clearly, a Gaussian of this form provides a reasonable fit to these data sets as well as essentially all of the independent yield distributions that have been measured. The quantity Z_p, which is the centroid of the Gaussian, is known as the *most probable charge of an isobar*. If you compare the ratios of N/Z for the three chains at their fitted values of Z_p, you discover another remarkable point. The ratio of N/Z in each

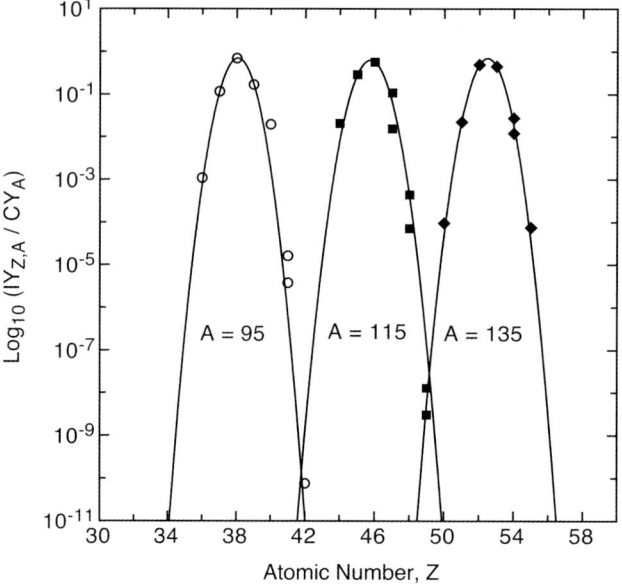

Fig. 12.9 Independent yields of the isobars of A = 95, 115, and 135 in the thermal fission of $^{235}_{92}U$. The errors on the data points (± 1σ) are roughly 1–2 times the size of the points. The curves shown are Gaussian fits where the parameters A, Z_p, and σ of Eq. (12.10) were all allowed to vary in each fitting procedure.

case is quite close to that for $^{236}_{92}U$ itself, namely 1.56. The fission process does not involve a significant redistribution of neutrons and protons in the fission fragments from that found in the parent nucleus.

If a Gaussian distribution is a universal descriptor of the independent fission yield data, then it should not matter what the actual chain yield is. A simple way to examine how "universal" a single distribution is in fitting the yield data is to normalize each of the distributions relative to its chain yield and relative to the difference between the nuclear charge Z and the centroid Z_p. That is, a function of the form

$$G(Z - Z_p) = A e^{-\frac{(Z-Z_p)^2}{2\sigma^2}} \qquad (12.11)$$

where $A = A'/CY_A$ would be expected to represent the fractional distribution in the chain yield among the isobars at each mass number.

To see how well this assumption describes experiment, the independent yield data in Fig. 12.9 were each divided by the corresponding chain yield and the quantities $Z - Z_p$ were calculated with the values of Z_p obtained from the fits shown in the figure. The entire set of data reduced in this way is shown in Fig. 12.10 along with a Gaussian fit of the form of Eq. (12.11). The overall fit is seen to be quite good.

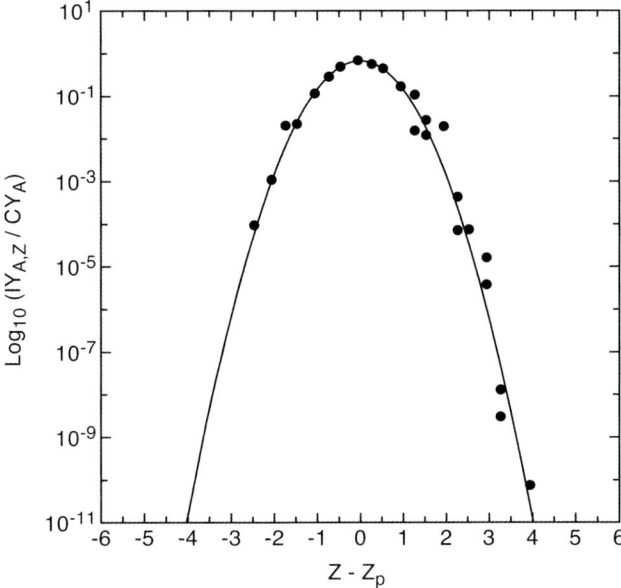

Fig. 12.10 Gaussian fit to the reduced independent yield data sets for A = 95, 115, and 135.

The standard deviation of the Gaussian is σ = 0.570 ± 0.020 and corresponds to a full width at half maximum of about 1.34 atomic numbers. The distribution is clearly quite narrow and about 76% of the chain yield will be found in but 1–2 isobars. One can perform such calculations for other mass chains and quite similar results will be found. In fact, more sophisticated approaches that take into account differences between odd-A and (even, even) nuclides demonstrate that the idea that the division of charge among the fission products is, on the average, about the same as in the fissioning nucleus and is an excellent description of the charge distribution.

A Gaussian distribution is often found to describe the probability distribution of random variables. The Central Limit Theorem of statistics states that, regardless of the actual forms of the distributions involved, the probability distribution of a function of independent randomly-distributed variables will, in the limit of a large number of such variables, be approximately Gaussian in nature. We can then conclude from fits of the type shown in Fig. 12.10 that the charge distribution in fission is governed by statistical processes. All of the data presented so far then infer that both nuclear structure and statistical effects play strong roles in nuclear fission.

The charge distributions provide the basis for understanding the very large number of nuclides produced in fission and the very complex decay properties of the fission products taken as a whole. If one fits the experimental data on charge distributions for each mass number A, one can obtain the most probable charges Z_p and can compare these to the charges of the hypothetical most stable isobars, Z_A,

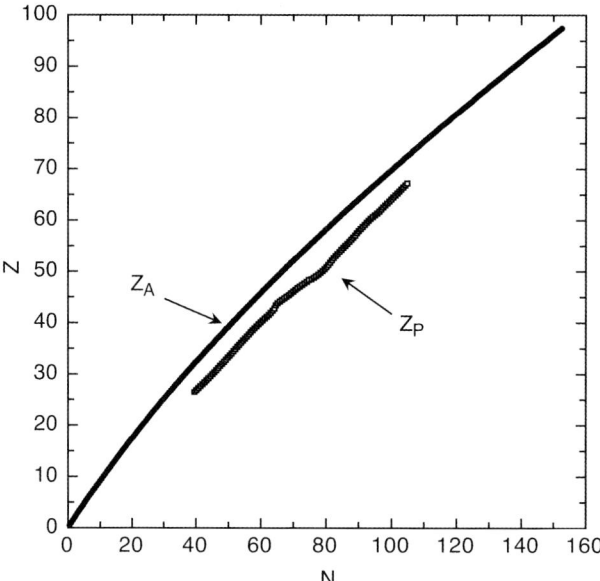

Fig. 12.11 The charge of the most stable isobar, Z_A, and the most probable charge Z_p of the products from thermal fission of $^{235}_{92}U$.

that define the valley of β stability. In Fig. 12.11 these data are shown in a plot of Z versus N. Over the entire mass range of the fission products, Z_p is roughly equidistant from Z_A and the average value of $Z_A - Z_p$ is 3.74 ± 0.39. Taking into account the half-width of the charge distribution in Fig. 12.10, we immediately have the result that at each mass number a β⁻ decay chain containing 4–5 isobars will be produced in fission. With measurable mass yields over the range 66 ≤ A ≤ 172, we conclude that at least 530 nuclides are produced in the thermal fission of $^{235}_{92}U$.

A careful examination of the Z_p data shown in Fig. 12.11 reveals a small but definite departure from the general trend in the immediate vicinity of Z = 50, N = 82 that corresponds to nuclides near the doubly magic $^{132}_{50}Sn$. This is the first small but definite indication that *spherical shell structure* exerts an influence on the fission process. A much more dramatic indication of the importance of shell structure on the fission process is presented in Fig. 12.12 where the mass-yield distributions from fission of $^{235}_{92}U$ and $^{239}_{94}Pu$ with thermal neutrons, and from the spontaneous fission of $^{252}_{98}Cf$, $^{254}_{100}Fm$, and $^{256}_{100}Fm$ are shown together. The striking feature of the distributions is the essential constancy of the location and width of the mass-yield distributions in the vicinity of the heavy mass peak. There is an immediate and sharp increase in yield near A = 128 in each distribution and the peak is located between A = 132 and 135. In contrast, the light mass peak is clearly seen to move progressively higher in mass as the mass and charge of the fissioning system increases. The valley between the light- and heavy-massed peaks becomes

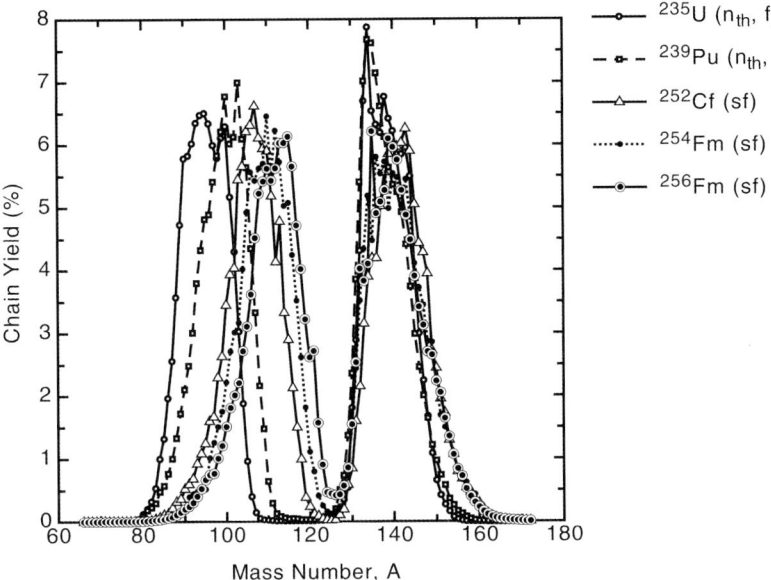

Fig. 12.12 The mass-yield distributions for thermal neutron-induced fission of $^{235}_{92}$U and $^{239}_{94}$Pu, and the spontaneous fission of $^{252}_{98}$Cf, $^{254}_{100}$Fm, and $^{256}_{100}$Fm.

narrower and narrower. At $^{256}_{100}$Fm, symmetric fission would lead to nuclei with Z = 50 and N = 78, just four neutrons below the doubly magic $^{132}_{50}$Sn. Although much less extensive data are available on mass yields of even heavier spontaneous fissioning nuclides, they suggest that as symmetric fission to produce $^{132}_{50}$Sn is approached, the probability for asymmetric fission decreases dramatically. As an example, the mass-yield distribution from spontaneous fission of $^{258}_{100}$Fm is shown in Fig. 12.13 [3]. A single relatively narrow mass peak is found. Only symmetric fission occurs with high probability and this demonstrates the extraordinary influence of the structure of the doubly magic Z = 50, N = 82 on the fission process.

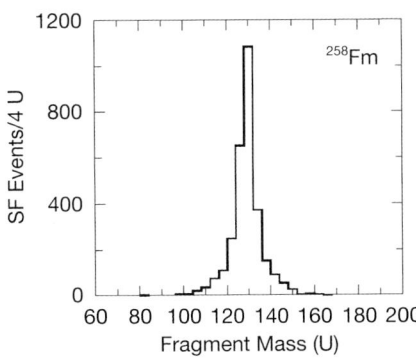

Fig. 12.13 The mass-yield distribution for spontaneous fission of $^{258}_{100}$Fm. The data were taken with semiconductor detectors and are of much lower mass resolution than the data sets shown in Fig. 12.12.

All experimental evidence indicates that both statistical and structural issues play strong roles in nuclear fission. The theory of fission, which greatly advanced over the past 50 years, is by no means complete. Both experimental and theoretical investigations are continuing to yield new and surprising information.

12.5
Energy Release in Fission

12.5.1
Fission Fragment Kinetic Energy

The principal part of the energy release in fission is contained in the kinetic energy of the fission fragments. Neglecting the small effects from emission of a few neutrons and prompt γ-rays, conservation of momentum requires that the fission fragments have equal and opposite momenta when the fissioning nucleus is at rest in the laboratory, i.e.,

$$p_L = -p_H \qquad (12.12)$$

Even though the total kinetic energy is sizeable, relativistic effects are negligible, and we can therefore write

$$T_L = \frac{p_L^2}{2m_L} = \frac{m_H}{m_L}\frac{p_H^2}{2m_H} = \frac{m_H}{m_L}T_H \qquad (12.13)$$

where the subscripts L and H refer to the light and heavy fragment, respectively. The energy release will obviously depend on the fissioning system and the actual mass division considered. But we can get a useful estimate from the mass difference between $^{236}_{92}U$ and the fragment masses corresponding to one of the more probable mass divisions. Using the nuclei $^{95}_{38}Sr$ and $^{141}_{54}Xe$ as an example, the mass difference between the initial and final states is calculated to be about 185 MeV. This would imply that the maximum kinetic energies of the fission fragments are about 110 MeV and 75 MeV, respectively, because we have neglected the excitation energy of the fragments which is contained in the prompt neutrons and γ-rays.

Many measurements of the kinetic energies of fission fragments have been performed. As an example of such data, the average total fragment kinetic energy in the fission of $^{235}_{92}U$ with thermal neutrons as a function of fragment mass number is shown in Fig. 12.14 [4]. The vertical lines shown in the figure represent ± 1σ limits on the uncertainties in the energies. The total kinetic energy is seen to increase roughly linearly with mass number among the light fragments until a maximum of about 176 MeV is reached near A = 102. Although only one datum is reported near symmetric fission, it is clear that the total kinetic energy release, on the average, is significantly larger than that found near A = 102. The variation of

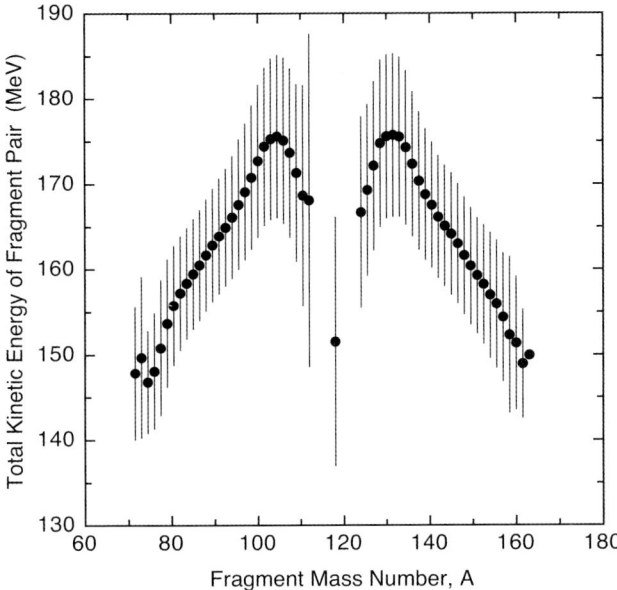

Fig. 12.14 The average total kinetic energy of fission fragment pairs as a function of fragment mass number, A. The data were taken for fission of $^{235}_{92}$U with thermal neutrons.

kinetic energy with mass number of the heavy fragment is very nearly the "mirror image" of that for the light fragments.

We can use these data to estimate the average kinetic energy of an individual fission fragment by use of Eq. (12.13). For simplicity, we will neglect emission of prompt neutrons. In Fig. 12.15, the average kinetic energies of individual fragments corresponding to the data in Fig. 12.14 are shown as a function of mass number along with the linear variation expected for a constant total fragment kinetic energy of 165 MeV for reference. Although the overall trend is roughly that expected for a constant total fragment kinetic energy, significant departures are evident. The lightest fragments have roughly constant kinetic energy, whereas the energies of the heavy fragments decrease at a rate greater than expected in the limit of constant total fragment kinetic energy. The average kinetic energies of fragments in the vicinity of the light and heavy mass peaks are significantly larger than for adjacent lighter or heavier fragments.

The data in Fig. 12.15 can also be used to estimate the actual spectrum of fission fragment energies with the aid of the mass-yield distribution shown in Fig. 12.8 that gives the relative probability that a given fragment mass will be produced in fission. The calculated energy spectrum is shown in Fig. 12.16. The predicted spectrum has the well-defined peaks represented by the mass-yield distribution. The average total fragment kinetic energy is calculated to be 165.0 MeV, quite close to the accepted value of 169.12 ± 0.49 MeV. Given experimental uncertainties and

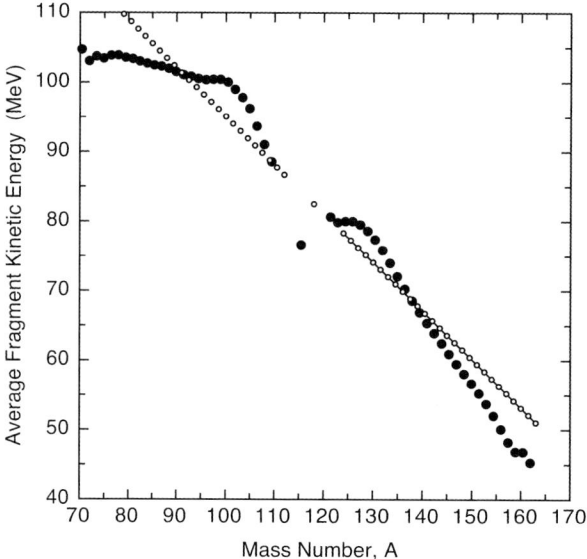

Fig. 12.15 The average kinetic energies of fission fragments from fission of $^{235}_{92}U$ with thermal neutrons estimated from the data of Zakharova, et al [4]. The small open circles represent the expected energies for a constant total fragment energy of 165 MeV.

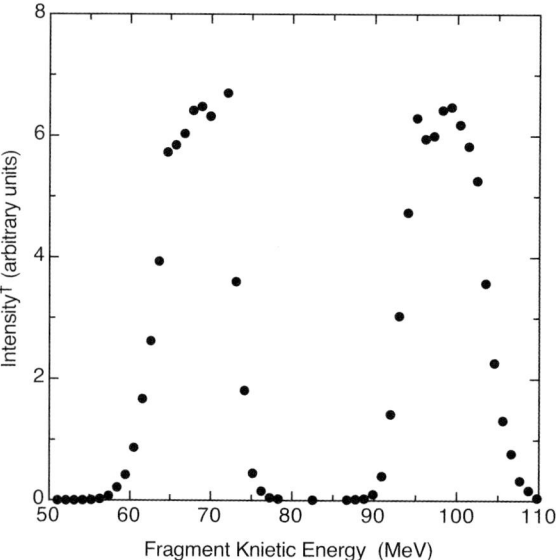

Fig. 12.16 The spectrum of fission fragment kinetic energies for the thermal fission of $^{235}_{92}U$ estimated from the data in Figs 12.8 and 12.15.

the simplifications we have made, our result can be considered to be in quite good agreement with evaluated data.

12.5.2
Kinetic Energy of Prompt Neutrons

Thermal neutron-induced fission of the lighter actinides is accompanied by the emission, on the average, of 2–4 prompt neutrons. Similar numbers are emitted in spontaneous fission as well. The energy spectrum of the prompt neutrons from fission of $^{235}_{92}$U with thermal neutrons is shown in Fig. 12.17 [5]. Although the measurements extend only over the range $0.01 \leq E_n \leq 10$ MeV, neutrons with both lower and higher energies are known to arise with very low abundances. The abscissa in the figure has been given in logarithmic form to allow easy visualization of both the shape of the spectrum at low energies and the uncertainties in the data. Two neutron detectors that depend upon very different detection mechanisms were used in the experiment and the data from these are indicated by the open and closed circles. Although the data overlap within errors, there is a clear systematic difference between the two data sets below about 0.8 MeV. (Measurements of low-energy neutrons are notoriously difficult, and they are usually biased by the detection of scattered neutrons.) The neutron spectrum is seen to have a most probable

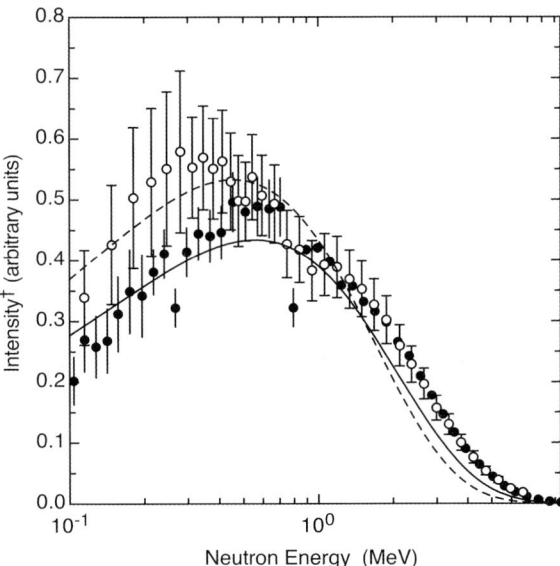

Fig. 12.17 The neutron spectrum in the energy range $0.01 \leq E_n \leq 10$ MeV from fission of $^{235}_{92}$U with thermal neutrons. The open and closed circles represent data taken with two detectors of different types. The dashed and solid lines in the figure represent fits to the two data sets with a simple Maxwell–Boltzmann distribution function.

energy in the vicinity of 0.6–0.7 MeV and the average energy derived from the data is about 1.94 MeV, quite close to the accepted value of 1.98 MeV. The spectrum is seen to vary smoothly and any quantum effects appear to be rather small. In fact, the spectrum is not too different from that found in the simple evaporation of molecules from an ordinary liquid. For the latter, the energy spectrum of the evaporated molecules is described accurately by the Maxwell–Boltzmann energy distribution that can be written as

$$N(E) = \frac{2\pi}{(\pi kT)^{3/2}} E^{1/2} e^{-E/kT} \tag{12.14}$$

where E is the energy of a molecule and k is the Boltzmann constant. For comparison, the two curves shown in Fig. 12.18 represent fits to each of the two data sets with Eq. (12.14). Now the energy spectrum of neutrons emitted from excited nuclei is not described by a simple Maxwell–Boltzmann function but it is described by what can be considered the quantum-equivalent of the evaporation of molecules from a classical liquid. The prompt fission neutrons appear to be emitted by "evaporation" from the excited fission fragments.

The stage of the fission process during which prompt neutrons are emitted can be understood with a simple semiclassical description of the fission process in a dynamical sense. We picture the fission process as proceeding through the shape changes shown in Fig. 12.18. The original nucleus in its ground state (spontaneous fission) or in its "equilibrium" shape after neutron capture (neutron-induced fission) is assumed to be the deformed prolate spheroid shown as the second object from the top of the figure. It is possible that this object can undergo shape changes that would lead to either more spherical or more deformed shapes. Classically, we can think of this as occurring through a vibration. A vibration that led to a more spherical shape (upward in Fig. 12.18) corresponds to an increase in potential energy (Fig. 12.7) at the expense of internal kinetic energy as does a small vibration that led to a more deformed shape. However, if the distortion proceeds far enough, a point will be reached where the separation of the protons is sufficiently large on the average, that the slightest increase in deformation would allow for the gross change in shape shown as the second object from the bottom. The fragments begin to look like spherical cores connected by a "neck" of nuclear matter. The cores will accelerate outward because of the Coulomb potential between them and fragment separation will then occur.

What we are picturing here are cartoons of the shape changes that we might envision as the nucleus moves along a potential energy curve such as those shown in Fig. 12.7. The shape of the original nucleus in its ground state corresponds to the minimum in the potential well. The point at which the maximum in the potential barrier occurs corresponds to a shape perhaps similar to the third or fourth object from the top in Fig. 12.18. This point is referred to in fission theory as the "saddle point". For uranium, the saddle point is reached before a neck develops, while for lighter nuclei it is found with a shape closer to that of the fourth object in the diagram.

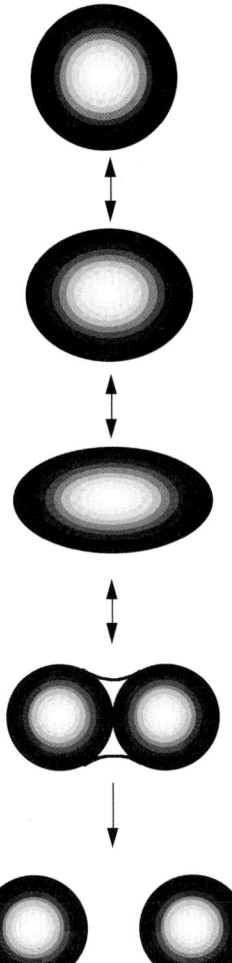

Fig. 12.18 Pictorial diagram of the shape changes during the fission process. The equilibrium shape of the fissioning nucleus is assumed to be the distorted spheroid that is the second object from the top of the diagram.

In keeping with this simple classical picture, and remembering that an increase in potential energy must come at the expense of internal kinetic energy, anything that would reduce the available internal energy will make it harder to reach the saddle point. The effective height of the fission barrier will have been increased if the internal energy is decreased. Therefore, we can conclude that neutron emission must take place *after* the saddle point is reached. The neutron emission must take place after the nucleus has achieved a configuration that "commits it" to fission. All available evidence suggests that the majority of the neutron emission takes place from the separated fission fragments themselves.

The same general argument suggests that the principal part of the prompt γ-rays emitted in fission must also come from the separated fragments. The average fragment must be born with rather high excitation energy. With an average neutron kinetic energy of about 2 MeV (Fig. 12.17) and a neutron binding energy of perhaps 7 MeV, the average fragment must be born with roughly 9–13.5 MeV of excitation energy. As such, fission fragments are indeed analogous to hot liquid droplets which can evaporate nucleons. Although proton emission is energetically possible, we know that it will be highly hindered relative to neutron emission because of the presence of a Coulomb barrier. Apart from possible angular momentum effects, there is no barrier for neutron emission.

The theoretical calculation of neutron emission from a highly-excited nucleus is rather complicated and goes beyond the level of this text. Nevertheless, the ideas presented above are essentially correct. Neutron emission from fission is reasonably described as the evaporation of neutrons from the excited fragments soon after the point of scission. The spectrum of particle evaporation from a "hot" nucleus is the quantum-equivalent of the Maxwell–Boltzmann spectrum of evaporated molecules from an ordinary liquid.

Experiments have been performed to examine the average number of neutrons emitted from fragments of different mass. In Fig. 12.19 is a summary of the results from two such measurements [6]. For the high-yield nuclides, the data from the two experiments are in quite good agreement, with the exception of nuclides with mass numbers near 138–150. Although the scatter in the data is rather large for the very

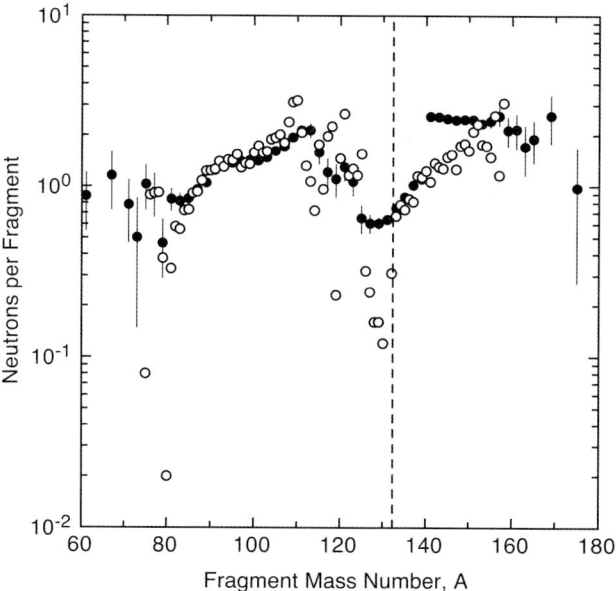

Fig. 12.19 The number of neutrons emitted per fragment when a fragment of mass number A is detected, following fission of $^{235}_{92}U$ with thermal neutrons. The closed and open circles represent measurements from different experiments. The error bars from only one of the experiments are shown for simplicity. The dashed vertical line is at A = 132.

low-yield nuclides, the same general trend is seen in both data sets. At low mass numbers, neutron emission per fragment tends to increase smoothly until a peak is reached at a mass number A = 110–112. Following this, the neutron emission decreases to a minimum at a mass number of about A = 128. The variation in the number of neutrons emitted when heavy fragments are detected roughly mirrors the behavior just described.

Neutron emission from a fragment will obviously depend upon its excitation energy and its neutron binding energy. If a neutron is emitted, the nuclide formed may also have an excitation energy greater than the its neutron binding energy and the same considerations apply to it, and so on. We can combine the information given in Fig. 12.19 with our knowledge of nuclear shell structure to make an additional inference with respect to the influence of closed shells on the fission process. Nuclei in the vicinity of the doubly magic $^{132}_{50}$Sn have about the lowest probability for prompt neutron emission of any of the fragments. This must imply that these fragments tend to be born with relatively low excitation energies. The fragments complementary to these nuclides will lie near Z = 42, A = 104 and it is seen that the prompt neutron emission from these is significantly larger. The inference is that the sharing of excitation energy by the two fragments must be affected by the shell structure during the actual nuclear division itself.

12.5.3
The Spectrum of Prompt γ-Rays

As discussed above, the prompt γ-rays from fission come primarily from decay of the excited fission fragments soon after scission. Although nuclei with excitation energies exceeding the neutron binding energy can decay by both γ-ray and neutron emission, photon emission is relatively unlikely because the decay probabilities for neutron emission are generally much larger. Thus neutron emission will tend to dominate the initial decay of an excited fragment if sufficient energy is available. If, after neutron emission, the residual nucleus of mass number A − 1 also has an excitation energy that exceeds its neutron binding energy, then a second neutron will most likely be emitted, etc. However, as soon as the excitation energy becomes less than the neutron binding energy, only photon emission can occur. Based on this analysis one expects, and indeed finds that the major part of the prompt γ-ray spectrum will have energies less than some 7–8 MeV. In Fig. 12.20 is the evaluated prompt photon spectrum following fission of $^{235}_{92}$U with thermal neutrons. The principal portion of the total intensity lies between about 0.1 and 2 MeV and very little intensity is found at energies comparable to neutron binding energies.

12.5.4
Summary of the Sources of Energy Release in Fission

From a theoretical viewpoint, the energy release in fission is just the sum of the kinetic energies of the fission fragments after neutron emission, the kinetic energy of the prompt neutrons and the energies of the prompt γ-rays. These, of course, will

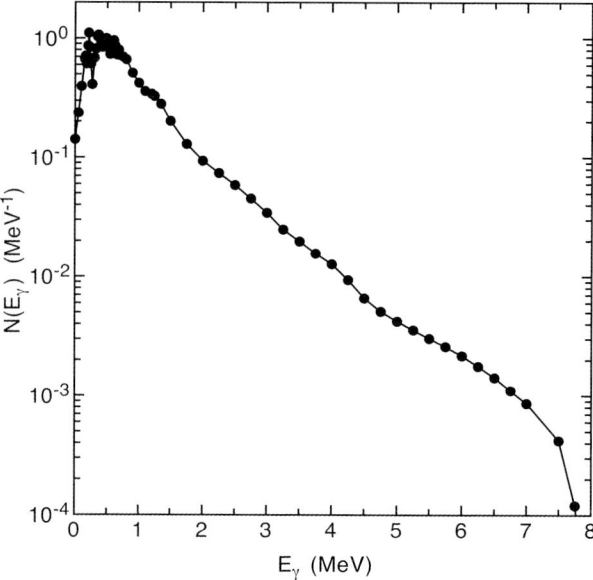

Fig. 12.20 The evaluated prompt photon spectrum following fission of $^{235}_{92}$U with thermal neutrons. Prompt photons with energies as high as at least 20 MeV are known to be emitted, but at considerably reduced intensity.

vary with each mass division and partition of excitation energy between the two fragments. From a practical viewpoint for many applications, it is the average value of these quantities that is most useful. With respect to the total energy release in nuclear power reactors, however, there are additional sources of energy that must be considered. The reactor's fuel will contain all of the fission products and thus the energy from decay of these, primarily the energy of the β particles and γ-rays emitted in their decay, must be considered as well. One must also account for the energies of delayed neutrons.

As examples of average energies from all of these sources, Table 12.1 contains the evaluated data for fission of $^{235}_{92}$U and $^{239}_{94}$Pu with thermal neutrons which are responsible for the majority of the energy release during operation of most power reactors. As seen in both cases, the total kinetic energy of the fission fragments amounts to roughly 93% of the average energy actually released in fission, 180–190 MeV. The average energy of the β particles and γ-rays from fission product decay are roughly the same and total about 10–12 MeV. The average energy of the antineutrinos is somewhat larger than the average energy of the β particles but is not recovered. Summing all of these sources, the actual average total energy release per fission is over 200 MeV but the recoverable energy is smaller by about 7–9 MeV.

Table 12.1 Averages of the different energy sources in the fission of $^{235}_{92}$U and $^{239}_{94}$Pu with thermal neutrons.

Energy Source in Thermal Fission	$^{235}_{92}$U (MeV)	$^{239}_{94}$Pu (MeV)
Average kinetic energy of fission fragments	169.12 ± 0.49	175.78 ± 0.10
Kinetic energy of prompt neutrons	4.79 ± 0.07	5.90 ± 0.10
Total energy of prompt gamma-rays	6.97 ± 0.5	7.76 ± 0.22
AVERAGE ENERGY FROM FISSION	180.88 ± 0.70	189.44 ± 0.26
Kinetic energy of delayed neutrons	$(7.4 \pm 1.1) \times 10^{-3}$	$(2.8 \pm 0.42) \times 10^{-3}$
Total energy of delayed γ-rays	6.33 ± 0.05	5.17 ± 0.06
Total energy release from β-rays	6.50 ± 0.05	5.31 ± 0.06
AVERAGE TOTAL RECOVERABLE ENERGY	193.7 ± 0.15	199.9 ± 0.23
Energy carried away by neutrinos	8.75 ± 0.07	7.14 ± 0.09
AVERAGE TOTAL ENERGY RELEASE PER FISSION	202.47 ± 0.13	207.06 ± 0.21

12.6
Fission Barriers and Fission Probabilities

The discussion of the changes in potential energy that accompany the distortion of a classical uniformly charged sphere have demonstrated that, for nuclides with Z^2/A somewhat smaller than the critical value of about 50, a potential barrier exists, which, if surmounted, would permit the nucleus to undergo fission. The classical model suggested that the magnitude of the fission barrier was on the order of 1 MeV or so for nuclei with Z^2/A of about 45. However, we are primarily interested in the barriers for nuclei such as those of uranium and plutonium with much lower fission parameters. Without resorting to detailed theory, there is a relatively easy way to estimate what the magnitudes of barriers must be in this region. This is related to the following observations.

Nuclides such as $^{235}_{92}$U and $^{239}_{94}$Pu with almost vanishingly small probabilities for spontaneous fission, undergo fission with high probability, upon capture of thermal neutrons. A number of lines of evidence suggest that within some 10^{-19} to 10^{-14} s after the neutron capture, the excited $^{236}_{92}$U and $^{240}_{94}$Pu nuclei so formed undergo fission, the emission of γ-rays or the re-emission of the neutron itself. This suggests that the excitation energies produced by neutron capture must be comparable to or greater than the magnitude of the fission barrier. On the other hand, the capture of a thermal neutron by the neighboring nuclides $^{238}_{92}$U and $^{240}_{94}$Pu does not lead to fission of the excited $^{239}_{92}$U and $^{241}_{94}$Pu formed with any significant probability. The overwhelming fate of these nuclei is decay by either the emission of γ-rays or the re-emission of a neutron.

Because the ratios of Z^2/A for the four nuclides formed by neutron capture are in the range 36–37, the liquid-drop model would suggest that their fission barriers ought to be quite similar. This in turn implies that the fission barrier in the region of these nuclides might be bounded by the excitation energies of the nuclides formed by neutron capture. The excitation energies for $^{236}_{92}$U and $^{240}_{94}$Pu must be larger than the fission barrier, whereas the excitation energies of the excited $^{239}_{92}$U and $^{241}_{94}$Pu formed by neutron capture must be less than the barrier height.

We can use the masses of the neutron and the uranium and plutonium nuclides involved to calculate the excitation energies of nuclides formed by neutron capture and these are given in Table 12.2. These data provide a rather clear picture. The fission barriers in the vicinity of these nuclides must be between about 5.3 and 6.5 MeV. From a number of different approaches, the fission barriers for the four capture products shown in the table are known to be about 5.9, 5.9, 6.2 and 5.9 MeV, respectively. The (even,even) capture products have excitation energies somewhat greater than the barriers but the (even, odd) capture products have excitation energies that are 0.65–1.4 MeV below the barrier height. The magnitudes of the barriers can be understood theoretically within the context of the liquid-drop model with a more sophisticated approach than we have considered. They can also be derived from analysis of reaction cross sections that are discussed in the next chapter.

Note that the difference in the excitation energies between the (even, even) nuclides $^{236}_{92}$U and $^{240}_{94}$Pu, and the (even, odd) nuclides $^{239}_{92}$U and $^{241}_{94}$Pu is in the range 1.3–1.75 MeV. We can easily understand this difference by consideration of the pairing energy. In the capture of a neutron by an odd-N target nuclide, the excitation energy of the (even, even) product will be given by the sum of the binding energy of the neutron and its pairing energy with the odd neutron in the target. On

Table 12.2 The excitation energies of the nuclides formed by capture of a zero-energy neutron in $^{235}_{92}$U, $^{239}_{94}$Pu, $^{238}_{92}$U and $^{240}_{94}$Pu.

Target	Excited capture product	Excitation energy (MeV)
$^{235}_{92}$U	$^{236}_{92}$U	6.545
$^{239}_{94}$Pu	$^{240}_{94}$Pu	6.533
$^{238}_{92}$U	$^{239}_{92}$U	4.806
$^{240}_{94}$Pu	$^{241}_{94}$Pu	5.242

the other hand, the capture of a neutron by an (even, even) target will produce an excitation energy that is just the binding energy of the last neutron. Therefore the excited states of (even, even) nuclides formed by neutron capture will have excitation energies that are greater than the adjacent (even, odd) nuclides formed by neutron capture by the pairing energy. The fissionability of odd-N nuclides by capture of low-energy neutrons and the lack of fissionability of even-N nuclides in the same reaction is commonplace among the actinides.

During the past decades fission has continued to be a topic that has generated considerable study, theoretically and experimentally. The complications that arise as a result of the shape changes which occur during fission and the influence of nuclear structure in both the deformed parent and the fission fragments themselves do not permit the development of a simple model that contains enough of reality and is tractable at the level of our study. As a result we will leave further investigation to more advanced study and turn our attention to the topic of low-energy nuclear reactions.

References

1. The principal article that announced the discovery of fission is O. Hahn and F. Strassmann, "Über den Nachweis und das Verhalten der bei der Bestrahlung des Urans mittels Neutronen entstehenden Erdalkalimetalle", Naturwissenschaften 27 (1939) 89.
2. A simple discussion of the work that produced the first chain reaction can be found in the publication written for the fortieth anniversary, "The First Reactor", DOE-NE 0046 (1982).
3. E.K. Hulet, J.F. Wild, R.J. Dougan, R.W. Lougheed, J.H. Landrum, A.D. Dougan, P.A. Baisden, C.M. Henderson, R.J. Dupzyk, R.L. Hahn, M. Schädel, K. Sümmer and G.R. Bethune, "Spontaneous fission properties of 258Fm, 259Md, 260Md, 258No, and 260[104]: Bimodal fission', Phys. Rev. C 40 (1989) 770.
4. V.P. Zakharova, D.K. Ryazanov, B.G. Basova, A.D. Rabinovich and V.A.Korostylev, Investigation Of Uranium-235 Fission By Thermal Neutrons, Yadernaya Fizika 16 (1972) 649
5. H.Werle, H.Bluhm, Fission Neutron Spectra Measurements Of ^{235}U, ^{239}Pu and ^{252}Cf, J. Nucl. Energy 26 (165) 1972.
6. E.E. Maslin, A.L. Rodgers and W.G.F. Core, Phys. Rev. 164 (1967) 1520, and references therein.

General References

For those interested in the history of the discovery of nuclear fission and the missed opportunities that were experienced by some researchers, a very nice review is found in G. Herrmann, "Discovery and Confirmation of Fission", Nuclear Physics A *502* (1989) 141 and the further publication by the same author "The discovery of nuclear fission – good solid chemistry got things on the right track", Radiochimica Acta 70–71 (1995) 51.

The theory of nuclear fission is described in many texts and articles. But there is perhaps no better place for the interested student to start than the paper N. Bohr and J.A. Wheeler, "The Mechanism of Fission", Physical Review 56 (1939) 426.

13
Low-Energy Nuclear Reactions

13.1
Introduction

The term nuclear reaction generally refers to the interaction of any particle or radiation with a nucleon or complex nucleus. In the broadest sense, it covers the interaction of nuclei with the very intense photon fields produced with high-intensity lasers as well as with highly relativistic particles. For most applications, we deal with the interaction of photons, electrons, individual nucleons or light complex nuclei, with other nuclei. In the majority of cases, we are interested in reactions for which the particles present initially – protons, neutrons and electrons – are the particles present after the interaction. That is, no "new" particles are created. We will restrict our attention to such reactions where the available energy is typically below 100–200 MeV. Notwithstanding the rather large energies involved, the velocities of many particles are generally small enough for relativistic effects to be neglected.

The most common reaction encountered involves the scattering of one particle on another, the familiar case of elastic scattering. For such reactions, there is no net change in internal energy of either of the particles. The kinematics of elastic scattering are just those of the familiar ideal scattering of billiard balls. The elastic scattering of neutrons with protons in water molecules is the means by which the high-energy prompt fission neutrons lose kinetic energy in most nuclear power reactors and become "thermalized". Thermal neutrons are much more likely to be absorbed in nuclei such as ^{235}U or ^{239}Pu to produce fission.

The interaction of protons and neutrons with complex nuclei can lead to the transfer of some kinetic energy to internal excitation of the reaction partner. This energy may later be emitted in the form of γ-rays. Such a reaction is referred to as inelastic scattering. The reaction partners are the "same" both before and after the reaction, but one or both has a different total internal excitation energy. The inelastic scattering of high-energy neutrons on nuclei of natural iron is a means by which the loss of kinetic energy by neutrons, referred to in the trade as "slowing down", can be accomplished with rather high efficiency.

Apart from these, all other reactions will result in reaction products with different atomic and/or mass numbers from those of the reaction partners. We have

Nuclear Physics for Applications. Stanley G. Prussin
Copyright © 2007 WILEY-VCH Verlag GmbH & Co., Weinheim
ISBN: 978-3-527-40700-2

already encountered the neutron-induced fission reaction in Chapter XII. In this case the reaction partners are the neutron and, say, a nucleus of $^{235}_{92}U$. After the reaction, only the excited state of $^{236}_{92}U$ exists, that later undergoes decay.

During the early studies of fission, the only readily available source of neutrons was produced by mixing $^{226}_{88}Ra$ with a light element such as beryllium, and neutrons were produced via reactions of the type

$$^{4}_{2}He + ^{9}_{4}Be \to n + ^{12}_{6}C \tag{13.1}$$

Fermi and others found that if the source was surrounded by paraffin, the probability that the neutrons would interact with stable nuclei to produce new β^- radioactive isotopes was generally greatly increased. The radioactivities were produced primarily by the absorption of a neutron in the target nucleus with the emission of the binding energy as prompt γ-rays. Reactions of this type are commonly referred to as *neutron-capture* reactions.

The study of the interaction of stable nuclides with nuclei of light elements that have been accelerated to high energies in cyclotrons or other accelerators has been one of the prime means by which information on nuclear structure has been obtained. Reactions produced in this way are also important sources of some of the radionuclides that are useful in practical applications. For example, one of the isotopes commonly employed for the diagnostic measurement of blood flow in the heart is $^{201}_{81}Tl$ with a half-life of 73 hr. Its use is based on the fact that thallium behaves quite similarly to potassium in the human body. $^{201}_{81}Tl$ is most commonly produced by the reaction

$$p + ^{203}_{81}Tl \to ^{201}_{82}Pb + 3n \tag{13.2}$$

followed by the electron capture decay of $^{201}_{82}Pb$ to $^{201}_{81}Tl$.

Over the years, a more-or-less standard shorthand notation has been developed to write down nuclear reactions and some common names have been attached to some of them. The typical shorthand notation for reactions is of the form X(x,y)Y, where X is the target, usually at rest in the laboratory, x is the projectile (a charged particle from an accelerator, a neutron in a beam emerging from a nuclear reactor, etc.), and y and Y are the light and heavy reaction products, respectively. If more than one light product is produced, it is commonly included in y. Thus Eq. (13.2) will frequently be written as $^{203}_{81}Tl(p,3n)^{201}_{82}Pb$ and one will hear that $^{201}_{82}Pb$ is produced by the (p,3n) reaction on $^{203}_{81}Tl$. The elastic scattering of neutrons on any target is described as the (n,n) reaction, whereas the inelastic scattering of neutrons will be described as the (n,n′) reaction. In Table 13.1 are the names and notations for many of the more common nuclear reactions of interest in applications.

Three fundamental characteristics are of most interest in the study of nuclear reactions; energetics, kinematics and the reaction probabilities, or cross sections. The energetics are easily handled using atomic masses in the manner outlined in Chapter II. For the present, we will be concerned primarily with the kinematics and reaction cross sections. We begin this chapter by an examination of the kinematics

Table 13.1 Some common nuclear reactions.

Name	Shorthand notation	Example
Elastic scattering	(x,x)	(p,p)
Inelastic scattering	(x,x′)	(p,p′)
Neutron capture or neutron absorption	(n,γ)	$^{197}_{79}\mathrm{Au}(n,\gamma)^{198}_{79}\mathrm{Au}$
Neutron-induced fission	(n,f)	$^{239}_{94}\mathrm{Pu}(n,f)$
Proton in, x neutrons out	(p,xn)	$^{203}_{81}\mathrm{Tl}(p,3n)^{201}_{82}\mathrm{Pb}$
Stripping reaction	(d,p), etc.	$^{88}_{38}\mathrm{Sr}(d,p)^{89}_{38}\mathrm{Sr}$
Pickup reaction	(d, $^3_2$He), etc.	$^{88}_{38}\mathrm{Sr}(d,^3_2\mathrm{He})^{87}_{37}\mathrm{Rb}$
Light element fusion		$^2_1\mathrm{H}(^3_1\mathrm{H},n)^4_2\mathrm{He}$

of binary reactions and a discussion of reactions that are *endoergic* ($Q < 0$) and *exoergic* ($Q > 0$). Following this we will examine cross sections in some detail, concentrating on neutron cross sections for simplicity.

13.2
Kinematics of Nonrelativistic Reactions

The kinematics of nuclear reactions are really quite simple in the nonrelativistic limit. As with all interactions, we need only consider the conservation of energy and momentum. The case of elastic scattering of nuclear particles is no more difficult than the ideal elastic scattering of billiard balls, because the reaction Q value is zero. Inelastic scattering or nuclear reactions in general are a bit more complex because of nonzero Q values, but the complications involve just a little more algebra. In the real world, we experience reactions in the "laboratory" coordinate system. Unfortunately, the laboratory is not the easiest place to examine what is really going on, especially for reactions with $Q < 0$. You will recall from classical mechanics that the motion of the center of mass is conserved and that means that some of the energy and momentum available in the laboratory is not available for reaction. This is not such a problem if we deal solely with kinematics, but it does pose a significant problem in the description of reaction cross sections. As a result, we will first consider the kinematics of interactions in the laboratory system and then recast the problems in the center of mass coordinate system, to point out some simplifications and insights into the effect of different Q values. We will then consider the cross section for simple elastic scattering of hard spheres and use it as

13.2.1
Kinematics of Elastic Scattering in the Laboratory Coordinate System

We consider the case of the elastic scattering of two bodies in the laboratory coordinate system under the action of conservative forces as shown in the general schematic of Fig. 13.1. Initially, a projectile of mass m_1 moving with velocity v_1 interacts with a target of mass m_2 that is at rest. Elastic scattering takes place and the projectile then moves with velocity v'_1 at the angle θ with respect to its initial trajectory. The target moves with velocity v'_2 at the angle ϕ after the interaction.

Conservation of linear momentum and energy are given by

$$\mathbf{p}_1 = \mathbf{p}'_1 + \mathbf{p}'_2 \tag{13.3}$$
$$T_1 = T'_1 + T'_2$$

We will assume that the characteristics of the projectile are of primary interest. That means that we want to eliminate the characteristics of the target after the interaction and this is easily accomplished by solving the first of Eqs (13.3) for $\mathbf{p}'_2$ and squaring. Thus,

$$p'^2_2 = \mathbf{p}'_2 \cdot \mathbf{p}'_2 = p_1^2 + p_1'^2 - 2\mathbf{p}_1 \cdot \mathbf{p}'_1 = p_1^2 + p_1'^2 - 2 p_1 p'_1 \cos\theta \tag{13.4}$$

In the nonrelativistic limit, the kinetic energy is just $T = \frac{1}{2}mv^2 = p^2/2m$. Dividing (13.4) by $2m_2$ gives

$$\frac{p'^2_2}{2m_2} = \frac{1}{2m_2}\left[2m_1\frac{p_1^2}{2m_1} + 2m_1\frac{p'^2_1}{2m_1} - 2\sqrt{2m_1}\sqrt{\frac{p_1^2}{2m_1}}\sqrt{2m_1}\sqrt{\frac{p'^2_1}{2m_1}}\cos\theta\right] \tag{13.5}$$

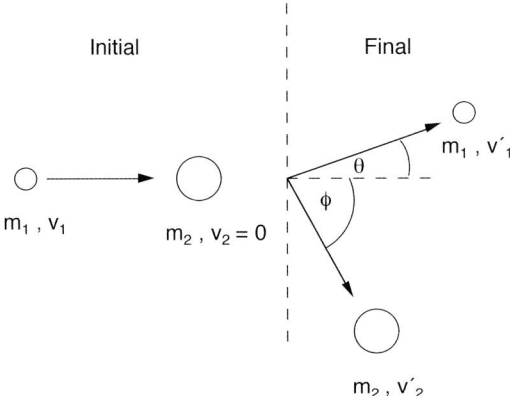

Fig. 13.1 Schematic diagram of elastic scattering in the laboratory coordinate system.

or

$$T_2' = \frac{m_1}{m_2}T_1 + \frac{m_1}{m_2}T_1' - \frac{2m_1}{m_2}\sqrt{T_1}\sqrt{T_1'}\cos\theta \tag{13.6}$$

Substitution of this expression into the second of Eqs (13.3) with a little additional algebra then gives

$$T_1'\left(\frac{m_1+m_2}{m_2}\right) - \frac{2m_1}{m_2}\sqrt{T_1}\sqrt{T_1'}\cos\theta + T_1\left(\frac{m_1-m_2}{m_2}\right) = 0 \tag{13.7}$$

This expression can be viewed as a simple quadratic equation in the square-root of either T_1 or T_1'. Because we are interested in the properties of the projectile after the scattering, we take it as a quadratic in $\sqrt{T_1'}$. Using the relation

$$x = \frac{-b \pm \sqrt{b^2 - 4ac}}{2a} \tag{13.8}$$

with $a = \left(\dfrac{m_1+m_2}{m_2}\right)$, $b = -\dfrac{2m_1}{m_2}\sqrt{T_1}\cos\theta$, and $c = T_1\left(\dfrac{m_1-m_2}{m_2}\right)$

a little algebra gives the final result as

$$\sqrt{T_1'} = \left(\frac{m_1}{m_1+m_2}\right)\sqrt{T_1}\cos\theta\left\{1 \pm \left[1 - \frac{(m_1^2 - m_2^2)}{m_1^2\cos^2\theta}\right]^{1/2}\right\} \tag{13.9}$$

Now there are two physical requirements that we must consider in the interpretation of this result. The first is that the kinetic energy is a real quantity and so must be any of its roots. In order for the square-root of the kinetic energy to be real, we must demand that

$$1 - \frac{m_1^2 - m_2^2}{m_1^2\cos^2\theta} \geq 0, \text{ or} \tag{13.10}$$

$$\cos^2\theta \geq \left(1 - \frac{m_2^2}{m_1^2}\right) \tag{13.11}$$

In the present problem, $0 \leq \theta \leq \pi$ and $0 \leq \cos^2\theta \leq 1$. So long as $m_2^2/m_1^2 > 1$, the right-hand side of the inequality in (13.11) will be a negative number and $\cos^2\theta$ will satisfy the inequality for all angles θ. In other words, the projectile can scatter through all possible angles. This is the situation most commonly found in nuclear reactions used to produce radioactive isotopes for applications.

If $m_1 = m_2$, the inequality says that $\cos^2\theta \geq 0$ and therefore $\cos\theta \geq 0$. Reference to Eq. (13.9) shows that the kinetic energy of the scattered particle goes to zero at $\theta = \pi/2$ in this case. Thus the scattering angle must lie in the range $0 \leq \theta \leq \pi/2$. The scattered projectile will always be found in the forward direction in the laboratory. This result has quite practical significance because the scattering of electrons on electrons and, to an excellent approximation, neutrons on protons, is frequently encountered.

If $m_1 > m_2$, life gets a little more interesting. Remembering that $\cos\theta$ decreases with increasing angle in the range $0 \le \theta \le \pi/2$, the inequality of (13.11) says that all scattering angles will be possible up to a maximum value θ_{max} given by expression

$$\cos^2\theta_{max} = 1 - \frac{m_2^2}{m_1^2} \qquad (13.12)$$

For example, if $m_1 = 1.01 m_2$, $\cos^2\theta_{max} = 0.0197$ and $\theta_{max} = 88.87°$. The scattered projectile will be found at almost all forward angles. But clearly, as the mass ratio m_1/m_2 increases, the maximum value of $\cos^2\theta$ will continuously increase and the maximum scattering angle of the projectile will progressively become smaller and smaller. In the limit that $m_1/m_2 \to \infty$, the projectile will be found only in the direction of its initial trajectory. This result also has important practical implications. In the next chapter we will examine the interaction of radiation with matter. We will find that the principal means by which α particles and other heavy charged particles interact with matter is via Coulomb scattering with atomic electrons. In an elastic scattering event with an electron, it is not difficult to show that the maximum scattering angle for the α particle is about 1.07×10^{-6} degrees. This is trivially small, and we can predict that the trajectories of α particles as they penetrate matter will be linear to a very good approximation. Now it turns out that β particles interact primarily by the same mechanism. If you have been following along, you can see that the electron trajectories will tend not to be linear.

The second physical requirement is that the kinetic energy must be equal to or greater than zero. You will probably recall that we encountered a negative kinetic energy for a particle tunneling under a barrier in quantum mechanical barrier penetration. But after an interaction, when the particles are no longer experiencing any mutual interaction, the kinetic energies must always be equal to or greater than zero. This is precisely the condition under which the conservation laws apply and the kinematics are described by Eqs (13.9).

First consider the case where $m_1 < m_2$ and all scattering angles are possible. The requirement of positive kinetic energies will be satisfied if and only if the positive sign in braces in (13.9) is chosen. With this, and by multiplying through by $\cos\theta$, Eq. (13.9) becomes

$$\sqrt{T_1'} = \left(\frac{m_1}{m_1 + m_2}\right)\sqrt{T_1}\left\{\cos\theta + \left[\cos^2\theta - \frac{(m_1^2 - m_2^2)}{m_1^2}\right]^{1/2}\right\} \qquad (13.13)$$

It will prove useful to consider the limits of the general expression given above. In the case that $\theta = 0$, a so-called "grazing" collision for which the impact parameter, the distance of closest approach that would result for the two particles if the particle trajectories were undeflected by the collision, is just equal to the sum of the radii of the collision partners, the energy transfer must go to zero, and $T'_1 = T_1$. It is not difficult to show that the quantity in braces reduces to $(m_1 + m_2)/m_1$ and we indeed have the expected result.

The maximum energy loss occurs in a "head-on" collision in which case $\theta = \pi$. Considering just this case, it is not difficult to show that (13.13) reduces to

$$T'_1 = \left(\frac{m_2 - m_1}{m_1 + m_2}\right)^2 T_1 \tag{13.14}$$

This simple result can be combined with the discussion that followed the limitation that the kinetic energy be real, to examine some practical applications. First, the expression shows that in the limit $m_1 = m_2$, the kinetic energy of the projectile in a head-on collision will be reduced to zero. As shown above, however, the range of scattering angles is reduced to $0 \leq \theta \leq \pi/2$. Head-on collisions of electrons on electrons, protons on protons and, to an excellent approximation, neutrons on protons, will result in total energy loss. This result is one of the principal reasons why most nuclear reactors use normal water for the purpose of reducing the energies of fission neutrons from the order of 2 MeV to thermal energies. It takes the fewest number of collisions, on the average, to carry out the reduction in energy. If the mass of the target is twice that of the projectile, the maximum energy loss in a single head-on collision is 8/9 of the initial kinetic energy. This means that only a few more collisions will be needed to obtain the same average energy loss as found in the elastic scattering of neutrons on protons, and "heavy water", $^2_1H_2O$ or D_2O is almost as good for thermalizing neutrons. Because of its lower neutron absorption cross section, the use of D_2O permits construction of a nuclear fission reactor with naturally-occurring uranium. When normal water is used, reactors must be fueled with uranium with a higher abundance of ^{235}U than is found in nature. The maximum energy loss per collision is seen to rapidly decrease with increasing m_2/m_1. Very little energy loss of a neutron takes place by elastic scattering on a uranium nucleus.

If $m_1 \geq m_2$, we must be aware of the limit on the scattering angle found above. The scattered projectile will be restricted to forward angles that are defined by Eq. (13.12). Nevertheless, you can demonstrate that the relation given in (13.14) holds regardless of the relative mass of the projectile and target. If we consider the case of the scattering of α particles on electrons, the ratio of the final to initial kinetic energy is about 0.99945, the maximum energy lost to the electron amounting to about 5.5×10^{-4} of the initial energy. Combined with the very small maximum scattering angle calculated above, we conclude that literally tens of thousands of collisions will be required for complete energy loss of the α particles emitted in radioactive decay as they traverse an essentially linear path through matter.

13.2.2
Kinematics of Elastic Scattering in the Center of Mass Coordinate System

The usefulness of treating reactions in the center of mass coordinate system is perhaps most easily illustrated by considering a reaction for which Q < 0. Such a reaction can proceed only by addition of energy to the system. For a binary reaction with a target that is at rest in the laboratory, the additional energy must be provided by the kinetic energy of the projectile. If the projectile kinetic energy was just equal to the absolute magnitude of the Q value, the total energy of the system would not be sufficient for the reaction to take place because of the conservation of the motion of the center of mass of the system. Similarly, if we wish to understand the energy-dependence of a reaction probability, we need to refer it directly to the energy actually available for the reaction. Because it is so easy to apply, we will discuss the relation between momentum and energy in the laboratory and center of mass coordinate systems first for the case of elastic scattering and relate it to the observations we have pointed out in the previous section. We again consider only the case of a projectile interacting with a target at rest in the laboratory coordinate system.

In developing the relation between dynamical variables in the two coordinate systems, one must pay close attention to notation. For our two-body problem, we will use the notation "1" and "2" to denote variables in the laboratory system and "1,cm" and "2,cm" to denote variables in the center of mass system. We also must distinguish variables before and after the interaction and we will use unprimed quantities for variables before the interaction and primed quantities for variables after the interaction has taken place.

The center of mass coordinate system is defined as that system in which the total linear momentum of the reaction partners is identically zero. It will be moving in the laboratory with some constant velocity $\mathbf{V}_{CM}$ and therefore will have some constant linear momentum $\mathbf{P}_{CM}$. If the reaction partners have masses m_1 and m_2 they will have velocities $\mathbf{v}_{1,cm}$ and $\mathbf{v}_{2,cm}$ in the center of mass system such that

$$m_1 \mathbf{v}_{1,cm} + m_2 \mathbf{v}_{2,cm} = 0 \tag{13.15}$$

or

$$\mathbf{p}_{1,cm} = -\mathbf{p}_{2,cm} \tag{13.16}$$

The momentum of the center of mass in the laboratory coordinate system is defined by the relation

$$(m_1 + m_2)\mathbf{V}_{CM} = m_1 \mathbf{v}_1 + m_2 \mathbf{v}_2 \quad \text{or} \tag{13.17}$$

$$\mathbf{P}_{CM} = \mathbf{p}_1 + \mathbf{p}_2 \tag{13.18}$$

The momentum of the center of mass in the laboratory system is just equal to the total initial momentum of the system. Its velocity is that of a single body with mass equal to the sum of the masses of the reaction partners and with velocity $\mathbf{V}_{CM}$ given by

$$\mathbf{V}_{CM} = \frac{m_1 \mathbf{v}_1 + m_2 \mathbf{v}_2}{(m_1 + m_2)} \tag{13.19}$$

We can relate the magnitudes of variables in the laboratory and center of mass systems by noting that they are vector quantities. Thus the velocity of a particle in the laboratory must be equal to the sum of its velocity in the center of mass system, plus the velocity of the center of mass itself, i.e.,

$$\mathbf{v}_1 = \mathbf{v}_{1,cm} + \mathbf{V}_{CM}$$
$$\mathbf{v}_2 = \mathbf{v}_{2,cm} + \mathbf{V}_{CM} \tag{13.20}$$

These relations also provide the relations between the momentum vectors and kinetic energies in the two coordinate systems.

We can now use the relations above to derive relations between parameters in the laboratory and center of mass coordinate systems for elastic scattering. First, because the target is at rest in the laboratory, the velocity of the center of mass in the laboratory is

$$\mathbf{V}_{CM} = \frac{m_1 \mathbf{v}_1}{(m_1 + m_2)} \tag{13.21}$$

The velocities of the projectile and target before scattering are then related by

$$\mathbf{v}_{1,cm} = \mathbf{v}_1 - \mathbf{V}_{CM} = \mathbf{v}_1 \left(\frac{m_2}{m_1 + m_2} \right)$$
$$\mathbf{v}_{2,cm} = \mathbf{v}_2 - \mathbf{V}_{CM} = -\mathbf{v}_1 \left(\frac{m_1}{m_1 + m_2} \right) \tag{13.22}$$

and the momenta of the two particles in the center of mass are then

$$\mathbf{p}_{1,cm} = \mathbf{v}_1 \left(\frac{m_1 m_2}{m_1 + m_2} \right) = \mu \mathbf{v}_1 ,$$
$$\mathbf{p}_{2,cm} = -\mathbf{v}_1 \left(\frac{m_1 m_2}{m_1 + m_2} \right) = -\mu \mathbf{v}_1 \tag{13.23}$$

equal and opposite as they should be. In these equations, μ is just the reduced mass of the system as normally defined.

The total kinetic energy in the laboratory is just that of the projectile, i.e.,

$$T_{tot} = T_1 + T_2 = \frac{1}{2} m_1 v_1^2 \tag{13.24}$$

The total kinetic energy in the center of mass coordinate system is

$$T_{tot,cm} = T_{1,cm} + T_{2,cm} = \frac{1}{2}m_1\left(v_1\frac{m_2}{m_1+m_2}\right)^2 + \frac{1}{2}m_2\left(-v_1\frac{m_1}{m_1+m_2}\right)^2$$
$$= \frac{1}{2}\left(\frac{m_1 m_2}{m_1+m_2}\right)v_1^2 = \frac{1}{2}\mu v_1^2 \quad (13.25)$$

This is just the kinetic energy of a particle with the reduced mass of the system moving with the velocity of the projectile in the laboratory. It is smaller than the kinetic energy available in the laboratory coordinate system by the kinetic energy of the center of mass, i.e.,

$$T_{CM} = \frac{1}{2}(m_1+m_2)V_{CM}^2 = \frac{1}{2}m_1 v_1^2 - \frac{1}{2}\mu v_1^2$$
$$= \frac{1}{2}\left(\frac{m_1^2 v_1^2}{m_1+m_2}\right) \quad (13.26)$$

If the mass of the projectile is large compared to the mass of the target, the kinetic energy of the center of mass will also be large and the kinetic energy available in the center of mass coordinate system will be small and vice versa. In most practical cases, the kinetic energy available in the laboratory is significantly less than the kinetic energy of the projectile, and this will have a major effect on nuclear reactions as we will see shortly.

The momentum of the center of mass in the laboratory is given directly by the definition in Eq. (13.18) as

$$P_{CM} = p_1 + p_2$$
$$= m_1 v_1 \quad (13.27)$$
$$= (m_1+m_2)V_{CM}$$

Eqs (13.23) indicate that the momenta of the projectile and target are equal and opposite to one another before the scattering and this must hold after the scattering as well. An observer at rest in the center of mass reference frame will see the two scattering partners approach one another with equal and opposite momenta, interact and then recede from one another with equal and opposite momenta. Because the momenta are always the same, the kinetic energies of the scattering partners will also be the same before and after the scattering. Elastic scattering in the center of mass system is a fairly boring affair. The only thing that can change as a result of the scattering is the angle with which the scattering partners recede relative to the trajectory of the incident projectile.

In Fig. 13.2 are shown schematics of the elastic scattering before and after the interaction in both the laboratory and center of mass coordinate systems. Having solved the kinematic equations by a purely algebraic approach, it will prove useful to consider some of the geometric relations that can be seen easily with use of the center of mass properties. For example, the constancy of the momentum of the

13.2 Kinematics of Nonrelativistic Reactions

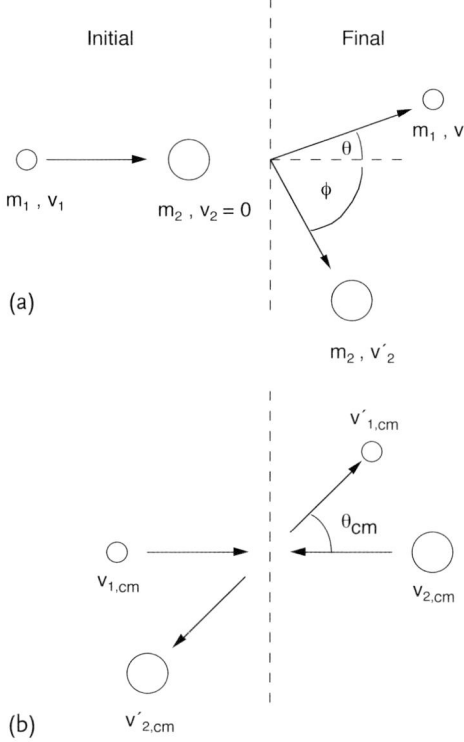

Fig. 13.2 Kinematics of elastic scattering in the laboratory (a) and center of mass (b) coordinate systems.

center of mass and the relation between a particle's velocity in the laboratory and center of mass coordinate systems can be used to examine the angular ranges in the laboratory available for a scattered particle. We first consider the case where the projectile and the target have identical masses. In this case, Eq. (13.21) gives the center of mass velocity in the laboratory as

$$\mathbf{V}_{CM} = \frac{m_1}{m_1 + m_2}\mathbf{v}_1 = \frac{1}{2}\mathbf{v}_1$$

and (13.22) gives the velocity of the projectile in the center of mass system as

$$\mathbf{v}_{1,\,cm} = \mathbf{v}_1\left(\frac{m_2}{m_1 + m_2}\right) = \frac{1}{2}\mathbf{v}_1$$

Because the two vectors are identical in length and can be oriented at any angle relative to one another in the cases of interest to us, we can easily see that if $\theta_{CM} = 0$, $\mathbf{v}'_1 = \mathbf{v}_{1,\,cm} + \mathbf{V}_{CM} = \mathbf{v}_1$. Scattering at $0°$ in the center of mass corresponds to

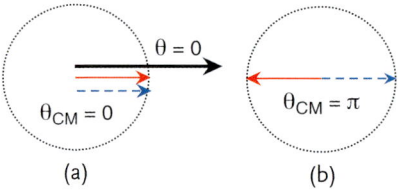
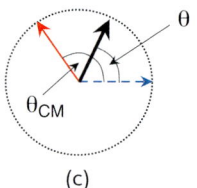

Fig. 13.3 Schematic diagram of the velocity vectors of the projectile in the center of mass frame (red), the center of mass (blue dashed) and the projectile in the laboratory coordinate system (black) for elastic scattering in the case that $m_1 = m_2$. The circle shown has a radius $|\mathbf{v}_{1,cm}|$. The projectile is scattered at $\theta_{CM} = 0$ (a), $\theta_{CM} = \pi$ (b), and at arbitrary scattering angle (c).

scattering at 0° in the laboratory. This is a grazing collision where no energy transfer takes place between the projectile and the target. If $\theta_{CM} = \pi$, $\mathbf{v}'_1 = \mathbf{V}_{CM} - |\mathbf{v}_{1,cm}| = 0$ and complete energy transfer to the target takes place. Because the vectors have the same lengths, it is seen that any angle $0 < \theta_{CM} < \pi$ will lead to the projectile appearing at an angle $\leq \pi/2$ in the laboratory coordinate system, and by a limiting procedure, $\theta_{CM} = \pi$ must translate into $\theta = \pi/2$. The relation between the center of mass velocity, the projectile in the center of mass and the projectile in the laboratory coordinate system are shown in the vector diagrams of Fig. 13.3 for the cases $\theta_{CM} = 0$, $\theta_{CM} = \pi$ and for an arbitrary scattering angle.

In the case that $m_1 < m_2$,

$$\frac{V_{CM}}{v_{1,cm}} = \frac{m_1}{m_1+m_2} v_1 \cdot \frac{m_1+m_2}{m_2 v_1} = \frac{m_1}{m_2} < 1$$

The length of $\mathbf{V}_{CM}$ is less than that of $\mathbf{v}_{1,cm}$. Thus summing the vectors $\mathbf{v}_{1,cm}$ and $\mathbf{V}_{CM}$ over all possible θ_{CM} will clearly allow the projectile to be found at all angles in the laboratory coordinate system (Fig. 13.4).

When $m_1 > m_2$, things get rather interesting. In this case, $V_{CM}/v_{1,cm} > 1$, and it should be clear that the projectile will only be observed at forward angles ($\theta > \pi/2$) in the laboratory system, regardless of the angle of scattering in the center of mass system. As seen in Fig. 13.5, projectiles scattered at both $\theta_{CM} = 0$ and $\theta_{CM} = \pi$ will be observed at $\theta = 0$ in the laboratory, and this means that projectiles with *two different energies* will be seen in the forward direction. By extension, particles with two different energies will be seen for a range of laboratory angles that depends inversely on the ratio of m_1/m_2. As the projectile gets more and more massive in comparison to the target, the range in laboratory angles where projectiles can be found will progressively decrease. With mass ratios as large as that found for the α particle and electron, it is easy to see from the vector diagrams that such projectiles must all be emitted very nearly at $\theta = 0$ in the laboratory.

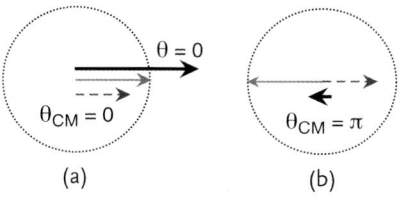

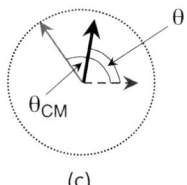

Fig. 13.4 Schematic diagram of the velocity vectors of the projectile in the center of mass frame (red), the center of mass (blue dashed) and the projectile in the laboratory coordinate system (black) for elastic scattering in the case that $m_1 < m_2$. See caption to Fig. 13.3.

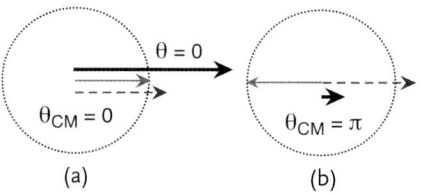

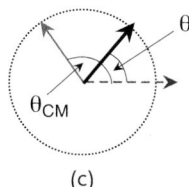

Fig. 13.5 Schematic diagram of the velocity vectors of the projectile in the center of mass frame (red), the center of mass (blue dashed) and the projectile in the laboratory coordinate system (black) for elastic scattering in the case that $m_1 > m_2$. See caption to Fig. 13.3.

The quantitative relation between the scattering angle in the center of mass and laboratory coordinate systems is easily derived from the vector diagrams. In Fig. 13.6, the relevant relations are shown for the velocity vectors of both the target and the projectile. We begin first with the geometrical relations for the projectile vectors. Directly from the diagram one has

$$v'_{1,\text{cm}} \sin\theta_{\text{CM}} = v'_1 \sin\theta$$
$$v'_{1,\text{cm}} \cos\theta_{\text{CM}} + V_{\text{CM}} = v'_1 \cos\theta \tag{13.28}$$

Taking the ratio of Eqs (13.28) and using the relations given in Eqs (13.21) and (13.22) we have

$$\tan\theta = \frac{\sin\theta_{\text{CM}}}{\cos\theta_{\text{CM}} + \dfrac{V_{\text{CM}}}{v'_{1,\text{cm}}}} = \frac{\sin\theta_{\text{CM}}}{\cos\theta_{\text{CM}} + \dfrac{V_{\text{CM}}}{v'_{1,\text{cm}}}} = \frac{\sin\theta_{\text{CM}}}{\cos\theta_{\text{CM}} + k} \tag{13.29}$$

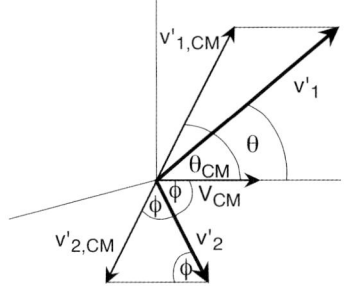

Fig. 13.6 Angular relations between laboratory and center of mass coordinates.

where $k = V_{CM}/v_{1,cm} = m_1/m_2$. This establishes the relation between the scattering angles in the two coordinate systems.

What remains to be found is the relation between the scattering angle of the projectile, θ, and the *recoil* angle of the target, ϕ, in the laboratory coordinate system. To do this we consider the triangle with sides of $v'_{2,cm}$, v'_2 and V_{CM} in Fig. 13.6. First, recall that $v'_{2,cm} = v_{2,cm}$; there is no change in the magnitude of velocity of a particle in the center of mass coordinate system as a result of elastic scattering. Then, from Eqs (13.21) and (13.22), $v_{2,cm} = V_{CM}$. Taken together, these relations indicate that the triangle is an isosceles triangle and the two angles opposite the sides of length $v'_{2,cm}$ and V_{CM} must be equal to one another and to the recoil angle ϕ. Thus, $\theta_{CM} = \pi - 2\phi$. If we substitute this into (13.29) and use the trigonometric relations

$$\sin(\alpha - \beta) = \sin\alpha\cos\beta - \cos\alpha\sin\beta$$
$$\cos(\alpha - \beta) = \cos\alpha\cos\beta + \sin\alpha\sin\beta \qquad (13.30)$$

we immediately arrive at the result

$$\tan\theta = \frac{\sin 2\phi}{k - \cos 2\phi} \qquad (13.31)$$

We now have all of the relations that are needed for a complete examination of the kinematics of nonrelativistic elastic scattering. All of these relations can be carried over and applied to other nuclear reactions with small modifications, as we will now see.

13.2.3
Kinematics of General Nonrelativistic Nuclear Reactions

The principal difference between elastic scattering and all other nonrelativistic reactions is the Q value. For elastic scattering, the energy available in the laboratory coordinate system is just the kinetic energy of the projectile. In all other reactions energy will either be produced as a result of the reaction, *exoergic* reactions with Q > 0, or energy will be converted into mass when Q < 0 (*endoergic* reactions). Now it is also true that the total rest mass of the target and projectile cannot be identical to the total rest mass of reaction products if $Q \neq 0$. However, the difference in mass is so small that, except for the most exacting calculations, it can be completely neglected in considering the reaction kinematics. We will make this approximation here.

In order to handle variable Q values in the general case of a projectile interacting with a stationary target, we make the following observation. If the reaction has $Q > 0$, the total kinetic energy of the reaction products will be the sum of the kinetic energy of the projectile and Q. If the reaction is endoergic, the kinetic energy of the reaction products will be the difference between the kinetic energy of the projectile and the absolute value of Q. A schematic of a general reaction is shown in Fig. 13.7. A projectile of mass m_1 is incident on a target of mass m_2 that is at rest in the laboratory. As a result of the reaction, a light product of mass m_3 is produced at an angle θ with respect to the direction of the incident projectile, and a heavy product of mass m_4 appears at the angle ϕ. The total energy available appears as kinetic energy of the reaction products. Thus

$$T_1 + Q = T_3 + T_4 \tag{13.32}$$

represents conservation of energy, and conservation of linear momentum is given by

$$\mathbf{p}_1 = \mathbf{p}_3 + \mathbf{p}_4 \tag{13.33}$$

We now follow a development exactly similar to that for elastic scattering. Assuming we wish to solve for the characteristics of the light product m_3, we solve (13.33) for $\mathbf{p}_4$, square the result, divide by $2m_4$ and obtain

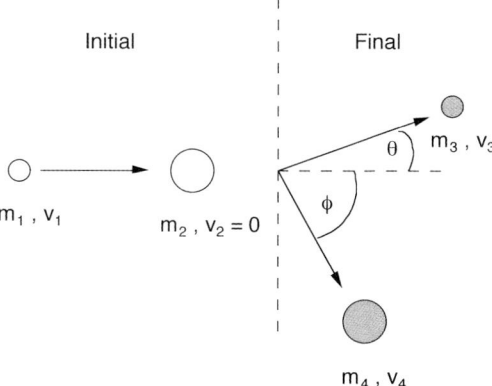

Fig. 13.7 Schematic diagram for a general binary nuclear reaction with a projectile incident upon a target nucleus at rest in the laboratory coordinate system.

$$T_4 = \frac{p_4^2}{2m_4} \tag{13.34}$$

$$= \frac{1}{2m_4}\left(2m_1\frac{p_1^2}{2m_1} + 2m_3\frac{p_3^2}{2m_3} - \frac{\sqrt{2m_1}}{\sqrt{2m_1}}\frac{\sqrt{2m_3}}{\sqrt{2m_3}}2p_1p_3\cos\theta\right) = T_1 + Q - T_3$$

After some algebra, this expression becomes

$$T_1\left(\frac{m_1}{m_4} - 1\right) + T_3\left(\frac{m_3}{m_4} + 1\right) - \frac{2\sqrt{m_1m_3}}{m_4}T_1^{1/2}T_3^{1/2}\cos\theta - Q = 0 \tag{13.35}$$

Treating the equation as a quadratic in $T_3^{1/2}$ and defining

$$a = \left(\frac{m_3}{m_4} + 1\right)$$

$$b = \frac{-2\sqrt{m_1m_3}}{m_4}T_1^{1/2}\cos\theta \tag{13.36}$$

$$c = T_1\left(\frac{m_1}{m_4} - 1\right) - Q$$

we obtain the solution as

$$T_3^{1/2} = \frac{-b}{2a} \pm \left(\frac{b^2}{4a^2} - \frac{c}{a}\right)^{1/2} \tag{13.37}$$

where

$$\frac{-b}{2a} = \frac{\sqrt{m_1m_3}}{(m_3 + m_4)}T_1^{1/2}\cos\theta$$

$$\frac{b^2}{4a^2} = \frac{m_1m_3}{(m_3 + m_4)^2}T_1\cos^2\theta \tag{13.38}$$

$$-\frac{c}{a} = \frac{m_4Q - T_1(m_1 - m_4)}{m_3 + m_4}$$

We now examine the solutions to Eq. (13.37) for different ranges of Q.

Q > 0
By assumption, $m_1 \le m_4$. Because $T_1 \ge 0$, $-c/a \ge 0$ and the quantity

$$\left(\frac{b^2}{4a^2} - \frac{c}{a}\right)^{1/2}$$

will then be real for all possible θ. Now we must remember that $T_3^{1/2} \ge 0$, and to ensure this for all choices of masses, we require that the positive root of

$$\left(\frac{b^2}{4a^2} - \frac{c}{a}\right)^{1/2}$$

be taken. Therefore, if $Q > 0$,

$$T_3^{1/2} = \frac{-b}{2a} + \left(\frac{b^2}{4a^2} - \frac{c}{a}\right)^{1/2} \tag{13.39}$$

A particularly important case that arises frequently is that where Q is very much larger than T_1. In the limit that $T_1 \to 0$, $b/2a \to 0$, $-c/a \to m_4 Q/(m_3 + m_4)$ and $T_3 \to m_4 Q/(m_3 + m_4)$. The kinetic energy of the light product is constant and independent of angle. The light product is emitted isotropically in the laboratory system. A simple example is the fission of $^{235}_{92}U$ with thermal neutrons. A second example is the fusion reaction $^2H(^2H, n)^3He$ with $Q_{^2H,n} = 3.269$ MeV. If the reaction takes place at "thermal" energies, such as in thermonuclear reactions, the kinetic energy of the neutrons would be 2.45 MeV, independent of the angle of emission in the laboratory. This reaction is a common means of producing neutrons in the laboratory by accelerating deuterium nuclei to energies of 0.2–0.4 MeV in a small linear accelerator and allowing them to interact with a target rich in deuterium nuclei. Although the kinetic energy of the deuterons is not completely negligible, it is approximately so and the neutrons produced have energies that are still roughly independent of angle in the laboratory.

$Q < 0$

When $Q < 0$, the reaction simply cannot proceed without sufficient projectile kinetic energy. Because of the requirement of conservation of the momentum of the center of mass, the reaction will not take place even if the projectile kinetic energy is equal to the absolute value of Q. Reactions with negative Q will therefore possess a *threshold* value, the minimum projectile kinetic energy that is required to just permit the reaction to occur in the center of mass coordinate system. If the kinetic energy of the projectile is just sufficient for the reaction to occur, the reaction products will have no kinetic energy in this system and therefore must appear to be moving together in the laboratory in the direction of the center of mass motion, the direction of the initial trajectory of the projectile. At threshold, both reaction products appear at $\theta = 0$ with the momentum of the center of mass.

These qualitative arguments can be put into a quantitative framework with relative ease. From Eq. (13.26) we have

$$T_{CM} = \frac{1}{2}\left(\frac{m_1^2 v_1^2}{m_1 + m_2}\right) = \left(\frac{m_1}{m_1 + m_2}\right) T_1 \tag{13.40}$$

This means that the kinetic energy available for the reaction in the center of mass coordinate system is

$$T_1 - \left(\frac{m_1}{m_1+m_2}\right)T_1 = \left(\frac{m_2}{m_1+m_2}\right)T_1 \tag{13.41}$$

and it is this energy that must be equal to or greater than $|Q|$ for the reaction to proceed. Thus, when $Q < 0$, there will be a threshold energy E_{th} in the laboratory coordinate system given by

$$E_{th} = (T_1)_{min} = \left(\frac{m_1+m_2}{m_2}\right)|Q| \tag{13.42}$$

Only if m_1/m_2 is small will we find $E_{th} \approx |Q|$.

At threshold, the reaction products appear at $\theta = 0$ in the laboratory. As T_1 increases, the products can appear at larger angles. The angular range available for any T_1 is found by analysis of the terms in Eqs (13.37). In order for the kinetic energy of the light product to be real, we must have

$$\frac{b^2}{4a^2} - \frac{c}{a} = \frac{m_1 m_3}{(m_3+m_4)^2}T_1\cos^2\theta + \frac{m_4 Q - T_1(m_1 - m_4)}{(m_3+m_4)} \geq 0 \tag{13.43}$$

and solving for $\cos^2\theta$ we have

$$\cos^2\theta \geq \frac{(m_3+m_4)^2}{m_1 m_3 T_1}\left[\frac{T_1(m_1-m_4) - m_4 Q}{(m_3+m_4)}\right] \tag{13.44}$$

Direct substitution of the relevant masses and energies into (13.44) will give the precise conditions for the angular range available to the reaction products in any specific case. However, with a bit of additional analysis we can obtain a more compact form that is somewhat more transparent. For this purpose, let

$$T_1 = \kappa E_{th} = \kappa\left(\frac{m_1+m_2}{m_2}\right)|Q| \tag{13.45}$$

where $\kappa \geq 1$. Substitution of this into (13.44) and recognizing that $(m_3+m_4) = (m_1+m_2)$ to a very good approximation, Eq. (13.44) can be reduced to

$$\cos^2\theta \geq \frac{1}{\kappa m_1 m_3}[m_2 m_4 - \kappa(m_4 - m_1)(m_1+m_2)] \tag{13.46}$$

Expansion of the factors in brackets with the same approximation then gives the final expression

$$\cos^2\theta \geq \left[1 + \frac{m_2 m_4}{m_1 m_3}\left(\frac{1-\kappa}{\kappa}\right)\right] \qquad \kappa \geq 1 \tag{13.47}$$

For any given endoergic reaction, the expression in (13.47) shows that the reaction products first appear at $\theta = 0$ in the laboratory, and as κ increases, the angular range increases continuously. The ratio of the kinetic energy to the threshold energy, κ, when the light product can first appear at θ = π/2 is readily found to be

$$\kappa = \frac{1}{1 - \frac{m_1 m_3}{m_2 m_4}} \tag{13.48}$$

For all $1 \leq \kappa \leq 1/(1 - (m_1 m_3/m_2 m_4))$, the kinetic energies are obviously real. Because essentially all realistic forces allow for scattering at all angles in the center of mass coordinate system, it must then be true that particles emitted at $\theta_{CM} > \pi/2$ will be forced to appear at *forward* angles in the laboratory. As a result, for energies near the reaction threshold, the kinetic energies of the particles seen at any allowed angle in the range $0 \leq \theta \leq \pi/2$ will have two values, one corresponding to particles emitted in the forward direction in the center of mass and one corresponding to particles emitted at a backward angle that corresponds to the same forward angle in the laboratory. So long as $\kappa \leq 1/(1 - (m_1 m_3/m_2 m_4))$, the solution to Eq. (13.47) provides a value for θ_{max} that represents the half angle of a cone in which all of the light reaction products must appear in the laboratory.

The general situation is shown schematically in Fig. 13.8. This rather odd situation actually has some practical applications. One case of particular interest is the reaction $^7_3\text{Li}(p, n)^7_4\text{Be}$ that has $Q_{p,n} = -1.6443$ MeV. It is a very useful reaction for producing monoenergetic neutrons for calibration of neutron spectrometers. At energies just above the reaction threshold, neutrons with two different energies are found at forward angles, making it very handy for measuring the efficiency and energy resolution of a detector at low energies. A second interesting application are cases where a beam of neutrons is desired for interrogating the content of a specimen by neutron-induced reactions. Whereas the neutrons from the $^2\text{H}(^2\text{H}_{th}, n)^3\text{He}$ reaction are emitted at all angles in the laboratory, low-energy neutrons from the $^7_3\text{Li}(p, n)^7_4\text{Be}$ reaction are concentrated in the forward direction. Thus, per unit neutron produced, a much more efficient use of the neutrons can result.

The relations given above provide an essentially complete means for determining the kinematics of most nuclear reactions of practical interest. With this completed, we now turn our attention to the general problem of reaction cross sections.

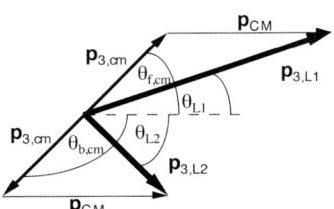

Fig. 13.8 For reactions with Q < 0 and in the vicinity of the threshold energy, the momentum of the center of mass is sufficiently large compared to the momentum of the light product in the center of mass system, that all particles will be found only at forward angles in the laboratory. In the case illustrated, light products emitted at angles $\theta_{f,cm}$ and $\theta_{b,cm}$ in the center of mass appear at angles θ_{L1} and θ_{L2} in the laboratory, respectively.

13.3
Cross Sections for Nuclear Reactions from First-Order Perturbation Theory

We will concentrate attention on the general properties of low-energy nuclear reactions taking the most common neutron reactions as our primary examples. Apart from the Coulomb barriers present in charged-particle reactions, there is basically no difference between the two reaction types except in rather specialized cases.

In Chapter III we introduced the general definition of the reaction cross section as the reaction probability per target nucleus per unit incident projectile flux. While we will not delve into real details, we do need to consider the properties of cross sections in the framework of quantum mechanics, and that means we must relate a cross section to a transition probability. The simplest way to accomplish this is via the Fermi Golden Rule. You will recall that in the limit of first-order perturbation theory, the transition probability from a specific initial state to a specific final state can be written as

$$w = \frac{2\pi}{\hbar}|H'|^2 \rho(E_f) \qquad (13.49)$$

where $|H'|$ represents the matrix element between the initial and final states for the perturbation potential H' that produces the transition and $\rho(E_f)$ is the density of final states available to the products emitted into the continuum. To apply this in the present case, we assume that the reactions are considered in the center of mass coordinate system, and we envision a general reaction of the form

$$a + A \rightarrow b + B \qquad (13.50)$$

where a and A are the projectile and target, respectively, and b and B are the light and heavy reaction products, respectively. We will use the subscript a,b on H' to represent the specific interaction we are considering. Now it is the nuclear potential, or at least part of it, that is responsible for the reaction and thus H' is far from a very small quantity. Nevertheless it is sufficiently small in most cases that the first-order approximation is reasonable.

With Eq. (13.49), the reaction cross section can now be written as

$$\sigma_{a,b} = \frac{w}{\phi_{a,\text{unit}}} \qquad (13.51)$$

where $\phi_{a,\text{unit}}$ represents a unit incident monoenergetic projectile flux. So long as it is uniform, a monoenergetic flux of projectiles ϕ_a can always be written as the product of the projectile density, n_a, and the projectile velocity, v_a, i.e.,

$$\phi_a = n_a v_a \text{ cm}^{-2}\text{s}^{-1} \qquad (13.52)$$

in common units. A unit density is then $n_a = 1 \text{ cm}^{-3}$, and a unit flux is more generally given by $\phi_a = v_a/V$ where V represents the unit volume. Using this with (13.51) gives the reaction cross section in the limit of first-order perturbation theory as

$$\sigma_{a,b} = \frac{W}{v_a} V \tag{13.53}$$

The initial state of the system that we consider is a single nucleus and a unit incident projectile flux. The target nucleus will have a total angular momentum quantum number J_A and the projectile will have a total angular momentum quantum number j_a. In the most common case, neither the target nor the projectile is polarized and we must consider that all possible orientations of these angular momentum vectors will be present. There will then be an equal probability that the target and projectile will interact through each of these combinations, subject to any requirements imposed by conservation of angular momentum. All we will observe is the average cross section over all orientations of the angular momentum vectors. We will assume that this averaging is implied in the evaluation of the matrix element in Eq. (13.49).

The final state of the system will be the heavy reaction product B and light product b, each with a specific angular momentum quantum number J_B and j_b, respectively. The density of final states represents the density of continuum states available to b. We assume that the wave function of b is normalized in a box of volume V and thus the number of states available with momenta p_b to $p_b + dp_b$ is just (Eq. (10.24))

$$dn_b' = \frac{4\pi V}{h^3} p_b^2 dp_b \tag{13.54}$$

As in the case of β decay, we do not consider the degeneracy with respect to spin. This will be taken care of in the evaluation of the matrix elements.

The density of states given in Eq. (13.54) implicitly assumes that there is only one combination of the orientations of the angular momentum vectors of the reaction products. But for angular momentum quantum numbers J_B and j_b, the total number of such combinations is the product of the number of magnetic substates for each vector, $(2J_B + 1)(2j_b + 1)$. Thus, the total density of final states with momenta p_b to $p_b + dp_b$ is

$$\frac{dn_b}{dp_b} = (2J_B + 1)(2j_b + 1)\frac{4\pi V}{h^3} p_b^2 \tag{13.55}$$

With

$$\frac{dn}{dp} = \frac{dn}{dE}\left(\frac{dp}{dE}\right)^{-1} \text{ and } \frac{dE}{dp} = \frac{p}{m} = v \tag{13.56}$$

the density of final states can be written

$$\rho(E_f) = \frac{dn}{dE_f} = \frac{4\pi}{h^3}(2J_B + 1)(2j_b + 1)\frac{p_b^2}{v_b}V \qquad (13.57)$$

Direct substitution of this result and the expression for w given in (13.49) into Eq. (13.53) gives the cross section as

$$\sigma_{a,b} = \frac{V^2}{\pi\hbar^4}|H_{a,b}|^2(2J_B + 1)(2j_b + 1)\frac{p_b^2}{v_a v_b} \qquad (13.58)$$

The energy dependence of the reaction cross section is contained in the matrix element $|H_{a,b}|$ and the ratio $p_b^2/v_a v_b$. Considering just the latter, cross sections will tend to increase with *decreasing* projectile energy and with *increasing* velocity (energy) of the light product, i.e., increasing Q values.

The expression for the cross section in (13.58) is quite general. If applied to reactions involving charged particles, the matrix element must contain transmission coefficients for each of the charged particles involved. If we considered a (p,α) reaction, for example, we would have to consider the Coulomb barrier between the incident proton and the target nucleus as well as the transmission coefficient involved with the emission of the α particle. Regardless of whether the particles are charged or neutral, reactions involving angular momentum different from zero will also require consideration of the centrifugal barriers that will be present. In most cases, the matrix elements are quite complicated functions and we will not deal with them explicitly. But we can use Eq. (13.58) to understand some qualitative characteristics of a number of simple reactions.

As a first example, consider the problem of the elastic scattering of very low-energy neutrons on nuclei. As we will show later, this takes place with zero orbital angular momentum and therefore no centrifugal barrier is involved. Now the perturbation potential is the strong interaction between the neutron and the target nucleus and we know that this must involve potentials of at least some 10 MeV or so. If this is the case, and we consider neutron kinetic energies in the center of mass of a few keV or less, it is reasonable to assume that the interaction should not be affected appreciably by small variations in the neutron energy. That is, $|H_{n,n}|$ should be roughly constant. If we make this assumption, the only variables in the cross section are those contained in the factors $p_b^2/v_a v_b$. In the center of mass system, there is no change in the kinetic energy of the neutron before and after the scattering. Therefore $v_a = v_b = v_n$ and $p_b^2/v_a v_b = (m_n v_n)^2/v_n^2 = m_n^2$. This indicates that the elastic scattering cross section of low-energy neutrons ought to be constant and independent of energy.

A simple test of this prediction can be obtained by examining the cross section for elastic scattering of neutrons on protons. The experimental data are shown in Fig. 13.9. The data have been taken from evaluated data files and therefore experimental errors are not shown. Also, and this is very important to remember, because of the practical importance of neutron cross sections in nuclear engineering, it has

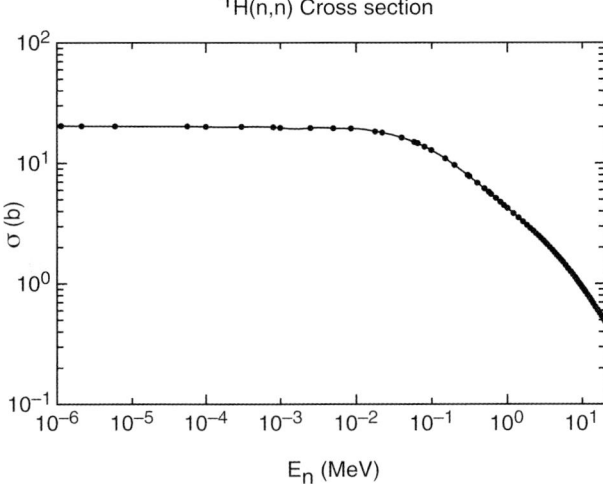

Fig. 13.9 Evaluated cross sections for elastic scattering of neutrons on protons. Note that the neutrons energies are quoted in the *laboratory* coordinate system. Cross sections are given in barns (1b = 10^{-24} cm^2).

become the *standard* in most data files to quote the cross section versus neutron kinetic energy in the *laboratory* coordinate system, not in the center of mass system. For the interaction of neutrons with protons, the kinetic energy in the center of mass system is almost exactly one-half that in the laboratory system. The evaluated data shown in the figure indicate that the elastic scattering cross section is indeed quite constant at about 21 barns up to a laboratory kinetic energy of about 10^4 eV.

It is important to realize that this is not the *total* reaction cross section, but just that for elastic scattering. Indeed, ^{1}H captures neutrons to produce ^{2}H and the cross section for this reaction is shown in Fig. 13.10. The cross section for this reaction is know as the *neutron absorption* cross section and the reaction is simply called the (n,γ) reaction. The strong energy dependence of the (n,γ) reaction is discussed in detail below.

The constancy of the scattering cross section at low energies is seen throughout the chart of the nuclides. A second example of technological importance is that of the reaction $^2_1$H(n, n)$^2_1$H. The total reaction cross section for neutrons on $^2_1$H, which is due almost entirely to elastic scattering, is shown in Fig. 13.11. Once again, the scattering cross section is essentially independent of energy to over 10 keV.

A final example of the low-energy elastic scattering cross section is that for the interaction of neutrons on the doubly magic $^{208}_{82}$Pb shown in Fig. 13.12. You will note that the cross section is again essentially constant to over 10 keV. But at energies above about 50 keV, there are two very sharp and narrow peaks in the cross section. These are termed *resonances* and are commonly observed in neutron cross sections for almost all nuclides in nature. We will discuss these in some detail later in this chapter.

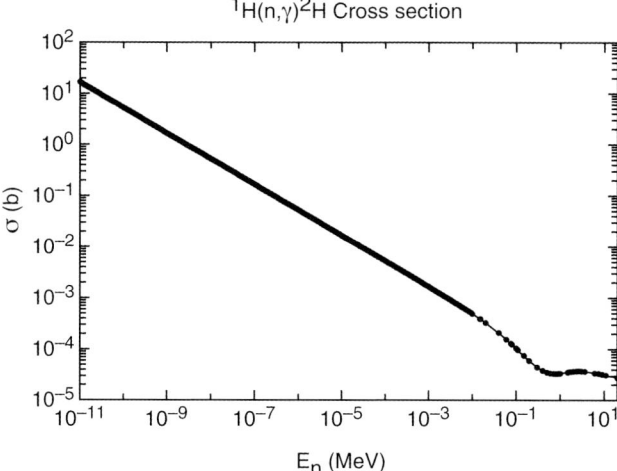

Fig. 13.10 Evaluated (n, γ) cross section for the interaction of neutrons on protons. Note that the neutrons energies are quoted in the *laboratory* coordinate system.

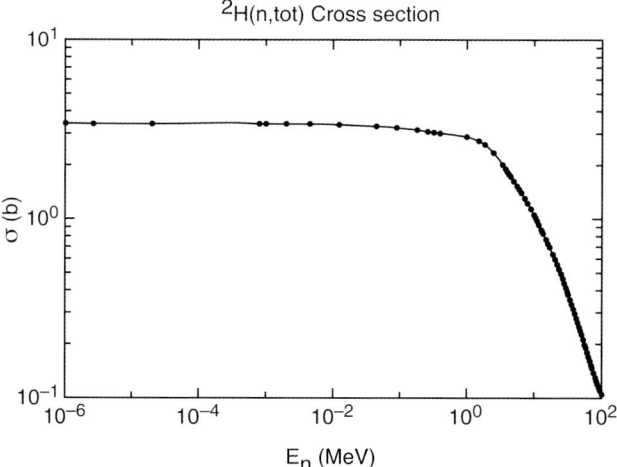

Fig. 13.11 Evaluated total cross section for elastic scattering of neutron on $^{2}_{1}$H. Note that the neutrons energies are quoted in the *laboratory* coordinate system.

The schematic form of the reaction cross section given in (13.58) can also be used to understand the qualitative characteristics of reactions with large positive Q values. The most common of these at low energies is the neutron absorption or (n,γ) reaction. We know from the semi-empirical mass formula, that the average neutron binding energy near the valley of β-stability is 7–8.5 MeV for most of the nuclides found in nature. If a neutron is captured, the excitation energy in the

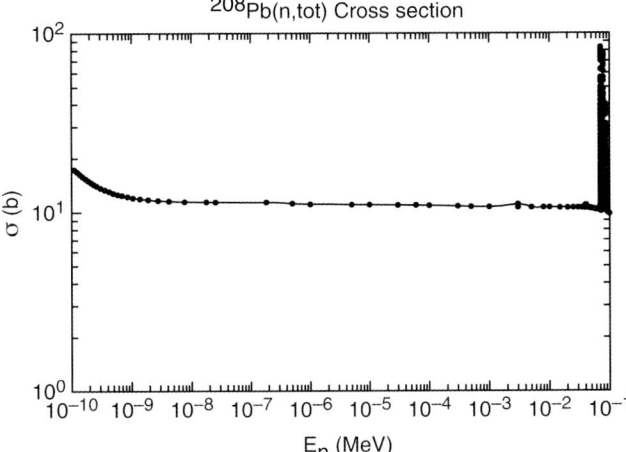

Fig. 13.12 Evaluated total cross section for elastic scattering of neutrons on $^{208}_{82}\text{Pb}$. Note that the neutrons energies are quoted in the *laboratory* coordinate system.

capture product will be roughly in this range and decay by photon emission will then take place with $Q_{n,\gamma} \approx 7 - 8\,\text{MeV}$. If we consider the cross sections of reactions where the center of mass energy of the neutron is no more than a few tens of keV, the total energy available is hardly affected. It then still seems reasonable to expect that the matrix element $|H_{a,b}| = |H_{n,\gamma}|$ will be rather insensitive to changes in total energy of this magnitude. If, in fact, $|H_{n,\gamma}| \approx$ constant and $p_b^2/v_b \approx$ constant, the only variable is the velocity of the projectile and we then predict that

$$\sigma_{a,b}\big|_{Q \gg 0} \propto \frac{1}{v_a} \propto \frac{1}{\sqrt{E_a}} \tag{13.59}$$

At low energies, the cross sections for reactions with large positive Q should be proportional to the inverse of the projectile velocity.

In Fig. 13.13 are the evaluated data on the (n,γ) cross section for low-energy neutrons on ^{1}H. The solid line shown in the figure represents a least-squares fit with an equation of the form

$$\sigma_{n,\gamma}(E) = \frac{\sigma_o}{\sqrt{E}} \tag{13.60}$$

that accurately describes the data to about 5–6 keV. A similar fit to the low-energy data for the reaction $^{197}_{79}\text{Au}(n,\gamma)^{198}_{79}\text{Au}$ is shown in Fig. 13.14. Again, the fit is excellent up to an energy of about 0.1 eV. At higher energies the cross section is dominated by resonances and the one centered at about 5 eV is truly spectacular. The cross section at the peak of the resonance is more than 10^4 b! To judge what

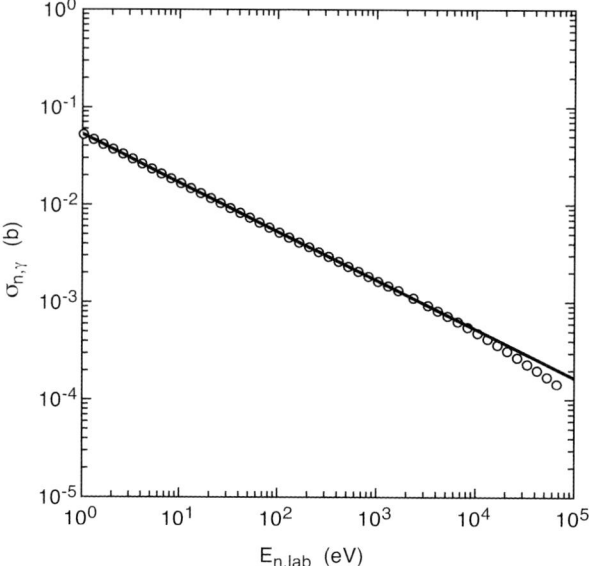

Fig. 13.13 The evaluated low-energy cross section for the (n,γ) reaction on ^{1}H. The open circles represent the evaluated data and the fit to the data represents a function of the form $\sigma_{n,\gamma}(E) = \sigma_0/\sqrt{E}$.

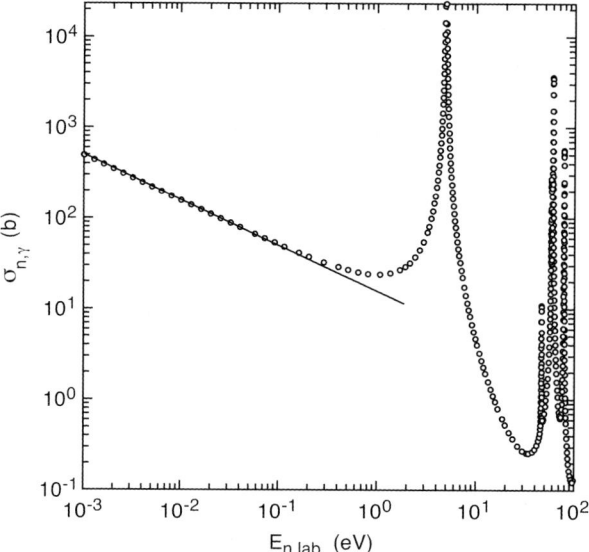

Fig. 13.14 The evaluated low-energy cross section for the (n,γ) reaction on ^{197}Au. The open circles represent the evaluated data and the fit to the data represents a function of the form $\sigma_{n,\gamma}(E) = \sigma_0/\sqrt{E}$.

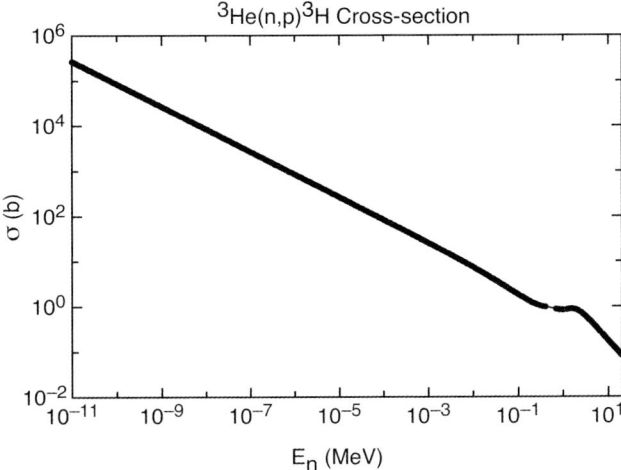

Fig. 13.15 The evaluated low-energy cross section for the (n,p) reaction on ^{3}He. The neutron energies are given in the laboratory coordinate system.

this implies, consider the cross sectional area of a ^{197}Au nucleus in the hard-sphere approximation. With $r_o = 1.25$ fm, the radius is calculated to be 7.27 fm leading to a cross sectional area of about 166 fm^2, or about 1.7 b! At the resonance peak, the neutron–nucleus combination appears to have a cross section about 6000 times the area of the nucleus. While this might seem very surprising in light of the fact that the range of the nuclear potential is so small, the wavelength of a neutron with a kinetic energy of 5 eV is about 1.28×10^4 fm!

The $1/\sqrt{E_n}$ variation of cross sections at low energies is not restricted to (n,γ) reactions but is found whenever the Q value is sufficiently large. A particularly interesting case is the reaction $^3_2$He(n, p)$^3_1$H that has a Q value of about 0.764 MeV. This is sufficiently large compared to the Coulomb barrier between the proton and the tritium nucleus that the latter does not affect the low-energy behavior of the cross section appreciably. The evaluated data up to a laboratory neutron energy of 3.0 MeV are shown in Fig. 13.15. In this case the $1/\sqrt{E_n}$ behavior persists to a laboratory neutron kinetic energy of over 1 keV. Note that at an energy of about 2 MeV one can see a broad "bump" in the cross section. This is also a resonance, but it is very much broader in energy than those seen in the data for Au and Pb given above. The widths of resonances vary greatly with mass number.

Because of the energy dependence of reactions with large Q values, it has become standard practice to quote cross sections for reactions such as (n,γ), (n,f), etc., for low-energy neutrons by reference to the interaction of neutrons with a laboratory velocity of 2200 m s^{-1}. This rather funny velocity corresponds to neutrons with an energy of $E_n = 0.0253$ eV. Back in the days when neutron spectra were being studied intensely, it was relatively easy for most laboratories in the business to discriminate low-energy neutrons on the basis of their velocities. A quasi-Maxwellian neutron beam coming from a reactor or other source of "thermal" neutrons was passed through a velocity selector and the resulting nearly monoenergetic particles were used to study the interaction of the neutrons with different materials. Since the neutrons were roughly in thermal equilibrium with a material that was near "room temperature", it was natural to select as a standard a neutron velocity corresponding to a relatively intense component of the beam. Classical particles in thermal equilibrium at some temperature T have an energy distribution that is described by the Maxwell–Boltzmann distribution function

$$n(E) = \frac{2\pi}{(\pi kT)^{3/2}} E^{1/2} e^{-E/(kT)} \quad (13.61)$$

The average kinetic energy of particles in this distribution is

$$\langle E \rangle = \frac{3}{2} kT \quad (13.62)$$

At a temperature of 25°C, the average energy is 0.03854 eV and kT = 0.02559 eV. As a result, neutrons with velocities of 2200 m s^{-1} have very nearly the magnitude of kT corresponding to a gas at thermal equilibrium with a temperature of 25° C. Cross sections corresponding to the interaction of 2200 m s^{-1} neutrons with matter are referred to as *thermal* neutron cross sections.

With this as background, we now note that the elastic scattering cross section for thermal neutrons on normal hydrogen is 20.491 ± 0.014 b and is roughly independent of energy up to about 10 keV. The thermal neutron cross section for $^{1}_{1}H$, called the absorption cross section, is found to be 0.3326 ± 0.0007 b, or only about 1.6% that of the elastic scattering cross section. It is because of the small magnitude of the absorption cross section that normal water can be used to thermalize neutrons in a nuclear fission reactor without serious loss by capture in the water itself. On the other hand, the thermal neutron absorption cross section for $^{2}_{1}H$ is only 0.519 ± 0.007 mb, or about 1.6×10^{-3} that of $^{1}_{1}H$. Hence, even fewer neutrons will be lost to capture in heavy water. As we noted earlier, it is essentially impossible to make a fission chain reaction with natural uranium when normal water is the medium for reducing the energy of fission neutrons, and reactors that use normal water must use uranium fuel that has a higher isotopic abundance of ^{235}U than exists in nature. But you can get a chain reaction to occur if heavy water is used with natural uranium, and the reactors developed in Canada are based on this design.

The last simple application of the schematic cross section given in Eq. (13.58) that we will consider is the case of inelastic neutron scattering, (n,n′), which obviously has Q < 0. In the center of mass system, no reaction occurs until the neutron has an energy $E_{n,cm} = |Q|$ that corresponds to the threshold energy in the laboratory. At threshold, the kinetic energy of the scattered neutron will be zero. Clearly, a small increase in the energy of the incident neutron above threshold will result in a rather large *fractional* change in the energy of the scattered neutron. If we assume that the matrix element does not change appreciably over small changes

13.3 Cross Sections for Nuclear Reactions from First-Order Perturbation Theory

in the incident energy, the energy dependence of the cross section will once again be due to the energies of the incident and scattered neutron.

If $E'_{n,\,cm}$ represents the energy of the scattered neutron in the center of mass system,

$$E'_{n,\,cm} = E_{n,\,cm} - |Q| \tag{13.63}$$

or

$$(v'_{n,\,cm})^2 = (v_{n,\,cm})^2 - \frac{2|Q|}{m_n} \tag{13.64}$$

Remembering that the subscripts a and b in (13.58) refer to the incident and scattered neutron, respectively,

$$\frac{p_b^2}{v_a v_b} = \frac{(p'_{n,\,cm})^2}{v_{n,\,cm} v'_{n,\,cm}} = m_n^2 \frac{v'_{n,\,cm}}{v_{n,\,cm}} = m_n^2 \left(1 - \frac{2|Q|}{m_n v_{n,\,cm}^2}\right)^{1/2} \tag{13.65}$$

and we then have

$$\frac{p_b^2}{v_a v_b} = \frac{m_n^2}{\sqrt{E_{n,\,cm}}}(E_{n,\,cm} - |Q|)^{1/2} \tag{13.66}$$

Just above the threshold, the factor $1/\sqrt{E_{n,\,cm}}$ will vary little with small changes in $E_{n,\,cm}$ if $|Q|$ is reasonable large. However, the relative change in the factor $(E_{n,\,cm} - |Q|)^{1/2}$ will be quite large. Therefore, we expect to find

$$\sigma_{n,\,n'} \sim (E_{n,\,cm} - |Q|)^{1/2} \tag{13.67}$$

As an example of the energy dependence of an inelastic scattering cross section, we show in Fig. 13.16 the evaluated data for ^{60}Ni. To interpret the data we need to consider the low-lying level structure of ^{60}Ni as shown in Fig. 13.17. The first-excited state of this nuclide is found at 1.342 MeV. The second through fifth excited states are clustered together in the energy range of about 2.17–2.64 MeV. The inelastic cross section will be zero until the center of mass neutron kinetic energy is equal to 1.342 MeV and no other states can be excited until a center of mass energy of 2.168 MeV is reached. At this energy, both the first and second-excited states can be involved in the reaction. Examination of the sparse data points in Fig. 13.16 shows a smooth increase in the cross section up to something near 2.2 MeV, just about where we can expect inelastic scattering to the second-excited state to be possible. At this energy there is a relatively abrupt increase in the cross section as would be expected from inelastic scattering involving both the first and second-excited states. The solid curve in Fig. 13.16 represents a least-squares fit to the first four data points shown. In the fit, both the Q value and amplitude of the function were free parameters. Although the uncertainties must be large owing to the very few data points, the functional form of Eq. (13.67) reproduces the data quite well.

1.5
1.2
1.0
0.8
0.5
0.2
0.0
1.0
1.5
2.0
2.5
3.0
σn,n′ (b)
En,cm (MeV)

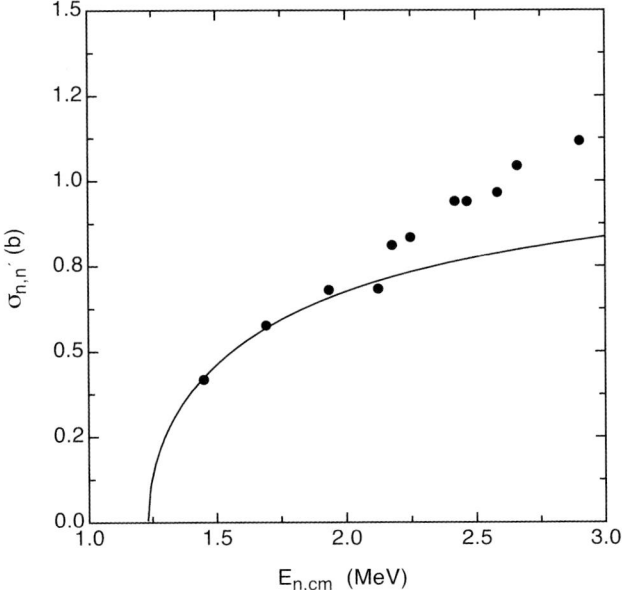

Fig. 13.16 Evaluated inelastic scattering cross section for neutrons on ^{60}Ni (closed circles). The solid curve represents the fit to the first four data points with an equation of the form of (13.67).

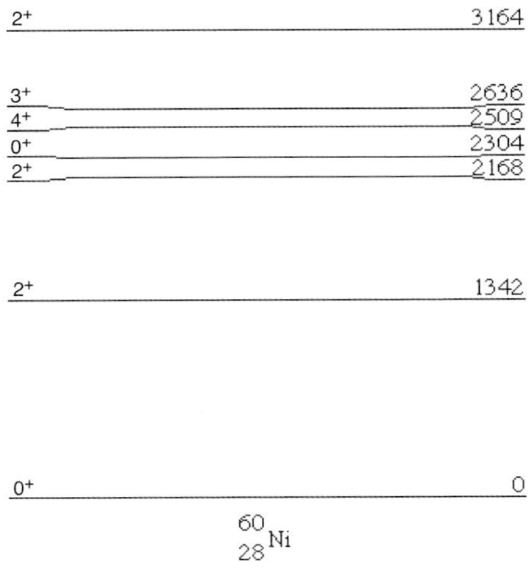

2+ 3164
3+ 2636
4+ 2509
0+ 2304
2+ 2168
2+ 1342
0+ 0
60 28 Ni

Fig. 13.17 The low-lying levels of ^{60}Ni as measured by inelastic neutron scattering.

Other reactions can be used as examples that satisfy the conditions we have considered, but the above is sufficient to give a qualitative understanding of some of the simple features of various cross sections. It should come as no surprise that calculations of matrix elements are, for the most part, very difficult to accomplish with realism and the details of these calculations go well beyond the intent of this text. We will be content with developing an understanding of the general properties of cross sections that can be obtained with relatively simple models. In this spirit, we now turn our attention to the application of the idea of microscopic reversibility to reaction cross sections.

13.4
The Reciprocity Theorem

The cross section for a reaction of the form A(a,b)B in the limit of first-order perturbation theory is given by (13.58) as

$$\sigma_{a,b} = \frac{V^2}{\pi \hbar^4}|H_{a,b}|^2(2J_B + 1)(2j_b + 1)\frac{p_b^2}{v_a v_b}$$

Suppose that we now consider the exact reverse of this equation, namely B(b,a)A. In an exactly similar fashion we can write the cross section as

$$\sigma_{b,a} = \frac{V^2}{\pi \hbar^4}|H_{b,a}|^2(2J_A + 1)(2j_a + 1)\frac{p_a^2}{v_b v_a} \quad (13.68)$$

If we take the ratio of the two cross sections, we have

$$\frac{\sigma_{a,b}}{\sigma_{b,a}} = \frac{|H_{a,b}|^2(2J_B + 1)(2j_b + 1)p_b^2}{|H_{b,a}|^2(2J_A + 1)(2j_a + 1)p_a^2} \quad (13.69)$$

Apart from the squares of the matrix elements, the two cross sections differ only by the density of states that are available to the light particles emitted in the two reactions. This makes sense and it is not too difficult to see why. Suppose we considered running the reactions such that the same total kinetic energy was available in the center of mass. If the Q value of the forward reaction was positive, that for the reverse reaction would be negative by an identical amount, and vice versa. Assuming that the available kinetic energy was greater than the absolute value of Q, both reactions could take place but the kinetic energy available to the reaction products would obviously be quite different. And, of course, the angular momenta of the states of the products formed will in general be different. The crux of the matter in understanding the cross section ratio is really the nature of the matrix elements themselves. No matter how complex they may be to evaluate, they are nothing more than integrals over the potential operating on the initial state to produce the final state.

Suppose we start with projectiles and target separated by a very large distance and map out the potential experienced by a projectile as it approaches and interacts with the target, and continue to map the potential as the light reaction product leaves the vicinity of the interaction. If we do this for the forward reaction and then repeat it for the reverse reaction at exactly the same total energy in the center of mass system, it should be clear that we will map out *exactly the same potential function*. This is no different from comparing the potential experienced by an α particle emitted in α decay and an α particle that approaches the α-decay daughter with exactly the same energy in the center of mass system. The very same potential function will be mapped in both cases.

In fact, we can view the forward and reverse reactions as the same reactions running in different senses of time. That means that the matrix elements must be *invariant* with respect to *time reversal*. This is the *principle* of *microscopic reversibility*. It applies not only to nuclear reactions but to chemical reactions as well. So long as we consider identical energies in the center of mass coordinate system, $|H_{a,b}| = |H_{b,a}|$. We then have that the ratio of the cross sections of the forward and reverse reactions must be given by

$$\frac{\sigma_{a,b}}{\sigma_{b,a}} = \frac{(2J_B + 1)(2j_b + 1)p_b^2}{(2J_A + 1)(2j_a + 1)p_a^2} \tag{13.70}$$

at the same total energy in the center of mass system. The only factor that distinguishes the cross sections is the density of states available for the light products. This is the *reciprocity theorem*.

The reciprocity theorem is quite general. It must apply not only to the absolute value of cross sections but to any other cross section properties as well. In particular, because the cross sections will vary with the angle of emission of the light product, it must apply to the *angular distributions* of the cross sections as well.

A number of experiments have been designed to investigate the applicability of the reciprocity theorem to nuclear reactions. Perhaps the most famous of these were the experiments on reaction $^{12}_{6}C(\alpha, d)^{14}_{7}N$ and its inverse, $^{14}_{7}N(d, \alpha)^{12}_{6}C$ at laboratory bombarding energies of E_α = 41.7 MeV and E_d = 20.0 MeV, respectively [1]. These energies ensured that the same total energy was available in the center of mass system in both reactions in the limit of no experimental energy errors. The test of the reciprocity theorem rested upon the agreement of the angular distributions of the forward and reverse reactions at the same energy in the center of mass coordinates. The experimental data are shown in Fig. 13.18 and they are truly remarkable. Within experimental errors, the cross sections are identical over the full range of angles measured.

A more practical example is represented by the reaction $^{3}_{2}He(n, p)^{3}_{1}H$ with a Q value of $Q_{n,p}$ = 0.764 MeV. This reaction is one of the most common means of detecting low-energy neutrons. High-pressure detectors containing $^{3}_{2}He$ are relatively easy to make and, if designed properly, can provide relatively high energy resolution with good efficiency. The thermal cross section is about 5330 barns. The experimental data on the (n,p) cross section by direct measurement, are collected

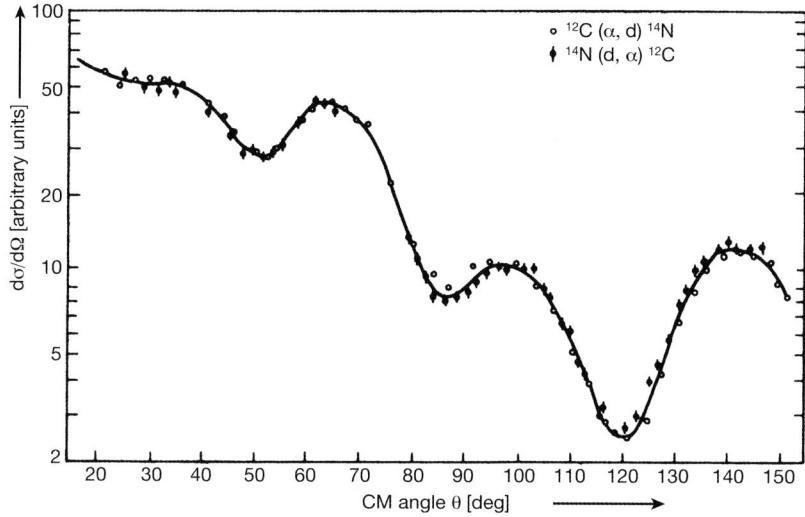

Fig. 13.18 The cross sections for the reactions $^{12}_{6}C(\alpha, d)^{14}_{7}N$ (open circles) and $^{14}_{7}N(d, \alpha)^{12}_{6}C$ (closed circles with error bars) at laboratory bombarding energies of $E_\alpha = 41.7$ MeV and $E_d = 20.0$ MeV, respectively, as a function of the center of mass angle of the light final product [1].

in Fig. 13.19. The data show an approximate $1/\sqrt{E_n}$ variation from thermal energies to well over 10 keV. A prominent "bump" is seen at energies near 1–10 MeV and this represents a resonance. Although high-quality monoenergetic neutron beams at low energies are now readily made, it was not always so. Thus, among others, a strong reason for studying the reverse reaction, $^{3}_{1}H(p, n)^{3}_{2}He$, was to obtain the cross section of the $^{3}_{2}He(n, p)^{3}_{1}H$ with use of the reciprocity theorem. Many measurements of the $^{3}_{1}H(p, n)^{3}_{2}He$ reaction cross section have been performed and a select set of experimental data is shown in Fig. 13.20. The reaction displays a sharp threshold characteristic of a reaction with Q < 0.

We now use the reciprocity theorem to generate cross sections for the $^{3}_{2}He(n, p)^{3}_{1}H$ reaction from the data given in Fig. 13.20. We first recognize that the ground-state spins and parities of the four particles involved in each reaction are identical. Second, none of the particles has a bound state in addition to the ground state. This guarantees that the reaction is not complicated by the possibility that different final states need be considered. Now it is certainly possible that nonzero orbital angular momentum can be carried by both the projectile and the light product. But so long as we consider the same energies in the center of mass for the forward and reverse reactions, these will automatically be accounted for. Hence, in the present case, the reciprocity relation implies that

$$\frac{\sigma_{a,b}}{\sigma_{b,a}} = \frac{p_b^2}{p_a^2} \qquad (13.71)$$

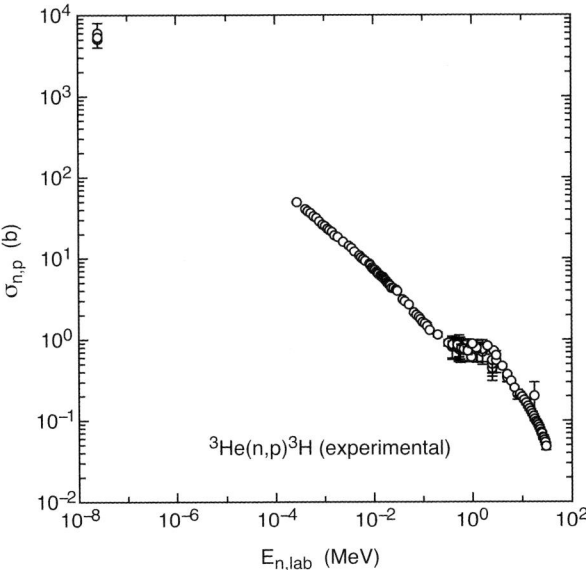

Fig. 13.19 Experimental data on the cross section of the $^{3}_{2}\text{He}(n, p)^{3}_{1}\text{H}$ in the laboratory coordinate system.

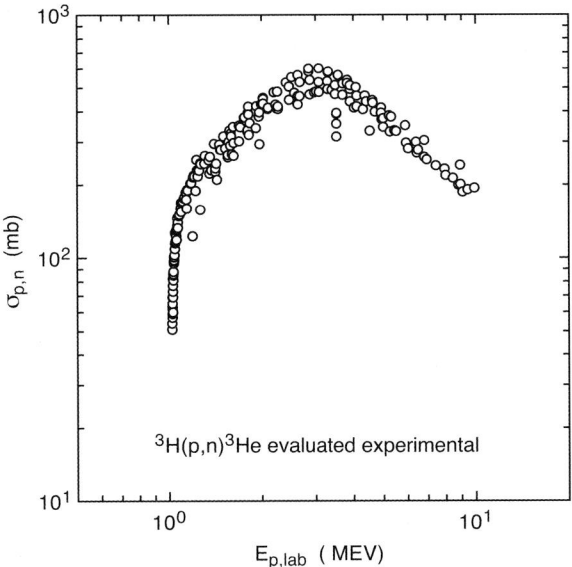

Fig. 13.20 Experimental data on the cross section of the $^{3}_{1}\text{H}(p, n)^{3}_{2}\text{He}$. The data are from a number of independent measurements.

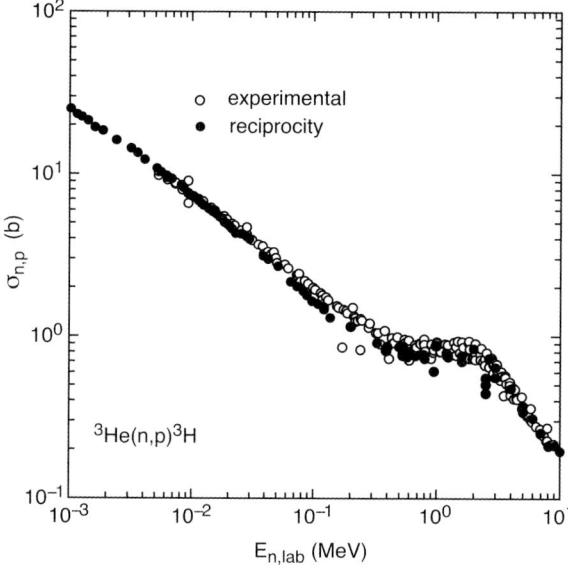

Fig. 13.21 Experimental cross section measurements on the reaction $^{3}_{2}\text{He}(n, p)^{3}_{1}\text{H}$ (open circles) and cross sections predicted with the data of Fig. 13.20 and the reciprocity theorem.

The cross sections at the same total energy in the center of mass system are just proportional to the squares of the momenta of the emitted light products in each case. Further, because of the near equivalence of the masses of the neutron and proton, the cross sections will be in the ratio of the center of mass kinetic energies of the light products. A straightforward application of the relations between the laboratory and center of mass energies given earlier in this chapter permits the calculation of the cross section ratio.

The cross sections calculated with the reciprocity theorem are shown in Fig. 13.21 along with the direct cross section measurements shown in Fig. 13.19. The agreement between the predictions and the experimental measurements is very good, especially considering the uncertainties in the data. There is very little doubt concerning the validity of the reciprocity theorem.

Simple considerations of the energy dependence of the reaction cross section and the principle of microscopic reversibility have given us a sound introduction to some of the characteristics of cross sections at low energies. But to understand the striking resonance structure that is found in many data sets, a more detailed knowledge of the matrix elements involved in the reactions is required. To proceed further it is necessary to consider quantum effects in greater depth and gain some understanding of the mechanism or mechanisms that are implied in both scattering and other types of reactions. We will first proceed by a qualitative examination of several general reaction mechanisms that make sense and are actually found to account quite well for many reactions. It should come as no surprise that we will need to consider some properties of nuclear structure, especially with respect to the properties of levels that can decay. The time dependence of such states naturally leads to an understanding of the general characteristics of the resonances seen in low-energy cross section data. This, coupled with a general formulation of quantum mechanical scattering and its extension to other reaction types, will provide us with enough information to understand the basics and to correlate the majority of reactions encountered in practice.

13.5
Qualitative Considerations of the Mechanisms of Low-Energy Nuclear Reactions

13.5.1
Potential Scattering

The simplest reactions we can think of are the interactions of protons and neutrons with one another or with a second proton or neutron. Using the scattering of neutrons on protons for simplicity, it is easy to see that the only reactions that need be considered are those of elastic scattering and neutron capture. The potential for these reactions is the strong interaction itself. Neglecting, for the moment, the spin dependence of the strong interaction, the reactions represent the interaction of two particles in a mutually attractive potential. So long as the energy is large enough, we can neglect neutron capture and consider elastic scattering alone. The situation is fundamentally the same as the scattering of electrons on protons. So long as the electron energy is sufficiently large, capture of the electron to form the hydrogen atom can be neglected to an excellent approximation. The two particles scatter in the potential field of the mutual interaction. There is no "intermediate" entity that is formed. This process is referred to as *potential* scattering.

As we have seen in the cross section data for neutron–proton scattering in Fig. 13.9, the elastic scattering cross section is essentially constant at low energies and then begins to decrease at energies approaching 10 keV in the laboratory. That the cross section ought to be energy dependent can be made plausible by a very simple argument. In a relative coordinate system with the proton at its origin, the neutron will approach with a velocity equal to that of its laboratory velocity. Now we know from the De Broglie relation, that a neutron with momentum p_n will have a wavelength $\lambda_n = h/p_n = h/\sqrt{2m_n E_n}$, and as the energy increases, the wavelength will decrease. Now the De Broglie wavelength is a measure of the effective size of a particle, and this implies the dimension over which the particle can experience the force of interaction with another particle or system of particles. For example, high-energy electrons tend to scatter from individual electrons in matter. If however, the electron energy is reduced such that its wavelength is comparable to the distance between atoms in a crystalline lattice, the electron will undergo diffraction in much the same way as x-rays. The same is true of neutrons. It is then reasonable to expect that the interaction of the neutron and proton will also vary with wavelength. Crudely, the neutron in the relative coordinate system will appear to have a "size" or "cross section" on the order of $\pi\lambda_n^2 = \pi h^2/2m_n E_n$, and for energies of 0.1, 1 and 10 MeV, this will be on the order of 257, 25.7 and 2.57 fm^2, respectively. To give this some perspective, the "cross section" corresponding to the hard-sphere radius of a nucleon, 1.25 fm, is about 5 fm^2. Viewed in this way, the probability that a neutron will interact with a proton when it is directed with a random impact parameter ought to decrease with increasing energy until its "cross section" becomes small compared to that of the target. Why the cross section becomes constant at very low energies will emerge once again when we consider partial wave analysis in Section 13.8.

If we consider the interaction of a neutron with a complex nucleus, we ought to expect that potential scattering can also occur and indeed it does. That the low-energy scattering behavior is very similar to that of neutrons on protons has already been demonstrated in Figs 13.11 and 13.12. However, this is not the only mechanism for scattering because resonances are seen in the elastic scattering cross section for neutrons on ^{208}Pb (Fig. 13.12).

13.5.2
The Compound Nucleus

One of the reaction mechanisms we can conceive of is one in which the projectile simply gets absorbed by the target nucleus A to form an intermediate a+A as shown schematically in Fig. 13.22. The intermediate state would be formed with an excitation energy given by the sum of the projectile kinetic energy in the center of mass system plus the binding energy of the projectile. The total energy of the target plus projectile in the center of mass system is $E_{a,cm} + (m_a + M_A)c^2$. The ground state mass of the intermediate is M_{a+A} and its total excitation energy upon

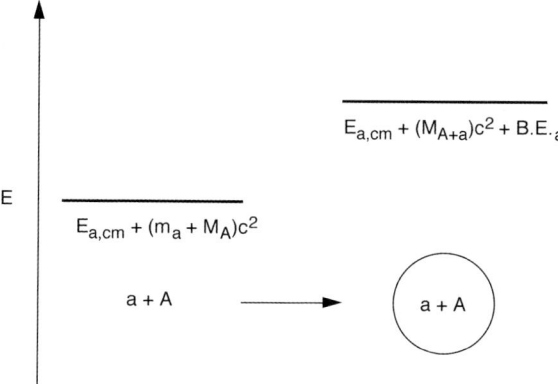

Fig. 13.22 Schematic diagram of the absorption of a projectile a by a target nucleus A in the center of mass coordinate system. The quantities m_a and m_A represent the rest masses of the projectile and target, respectively. After absorption of the projectile, the product nucleus A + a has an excitation energy above its ground state of $E_{a,cm} + (B.E.)_a$.

absorption of the projectile is $E_{a,cm} + (B.E.)_a$. If the projectile were a neutron, the binding energy might be on the order of 7–8 MeV. Clearly, there is more than enough energy for the projectile to be re-emitted, and depending upon $E_{a,cm}$, there might be sufficient energy for other particles or particle combinations to be emitted as well. Regardless of what is emitted, the excitation energy of the intermediate is high and we expect it to decay with a rather short half-life.

The question of the half-life of the intermediate state is quite important, for it determines whether it can really be considered a well-defined state of a nucleus or one that is of such a transitory nature that its formation and decay take place essentially "instantly". The description of the reactions in these limiting cases will be quite different. A natural reference for lifetimes is the time required for a bound nucleon to move about the nucleus. We know that the average binding energy of a nucleon in a nucleus near stability is some 7–8.5 MeV. We also know that the depth of the central potential well in a medium-massed nucleus is of order 45 MeV. If we make the crude approximation that the energy levels in the well are equally spaced, then, on the average, the kinetic energy of a particle is about 18 MeV. For a nucleus of A = 100, with a hard-sphere radius of about 6 fm, the time required by the average nucleon to move a distance equal to the nuclear diameter is about 10^{-22} s. If a collection of nucleons exists for times long compared to 10^{-22} s we have a nuclear system that will have characteristics very nearly those of a time-independent system. It will be a quantized system with a rather well-defined level structure. If, on the other hand, the lifetime is comparable to or shorter than 10^{-22} s, no such structure will exist and the dynamics of the system will not permit an examination of any structure in the intermediate state.

There are many lines of evidence that demonstrate that, for many low-energy nuclear reactions, the lifetime of the intermediate is very much longer than 10^{-22} s. A simple means of establishing this is by consideration of neutron capture. Following capture, the intermediate can and does decay by emission of one or more γ-rays to known lower-lying levels in the intermediate nuclide. From our discussions in Chapter XI, we know that photon emission is a relatively slow process that is due to the electromagnetic interaction, and the fastest transitions are of the E1 type. From Table 11.1 the single-particle estimate for the half-life of a level that decayed solely by a single E1 transition, is

$$t_{1/2} = \frac{6.76 \times 10^{-6}}{E_\gamma^3 A^{2/3}} \text{ s} \tag{13.72}$$

Suppose we consider a γ-ray transition from the excited state of the newly formed intermediate to its ground state. Taking an excitation energy of 10 MeV, the half-life of the intermediate is estimated to be 3×10^{-19} s. While this seems trivially small, it is some 3000 times longer than the time required for the average nucleon to traverse the nuclear dimension. Further, we know that the real lifetimes of E1 transitions are generally much longer than the Weisskopf estimates, at least at lower energies. The conclusion we come to is that the intermediate must have a very long lifetime compared to characteristic nuclear times if photon emission is to

13.5 Qualitative Considerations of the Mechanisms of Low-Energy Nuclear Reactions

take place with reasonable probability. The lifetimes of many intermediates formed in nuclear reactions are known very well. Although they are short and usually cannot be directly measured, they can be inferred accurately by indirect means, one of which we will develop shortly. Such measurements provide lifetimes typically in the range of 10^{-19}–10^{-14} s.

The long-lived nature of the intermediate formed by absorption of a projectile was postulated long ago by Niels Bohr. The intermediate is called the *compound nucleus*. Because it is so long lived, it has the characteristic quantum properties of any nucleus, including the fact that it has states with well-defined spins and parities. There is nothing magic about these states. They are, in fact, nothing more than excited states of the nucleus represented by the combination of the target and projectile in its ground state. In principle, each can be formed by direct excitation of the ground state.

Why the compound nucleus has such a long lifetime is worth a little thought. While nuclei at low energies have relatively large energy spacing between adjacent levels, we know that the level densities increase rapidly with excitation energy. This will become quite evident later in this chapter when we consider more empirical data on the cross sections for neutron reactions. When level densities are large, it takes but a small energy to promote one nucleon from one state to another. Further, it will be possible to promote many particles into excited states and thus the nuclear excitations will generally be quite complex in nature.

Remembering that the level spacing in the vicinity of the ground state of an odd-A nuclide of $A \geq 100$ is on the order 0.1–0.5 MeV, if 10 MeV is placed into the nucleus it is clearly possible for a number of nucleons to be excited from the states they occupy when the nuclide is in its ground state, and the possible combinations of "many-particle" excitations that can exist will be very much larger than the number of single particle states in the nucleus.

If we use the Fermi gas model as an approximation, a simple mechanical picture shows that, once formed, the compound nucleus must have a relatively long lifetime. Consider the absorption of a neutron by a nucleus. Once it is absorbed, the neutron can, through a collision with another particle, excite it. The excitation energy of the nucleus is now shared by two nucleons. Because the nucleons can be excited with energies only on the order of 0.1–0.5 MeV, two nucleons that share an excitation energy of 10 MeV can, through subsequent collisions, excite many more particles. In this picture, the absorption of a particle by the target will, through collisional mechanisms, quickly cause the total excitation energy to be shared by a relatively large number of particles, no one of which has a large fraction of the total excitation energy. If the nucleus were indeed a gas, it is not difficult to see that, after a very short time, the energy will be shared by a large number of particles and approach a thermal equilibrium with each of the excited particles possessing a small fraction of the total excitation energy. Although the nucleus has more than enough energy to re-emit a neutron, a proton, or perhaps some combination of particles, no single nucleon, on the average, has sufficient energy to be re-emitted. In this sense, the compound nucleus can be viewed as a "bound state in the continuum".

The long lifetime of the compound nucleus should now be understood. The sharing of the excitation energy among a number of particles means that the nucleus cannot decay by particle emission until, by chance, the collisional process reconcentrates sufficient energy on a single nucleon for this to happen. Of course, before this happens, other decay modes, even though they are inherently slow, may take place. This is the case with photon emission and the model provides a plausible explanation for the relatively high probability for the (n,γ) reaction after absorption of low-energy neutrons.

A key consequence of this picture of the compound nucleus is that, so long as some sort of equilibrium state is produced before decay occurs, the decay of the compound nucleus will be *completely independent* of the means by which it is formed.

This simple picture is essentially correct. In quantum mechanical terms, the wave function of the compound nucleus will be extremely complex. In terms of the independent particle model, the wave function will contain a very large number of terms, each of which represents one of the specific particle configurations that is energetically allowed and possesses the spin and parity of the compound nucleus. Each such term will have such a small amplitude on the average that, by its square, represents a very small probability of finding the nucleus in that particular configuration. One of these terms would, for example, represent the target nucleus in its ground state and a neutron with an energy representing the entire excitation energy of the compound nucleus. It is this term that can give rise to the re-emission of a neutron to produce the initial state in the center of mass coordinate system. Such a process is indeed an elastic scattering process. It is *compound* elastic scattering and it is this mechanism that gives rise to the resonances seen in the elastic scattering cross section for neutrons on ^{208}Pb shown in Fig. 13.12. If sufficient excitation energy were available, a second term in the wave function might represent the target nucleus in its first excited state with the neutron possessing the remaining excitation energy of the compound nucleus. This term can give rise to the re-emission of a neutron leaving the target in its first-excited state. This is the process of compound *inelastic* scattering. Other reactions that can be postulated would also occur by the presence of small terms in the very complex wave function of the compound nucleus.

The formation and decay of the compound nucleus can be represented schematically as shown in Fig. 13.23. The projectile is absorbed by the target to produce the

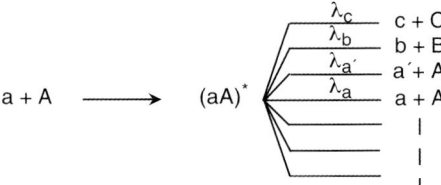

Fig. 13.23 Schematic diagram of the formation and decay of the compound nucleus (aA)*.

compound nucleus (aA)*. Once formed, decay will take place by any and all modes that are energetically allowed, such as elastic scattering (a + A), inelastic scattering (a′ + A), or other particle emissions (b + B, c + C, etc.). The probability for decay by any mode is an independent property of the compound nuclear state involved, just as for any other excited state of a nucleus. The lifetime of the state is determined by its total decay constant

$$\lambda_{tot} = \sum_{\text{all i modes}} \lambda_i \qquad (13.73)$$

where i is an index representing all possible decay modes of the state.

Because of the long lifetime of the compound nucleus, a reaction that takes place by this mechanism can rightly be viewed as occurring in two steps, the first being the formation of the compound nucleus and the second its decay at a later time. Therefore, we can write the cross section for a reaction (a,b) that takes place by this mechanism as

$$\sigma_{a,b} = \sigma_{(aA)^*} \frac{\lambda_b}{\lambda_{tot}} \qquad (13.74)$$

where $\sigma_{(aA)^*}$ is the cross section for forming the compound nucleus and λ_b/λ_{tot} is the fraction of decays that result in the emission of b.

The case of compound elastic scattering at an isolated resonance is particularly interesting and illustrates an important concept. For this interaction to occur, there must obviously be a nonzero value of λ_a, which, for the moment we write as $\lambda_{(aA)^* \to a}$, as well as a nonzero probability $\lambda_{a \to (aA)^*}$ for the projectile to enter into the target in the first place. The relation between these two is simply arrived at by consideration of the principle of microscopic reversibility. They must be identical. You can extend this argument to any of the other partial decay constants of the compound nucleus. If a decay mode has a nonzero decay constant, the reciprocity theorem (Eq. (13.69)) indicates that the reverse reaction must occur with the same probability so long as the same density of states is accounted for in the two reactions. This must be true for elastic scattering because there is no change in the spins and parities of the reactants and products nor is there any change in the center of mass momentum of the projectile as a result of the scattering. We then arrive at an important conclusion. We already know that potential scattering will always occur with some probability. But we can now add that if any reaction between a target and projectile takes place via compound nucleus formation, there must be a contribution from compound elastic scattering as well. This has a direct effect on the energy-dependence of the elastic scattering cross section and is an observable that can be used to determine properties of certain resonances. The equality of $\lambda_{(aA)^* \to a}$ and $\lambda_{a \to (aA)^*}$ has important implications to a physical development of the form of the reaction cross section through an isolated resonance as will be shown in Section 13.7.

13.5.3
Direct Reactions

The mechanisms of potential scattering and compound nucleus formation are by far the main contributors to the total cross section for low-energy nuclear reactions throughout the chart of the nuclides. But there is one other reaction mechanism that must be included because of its great importance for uncovering the detailed properties of the low-energy levels of nuclei, and that is the mechanism of so-called *direct* reactions.

Suppose that a projectile passes so close to the target nucleus that it experiences the strong interaction due to the nuclear force. The projectile can simply scatter in the potential field as in potential scattering. But there is also some probability that it can transfer one or more nucleons to the target or capture one or more nucleons from the target. The transfer reactions are referred to as *stripping* reactions and those that capture particles from the target are referred to simply as *pickup* reactions.

A rather simple example is that of the $(^2_1H, p)$ or (d, p) reaction. The deuteron, as you will recall, has the smallest binding energy of any nucleus found in nature, only about 2.22 MeV. If the deuteron approaches a target closely, it is not difficult to see that there will be a natural tendency for the deuteron to align itself such that the proton is farthest from the target because of Coulomb repulsion. If the approach is sufficiently close that the nuclear potential between the deuteron and target is significant, the neutron can be stripped from the deuteron with the proton simply scattering away. The time required for such reactions will obviously be the time during which the mutual interaction of the target and projectile is large. Because of the very short range of the nuclear force, this time will be on the order of the time for the projectile to travel a distance comparable to the nuclear diameter. For projectiles with kinetic energies of some tens of MeV per nucleon, it is readily found that this time is on the order of $2-5 \times 10^{-22}$ s for a nucleus with A = 100. This is comparable to the characteristic time for nucleon motion in the nucleus. In the direct reaction, there is no intermediate state at all.

The beauty of direct reactions is that a particle stripped from a projectile frequently is captured into a specific low-lying level of the product nucleus. It can be captured to directly produce the ground state, first excited state, etc. When this occurs, all available energy resides in the remaining projectile fragment. So, for example, measurement of the energies of the protons arising by direct (d,p) reactions on a target will reveal the energies of the excited states of the product nuclide. Not only that, any angular momentum transfer that takes place is reflected in the waves representing the proton in the final state of the system. Thus measurements of the angular distributions of the protons can reveal the angular momentum of the state into which the neutron was captured. If one wants to get fancy about it, the use of polarized targets and projectiles also permits the determination of the change in the spin orientation of the captured particle. This is very powerful stuff. Over the years very sophisticated theories have been developed to analyze data obtained in studies of direct-reaction cross sections, such that the magnitudes can be related to the single particle character of states in the product nucleus. Taken together, the information gleaned from the study of direct reactions has been instrumental in understanding the nature of excited states in nuclei and thus in our overall understanding of nuclear structure.

13.6
The Properties of Time-Dependent States

During our discussions of α, β, γ decay and fission, we have alluded to the fact that states that can undergo decay are not truly stationary states of the time-independent Schrödinger equation. Based on our scheme of separation of variables, we demonstrated in Chapter VII that the time-dependent part of the equation has the general solution (Eq. (7.6)) $T = Ce^{-\frac{iEt}{\hbar}}$. Therefore if ψ_o is the solution to the stationary equation, the complete solution for a state with a time-dependent wave function is of the form

$$\Psi = \psi_o e^{-\frac{iEt}{\hbar}} \tag{13.75}$$

Assuming proper normalization, if the state were stationary, we would have

$$\langle \psi_o | \psi_o \rangle = \int_{\text{all space}} \psi_o^* \psi_o d\tau = 1 \tag{13.76}$$

where τ represents the differentials of all space and spin variables of the wave function, not the mean level lifetime. The system would have to be found somewhere at all times. However, if the state can decay, it will have a well-defined decay constant λ, and thus the probability of finding it anywhere at time t after it was created would be

$$\langle \Psi | \Psi \rangle = e^{-\lambda t} \tag{13.77}$$

Examination of Eq. (13.75) immediately indicates that, if the energy E were purely real, the requirement expressed in (13.77) could not be satisfied. We are then led to require that the energy of a state that can undergo decay be *complex*. If we assume, for example, that $E = E_o + i\beta$ direct substitution of (13.75) into (13.77) gives

$$\langle \Psi | \Psi \rangle = \langle \psi_o | \psi_o \rangle e^{\frac{-i}{\hbar}(E_o + i\beta)t} e^{\frac{i}{\hbar}(E_o - i\beta)t}$$
$$= e^{\frac{2\beta}{\hbar}t} = e^{-\lambda t} \tag{13.78}$$

That is, $\lambda = -2\beta/\hbar$ and the quantity $-\hbar/2\beta$ must represent the mean lifetime of the state.

In the literature, the quantity -2β is given the symbol Γ. We then write

$$\lambda = \frac{\Gamma}{\hbar} = \frac{1}{\tau} \quad \text{or} \tag{13.79}$$

$$\Gamma\tau = \hbar \tag{13.80}$$

The energy Γ is directly proportional to the decay constant. Levels with very short half-lives will have large Γ and vice versa. This last expression is a statement of the Heisenberg Uncertainty Principle, $\Delta E \Delta t \geq \hbar$, as it applies to states that can decay. The time-dependent wave function is then written as

$$\Psi = \psi_o e^{\frac{-i}{\hbar}\left(E_o - \frac{i\Gamma}{2}\right)t} \tag{13.81}$$

The uncertainty principle indicates that the energy of a state that can decay is not "sharp", i.e., it is not represented by a δ-function. What can we say about this energy? In order to be quantitative it is useful to remember the Fourier transform that allows for the transformation of a function of time into an equivalent function of frequency. In general, if $f(t)$ is an arbitrary function of time, it is possible to relate it to a function $g(\omega)$ of the frequency ω by the *definition*

$$f(t) = \frac{1}{\sqrt{2\pi}} \int_{-\infty}^{\infty} g(\omega) e^{-i\omega t} d\omega \tag{13.82}$$

and the frequency function can be obtained by the inverse relation

$$g(\omega) = \frac{1}{\sqrt{2\pi}} \int_{-\infty}^{\infty} f(t) e^{i\omega t} dt \tag{13.83}$$

If we apply this to the wave function $\Psi(t)$, that is, take $f(t) = \Psi(t) = \psi_o e^{\frac{-i}{\hbar}\left(E_o - \frac{i\Gamma}{2}\right)t}$, we then have

$$g(\omega) = \frac{1}{\sqrt{2\pi}} \int_{-\infty}^{\infty} \psi_o e^{\frac{-i}{\hbar}\left(E_o - \frac{i\Gamma}{2}\right)t} e^{i\omega t} dt$$

$$= \frac{1}{\sqrt{2\pi}} \int_{-\infty}^{\infty} \psi_o e^{i\left(\omega - \frac{E_o}{\hbar}\right)t} e^{-\frac{\Gamma}{2\hbar}t} dt \tag{13.84}$$

The timescale deserves some attention. A state that can decay cannot exist from $t = -\infty$ to $t = +\infty$. It then makes sense to consider the time range over which it can exist. For this purpose we assume that the state is formed at $t = 0$ and the lower limit of the integral in (13.84) is set to zero. We then have

13.6 The Properties of Time-Dependent States

$$g(\omega) = \frac{1}{\sqrt{2\pi}} \int_0^\infty \psi_o e^{\left\{i\left(\omega - \frac{E_o}{\hbar}\right) - \frac{\Gamma}{2\hbar}\right\}t} dt$$

$$= \frac{\psi_o}{\sqrt{2\pi}} \left. \frac{e^{\left\{i\left(\omega - \frac{E_o}{\hbar}\right) - \frac{\Gamma}{2\hbar}\right\}t}}{i\left(\omega - \frac{E_o}{\hbar}\right) - \frac{\Gamma}{2\hbar}} \right|_0^\infty \qquad (13.85)$$

$$= \frac{\psi_o}{\sqrt{2\pi}} \frac{-1}{i\left(\omega - \frac{E_o}{\hbar}\right) - \frac{\Gamma}{2\hbar}}$$

and

$$g(\omega) = \frac{\psi_o}{\sqrt{2\pi}} \frac{i\hbar}{(\hbar\omega - E_o) + \frac{i\Gamma}{2}} \qquad (13.86)$$

Now we are close to understanding the energy dependence of a time-dependent state. Just as there must be a probability for finding a system in space, there must be a probability for finding a state in *energy* or with a corresponding *frequency*. Recognizing that $E = \hbar\omega$, there is a one-to-one correspondence between energy and frequency. Because $g(\omega)$ represents the wave function in frequency "space" (see Eq. (13.83)), the probability for finding the state with frequency ω to $\omega + d\omega$ must be $P(\omega)d\omega = |g(\omega)|^2 d\omega$, and because of the one-to-one correspondence between energy and frequency, we can obviously write

$$g(\omega)d\omega = g(E)dE \qquad \text{or} \qquad (13.87)$$

$$g(E) = g(\omega)\frac{d\omega}{dE} = \frac{1}{\hbar}g(\omega) \qquad (13.88)$$

We can now express the probability distribution for the level in terms of energy as

$$P(E) = |g(E)|^2 = \frac{|\psi_o|^2}{2\pi} \left[\frac{i}{(E - E_o) + \frac{i\Gamma}{2}}\right]\left[\frac{-i}{(E - E_o) - \frac{i\Gamma}{2}}\right]$$

$$= \frac{|\psi_o|^2}{2\pi} \frac{1}{(E - E_o)^2 + \frac{\Gamma^2}{4}} \qquad (13.89)$$

The probability for finding the state with any energy must be given by

$$\int_{-\infty}^\infty P(E)dE = 1 \qquad (13.90)$$

Substitution of the last relation in (13.89) for P(E) then gives

$$\int_{-\infty}^{\infty} \frac{|\psi_o|^2}{2\pi} \frac{dE}{(E-E_o)^2 + \frac{\Gamma^2}{4}} = \frac{|\psi_o|^2}{2\pi} \int_{-\infty}^{\infty} \frac{dx}{x^2 + b^2} = \frac{|\psi_o|^2}{2\pi} \frac{1}{b} \tan^{-1}\left(\frac{x}{b}\right)\Big|_{-\infty}^{\infty} \quad (13.91)$$

$$= \frac{|\psi_o|^2 \pi}{2\pi \; b} = \frac{|\psi_o|^2}{\Gamma} = 1$$

In the derivation we have used the substitutions $x = (E_o - E)$ and $b = \Gamma/2$ for simplicity. With the normalization constant from (13.91), the probability distribution function for finding a time-dependent state in energy is (Eq. (13.89))

$$P(E) = \frac{\Gamma}{2\pi} \frac{1}{(E-E_o)^2 + \Gamma^2/4} \quad (13.92)$$

We now have a direct relation between the decay constant λ, as represented by Γ, and the probability distribution in energy for a time-dependent state. For any given Γ, the probability distribution is seen to be a peaked about the energy E_o. For large values of $E - E_o$, the probability for finding the state is very small. As E approaches E_o, the probability becomes a maximum. At $E = E_o$, the probability is given by

$$P(E_o) = \frac{\Gamma}{2\pi}\left(\frac{4}{\Gamma^2}\right) = \frac{2}{\pi \Gamma} \quad (13.93)$$

Further, the probability is equal to $P(E_o)/2$ when

$$\frac{1}{\pi \Gamma} = \frac{\Gamma}{2\pi} \frac{1}{(E-E_o)^2 + \Gamma^2/4} \quad (13.94)$$

or

$$(E - E_o) = \pm \frac{\Gamma}{2} \quad (13.95)$$

In Fig. 13.24 we show a schematic plot of $P(E) / P(E_o)$ as a function of E. For the choice of $\Gamma = 0.2$ units in $E - E_o$, the distribution is rather sharply peaked and symmetric about E_o. Γ is seen to be the full width of the distribution at half its maximum height. The quantity Γ is referred to as the *total width* of the level.

The derivation we have performed is very general. It applies to the levels of any quantized system that are "isolated" in the sense that neighboring levels are separated from it by energies that are large compared to their total widths. For low-lying levels in nuclei that decay predominantly by photon emission, β decay, etc., and have relatively long half-lives, level widths are very, very small. A level with a half-life of 1 s has a width Γ of only about 4.6×10^{-16} eV! A level with a half-life of 10^{-14} s has a width of about 0.046 eV. But a level that lives about as long as the characteristic nuclear time of about 10^{-22} s would have a width of about 4.6 MeV.

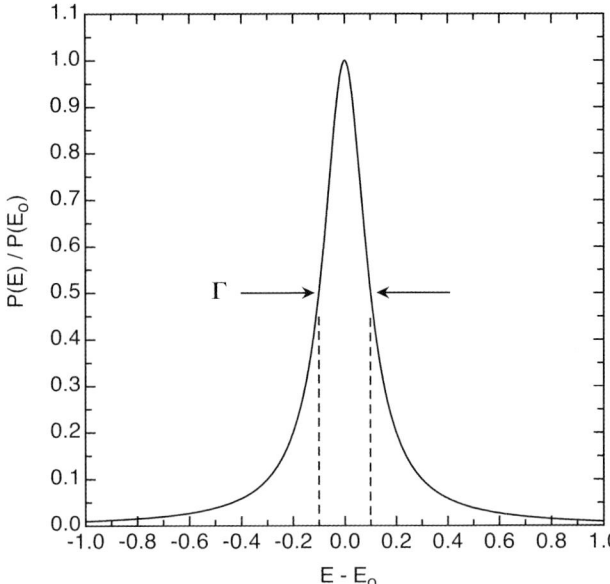

Fig. 13.24 The Lorentzian probability distribution in energy for a time-dependent state. The curve was calculated for $\Gamma = 0.2$ and is normalized to the magnitude of the probability function at $E = E_o$.

The total decay probability of any level is the sum of the decay probabilities for all allowed decay modes. Thus the total level width must be the sum of the *partial* widths for decay by all possible modes, i.e., $\Gamma = \sum_i \Gamma_i$, where i is the index of the decay modes of a level.

The connection between the developments in this section and the resonances seen in reaction cross sections at low energies should now be clear. The resonances represent the formation of a compound nucleus in specific individual states. Although the general derivation of the reaction cross section is somewhat lengthy and complex, a simple physical approach can quickly give us the general form of the cross section and that is the next topic that we will consider.

13.7
A Physical Approach to the Form of Cross Sections for Compound Nucleus Reactions: The Breit–Wigner Single-Level Formula

If you have been following along, you have probably begun to see the main features of an approach to understanding how one might express the cross section for a reaction that occurs through a single state in the compound nucleus. Four factors must be considered. First, the projectile must be able to interact with the target. If the target and projectile were classical ideal hard spheres, the distance between

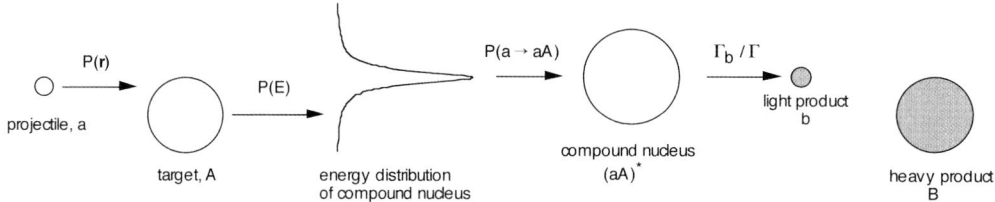

Fig. 13.25 Schematic drawing representing the spatial, energetic, formation and decay factors for a reaction proceeding through a single isolated level in the compound nucleus.

their centers at the point of closest approach must be no larger than the sum of their radii. In the more general case, it should be clear that the distance of closest approach can be no more than the range of the potential of their mutual interaction. In short, the projectile must "find" the target in space. Second, even if the two reaction partners find one another, there must be some probability that the projectile can enter the target. If this probability was zero, the result of an interaction would be simple potential scattering. Third, if the compound nucleus is formed through a single state that is narrow in energy and well-separated energetically from all other states of the system, the total energy in the center of mass must be very close to the energy of that state. That is, the projectile–target combination must "find" themselves within the energy distribution of the level as well. If the level is very narrow in energy, as most are at low energies, and we do not get close to the centroid, the probability for forming the compound nucleus can be expected to be small in most cases[1]. Finally, if we do form the compound nucleus, the probability for any specific reaction will be dependent upon the fraction of decays of the compound nucleus that take place in the desired way.

The situation is shown pictorially in Fig. 13.25. Taking into account the discussions in the preceding paragraph, we are tempted to write for the general reaction

$$a + A \rightarrow (aA)^*\big|_{res} \rightarrow b + B$$

proceeding through a single excited state, or resonance, as

$$\sigma_{a,b} \propto P(r)P(E)P(a \rightarrow aA)\frac{\Gamma_b}{\Gamma} \quad (13.96)$$

In this expression, P(r) is the probability for the projectile and target to interact in space and P(E) is the probability that the resonance through which the reaction is assumed to proceed will be found when the kinetic energy in the center of mass is E. The probability that the projectile will enter the target and form the compound

---

1) As it turns out, this is not necessarily the case for all resonances. The exceptions will be seen to arise naturally from our formulation and this assumption does not invalidate our approach.

13.7 The Breit–Wigner Single-Level Formula

nucleus $(aA)^*$ is represented by $P(a \to aA)$ and the ratio Γ_b/Γ gives the probability that the compound nucleus will decay by emission of the light product b.

From the discussion in Section 13.5.1, the range over which the nuclear interaction will be effective will depend upon the relative wavelength of the projectile–target combination in the center of mass coordinate system. In this system, the wavelength of the projectile is $\lambda_a(E) = h/p_a$ and thus the effective cross section of the incoming projectile will be on the order of $\pi\lambda_a^2(E)$. For projectiles randomly incident on the target, we take this as the measure of $P(r)$. The probability for finding the single level in the compound nucleus in energy is just the $P(E)$ given in Eq. (13.92).

The probability $P(a \to aA)$ for the projectile entering the target and forming the compound nucleus can be inferred by use of the reciprocity theorem. The width of the resonance for decay by emission of the projectile to yield the products a + A, i.e., compound elastic scattering, is just Γ_a. Thus the probability for forming the compound nucleus in the first place at the same total energy in the center of mass coordinate system must also be proportional to Γ_a. It is then plausible to take $P(a \to aA) \sim \Gamma_a$.

Combining these three factors with the ratio Γ_b/Γ that gives the fraction of decays of the compound nucleus that occur by emission of b leads to the expectation that the cross section for the reaction $a + A \to (aA)^*|_{res} \to b + B$ through a single isolated resonance will be proportional to

$$\sigma_{a,b}(E) \propto P(r)P(a \to aA)P(E)\frac{\Gamma_b}{\Gamma} = \pi\lambda_a^2 \cdot \Gamma_a \cdot \frac{\Gamma}{2\pi} \frac{1}{E - E_o^2 + \Gamma^2/4} \cdot \frac{\Gamma_b}{\Gamma}$$

$$= \frac{\lambda_a^2}{2} \frac{\Gamma_a \Gamma_b}{(E - E_o)^2 + \Gamma^2/4} \quad (13.97)$$

This proportionality gives the correct dependence on the resonance properties and the projectile wavelength and thus contains the essential physics of a reaction cross section that proceeds through a single isolated resonance.

To arrive at a quantitative result, we must consider two additional factors. First, we have neglected the problem of the angular momenta of the projectile and target and the fact that the level in the compound nucleus has a definite angular momentum and parity. Referring to the diagram in Fig. 13.26, we must have that

$$I_a + I_A + l_a = J_C \quad (13.98)$$

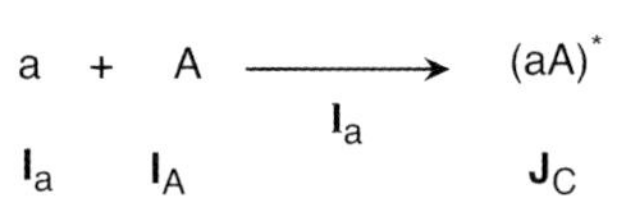

Fig. 13.26 The angular momenta that must be considered in the formation of the compound nucleus. The quantities I_a, I_A and J_C are the ground state angular momenta of the projectile and target and the angular momentum of the resonance, respectively. The quantity l_a is the orbital angular momentum of the projectile in the center of mass system.

Remembering that each of the vectors I_a, I_A and l_a has magnetic substates and that the z component of angular momentum must be conserved, not all possible combinations of the vector additions may lead to one of the $(2J_C + 1)$ values of the z components of angular momentum of the resonance in the compound nucleus when the target and projectile are not polarized. The total number of possible combinations is $(2I_a + 1)(2I_A + 1)(2l_a + 1)$. Hence, the fraction of all possible combinations that can produce a z component of angular momentum of the compound nucleus is simply

$$\frac{(2J_C + 1)}{(2I_a + 1)(2I_A + 1)(2l_a + 1)}$$

This "statistical" factor should multiply the expression we have developed in Eq. (13.97).

The second factor that is missing is not so easily uncovered. To obtain it, one must go through the complete theoretical development of the cross section expression. The principal problem is that we have treated the incoming projectile as though it were a classical point particle. If this were the case, it would have a definite orbital angular momentum defined by its impact parameter. As a wave, however, the projectile really has a finite extent and thus a range of angular momenta relative to the target. For the moment, we will simply note that, for each possible angular momentum of the projectile in the center of mass system, the cross section given in Eq. (13.97) must be multiplied by the factor $(2l_a + 1)$. This leads to a statistical factor, usually symbolized by g, of the form

$$g = \frac{(2J_C + 1)}{(2I_a + 1)(2I_A + 1)} \tag{13.99}$$

The actual expression for the reaction cross section through a single isolated resonance is given by

$$\sigma_{a,b}(E) = \pi g \lambda_a^2 \frac{\Gamma_a \Gamma_b}{(E - E_o)^2 + \Gamma^2/4} \tag{13.100}$$

where $\lambda_a = \lambda_a/2\pi$. Our result from simple physical arguments differs from this by a factor of 2π. Not too shabby.

The result given in Eq. (13.100) is known as the Breit–Wigner single-level formula. It represents prominent features of the cross sections of all low-energy reactions that proceed through the formation of a compound nucleus. Its characteristics are especially evident in the low-energy behavior of the cross sections of neutron reactions. Indeed, the resonances in the $^{197}_{79}\text{Au}(n, \gamma)^{198}_{79}\text{Au}$ reaction shown in Fig. 13.14 are typical of what is found in such reactions throughout the chart of the nuclides. We will discuss some of the details of these further below. But for the present, it is worthwhile examining just how well this form describes experimental data. To do this, we consider the large resonance in the $^{197}_{79}\text{Au}(n, \gamma)^{198}_{79}\text{Au}$ reaction

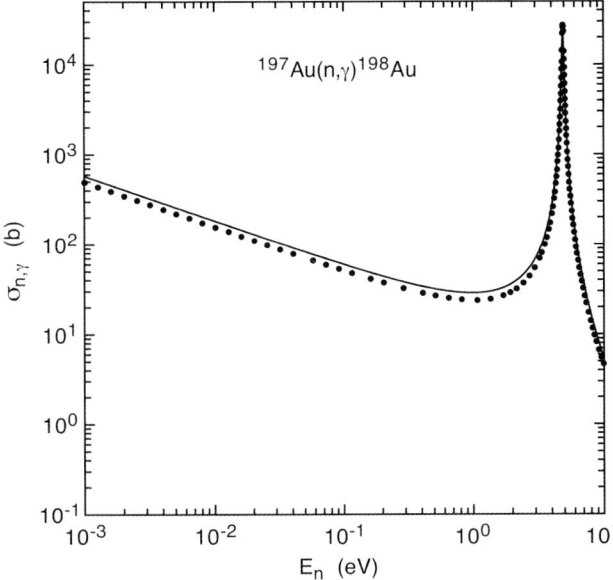

Fig. 13.27 The cross section data for the $^{197}_{79}\text{Au}(n, \gamma)^{198}_{79}\text{Au}$ (data points) and a fit with a Breit–Wigner function neglecting the contribution to the cross section from higher-lying resonances (solid curve).

cross section at an energy of about 5 eV. This resonance is so much larger than the others that we will make the assumption that the higher-energy resonances can be completely neglected. A least-squares fit to the cross section data with a single resonance of the form given in (13.100) is shown in Fig. 13.27. The fit, shown as a solid line in the figure, is really excellent. The overestimation of the data below the peak is due to the neglect of the contributions from the "tails" of higher lying resonances. But there can be no doubt that the Breit–Wigner form is an accurate description of the cross section for an isolated resonance in a compound nucleus.

In Section 13.3, we showed how the general formulation of the reaction cross section, based on first-order perturbation theory predicted that the low-energy cross section of a reaction with large positive Q value, should vary inversely with the velocity of the projectile. Fig. 13.27 demonstrates that this behavior arises naturally from the Breit–Wigner expression. The conclusion is that the $1/v_{\text{projectile}}$ behavior of the cross sections is due to the summation of contributions from the tails of all of the resonances that are found in the cross section. This result can be obtained from the Breit–Wigner form of Eq. (13.100) with a little bit of work. The first thing we must consider is the energy dependence of the widths. The total width Γ is the sum of the partial widths. If all of these are constants, the total width will also be constant. For the present, we restrict attention to the (n,γ) reaction at low energies in nuclides where proton, α particle, etc., emission can be neglected. In such cases, the total width is, to an excellent approximation, just the sum of Γ_n and Γ_γ. The γ

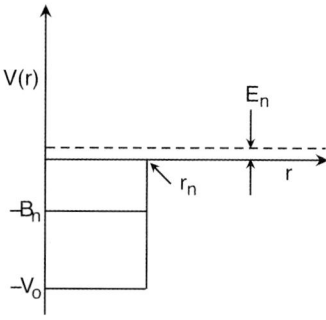

Fig. 13.28 Schematic drawing of the potential function experienced by a neutron with E_n slightly greater than zero in the presence of a nucleus represented by a spherical potential well of depth $-V_o$.

width depends upon the energies of all of the possible photons that can be emitted in the decay of the resonance. Most resonances decay to a large number of levels because of their high excitation energies. Nevertheless, the energies of photons emitted with appreciable probability are typically $E_\gamma \geq 0.5$ MeV and they tend to be of E1 multipolarity. In the case of low-energy resonances of the type shown in Fig. 13.27, we are dealing with compound nuclear levels within some tens of eV of the neutron binding energy. The change in the decay probability of a 0.5 MeV γ-ray by addition of, say 50 eV is no more than about three parts in 10^4. Clearly, to an excellent approximation, we can consider the probability for photon emission over such energy ranges to be constant, and hence Γ_γ is essentially a constant.

The same cannot be said for the neutron width. Over the energy range $0 < E_n \leq 50$ eV, the velocity of the neutron changes appreciably. Remembering that the neutron width is a measure of the probability that the compound nucleus will decay by neutron emission, we can use the simple ideas of barrier penetration to show that the width will be proportional to the square-root of E_n itself.

Suppose that the compound nucleus is formed at an excitation energy just slightly larger than the neutron binding energy – this is the real situation for low-energy resonances. We can picture the problem of neutron emission as shown in the schematic of Fig. 13.28. A neutron with an energy slightly higher than that of the neutron binding energy will have E_n just slightly greater than zero. Outside the range of the nuclear potential this will be its kinetic energy. Taking the nuclear potential as that of a spherical potential well for simplicity, the kinetic energy of the neutron at radial dimensions $r \leq r_n$ will be equal to $E_n + V_o \gg E_n$. Small changes in the total neutron energy will produce large changes in kinetic energy outside the region of the potential well, but will result in negligible changes in the kinetic energy when the neutron has radial dimensions $r \leq r_n$.

Notwithstanding the fact that the total energy is positive and, in the absence of orbital angular momentum, there is no potential barrier through which the neutron must penetrate, there will nevertheless be some reflection that takes place at the potential discontinuity at $r = r_n$. This is easily demonstrated by analysis of a simple one-dimensional barrier of the type that was analyzed in Chapter IX where we developed the fundamental properties of barrier penetration that were then applied to α decay. The transmission coefficient T we recall is of the form

$$T = \left.\frac{|\psi_{r>r_n}|^2 v_{r>r_n}}{|\psi_{r\le r_n}|^2 v_{r\le r_n}}\right|_{r=r_n} \qquad (13.101)$$

where $\psi_{r>r_n}$ is the wave function of the incident projectile, $\psi_{r\le r_n}$ is the wave function of the bound state, the numerator represents the flux of neutrons moving outward at $r \ge r_n$ and the denominator represents the flux of a neutron that is moving outward in the region of $r \le r_n$. Now $\psi_{r\le r_n}$ is the wave function of a neutron with total energy $E_n + V_o$ and the argument given above means that the properties of the neutron within the well will be essentially independent of E_n. As a result, the energy dependence of the transmission coefficient will be dominated by the velocity of the neutron after emission. Because the decay probability must be proportional to the transmission coefficient, we arrive at the conclusion that the neutron width must be proportional to $v_{r>r_n}$ or $\Gamma_n \propto \sqrt{E_n}$.

We now apply this conclusion to the Breit–Wigner expression of Eq. (13.100) in the limit that $E_n \to 0$.

$$\begin{aligned}\sigma_{n,\gamma}(E_n) &= \pi g \lambdabar_n^2 \frac{\Gamma_n \Gamma_\gamma}{(E_n - E_o)^2 + \Gamma^2/4} \\ &= \left.\pi g \left(\frac{\hbar^2}{2 m_n E_n}\right)\frac{(\text{const}\sqrt{E_n})\Gamma_\gamma}{(E_n - E_o)^2 + \Gamma^2/4}\right|_{E_n \to 0} \to \text{Const}\frac{1}{\sqrt{E_n}}\end{aligned} \qquad (13.102)$$

The $1/v$ behavior of the reaction cross section is not "exact" but it is an excellent approximation in practical cases. The transmission coefficient is not entirely independent of the change in neutron energy within the nucleus, but it is very nearly so. The total width of a level is also not a constant if any of the partial widths is not a constant. However, only in cases of resonances in very light nuclei, where the widths are quite large, must these complications be considered. We will have more to say about resonances in reactions later, but some of their features cannot be understood without an examination of the general formulations of reaction cross sections to which we now turn.

13.8
Scattering in Quantum Mechanics: Partial Wave Analysis

The simplest reaction we can consider is that of elastic scattering. The treatment in the preceding sections has considered the kinematics and some of the properties of elastic scattering that can be uncovered by examination of the density of states available in the process. In the present section, we want to concentrate on the description of the scattering process in terms of the wave properties of the neutron. If we were to try to do this problem completely, we would have to worry about the details of the interaction of the neutron with an arbitrary nucleus and all of the complications that this most assuredly entails. But there is a much simpler approach that is very general and provides a great deal of insight into the characteristics of scattering cross sections. This approach is the quantum mechanical equivalent of the classical kinematics calculations we have performed earlier. For kinematics, we do not care about the details of an interaction so long as it takes place via conservative forces. The conservation of energy, momentum and angular momentum are guaranteed for such cases so long as we consider the initial state in the limit that the scattering partners have not yet begun to interact, and consider the final state sufficiently long after the interaction that the scattering partners are again no longer interacting.

We can apply the same conditions to the scattering of neutrons on nuclei in terms of their wave properties. Well before a neutron is within range of the nuclear force of the target, it is a free particle, and well after the interaction, when it is again separated from the target by a dimension large compared to the range of the nuclear interaction, it is again a free particle. The wave functions before and after the interaction must express the conservation of energy, linear and angular momentum. We know that the only effect that can take place is a change in the angle of the scattered particle relative to the trajectory of the incident neutron. As we have alluded to above, the incoming wave can possess a range of angular momenta and a wave description must account for this. In the classical case, if the projectile is incident with a define impact parameter, the scattering will preserve this angular momentum. We expect the same conservation to be found in the quantum mechanical scattering. Each angular momentum component contained in the wave function of the projectile will be represented by a definite amplitude or probability, and the conservation of angular momentum implies that each of these amplitudes will be conserved individually. However complicated the wave functions are during the interaction itself, the wave functions in the *asymptotic* limits before and after the scattering must be calculable in a straightforward way because they reflect only the requirements of the conservation laws.

The development that we will then embark upon is the development of wave functions that reflect the quantized nature of orbital angular momentum and the general form of the wave functions after the scattering. This will contain a function that represents the probability that scattering will result in the projectile appearing at a specific angle θ with respect to its initial trajectory. It defines the *angular distribution* of the scattered particles. We will not calculate this function explicitly. Its form is determined completely by the potential function. (This is, by the way, why the study of nuclear reactions is so important fundamentally. It is through studies of the angular distributions that we can infer the properties of the nuclear interaction.)

The elastic scattering of a neutron by an arbitrary nucleus is shown schematically in Fig. 13.29. Only the wave functions of the neutron outside of the shaded area that signifies the range of the nuclear force between the neutron and the target nucleus will be considered. As usual, we will assume that we treat the scattering in the center of mass coordinate system and that the system is unpolarized so that scattering will have cylindrical symmetry about the direction of incidence of the neutron. Outside the range of the nuclear interaction, the incident neutron is a free

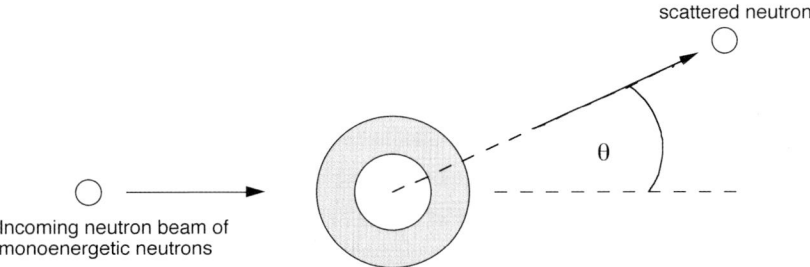

Fig. 13.29 Schematic diagram of the elastic scattering of a neutron on an arbitrary nucleus. The shaded area is meant to represent the region outside of the hard sphere radius of the target in which the nuclear potential is appreciably different from zero.

particle and can be described as a plane wave. If the neutron moves along the z-axis in the positive direction, its wave function is of the form $\psi_o = Ae^{ikz}$ where $k = p/\hbar = 1/\lambdabar$. If, as a result of the scattering, the outgoing neutron had an equal probability of being scattered into any solid angle element, i.e., the scattering was isotropic, the neutron would be described as a spherical wave emanating from the target. Because it has the same kinetic energy as the incoming neutron, the wave function might be guessed by analogy as being proportional to e^{ikr}. While this is certainly spherically symmetric about an origin taken as the center of the target, it clearly cannot be normalized. However a function of the form e^{ikr}/r does have the desired properties and indeed represents the form of an outgoing spherical wave. Thus, in the general case where the scattering is not isotropic, we take the wave function of the scattered neutron in the form $\psi_1 = Af(\theta)e^{ikr}/r$. Note that the total amplitude of the incoming and outgoing waves must be the same.

The total wave function of the system as a whole, reckoned outside of the range of the nuclear interaction, will then be the sum of the incoming and outgoing waves and we can write

$$\psi_{tot} = \psi_o + \psi_1 = A\left(e^{ikz} + f(\theta)\frac{e^{ikr}}{r}\right) \quad (13.103)$$

We obviously need to examine the scattering cross section as a function of the scattering angle. For this purpose, and in complete analogy with the definition of the total cross section, we define the *differential elastic scattering cross section* for unpolarized systems as

$$\sigma(\theta) = \frac{d\sigma}{d\Omega_\theta} \quad (13.104)$$

where σ is the total cross section and Ω_θ is the solid angle between θ and $\theta + d\theta$ as shown in Fig. 13.30. The spherical surface area swept out by variation of θ over the range θ to $\theta + d\theta$ and ϕ over the range ϕ to $\phi + d\phi$ is $dS = r^2\sin\theta d\theta d\phi$. The area

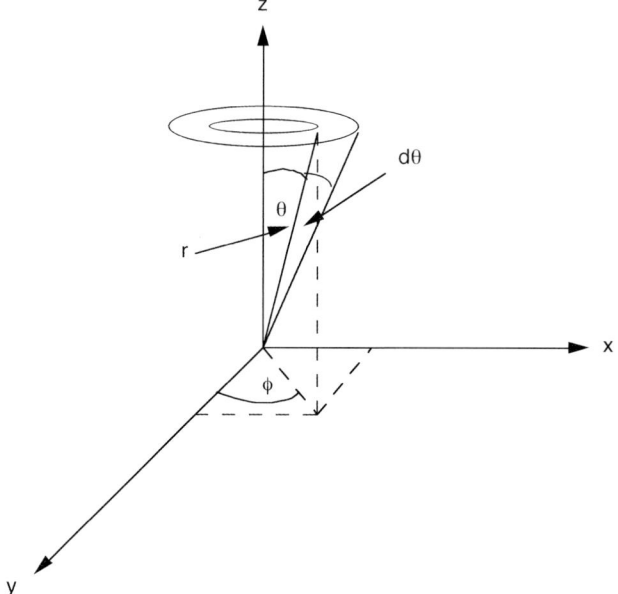

Fig. 13.30 Relations between Cartesian and spherical polar coordinates in the calculation of the solid angle between θ and $\theta + d\theta$.

swept out as ϕ goes from 0 to 2π is then $dS_\theta = 2\pi r^2 \sin\theta d\theta$. By definition, the solid angle subtended by dS_θ and the origin, $d\Omega_\theta$, is

$$d\Omega_\theta = \frac{dS_\theta}{r^2} = 2\pi \sin\theta d\theta \tag{13.105}$$

The differential scattering cross section is then seen to be the probability for scattering of particles at angles θ to $\theta + d\theta$ per unit incident neutron flux.

Now we can use the wave function for the incident and scattered neutrons to relate them directly to $\sigma(\theta)$. First, the incident flux is just $|\psi_o|^2 v$ as we have seen before. The probability for neutrons to be scattered at θ to $\theta + d\theta$ per unit incident flux must be $|\psi_1|^2 v$ times the area of the spherical surface $dS_\theta = r^2 d\Omega_\theta$. Therefore

$$\sigma(\theta) d\Omega_\theta = \frac{|\psi_1|^2 v r^2 d\Omega_\theta}{|\psi_o|^2 v} = \frac{A^2 f^2(\theta)}{r^2}\left(\frac{1}{A^2}\right) r^2 d\Omega_\theta = |f(\theta)|^2 d\Omega_\theta \tag{13.106}$$

and we find that

$$\sigma(\theta) = |f(\theta)|^2 \tag{13.107}$$

The differential scattering cross section, and hence the total scattering cross section which is obtained by integration over σ(θ), is defined completely by the angular distribution function that is itself defined completely by the forces that act in the scattering interaction.

To understand the behavior of the scattering, we will have to obtain expressions for the wave functions that contain angular momentum. In texts on mathematical physics, it is shown that the function e^{ikz} can be expressed in terms of our old friends the spherical Bessel functions and the spherical harmonics. The relation is

$$e^{ikz} = \sum_{l=0}^{\infty} (2l+1) i^l j_l(kr) Y_{l,0}(\theta, \phi) \qquad (13.108)$$

(It's amazing how these functions keep cropping up.) We know the spherical Bessel functions from the solution to the Schrödinger equation for a spherically symmetric potential well and from the description of the wave functions of the leptons emitted in β decay. The spherical harmonics will be remembered to be the universal eigenfunctions of the angular coordinates for any spherically symmetric potential. Eq. (13.108) then indicates that a plane wave is just the superposition of wave functions for a spherical symmetric potential, the superposition containing contributions from *all* possible orbital angular momenta. Remember that the spherical harmonics of m = 0 are just the Legendre polynomials $P_l(\cos \theta)$.

This handy relation can be simplified greatly for the present purposes, and that lies in the fact that we are considering only the asymptotic wave functions where the nuclear potential is negligible. In this range, we can take our wave functions at any point we like; they will be the wave functions of free particles. If so, we might just as well consider the wave functions at very large values of r. And the reason for this is that for arguments $kr \to \infty$, the spherical Bessel functions have the limiting solutions

$$j_l(kr)|_{kr \gg 1} \to \frac{1}{kr} \sin\left(kr - \frac{l\pi}{2}\right) \qquad (13.109)$$

They just reduce to simple sinusoidal functions. If we take this limit and substitute it into Eq. (13.108), expand $\sin(kr - (l\pi/2))$ in terms of its representation in the functions $e^{\pm i(kr - \frac{l\pi}{2})}$, and use the relations $e^{il\pi/2} = i^l$ and $e^{il\pi} = (-1)^l$, we obtain the result

$$e^{ikz}|_{r \to \infty} = \sum_{l=0}^{\infty} \frac{(2l+1)}{2ikr} [e^{ikr} - (-1)^l e^{-ikr}] P_l(\cos \theta) \qquad (13.110)$$

This is a remarkable result. If you multiply through by 1/kr, you will notice that the two terms in brackets are just outgoing (e^{ikr}/r) and incoming (e^{-ikr}/r) spherical waves. At large radial dimensions, a plane wave is just the superposition of incoming and outgoing spherical waves of definite orbital angular momentum modulated by the Legendre polynomials. And because the wave function of the scattered neutron is also described as an outgoing spherical wave modulated by the angular distribution function f(θ), the entire wave disturbance is just the summation of angle-modulated incoming and outgoing spherical waves.

Considering the conservation laws, how can the incoming and outgoing spherical waves differ from one another? In elastic scattering nothing really changes, except the angle of the scattered particle relative to the trajectory of the projectile. The linear momentum and kinetic energy remain unchanged and so does the angular momentum. This must apply individually to each of the angular momentum components in the wave function of the projectile. The only difference between the incoming and outgoing waves for a given l is a possible change in sign, a so-called phase factor and, of course, an angular variation due to the scattering. The total wave disturbance should then be described quite generally as the sum of incoming and outgoing waves of the form given in (13.110) multiplied by a factor $\alpha_l e^{i\delta_l}$. Quite generally, then, we write the total wave disturbance as

$$\psi_{tot}|_{r \to \infty} = \alpha_l \left\{ A \sum_{l=0}^{\infty} \frac{(2l+1)}{2ikr} [e^{i(kr+\delta_l)} - (-1)^l e^{-i(kr+\delta_l)}] P_l \cos(\theta) \right\} \quad (13.111)$$

This must be equal to the original expression we wrote down for the total wave disturbance in Eq. (13.103). Rewriting this using the asymptotic expression for the incident plane wave, we have

$$\psi_{tot}|_{r \to \infty} = A \left\{ \sum_{l=0}^{\infty} \frac{(2l+1)}{2ikr} [e^{ikr} - (-1)^l e^{-ikr}] P_l(\cos\theta) \right\} + Af(\theta)\frac{e^{ikr}}{r} \quad (13.112)$$

If we equate these two expressions, we can solve to obtain an expression for $f(\theta)e^{ikr}/r$, the wave function of the scattered neutron in terms of α and δ_l. Immediately one sees that

$$f(\theta)\frac{e^{ikr}}{r} = \sum_{l=0}^{\infty} \alpha_l \frac{(2l+1)}{2ikr} [e^{i(kr+\delta_l)} - (-1)^l e^{-i(kr+\delta_l)}] P_l(\cos\theta)$$

$$-\sum_{l=0}^{\infty} \frac{(2l+1)}{2ikr} [e^{ikr} - (-1)^l e^{-ikr}] P_l(\cos\theta) \quad (13.113)$$

Now $f(\theta)e^{ikr}/r$ represents a purely outgoing wave, and that means the total amplitude of each of the incoming waves of each l on the right-hand side must be zero. That is, we must have

$$-\alpha_l \frac{(2l+1)}{2ikr}(-1)^l e^{-i(kr+\delta_l)} + \frac{(2l+1)}{2ikr}(-1)^l e^{-ikr} = 0 \quad \text{or} \quad (13.114)$$

$$\alpha_l = e^{i\delta_l} \tag{13.115}$$

The conclusion is that, as one should expect, only a single parameter is needed to describe the differences between the incoming and outgoing waves produced through elastic scattering. With the use of (13.115), we can now use (13.113) to solve for f(θ) directly as

$$f(\theta) = \sum_{l=0}^{\infty} \frac{(2l+1)}{2ik}(e^{2i\delta_l} - 1)P_l(\cos\theta) \tag{13.116}$$

We can manipulate this a bit by use of the relation $\sin\theta = (1/2i)(e^{i\theta} - e^{-i\theta})$. Noting that $e^{i\theta}\sin\theta = (1/2i)(e^{2i\theta} - 1)$, we rewrite (13.116) as

$$f(\theta) = \sum_{l=0}^{\infty} \frac{(2l+1)}{k} \sin\delta_l e^{i\delta_l} P_l(\cos\theta) \tag{13.117}$$

Finally, the substitution of this expression into (13.107) gives the differential scattering cross section as

$$\sigma(\theta) = \frac{d\sigma}{d\Omega_\theta} = \left| \sum_{l=0}^{\infty} \frac{(2l+1)}{k} \sin\delta_l e^{i\delta_l} P_l(\cos\theta) \right|^2 \tag{13.118}$$

All of the effects of the scattering potential are contained in the phase shifts. For each l, the phase shift determines the magnitude of the cross section while the angular momentum determines the relative probability for scattering at each θ to θ + dθ through the Legendre polynomial. If l = 0, $P_0(\cos\theta) = 1$, and if this is the only angular momentum involved, the scattering will be isotropic in the center of mass. If l = 1, $P_1(\cos\theta) = \cos\theta$, and so forth. Measurement of the differential cross section will then give information concerning the angular momentum involved in the scattering. In the general case, where the scattering involves a range of angular momenta, we will observe the square of the sum of the contributions from each. There will be *interference* between the different scattering distributions and the shape will give information on both the angular momenta involved as well as the relative phase shifts for each l.

The total scattering cross section is obtained by integrating Eq. (13.118) over all θ. Thus

$$\sigma = \int \left(\frac{d\sigma}{d\Omega_\theta}\right) d\Omega_\theta$$

$$= 2\pi \int_0^\pi \left(\frac{d\sigma}{d\Omega_\theta}\right) \sin\theta \, d\theta \tag{13.119}$$

$$= 2\pi \int_{-1}^1 |f(\theta)|^2 d(\cos\theta)$$

At first sight, this integration looks messy. But the orthogonality of the Legendre polynomials actually makes life simple. In texts on applied mathematics, you can find the proof of the orthogonality relations

$$\int_{-1}^1 P_l(\cos\theta) P_{l'}(\cos\theta) d(\cos\theta) = \begin{cases} 0 & \text{if } l \neq l' \\ \dfrac{2}{2l+1} & \text{if } l = l' \end{cases} \tag{13.120}$$

Of all the terms in the square of the right-hand side of Eq. (13.119), the only ones that will not give vanishing integrals are the terms where $l = l'$. Therefore, the differential cross section is easily found to be

$$\sigma_{el} = 2\pi \sum_{l=0}^\infty \frac{(2l+1)^2}{k^2} \sin^2\delta_l \left(\frac{2}{2l+1}\right) \quad \text{or} \tag{13.121}$$

$$= \frac{4\pi}{k^2} \sum_{l=0}^\infty (2l+1) \sin^2\delta_l$$

$$\sigma_{el} = 4\pi\lambdabar^2 \sum_{l=0}^\infty (2l+1) \sin^2\delta_l \tag{13.122}$$

The total cross section is dependent solely on the wavelength of the neutron in the center of mass coordinate system and the phase shifts. The cross section is the *sum* of the *partial* cross sections for each l. Because the maximum value of the sine is unity, the maximum value of a partial cross section is just $4\pi\lambdabar^2(2l+1)$.

At very low energies it is not difficult to demonstrate that the only scattering of significance will be s-wave, i.e., $l = 0$, scattering. To do this we need only remember that the waves making up a plane wave are proportional to $j_l(kr)$ (Eq. (13.108)). For small values of the argument, $j_l(kr)$ has the limiting form

$$j_l(kr)|_{kr \to 0} \to \frac{(kr)^l}{1 \cdot 2 \cdot 5 \cdots (2l+1)} \tag{13.123}$$

13.8 Scattering in Quantum Mechanics: Partial Wave Analysis

Therefore, for sufficiently small kr, the only term of appreciable size in the expansion of e^{ikz} will be that for $l = 0$. What "sufficiently small" means can be estimated by a simple calculation. From (13.123),

$$j_0(kr)/j_1(kr)/j_2(kr)\ldots = 1/\frac{kr}{3}/\frac{(kr)^2}{15}\ldots$$

So long as $kr \ll 0.01$, terms in the expansion with $l > 0$ will be negligible in comparison to that for $l = 0$. In the hard-sphere approximation, the scattering will occur when the radial separation between the target and the neutron is about equal to the target radius r_n. At this separation,

$$kr_n = \frac{p}{\hbar}r_n = \frac{(2m_nc^2)^{1/2}\sqrt{E_n}}{\hbar c}r_oA^{1/3} \approx 0.27A^{1/3}\sqrt{E_n} \quad (13.124)$$

with energy in units of MeV. We can then say that so long as

$$E_n \ll \frac{1.3 \times 10^{-3}}{A^{2/3}} \text{MeV} \quad (13.125)$$

only s-wave scattering is likely to occur. For neutron–proton scattering, this means that kinetic energies in the center of mass of less than 1 keV can be considered, whereas for A = 200, kinetic energies of less than 0.04 keV can be considered. For elastic scattering of neutrons with kinetic energies of less than a few tens of eV, only s-wave scattering need be considered for any target mass.

If we restrict attention to very low energy neutrons, we then have, to a good approximation,

$$\sigma_{el} \approx \sigma_{el}(0) = 4\pi\lambda^2\sin^2\delta_o \quad (13.126)$$

The low-energy behavior of the scattering cross sections of neutrons on ^{1}H, ^{2}H and ^{208}Pb have already been discussed in Section 13.3. They all demonstrate constant cross sections at low energies where s-wave scattering must dominate. For this to be true, Eq. (13.126) indicates that the phase shift must be energy dependent to compensate for the variation in the neutron wavelength. Indeed, it can be shown that, for the spherical potential well at low energies, $\delta_o \propto 1/\lambda$.

The dependence of the differential scattering cross section on the neutron energy in the center of mass coordinate system is illustrative of the importance of different l-values on the scattering process. As an example, we consider the elastic scattering of neutrons on the nucleus $^{56}_{26}$Fe. In Fig. 13.31 are the evaluated differential scattering cross sections for neutron energies of thermal, 50.9 keV and 802 keV. The data plotted are the differential scattering cross sections per steradian divided by the total cross section at each energy. Hence

$$\int \frac{\sigma(E_n, \theta)}{\sigma(E_n)} d\Omega_\theta = 1$$

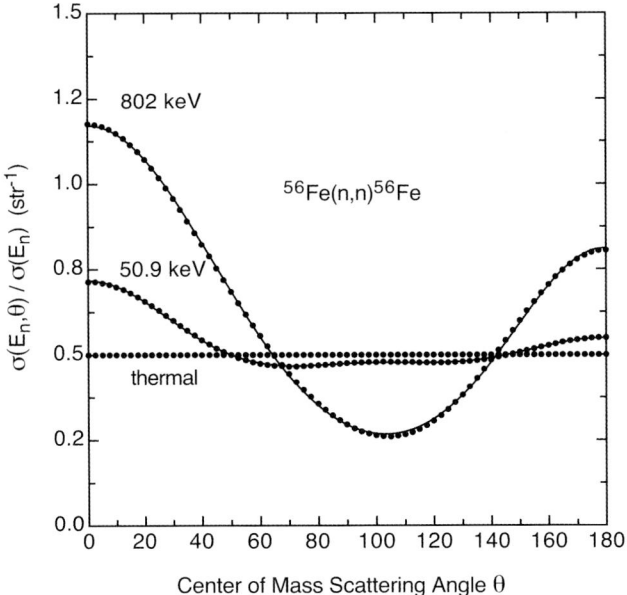

Fig. 13.31 The evaluated differential scattering cross sections for neutrons on $^{56}_{26}\text{Fe}$ at energies of thermal, 50.9 keV and 802 keV, each divided by the total cross section at that energy.

The differential cross section at thermal energies is isotropic and thus is shown as a constant with magnitude of 0.5 str^{-1}. This is consistent with all known information. The differential cross sections at 50.9 and 802 keV show distinct departures from that expected for l = 0 scattering. Both have higher probabilities for scattering at small angles than at angles near 180°. These indicate that significant scattering at l > 0 must be taking place. According to (13.118), the angular distribution can be written as

$$\sigma(\theta) = \left| \sum_{l=0}^{\infty} a_l P_l(\cos\theta) \right|^2 \tag{13.127}$$

where

$$a_l = \frac{(2l+1)}{k} \sin\delta_l e^{i\delta_l}$$

Hence, a fit of the data in Fig. 13.31 with (13.127) will yield

$$\sigma_{el} = \left| \sum_{l=0}^{\infty} \frac{2l+1}{k} \sin\delta_l e^{i\delta_l} P_l \cos\theta \right|^2$$

$$= 2\pi \sum_{l=0}^{\infty} \frac{(2l+1)^2}{k^2} \sin^2\delta_l \left(\frac{2}{2l+1} \right) \quad (13.128)$$

$$= 2\pi \sum_{l=0}^{\infty} a_l^2 \frac{2}{2l+1}$$

Thus the $a_l^2/(2l+1)$ will provide the relative contribution of the scattering at each l to the total scattering cross section.

The solid lines in Fig. 13.31 represent fits to the evaluated data with Eq. (13.128) and the representations are seen to be excellent. The coefficients a_l obtained from the fits are shown in Table 13.2 along with relative values of $a_l^2/(2l+1)$ for the scattering data at 50.9 and 802 keV. Notwithstanding the marked differences from an isotropic distribution, the scattering at both energies is seen to be dominated by s-wave scattering. Why this is true is, of course, dependent upon the details of the nuclear interaction between the neutron and the $^{56}_{26}$Fe nucleus, and by just how much angular momentum the neutron can "bring in" during the scattering. This can be estimated crudely with a simple mechanical picture. Treating the neutron and target as hard spheres, the maximum angular momentum that is brought in by the neutron is given by $l\hbar = rp = (r_{Fe} + r_n)\sqrt{2m_n E_n}$. For a 1 MeV neutron, this estimate gives l = 1.32. It is then not unreasonable to expect that s-wave scattering will be a major contributor to the scattering cross section at such energies.

Table 13.2 The coefficients obtained from fits to the data in Fig. 13.31 and the relative contribution of scattering at each angular momentum to the total scattering cross section.

l	a_l (50.9 keV)	$\left(\frac{a_l^2(50.9\,keV)}{2l+1}\right)_{rel}$	a_l (802 keV)	$\left(\frac{a_l^2(802\,keV)}{2l+1}\right)_{rel}$
0	7.063×10^{-1}	1.000	6.896×10^{-1}	1.000
1	2.417×10^{-2}	3.903×10^{-4}	1.406×10^{-1}	1.386×10^{-2}
2	5.735×10^{-2}	1.319×10^{-3}	2.970×10^{-1}	3.710×10^{-2}
3	2.857×10^{-2}	2.338×10^{-4}	-5.005×10^{-2}	7.524×10^{-4}
4	3.010×10^{-2}	2.108×10^{-4}	5.006×10^{-3}	5.855×10^{-6}

13.9
Extension of the Partial Wave Analysis to Nuclear Reactions

The formalism presented above for elastic scattering can be extended in a relatively simple way to all other nuclear reactions. Because the interaction causing the reactions is the same as that causing the scattering, it must be true that the equations describing the differential and total cross sections for elastic scattering and other reactions will be closely related. Regardless of the reaction type, the principal difference between these is that the "original" neutrons are lost as a result of a reaction. That is, relative to elastic scattering, the intensity of neutrons appearing at any angle in the center of mass must be reduced from that calculated in the previous section. For the present purposes, we define the total reaction cross section as the cross section that results in the disappearance of neutrons with energies equal to the initial energy. With this, we not only can have a phase shift, but a loss in the amplitude of waves corresponding to a given orbital angular momentum. The total cross section is then the sum of the total reaction and elastic scattering cross sections.

In the previous section, we showed that the total wave function for the case of purely elastic scattering could be written as,

$$\psi_{tot}\big|_{r \to \infty} = A \sum_{l=0}^{\infty} \frac{(2l+1)}{2ikr} [e^{i(kr+2\delta_l)} - (-1)^l e^{-ikr}] P_l \cos(\theta) \tag{13.129}$$

where we have used the relation $\alpha_l = e^{i\delta_l}$. The only effect of the elastic scattering is the production of the phase shift δ_l. To account for the loss in amplitude that accompanies reactions, we introduce a factor $\widehat{\eta}_l = |\eta_l| e^{2i\delta_l}$ that expresses both the phase shift and amplitude change in the outgoing waves. If we use this, Eq. (13.129) becomes

$$\psi_{tot}\big|_{r \to \infty} = A \sum_{l=0}^{\infty} \frac{(2l+1)}{2ikr} [\widehat{\eta}_l e^{ikr} - (-1)^l e^{-ikr}] P_l \cos(\theta) \tag{13.130}$$

If we equate this to the general expression (Eq. (13.103))
$\psi_{tot} = A(e^{ikz} + f(\theta)e^{ikr}/r)$, it is easy to show that

$$f(\theta) = \sum_{l=0}^{\infty} \frac{(2l+1)}{2ikr} (\widehat{\eta}_l - 1) P_l \cos(\theta) \tag{13.131}$$

in complete analogy with the development shown previously for pure elastic scattering.

Our problem is now to relate the amplitude of the outgoing neutron waves in the reactions to those of the incoming waves for each l. To do this we need to resort to

13.9 Extension of the Partial Wave Analysis to Nuclear Reactions

the quantum mechanical expression for a current density. In texts on quantum mechanics, it is shown that the current density **j** corresponding to particles of mass m is given generally by

$$\mathbf{j} = \frac{-i\hbar}{2m}(\psi^*\nabla\psi - \psi\nabla\psi^*) \tag{13.132}$$

where ψ is the wave function of the particle. For an incoming neutron represented by the plane wave $\psi_o = Ae^{ikz}$, direct substitution into 13.132 gives

$$\mathbf{j}_o = A^2\frac{\hbar k}{m} = A^2 v \tag{13.133}$$

just what you expect, remembering that the wave function must be normalized in some arbitrarily large volume V and $A = V^{-1/2}$ as we have found before. For the flux of outgoing *scattered* particles, we can use the general form $\psi_1 = Af(\theta)e^{ikr}/r$ and obtain

$$\mathbf{j}_1 = A^2\frac{\hbar k}{mr^2}|f(\theta)|^2 = A^2\frac{v}{r^2}|f(\theta)|^2 \tag{13.134}$$

The *number* of scattered particles can be obtained by integrating the current density j_1 over some spherical surface with radius r_o that is so large that the asymptotic solutions are valid. If N is the number of scattered particles, we have, in general,

$$N = \int_{\text{sphere}} \mathbf{j}_1 \cdot d\mathbf{S} = \int A^2\frac{v}{r_o^2}|f(\theta)|^2 r_o^2 d\Omega_\theta = A^2 v \int |f(\theta)|^2 d\Omega_\theta \tag{13.135}$$

The differential scattering cross section is just

$$\frac{d\sigma}{d\Omega_\theta} = \frac{N(\theta)}{j_o} = |f(\theta)|^2 \tag{13.136}$$

or, from Eq. (13.131),

$$\frac{d\sigma}{d\Omega_\theta} = \left|\sum_{l=0}^{\infty}\frac{(2l+1)}{2ik}(\widehat{\eta}_l - 1)P_l(\cos\theta)\right|^2 \tag{13.137}$$

Further, following the same procedure used in the last section to obtain the total cross section by integration, we obtain

$$\sigma_{el} = \frac{\pi}{k^2}\sum_{l=0}^{\infty}(2l+1)\left|\widehat{\eta}_l - 1\right|^2 \tag{13.138}$$

At this point it is worthwhile to write down again the equivalent expressions derived for elastic scattering in the last section (Eqs (13.118) and (13.122)) for comparison.

$$\sigma(\theta) = \frac{d\sigma}{d\Omega_\theta} = \left| \sum_{l=0}^{\infty} \frac{(2l+1)}{k} \sin\delta_l e^{i\delta_l} P_l(\cos\theta) \right|^2$$

$$\sigma_{el} = 4\pi\lambda^2 \sum_{l=0}^{\infty} (2l+1)\sin^2\delta_l$$

If $|\eta_l| = 1$, so that $\widehat{\eta}_l = e^{2i\delta_l}$, then the quantity $(1/2i)(\widehat{\eta}_l - 1)$ in (13.137) is just $(1/2i)(e^{2i\delta_l} - 1) = \sin\delta_l e^{i\delta_l}$ and the two expressions for the differential scattering cross sections are identical. Each neutron in the incoming wave appears as a scattered neutron in the outgoing wave and the total cross section is due solely to elastic scattering. However, if $|\eta_l| < 1$, the intensity of the scattered neutrons is less than that in the incoming wave and the difference between the two must be due to the sum of all other reactions. Because the probability function for the neutron is proportional to $|\eta_l|^2$, the fraction of the incoming amplitude that is associated with reactions must be proportional to $(1 - |\eta_l|^2)$.

We now formally define the total cross section as

$$\sigma_{tot} = \sigma_{el} + \sigma_{react} \tag{13.139}$$

where

$$\sigma_{react} = \sum_i \sigma_{i, reac} \tag{13.140}$$

summed over all i interactions *not* including elastic scattering. From Eq. (13.138), we then have

$$\sigma_{el} = \frac{\pi}{k^2} \sum_{l=0}^{\infty} (2l+1) |\widehat{\eta}_l - 1|^2$$

$$\sigma_{react} = \frac{\pi}{k^2} \sum_{l=0}^{\infty} (2l+1)(1 - |\widehat{\eta}_l|^2) \tag{13.141}$$

The total cross section is the sum of these two. Writing

$$|\widehat{\eta}_l - 1|^2 + 1 - |\widehat{\eta}_l|^2 = (|\eta_l| - 1)(|\eta_l| - 1) + 1 - |\widehat{\eta}_l|^2$$
$$= |\eta_l|^2 + 1 - 2|\eta_l| + 1 - |\eta_l|^2 = 2(1 - \text{Re}(\widehat{\eta}_l)) \tag{13.142}$$

the total cross section can now be written as

$$\sigma_{tot} = \frac{2\pi}{k^2} \sum_{l=0}^{\infty} (2l+1)(1 - \text{Re}(\widehat{\eta}_l))$$

(13.143)

$$= 2\pi\lambda^2 \sum_{l=0}^{\infty} (2l+1)(1 - \text{Re}(\widehat{\eta}_l))$$

The relations given in Eqs (13.141) and (13.143) give the limits on the cross sections that we can expect. To be sure, we cannot obtain quantitative results without evaluating the phase shifts from a nuclear model. But the relations do permit us to understand the relative magnitudes of elastic scattering and all other interactions. First, let us assume that the $|\eta_l|$ is unity. The elastic scattering cross section is then given by (13.122). Because the maximum value of $\sin^2\delta_l$ is unity, the theoretical maximum value that the elastic scattering cross section can have is

$$(\sigma_{el})_{max} = 4\pi\lambda^2 \sum_{l=0}^{\infty} (2l+1)$$

(13.144)

But if $|\eta_l| = 1$, (13.141) indicates that the total reaction cross section must be zero. This is just a check on the assumptions we have already made in our previous derivations but it serves to emphasize the point that it is completely possible for an interaction to take place by nothing but elastic scattering. If $\widehat{\eta}_l = |\widehat{\eta}_l|e^{2i\delta_l} = 1$, Eq. (13.141) shows that the elastic scattering cross section vanishes. In this case, the phase shift is zero. There can be no interaction at all if the phase shift is zero. That is, in (13.143), $(1 - \text{Re}(\widehat{\eta}_l)) = 0$.

Eq. (13.141) indicates us that the reaction cross section will be a maximum when $|\eta_l|^2 = 0$, and

$$(\sigma_{react})_{max} = \frac{\pi}{k^2} \sum_{l=0}^{\infty} (2l+1) = \pi\lambda^2 \sum_{l=0}^{\infty} (2l+1)$$

(13.145)

But in this case, $\left|\widehat{\eta}_l - 1\right|^2 = \left|\widehat{\eta}_l\right|^2 - \widehat{\eta}_l - \widehat{\eta}_l^* + 1 = 1$ and

$$\sigma_{el}\Big|_{|\widehat{\eta}_l|^2 = 0} = \pi\lambda^2 \sum_{l=0}^{\infty} (2l+1)$$

(13.146)

When the reaction cross section has achieved its maximum theoretical value, the elastic scattering cross section has the same magnitude.

It should be remembered that, in order for reactions to occur, the projectile must in some way interact with the target beyond simple potential scattering. Within the models we have discussed, perhaps the simplest to consider is the case of compound nucleus formation wherein the neutron is actually absorbed into the target. But if absorption is to take place, it must be true that the probability will depend not only upon the projectile energy but also upon the absorptive properties of the nucleus for the particle waves themselves. In this regard, an analogy between the classical scattering and absorption of light can be made. Light waves incident upon a transparent medium will undergo both reflection (scattering) and refraction (absorption). The relative intensity of the light that is scattered or refracted depends upon the properties of the medium and the angle of incidence. We should therefore expect that the absorption of the projectile wave will also depend upon the geometrical relations between the wave and the surface of the target nucleus.

The presence of both potential and compound elastic scattering should both be seen in the cross sections. Considering only low-energy neutrons and an energy region far from any resonances, we expect that there cannot be any significant compound elastic scattering, and the constant cross section predicted by simple potential scattering ought to be observed. However, in regions near resonances, both types of scattering take place and the scattering cross section must reflect the contributions from both mechanisms.

Without considering some form for the nuclear potential it is difficult to proceed further. However, if we can be content with considering the simple spherical-well approximation and restrict attention to s-wave neutrons only, we can gain a great deal of insight into the forms of the scattering and reaction cross sections and show quantitatively how the Breit–Wigner expressions arise. This is the object of the next section.

13.10
S-Wave Scattering and Reactions in the Limit of the Spherical Potential Well Model

We consider the case of low-energy s-wave neutrons interacting with a spherical nucleus that is represented by a potential well of the form

$$V(r) = \begin{cases} -V_o, & r < R_n \\ 0, & r \geq R_n \end{cases} \quad (13.147)$$

where R_n is the radius of the target. The problem is, as usual, posed in the center of mass coordinate system. We treat the case where potential scattering, compound scattering and reactions can occur. If a compound nucleus is formed, there will, of necessity, be a wave function for the neutron within the nucleus and in the region outside where no potential exists. As it turns out, we will not have to solve the problem completely, but we will have to consider the continuity conditions between these two wave functions at the radial dimension $r = R_n$.

We suppose that within the nucleus, the wave function of a neutron that is absorbed is given by ψ_i. The wave function of the neutron at $r > R_n$ is taken as ψ_o.

13.10 S-Wave Scattering and Reactions in the Limit of the Spherical Potential Well Model

Because the potential is zero here and the asymptotic expressions we have introduced in the previous sections apply explicitly to regions where the nuclear interaction can be neglected, ψ_o must be describable by an asymptotic form. This is a very major simplification. From Eq. (13.130) the total wave function for an s-wave is simply

$$\psi_o = A\frac{1}{2ikr}[\widehat{\eta}_o e^{ikr} - e^{-ikr}] \tag{13.148}$$

It is common practice in spherically symmetric problems of this type to make the substitution $u = r\psi$ and, with this, the continuity conditions can be written as

$$\begin{aligned} u_o(R_n) &= u_i(R_n) \\ \left.\frac{du_o}{dr}\right|_{r=R_n} &= \left.\frac{du_i}{dr}\right|_{r=R_n} \end{aligned} \tag{13.149}$$

These two requirements can be combined into the single condition

$$\left.\frac{1}{u_o}\frac{du_o}{dr}\right|_{r=R_n} = \left.\frac{1}{u_i}\frac{du_i}{dr}\right|_{r=R_n} \tag{13.150}$$

If this is multiplied through by R_n, the equation is completely dimensionless. Further, we can define the quantity

$$f = \lim_{r \to R_n}\left(\frac{R_n}{u}\frac{du}{dr}\right) \tag{13.151}$$

and the combined continuity conditions in Eq. (13.150) can be written simply as

$$f_o = f_i \tag{13.152}$$

A physical picture of the meaning of f is worthwhile. Suppose we consider f_o. Regardless of the magnitude of u_o at $r = R_n$, $f_o = 0$ if the slope of the wave function at $r = R_n$ is zero. Similarly, $f_o \to \infty$ if the slope of the wave function becomes infinite at $r \to R_n$. We will see shortly that this picture has a great deal to do with whether the incident neutron can penetrate and form a compound nucleus.

Now comes quite a bit of algebra and mathematical details will be written out explicitly where significant complications can arise. We first want to obtain an expression for f_o using the definition in (13.151). We use the expression for ψ_o given in (13.148). Recognizing that $\widehat{\eta}_o$ is not a function of r, the differentiation is simple and the result is

$$f_o = ikR_n\left(\frac{\widehat{\eta}_o e^{ikR_n} + e^{-ikR_n}}{\widehat{\eta}_o e^{ikR_n} - e^{-ikR_n}}\right) \tag{13.153}$$

We have now related the scattering function $\widehat{\eta}_o$ to the continuity conditions at $r = R_n$. Eq. (13.153) can be "inverted" to give $\widehat{\eta}_o$ in terms of f_o, and the result is

$$\widehat{\eta}_o = e^{-2ikR_n}\left(\frac{f_o + ikR_n}{f_o - ikR_n}\right) \tag{13.154}$$

This is what we are after, because with it, we can write down the equations for the elastic scattering and reaction cross sections in terms of the continuity conditions.

We first evaluate $|\widehat{\eta}_o - 1|^2$. Direct substitution of (13.154) and combination of the two terms gives

$$|\widehat{\eta}_o - 1|^2 = \left|\frac{f_o(e^{-2ikR_n} - 1) + ikR_n(e^{-2ikR_n} + 1)}{f_o - ikR_n}\right|^2 \tag{13.155}$$

We now replace the two factors in parentheses by

$$(e^{-2ikR_n} - 1) = -2ie^{-ikR_n}\sin kR_n$$
$$(e^{-2ikR_n} + 1) = 2e^{-ikR_n}\cos kR_n \quad,$$

factor out the quantities $2ie^{-ikR_n}$ and square to obtain

$$|\widehat{\eta}_o - 1|^2 = 4\left|\frac{kR_n\cos kR_n - f_o\sin kR_n}{f_o - ikR_n}\right|^2 \tag{13.156}$$

To proceed further, we need to recognize that the quantity f_o is, in general, complex (see Eq. (13.153)). To ensure that we handle this properly, we make its real and imaginary parts explicit by writing

$$f_o = kR_n(g + ih) \tag{13.157}$$

where both g and h are purely real. Substitution of this into (13.156) and careful algebraic manipulation gives the result

$$|\widehat{\eta}_o - 1|^2 \tag{13.158}$$
$$= 4\left[\sin^2 kR_n + 2\sin kR_n\frac{(h-1)\sin kR_n - g\cos kR_n}{g^2 + (h-1)^2} + \frac{1}{g^2 + (h-1)^2}\right]$$

A similar calculation gives the result

$$(1 - |\widehat{\eta}_o|^2) = \frac{-4h}{g^2 + (h-1)^2} \tag{13.159}$$

13.10 S-Wave Scattering and Reactions in the Limit of the Spherical Potential Well Model

Finally, substitution of the expressions in Eqs (13.158) and (13.159) into Eqs. (13.141) for the single term corresponding to l = 0, gives the results

$$\sigma_{es}(0) \tag{13.160}$$

$$= 4\pi\lambda^2 \left[\sin^2 k R_n + 2 \sin k R_n \frac{(h-1)\sin k R_n - g\cos k R_n}{g^2 + (h-1)^2} + \frac{1}{g^2 + (h-1)^2} \right]$$

$$\sigma_{react}(0) = 4\pi\lambda^2 \left[\frac{-h}{g^2 + (h-1)^2} \right]$$

We now have general expressions for the elastic scattering and reaction cross sections in terms of parameters associated with the continuity conditions of the neutron wave function at the nuclear radius. These expressions are beginning to demonstrate a character that is very similar to the expression (Eq. (13.97)) which we obtained above for the Breit–Wigner single-level formula by simple physical arguments.

Consider first the second of Eqs (13.160) representing the total reaction cross section. It is seen to possess a resonant form. It will be maximized when $h - 1$ is minimized. Because the reaction cross section must be equal to or greater than zero, h must be less than zero. Further, if h = 0, the reaction cross section vanishes. From the definition in Eq. (13.157), it is seen that f_o must be complex if any reaction is to take place. If we expand the denominator in the second of Eqs (13.160), we can write

$$g^2 + (h-1)^2 = g^2 + h^2 + (1-2h) = \frac{|f_o|^2}{(kR_n)^2} + (1-2h) \tag{13.161}$$

This indicates that the maximum reaction cross section will be found when $|f_o|^2$ is minimized for any given value of h. But from the definition

$$f_o = \lim_{r \to R_n} \left(\frac{R_n du}{u \, dr} \right)$$

$|f_o|^2$ will be minimized when the slope of u_o tends toward zero. On the other hand, when $f_o \to \infty$, the cross section goes to zero and that implies that the neutron cannot penetrate into the nucleus at all. If this is the case, the only interaction that can take place is that of potential scattering. Examination of the first of Eqs (13.160) shows that this is precisely what happens. If $f_o \to \infty$, the denominators in the second and third terms become infinite and the cross section goes to $\sigma_{es}(0) = 4\pi\lambda^2 \sin^2 k R_n$, exactly the result from (13.126) if $\delta_o = kR_n$. Although we have not shown it, you can show that this result is correct to an excellent approximation.

The cross section for elastic scattering has three terms. We have identified the first as representing potential scattering and it should be clear from its form that the third term must represent compound elastic scattering. It has the same resonant form as seen in the expression for the total reaction cross section. The second term has characteristics of both of the first and third terms and represents the *interference* between potential and compound elastic scattering. This interference is characteristic of wave phenomena and, as we shall see, it represents a very distinctive feature of resonances for which the reactions are dominated by elastic scattering of the $l = 0$ type.

What now remains is development of the relation between the physical parameters of a resonance and the continuity parameters g and h. We can get the desired result by use of an expansion of $f_o(E)$ about the energy corresponding to the resonance peak, E_R. Using a Taylor series, we can write

$$f_o(E) = f_o(E_R) + \left.\frac{df_o}{dE}\right|_{E=E_o} (E - E_R) + \left.\frac{d^2 f_o}{dE^2}\right|_{E=E_o} \frac{(E-E_R)^2}{2} + \ldots \quad (13.162)$$

Near the resonance center, the change in f_o with variation in E is expected to be relatively slow. Making this assumption, we can use just the first two terms in the expansion. At the resonance center we have $f_o(E_R) = 0$ and therefore only the second term remains. We know that the energy must be complex if a reaction is to be possible and, in analogy with our discussion of the wave functions of time-dependent states in Section 13.6, we write the energy as $E_R = E_o - \frac{i\Gamma_o}{2}$, where E_o is the real part of the energy at the centroid of the resonance and Γ_o must be some part of the total width of the resonance. We then have the approximation

$$f_o(E) = kR_n(g + ih) = \left.\frac{df_o}{dE}\right|_{E=E_o} \left(E - E_o + \frac{i\Gamma_o}{2}\right) \quad (13.163)$$

Letting $\alpha = \left.\frac{df_o}{dE}\right|_{E=E_o}$, we can write

$$\begin{aligned} g &= \frac{\alpha}{kR_n}(E - E_o) \\ h &= \frac{\alpha}{kR_n}\frac{\Gamma_o}{2} \end{aligned} \quad (13.164)$$

and direct substitution into the second of Eqs (13.160) gives

$$\sigma_{react}(0) = 4\pi\lambda^2 \left[\frac{-\left(\frac{\alpha}{kR_n}\frac{\Gamma_o}{2}\right)}{\left(\frac{\alpha}{kR_n}\right)^2 (E-E_o)^2 + \left(\frac{\alpha}{kR_n}\frac{\Gamma_o}{2} - 1\right)^2}\right] \quad (13.165)$$

$$= 4\pi\lambda^2 \frac{-(kR_n/\alpha)(\Gamma_o/2)}{(E-E_o)^2 + (\Gamma_o/2 - kR_n/\alpha)^2}$$

13.10 S-Wave Scattering and Reactions in the Limit of the Spherical Potential Well Model

Similarly, substitution into the resonance term of the expression for the elastic scattering cross section gives

$$\sigma_{es,res}(0) = 4\pi\lambda^2 \frac{(kR_n/\alpha)^2}{(E-E_o)^2 + (\Gamma_o/2 - kR_n/\alpha)^2} \tag{13.166}$$

What now remains is the relation between the parameters in the preceding equations and the observable width parameters of the resonance. To do this, we first examine the denominators of (13.165) and (13.166). They are identical resonance forms and, from (13.92), must represent the probability of finding the compound nucleus in energy. We then conclude that

$$\frac{\Gamma_{tot}}{2} = \frac{\Gamma_o}{2} - \frac{kR_n}{\alpha} = \frac{\Gamma_n}{2} + \frac{\Gamma_{reac}}{2} \tag{13.167}$$

Elastic scattering is proportional to the probability of forming the compound nucleus by neutron absorption times the probability for the compound system to decay by neutron emission. Thus we can conclude from the numerator of (13.166) that

$$\frac{\Gamma_n}{2} = -\frac{kR_n}{\alpha} \tag{13.168}$$

and Γ_o must represent the total reaction width. The resonance term of the elastic scattering cross section and the total reaction cross section, including the statistical factor g (Eq. (13.99)), are then

$$\sigma_{es,res}(0) = 4\pi g\lambda^2 \frac{(\Gamma_n/2)^2}{(E-E_o)^2 + (\Gamma_{tot}/2)^2}$$

$$\sigma_{reac,res}(0) = 4\pi g\lambda^2 \frac{(\Gamma_n/2)(\Gamma_{reac}/2)}{(E-E_o)^2 + (\Gamma_{tot}/2)^2} \tag{13.169}$$

where $\Gamma_{tot} = \Gamma_n + \Gamma_{reac}$. In the absence of potential scattering, the sum of these two cross sections would represent the total cross section of an isolated resonance with s-wave neutrons. The total elastic scattering cross section is, however, given by the first of Eqs (13.160). Using the relations given in (13.164), the total elastic scattering cross section is given by

$$\sigma_{es}(0) = 4\pi\lambda^2 \left[\sin^2 kR_n + 2\sin kR_n \frac{\left(-\frac{\Gamma_n\Gamma_{tot}}{4}\right)\sin kR_n + \left(\frac{\Gamma_n}{2}\right)(E-E_o)\cos kR_n}{(E-E_o)^2 + (\Gamma_{tot}/2)^2} \right.$$

$$\left. + \frac{(\Gamma_n/2)^2}{(E-E_o)^2 + (\Gamma_{tot}/2)^2} \right] \tag{13.170}$$

The expressions we have derived for the Breit–Wigner single-level cross sections are correct in the case of pure s-wave scattering. The results are not really dependent upon our choice of potential, but are dependent upon the assumption that the interaction occurs with l = 0. If l ≠ 0, there is a slight difference between the correct result and what we have found here but this difference is not terribly important from a practical viewpoint. The essential physics has been presented. What now remains is to understand how well our results agree with experiment and that is the subject of the next section.

13.11
The Breit–Wigner Single-Level Formula and Experimental Cross Sections

The cross section relations derived in the previous section should represent the behavior of cross sections in the vicinity of isolated resonances. In order to determine just how well the theory represents experimental fact, it is perhaps most reasonable to begin by examination of elastic scattering as it should provide a very stringent test, not only of the general resonance form, but also of the interference between potential and resonant scattering.

In order to understand the relation of the various terms in Eq. (13.170) to the behavior of the scattering cross section, we will assume that we have a nucleus of mass number A = 100 for which a single dominant resonance exists with a neutron width of $\Gamma_n = 0.1\,\text{keV}$ that is equal to the total width Γ_{tot}, i.e., only elastic scattering occurs. We also assume that the centroid is located at $E_o = 1.0\,\text{keV}$. In Fig. 13.32 are plots of the total elastic scattering cross section, along with the contributions from the various terms of which it is composed. The total cross section is seen to display a marked signature from the interference between the potential and resonance scattering. The effect is to decrease the cross section below the resonance centroid and increase it just after the resonance. The potential scattering term is essentially constant. This can be understood by considering the magnitude of kR_n. In this case,

$$kR_n = \frac{r_o A^{1/3}\sqrt{2m_n c^2 E_n}}{\hbar c} \leq 0.127$$

for all energies less than 10 keV. Hence,

$$4\pi\lambda^2\sin^2 kR_n \approx 4\pi\lambda^2(kR_n)^2 = 4\pi R_n^2 \tag{13.171}$$

This result is quite general. At low energies the potential scattering cross section is equal to the surface area of the target nucleus in the hard-sphere approximation.

The interference term is negative below the resonance and rises sharply just at the resonance center. The resonance term rises slowly at low energies but decreases rapidly above the resonance. Because the resonances of general importance occur at low energies, the majority of those observed will correspond to s-wave resonanc-

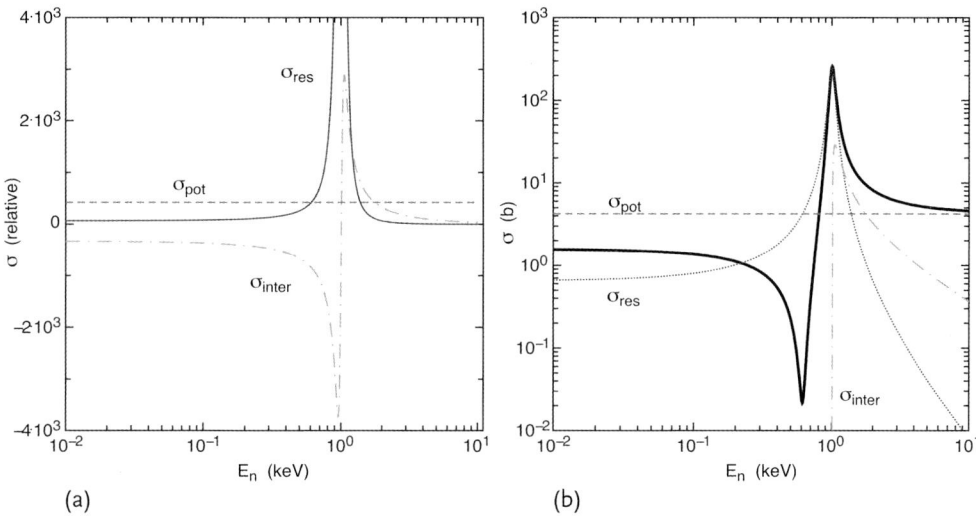

Fig. 13.32 Hypothetical elastic scattering cross section at an isolated resonance in a nucleus of A = 100 with parameters $E_0 = 1.0$ keV, $\Gamma_n = \Gamma_{tot} = 0.1$ keV. The curves labeled σ_{pot}, σ_{inter} and σ_{res} represent the contributions from the first, second and third terms in (13.170), respectively. (a) Relative linear scale. (b) log-log plot to show the total scattering cross section (black) over the full energy range.

es. Hence, so long as elastic scattering represents a large portion of the total cross section, we expect to observe the signature shape of that shown in Fig. 13.32.

To demonstrate the characteristics of neutron reaction cross sections, we turn to experimental data for a few nuclides that represent typical n + nucleus reactions throughout the chart of the nuclides. The first is the cross section for the reaction $^{6}_{3}\text{Li}(n, ^{3}_{1}\text{H})\alpha$ (Q = 4.784 MeV) shown in Fig. 13.33. Over the energy range 0.01 eV to 1 MeV, the cross section is dominated by a single resonance centered near 0.25 MeV and its low-energy tail that accounts for almost all of the $1/v$ cross section seen at low energies. The Q value is so large that Coulomb effects on barrier penetration are negligible. Further, the reaction width is so large that the reaction cross section dominates the total cross section, much as in the case of the $^{3}_{2}\text{He}(n, p)^{3}_{1}\text{H}$ reaction cross section shown in Fig. 13.21.

As a second example we show the total cross section for neutrons on $^{207}_{82}\text{Pb}$ in the range 10–70 KeV (Fig. 13.34). The principal features in this region are a number of very sharp, narrow resonances and the resonance centered at 41.3 keV is a beautiful example of an s-wave resonance that is dominated by elastic scattering. In this case $\Gamma_n/\Gamma_\gamma \approx 366$. The other resonances shown correspond to s- and p-wave resonances that have much smaller overall widths.

As a third example we show the (n,γ) cross section for $^{135}_{54}\text{Xe}$ (Fig. 13.35). This reaction is both a curiosity and an extremely important technological issue. $^{135}_{54}\text{Xe}$ is a fission product with a half-life of 9.104 hr. If you look carefully at the abscissa,

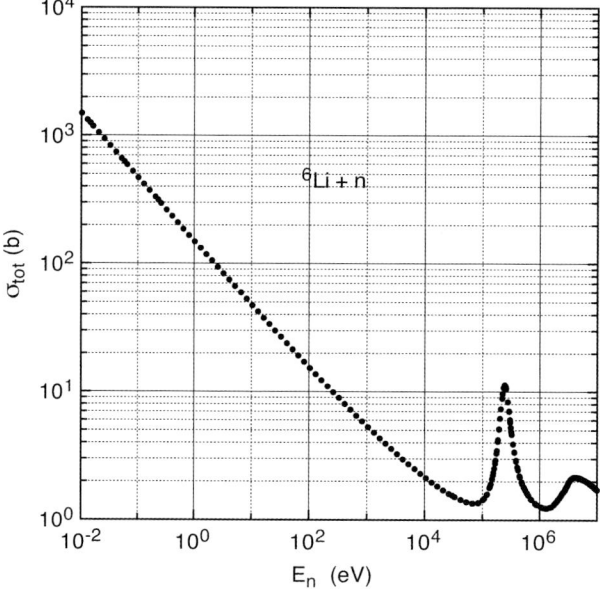

Fig. 13.33 The cross section for the reaction ${}^{6}_{3}\text{Li}(n, {}^{3}_{1}\text{H})\alpha$.

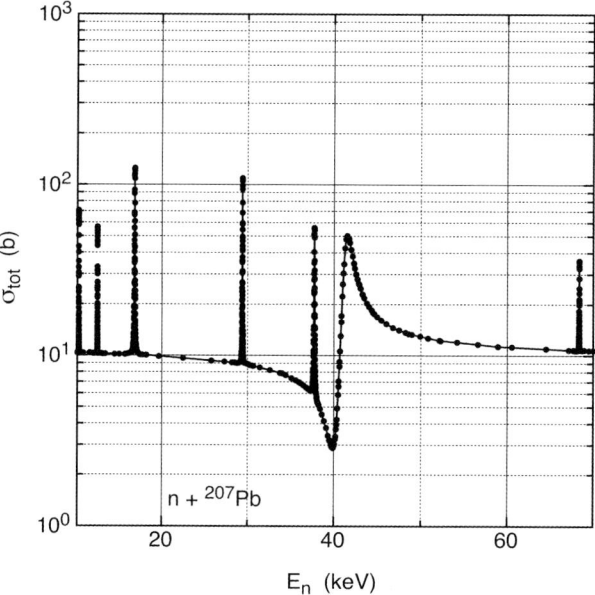

Fig. 13.34 The total cross section for the reaction $n + {}^{207}_{82}\text{Pb}$.

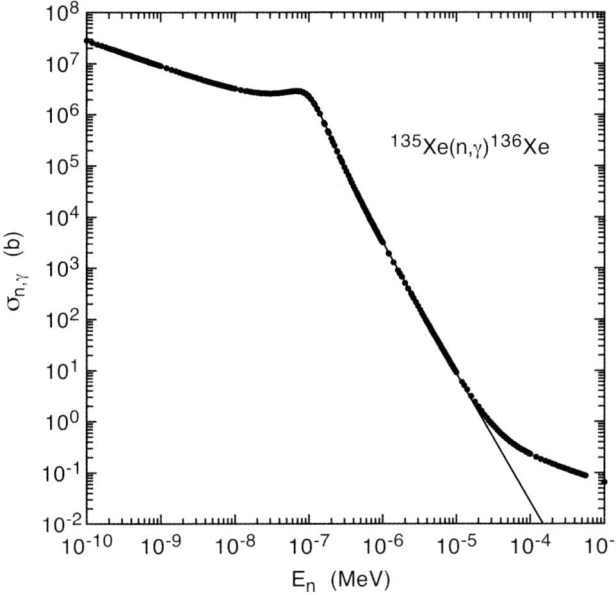

Fig. 13.35 The cross section for the reaction $^{135}_{54}\text{Xe}(n,\gamma)^{136}_{54}\text{Xe}$. The solid line shown in the figure represents a fit with the Breit–Wigner single-level formula.

you find that the cross section at thermal energies is on the order of 3×10^6 b! Absolutely enormous. Although it does not look like the nice simple resonances displayed in the previous figures, the cross section is dominated by a single resonance located at about 8.13×10^{-2} eV and has a ratio of $\Gamma_n/\Gamma_\gamma \approx 0.20$. Again, ones sees a 1/v energy variation at lower energies. The technological significance of this reaction arises because of the enormous cross section and the fact that $^{135}_{54}\text{Xe}$ has a large cumulative yield in the fission of $^{235}_{92}\text{U}$. In fact, the yield and cross section are so large that a few percent of all neutrons produced in fission are captured in $^{135}_{54}\text{Xe}$ during the operation of nuclear power reactors and are therefore unavailable to the fission process. $^{135}_{54}\text{Xe}$ is known as a "reactor poison" and is so significant that calculations of reactor dynamics and power levels must take account of its presence.

Having examined a number of partial reactions to demonstrate how our simple models pertain to reality, it is worthwhile to examine some neutron cross sections in a more global sense. In Figs 13.36–13.40 are reaction cross sections in the range 10^{-3} eV to over 20 MeV for selected nuclides representing examples throughout the chart of the nuclides.

The total cross section for the interaction of neutrons with ^{40}Ca is shown in Fig. 13.36. The total cross section for thermal neutrons ($v_n = 2200$ m s^{-1}) is due to elastic scattering (3.01 ± 0.08 b) and capture (0.41 ± 0.02 b). Because of the small capture cross section, the 1/v dependence at low energies is relatively weak. No resonances are seen below a neutron energy of 1 keV and the first prominent ones

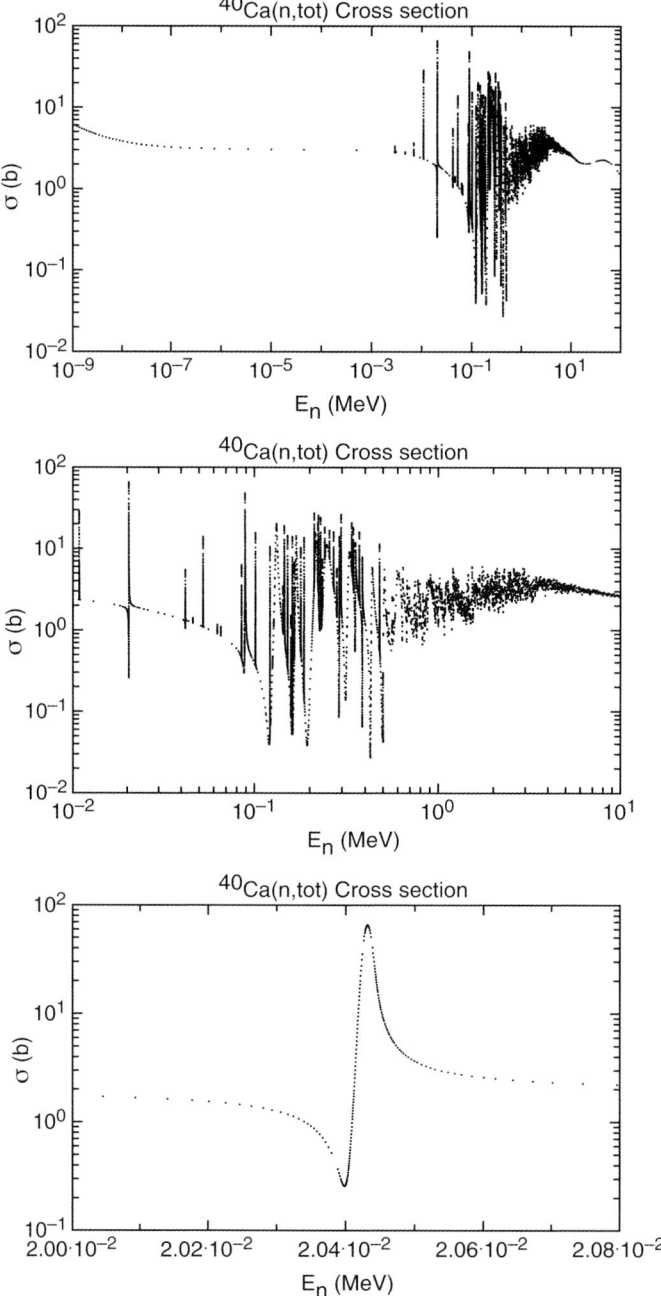

Fig. 13.36 The total cross section for neutrons interacting with ^{40}Ca.

found in the energy range 10–100 keV. The average energy separation between these is about 15 keV. It should be remembered that the resonances represent levels in the compound nucleus ^{41}Ca just above the binding energy of the last neutron, which is about 8.36 MeV. Resonances that are predominantly capture and predominantly scattering are seen; the latter by the characteristic interference pattern between resonant and potential scattering. The details of the resonance near 20.4 keV, shown in expanded form in the last panel of the figure, illustrates the interference effects quite beautifully.

Fig. 13.37 shows the total cross section for the interaction of neutrons with ^{138}Ba. The thermal capture cross section is only 0.360 ± 0.036 b, quite small compared to that for potential scattering. The first sizable resonance is seen below 1 keV and the spacing between the lowest-energy resonances is on the order of 3.5 keV, a factor of 4–5 smaller than seen in Fig. 13.36. The binding energy of the last neutron in the compound nucleus is only about 4.72 MeV and thus we are observing levels at a significantly lower excitation energy than in the ^{40}Ca + n

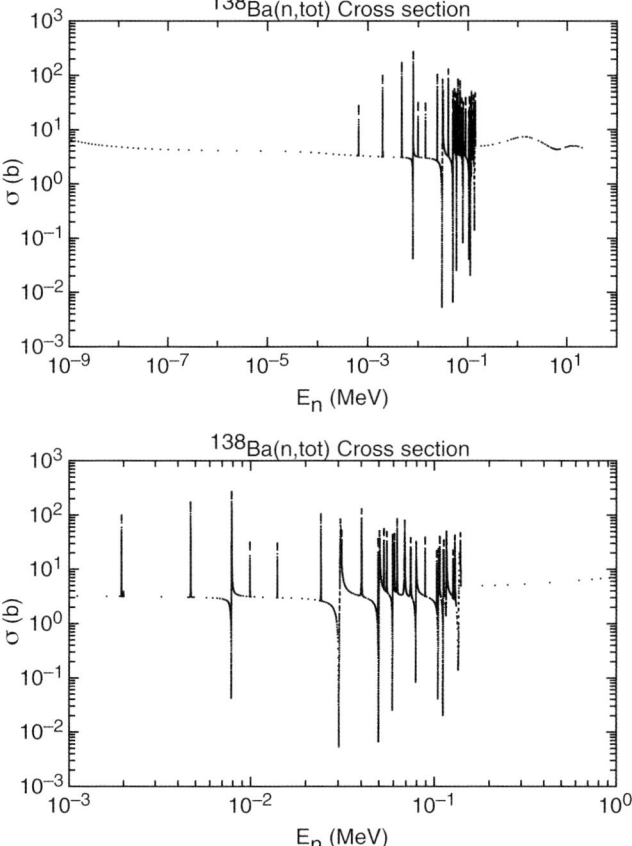

Fig. 13.37 The total cross section for neutrons interacting with ^{138}Ba.

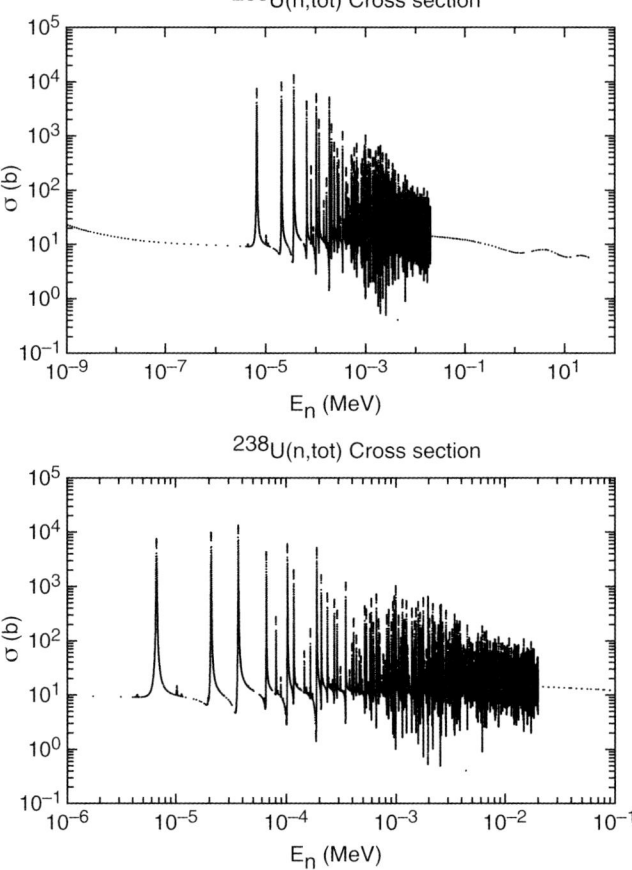

Fig. 13.38 The total cross section for neutrons interacting with ^{238}U.

reaction. Clearly the level density must be increasing with increasing mass number. Above about 200 keV, the spacing between resonances becomes very small and the cross section tends to vary smoothly with energy.

The trend of increasing resonance density with mass number is seen in the total cross section for neutrons interacting with ^{238}U as shown in Fig. 13.38. The thermal scattering cross section is 9.38 ± 0.09 b and the capture cross section is 2.680 ± 0.019 b. The lowest-energy resonance is now found near a neutron energy of 7 eV and is predominantly due to capture. But the majority of the other resonances are dominated by elastic scattering as seen by the interference patterns. The density of the resonances is very high and the separation between the lowest-energy resonances is now only about 12–13 eV. Above about 10 keV, individual resonances become hardly distinguishable.

The general picture emerging is that the density of excited states just above the neutron binding energy is increasing dramatically with increasing mass number.

Even though the majority of the isolated resonances in neutron reactions are s-wave, there is evidence from a number of sources that the increase in level density occurs for all wave types.

We have purposely chosen (even, even) nuclei as our examples so that the issue of pairing energy need not be considered. If we examined the density of resonances for the interaction of neutrons on odd-N nuclides, we can expect to see even higher resonance densities because of the higher excitation energies of the compound nuclei as a result of pairing. As an example, the cross section for the interaction of neutrons with ^{235}U is shown in Fig. 13.39. The thermal cross sections for the elastic scattering, capture and fission reactions are 14.3 ± 0.5, 98.3 ± 0.8 and 582.6 ± 1.1 b, respectively. At the lowest energies, an intense 1/v variation in the cross

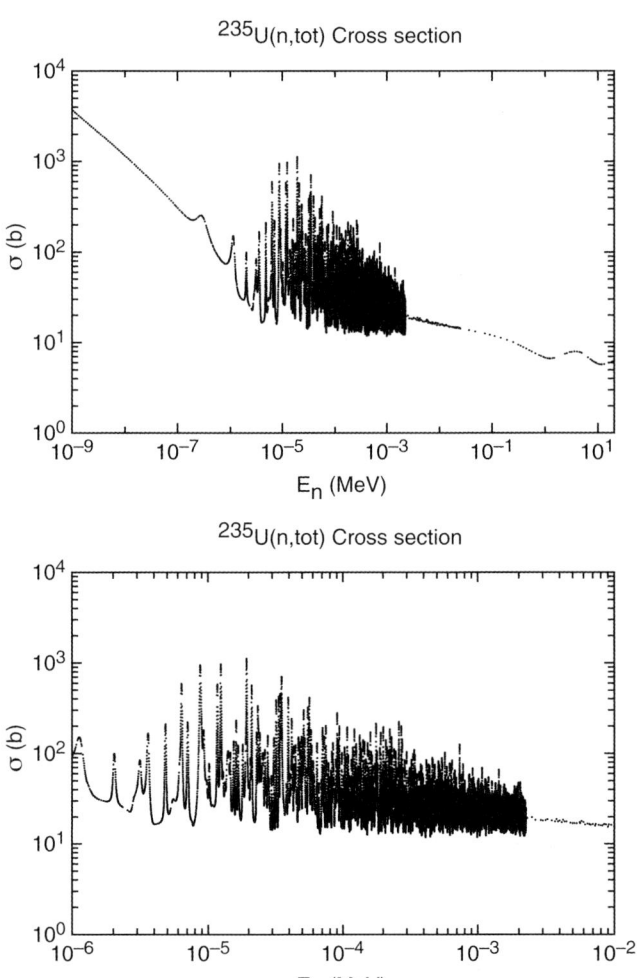

Fig. 13.39 The total cross section for neutrons interacting with ^{235}U.

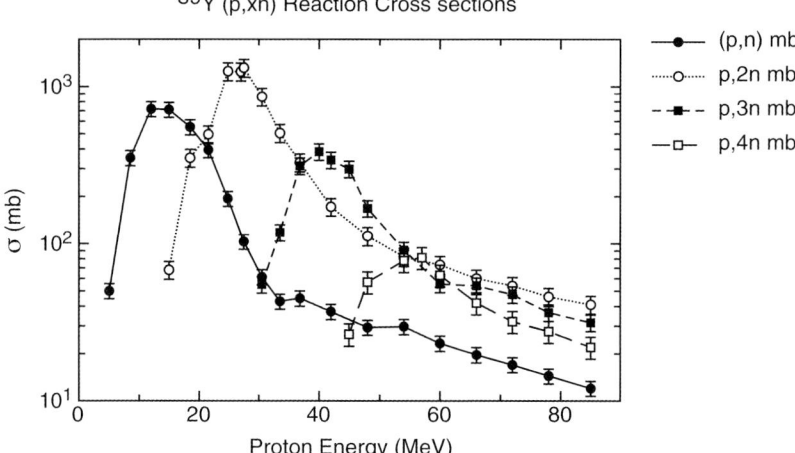

Fig. 13.40 Cross sections for (p, xn) reactions on ^{89}Y.

section is seen but it is somewhat distorted because of the strong resonances seen at energies as low as 0.3 eV. The resonances are so numerous that many overlap with one another even at the lowest energies. Nevertheless, the dominance of the capture and fission cross sections is reflected in the fact that the interference between potential and compound elastic scattering is not strongly evident anywhere. The mean spacing between the lowest energy resonances is only about 1 eV, indicating a level density more than an order of magnitude greater than that seen in the cross section for neutrons interacting with ^{238}U. At an energy of only a few keV, one can no longer distinguish individual resonances.

The characteristics of resonances in neutron reactions are very important to nuclear technology and, in some cases, to the production of radionuclides for application purposes. Resonances in the reaction of neutrons with ^{238}U have strong implications with respect to the design and operation of nuclear power reactors and to the production of plutonium isotopes. The strong capture resonances in ^{238}U not only are the principal source of ^{239}Pu production through decay of ^{239}Np, but they also give rise to a very sizeable radial dependence in the production of plutonium within a reactor fuel element.

Isolated resonances are certainly of importance in some charged-particle reactions as well, but these are restricted to reactions on targets of low atomic number. Except for these, the Coulomb barrier that must be overcome for compound nucleus formation is so large that one no longer probes levels near the particle binding energy.

As a specific example, the cross sections for the (p,n), (p,2n), (p,3n) and (p,4n) reactions on ^{89}Y are shown in Fig. 13.40 for incident proton energies up to about 82 MeV [2]. The cross section for the (p,n) reaction rises very steeply from below about 5 MeV and reaches a maximum near 15 MeV. The Q value for the reaction $^{89}_{39}Y + p = {}^{90}_{40}Zr^*$ is about 8.87 MeV and the binding energy of the last neutron in

^{90}Zr is about 12.0 MeV. Thus, to observe neutron emission from individual resonances in the compound nucleus, the kinetic energy of the incident proton would have to be about 3 MeV. Now the Coulomb barrier for protons on ^{89}Y is about 8.5 MeV, and the very small probability for transmission of protons through the Coulomb barrier with energies near 3 MeV means that cross sections at such energies will be very small indeed. Under normal conditions such small cross sections simply go undetected.

The data in Fig. 13.40 point out one additional important characteristic that arises when the compound nucleus has high excitation energies, namely, the sequential emission of particles from the daughter products of reactions. In the present case, note that the (p,2n) reaction cross section is observed to rise sharply and reach a maximum near 23 MeV, about 10 MeV above the threshold value of about 13 MeV. Similar characteristics are seen with the (p,3n) reaction, with a threshold value of about 25 MeV, and with the (p,4n) reaction with a threshold of about 35 MeV. Now all experimental evidence shows that the neutrons are not emitted simultaneously but arise sequentially from successive and different compound nuclei. So, for example, the data indicate that at an incident proton energy of about 35 MeV, the compound nucleus formed by proton absorption, ^{90}Zr, can emit a neutron leaving the daughter ^{89}Zr with sufficient excitation energy that it too can emit a neutron to yield ^{89}Zr that also has sufficient excitation energy to emit a neutron.

Such sequential reactions are quite the norm when sufficient excitation energy is available. As a gross approximation, the average kinetic energy of a neutron that is "evaporated" from the compound nucleus is about 2–3 MeV. Thus, with the addition of about 10–11 MeV of excitation energy above that required for a single-neutron emission, a second emission is likely from the daughter nucleus formed. Examination of the data in the figure shows that this is just about the energy differences between the peaks in the cross sections for the different reactions.

Finally, it is evident that the total cross section for the reactions shown in Fig. 13.40, commonly called the (p,xn) reaction cross section, decreases significantly with increasing proton energy. In fact, the thresholds for some 44 reactions lie below 25 MeV and there are some 300 reactions that have thresholds below 50 MeV. Hence the (p,xn) cross section represents only a fraction of the total cross section for the p + ^{89}Y reaction and the decrease seen with increasing excitation energy in the compound nucleus simply reflects the increased competition from additional decay modes.

13.12
About Fission Cross Sections

Neutron-induced fission cross sections are so important technologically that we will end this chapter with a brief discussion of their general properties using the reactions $^{235}_{92}$U(n, f) and $^{238}_{92}$U(n, f) as our examples. In Fig. 13.41 are the evaluated cross section data over the energy range $10^{-5} - 2 \times 10^7$ eV. One immediately sees that at energies less than about 1 MeV, the fission cross section for $^{235}_{92}$U(n, f) is

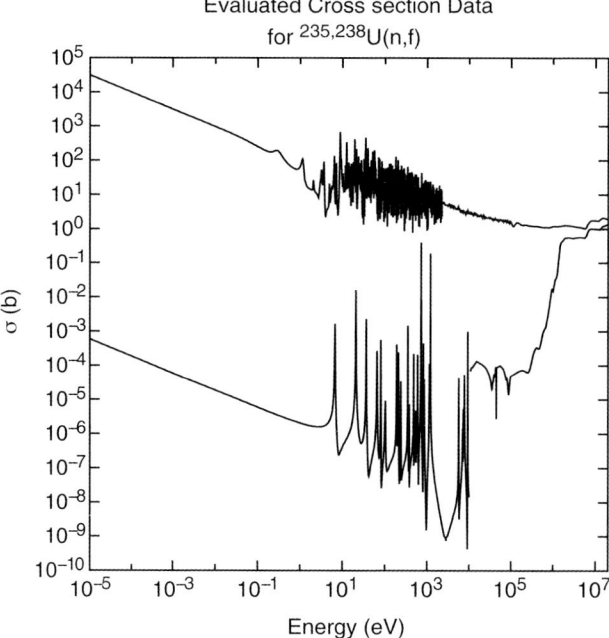

Fig. 13.41 Evaluated experimental fission cross sections for the reactions $^{235}_{92}U(n, f)$ and $^{238}_{92}U(n, f)$.

everywhere 4–7 orders of magnitude larger than for the $^{238}_{92}U(n, f)$ reaction. This is typical of the fission cross sections for adjacent (even, even) and (even, odd) nuclides. You will recall from our discussion of fission barriers and excitation energies in Chapter XII that capture of a thermal neutron in ^{235}U produces an excitation energy of about 6.545 MeV (Table 12.2) that is larger than the fission barrier of about 5.9 MeV. As a result, we expect and indeed find, that fission takes place with thermal neutrons and because of the many resonances with large fission widths, we see strong $1/v_n$ variation in the cross section followed by a region of very dense resonances. The majority of these are the same as those shown with much higher resolution in Fig. 13.39. The resonances get so dense above about 1 keV that, within the experimental energy resolution, the cross section tends to become essentially continuous. At higher energies, the cross section decreases continuously until an energy of about 1 MeV where it tends to increase.

The fission cross section for the reaction $^{238}_{92}U(n, f)$ is negligibly small below an energy of about 1 MeV, and this reflects the fact that the excitation energy produced by capture of a thermal neutron, 4.806 MeV, is smaller than the fission barrier by just about 1 MeV. Below about 1 MeV, the cross section is dominated by elastic scattering and neutron capture. Once sufficient kinetic energy is brought in by the captured neutron, the cross section rises very abruptly as the fission reaction can now occur with high probability.

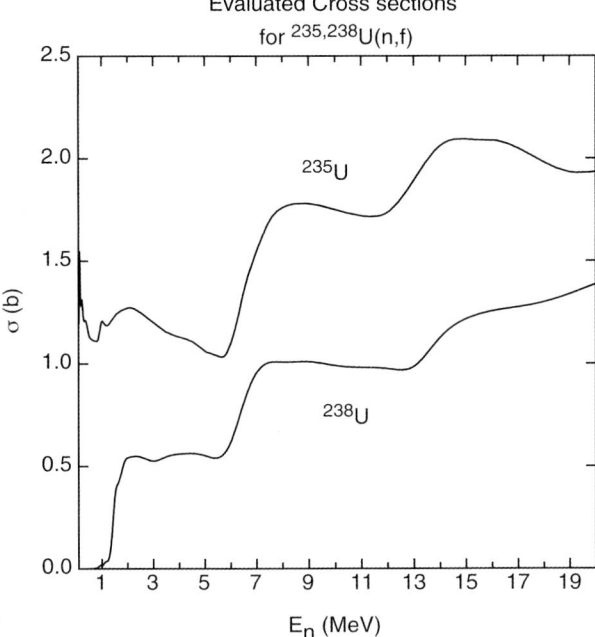

Fig. 13.42 Evaluated experimental fission cross sections for the reactions $^{235}_{92}U(n, f)$ and $^{238}_{92}U(n, f)$ at high energies.

The fission cross sections above about 1 MeV for both reactions are shown in expanded form in Fig. 13.42. In both reactions there is an increase in the fission cross section above about 5 MeV. Indeed, the behavior of the fission cross sections above about 1 MeV is remarkably similar. Because of the very large level densities at such energies, this increase cannot be due to individual resonances but must be due to the onset of some additional effect. In fact, the increase in the cross section is due to what is referred to as *second-chance* fission. This results from the same general effect as that discussed above for the $^{89}_{39}Y(p, xn)$ reaction.

At any kinetic energy of the captured neutron, the compound nuclei $^{236}U^*$ and $^{239}U^*$ can decay by neutron emission, neutron capture or fission. But because photon emission is such a slow process, almost all decay will occur by neutron emission or fission so long as the excitation energy is above the fission barrier. If neutron emission takes place and the neutron daughter has an excitation energy below the fission barrier, γ-ray emission will take place. But if the excitation energy is still above the fission barrier, neutron emission and fission can take place. Thus there is a second chance for fission to occur even if the initial compound nucleus decayed by neutron emission. The fission cross section must increase. This is exactly what is reflected in the step increases in the fission cross sections shown above about 6 MeV in Fig. 13.42.

References

1. D. Bodansky, S.F. Eccles, G.W. Farwell, M.E. Rickey, and P.C. Robison, Time Reversal Invariance and the Inverse Reactions $C^{12} + \alpha \Leftrightarrow N^{14} + d^*$, Phys. Rev. Letters 2 (1959) 101.
2. G.B. Saha, N.T. Porile and L. Yaffe, "(p,xn) and (p,pxn) Reactions of Yttrium-89 with 5-85-MeV Protons", Phys. Rev. 144 (1966) 962.

General References

More extensive discussions of nuclear reactions at a somewhat higher level than given here can be found in P. Marmier and E. Sheldon, "Physics of Nuclei and Particles", Academic Press, volume I, New York (1969), and in K.S. Krane, "Introductory Nuclear Physics", John Wiley and Sons, New York (1987). Practical applications of neutron reactions can be found in essentially all introductory nuclear engineering texts.

Problems

1. A very thin foil of pure ^{63}Cu is bombarded with a beam of 11 MeV protons and 38.1 min ^{63}Zn is produced by the (p,n) reaction. The bombardment takes place for 1.00 hr and the beam intensity is 1μA. The cross section for this reaction at 11 MeV is 415mb.
(a) What is the Q value for the (p,n) reaction?
(b) Assuming that the reaction takes place by compound nucleus formation, calculate the excitation energy of the compound nucleus.
(c) Calculate the kinetic energy of the compound nucleus. You may assume that the nuclear masses involved are directly proportional to their mass numbers.
(d) Estimate the Coulomb barrier for this reaction, assuming the equivalent hard-sphere radius of the proton is 1.05 fm.
(e) Suppose that the energy available in the center of mass coordinates was |Q| + 1.0 MeV. Discuss qualitatively how large the cross section for the reaction would be in comparison to the cross section if the excitation energy of the compound nucleus were 11.0 MeV.
(f) Neglecting the energy loss of the protons in the target, calculate the activity of ^{63}Zn (Ci) at the end of irradiation, assuming that the target thickness was 10μm. Assume the mass density of ^{63}Cu to be 8.843 g cm^{-3}.

2. The reaction ^{239}Pu + n displays a resonance at a neutron energy of 10.93 eV with the widths Γ_n = 0.00187 eV, Γ_γ = 0.040 eV and Γ_f = 0.144 eV. The total angular momentum quantum number of the resonance is J = 1.
(a) Calculate the *half-life* of the compound nucleus represented by the resonance.
(b) Calculate the cross section ratios $\sigma_{n,f}/\sigma_{n,\gamma}/\sigma_{n,n}$ at the resonance centroid.

(c) Calculate the fission cross section of ^{239}Pu at a neutron energy of 0.025 eV if fission at this energy were due *solely* to this resonance.

3. Endoergic reactions have the property that at and just above the threshold for a reaction, particle emission is restricted to a small range in angles. Consider the reaction $^{7}_{3}$Li(p, n)$^{7}_{4}$Be that has $Q_{p,n} = -1.6443$ MeV.
(a) Calculate the threshold energy for the reaction.
(b) For a proton energy of $E_{th} + 0.01$ MeV in the laboratory, calculate the maximum angle with respect to the incident beam direction at which neutrons will be observed.
(c) Calculate the energies of the emitted neutrons at $0°$, the maximum angle found in part b., and nine evenly spaced angles between these limits.

4. The first excited state of ^{7}Li lies at an excitation energy of 429.1 keV above the ground state.
(a) What is the threshold energy for exciting this state in the p + ^{7}Li reaction?
(b) If 4.0 MeV protons undergo inelastic scattering and leave ^{7}Li in the first excited state, what is the energy of the protons that will appear at an angle of $20°$ with respect to the beam direction?

5. A target containing deuterium is bombarded with a beam of photons in order to produce neutrons by the (γ,n) reaction. Use a first-order approximation to estimate the minimum difference between the photon energy and the deuteron binding energy that will permit the reaction to occur.

6. If the Q value for the reaction ^{3}He(n,p)^{3}H is 0.764 MeV, what is the threshold for the reaction ^{3}H(p,n)^{3}He?

7. A common method of making a portable neutron source is to mix ^{241}Am with natural Be (100% ^{9}Be) on the atomic scale. Neutrons are then produced by the (α,n) reaction on Be. Assuming that the maximum particle energy is 5544.5 keV, calculate the maximum energy of the emitted neutrons.

8. In Chapter IX, in our discussion of a simple barrier penetration model, we discussed the fact that the angular momentum effect on α emission was relatively small in comparison to that of the Coulomb barrier. But this is not the case when neutrons are incident upon a nucleus. As an example, consider the case of p-wave neutrons of energy 0.025eV incident on a nucleus of ^{208}Pb. Use the simple barrier penetration model from Chapter IX with the centrifugal barrier in the potential to obtain a crude estimate of the transmission coefficient. For this estimate, neglect the small recoil energy of ^{208}Pb and take the range over which integration must be performed to be the dimension between the hard-sphere radius of ^{208}Pb and the dimension at which the kinetic energy of the neutron is equal to the centrifugal potential. Use the hard-sphere radius of the neutron. What do you conclude?

A useful standard integral for this problem is

$$\int \frac{(a^2 - x^2)^{1/2}}{x} dx = (a^2 - x^2)^{1/2} - a\ln\left(\frac{a + (a^2 - x^2)^{1/2}}{x}\right)$$

9. Consider the first of the two equations given in Eq. (13.160). Rewrite this equation such that it does not contain the parameters g and h but does contain parameters that can be measured experimentally.

10. Show that the Breit–Wigner single-level resonance formula for neutron reactions can be written in the form

$$\sigma(E) = \sigma_0 \left(\frac{E_0}{E}\right)^{1/2} \left(\frac{\Gamma^2}{4(E - E_0)^2 + \Gamma^2}\right)$$

where σ_0 is the cross section at the resonance center, E_0, and Γ is the total width of the resonance.

11. Use the expression given in Eq. (13.130) with Eq. (13.103) to derive Eq. (13.131) explicitly.

12. From the second panel of Fig. 13.38, it is seem that the total cross section for the reaction ^{238}U + n far from resonance centroids is approximately 10 b. Assuming that this represents only elastic scattering, use this to estimate the radius parameter r_0.

14
The Interaction of Ionizing Radiation with Matter

14.1
Introduction

No text on low-energy nuclear physics is complete without at least an introduction to the interaction of radiation with matter. The means by which we observe radioactive decay and nuclear reactions is via the interaction of the radiations involved with material objects designed to distinguish one radiation type from another, to determine the energy of the radiation and, in some cases, the direction in space of the trajectory of the radiation. The interaction of radiation with matter causes disruption of the structure of the medium that can be both beneficial and deleterious. Radiation is frequently used to change the electrical properties of the surface or bulk of solid materials, especially semiconductors. The irradiation of the fuel in a nuclear reactor by the fission fragments produced in fission, by the α, β and γ-rays emitted by the radionuclides produced by fission and neutron capture and, of course, by neutrons themselves, cause a significant and continuous disruption in the structure of the fuel that has a marked effect on its properties. The interaction of radiation with living matter produces a wide range of effects that can affect both the short- and long-term health of the cells and tissues involved.

We use the term ionizing radiation to indicate the range of energies that must be considered. The lowest electron binding energies in atoms and molecules are typically on the order of 5–10 eV, and thus represent the lower limit of the energy of an incident radiation that must be considered. In practice, we usually consider the lower limit for charged particles and photons to be on the order of 1–10 keV, while for neutrons we need to consider energies as low as thermal (≥ 0.01 eV) in order to account for the ionizing radiation that is produced subsequent to the interactions of neutrons with matter.

Because of the involvement of electrons bound in atoms and molecules, the interaction of radiation with matter is extremely complex. It is fair to say that, after more than a century of use and study, a complete picture of the details of all of the interaction processes is still lacking. We clearly understand the principal features of the primary interactions and some of the features of the subsequent interactions, but certainly not all. As important as it is, our knowledge of the effects of ionizing

Nuclear Physics for Applications. Stanley G. Prussin
Copyright © 2007 WILEY-VCH Verlag GmbH & Co., Weinheim
ISBN: 978-3-527-40700-2

radiation on living matter is still rudimentary in many respects. This problem is complicated both by the complexity of the interactions themselves and the complicated response of living system to the effects produced by the interactions.

The purpose of the present chapter is to provide an introduction to the fundamentals of the interaction of radiation with matter. It is far from exhaustive and, in keeping with the tenor of the preceding chapters, we will use the simplest possible models to understand the main features of the processes involved. We will begin with an examination of the principal mechanisms by which photons interact with matter. The classical model of the elastic scattering of photons on electrons will be examined first, both because of the importance of the concepts involved and because the cross section for the interaction provides a natural measure for the magnitudes of the other interactions which we will consider. We then proceed to examine the inelastic scattering of photons on electrons where the kinematics is easily described in the limit that the electron is free. However, the cross sections for these interactions cannot be approached by simple means and we will have to be satisfied with a presentation of the results of quantum theory. The details of elastic and inelastic scattering of photons with bound electrons are even more complex and we will only present a qualitative description of the processes and an examination of the results of sophisticated calculations. We will then delve into qualitative discussions of the photoelectric effect and the process of pair production, where the electromagnetic interaction results in the transformation of the energy of a photon into the creation of an electron–positron pair. Once the fundamentals have been covered, we will look at the means by which the interactions are handled on a "macroscopic" scale for general applications.

We then turn to the interaction of charged particles with matter. Regardless of the real complexities involved, the main aspects of the interactions can be uncovered with a simple classical model that considers just the scattering of the charged particles on atomic electrons as a result of the Coulomb force alone. While simple, the classical result is very nearly that from a detailed quantum mechanical treatment, which we will then write down and discuss in some detail. Finally, when low-energy charged particles are considered, a significant fraction of the energy loss is due to the interaction of the incident particle with atoms or ions as a whole and we will consider some of the general aspects of this so-called "nuclear" stopping.

The interaction of neutrons with matter will not be addressed directly. Neutrons will interact by all of the reactions we have discussed throughout this text, especially in Chapter XIII. The fact is that it is the products of the neutron interactions that affect matter, including the charged particles and photons produced by elastic and inelastic scattering and the charged particles and photons produced subsequent to other reactions such as neutron capture, fission, etc.

14.2
The Interaction of Photons with Matter

The interaction of photons with matter occurs via four main mechanisms:

Elastic scattering
The classical interaction of a photon with a single bound electron is treated as an adiabatic process where the energy absorbed by the electron is re-emitted without loss. This process is referred to as *Thompson scattering*. The only difference between the incident and scattered photon is a change in the photon's trajectory. The similar, but more complex, process where the photon interacts simultaneously with all of the electrons bound in an atom is called *Raleigh* or *coherent scattering*. Thompson or Raleigh scattering is relatively unimportant except in the case of very low-energy photons such as are found in the imaging of human tissues with x-rays.

Inelastic Scattering
Photons interacting with electrons can undergo an inelastic scattering process wherein a fraction of the energy of the incident photon is transferred to the electron and a lower-energy scattered photon is produced. As in all such processes, the kinematics can be understood by the simple application of conservation of energy and linear momentum. It is not difficult to anticipate that the inelastic scattering process will give rise to photons and electrons with a range of energies. In the limit that the electron can be considered to be free (unbound), the scattering process is referred to as *Compton scattering*. When electron binding is considered, the scattering is referred to as *incoherent scattering*. Inelastic scattering is the most probable interaction for photons with energies greater than a few tenths of an MeV for all but the heaviest elements.

Photoelectric Absorption
Photons can interact with electrons bound in atoms by a process in which the energy of the incident photon is completely transferred to the electron, ejecting it from the atom. This is the *photoelectric* interaction or photoelectric effect. The photoelectric effect is especially important for photons with energies below about 200–300 keV interacting with all but the lighter elements. The ion produced after electron ejection will then decay by emission of x-rays or Auger electron ejection, as has been discussed previously in Chapter X.

Pair Production
When the energy of a photon exceeds twice the rest mass of an electron, the photon can interact with the electromagnetic field of a nucleus to create an electron–positron pair. The photon disappears completely and its energy is transformed into the rest mass and kinetic energies of the two particles. Pair production is especially important for photons with energies above about 4–6 MeV. Pair production can and does take place in the field of electrons. However, the mini-

mum energy for pair creation in this case can be shown to be four times the rest mass of an electron.

14.2.1
Elastic Scattering of Photons on Unbound Electrons

As an introduction to the interaction of photons with matter, we first consider the classical process of Thompson scattering. The fundamental ideas of this model are rather simple. We suppose that an electron is bound at the origin of our coordinate system, its equilibrium position. Although bound, the electron is assumed free to oscillate if driven by some external force. An incident photon represents an oscillating electric field and its interaction with the electron will force the oscillation at the frequency $v = E_o/h$ where E_o is the photon's energy and h is Planck's constant. Now an oscillating electron is obviously undergoing acceleration and, classically, an accelerating electron will radiate energy in proportion to its acceleration. In Thompson scattering it is assumed that the re-radiation of the incident photon energy occurs *adiabatically* – the energy radiated is equal to the energy absorbed. The only change possible between the incident and scattered photon is the angle of the latter with respect to the direction of the incident photon, its direction of propagation. To make the problem as simple as possible, and to represent the most common case of interest, we will assume that the incident photons are *unpolarized*. That is, there is an equal probability of finding the electric vector of the incident photons pointing at any azimuthal angle about the direction of propagation of the photon.

The calculations will be performed with reference to the schematic diagram shown in Fig. 14.1. The direction of propagation of the incident photon is taken to be the positive y-axis. The photon has an electric vector given by

$$\mathbf{E} = E_o \sin \omega t \tag{14.1}$$

and we will assume for the moment that it is parallel to the z-axis. The energy of the photon per unit area normal to the direction of its propagation is given by the *Poynting* vector

$$\mathbf{S}_{in} = \varepsilon_o c^2 (\mathbf{E} \times \mathbf{B}) \tag{14.2}$$

where **B** is the photon's magnetic vector that is normal to both **E** and the direction of proagation. For a plane monochromatic wave, there are equal contributions from both the electric and magnetic vectors and thus the absolute value of the Poynting vector can be written as

$$|\mathbf{S}_{in}| = \varepsilon_o c |\mathbf{E}|^2 \tag{14.3}$$

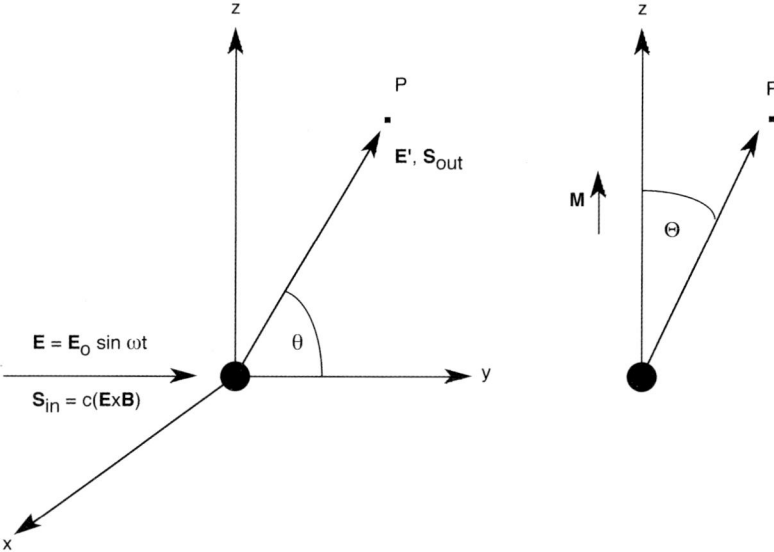

Fig. 14.1 Schematic diagram of the interaction of an incident photon with an electron (filled circle), bound at the origin, but free to oscillate adiabatically.

The interaction of the electric field of the photon with the electron produces a force

$$\mathbf{F} = e\mathbf{E} \tag{14.4}$$

The oscillating electron re-emits the absorbed energy and this results in an electric field of $\mathbf{E'}$ and a Poynting vector $\mathbf{S}_{out}$ at the point P. The vector from the origin to P makes the angle θ with the y-axis and the angle Θ with the z-axis.

The oscillation of the electron is in the direction of the electric field $\mathbf{E}$. As a result, it produces an electric dipole moment of magnitude

$$\mathbf{M} = e\mathbf{z} \tag{14.5}$$

that is also directed along the z axis. The equation of motion of the electron is given by

$$m_e \frac{d^2z}{dt^2} = eE_o \sin\omega t \tag{14.6}$$

We can use Eqs (14.1) and (14.3) to write the rate at which energy is absorbed by the electron per unit area normal to the z direction, averaged over a cycle in time as

$$\langle |\mathbf{S}_{in}| \rangle = \varepsilon_o c |\mathbf{E}_o|^2 \langle \sin^2 \omega t \rangle$$

$$= \varepsilon_o c |\mathbf{E}_o|^2 \frac{\int_{\omega t = 0}^{\pi} \sin^2 \omega t \, d(\omega t)}{\int_{\omega t = 0}^{\pi} d(\omega t)} \tag{14.7}$$

$$= \frac{\varepsilon_o c}{2} |\mathbf{E}_o|^2$$

Having established the energy absorption rate, we now consider its re-emission by the oscillating electron. From classical electricity and magnetism, the electric field of an oscillating electric dipole at the point P and angle Θ can be written

$$\mathbf{E}' = \frac{1}{c^2 r} \frac{d^2 \mathbf{M}}{dt^2} \sin \Theta \tag{14.8}$$

where r is the radial distance between the dipole and the point P. We now use this and Eqs. (14.5) and (14.6) to obtain the rate of energy radiated at P per unit area normal to r as

$$\langle |\mathbf{S}_{out}| \rangle = \varepsilon_o c \left| \frac{e^2 \mathbf{E}_o}{4 \pi \varepsilon_o m_e c^2 r} \sin \omega t \sin \Theta \right|^2$$

$$= \frac{e^4}{16 \pi^2 \varepsilon_o m_e^2 c^3 r^2} |\mathbf{E}_o \sin \omega t \sin \Theta|^2 \tag{14.9}$$

The averaging implied in these equations is relatively easy to evaluate because we consider only the case of unpolarized incident radiation. Because all azimuthal angles of the incident photon's electric vector are equally possible, we have axial symmetry about the y-axis. As a result, we can rotate the electric vector of the incident photon about the y-axis without changing the nature of the problem and we might just as well rotate it into the x–y plane. In this case, both θ and Θ also lie in this plane. We then have the situation shown schematically in Fig. 14.2. (a) shows that the radius vector to the point P now lies in the x–y plane and thus the angle between $\mathbf{E}_o$ and the z–axis is $\pi/2$. (b) is the view down the z-axis and shows that $\Theta = \pi/2 - \theta$.

The evaluation of $\overline{|\mathbf{E}_o \sin \omega t \sin \Theta|^2}$ can now be accomplished easily. First we recognize that the angles ωt and θ are completely independent. Therefore we can write

$$\overline{|\mathbf{E}_o \sin \omega t \sin \Theta|^2} = \langle \sin^2 \omega t \rangle \langle (\mathbf{E}_o \sin \Theta)^2 \rangle$$

$$= \frac{1}{2} \langle (\mathbf{E}_o \sin \Theta)^2 \rangle \tag{14.10}$$

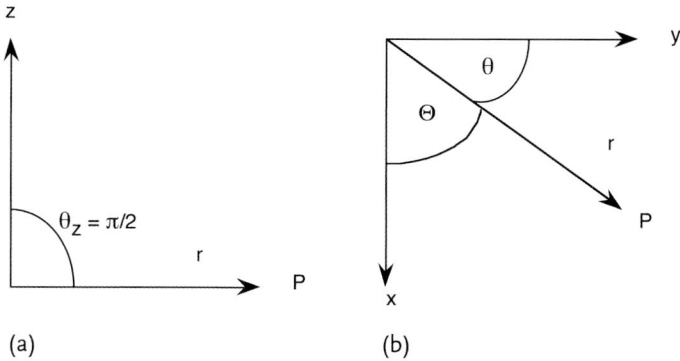

Fig. 14.2 Relations between the radius vector to the point P and the coordinate axes when the electric vector of the incident photon is rotated into the x–y plane.

The electric vector $\mathbf{E}_o$ is always in the x–z plane. The average lengths of the projection of $\mathbf{E}_o$ on both the x and z-axes are obviously the same and thus the average value of the squares of the components must be equal to $\overline{|\mathbf{E}_o|^2}_x = \overline{|\mathbf{E}_o|^2}_z = 1/2|\mathbf{E}_o|^2$. The average value of $(\mathbf{E}_o \sin\Theta)^2$ at the point P can now be calculated by summing the contributions these two averages make at the point P. $\overline{|\mathbf{E}_o|^2}_z$ is normal to r and, from Fig. 14.2 (b),

$$\overline{|(\mathbf{E}_o \sin\Theta)^2|}_x = \frac{1}{2}|\mathbf{E}_o|^2 \sin^2\left(\frac{\pi}{2} - \theta\right) \tag{14.11}$$

$$= \frac{1}{2}|\mathbf{E}_o|^2 \cos^2\theta$$

Therefore

$$\overline{|\mathbf{E}_o \sin\omega t \sin\Theta|}^2 = \frac{1}{4}|\mathbf{E}_o|^2 + \frac{1}{4}|\mathbf{E}_o|^2 \cos^2\theta \tag{14.12}$$

We can now calculate the energy per unit area scattered at P due to the photon incident at the origin by substitution of this result into (14.9). Thus,

$$\langle |\mathbf{S}_{out}| \rangle = \frac{e^4 |\mathbf{E}_o|^2}{64\pi^2 \varepsilon_o m_e^2 c^3 r^2}(1 + \cos^2\theta) \tag{14.13}$$

The expressions in Eqs (14.7) and (14.13) permit us now to determine the differential Thompson scattering cross section, $\sigma_{Th}(\theta)$. By definition,

$$\sigma(\theta) = \frac{d\sigma}{d\Omega_\theta} \tag{14.14}$$

and $\sigma(\theta)d\Omega_\theta$ gives the scattering of the incident radiation into the solid angle between θ to $\theta + d\theta$ per unit incident flux. In the present case the incident flux is represented by the energy of the incident photon per unit area normal to its direction of propagation, $|\mathbf{S}_{in}|$. The energy scattered into an area element dA at the point P is just $|\mathbf{S}_{out}|dA$. With $dA = r^2 d\Omega_\theta$, we then have

$$\sigma_{Th}(\theta)d\Omega_\theta = \frac{\langle|\mathbf{S}_{out}|\rangle r^2 d\Omega_\theta}{\langle|\mathbf{S}_{in}|\rangle}$$

$$= \frac{\frac{e^4|E_o|^2}{64\pi^2\varepsilon_o m_e^2 c^3 r^2}(1+\cos^2\theta)r^2 d\Omega_\theta}{\frac{\varepsilon_o c}{2}|E_o|^2} \quad (14.15)$$

and thus

$$\sigma_{Th}(\theta) = \frac{k_C^2 e^4}{2m_e^2 c^4}(1+\cos^2\theta) \quad (14.16)$$

The expression in (14.16) is simplified a bit more by use of the classical electron radius, $r_e = k_C e^2/m_e c^2$,

$$\sigma_{Th}(\theta) = \frac{r_e^2}{2}(1+\cos^2\theta) \quad (14.17)$$

The total scattering cross section is obtained by integration as

$$\sigma_{Th} = \int_0^\pi \sigma_{Th}(\theta)d\Omega_\theta = \pi r_e^2 \int_0^\pi (1+\cos^2\theta)\sin\theta d\theta = \frac{8}{3}\pi r_e^2 \quad (14.18)$$

The total cross section for Thompson scattering per electron is found to be about 0.665b. Because the electron radiates all of the incident energy and behaves as if it had infinite mass, photons can be scattered elastically at all angles. The differential Thompson scattering cross section is shown in Fig. 14.3 as a function of the scattering angle θ in units of mb str^{-1}. (Remember that the total solid angle subtending a point is 4π steradians.) It is symmetric about $\pi/2$ and has the limiting values of 79.5 mb str^{-1} and 39.7 mb str^{-1} at the angles 0 and π, and $\pi/2$, respectively.

The total cross section for scattering at angles θ to $\theta + d\theta$ in the laboratory will, of course, be proportional to the solid angle in this range. The quantity $\frac{d\sigma_{Th}}{d\theta}$ is shown in Fig. 14.4. The probability for scattering at very forward and very back angles is strongly suppressed and there are small peaks in the scattering distribution at angles slightly greater than and slightly less than $\pi/4$ and $3\pi/4$.

It is important to note that Thompson scattering is independent of the photon energy. To the extent that all of the electrons in an atom can be treated independently, the scattering per atom is just $Z\sigma_{Th}$. One additional point to note is the depen-

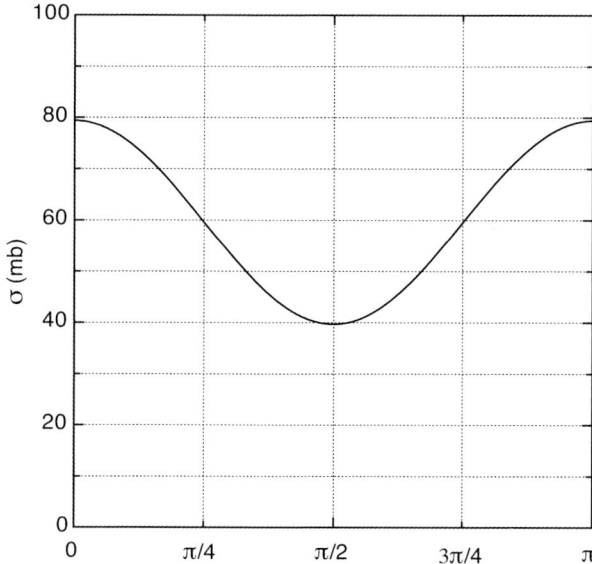

Fig. 14.3 The differential Thompson scattering cross section.

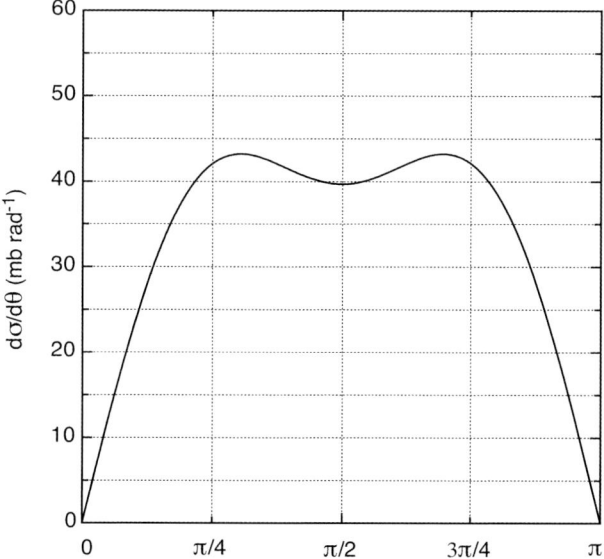

Fig. 14.4 The differential of the Thompson cross section with respect to the scattering angle in the laboratory.

dence of the cross section on the inverse square of the electron mass. We can use this to infer that scattering of photons on nuclei, the equivalent of Thompson scattering on electrons, will have a very much smaller cross section and can generally be neglected.

Thompson scattering is really an ideal approximation. Electrons are generally bound in atoms and the scattering cannot really take place without some change in the photon energy because of the requirements of energy and momentum conservation. But this effect is quite negligible in almost all cases. What is not negligible is the fact the electrons in a material are not at rest and the binding of the electrons in different atomic orbitals means that the scattering will be dependent upon the energies and momenta of the different electrons. The elastic scattering from atomic electrons is referred to as *coherent* scattering. The theory of coherent scattering is well beyond the level of this text and we will be content with comparing the results of detailed calculations with the simple Thompson model later in this chapter.

14.2.2
Compton Scattering

Kinematics

The kinematics of Compton scattering is easily developed by application of the laws of conservation of energy and momentum. The only complication is the fact that both the photon and electron must be considered as relativistic particles. The scattering process is shown schematically in Fig. 14.5. A photon of energy E_o is assumed incident on a free unbound electron at rest in the laboratory. After the interaction, the electron is found moving at the *recoil* angle ϕ with a kinetic energy T_e and a photon with the lower energy E' is found moving at the *scattering* angle θ.

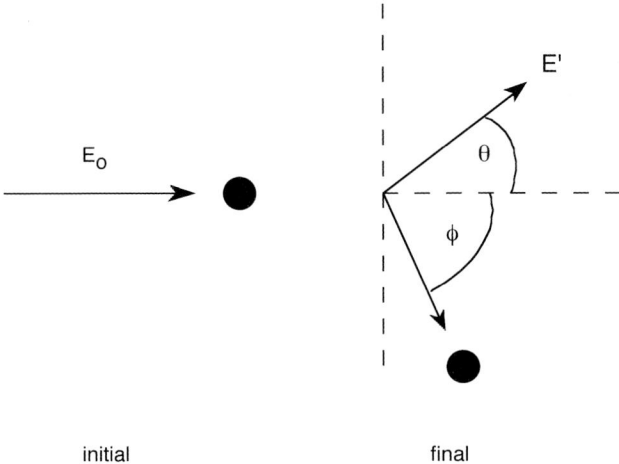

Fig. 14.5 Schematic diagram of Compton scattering.

Conservation of energy and momentum can be written as

$$E_o = E' + T_e \tag{14.19}$$
$$\mathbf{p}_o = \mathbf{p}' + \mathbf{p}_e$$

We will solve these equations to determine the characteristics of the scattered photon as a function of the properties of the incident photon. We start by solving the second equation for the electron momentum and squaring the result;

$$p_e^2 = (\mathbf{p}_o - \mathbf{p}')(\mathbf{p}_o - \mathbf{p}') = p_o^2 + p'^2 - 2|\mathbf{p}_o||\mathbf{p}'|\cos\theta \tag{14.20}$$

The electron momentum is related to the total electron energy E_e by

$$p_e^2 c^2 = E_e^2 - (m_e c^2)^2 \tag{14.21}$$

while the energy of a photon is related to its momentum by $E = pc$. Substituting these into Eq. (14.20) then gives

$$E_e^2 - (m_e c^2)^2 = E_o^2 + E'^2 - 2 E_o E' \cos\theta \tag{14.22}$$

The total energy of the electron is just the sum of the energy-equivalent of its rest mass and its kinetic energy. We can use this with Eq. (14.19) to write

$$E_o - E' = E_e - m_e c^2 \tag{14.23}$$

Solving for E_e, squaring and substituting the result into (14.22) then gives

$$\tag{14.24}$$
$$(E_o - E')^2 + (m_e c^2)^2 + 2 m_e c^2 (E_o - E') - (m_e c^2)^2 = E_o^2 + E'^2 - 2 E_o E' \cos\theta$$

Expanding and combining terms gives the final result that

$$E' = \frac{E_o}{1 + \dfrac{E_o}{m_e c^2}(1 - \cos\theta)} \tag{14.25}$$

One often sees this written as

$$E' = \frac{E_o}{1 + \alpha(1 - \cos\theta)} \tag{14.26}$$

where $\alpha = E_o/m_e c^2$ is the incident photon energy in units of the energy-equivalent of the electron rest mass, or the *reduced* energy of the photon.

The energy of the scattered photon generally varies over a considerable range but it does not go to zero unless the energy of the incident photon goes to zero. Scattering at $\theta = 0$ leaves the photon energy unchanged while scattering at $\theta = \pi$ results in the lowest energy. Thus

$$\frac{E_o}{1+2\alpha} \leq E' \leq E_o \qquad (14.27)$$

Compton scattering at low energies ($\alpha \ll 1$) results in very little change in the photon energy and in the limit that $\alpha \to 0$, only elastic scattering occurs. If we divide the numerator and denominator on the right-hand side of (14.26) by E_o, we see that, in the limit $E_o \to \infty$, the energy of the scattered photon appearing at $\theta = \pi$ has the limiting value of $E' \to 1/2 m_e c^2$.

The energy difference between the incident and scattered photon is just the kinetic energy of the Compton electron;

$$E_o - E' = E_o \left[\frac{\alpha(1-\cos\theta)}{1+\alpha(1-\cos\theta)} \right] = T_e \qquad (14.28)$$

Clearly, Compton scattering at forward angles produces low electron kinetic energies and the maximum electron kinetic energy is achieved at $\theta = \pi$.

The relation between the scattering angle of the photon and the recoil angle of the electron can be obtained by reference to Fig. 14.5. In the initial state there is no momentum normal to the direction of propagation of the incident photon and therefore we must have, after the interaction,

$$p_e \sin\phi = p' \sin\theta = \frac{E'}{c} \sin\theta \qquad (14.29)$$

Further, the momentum balance in the direction of the incident photon gives

$$p_o = \frac{E'}{c} \cos\theta + p_e \cos\phi \qquad (14.30)$$

These two equations can be combined to yield the result

$$\frac{E_o}{E'} = \cos\theta + \sin\theta \cot\phi \qquad (14.31)$$

Equating this to the ratio E_o/E' obtained from (14.26) leads to the final result

$$\cot\phi = (1+\alpha)\tan\frac{\theta}{2} \qquad (14.32)$$

This relation provides some important insights into the scattering process. When $\theta = 0$, $\cot\phi|_{\theta=0} = (1+\alpha)\tan 0 = 0$. Therefore, when the incident photon produces the minimum energy transfer to the electron, $\phi|_{\theta=0} = \pi/2$. On the other hand, $\cot\phi|_{\theta=\pi} = (1+\alpha)\tan(\pi/2) = \infty$ and therefore $\phi|_{\theta=\pi} = 0$. When the maximum energy transfer to the electron takes place, the Compton electron appears in the direction of the incident photon. We then conclude that Compton scattering results in electrons being found only at forward angles.

The Klein–Nishina Cross Section

The calculation of the differential cross section for Compton scattering is fairly complex. It involves relativistic quantum mechanics and will not be treated here. We will be satisfied by writing down and analyzing the result found for the *differential Compton collision* cross section $_e\sigma_C(\theta) = \frac{d}{d\Omega_\theta}\,_e\sigma_C$ that represents the probability for an incident photon with frequency $v_o = E_o/h$, to produce a scattered photon of frequency $v' = E/h$ at the angle θ to $\theta + d\theta$ to the trajectory of the incident photon when unpolarized photons are scattered from an electron initially at rest in the laboratory. The expression is

$$_e\sigma_C(\theta) = \frac{r_e^2}{2}\left(\frac{v'}{v_o}\right)^2\left(\frac{v_o}{v'} + \frac{v'}{v_o} - \sin^2\theta\right) \text{ cm}^2 \text{ str}^{-1} \text{ electron}^{-1} \qquad (14.33)$$

Using Eq. (14.26), the result can be written in the equivalent form

$$_e\sigma_C(\theta) = \frac{r_e^2}{2}\{[1 + \alpha(1-\cos\theta)]^{-3} \qquad (14.34)$$

$$[-\alpha\cos^3\theta + (\alpha^2 + \alpha + 1)(1 + \cos^2\theta) - \alpha(2\alpha + 1)\cos\theta]\}$$

The result is the differential Klein–Nishina formula for unpolarized radiation. It can be integrated to yield the total cross section:

$$_e\sigma_C = 2\pi r_e^2 \left\{ \frac{1+\alpha}{\alpha^2}\left[\frac{2(1+\alpha)}{1+2\alpha} - \frac{1}{\alpha}\ln(1+2\alpha)\right] \right. \qquad (14.35)$$

$$\left. + \frac{1}{2\alpha}\ln(1+2\alpha) - \frac{1+3\alpha}{(1+2\alpha)^2} \right\} \text{ cm}^2 \text{ electron}^{-1}$$

We begin our analysis of the cross section by noting that, in the limit $\alpha \to 0$, $_e\sigma_C(\theta) \to (r_e^2/2)(1 + \cos^2\theta)$. The Klein–Nishina expression reduces to the differential cross section for Thompson scattering, the latter providing a natural scale for the magnitude and angular variation of the cross section for Compton scattering.

The differential Compton collision cross sections are shown in Fig. 14.6 for reduced photon energies in the range $0.01 \le \alpha \le 10$. For $\alpha = 0.01$ ($E_o = 5.11\,\text{keV}$) the cross section is almost identical to that shown in Fig. 14.3 for Thompson scattering. As the energy increases, especially for $\alpha > 0.1$, the distribution becomes more and more forward peaked. The differential cross section for scattering at angles greater than about $\pi/2$ becomes small for incident photon energies greater than several MeV. This is seen quite clearly in Fig. 14.7 where we show the differential of the total cross section with respect to the scattering angle, i.e., $\frac{d}{d\theta}\,_e\sigma_C = 2\pi\,_e\sigma_C \sin\theta\, d\theta$. The variation in the solid angle with θ exerts a very strong effect on the total probability for scattering at any angle, suppressing the probability at forward and backward angles. As the energy of the incident photon increases, the total cross section decreases and the probability for scattering becomes more forward peaked.

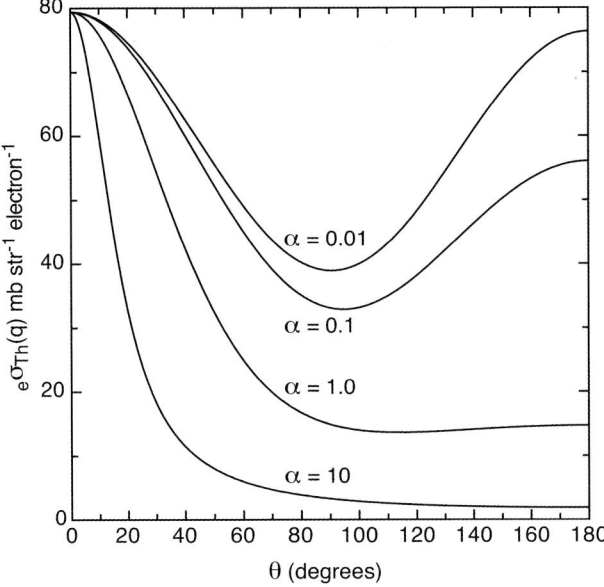

Fig. 14.6 The differential Klein–Nishina cross section for reduced photon energies in the range $0.01 \leq \alpha \leq 10$.

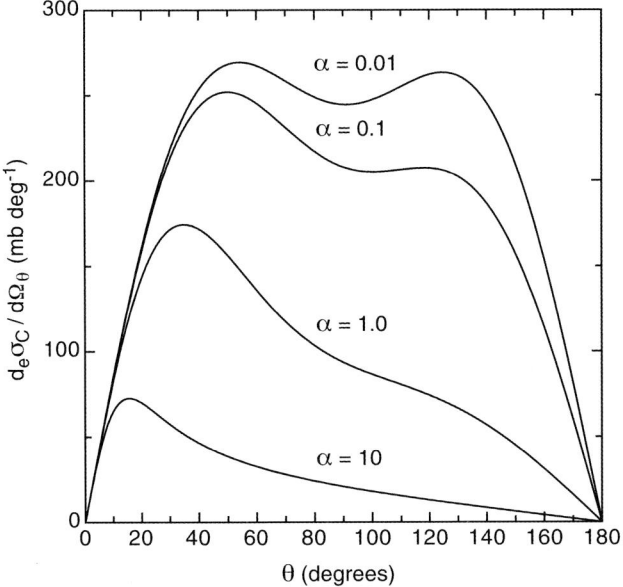

Fig. 14.7 The differential Klein–Nishina cross section with respect to the scattering angle of the photon.

Applications of the Klein–Nishina Cross Section

The expressions we have developed can be used to derive expressions for the *energy spectra* of Compton scattered photons and electrons. All we need recognize is that: (a) the differential scattering cross section provides a means of determining the number of scattering events that occur at all angles; and (b) there is a one-to-one correspondence between the scattering angle and the energies of the scattered photons and electrons. Given a single electron and a flux ϕ_o of incident photons, the number of photons scattered at the angle θ to $\theta + d\theta$ is

$$dN = \phi_o \sigma_C(\theta) d\Omega = 2\pi \phi_o \sigma_C(\theta) \sin\theta \, d\theta \tag{14.36}$$

The quantity

$$\frac{1}{\phi_o}\frac{dN}{d\theta} = N(\theta)$$

is called the *distribution function* or *spectral function* for scattering at the angle θ.

In general, the number of photons scattered at θ to $\theta + d\theta$ per target electron per unit incident photon flux is just $N(\theta)d\theta$. Each of these will have an energy given by (14.26). Because of the one-to-one correspondence between energy and angle, it must be true that

$$N(E')dE' = N(\theta)d\theta \tag{14.37}$$

or

$$N(E') = N(\theta)\frac{d\theta}{dE'} \tag{14.38}$$

where $N(E')$ is the *spectral function* or the *energy spectrum* of the scattered photons. From (14.26) we can show that

$$\frac{dE'}{d\theta} = \frac{-E_o \alpha \sin\theta}{[1 + \alpha(1 - \cos\theta)]^2} \tag{14.39}$$

the negative sign indicating that the photon energy decreases with increasing scattering angle. We can now combine Eq. (14.38) and (14.39) to obtain the energy spectrum (absolute value) of the scattered photons as

$$N(E') = \left(\frac{2\pi}{\alpha E_o}\right)_e\sigma_C(\theta)[1 + \alpha(1 - \cos\theta)]^2 \tag{14.40}$$

In Fig. 14.8 are shown energy spectra of scattered photons for E_o in the range $1 \leq \alpha \leq 10$. The maximum energy in each case is just the incident photon energy. The minimum energy rapidly approaches the limiting value of $1/2 m_e c^2$ for $E_o > 2 m_e c^2$. Except for the lowest energies, the shape of the spectra above an MeV

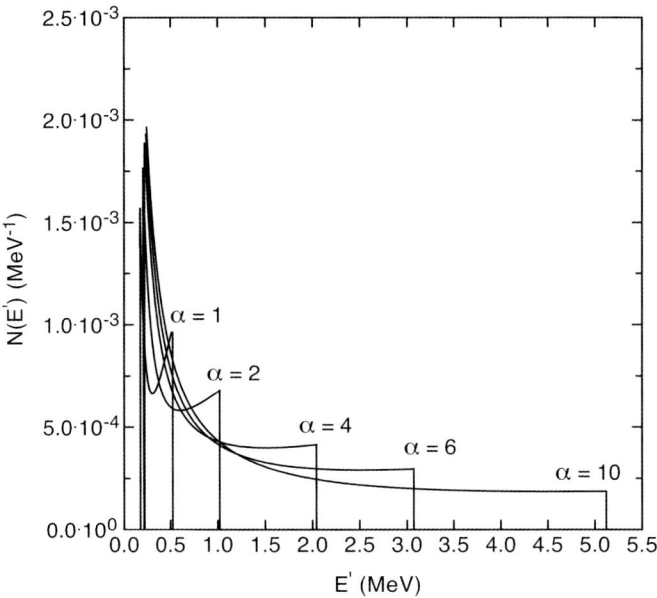

Fig. 14.8 The spectra of Compton-scattered photons for E_o in the range $1 \leq \alpha \leq 10$.

or so are very similar, and at the higher energies, the probability for finding any photon energy is roughly constant.

In Fig. 14.9 are the kinetic energy spectra of the Compton electrons corresponding to the scattered photon spectra shown in Fig. 14.8. In each case the maximum energy corresponds to Compton scattering at $\theta = \pi$ and rapidly approaches the limiting energy of $E_o - 1/2 m_e c^2$. Reflecting the distributions shown in Fig. 14.8, the probability for finding small kinetic energies becomes roughly constant when the incident photon energy is high.

These results have rather immediate applications in radiation detection. Many photon detectors have low detection efficiencies. That is, the probability that an incident photon will interact in them is much less than unity. In such cases if a Compton event occurs, the probability that the scattered photon will interact is also very small. Because the Compton electrons generally deposit their energy over small linear dimensions, it is common to find that the spectrum of energy deposited in the detector approximates just the spectrum of the Compton electrons. One will observe an energy spectrum from Compton interactions that is very nearly that shown in Fig. 14.9. Even for fairly efficient photon detectors where the scattered photons have reasonable detection efficiencies, one typically sees a "continuum" of energies that is still quite similar to the shape of the Compton electron spectrum as the maximum electron energy is approached. The maximum in the Compton electron distribution is commonly referred to as the "Compton edge".

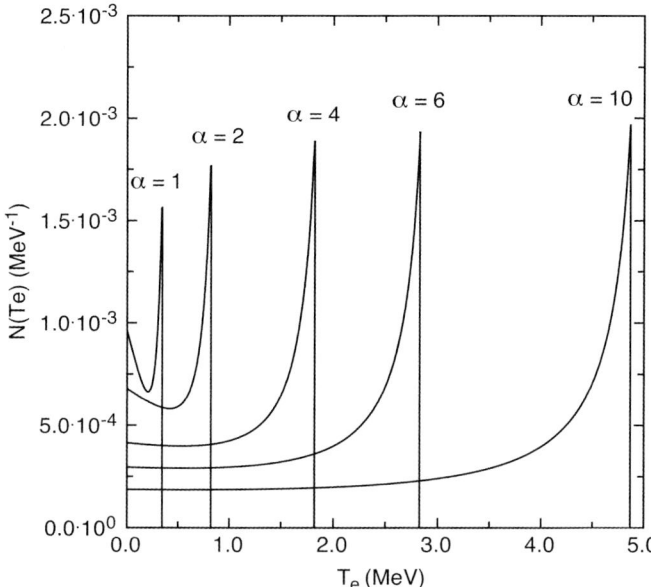

Fig. 14.9 The energy spectra of Compton electrons for photon energies in the range $1 \leq \alpha \leq 10$.

In the engineering literature, one often finds analytical models or computer codes that are used to estimate the transport and energy deposition of photons in various materials. These make use of two more cross sections that are related to the Compton collision cross section. The first of these is the differential Compton *energy-scattering* cross section. It is defined as the differential Compton collision cross section multiplied by the fraction of the incident photon energy represented by the scattered photon. That is,

$$_e\sigma_s(\theta) = \frac{E'}{E_o} {}_e\sigma_C(\theta) \tag{14.41}$$

This can be viewed as the differential cross section for transfer of the incident photon energy to the scattered photon. Under most conditions where one is interested in shielding against photon fields or for determining where energy is deposited on a macroscopic scale, it is often a good approximation to assume that the kinetic energy of a Compton electron is deposited at the site of the interaction while the scattered photon will be transported far from that site. This derives from the relatively short range of electrons in matter compared to the distance required for the photons to undergo a second interaction.

The total energy scattering cross section can be obtained by integrating Eq. (14.41) to give

$$_e\sigma_s = \int_0^\pi 2\pi\,_e\sigma_s(\theta)\sin\theta\,d\theta \tag{14.42}$$

$$= 2\pi r_e^2\left[\frac{4\alpha^2}{3(1+2\alpha)^3} - \frac{(1+\alpha)}{\alpha^2(1+2\alpha)^2}(1+2\alpha-2\alpha^2) + \frac{1}{2\alpha^3}\ln(1+2\alpha)\right]$$

The second cross section is the *energy-absorption* cross section, and it is given naturally by the relation

$$_e\sigma_a = {}_e\sigma_C - {}_e\sigma_s \tag{14.43}$$

It represents the cross section for transfer of the incident photon energy to the Compton electron.

Coherent and Incoherent Scattering

The majority of the electrons in matter are bound within atoms and molecules. As such, the processes of Thompson and Compton scattering are approximations that must be modified by consideration of the effects of electron binding on the scattering process. Thompson scattering is a good approximation in the limit of very low photon energies while Compton scattering is a very good approximation in the limit of very high photon energies. The theoretical development of the scattering probabilities when electron binding is considered explicitly goes well beyond the purpose of this text. Nevertheless, is not difficult to understand qualitatively how binding affects the scattering. We will present qualitative discussions and then show how they are reflected in the results of detailed theoretical calculations. The principal purpose of this discussion is to provide the reader with some feeling for the importance of electron binding and the conditions under which the simple models discussed above must be viewed with care.

We are all familiar with the simple elastic scattering of low-energy photons with mirrors. The classical picture of the scattering, where the angle of incidence of a light wave is equal to the angle of reflection is obtained without regard to the details of the structure of the mirror itself. Implicit in the classical model is the underlying assumption that the incident light wave will interact with the medium in such a way that the scattering can be considered to take place with a single entity – the mirror surface. From a microscopic point of view, this implies that the electrons in the mirror with which the photon interacts act in a way so as to scatter the wave as though it interacted with a single electron alone. In this sense, the scattering from many electrons is said to take place *coherently*. The different amplitudes of the wave scattered by individual electrons are added in a way that produces the linear result of the classical picture. This is actually a very complex process.

Consider a photon incident on an isolated neutral atom. The photon will have its momentum vector **p**, or equivalently its propagation vector **k**, directly along its trajectory. The electrons in the atom are each bound in well-defined orbitals and each has a kinetic energy characteristic of that orbital. A photon interacting with one such electron must see a moving particle that not only has a probability distribution in space but also with a probability distribution for the vector direction of its linear momentum **p**. It is not difficult to see that the relative energy of the electron sensed by the photon will vary over a considerable range. Thus the momentum transfer that is possible will vary and this must be accounted for in the calculation of the scattering probability. The scattering of the photon from the atom must account explicitly for the spatial distributions and momentum distributions of all of the electrons.

When dealing with bound electrons, it is not difficult to see that elastic scattering – coherent scattering – will have a much smaller probability of occurring at angles very different from zero except for very small photon energies or very large electron binding energies. Consider, for example, the case of photons with energies of 10 keV interacting with atoms of aluminum (Z = 13). The ionization energy of aluminum is about 6 eV and the K- and average L-electron binding energies are 1.5596 and about 0.1 keV, respectively. From Compton-scattering kinematics, the scattering angle that would result in transfer of 6 eV to a free electron is 14.22°. Scattering at smaller angles simply cannot ionize the atom but scattering at larger angles can result in ionization. Because the maximum possible energy transfer to a free electron is 0.377 keV, all but the K electrons could be ejected from the atom at larger scattering angles. Thus we can project that except for very small angles, coherent scattering will be suppressed in comparison to incoherent scattering and the number of bound electrons for which incoherent scattering is possible will vary with scattering angle. If we now consider a photon with an energy of 100 keV, the angle at which 6 eV could be transferred to a free electron is only 1.419° and the maximum energy transfer of 28.1 keV is more than sufficient to eject the K electrons. Elastic scattering will be the only possible process at only very small angles indeed. Furthermore, elastic scattering will be reduced even if only the excitation of the atom occurs. If we consider that the most weakly bound electrons can be excited by energy transfers as small as 5 eV or less, the probability for elastic scattering at all but the smallest angles is seen to be rather unlikely. The implication of these observations is that the probability of elastic scattering from bound electrons will be largest at forward angles and will decrease rapidly at large angles as the photon energy increases.

The cross section for coherent scattering is usually given in the form

$$\sigma_{coh}(\theta) = {}_e\sigma_T(\theta)[F(x, Z)]^2 \qquad (14.44)$$

Here, ${}_e\sigma_T(\theta)$ is just the differential Thompson scattering cross section discussed above. All of the modifying effects of electron binding on elastic scattering are formally accounted for by introduction of the *atomic form factor* F(x, Z) where

$$x = \left(\sin\frac{\theta}{2}\right)/\lambda, \qquad (14.45)$$

λ is the wavelength of the photon and Z is the atomic number of the atom. In the limit that electron binding energies were infinite, so that all electrons remained bound and yet were free to oscillate and radiate coherently, the quantity $[F(x, Z)]$ reduces to the atomic number of the atom, Z. On the other hand, when binding energies are small, the electron appears to be "quasi" free and it cannot satisfy the requirements for adiabatic energy absorption and re-emission.

In Fig. 14.10 are plots of theoretical form factors for 10 and 100 keV photons interacting with aluminum atoms [1]. The form factors are seen to have the limiting value of 13 at $\theta = 0$. With increasing angle, the form factor decreases rapidly. Considering that the cross section is proportional to the square of the form factor, it becomes evident that coherent scattering will be strongly forward peaked for most photon energies of interest. In Fig. 14.11 are shown the differential cross sections for coherent scattering at these energies along with the differential Thompson scattering cross section multiplied by a factor of 13. The effects of the electron binding are indeed dramatic. Even at 10 keV, the cross section is strong only at small angles and is strongly depressed as $\theta \to \pi$. For photons of 100 keV, coherent scattering beyond an angle of about 20° is negligible.

Such effects are seen throughout the periodic table of the elements. In Fig. 14.12 are shown the corresponding differential cross sections for the element lead (Z = 82). At all but the lowest energies, elastic scattering cross sections are strongly enhanced in the forward direction and very strongly depressed beyond some tens of degrees. The general result is that coherent scattering from bound electrons is strongly directed in the forward direction, except at relatively low energies.

The effect of electron binding on inelastic or *incoherent* scattering is rather easy to see qualitatively from the preceding discussion. For inelastic scattering to occur,

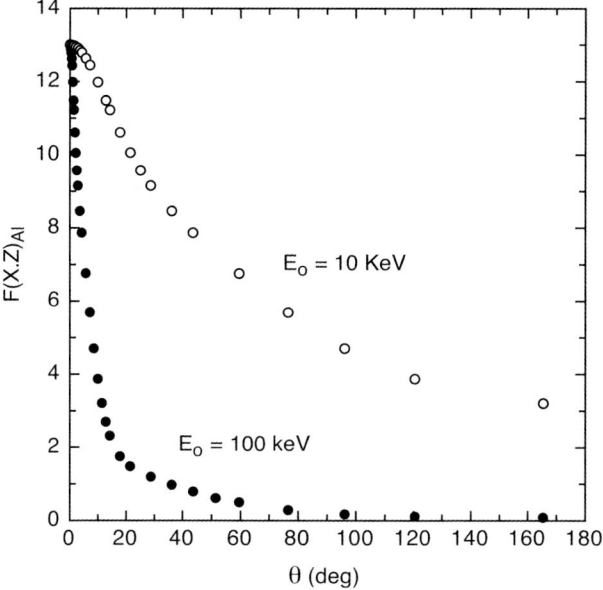

Fig. 14.10 The atomic form factors for 10 and 100 keV photons interacting with aluminum atoms [1].

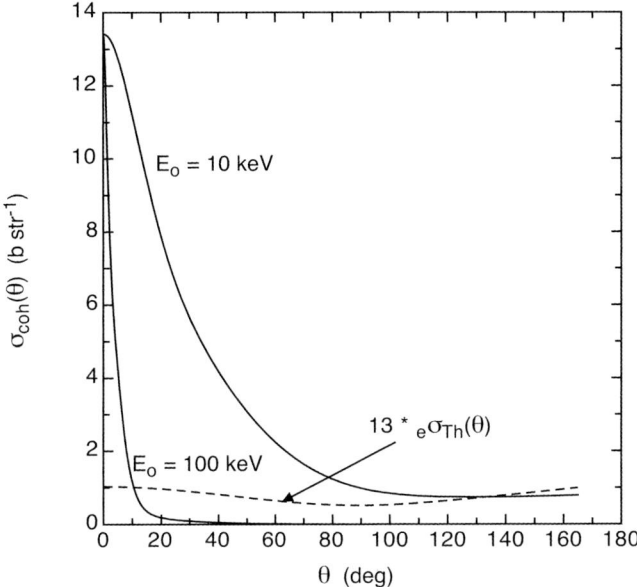

Fig. 14.11 The differential coherent scattering cross sections for 10 and 100 keV photons interacting with aluminum atoms. These distributions were calculated with relativistic quantum mechanical calculations to correct for electron binding. For reference, the dashed curve represents 13 times the Thompson differential cross section [1].

we must have energy transfer to the electron. The cross section for incoherent scattering must then go to zero as $\theta \to 0$ because no energy transfer can occur at this angle. The more weakly bound electrons will easily be excited or ejected from the atom as the scattering angle increases. But as the binding energy of an electron increases, there must be an ever increasing minimum scattering angle for which incoherent scattering is possible. In total, we can then expect the incoherent scattering cross section to rise from zero at $\theta = 0$ and not approach $Z_e \sigma_C$ until the energy that can be transferred to the most tightly bound electrons is enough to eject it from the atom.

The differential incoherent scattering cross section from an atom is written as

$$\sigma_{inc}(\theta) = \sigma_{KN}(\theta) S(x, Z) \qquad (14.46)$$

where $\sigma_{KN}(\theta)$ is the differential Klein–Nishina cross section and $S(x,Z)$ is called the *incoherent scattering function*. The latter represents the correction to the Klein–Nishina cross section for the binding of all of the electrons in the atom. As for the atomic form factor, $S(x,Z)$ must be calculated by use of the best available wave functions for all electrons in a neutral atom. In Fig. 14.13 are the incoherent scattering functions for the interaction of 10 and 100 keV photons with aluminum

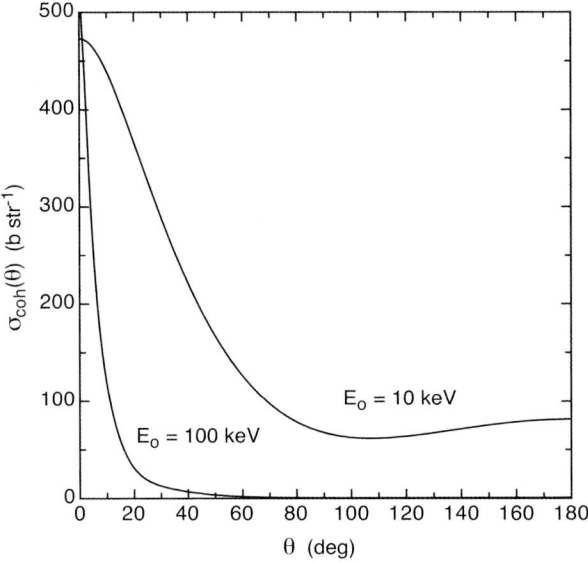

Fig. 14.12 The differential coherent scattering cross sections for 10 and 100 keV photons interacting with lead atoms. These distribution were calculated with relativistic quantum mechanical calculations to correct for electron binding [1].

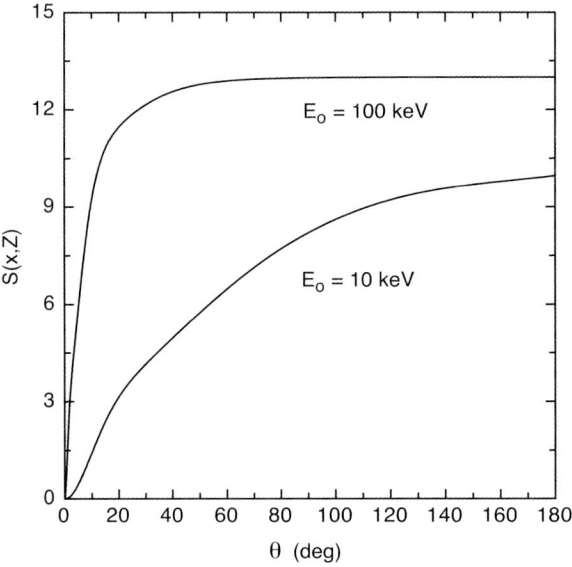

Fig. 14.13 The incoherent scattering functions for the interaction of 10 and 100 keV photons with aluminum atoms [1].

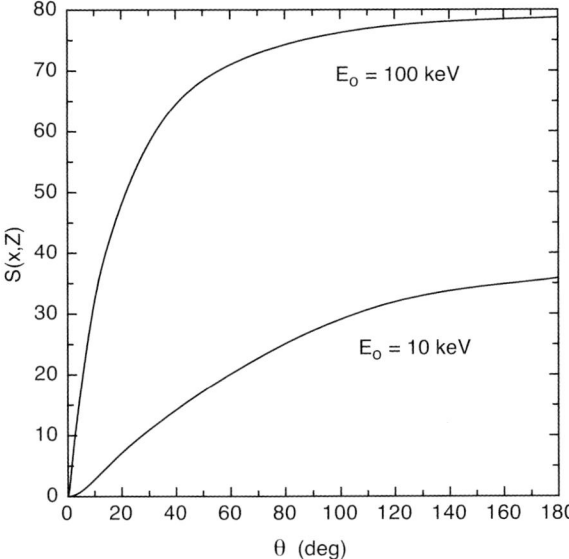

Fig. 14.14 The incoherent scattering functions for the interaction of 10 and 100 keV photons with lead atoms [1].

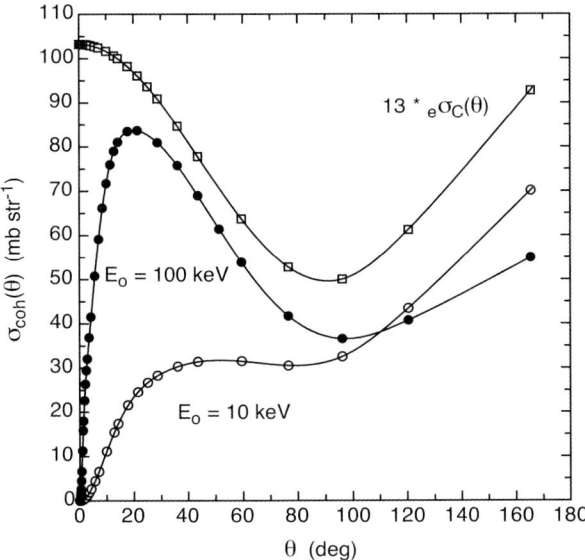

Fig. 14.15 The differential incoherent scattering cross sections for 10 and 100 keV photons interacting with aluminum atoms. For reference, the figure includes the Compton collision cross section at 10 keV multiplied by 13 to reflect the cross section expected in the case that binding energy effects are neglected [1].

atoms. For 10 keV photons, the scattering function indeed vanishes at $\theta = 0°$, increases strongly with increasing angle, but does not approach the limiting value of 13 even for scattering at $\theta = \pi$. Reference to Fig. 14.10 shows why this is so. Coherent scattering is significant at all angles at this photon energy. On the other hand, for a photon energy of 100 keV, the scattering function rises very rapidly and approaches the limiting value of 13 already at $\theta = \pi/2$.

Similar variations of $S(x,Z)$ with energy are seen throughout the periodic table. Fig. 14.14 shows the incoherent scattering functions for the interaction of 10 and 100 keV photons with lead atoms as an example of their behavior at high atomic number.

The differential incoherent cross sections for the interaction of 10 and 100 keV photons with aluminum are shown in Fig. 14.15 along with the Compton collision cross section multiplied by 13 to reflect the cross section expected in the absence of electron binding. The cross section is significantly suppressed everywhere in comparison to the free electron limit.

14.2.3
The Photoelectric Effect

The photoelectric interaction is the process in which a photon incident upon an atom is completely absorbed and a bound electron is ejected from the atom. The process is shown schematically in Fig. 14.16. To a good approximation, the total energy of the photon is represented by the sum of the electron kinetic energy and its binding energy in the atom. The recoil energy of the ion is generally negligibly small. Thus,

$$T_e = E_o - be_e \tag{14.47}$$

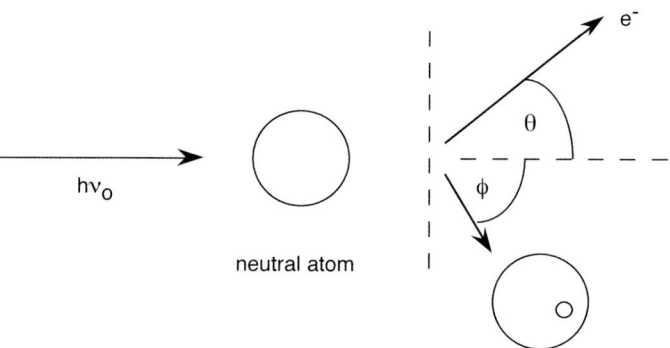

Fig. 14.16 Schematic diagram of a photoelectric interaction
The interaction of a photon with a neutral atom results in the ejection of an atomic electron and the production of an ion of electronic charge +1.

where be_e is the binding electron of the ejected electron. The spectrum of photoelectrons will then be composed of lines representing the ejection of electrons from the various shells in the atom.

As surprising as it may be at first examination, the probability for the photoelectric interaction is greatest for the most tightly bound electrons. So long as sufficient energy is available, the cross sections for photoelectric interactions in the various atomic electron shells will vary as K > L > M > ... The calculation of the cross section for the photoelectric interaction will not be given here. But we can present a simple argument that makes this sequence plausible. First, it is not difficult to show that conservation of momentum requires that the electron must be bound for the photoelectric effect to occur. The binding essentially means that there is a strong enough coupling between the electron and the atom that the latter has a high probability for "accepting" the recoil energy and momentum. If you think of two masses connected to one another by a spring, it is easy to see that, if the spring constant is large and one of the masses is subject to an impulse, momentum transfer through the spring will be effective and both masses will be subject to the motion induced by the impulse. But if the spring constant is very small and the spring can "stretch" to a large extent, the mass that is subject to the impulse will move essentially as if it were free. Very little momentum transfer will occur to the second mass. We can use this simple model to infer that, for the photoelectric effect to occur, there must be a high probability for momentum transfer to the ion that is produced upon ejection of an electron. The more tightly bound an electron, the more strongly "coupled" it is to the atom, and the higher will be the probability for effective momentum transfer.

Simplified models of the photoelectric effect generally make approximations that are applicable to the case that the photon energy is large compared to the binding energy of the ejected electron. Such models predict that the cross section for ejection of K electrons is approximately

$$\sigma_{pe, K} \approx \left(\frac{32}{\alpha^7}\right)^{1/2} \left(k_C \frac{e^2}{\hbar c}\right)^4 Z^5 {}_e\sigma_{Th} \text{ cm}^2 \text{atom}^{-1} \tag{14.48}$$

where $\alpha = E_o/m_e c^2$ and $k_C(e^2/\hbar c) \approx 1/137$ is the fine-structure constant. The cross sections for interaction with electrons in L, M, etc., shells will, of course, differ in magnitude and have somewhat different dependences on both Z and E_o.

One of the principal predictions of the model is that the cross section is proportional to $Z^5/E_o^{3.5}$. Thus, photoelectric absorption in the K shell should be most probable in the heavier elements and at low energies. As an example of empirical evaluated data, the energy dependence of the total photoelectric cross section for photons interacting with atoms of lead [2] is shown in Fig. 14.17. For clarity, the energy range has been restricted to permit easy visualization of the effect of the binding of the innermost electrons in the atom. In the energy range of 1–2.5 keV, the total cross section decreases monotonically with energy. Such photons have sufficient energy to eject electrons from all but the K-, L-, and M- electron shells. Electrons in the latter shell have binding energies between 2.484 and 3.851keV. As

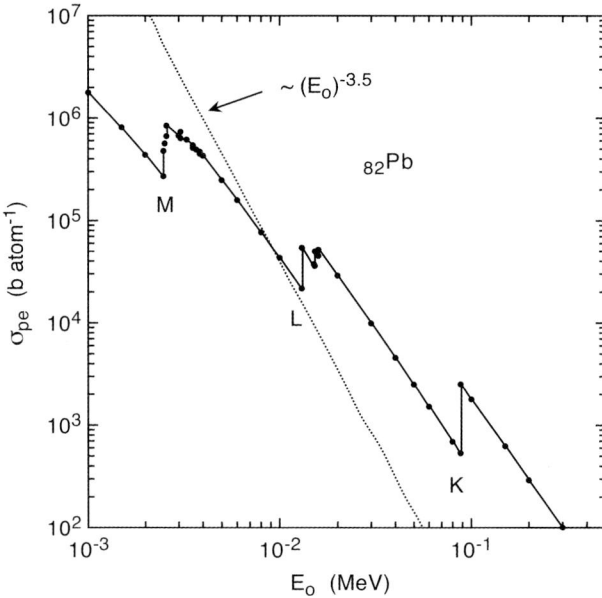

Fig. 14.17 The photoelectric cross section for low-energy photons interacting with atoms of lead. The cross sections were taken from the evaluated data given by the code XCOM [2]. The locations of the K-, L-, and M-edges are indicated.

soon as photons have energies just slightly greater than 2.484 keV, photoelectric absorption in the M shell can taken place and the cross section is seen to increase suddenly. As the energy is increased further, several abrupt increases in the cross section are found as electrons of ever-increasing binding energies can be ejected. Similar sudden increases are seen when the L and then K electrons can be ejected.

The photon energy above which electrons from the ith electron shell can be ejected is referred to as the energy of the ith *edge*. Thus one will hear that the K edge in Pb is at 88.005 keV, etc. A curve representing the energy variation $(E_o)^{-3.5}$ is shown in Fig. 14.17 for comparison with the experimental data. This prediction is in reasonable agreement with the energy variation of the cross section at energies above the L edge but at lower energies the cross section does not decrease as rapidly as predicted by the model.

The ion produced as a result of a photoelectric interaction generally has a vacancy in one of the inner electron shells. Just as in the case of the ion produced as a result of electron capture decay (see Sections 10.9.1 and 10.9.2), such vacancies will be filled by the emission of x-rays and/or Auger electrons. Thus, photoelectric interactions in the inner electron shells of high-Z elements will generally result in the production of a highly charged ion and, essentially, the total energy of the incident photon will be quickly released.

Photoelectric interactions are particularly important with respect to both photon detection and to energy deposition in materials. The fact that the total energy of a photon is deposited in a material means that a detector designed to measure a signal proportional to the energy deposited will see the full photon energy. If such a detector has sufficient energy resolution, a peak in the spectrum of the detected events will be produced that will allow the determination of the photon energies with relative ease. Most modern detectors for relatively low-energy photons are designed to meet this requirement.

14.2.4
Pair Production

When the energy of a photon exceeds $2m_ec^2$, pair production becomes possible. In this process, the photon, interacting with the electromagnetic field of a nucleus, is transformed into an electron–positron pair. The total kinetic energy of the pair is given by

$$T_{e^-} + T_{e^+} = E_o - 2m_ec^2 \tag{14.49}$$

The kinetic energy can be shared between the two particles in the same way as the energy released in β decay can be shared between the antineutrino and the β⁻ particle. Once formed, the electron and positron will lose energy primarily by Coulomb interactions with the atomic electrons in the medium in which the particles are moving. Once the positron has reached essentially thermal energies, it will *annihilate* with an electron. In this case, the annihilation almost always leads to the creation of two photons, each with an energy of m_ec^2, and with equal and opposite momentum. Thus, a signature of pair production is the subsequent observation of 0.511 MeV photons emitted at a relative angle of 180° with respect to one another.

In the extreme relativistic limit, the cross section for interaction with a "bare" nucleus, i.e., in the limit that screening of the nuclear charge by the atomic electrons is neglected, can be shown to be given by

$$_a\sigma_{pp} = \frac{2_e\sigma_{Th}}{k_C(e^2/\hbar c)} Z^2 \left(\frac{28}{9} \ln 2\alpha - \frac{218}{27} \right) \text{cm}^2\text{atom}^{-1} \tag{14.50}$$

while in the limit that the interaction takes place with the atom as a whole, the cross section is given by

$$_a\sigma_{pp} = \frac{2_e\sigma_{Th}}{k_C(e^2/\hbar c)} Z^2 \left[\frac{28}{9} \ln\left(\frac{183}{Z^{1/3}}\right) - \frac{2}{27} \right] \text{cm}^2\text{atom}^{-1} \tag{14.51}$$

Regardless of the approximation considered, the cross section per atom is predicted to be proportional to the square of the atomic number.

The cross sections discussed above were derived in the limit that the recoil energy of the nucleus or atom, in whose field the interaction took place, was negligible. Because the interaction that gives rise to the transformation is the

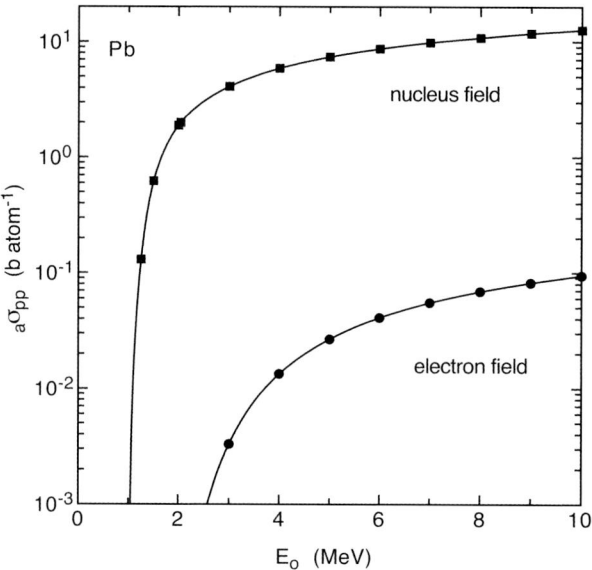

Fig. 14.18 The pair production cross section for low-energy photons interacting with atoms of lead. The upper curve represents the cross section for pair production in the field of the nucleus for a neutral atom. The lower curve represents pair production in the field of the electrons only. The cross sections were taken from the evaluated data given by the code XCOM [2].

electromagnetic interaction, and real media are filled with electrons, it is natural to enquire whether pair production takes place in the field of an electron. It does. But in this case, conservation of energy and momentum show that the threshold for pair production in the field of a free electron is $4m_e c^2$. Pair production in the field of electrons is really only significant for the very lightest elements.

In Fig. 14.18 are shown the pair production cross sections for photons interacting with lead nuclei and the cross section for photons interacting in the field of the electrons only. For atoms of high atomic number, the cross section for interaction in the field of the atomic electrons is relatively small. However, with atoms of low atomic number, the interaction in the electron field can be significant. The shape of the cross section near threshold for interaction in the field of the nucleus is always similar to that found for Pb. The cross section rises steeply just above threshold and becomes rather large at energies above 1.5 MeV.

14.2.5
Total Cross Sections and Attenuation Coefficients

The interaction of photons with matter will obviously take place by all possible mechanisms. From a macroscopic point of view, it is often the total interaction cross section or the total nonelastic cross section that is needed. The total cross section per atom is defined as the sum of the coherent scattering, incoherent scattering, photoelectric and pair production cross sections:

$$_a\sigma_{tot} = {_a\sigma_{coh}} + {_a\sigma_{incoh}} + {_a\sigma_{pe}} + {_a\sigma_{pp}} \qquad (14.52)$$

The total cross section and each of its components for photons interacting with lead are shown in Fig. 14.19. As can be seen, the total photoelectric cross section dominates at energies below about 0.1 MeV. The coherent scattering cross section is actually larger than that of incoherent scattering below about 0.1 MeV but it is at least an order of magnitude smaller than the photoelectric cross section. Incoherent scattering is the dominant cross section in the energy range of about 0.6–2.5 MeV. The relative contributions from the different interactions vary with atomic number because of their different dependence on Z.

As an example, Fig. 14.20 shows the cross sections for photons interacting with aluminum. Comparison with Fig. 14.19 shows immediately the dramatic reduction in the photoelectric cross section. The incoherent cross section is now dominant over a considerably larger energy range. To a good approximation, incoherent scattering is the only important interaction in the energy range of about 0.1–5 MeV, and this is found for all of the low-Z elements.

A number of different effects result from the interaction of photons with matter. Coherent scattering will result in a change in the trajectory of a photon but not its energy. Incoherent scattering will result in partial energy loss with the scattered

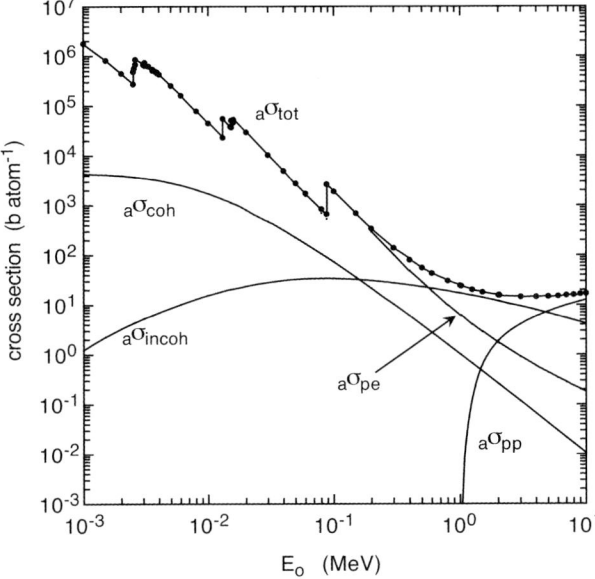

Fig. 14.19 The total cross section and its components for photons interacting with lead atoms. The data were taken from the evaluations of XCOM [2].

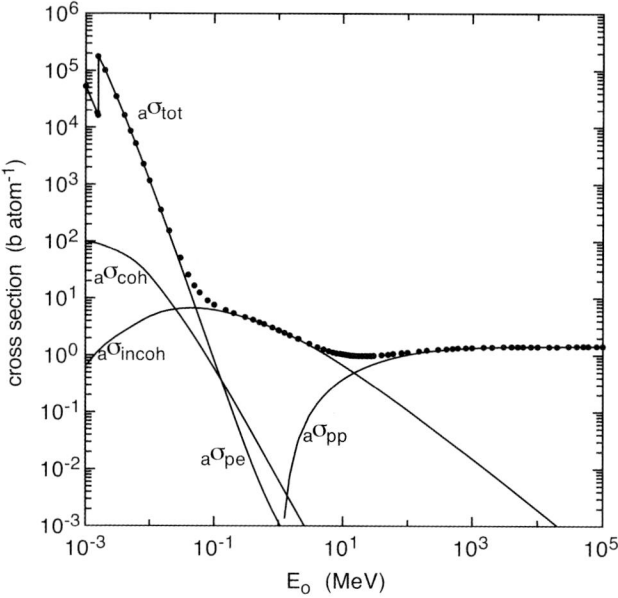

Fig. 14.20 The total cross section and its components for photons interacting with aluminum atoms. The data were taken from the evaluations of XCOM [2].

photon escaping the site of interaction but with the electron depositing its energy not far from the interaction site. The photoelectric effect will result in the loss of the kinetic energy of the electron near the interaction site. Pair production will also result in the loss of the electron and positron kinetic energy near the interaction site, but the subsequent annihilation of the positron will result in the two 511 keV photons being transported beyond that site. Clearly, the situation with respect to energy transport and deposition will be quite complicated and will vary significantly with the energy of the photon.

A detailed description of the energy deposition and its transport following the interaction of photons with matter on a macroscopic scale, is a very complicated proposition. Large computer codes that use a statistical approach are most often used for solving such problems. These "Monte Carlo" codes are capable of providing excellent estimates of energy transport and deposition for almost any problem that arises in practice. For general applications involving relatively simple geometries, or when a quick, rough estimate is needed, a much simpler approach is possible. To understand this, all that is necessary is an examination of an ideal experimental arrangement shown in the schematic of Fig. 14.21. We assume that a beam of monoenergetic photons is incident upon some medium called the *absorber*. The absorber is surrounded by a *collimator* constructed from some high-Z material, such as lead. On the opposite side of the absorber is a detector that has a very high probability of detecting any photons incident upon it. The thickness of the collimator is chosen such that, in the absence of any hole or penetration,

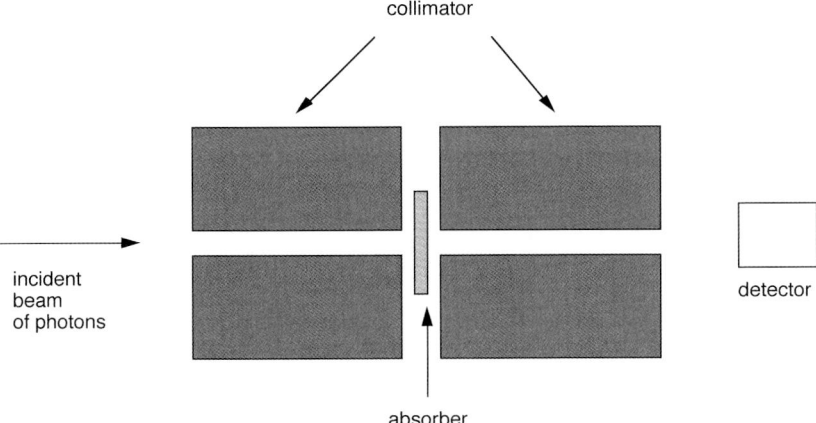

Fig. 14.21 Schematic diagram of a photon beam passing through a hole in a medium that strongly absorbs the incident photons. That part of the beam that passes through the opening in the absorber is incident upon a detector with a high efficiency.

essentially no photons would reach the detector. However, a small cylindrical hole is present that allows a portion of the beam to penetrate and reach the detector. Photons that undergo photoelectric or pair production interactions in the absorber are, of course, completely removed from the beam. Further, in the limit that the diameter of the opening in the collimator goes to zero, any photon suffering coherent or incoherent scattering in the absorber will change trajectory, interact with the collimator and thus not be able to reach the detector. Thus, only source photons that do not interact at all with the absorber will actually penetrate to and be registered in the detector.

As discussed in Chapter III, the photon intensity I (cm^{-2} s^{-1}) that penetrates an elemental absorber of thickness x when a beam of radiation with intensity I_o is incident on it is

$$I = I_o e^{-n_a \sigma_{tot} x} \qquad (14.53)$$

The quantity $n_a \sigma_{tot}$ with the usual dimensions of cm^{-1} represents the probability that an incident photon will be "removed" from the beam per unit thickness of absorber traversed. The sense of the word *removed* is that any interaction that causes the complete disappearance of the photon, a change in its energy, or even a change in its trajectory, represents a change from the source characteristics. These are just the changes that can be distinguished by the experimental arrangement discussed above. We use the word *attenuation* to describe this effect on the incident beam. It is important to recognize that attenuation does not necessarily imply the complete disappearance of a photon.

The quantity $n_a \sigma_{tot}$ is often referred to as the total *macroscopic* cross section and is written as

$$\Sigma_{tot} = n_a \sigma_{tot} \tag{14.54}$$

and, just as in the case where the inverse of the radioactive decay constant $1/\lambda$ represents the mean lifetime of a nuclide, the inverse of the macroscopic cross section represents the thickness of an absorber that results in a photon beam being attenuated to $1/e$ of its incident intensity. This thickness is referred to as the *mean free path* for attenuation. The quantity $n_a \sigma_{tot}$ is also referred to as the *linear attenuation coefficient*, μ_o.

A little care is warranted here. In most references, the linear attenuation coefficient is given as the sum of the incoherent, photoelectric and pair production cross sections. The coherent scattering cross section, which is usually quite small, is neglected. Nevertheless, by the definition of attenuation, any interaction that causes the disappearance, change in energy or change in trajectory of a photon, contributes to attenuation.

Frequently, attenuation coefficients will be given as the ratio of the linear attenuation coefficient to the normal density of a material. This is called the *mass attenuation coefficient* and is defined by

$$\mu = \frac{\mu_o}{\rho} \tag{14.55}$$

with the usual dimensions of $cm^2\, gm^{-1}$.

Given attenuation coefficients, it is a relatively easy matter to estimate shielding requirements for many photon sources, especially if high-Z materials are used. Further, with use of the cross sections for energy scattering and energy absorption from Compton interactions discussed in Section 14.2.2, rather accurate energy transport and energy deposition calculations can be accomplished for simple geometries.

14.3
The Interaction of Charged Particles with Matter

The interaction of charged particles with matter takes place predominantly with the atomic electrons of the medium through which the particle is traversing. To a good approximation, the interaction can be limited to the Coulomb interaction and it is this approximation that we will consider here. Although the interaction with atomic nuclei must be considered in a complete description, the energy loss by such interactions is entirely negligible in almost all practical cases. This is easy to understand when one considers that the Coulomb field of the nucleus is effectively shielded by the bound electrons and, with the exception of those trajectories that would bring the incident particle quite close to the nucleus, an incident particle will not "see" it at all. On the other hand, essentially all random trajectories will experience the fields from the electrons that occupy essentially all of the volume of ordinary matter.

It is convenient to consider the interaction of heavy and light charged particles separately, primarily because of the simple, essentially linear trajectories of the former. Heavy charged particles are all particles with masses much larger than the mass of an electron, including all ions of the chemical elements. For practical purposes, the term "light charged particles" includes just electrons and positrons. In the following, a relatively simple classical model will be developed that gives a clear physical picture of the main features of the energy loss of heavy charged particles. The quantum mechanical theory is quite complex and we will be content with presenting the results from such calculations and discussing their implications.

14.3.1
The Stopping of Heavy Charged Particles in Matter

A relatively simple model of the energy loss of heavy charged particles can be developed with the following assumptions. A charged particle with velocity $v \gg 0$ and mass $m_{HCP} \gg m_e$ is incident upon a homogeneous isotropic medium, the *stopping medium*. The only interactions that result in energy loss are those with electrons. According to Eq. (13.14), the maximum energy loss in an elastic collision of a heavy charged particle with a free electron is

$$\begin{aligned} \Delta E &= T_{HCP} - T'_{HCP} \\ &= T_{HCP}\left[1 - \left(\frac{m_{HCP} - m_e}{m_{HCP} + m_e}\right)^2\right] \\ &\approx \frac{4m_e}{m_{HCP}} T_{HCP} \end{aligned} \qquad (14.56)$$

Even for scattering of protons on electrons, the maximum fraction of the incident kinetic energy that can be transferred in a single collision is about 2×10^{-3}. Thus the large mass of the heavy charged particle guarantees that any single scattering interaction cannot result in the loss of a significant fraction of the particle's kinetic energy and cannot affect the momentum vector of the heavy charged particle very much at all. A very large number of collisions will be needed, on the average, for a heavy charged particle to come to rest in a medium.

Because of the small energy loss per collision and the assumption of an isotropic stopping medium, we can then make the assumption that the trajectory of a heavy charged particle is linear and that, in any single collision, the energy loss is so small that the velocity of the particle can be considered constant. Further, we will assume that the energy loss in a collision can be treated classically and account will be taken of the electron binding in an average way.

The classical model of the energy loss of a heavy charged particle proceeds with reference to the schematic diagram shown in Fig. 14.22. A heavy charged particle with velocity v and charge Z is moving in a medium of constant electron density. An electron located at a distance x and impact parameter b from its trajectory will experience a force of magnitude

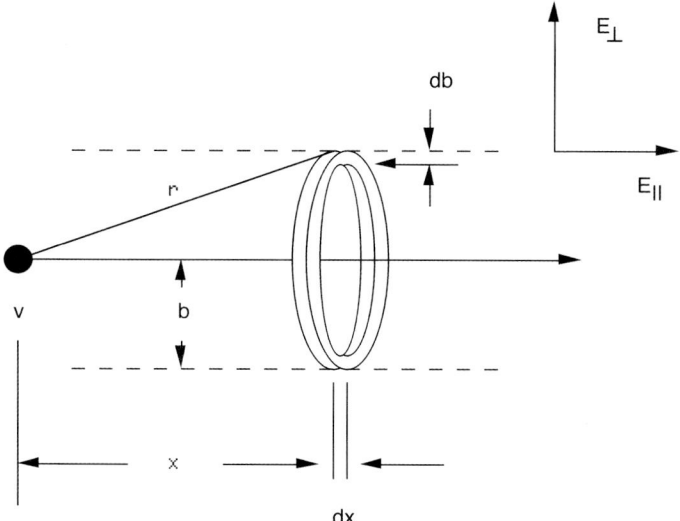

Fig. 14.22 Schematic diagram for the interaction of a heavy charged particle (shown in black) with electrons contained in the annulus of thickness dx extending from b to b + db.

$$F = -k_C \frac{Ze^2}{r^2} = |eE| = e(E_\perp^2 + E_\parallel^2)^{1/2} \tag{14.57}$$

If the electron were held fixed and the trajectory of the charged particle were maintained along the initial trajectory, the force experienced by the electron would be

$$F(x, b) = -k_C \frac{Ze^2}{(x^2 + b^2)} = -k_C \frac{Ze^2}{[(vt)^2 + b^2]}. \tag{14.58}$$

In Fig. 14.23 are shown the forces felt by the electron as a result of a proton moving along a linear trajectory at impact parameters of 2 and 20 Angstroms, respectively, as the proton moves over the range $-\infty \leq x \leq \infty$. For an impact parameter of 2 Å, the force is significant only over a dimension in x of about 30 Å while for b = 20 Å, the force is significant over a dimension of about 100 Å. In the latter case, however, the maximum force felt by the electron is only 0.01 that felt by the electron when the impact parameter is 2 Å.

In general, we will deal with charged particles moving with initial kinetic energies on the order of 1 MeV per nucleon or greater, i.e., 1 MeV protons, 4 MeV α particles, etc. The velocity of such a particle is about 1.5×10^9 cm s^{-1}. The times required for the particle to move a distance of 2 and 20 Å are then about 1.4×10^{-17} and 1.4×10^{-16} s, very short indeed. These very short times, during which the interaction is strong, suggests that a simple impulse approximation should suffice

14.3 The Interaction of Charged Particles with Matter

to estimate the energy transfer. That is, we can assume that during the time of interaction, the electron does not move significantly from its initial position. With this assumption, the momentum transfer to the electron is approximated by

$$\Delta p_e = \int F dt = e \int E dt = e \int (E_\perp \hat{j} + E_\parallel \hat{i}) dt = e \int E_\perp \hat{j} \frac{dx}{v} \quad (14.59)$$

where $\hat{i}$ and $\hat{j}$ are unit vectors parallel to and normal to the trajectory of the heavy charged particle, respectively. The last of the equalities in (14.59) is obtained when it is recognized that the contribution from integration of $E_\parallel dt$ over the range $-\infty \leq x \leq 0$ will exactly cancel the contribution from integration over the range $0 \leq x \leq \infty$.

The evaluation of the last integral in (14.59) is easily accomplished with use of Gauss's law. Namely, the integral of the scalar product of the electric field and an element of surface area that completely encloses an electric charge Ze is given by

$$\int_{\text{surface}} (\mathbf{E} \cdot d\mathbf{S}) = \frac{1}{\varepsilon_0} Ze \quad (14.60)$$

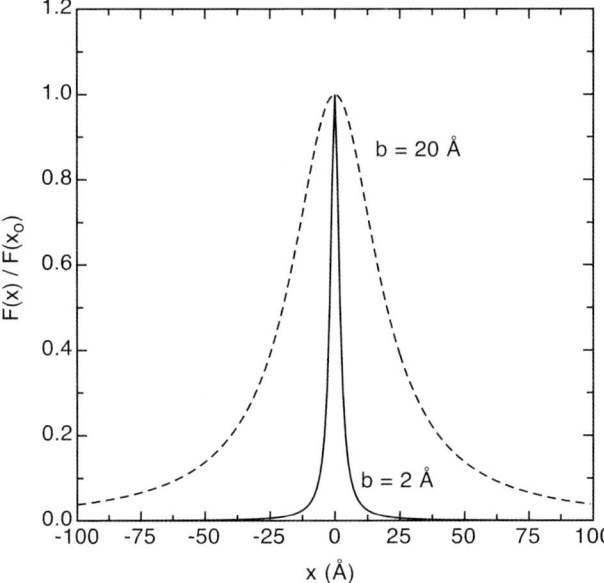

Fig. 14.23 The relative force $F(x)/F(x_o)$ felt by an electron held fixed when a proton passes along a linear trajectory with an impact parameter of 2 and 20 Å. $F(x_o)$ is the force at the electron location when the distance of separation between the proton and electron is the impact parameter b.

For simplicity, we choose a cylindrical surface of radius b whose axis is the trajectory of the ion. The only component of the electric field we need consider is $E_\perp$ and this is everywhere normal to the surface. Therefore

$$\int_{\text{surface}} (\mathbf{E} \cdot d\mathbf{S}) = \int_{\text{surface}} E_\perp \cdot 2\pi b \, dx \tag{14.61}$$

Combining the results of (14.60) and (14.61) we then have

$$\int_{\text{surface}} E_\perp \, dx = \frac{2Ze}{\varepsilon_o b} \tag{14.62}$$

and, from (14.59) we then have the result

$$\Delta p_e = k_C \frac{2Ze^2}{bv} \tag{14.63}$$

The kinetic energy transferred to the electron is then

$$T_e = k_C \frac{2Z^2 e^4}{m_e b^2 v^2} \tag{14.64}$$

This can be used as the estimate of the energy transfer to all electrons with the same impact parameter. If n_e is the electron density, the number of electrons with impact parameters b to b + db in a differential distance dx along the heavy ion trajectory is just $2\pi n_e b \, db dx$ and the total energy transfer to these is $2\pi T_e n_e b \, db dx$. We can take this as the energy loss, –dE, of the heavy charged particle in moving a distance dx. Thus

$$-\frac{dE}{dx}\bigg|_{b \text{ to } b+db} = 2\pi T_e n_e b \, db dx = k_C^2 \frac{4\pi n_e Z^2 e^4}{m_e v^2} \left(\frac{db}{b}\right) \tag{14.65}$$

The total energy loss per unit distance traversed is now obtained by integration over all possible impact parameters. Therefore,

$$-\frac{dE}{dx} = k_C^2 \frac{4\pi n_e Z^2 e^4}{m_e v^2} \int \frac{db}{b} \tag{14.66}$$

The integration over the impact parameter must be handled with some care. Integration over all possible impact parameters $0 \le b \le \infty$ will clearly get us into trouble. Further, it is not physically reasonable to integrate over this range. At b = 0, the energy loss rate would be infinite. But we know physically that there is a maximum energy that can be transferred per collision and that is the energy transfer allowed in a head-on collision. In such a collision it is not difficult to show that the momentum transfer is just

$$\Delta p_e = 2 m_e v \tag{14.67}$$

in the limit that the mass of the ion is very much larger than the mass of the electron. As a result, the maximum kinetic energy transfer is $(T_e)_{max} = 2 m_e v^2$. Within the spirit of our computation, this maximum energy transfer can be taken to represent the minimum value of a physically acceptable impact parameter. Equating this result to the expression given in (14.64) then gives

$$b_{min} = k_C \frac{Ze^2}{m_e v^2} \tag{14.68}$$

Up to the present, our derivation of the energy loss has treated the electrons as though they were free. While this is a reasonable approximation for the valence electrons in atoms, the most weakly bound electrons in molecules and the conduction electrons in metals, it is certainly not reasonable when considering the most tightly bound electrons, especially the K and L electrons in the high-Z elements. Somehow, the free electron approximation must be modified to take electron binding into account. We can do this in a schematic way by defining an *effective ionization constant*, I, for each element in the periodic table. In principle, this constant could be calculated theoretically. But it is usually taken as an empirical constant derived by fitting the energy loss expression to experimental data. Given I, energy transfers corresponding to $T_e < I$ are simply forbidden and impact parameters that are sufficiently large that they would produce energy transfers less than I will not contribute to the integral in Eq. (14.66). Hence, we can define an effective maximum impact parameter b_{max} by setting

$$I = k_C^2 \frac{2 Z^2 e^4}{m_e b_{max}^2 v^2} \tag{14.69}$$

or,

$$b_{max} = k_C \frac{Ze^2}{v} \left(\frac{2}{m_e I} \right)^{1/2} \tag{14.70}$$

If we now use these limits in (14.66), we obtain

$$-\left(\frac{dE}{dx}\right)_{class} = k_C^2 \frac{4\pi n_e Z^2 e^4}{m_e v^2} \ln \left(\frac{2 m_e v^2}{I} \right)^{1/2} \tag{14.71}$$

We have used the subscript "class" to indicate that the energy loss expression was obtained by a simple classical model.

A more rigorous approximation that is obtained with use of quantum mechanics that takes into account relativistic effects at high energies gives

$$-\left(\frac{dE}{dx}\right)_{QM} = k_C^2 \frac{4\pi n_e Z^2 e^4}{m_e v^2} \ln \left[\frac{2 m_e v^2}{I(1-\beta^2)} - \beta^2 \right] \tag{14.72}$$

A comparison of (14.71) and (14.72) shows that the classical model differs from the quantum mechanical result by a factor of 2 in the limit of small velocities. Thus, the physical ideas presented in the classical model are essentially correct and the insight gained in the derivation was well worth the effort.

The energy loss given by (14.72) is referred to as the *stopping power* due to ionization. In essentially all practical cases it can be taken as the total stopping power for heavy ions. Apart from problems encountered at low velocities of an ion, the stopping power expression has been shown to represent the energy loss very well, even for highly relativistic particles.

The stopping power expression is quite remarkable because the only properties of the heavy ion that are included are its velocity and ionic charge. Thus, regardless of their masses, all heavy ions with the same velocity are predicted to suffer an energy loss per unit distance traversed that is directly proportional to the squares of their ionic charges. In principle, then, a measurement of the stopping power of protons in a material can be used to estimate the stopping power for all other heavy ions that one might wish to consider. Again, neglecting very low velocities, the stopping power of all heavy ions are found to be approximately Z^2 times the stopping power of protons of the same velocity moving in the same material. In this sense, we can consider the result given in (14.72) as a universal function, at least in first order. As examples, the stopping powers for protons traversing aluminum and gold at their normal densities calculated with (14.72) are shown in Fig. 14.24 in a log-log representation to illustrate the predicted low- and high-energy variations.

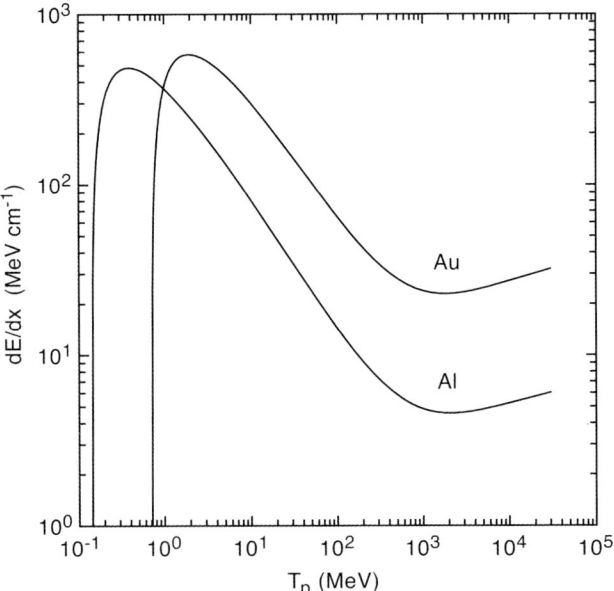

Fig. 14.24 The estimated stopping powers for protons in aluminum and gold when these elements are at their normal densities.

For these calculations, the effective ionization constant was taken from the empirical relation [3]

$$I \approx 9.1Z_o(1 + 1.9Z_o^{-2/3}) \text{ eV} \tag{14.73}$$

where Z_o is the atomic number of the stopping medium.

The stopping power is predicted to increase from zero when $2m_ev^2 = I$, go through maxima at energies in the range of about 0.3 and 2 MeV, respectively, decrease smoothly until the kinetic energy is about 1–3 m_pc^2 and then begin a monotonic increase, the latter due to relativistic effects. In the vicinity of the peaks, the stopping power is roughly 500 MeV cm^{-1}, or about 0.05 MeV μ^{-1}. We can therefore conclude that low-energy protons will penetrate only some tens of microns before they lose all of their kinetic energy. For α particles of the same total kinetic energy, we expect even shorter distances traversed before their kinetic energies fall to zero.

Of particular importance in a variety of applications is the distance penetrated by a heavy ion before it is "stopped", and the energy deposited along its trajectory. We can get both of these from the stopping power expression of Eq. (14.72). For reasons that will become clear later, the distance traversed by a heavy charged particle that is calculated by use of the stopping power expression is known as the *average range* of the particle in the medium in question. We can calculate our estimate of this average range in the following way. Formally, the range is just the sum of the differential path lengths, dx, traversed by the ion as it loses energy. That is,

$$R = \int_{x=0}^{x=R} dx \tag{14.74}$$

We can relate this to the stopping power by noting that the inverse of the latter multiplied by a differential in energy is just dx, i.e.,

$$R = \int_{E_o}^{0} \left(\frac{dE}{dx}\right)^{-1} dE \tag{14.75}$$

The initial energy of the ion, the energy at $x = 0$, is defined as E_o. The final energy is, of course, zero when the ion has traversed the distance $x = R$. To get an explicit expression for the integrand we need to obtain the differential of the energy. Although we have used $-\frac{dE}{dx}$ to represent the stopping power and be consistent with common usage, we must remember that it is really the kinetic energy T with which we are concerned. As shown in Eq. (2.11), the kinetic energy of a relativistic particle is related to the rest mass by

$$T = (\gamma - 1)m_oc^2 = \left(\frac{1}{\sqrt{1-\beta^2}} - 1\right)m_oc^2 \tag{14.76}$$

and differentiation with respect to β gives

$$\frac{dT}{d\beta} = \frac{m_o c^2 \beta}{(1-\beta^2)^{3/2}} \tag{14.77}$$

We can now use Eqs (14.72) and (14.77) in (14.75) to obtain the explicit expression

$$R = \int_{\beta(E_o)}^{0} \left\{ k_C^2 \frac{4\pi n_e Z^2 e^4}{m_e v^2} \ln\left[\frac{2 m_e v^2}{I(1-\beta^2)} - \beta^2\right] \right\}^{-1} \frac{m_o c^2 \beta}{(1-\beta^2)^{3/2}} d\beta$$

$$= \frac{(m_o c^2)(m_e c^2)}{k_C^2 4\pi Z^2 e^4} \int_{\beta(E_o)}^{0} \frac{\beta^3}{n_e(1-\beta^2)^{3/2} \ln\left[\frac{2 m_e c^2 \beta^2}{I(1-\beta^2)} - \beta^2\right]} d\beta \tag{14.78}$$

This rather messy looking equation has a very simple structure that points out a very important generality. The integral is a function only of the velocity of the heavy ion and the properties of the stopping medium as expressed in n_e and I. It contains no reference to the mass or charge of the ion. Hence, it is a universal function for all ions traversing a given medium. It can be calculated once and for all. We therefore write

$$R = \frac{(m_o c^2)(m_e c^2)}{k_C^2 4\pi Z^2 e^4} \int_{\beta(E_o)}^{0} f(\beta) d\beta \tag{14.79}$$

where,

$$f(\beta) d\beta = \frac{\beta^3}{n_e(1-\beta^2)^{3/2} \ln\left[\frac{2 m_e c^2 \beta^2}{I(1-\beta^2)} - \beta^2\right]} d\beta \tag{14.80}$$

Once the universal integrals have been performed, we can get the range of any heavy ion in the medium by adjusting the mass and charge only.

Suppose we have calculated the ranges of protons of all practical energies. If we take the ratio of the range for an arbitrary heavy ion to that of a proton with the same initial velocity, β, we have from Eq. (14.79),

$$\left.\frac{R_{HI}}{R_p}\right|_{\beta_p = \beta_{HI}} = \frac{m_{o,HI} Z_p^2}{m_{o,p} Z_{HI}^2} \approx \frac{A_{o,HI}}{Z_{HI}^2} \tag{14.81}$$

where $A_{o,HI}$ is the mass number of the heavy ion. This result is referred to as the *range-scaling* law. It says that the range of any heavy ion can be obtained from the range of a proton (or any other heavy ion for that matter) that has the same velocity. This is a very nice simplification for estimating the ranges of ions in matter. Thus, a proton and an α particle with the same velocity should have the same range in a

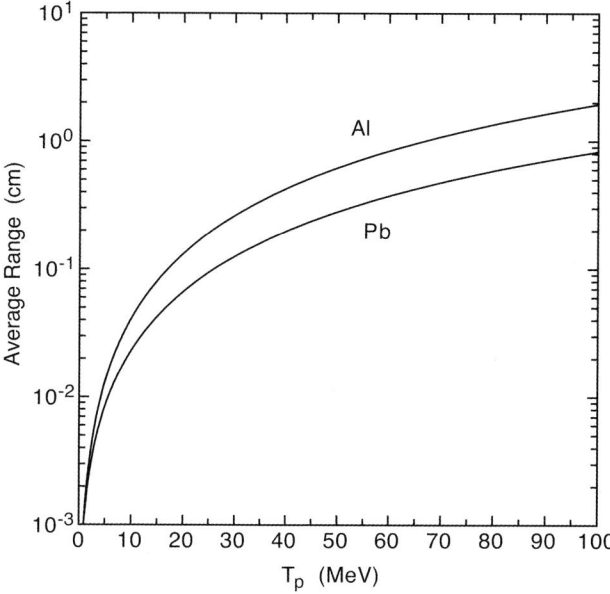

Fig. 14.25 The ranges of protons in aluminum and lead at normal densities.

given material. These two particles will have the same velocity if the energy of the α particle is about four times that of the proton.

To set the scale for ranges of ions in condensed matter and to demonstrate the expected dependence of range on initial energy, Eq. (14.80) has been used to calculate the range of protons in aluminum and lead as shown in Fig. 14.25. As can be seen, protons with energies of a few tens of MeV or less will lose their kinetic energies in linear dimensions of a small fraction of a centimeter. We can clearly infer that α particles with energies of 5–8 MeV, the energies commonly found in α decay, will have ranges on the order of 1–10 μm in these materials. Ions with larger ratios of $Z^2/m_{o,HI}$ can be expected to have even smaller ranges. The simple fact that can be learned from these model calculations is that the range of heavy charged particles in matter is generally quite small, owing to the very large number of collisions that take place with the electrons.

The calculations presented above are based on a rather simple model. While we can expect it to provide the correct order of magnitude of the stopping power and ranges, we should not expect it to provide quantitative information with high accuracy. The principal deficiencies in the model are rather easy to see. First, the ionization potential is essentially an empirical average that cannot be expected to be very accurate at low energies, especially in higher-Z materials where K and L binding energies are significant. Second, the model implicitly assumes that the heavy ion is traveling as a fully ionized atom, completely stripped of all atomic electrons. We know that this cannot be true as the ion velocity goes to zero. In fact, only at very high energies can we expect the ionic charge of all but the lightest ions

$$HI^{+Z} + SM \;\rightleftarrows\; HI^{+(Z-1)} + SM^{+1} \;\rightleftarrows\; HI^{+(Z-2)} + SM^{+2} \;\rightleftarrows\; \ldots\ldots$$

Fig. 14.26 Schematic diagram of charge exchange between a heavy ion and the stopping medium. The dots to the right of the figure are meant to indicate that the exchange reactions can proceed to lower ionic charges on the heavy ion.

to be Ze. For any arbitrary ion, we can expect that it will bind an electron and ionize the medium itself so long as this will result in a lower energy for the system. As the heavy ion moves through matter there will be some probability that it will "pick up" an electron and even some probability that an electron bound to it can be taken up, once again, by the stopping medium. The process is statistical in nature. We can picture it schematically as shown in Fig. 14.26. We suppose that the heavy ion is moving initially with sufficiently high speed that is fully ionized. As it slows down the process of *charge exchange* with the stopping medium begins and the ion captures an electron from the stopping medium. As it continues to move, the process of charge exchange can cause loss of the captured electron or capture of additional electrons. The probability for capture or release depends upon the binding energies of the electrons on the ion and the energies of the electrons in the stopping medium. On the average, especially for ions of high atomic number, there will be a continuous reduction in the ionic charge as it slows down. Because the stopping power depends upon the square of the ion's charge, the process of charge exchange will lead to significantly larger ranges than estimated by Eq. (14.72) when initial ion velocities are small.

The third effect not contained in the model is the interaction of low-energy ions of low ionic charge with the atoms in the stopping medium. At low energies, there is a significant probability that ion–atom and ion–ion collisions will be effective in the transfer of energy from the heavy ion to the stopping medium. This mechanism is referred to as *nuclear stopping* as opposed to the electronic stopping that we have considered up to now. Such collisions involve partners of heavy mass, sufficiently large that they can actually cause an ion in a lattice to be ejected from its normal site. Nuclear stopping is especially important for heavier ions and it reduces the range of an ion relative to that given by our model that considers only the interaction with electrons.

It should be clear that the interaction of heavy ions with matter is an extremely complicated process. It is fair to say that no complete theory has yet been developed. Rather, a combination of empirical data and relatively simple theory has been used to develop models that can predict the range and energy loss of heavy ions quite accurately in essentially all materials. There are, as you may suspect, many approaches that can be taken in this effort. Perhaps the most successful and easiest to understand in principle makes use of the fact that as far as we can tell, protons remain completely stripped of their orbital electrons at all but such low energies that the error in assuming that they are stripped is negligible. This suggests that measurements of the stopping power and range of protons in matter can serve as a reference for the stopping of all other ions. Because the only difference between the stopping of protons and any other heavy ion in a given material is the ionic charge of the ion for the same velocities, we can use this reference and empirical stopping powers of some heavy ions to develop a model that predicts the ionic charge of the latter as a function of their velocity. If the model is reasonable, we can obtain a rather accurate means of estimating stopping power and ranges of any and all heavy ions in matter.

Stopping powers and ranges have been calculated from such approaches and can be found in a number of references. To illustrate the results of one such model, we show in Fig. 14.27 the stopping powers and in Fig. 14.28 the ranges of a number of ions in different materials that have been calculated with the model developed by Ziegler and co-workers that is described in detail in the references [4].

In keeping with standard practice, the stopping powers shown in the figure are expressed in MeV cm^{-1} divided by the normal density ρ of the element in question, $-\frac{dE}{d\rho x}$, and thus have the dimensions of MeV (mg cm^{-2})$^{-1}$. The quantity ρx is known as the *areal density* and the stopping powers are properly referred to the energy loss per unit areal density. This form is regularly used because it represents the stopping power per unit of the physical material normal to the trajectory of the heavy charged particle, regardless of the physical state of the matter. Clearly, the stopping power of protons in gaseous or solid xenon ought to be the same in units of areal density to an excellent approximation. Further, one often encounters mixtures or compounds as the stopping medium. In that case one can, to an excellent approximation, determine the total stopping power by summing the stopping powers for each with areal densities easily calculated from the known density and composition of the compound or mixture.

In Fig. 14.27 it is clear that the stopping powers of protons in different materials have very different magnitudes especially at energies below about 100 keV. Even greater differences will be found when we consider heavier ions where significant differences in nuclear stopping will be found.

Although we have alluded to some of the statistical aspects involved, we have to this point considered the stopping power calculation as one that is fundamentally deterministic. But this is hardly the case. A more fundamental approach would directly consider the cross sections for energy transfer to electrons. Given the random trajectory of the heavy ion, the energy-dependent cross section and the fact that the electron density can be expected to fluctuate somewhat about its mean, the stopping power of a heavy charged particle is subject to considerable statistical fluctuation about its mean. This implies that the ranges of the particles in a beam will not all be the same nor will they all have exactly linear trajectories. There will be some uncertainty in the actual distance penetrated into a material. This effect is referred to as *straggling*. It is not very important for high-energy ions but can be quite important for low-energy ions. While we will not delve into the matter further, it is important to recognize that the calculations we have presented represent the mean distance *traveled* by a heavy ion until its kinetic energy is reduced to zero. As such the actual distance of penetration into a medium will tend to be somewhat less than calculated. The effect is not very great for heavy ions but is quite significant for the stopping of electrons and positrons to which we now turn.

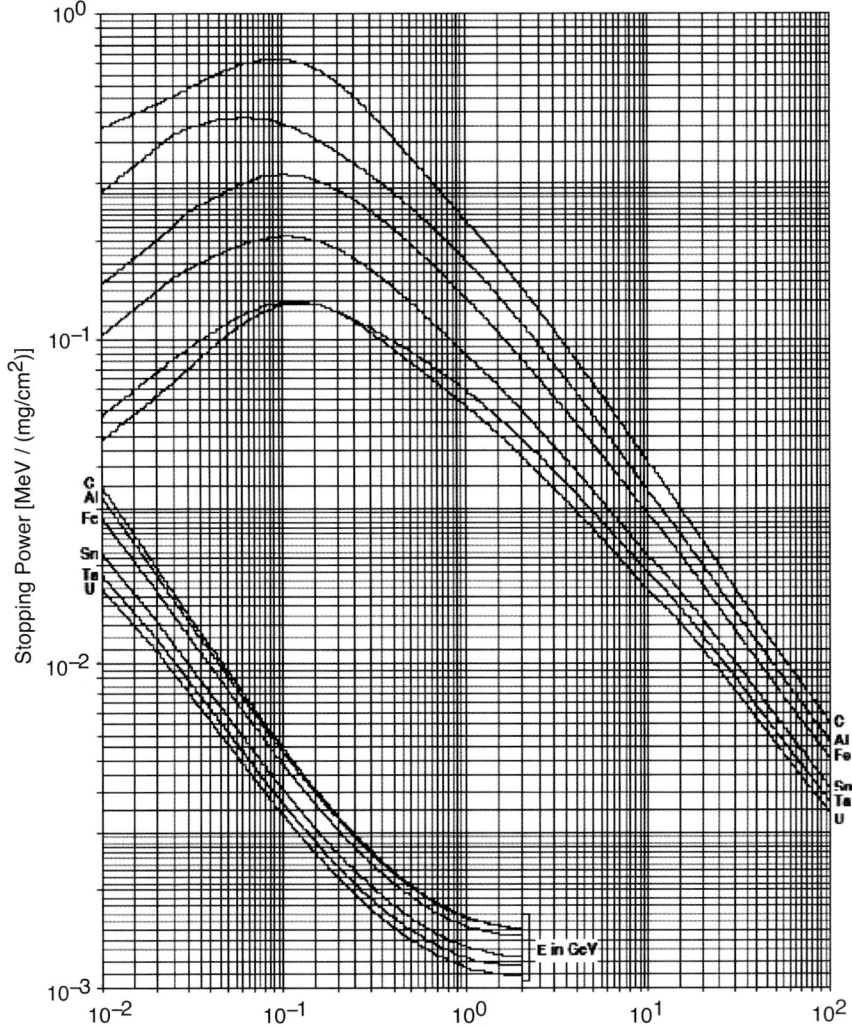

Fig. 14.27 The stopping power of protons in various materials over the energy range 10 keV - 2 GeV. The stopping powers were calculated with the model developed by Ziegler and co-workers. The ordinate is given in units of MeV (mg cm^{-2})$^{-1}$ for the elements at their normal densities and the abscissa is in units of MeV.

14.3 The Interaction of Charged Particles with Matter | 537

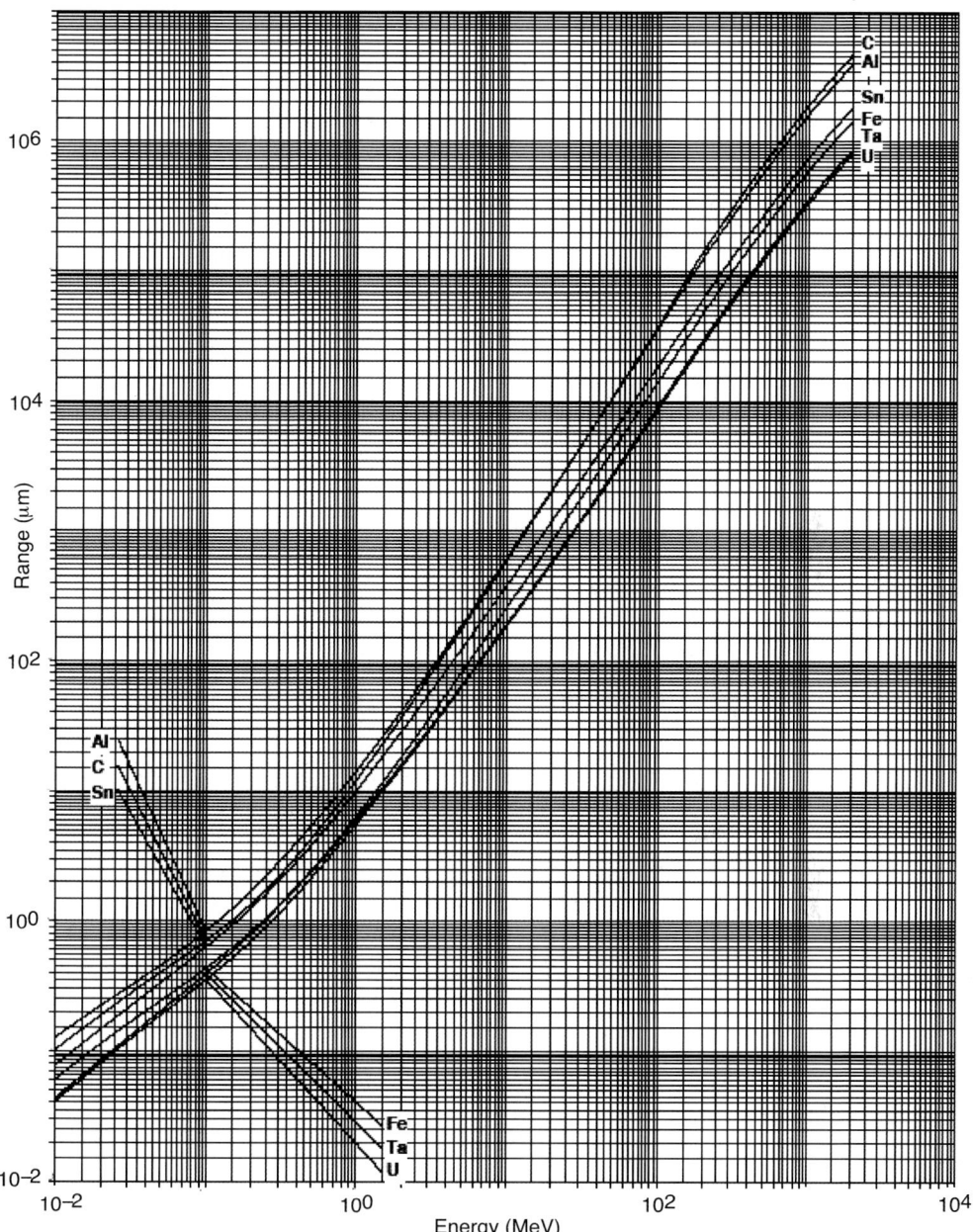

Fig. 14.28 The range of protons in different elements calculated with the model developed by Ziegler and co-workers.

14.3.2
The Stopping of Electrons and Positrons in Matter

The classical stopping power model that we discussed in the previous section was based on the assumptions that the trajectory of the heavy ion was linear and the time required for interaction with an electron is so short that an impulse approximation can be used to estimate the average energy transfer. While the latter will clearly apply to the interaction of high-energy electrons with the electrons in the stopping medium, the assumption of a linear trajectory cannot be satisfied. A single collision of a projectile with a target, when the two have identical masses, can result in the complete transfer of the projectile kinetic energy. As you will recall from our discussion of the kinematics of nuclear reactions, the scattering of particles of identical mass will result in both the target and projectile appearing at forward angles in the laboratory. Thus, a high-energy electron that scatters from an electron at rest can suffer rather wide angle scattering and therefore cannot be expected to have a truly linear trajectory.

The situation is further complicated by the fact that, in the scattering of electrons on electrons, we have no means of identifying which of the particles was the projectile and which was the target after the collision. A head-on collision, in which all of the kinetic energy of the projectile is transferred to the target electron, will appear to us as no collision at all! The "target" will simply move away from the site of the interaction with exactly the same kinetic energy as the projectile had.

Even with these complications, we can still use the basic ideas of the classical model to understand the stopping of electrons in matter because, as seen in Eq. (14.72), the stopping power is not dependent upon the mass of the charged projectile. It depends only on its velocity. Therefore, if we relax the notion of the stopping power describing the energy loss along the linear trajectory of the projectile to one that describes the energy loss along the *actual* trajectory, no matter how nonlinear it is, the same basic physics should apply. In fact, accounting for the identity of the projectile and target in the stopping of electrons in matter, the quantum mechanical result for the energy loss of electrons by ionization is very nearly the same as that for the stopping of protons in matter. The result for the energy loss of electrons by ionization can be written in the form

$$-\left(\frac{dE}{dx}\right)_{e,\,ion} = k_{\bar{C}}^2 \frac{4\pi n_e e^4}{m_e v^2} \left[\ln \frac{2m_e c^2}{I} + \ln(\gamma - 1) + \frac{1}{2}\ln(\gamma + 1) \right.$$
$$\left. -\left(3 + \frac{2}{\gamma} - \frac{1}{\gamma^2}\right) \ln 2^{1/2} + \frac{1}{16} - \frac{1}{8\gamma} + \frac{9}{16\gamma^2} \right] \quad (14.82)$$

where $\gamma = (1 - \beta^2)^{-1/2}$. A direct comparison with (14.72) shows that the ratio of the constants preceding the bracketed terms as well as the first term in brackets are identical if $Z = 1$. It is not difficult to show that the stopping powers of electrons and protons by ionization do not differ by more than about 13% over the energy range of 0.01–10 MeV. Therefore for simple estimates, one can take the stopping

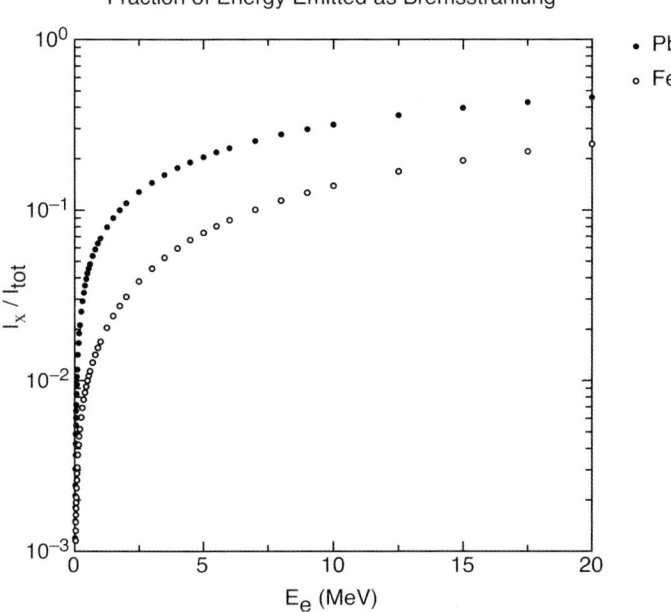

Fig. 14.29 The fraction of electron kinetic energy emitted as bremsstrahlung for electrons slowing down in iron and lead.

power of electrons by ionization as the same as that of protons with the same velocity.

There is, however, a rather significant difference between the total energy loss rate of electrons and heavy ions. As you will recall classically, a charge that is accelerating in an electric field will emit energy as electromagnetic radiation. The theory is a bit complex and will not be discussed here. But experiment is in agreement with theory that the probability for emission of electromagnetic energy is inversely proportional to the square of the mass of the particle. As a result, energy loss by emission of electromagnetic radiation is generally negligible for heavy ions such as protons, but can be significant in the case of the stopping of electrons and positrons. This radiation is known as *bremsstrahlung*, literally translated from German to English as "braking radiation". The probability for bremsstrahlung in the field of a nucleus is proportional to the square of the atomic number. As a result, it is generally small for electrons stopping in media composed of low-Z nuclides but can be significant in the stopping of electrons in high-Z elements such as lead. In Fig. 14.29 are shown calculations of the fraction of total energy loss due to bremsstrahlung when electrons slow down in iron and lead [5]. Clearly, bremsstrahlung radiation is only significant for electron energies above a few MeV. It is therefore not very important for energy loss of electrons emitted in most radioactive decays. It is, however, very important for providing a source of continuous radiation extending up to the electron kinetic energy for applications in radiography for both medical and industrial applications. Indeed, such sources are the mainstay of the common medical procedure of computed axial tomography (CAT scans) and are commonly employed for analysis of defects in jet engines, etc.

The problem of large-angle scattering makes the calculation of the range of electrons and positrons in matter very complicated and there is still considerable effort being expended to study this problem. For practical applications, a number of studies have combined experimental determinations of the range of monoenergetic electrons and electrons from β decay to develop useful correlations for range estimations. One of the more useful of these are the Katz–Penfold [6] relations. These represent empirical fits to experimental data on the range of electrons in aluminum at normal density. The range in g cm^{-2} is parameterized as

$$\bar{R} = 0.412 E^n, \quad n = 1.265 - 0.0954 \ln E \quad (0.01 < E < 3 \text{ MeV})$$
$$= 0.530 E - 0.106 \quad (2.5 < E < 20 \text{ MeV}) \quad (14.83)$$

One must exercise some care when using these range estimates because of the rather extensive straggling of electrons. Nevertheless, the predictions are in good agreement with the ranges measured for the maximum range of the continuous electrons from β decay and from ranges obtained from beams of monoeneregetic electrons. In Fig. 14.30 the Katz–Penfold correlations are shown graphically over the energy range 0.01–10 MeV.

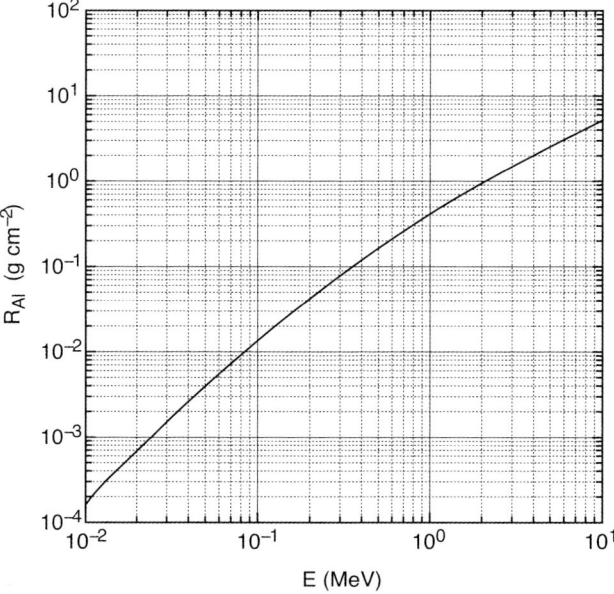

Fig. 14.30 Graphical display of the ranges of electrons in aluminum as given by the Katz–Penfold relations.

Given the ranges in aluminum, we can obtain good approximations for the ranges of electrons in other materials composed of the lighter elements. This relies on the fact that for all elements with $Z \leq 20$, the ratio N/Z of isotopes found in nature is approximately unity. If ρ is the mass density and A and Z are the mass and atomic numbers of the element in question, the electron density is given by

$$n_e = \frac{\rho}{A} Z N_o \approx \frac{\rho Z N_o}{2Z} = \frac{\rho N_o}{2} \qquad (14.84)$$

Therefore, any material composed of light elements will have the same areal electron density. The range of electrons in any material composed of light elements will be approximately the same as the range in aluminum.

References

1. J.H. Hubbell, W.J. Veigele, E.A. Briggs, R.T. Brown, D.T. Cromer, R.J. Howerton, "Atomic form factors, incoherent scattering functions, and photon scattering cross sections", J. Physical and Chemical Reference Data 4 (1975) 471.
2. National Institute of Standards and Technology, XCOM: Photon Cross Sections Database. Available on the worldwide web at http://www.physics.nist.gov/PhysRefData/Xcom.
3. See P. Marmier and E. Sheldon, "Physics of Nuclei and Particles", Volume I, Academic Press, New York (1969).
4. J.F. Ziegler, J.P. Biersack, and U. Littmark, The Stopping and Range of Ions in Solids, Vol. 1 of "The Stopping and Ranges of Ions in Matter", Pergamon Press, New York (1985), TRIM version 92.16, J.F. Ziegler and J.P. Biersack; updated version of a computer code for calculating stopping and ranges, described in reference 1, and H.H. Andersen and J.F. Ziegler, Hydrogen Stopping Powers and Ranges in All Elements, Vol. 2 of "The Stopping and Ranges of Ions in Matter", Pergamon Press, New York (1977).
5. National Institute of Standards and Technology, ESTAR Database Program, available on the worldwide web at http://www.physics.nist.gov/PhysRefData/Star/Text/ESTAR.html.
6. L. Katz and A. S. Penfold, "Range-Energy Relations for Electrons and the Determination of Beta-Ray End-Point Energies by Absorption", Rev. Mod. Phys. 24 (1952) 28.

General References

The best discussion of the interaction of radiation with matter at about the level of this text is given in R.D. Evans," The Atomic Nucleus", McGraw-Hill Book Co., New York (1955); reprinted by Krieger Publishing Co., Malabar, Florida (1982).

An excellent reference for the applications of the interaction of radiation with matter, as they relate to radiation detection and measurement, is given in G.E. Knoll, "Radiation Detection and Measurement", J. Wiley and Sons, New York (2000).

Problems

1. (a) Calculate the energy of the Compton electron when a 1.00 MeV photon undergoes Compton scattering and the scattered photon appears at an angle of 45° relative to the initial trajectory of the incident photon.
(b) Calculate the angle at which the Compton electron appears relative to the trajectory of the incident photon.

2. Monoenergetic photons undergo Compton scattering and the maximum energy of the Compton scattered electrons is 1.00% of the energy of the incident photons. What is the incident photon energy?

3. In the limit of the nonrelativistic form of Eq. (14.72), determine an expression for the maximum in the stopping power of a heavy charged particle.

4. Estimate the depth of penetration of an (a) 10 keV, (b) 100 keV and (c) 1.00 MeV electron in aluminum. Give your answer in cm. The density of aluminum is 2.7 g cm^{-3}.

5. Assume that a semi-infinite slab of homogeneous, isotropic material contains a uniform distribution of a monoenergetic α emitter. Assume further, that straggling can be neglected, so that the range of all α particles is identical. Derive an expression for the energy spectrum of the α particles that emerge from the surface of the medium.

6. Beams of light heavy ions have been used to deliver high-level irradiation of tumors deep within the body. The idea is to choose an incident energy such that the beam stops at the desired depth. Because of the strong peak in the stopping power near the end of the range of the charged particles, most of the energy loss will occur where desired and not in the healthy tissue through which the beam must penetrate.

Suppose that the soft tissue of the body can be approximated as water and that 10 cm of tissue must be penetrated to reach a tumor. A beam of ^{20}Ne ions is used to provide the energy loss at the tumor site. Estimate the kinetic energy required such that the range of the neon ions in water is 10 cm. You may assume that heavy-ion ranges in low-Z materials are inversely proportional to the materials' electron densities.

7. Consider Eq. (14.72) in the general form $-(dE/dx) = g(v)$, where v is the velocity of a heavy charged particle.
(a) Derive an equation for the time required for a heavy charged particle to suffer complete loss of kinetic energy.
(b) Use the information in Fig. 14.24 to obtain a crude estimate of the time required for a 2 MeV proton to lose 200 keV in gold (Au).

(c) Estimate the time required for complete loss of kinetic energy by a 2 MeV proton in Au by numerical integration of the expression in part (a). Discuss any limitations to the stopping power expression and whether this estimate is expected to be too small, too large or just about right.

Appendix 1
Atomic Masses

The data presented here represent a selection of the atomic masses presented in "The 1995 update to the atomic mass evaluation" by G.Audi and A.H.Wapstra, Nuclear Physics *A595* vol. 4 p.409–480, December 25, 1995. The masses represent the best evaluated data, i.e., *recommended* masses, and are given as the mass excesses in keV and masses in atomic mass units, u. Although no errors are given, they typically range from less than 1 keV to several hundred keV. When the highest accuracy is needed, the reader should refer to the original article or the downloadable forms from the National Nuclear Data Center available at http://www.nndc.bnl.gov.

The nuclides included in the table are those for which mass measurements have been made, for which the ground states have known half-lives and where the half-lives are are long compared to the characteristic nuclear time. The columns in the table represent

1. The atomic number Z.
2. The chemical symbol El.
3. The mass number A.
4. The mass excess in keV.
5. The mass in atomic mass units u.

0	n	1	8071.32	1.0086649	3	Li	6	14086.31	6.0151223
							7	14907.67	7.0160040
1	H	1	7288.97	1.0078250			8	20946.19	8.0224867
		2	13135.72	2.0141018			9	24953.90	9.0267891
		3	14949.79	3.0160493			10	33050.23	10.0354809
							11	40795.86	11.0437961
2	He	3	14931.20	3.0160293					
		4	2424.91	4.0026032	4	Be	7	15769.49	7.0169292
		5	11386.23	5.0122236			8	4941.66	8.0053051
		6	17594.12	6.0188881			9	11347.58	9.0121821
		7	26110.26	7.0280305			10	12606.58	10.0135337
		8	31597.98	8.0339218			11	20173.97	11.0216576

Nuclear Physics for Applications. Stanley G. Prussin
Copyright © 2007 WILEY-VCH Verlag GmbH & Co., Weinheim
ISBN: 978-3-527-40700-2

Appendix 1 Atomic Masses

Z	El	A	Δ	Mass
		12	25076.40	12.0269206
		14	39882.40	14.0428155
5	B	8	22921.00	8.0246067
		10	12050.76	10.0129370
		11	8667.98	11.0093055
		12	13368.90	12.0143521
		13	16562.21	13.0177803
		14	23663.73	14.0254041
		15	28966.94	15.0310973
		17	43716.31	17.0469314
6	C	9	28913.65	9.0310401
		10	15698.57	10.0168531
		11	10650.53	11.0114338
		12	0.00	12.0000000
		13	3125.01	13.0033548
		14	3019.89	14.0032420
		15	9873.14	15.0105993
		16	13694.12	16.0147012
		17	21036.59	17.0225837
		18	24924.04	18.0267570
		19	32833.38	19.0352481
		20	37560.06	20.0403224
7	N	12	17338.08	12.0186132
		13	5345.46	13.0057386
		14	2863.42	14.0030740
		15	101.44	15.0001089
		16	5683.43	16.0061014
		17	7870.82	17.0084497
		18	13117.14	18.0140818
		19	15860.45	19.0170269
		20	21766.49	20.0233673
		21	25231.91	21.0270876
		22	32080.89	22.0344402
8	O	13	23110.74	13.0248104
		14	8006.46	14.0085953
		15	2855.39	15.0030654
		16	−4737.00	15.9949146
		17	−809.00	16.9991315
		18	−782.06	17.9991604
		19	3333.57	19.0035787
		20	3796.91	20.0040761
		21	8061.74	21.0086546
		22	9284.35	22.0099672
		23	14616.37	23.0156913
		24	18974.46	24.0203699
9	F	17	1951.70	17.0020952
		18	873.43	18.0009377
		19	−1487.40	18.9984032
		20	−17.40	19.9999813
		21	−47.58	20.9999489
		22	2793.78	22.0029992
		23	3329.52	23.0035744
		24	7544.51	24.0080994
		25	11266.38	25.0120950
10	Ne	17	16485.17	17.0176976
		18	5306.78	18.0056971
		19	1751.06	19.0018798
		20	−7041.93	19.9924402
		21	−5731.72	20.9938467
		22	−8024.34	21.9913855
		23	−5153.64	22.9944673
		24	−5947.52	23.9936151
		25	−2058.70	24.9977899
		26	429.88	26.0004615
		27	7093.51	27.0076152
		28	11278.59	28.0121081
		29	18020.59	29.0193459
11	Na	20	6844.86	20.0073483
		21	−2184.26	20.9976551
		22	−5182.10	21.9944368
		23	−9529.49	22.9897697
		24	−8417.60	23.9909633
		25	−9357.46	24.9899544
		26	−6902.47	25.9925899
		27	−5580.86	26.9940087
		28	−1033.58	27.9988904
		29	2618.71	29.0028113
		30	8594.42	30.0092265
		31	12663.76	31.0135951
		32	18303.66	32.0196498
		33	25509.89	33.0273860
12	Mg	20	17570.53	20.0188627
		21	10911.68	21.0117142
		22	−396.77	21.9995741
		23	−5472.67	22.9941249
		24	−13933.38	23.9850419
		25	−13192.73	24.9858370
		26	−16214.48	25.9825930
		27	−14586.50	26.9843407
		28	−15018.75	27.9838767
		29	−10661.19	28.9885547
		30	−8882.23	29.9904645

		A	Mass excess	Atomic mass
		31	−3215.09	30.9965485
		32	−795.60	31.9991459
		33	5204.23	33.0055870
		34	8450.92	34.0090724
13	Al	23	6767.21	23.0072649
		24	−55.04	23.9999409
		25	−8915.74	24.9904286
		26	−12210.34	25.9868917
		27	−17196.83	26.9815384
		28	−16850.55	27.9819102
		29	−18215.50	28.9804449
		30	−15872.37	29.9829603
		31	−14954.18	30.9839460
		32	−11062.07	31.9881244
		33	−8504.92	32.9908696
		34	−2862.24	33.9969273
		35	−58.08	34.9999376
14	Si	24	10754.76	24.0115457
		25	3825.31	25.0041066
		26	−7144.62	25.9923299
		27	−12384.43	26.9867048
		28	−21492.79	27.9769265
		29	−21895.03	28.9764947
		30	−24432.88	29.9737702
		31	−22948.96	30.9753633
		32	−24080.86	31.9741481
		33	−20492.38	32.9780005
		34	−19956.56	33.9785758
		35	−14359.76	34.9845842
		36	−12400.64	35.9866874
15	P	27	−752.98	26.9991916
		28	−7161.02	27.9923123
		29	−16951.91	28.9818014
		30	−20200.56	29.9783138
		31	−24440.99	30.9737615
		32	−24305.32	31.9739072
		33	−26337.73	32.9717253
		34	−24557.55	33.9736364
		35	−24857.61	34.9733143
		36	−20250.84	35.9782598
		37	−18994.71	36.9796083
		38	−14466.10	37.9844700
		39	−12649.69	38.9864200
		40	−8336.87	39.9910500
		41	−4843.77	40.9948000
16	S	28	4073.11	28.0043727
		29	−3158.88	28.9966088
		30	−14062.81	29.9849030
		31	−19044.93	30.9795544
		32	−26015.98	31.9720707
		33	−26586.24	32.9714585
		34	−29931.85	33.9678668
		35	−28846.37	34.9690322
		36	−30663.96	35.9670809
		37	−26896.22	36.9711257
		38	−26861.08	37.9711635
		39	−23161.34	38.9751353
		40	−22849.54	39.9754700
		41	−18601.93	40.9800300
		42	−17241.95	41.9814900
		43	−12482.02	42.9866000
17	Cl	31	−7064.44	30.9924160
		32	−13330.69	31.9856889
		33	−21003.51	32.9774518
		34	−24440.57	33.9737620
		35	−29013.51	34.9688527
		36	−29521.89	35.9683070
		37	−31761.52	36.9659026
		38	−29797.98	37.9680106
		39	−29800.65	38.9680077
		40	−27557.73	39.9704156
		41	−27339.15	40.9706502
		42	−24987.33	41.9731750
		43	−24029.39	42.9742034
		44	−19991.06	43.9785387
		45	−18909.33	44.9797000
18	Ar	32	−2179.08	31.9976607
		33	−9381.34	32.9899287
		34	−18378.26	33.9802701
		35	−23048.21	34.9752567
		36	−30230.44	35.9675463
		37	−30948.03	36.9667759
		38	−34714.76	37.9627322
		39	−33241.84	38.9643134
		40	−35039.89	39.9623831
		41	−33067.26	40.9645008
		42	−34422.07	41.9630464
		43	−31977.53	42.9656707
		44	−32262.04	43.9653653
		45	−29719.33	44.9680950
		46	−29720.74	45.9680935
		47	−25908.35	46.9721863

19 K	35	−11167.11	34.9880116	22 Ti	40	−8850.21	39.9904989
	36	−17425.08	35.9812934		42	−25120.88	41.9730316
	37	−24799.24	36.9733769		43	−29320.31	42.9685234
	38	−28801.69	37.9690801		44	−37548.30	43.9596903
	39	−33806.84	38.9637069		45	−39006.91	44.9581244
	40	−33535.02	39.9639987		46	−44125.34	45.9526295
	41	−35558.87	40.9618260		47	−44931.73	46.9517638
	42	−35021.32	41.9624031		48	−48487.00	47.9479471
	43	−36593.04	42.9607158		49	−48558.04	48.9478708
	44	−35810.21	43.9615562		50	−51425.85	49.9447921
	45	−36608.03	44.9606997		51	−49726.85	50.9466160
	46	−35418.93	45.9619762		52	−49464.02	51.9468982
	47	−35696.89	46.9616778		53	−46824.60	52.9497317
	48	−32124.48	47.9655130		54	−45764.29	53.9508700
	49	−30320.05	48.9674501		55	−41805.44	54.9551200
	50	−25352.62	49.9727828		56	−39132.06	55.9579900
20 Ca	36	−6439.20	35.9930872	23 V	45	−31873.59	44.9657823
	37	−13160.61	36.9858715		46	−37073.93	45.9601995
	38	−22059.04	37.9763186		47	−42003.93	46.9549069
	39	−27276.26	38.9707177		48	−44474.66	47.9522545
	40	−34846.11	39.9625912		49	−47956.18	48.9485169
	41	−35137.49	40.9622784		50	−49217.54	49.9471628
	42	−38546.76	41.9586184		51	−52197.49	50.9439637
	43	−38408.44	42.9587669		52	−51437.41	51.9447797
	44	−41469.09	43.9554811		53	−51844.60	52.9443425
	45	−40812.53	44.9561860		54	−49886.73	53.9464444
	46	−43134.91	45.9536928		55	−49147.30	54.9472382
	47	−42339.69	46.9545465		56	−46239.36	55.9503600
	48	−44214.74	47.9525335		57	−44376.37	56.9523600
	49	−41290.05	48.9556733		58	−40380.26	57.9566500
	50	−39571.46	49.9575183		59	−37911.80	58.9593000
	51	−35886.51	50.9614743		60	−33068.03	59.9645000
	52	−32509.14	51.9651000	24 Cr	46	−29470.93	45.9683617
21 Sc	40	−20526.39	39.9779640		47	−34552.36	46.9629065
	41	−28642.21	40.9692513		48	−42815.31	47.9540359
	42	−32120.92	41.9655168		49	−45325.43	48.9513412
	43	−36187.62	42.9611510		50	−50254.46	49.9460496
	44	−37815.81	43.9594031		51	−51444.76	50.9447718
	45	−41069.34	44.9559103		52	−55412.80	51.9405119
	46	−41758.64	45.9551703		53	−55280.64	52.9406538
	47	−44331.63	46.9524081		54	−56928.32	53.9388850
	48	−44492.81	47.9522350		55	−55103.30	54.9408442
	49	−46552.28	48.9500241		56	−55288.60	55.9406453
	50	−44537.51	49.9521870		57	−52392.99	56.9437538
	51	−43218.80	50.9536027		58	−51930.78	57.9442500
	52	−40380.26	51.9566500		59	−47850.84	58.9486300
	54	−34465.27	53.9630000		60	−46826.20	59.9497300

Z	Element	A	Mass excess	Atomic mass
		61	−42764.88	60.9540900
		62	−41172.03	61.9558000
25	Mn	49	−37610.54	48.9596234
		50	−42621.47	49.9542440
		51	−48236.96	50.9482155
		52	−50701.14	51.9455701
		53	−54683.62	52.9412947
		54	−55551.27	53.9403633
		55	−57706.38	54.9380497
		56	−56905.55	55.9389094
		57	−57484.85	56.9382875
		58	−55902.25	57.9399865
		59	−55473.10	58.9404472
		60	−52914.44	59.9431940
		61	−51735.17	60.9444600
		62	−48465.63	61.9479700
		63	−46751.68	62.9498100
26	Fe	50	−34471.50	49.9629933
		51	−40217.31	50.9568250
		52	−48329.14	51.9481166
		53	−50941.27	52.9453123
		54	−56248.41	53.9396149
		55	−57475.01	54.9382981
		56	−60601.00	55.9349422
		57	−60175.71	56.9353987
		58	−62148.84	57.9332805
		59	−60658.42	58.9348805
		60	−61406.92	59.9340770
		61	−58917.49	60.9367495
		62	−58897.90	61.9367705
		63	−55779.30	62.9401185
		64	−55079.23	63.9408700
		65	−51288.05	64.9449400
		66	−50319.30	65.9459800
		67	−46574.69	66.9500000
27	Co	53	−42639.14	52.9542250
		54	−48005.33	53.9484642
		55	−54023.71	54.9420032
		56	−56035.00	55.9398440
		57	−59339.67	56.9362963
		58	−59841.43	57.9357576
		59	−62223.61	58.9332002
		60	−61644.22	59.9338222
		61	−62895.04	60.9324794
		62	−61428.10	61.9340542
		63	−61837.02	62.9336153
		64	−59789.31	63.9358136
		65	−59164.22	64.9364846
		66	−56052.26	65.9398254
		67	−55321.42	66.9406100
		68	−51828.32	67.9443600
		69	−51045.86	68.9452000
28	Ni	55	−45329.91	54.9513364
		56	−53899.64	55.9421364
		57	−56075.47	56.9398005
		58	−60223.01	57.9353480
		59	−61151.12	58.9343516
		60	−64468.10	59.9307907
		61	−64216.78	60.9310605
		62	−66742.69	61.9283488
		63	−65509.22	62.9296730
		64	−67095.90	63.9279696
		65	−65122.59	64.9300880
		66	−66028.73	65.9291153
		67	−63742.46	66.9315697
		68	−63486.03	67.9318450
		69	−60377.72	68.9351819
		71	−55889.63	70.9400000
		72	−54678.69	71.9413000
29	Cu	57	−47305.27	56.9492157
		58	−51659.97	57.9445408
		59	−56351.55	58.9395041
		60	−58341.21	59.9373682
		61	−61979.57	60.9334622
		62	−62794.52	61.9325873
		63	−65576.16	62.9296011
		64	−65420.80	63.9297679
		65	−67259.72	64.9277937
		66	−66254.33	65.9288731
		67	−67300.16	66.9277503
		68	−65541.89	67.9296379
		69	−65739.92	68.9294253
		70	−62960.33	69.9324093
		71	−62764.22	70.9326199
30	Zn	58	−42293.11	57.9545965
		59	−47257.41	58.9492671
		60	−54183.11	59.9418321
		61	−56342.43	60.9395139
		62	−61167.35	61.9343342
		63	−62209.29	62.9332156
		64	−65999.53	63.9291466
		65	−65907.77	64.9292451
		66	−68896.30	65.9260368
		67	−67877.16	66.9271309

		68	−70004.03	67.9248476			80	−69447.74	79.9254448
		69	−68414.93	68.9265536			81	−66302.74	80.9288211
		70	−69559.43	69.9253249			82	−65623.44	81.9295504
		71	−67321.67	70.9277272					
		72	−68128.42	71.9268612	33	As	67	−56643.75	66.9391904
		73	−65409.99	72.9297795			68	−58876.96	67.9367930
		74	−65709.20	73.9294583			69	−63080.62	68.9322802
		75	−62468.42	74.9329374			70	−64340.32	69.9309278
		76	−62042.89	75.9333942			71	−67892.19	70.9271148
		77	−58604.14	76.9370859			72	−68229.46	71.9267527
		78	−57222.06	77.9385696			73	−70956.28	72.9238253
		80	−51777.35	79.9444147			74	−70859.60	73.9239291
							75	−73032.46	74.9215965
31	Ga	62	−51996.35	61.9441796			76	−72289.57	75.9223940
		63	−56689.29	62.9391416			77	−73916.18	76.9206477
		64	−58834.73	63.9368383			78	−72816.20	77.9218286
		65	−62652.91	64.9327394			79	−73635.99	78.9209485
		66	−63721.30	65.9315924			80	−72117.97	79.9225782
		67	−66876.68	66.9282050			81	−72532.74	80.9221329
		68	−67082.93	67.9279835			82	−70323.44	81.9245047
		69	−69320.92	68.9255809			83	−69880.09	82.9249807
		70	−68904.71	69.9260278					
		71	−70136.82	70.9247050	34	Se	69	−56297.48	68.9395622
		72	−68586.50	71.9263694			72	−67894.43	71.9271123
		73	−69703.84	72.9251699			73	−68216.28	72.9267668
		74	−68054.01	73.9269410			74	−72212.61	73.9224766
		75	−68464.20	74.9265007			75	−72168.82	74.9225236
		76	−66202.89	75.9289283			76	−75251.56	75.9192141
		77	−65874.14	76.9292812			77	−74599.05	76.9199146
		78	−63662.06	77.9316560			78	−77025.67	77.9173096
		79	−62487.99	78.9329164			79	−75916.93	78.9184998
		80	−59067.75	79.9365882			80	−77759.40	79.9165219
		81	−57982.74	80.9377530			81	−76389.08	80.9179930
							82	−77593.44	81.9167000
32	Ge	64	−54424.73	63.9415727			83	−75340.09	82.9191191
		65	−56410.56	64.9394408			84	−75949.80	83.9184646
		66	−61621.30	65.9338468			85	−72428.60	84.9222447
		67	−62653.75	66.9327384			86	−70540.95	85.9242712
		68	−66976.96	67.9280973			87	−66582.48	86.9285208
		69	−67093.64	68.9279720			88	−63878.14	87.9314240
		70	−70560.32	69.9242504					
		71	−69904.90	70.9249540	35	Br	72	−59152.77	71.9364969
		72	−72585.56	71.9220762			73	−63532.64	72.9317949
		73	−71297.14	72.9234594			74	−65305.96	73.9298912
		74	−73422.01	73.9211783			75	−69138.82	74.9257764
		75	−71855.91	74.9228595			76	−70288.69	75.9245420
		76	−73212.89	75.9214028			77	−73233.93	76.9213802
		77	−71214.14	76.9235485			78	−73451.90	77.9211462
		78	−71862.06	77.9228529			79	−76067.98	78.9183377
		79	−69487.99	78.9254016			80	−75888.85	79.9185300

		A	Δ	Mass
		81	−77974.36	80.9162911
		82	−77495.94	81.9168047
		83	−79009.11	82.9151803
		84	−77776.30	83.9165037
		85	−78610.60	84.9156081
		86	−75639.95	85.9187972
		87	−73857.48	86.9207108
		88	−70732.14	87.9240659
		89	−68569.82	88.9263873
		90	−64613.08	89.9306350
		91	−61510.92	90.9339653
		92	−56583.35	91.9392553
36	Kr	72	−54112.77	71.9419076
		73	−56885.29	72.9389311
		74	−62169.55	73.9332583
		75	−64241.60	74.9310338
		76	−68978.70	75.9259483
		77	−70171.41	76.9246679
		78	−74159.70	77.9203863
		79	−74442.20	78.9200830
		80	−77893.34	79.9163781
		81	−77693.65	80.9165925
		82	−80588.56	81.9134846
		83	−79981.83	82.9141360
		84	−82431.03	83.9115067
		85	−81480.60	84.9125270
		86	−83265.95	85.9106104
		87	−80709.98	86.9133543
		88	−79692.14	87.9144470
		89	−76724.82	88.9176325
		90	−74963.08	89.9195238
		91	−71312.92	90.9234425
		92	−68788.26	91.9261528
		93	−64026.00	92.9312653
37	Rb	74	−51725.50	73.9444704
		75	−57222.41	74.9385692
		76	−60480.55	75.9350715
		77	−64825.83	76.9304066
		78	−66935.77	77.9281415
		79	−70796.59	78.9239968
		80	−72172.78	79.9225194
		81	−75456.44	80.9189942
		82	−76189.03	81.9182077
		83	−79072.70	82.9151120
		84	−79750.15	83.9143847
		85	−82167.69	84.9117894
		86	−82747.32	85.9111671
		87	−84595.04	86.9091835
		88	−82606.22	87.9113186
		89	−81710.70	88.9122800
		90	−79354.95	89.9148090
		91	−77747.92	90.9165342
		92	−74775.26	91.9197255
		93	−72626.00	92.9220328
		94	−68551.12	93.9264074
		95	−65838.68	94.9293193
		96	−61214.09	95.9342840
		97	−58364.74	96.9373429
		98	−54302.78	97.9417036
		99	−50840.36	98.9454206
		101	−43597.64	100.9531960
38	Sr	77	−57974.77	76.9377615
		78	−63174.51	77.9321794
		79	−65477.43	78.9297071
		80	−70304.88	79.9245246
		81	−71526.53	80.9232131
		82	−76008.73	81.9184013
		83	−76796.99	82.9175551
		84	−80644.29	83.9134248
		85	−81102.66	84.9129327
		86	−84521.56	85.9092624
		87	−84878.36	86.9088794
		88	−87919.66	87.9056144
		89	−86207.05	88.9074530
		90	−85941.86	89.9077376
		91	−83638.98	90.9102099
		92	−82875.11	91.9110299
		93	−80087.60	92.9140225
		94	−78841.77	93.9153599
		95	−75117.33	94.9193583
		96	−72954.16	95.9216805
		97	−68791.98	96.9261488
		98	−66628.66	97.9284712
		99	−62116.63	98.9333151
		100	−60219.47	99.9353518
		101	−55407.64	100.9405175
		102	−53077.64	101.9430188
39	Y	79	−58357.43	78.9373507
		81	−66016.16	80.9291288
		82	−68192.74	81.9267921
		83	−72328.10	82.9223526
		84	−74158.31	83.9203878
		85	−77847.66	84.9164271
		86	−79281.56	85.9148878
		87	−83016.75	86.9108779
		88	−84297.06	87.9095034

Appendix 1 Atomic Masses

		89	−87702.10	88.9058479			97	−85606.95	96.9080972
		90	−86487.86	89.9071515			98	−83526.41	97.9103307
		91	−86346.30	90.9073035			99	−82327.42	98.9116179
		92	−84815.47	91.9089469			100	−79939.47	99.9141815
		93	−84224.20	92.9095816			101	−78942.65	100.9152516
		94	−82349.64	93.9115941			102	−76347.64	101.9180375
		95	−81204.18	94.9128238			103	−75319.39	102.9191413
		96	−78340.70	95.9158978			104	−72228.53	103.9224595
		97	−76260.46	96.9181311			105	−70854.99	104.9239341
		98	−72452.04	97.9222196					
		99	−70202.28	98.9246348	42	Mo	86	−64557.03	85.9306952
		100	−67294.47	99.9277564			87	−67694.61	86.9273269
		101	−64912.64	100.9303134			88	−72700.55	87.9219528
		102	−61892.64	101.9335555			89	−75003.36	88.9194806
							90	−80167.94	89.9139362
40	Zr	81	−58856.16	80.9368153			91	−82203.63	90.9117508
		82	−64192.74	81.9310863			92	−86805.47	91.9068105
		83	−66460.10	82.9286522			93	−86803.85	92.9068123
		85	−73154.66	84.9214653			94	−88410.34	93.9050876
		86	−77805.03	85.9164729			95	−87708.08	94.9058415
		87	−79347.84	86.9148166			96	−88791.01	95.9046790
		88	−83623.76	87.9102262			97	−87540.83	96.9060211
		89	−84869.42	88.9088890			98	−88112.01	97.9054079
		90	−88767.94	89.9047037			99	−85966.08	98.9077116
		91	−87891.13	90.9056450			100	−86184.47	99.9074772
		92	−88454.56	91.9050402			101	−83511.65	100.9103466
		93	−87117.38	92.9064757			102	−83557.64	101.9102972
		94	−87266.29	93.9063158			103	−80849.39	102.9132046
		95	−85657.62	94.9080428			104	−80333.53	103.9137584
		96	−85440.65	95.9082757			105	−77339.99	104.9169721
		97	−82948.86	96.9109508			106	−76257.41	105.9181343
		98	−81276.22	97.9127464			107	−72940.88	106.9216948
		99	−77769.41	98.9165111					
		100	−76604.47	99.9177617	43	Tc	89	−67493.36	88.9275429
		101	−73457.65	100.9211400			90	−71207.27	89.9235559
		102	−71742.64	101.9229811			91	−75983.63	90.9184282
		103	−68374.39	102.9265971			92	−78935.11	91.9152597
							93	−83603.00	92.9102485
41	Nb	83	−58960.10	82.9367037			94	−84154.59	93.9096564
		85	−67154.66	84.9279065			95	−86017.45	94.9076565
		86	−69827.03	85.9250376			96	−85817.78	95.9078708
		87	−74182.84	86.9203615			97	−87220.57	96.9063649
		89	−80578.41	88.9134955			98	−86428.01	97.9072157
		90	−82656.94	89.9112642			99	−87323.31	98.9062546
		91	−86637.74	90.9069906			100	−86016.38	99.9076576
		92	−86448.95	91.9071933			101	−86336.09	100.9073144
		93	−87208.74	92.9063776			102	−84567.59	101.9092130
		94	−86364.89	93.9072835			103	−84599.39	102.9091788
		95	−86782.46	94.9068352			104	−82488.53	103.9114449
		96	−85604.21	95.9081001			105	−82289.99	104.9116581

		A	Δ	Mass
		106	−79777.41	105.9143554
		107	−79100.88	106.9150817
		108	−75935.40	107.9184800
44	Ru	91	−68578.97	90.9263775
		93	−77266.00	92.9170516
		94	−82568.02	93.9113596
		95	−83449.99	94.9104128
		96	−86072.19	95.9075977
		97	−86112.37	96.9075546
		98	−88224.47	97.9052872
		99	−87616.96	98.9059394
		100	−89218.79	99.9042197
		101	−87949.58	100.9055823
		102	−89097.85	101.9043495
		103	−87258.92	102.9063237
		104	−88091.24	103.9054302
		105	−85929.99	104.9077504
		106	−86324.41	105.9073270
		107	−83920.88	106.9099073
		108	−83655.40	107.9101923
		109	−80852.21	108.9132016
		110	−80139.97	109.9139662
45	Rh	95	−78339.99	94.9158986
		96	−79625.76	95.9145183
		97	−82589.37	96.9113367
		98	−83167.10	97.9107165
		99	−85574.38	98.9081321
		100	−85588.79	99.9081167
		101	−87408.10	100.9061636
		102	−86775.32	101.9068429
		103	−88022.27	102.9055042
		104	−86950.00	103.9066554
		105	−87846.91	104.9056925
		106	−86363.81	105.9072847
		107	−86861.30	106.9067506
		108	−85016.73	107.9087308
		109	−85012.21	108.9087357
		110	−82949.97	109.9109496
46	Pd	96	−76175.76	95.9182220
		97	−77799.37	96.9164790
		98	−81300.08	97.9127208
		99	−82187.79	98.9117678
		100	−85227.41	99.9085046
		101	−85428.10	100.9082892
		102	−87925.83	101.9056078
		103	−87479.19	102.9060873
		104	−89390.89	103.9040350
		105	−88413.63	104.9050841
		106	−89904.91	105.9034831
		107	−88372.26	106.9051285
		108	−89521.73	107.9038945
		109	−87603.71	108.9059536
		110	−88349.97	109.9051524
		111	−86029.09	110.9076440
		112	−86337.11	111.9073133
		113	−83693.47	112.9101514
		114	−83494.15	113.9103654
		115	−80403.37	114.9136835
		116	−79961.03	115.9141583
		118	−75465.99	117.9189840
47	Ag	98	−72880.08	97.9217600
		99	−76757.79	98.9175971
		100	−78180.85	99.9160694
		101	−81224.27	100.9128022
		102	−81971.46	101.9120000
		103	−84791.60	102.9089725
		104	−85112.24	103.9086283
		105	−87068.38	104.9065283
		106	−86939.65	105.9066665
		107	−88405.27	106.9050931
		108	−87603.55	107.9059538
		109	−88719.65	108.9047556
		110	−87457.53	109.9061105
		111	−88217.43	110.9052947
		112	−86625.08	111.9070042
		113	−87033.47	112.9065658
		114	−84944.88	113.9088080
		115	−84987.37	114.9087623
		116	−82568.03	115.9113596
		117	−82265.64	116.9116842
		118	−79565.99	117.9145824
		119	−78556.56	118.9156661
		120	−75647.91	119.9187886
		121	−74658.23	120.9198511
48	Cd	100	−74305.05	99.9202303
		101	−75747.74	100.9186815
		102	−79384.46	101.9147773
		103	−80649.71	102.9134190
		104	−83975.95	103.9098481
		105	−84330.17	104.9094679
		106	−87133.79	105.9064581
		107	−86988.27	106.9066143
		108	−89252.57	107.9041835
		109	−88505.36	108.9049856
		110	−90349.71	109.9030056

		A	Δ	Mass
		111	−89254.23	110.9041817
		112	−90581.05	111.9027573
		113	−89049.93	112.9044010
		114	−90021.32	113.9033582
		115	−88090.86	114.9054306
		116	−88719.73	115.9047555
		117	−86425.64	116.9072183
		118	−86708.90	117.9069142
		119	−83906.56	118.9099226
		120	−83972.91	119.9098514
		121	−81058.23	120.9129804
		123	−77310.57	122.9170037
		124	−76710.10	123.9176483
		125	−73357.78	124.9212472
		126	−72326.78	125.9223540
		127	−68525.51	126.9264349
		128	−67290.54	127.9277607
49	In	100	−64134.25	99.9311491
		102	−70134.46	101.9247076
		103	−74599.71	102.9199139
		104	−76067.26	103.9183385
		105	−79481.17	104.9146735
		106	−80610.42	105.9134612
		107	−83562.27	106.9102922
		108	−84095.56	107.9097197
		109	−86485.41	108.9071541
		110	−86471.71	109.9071688
		111	−88388.82	110.9051107
		112	−87995.12	111.9055334
		113	−89366.38	112.9040613
		114	−88569.46	113.9049168
		115	−89536.75	114.9038784
		116	−88249.73	115.9052600
		117	−88943.01	116.9045158
		118	−87230.09	117.9063547
		119	−87703.56	118.9058464
		120	−85733.29	119.9079616
		121	−85838.23	120.9078489
		122	−83576.33	121.9102771
		123	−83425.57	122.9104390
		124	−80876.10	123.9131760
		125	−80479.78	124.9136014
		126	−77812.78	125.9164646
		127	−76993.51	126.9173441
		128	−74360.54	127.9201707
		129	−72975.13	128.9216580
		130	−69997.16	129.9248550
		131	−68215.71	130.9267674
		132	−62485.54	131.9329190
50	Sn	104	−71552.26	103.9231855
		105	−73224.35	104.9213904
		106	−77425.33	105.9168805
		107	−78555.95	106.9156667
		108	−82003.75	107.9119654
		109	−82635.73	108.9112869
		110	−85834.66	109.9078527
		111	−85943.90	110.9077354
		112	−88658.83	111.9048209
		113	−88330.42	112.9051734
		114	−90558.14	113.9027819
		115	−90032.63	114.9033460
		116	−91524.72	115.9017442
		117	−90397.97	116.9029538
		118	−91653.10	117.9016064
		119	−90067.18	118.9033089
		120	−91103.29	119.9021966
		121	−89202.77	120.9042369
		122	−89944.92	121.9034402
		123	−87819.47	122.9057219
		124	−88236.10	123.9052747
		125	−85897.78	124.9077850
		126	−86019.78	125.9076540
		127	−83507.51	126.9103510
		128	−83336.14	127.9105350
		129	−80630.13	128.9134400
		130	−80246.16	129.9138522
		131	−77389.31	130.9169192
		132	−76620.54	131.9177445
		133	−70966.71	132.9238141
		134	−66635.74	133.9284636
51	Sb	105	−63780.69	104.9315286
		109	−76255.73	108.9181361
		112	−81603.85	111.9123947
		113	−84413.89	112.9093780
		114	−84676.63	113.9090959
		115	−87002.63	114.9065989
		116	−86817.80	115.9067973
		117	−88641.34	116.9048396
		118	−87996.47	117.9055319
		119	−89473.28	118.9039465
		120	−88422.69	119.9050744
		121	−89592.90	120.9038181
		122	−88328.52	121.9051755
		123	−89222.49	122.9042157
		124	−87618.62	123.9059376
		125	−88261.09	124.9052479
		126	−86397.78	125.9072482
		127	−86708.51	126.9069146

		128	−84610.07	127.9091674			126	−87914.96	125.9056194
		129	−84626.13	128.9091501			127	−88987.08	126.9044685
		130	−82393.93	129.9115465			128	−87741.83	127.9058053
		131	−82021.31	130.9119465			129	−88503.57	128.9049875
		132	−79723.54	131.9144133			130	−86932.58	129.9066741
		133	−78956.71	132.9152365			131	−87444.76	130.9061242
		134	−74005.74	133.9205516			132	−85702.54	131.9079946
		135	−69705.58	134.9251680			133	−85877.71	132.9078065
							134	−83949.44	133.9098766
52	Te	108	−65682.58	107.9294869			135	−83787.58	134.9100504
		109	−67573.84	108.9274565			136	−79498.25	135.9146551
		110	−72277.25	109.9224072			137	−76501.12	136.9178727
		111	−73475.69	110.9211206			138	−72299.14	137.9223837
		112	−77256.59	111.9170617			139	−68843.54	138.9260934
		115	−82363.97	114.9115787					
		116	−85305.97	115.9084203	54	Xe	112	−59927.33	111.9356654
		117	−85106.70	116.9086342			113	−62053.48	112.9333829
		118	−87723.26	117.9058252			117	−73993.82	116.9205644
		119	−87180.27	118.9064082			118	−77713.68	117.9165710
		120	−89404.88	119.9040199			119	−78660.66	118.9155543
		121	−88557.29	120.9049299			120	−81829.88	119.9121520
		122	−90311.06	121.9030471			121	−82542.93	120.9113865
		123	−89169.16	122.9042730			122	−85186.61	121.9085484
		124	−90523.07	123.9028195			123	−85258.94	122.9084708
		125	−89027.79	124.9044248			124	−87657.51	123.9058958
		126	−90070.29	125.9033056			125	−87189.47	124.9063983
		127	−88289.51	126.9052173			126	−89172.96	125.9042689
		128	−88993.63	127.9044614			127	−88324.64	126.9051796
		129	−87005.63	128.9065956			128	−89860.81	127.9035305
		130	−87352.93	129.9062228			129	−88697.35	128.9047795
		131	−85211.31	130.9085219			130	−89881.80	129.9035080
		132	−85209.54	131.9085238			131	−88415.61	130.9050820
		133	−82959.71	132.9109391			132	−89279.54	131.9041545
		134	−82399.44	133.9115406			133	−87648.30	132.9059057
		135	−77825.58	134.9164508			134	−88124.44	133.9053946
		136	−74423.42	135.9201032			135	−86435.65	134.9072075
		137	−69559.52	136.9253248			136	−86424.44	135.9072196
							137	−82378.58	136.9115630
53	I	109	−57574.09	108.9381917			138	−80119.14	137.9139886
		113	−71124.92	112.9236443			139	−75649.54	138.9187869
		116	−77560.82	115.9167351			140	−72995.90	139.9216357
		117	−80436.65	116.9136477			141	−68328.54	140.9266463
		118	−80690.44	117.9133753			142	−65481.24	141.9297030
		119	−83666.00	118.9101809					
		120	−83789.88	119.9100479	55	Cs	113	−51664.83	112.9445355
		121	−86287.94	120.9073661			116	−62490.06	115.9329142
		122	−86077.06	121.9075925			117	−66471.88	116.9286395
		123	−87934.94	122.9055980			118	−68413.68	117.9265549
		124	−87363.48	123.9062115			119	−72311.05	118.9223709
		125	−88842.02	124.9046242			120	−73887.75	119.9206783

		A	Δ	Mass			A	Δ	Mass
		121	−77142.93	120.9171837			142	−77828.01	141.9164482
		122	−78131.89	121.9161220			143	−73944.61	142.9206172
		123	−81049.12	122.9129902			144	−71780.48	143.9229405
		124	−81742.56	123.9122458			145	−68070.02	144.9269238
		125	−84090.73	124.9097249			146	−65105.23	145.9301067
		126	−84348.68	125.9094480			147	−61485.56	146.9339926
		127	−86239.94	126.9074176			148	−58048.48	147.9376824
		128	−85932.25	127.9077480					
		129	−87501.40	128.9060634	57	La	128	−78759.73	127.9154480
		130	−86902.64	129.9067062			129	−81349.84	128.9126674
		131	−88063.21	130.9054603			131	−83733.39	130.9101085
		132	−87160.07	131.9064298			132	−83731.61	131.9101104
		133	−88075.66	132.9054469			133	−85328.22	132.9083964
		134	−86895.88	133.9067135			134	−85241.37	133.9084897
		135	−87586.60	134.9059719			135	−86655.94	134.9069710
		136	−86344.13	135.9073058			136	−86022.36	135.9076512
		137	−86551.15	136.9070836			137	−87126.67	136.9064657
		138	−82893.14	137.9110106			138	−86529.42	137.9071069
		139	−80706.57	138.9133580			139	−87236.11	138.9063482
		140	−77055.90	139.9172771			140	−84325.76	139.9094726
		141	−74478.54	140.9200440			141	−82942.99	140.9109571
		142	−70521.24	141.9242924			142	−80039.09	141.9140745
		143	−67691.39	142.9273303			143	−78190.86	142.9160587
		144	−63316.08	143.9320274			144	−74899.87	143.9195917
		145	−60185.47	144.9353883			145	−72993.38	144.9216384
		146	−55738.70	145.9401621			146	−69209.86	145.9257002
		147	−52289.93	146.9438645			147	−67235.56	146.9278197
		148	−47599.77	147.9488996			148	−63163.48	147.9321912
56	Ba	119	−64224.71	118.9310520	58	Ce	131	−79713.39	130.9144242
		120	−68887.75	119.9260460			134	−84741.37	133.9090264
		121	−70340.91	120.9244859			135	−84630.36	134.9091456
		124	−79094.60	123.9150885			136	−86495.19	135.9071436
		125	−79530.73	124.9146203			137	−85904.57	136.9077777
		126	−82675.53	125.9112442			138	−87573.86	137.9059856
		127	−82789.94	126.9111214			139	−86958.11	138.9066467
		128	−85409.73	127.9083089			140	−88087.62	139.9054341
		129	−85069.84	128.9086738			141	−85444.90	140.9082711
		130	−87271.21	129.9063105			142	−84542.63	141.9092398
		131	−86693.39	130.9069308			143	−81616.41	142.9123812
		132	−88439.61	131.9050562			144	−80441.31	143.9136427
		133	−87558.22	132.9060024			145	−77101.73	144.9172279
		134	−88954.55	133.9045034			146	−75740.02	145.9186898
		135	−87855.94	134.9056828			147	−72180.56	146.9225110
		136	−88892.36	135.9045702			148	−70425.84	147.9243948
		137	−87726.77	136.9058215			149	−66798.16	148.9282892
		138	−88267.17	137.9052413			150	−64993.68	149.9302264
		139	−84919.28	138.9088354					
		140	−83276.03	139.9105995	59	Pr	131	−74463.39	130.9200603
		141	−79729.88	140.9144065			135	−80910.36	134.9131392

		136	−81368.84	135.9126470			150	−73607.13	149.9209795
		137	−83202.57	136.9106784			151	−73399.21	150.9212027
		138	−83136.86	137.9107489			152	−71268.08	151.9234906
		139	−84829.11	138.9089322			153	−70688.10	152.9241132
		140	−84699.62	139.9090712			154	−68421.00	153.9265471
		141	−86025.58	140.9076478			155	−66977.16	154.9280971
		142	−83797.31	141.9100399			156	−64216.85	155.9310604
		143	−83077.86	142.9108123					
		144	−80759.96	143.9133006	62	Sm	137	−67955.54	136.9270467
		145	−79636.30	144.9145069			139	−72375.21	138.9223020
		146	−76766.26	145.9175881			140	−75459.39	139.9189910
		147	−75470.56	146.9189790			141	−75946.08	140.9184686
		148	−72485.84	147.9221833			142	−78996.94	141.9151933
		149	−70988.16	148.9237911			143	−79527.63	142.9146236
		150	−68003.68	149.9269951			144	−81976.37	143.9119948
		151	−66855.30	150.9282279			145	−80662.14	144.9134057
							146	−81005.72	145.9130368
60	Nd	131	−67903.39	130.9271027			147	−79276.39	146.9148933
		136	−79157.84	135.9150206			148	−79346.59	147.9148180
		137	−79512.57	136.9146398			149	−77146.77	148.9171796
		139	−82042.11	138.9119242			150	−77061.13	149.9172715
		140	−84477.34	139.9093099			151	−74586.25	150.9199284
		141	−84202.57	140.9096048			152	−74772.65	151.9197283
		142	−85959.52	141.9077187			153	−72569.05	152.9220939
		143	−84011.78	142.9098097			154	−72465.28	153.9222053
		144	−83757.48	143.9100827			155	−70201.16	154.9246360
		145	−81441.58	144.9125689			156	−69371.85	155.9255263
		146	−80935.51	145.9131122			157	−66737.34	156.9283545
		147	−78156.25	146.9160958			158	−65215.81	157.9299880
		148	−77417.84	147.9168886					
		149	−74385.20	148.9201442	63	Eu	140	−66989.39	139.9280840
		150	−73693.68	149.9208866			141	−69968.35	140.9248859
		151	−70956.79	150.9238248			142	−71352.40	141.9234001
		152	−70157.86	151.9246825			143	−74252.51	142.9202867
		153	−67352.10	152.9276946			144	−75661.41	143.9187742
		154	−65685.88	153.9294833			145	−78002.10	144.9162613
		155	−62755.16	154.9326296			146	−77127.96	145.9171998
							147	−77555.05	146.9167412
61	Pm	136	−71307.84	135.9234479			148	−76239.26	147.9181538
		139	−77537.72	138.9167599			149	−76451.50	148.9179260
		140	−78430.25	139.9158017			150	−74800.55	149.9196983
		141	−80474.89	140.9136067			151	−74662.94	150.9198461
		142	−81085.85	141.9129508			152	−72898.34	151.9217404
		143	−82970.42	142.9109276			153	−73377.29	152.9212263
		144	−81425.82	143.9125858			154	−71747.96	153.9229754
		145	−81278.54	144.9127439			155	−71828.02	154.9228895
		146	−79463.72	145.9146922			156	−70094.12	155.9247509
		147	−79052.25	146.9151339			157	−69471.34	156.9254195
		148	−76878.25	147.9174678			158	−67214.81	157.9278420
		149	−76075.85	148.9183292			159	−66057.35	158.9290845

Z	El	A	Δ	Mass
64	Gd	143	−68242.51	142.9267387
		145	−72947.62	144.9216875
		146	−76098.07	145.9183054
		147	−75367.68	146.9190895
		148	−76280.24	147.9181098
		149	−75137.62	148.9193365
		150	−75771.94	149.9186555
		151	−74198.82	150.9203443
		152	−74717.10	151.9197879
		153	−72892.86	152.9217463
		154	−73716.31	153.9208623
		155	−72080.11	154.9226188
		156	−72545.16	155.9221196
		157	−70833.88	156.9239567
		158	−70699.89	157.9241006
		159	−68571.85	158.9263851
		160	−67951.90	159.9270507
		161	−65515.98	160.9296657
		162	−64290.58	161.9309812
65	Tb	146	−67830.80	145.9271807
		147	−70758.90	146.9240372
		148	−70515.36	147.9242987
		149	−71499.95	148.9232417
		150	−71115.68	149.9236542
		151	−71633.58	150.9230982
		152	−70727.10	151.9240714
		153	−71323.69	152.9234309
		154	−70154.31	153.9246863
		155	−71258.90	154.9235004
		156	−70100.74	155.9247438
		157	−70773.83	156.9240212
		158	−69479.88	157.9254103
		159	−69542.41	158.9253432
		160	−67846.27	159.9271641
		161	−67471.56	160.9275663
		162	−65684.47	161.9294848
		163	−64604.74	162.9306440
		164	−62086.62	163.9333473
66	Dy	146	−62670.80	145.9327202
		147	−64386.26	146.9308785
		148	−67833.36	147.9271779
		149	−67687.95	148.9273340
		150	−69322.03	149.9255798
		151	−68763.22	150.9261797
		152	−70128.56	151.9247139
		153	−69153.30	152.9257609
		154	−70400.40	153.9244221
		155	−69164.40	154.9257490
		156	−70534.32	155.9242783
		157	−69432.38	156.9254613
		158	−70416.62	157.9244047
		159	−69176.78	158.9257357
		160	−69681.59	159.9251938
		161	−68064.63	160.9269296
		162	−68190.26	161.9267948
		163	−66389.87	162.9287276
		164	−65976.62	163.9291712
		165	−63621.19	164.9316999
		166	−62593.37	165.9328033
		167	−59942.54	166.9356491
		169	−55606.79	168.9403037
67	Ho	149	−61674.26	148.9337900
		151	−63638.92	150.9316808
		152	−63583.21	151.9317406
		153	−65023.39	152.9301945
		154	−64649.15	153.9305963
		155	−66062.40	154.9290791
		157	−66892.38	156.9281881
		158	−66186.62	157.9289458
		159	−67339.05	158.9277086
		160	−66391.59	159.9287257
		161	−67205.73	160.9278517
		162	−66049.98	161.9290925
		163	−66387.30	162.9287303
		164	−64989.79	163.9302306
		165	−64907.27	164.9303192
		166	−63079.58	165.9322813
		167	−62292.54	166.9331262
		168	−60084.68	167.9354965
		169	−58806.79	168.9368683
		170	−56248.30	169.9396150
		171	−54528.51	170.9414613
68	Er	152	−60474.02	151.9350785
		153	−60460.36	152.9350932
		154	−62617.54	153.9327773
		155	−62219.81	154.9332043
		156	−64259.10	155.9310150
		157	−63392.33	156.9319456
		159	−64570.49	158.9306808
		160	−66062.55	159.9290790
		161	−65203.32	160.9300014
		162	−66345.72	161.9287750
		163	−65177.30	162.9300293
		164	−65952.56	163.9291970
		165	−64531.29	164.9307228
		166	−64934.46	165.9302900

Appendix 1 Atomic Masses

		A	Δ	Atomic mass
		167	−63299.25	166.9320455
		168	−62999.00	167.9323678
		169	−60930.80	168.9345881
		170	−60118.30	169.9354604
		171	−57728.51	170.9380259
		172	−56493.10	171.9393522
69	Tm	153	−54000.90	152.9420277
		155	−56642.69	154.9391916
		156	−56814.70	155.9390069
		157	−58911.33	156.9367561
		159	−60725.05	158.9348090
		160	−60462.55	159.9350908
		161	−62039.32	160.9333981
		162	−61506.40	161.9339702
		163	−62738.30	162.9326477
		164	−61990.01	163.9334510
		165	−62938.75	164.9324325
		166	−61894.85	165.9335532
		167	−62550.89	166.9328489
		168	−61319.89	167.9341704
		169	−61281.94	168.9342112
		170	−59803.88	169.9357979
		171	−59218.96	170.9364258
		172	−57383.64	171.9383961
		173	−56261.92	172.9396004
		174	−53873.30	173.9421646
		175	−52319.31	174.9438329
		176	−49377.17	175.9469914
70	Yb	156	−53237.57	155.9428471
		157	−53413.11	156.9426587
		158	−56022.00	157.9398579
		159	−55746.43	158.9401538
		163	−59368.30	162.9362655
		165	−60176.75	164.9353976
		166	−61590.72	165.9338797
		167	−60596.60	166.9349469
		168	−61576.90	167.9338945
		169	−60372.80	168.9351872
		170	−60771.92	169.9347587
		171	−59315.39	170.9363223
		172	−59263.79	171.9363777
		173	−57560.03	172.9382068
		174	−56953.30	173.9388581
		175	−54704.31	174.9412725
		176	−53497.17	175.9425684
		177	−50992.65	176.9452572
		178	−49701.35	177.9466434
71	Lu	157	−46480.11	156.9501016
		159	−49727.69	158.9466151
		163	−54768.30	162.9412038
		165	−56256.75	164.9396059
		166	−56110.72	165.9397627
		167	−57466.60	166.9383071
		168	−57101.90	167.9386986
		169	−58079.80	168.9376488
		170	−57312.78	169.9384722
		171	−57836.54	170.9379099
		172	−56744.52	171.9390823
		173	−56889.22	172.9389269
		174	−55578.96	173.9403336
		175	−55174.33	174.9407679
		176	−53390.99	175.9426824
		177	−52391.88	176.9437550
		178	−50345.97	177.9459514
		179	−49067.17	178.9473242
		180	−46686.50	179.9498800
72	Hf	160	−45909.99	159.9507136
		161	−46266.51	160.9503309
		162	−49180.10	161.9472030
		169	−54810.43	168.9411586
		172	−56394.52	171.9394580
		174	−55852.22	173.9400402
		175	−54489.61	174.9415030
		176	−54583.84	175.9414019
		177	−52890.21	176.9432200
		178	−52445.22	177.9436978
		179	−50472.93	178.9458151
		180	−49789.50	179.9465488
		181	−47413.85	180.9490991
		182	−46059.68	181.9505529
		183	−43285.58	182.9535310
		184	−41500.03	183.9554479
73	Ta	161	−38775.30	160.9583730
		163	−42553.76	162.9543167
		172	−51474.52	171.9447398
		174	−52007.22	173.9441680
		176	−51473.84	175.9447406
		177	−51724.21	176.9444718
		178	−50533.22	177.9457504
		179	−50362.04	178.9459341
		180	−48935.42	179.9474657
		181	−48441.08	180.9479964
		182	−46432.72	181.9501524
		183	−45295.58	182.9513732
		184	−42840.03	183.9540094

	185	−41396.44	184.9555591		192	−35882.03	191.9614791
	186	−38610.33	185.9585501		193	−33395.84	192.9641481
					194	−32435.26	193.9651793
74 W	164	−38206.33	163.9589838		195	−29692.40	194.9681239
	165	−38809.79	164.9583360		196	−28296.41	195.9696226
	166	−41898.69	165.9550199				
	178	−50441.92	177.9458484	77 Ir	169	−21991.76	168.9763909
	179	−49302.36	178.9470718		181	−39456.07	180.9576422
	180	−49643.28	179.9467058		182	−39003.80	181.9581277
	181	−48253.19	180.9481981		184	−39692.52	183.9573883
	182	−48246.24	181.9482055		186	−39168.29	185.9579511
	183	−46365.61	182.9502245		187	−39718.12	186.9573609
	184	−45706.03	183.9509326		188	−38329.14	187.9588520
	185	−43388.44	184.9534206		189	−38455.35	188.9587165
	186	−42511.33	185.9543622		190	−36708.03	189.9605923
	187	−39906.72	186.9571584		191	−36709.06	190.9605912
	188	−38669.15	187.9584870		192	−34835.82	191.9626022
	189	−35478.53	188.9619122		193	−34536.35	192.9629237
	190	−34298.03	189.9631796		194	−32531.85	193.9650756
					195	−31692.40	194.9659768
75 Re	165	−30692.47	164.9670503		196	−29453.92	195.9683799
	178	−45781.92	177.9508511		197	−28283.42	196.9696365
	179	−46592.36	178.9499811				
	180	−45840.97	179.9507877	78 Pt	172	−21073.99	171.9773761
	181	−46514.52	180.9500646		173	−21890.44	172.9764997
	182	−45446.24	181.9512115		174	−25326.13	173.9728113
	183	−45809.61	182.9508214		182	−36078.96	181.9612677
	184	−44223.33	183.9525243		185	−36557.61	184.9607538
	185	−43821.43	184.9529558		186	−37788.52	185.9594324
	186	−41929.77	185.9549866		188	−37822.66	187.9593957
	187	−41217.87	186.9557508		189	−36484.85	188.9608319
	188	−39018.15	187.9581123		190	−37324.89	189.9599301
	189	−37978.53	188.9592284		191	−35690.51	190.9616847
	190	−35568.03	189.9618162		192	−36295.51	191.9610352
	191	−34350.15	190.9631236		193	−34479.71	192.9629845
					194	−34778.64	193.9626636
76 Os	168	−29963.45	167.9678329		195	−32812.39	194.9647745
	169	−30668.31	168.9670762		196	−32662.94	195.9649349
	170	−33934.59	169.9635697		197	−30438.05	196.9673234
	178	−43455.84	177.9533482		198	−29923.30	197.9678760
	181	−43524.52	180.9532745		199	−27408.08	198.9705762
	182	−44538.24	181.9521862		200	−26618.47	199.9714239
	184	−44254.52	183.9524908		201	−23756.38	200.9744965
	185	−42808.64	184.9540430				
	186	−42999.29	185.9538384	79 Au	173	−12670.05	172.9863981
	187	−41220.53	186.9557479		185	−31850.61	184.9658070
	188	−41138.50	187.9558360		186	−31672.96	185.9659977
	189	−38987.80	188.9581449		190	−32882.89	189.9646988
	190	−38708.03	189.9584452		191	−33860.51	190.9636493
	191	−36395.37	190.9609280		192	−32779.17	191.9648101

		193	−33411.06	192.9641318			199	−25234.74	198.9729094
		194	−32286.61	193.9653389			200	−26253.63	199.9718156
		195	−32585.58	194.9650179			201	−25293.24	200.9728466
		196	−31157.24	195.9665513			202	−25947.89	201.9721438
		197	−31156.97	196.9665516			203	−24800.57	202.9733755
		198	−29597.99	197.9682253			204	−25123.54	203.9730288
		199	−29111.03	198.9687480			205	−23783.73	204.9744671
		200	−27276.11	199.9707179			206	−23800.60	205.9744490
		201	−26416.38	200.9716409			207	−22467.07	206.9758806
		202	−24415.92	201.9737884			208	−21763.56	207.9766359
		203	−23159.49	202.9751373			209	−17628.71	208.9810748
							210	−14742.63	209.9841731
80	Hg	176	−11724.48	175.9874133			211	−10496.56	210.9887315
		177	−12727.12	176.9863369			212	−7556.75	211.9918875
		178	−16323.19	177.9824763			214	−188.03	213.9997981
		186	−28447.80	185.9694600					
		191	−30680.51	190.9670631	83	Bi	197	−19622.58	196.9789343
		193	−31070.75	192.9666442			198	−19538.65	197.9790244
		194	−32246.61	193.9653818			199	−20886.93	198.9775770
		195	−31075.58	194.9666390			200	−20360.61	199.9781420
		196	−31843.26	195.9658149			201	−21451.63	200.9769707
		197	−30557.35	196.9671953			202	−20796.06	201.9776745
		198	−30970.47	197.9667518			203	−21547.21	202.9768681
		199	−29563.30	198.9682625			204	−20674.36	203.9778052
		200	−29520.23	199.9683087			205	−21075.34	204.9773747
		201	−27679.08	200.9702853			206	−20043.09	205.9784829
		202	−27362.07	201.9706256			207	−20068.83	206.9784552
		203	−25283.45	202.9728571			208	−18884.46	207.9797267
		204	−24707.28	203.9734757			209	−18272.89	208.9803833
		205	−22303.59	204.9760561			210	−14806.15	209.9841050
		206	−20959.85	205.9774987			211	−11868.97	210.9872581
		207	−16229.40	206.9825770			212	−8130.51	211.9912715
							213	−5239.81	212.9943748
81	Tl	197	−28376.84	196.9695362			214	−1212.19	213.9986987
		198	−27510.47	197.9704663			215	1706.82	215.0018323
		199	−28118.22	198.9698139					
		200	−27064.19	199.9709454	84	Po	194	−10913.28	193.9882841
		201	−27196.11	200.9708038			203	−17313.80	202.9814129
		202	−25997.46	201.9720906			204	−18343.80	203.9803071
		203	−25775.28	202.9723291			205	−17544.32	204.9811654
		204	−24359.83	203.9738487			206	−18196.51	205.9804653
		205	−23834.81	204.9744123			207	−17159.77	206.9815782
		206	−22267.06	205.9760953			208	−17483.15	207.9812311
		207	−21044.40	206.9774079			209	−16379.59	208.9824158
		208	−16762.56	207.9820047			210	−15968.23	209.9828574
		209	−13647.20	208.9853491			211	−12447.67	210.9866369
		210	−9253.86	209.9900656			212	−10384.52	211.9888518
							213	−6667.15	212.9928425
82	Pb	182	−6822.17	181.9926761			214	−4484.26	213.9951860
		190	−20325.19	189.9781800			215	−545.29	214.9994146

		216	1774.68	216.0019052			215	303.69	215.0003260
		218	8351.56	218.0089658			216	2969.48	216.0031879
85	At	201	−10724.37	200.9884869			217	4300.20	217.0046164
		202	−10760.03	201.9884486			218	7045.19	218.0075633
		203	−12251.74	202.9868472			219	8607.79	219.0092408
		204	−11865.78	203.9872616			220	11469.46	220.0123130
		205	−13007.05	204.9860364			221	13269.74	221.0142456
		206	−12482.72	205.9865992			222	16342.09	222.0175439
		207	−13249.70	206.9857759			223	18379.04	223.0197307
		208	−12498.31	207.9865825			224	21643.74	224.0232355
		209	−12893.11	208.9861587			225	23852.68	225.0256069
		210	−11987.11	209.9871313			226	27333.22	226.0293434
		211	−11661.55	210.9874808			227	29652.40	227.0318332
		212	−8630.61	211.9907347	88	Ra	211	832.75	211.0008940
		213	−6593.91	212.9929212			212	−201.68	211.9997835
		214	−3393.98	213.9963564			213	322.15	213.0003458
		215	−1265.67	214.9986412			214	84.90	214.0000911
		216	2243.82	216.0024088			215	2518.94	215.0027042
		217	4386.98	217.0047096			216	3277.37	216.0035184
		218	8086.73	218.0086815			217	5874.01	217.0063060
		219	10522.60	219.0112965			218	6635.91	218.0071239
							219	9379.01	219.0100688
86	Rn	198	−1136.44	197.9987800			220	10260.10	220.0110147
		207	−8637.91	206.9907268			221	12955.00	221.0139078
		208	−9658.44	207.9896312			222	14309.44	222.0153618
		209	−8964.11	208.9903766			223	17229.97	223.0184971
		210	−9613.15	209.9896799			224	18818.04	224.0202020
		211	−8769.63	210.9905854			225	21987.41	225.0236045
		212	−8673.23	211.9906889			226	23662.32	226.0254025
		213	−5711.59	212.9938684			227	27172.31	227.0291707
		214	−4334.92	213.9953463			228	28936.02	228.0310641
		215	−1183.75	214.9987292			229	32434.90	229.0348203
		216	240.47	216.0002582			230	34544.24	230.0370848
		217	3646.38	217.0039146					
		218	5203.62	218.0055863	89	Ac	209	8913.22	209.0095687
		219	8825.75	219.0094748			210	8622.65	210.0092568
		220	10604.26	220.0113841			211	7124.25	211.0076482
		222	16366.79	222.0175705			212	7276.31	212.0078114
							213	6123.35	213.0065737
87	Fr	205	−1244.51	204.9986640			214	6420.85	214.0068931
		206	−1409.46	205.9984869			215	6008.91	215.0064508
		207	−2925.46	206.9968594			216	8123.81	216.0087213
		208	−2669.80	207.9971339			217	8693.33	217.0093327
		209	−3804.76	208.9959154			218	10828.66	218.0116250
		210	−3354.94	209.9963983			219	11555.10	219.0124049
		211	−4164.40	210.9955293			220	13741.50	220.0147521
		212	−3544.35	211.9961950			221	14508.71	221.0155757
		213	−3563.11	212.9961748			222	16607.47	222.0178288
		214	−973.65	213.9989547			223	17815.78	223.0191260

		224	20221.27	224.0217084			232	35938.64	232.0385817
		225	21629.82	225.0232206			233	37483.53	233.0402402
		226	24302.53	226.0260898			234	40335.85	234.0433023
		227	25846.14	227.0277470			235	42324.06	235.0454367
		228	28890.12	228.0310148			236	45340.63	236.0486752
		229	30674.90	229.0329309			237	47636.07	237.0511394
		230	33557.20	230.0360251			238	50763.66	238.0544970
		231	35910.49	231.0385515					
		232	39143.68	232.0420225	92	U	219	23208.56	219.0249154
90	Th	215	10923.25	215.0117266			223	25823.76	223.0277229
		216	10293.90	216.0110510			224	25700.04	224.0275901
		217	12171.06	217.0130662			225	27371.36	225.0293844
		218	12358.82	218.0132677			226	27329.80	226.0293397
		219	14457.95	219.0155212			227	29006.78	227.0311401
		220	14655.31	220.0157331			228	29217.57	228.0313663
		221	16926.64	221.0181715			229	31201.45	229.0334961
		222	17189.91	222.0184541			230	31603.16	230.0339274
		223	19370.56	223.0207951			231	33803.13	231.0362891
		224	19989.16	224.0214592			232	34601.53	232.0371463
		225	22301.31	225.0239414			233	36913.42	233.0396282
		226	23185.52	226.0248907			234	38140.58	234.0409456
		227	25801.32	227.0276988			235	40914.06	235.0439230
		228	26763.08	228.0287313			236	42440.63	236.0455619
		229	29579.90	229.0317553			237	45386.07	237.0487239
		230	30857.20	230.0331266			238	47303.66	238.0507826
		231	33810.49	231.0362970			239	50568.73	239.0542878
		232	35443.68	232.0380503			240	52709.26	240.0565857
		233	38728.65	233.0415769	93	Np	225	31577.35	225.0338997
		234	40608.94	234.0435955			227	32563.41	227.0349582
		235	44250.08	235.0475044			229	33763.73	229.0362468
							230	35222.20	230.0378126
91	Pa	213	19732.03	213.0211832			231	35613.96	231.0382331
		214	19318.47	214.0207392			233	37941.93	233.0407323
		215	17789.31	215.0190976			234	39950.43	234.0428885
		216	17800.52	216.0191096			235	41037.78	235.0440559
		217	17035.69	217.0182886			236	43370.10	236.0465597
		218	18637.24	218.0200079			237	44867.50	237.0481672
		219	18518.42	219.0198803			238	47450.73	238.0509404
		220	20377.82	220.0218765			239	49305.27	239.0529314
		221	20365.94	221.0218637			240	52320.92	240.0561688
		223	22322.08	223.0239637			241	54256.04	241.0582462
		224	23860.08	224.0256148					
		225	24326.12	225.0261152	94	Pu	228	36074.60	228.0387277
		226	26019.19	226.0279327			229	37389.17	229.0401389
		227	26820.64	227.0287931			230	36929.64	230.0396456
		228	28910.72	228.0310369			232	38358.40	232.0411794
		229	29890.33	229.0320886			233	40042.66	233.0429875
		230	32166.87	230.0345325			234	40338.04	234.0433047
		231	33420.97	231.0358789			235	42179.44	235.0452815

		236	42893.51	236.0460481	98	Cf	242	59325.64	242.0636887
		237	45087.82	237.0484037			244	61469.64	244.0659904
		238	46158.69	238.0495534			246	64085.67	246.0687988
		239	48583.48	239.0521565			247	66128.65	247.0709920
		240	50121.32	240.0538074			248	67233.44	248.0721780
		241	52951.04	241.0568453			249	69719.35	249.0748468
		242	54713.01	242.0587368			250	71166.09	250.0763999
		243	57749.84	243.0619970			251	74128.33	251.0795800
		244	59799.72	244.0641976			252	76028.14	252.0816195
		245	63098.14	245.0677386			253	79295.08	253.0851267
		246	65389.41	246.0701984			254	81334.50	254.0873162
95	Am	237	46547.44	237.0499707	99	Es	251	74504.23	251.0799836
		238	48417.04	238.0519778			252	77288.14	252.0829722
		239	49386.39	239.0530185			253	79007.42	253.0848179
		240	51500.27	240.0552878			254	81986.39	254.0880160
		241	52930.22	241.0568229			255	84082.58	255.0902663
		242	55463.97	242.0595430					
		243	57168.28	243.0613727	100	Fm	246	70124.38	246.0752816
		244	59875.89	244.0642794			248	71896.80	248.0771844
		245	61893.48	245.0664454			250	74067.51	250.0795147
		246	64988.87	246.0697684			251	75978.66	251.0815664
							252	76811.05	252.0824600
96	Cm	238	49384.36	238.0530163			253	79341.16	253.0851762
		240	51715.65	240.0555190			254	80898.19	254.0868478
		241	53697.58	241.0576467			255	83792.96	255.0899554
		242	54799.16	242.0588293			256	85479.95	256.0917665
		243	57177.19	243.0613822			257	88583.79	257.0950986
		244	58447.84	244.0627463					
		245	60999.42	245.0654856	101	Md	255	84835.99	255.0910752
		246	62612.74	246.0672175			256	87609.57	256.0940527
		247	65527.62	247.0703468			257	88989.93	257.0955346
		248	67386.36	248.0723422			258	91682.58	258.0984253
		249	70744.22	249.0759470					
		250	72983.18	250.0783506	102	No	252	82871.20	252.0889659
		251	76641.33	251.0822778			254	84718.20	254.0909487
							255	86845.45	255.0932324
97	Bk	243	58685.58	243.0630015			256	87817.40	256.0942758
		244	60703.48	244.0651678			257	90217.77	257.0968527
		245	61809.64	245.0663554					
		246	63962.74	246.0686668	104	Rf	256	94248.15	256.1011795
		247	65482.65	247.0702985					
		249	69843.35	249.0749799	106	Sg	260	106595.92	260.1144354
		250	72945.78	250.0783105					
		251	75221.33	251.0807534	108	Hs	264	119611.50	264.1284082

Appendix 2
Nuclide Table

The following table presents the properties of a selected set of known radioactive nuclides as well as all nuclides that exist in nature. They have been complied from the data set contained in the code "The NuDat Program for Nuclear Data on the Web" available from the National Nuclear Data Center, http://www.nndc.bnl.gov/nndc. The present table contains data as quoted in the data tables. For brevity, experimental errors have not been included here. The reader is urged to examine NuDat or other authoritative sources for error estimates when high accuracy is needed.

Z	Element	A	Abund (%)	J^π	$t_{1/2}$	Decay mode	Intensity (%)
0	n	1		1/2+	10.4 m	β^-	100
1	H	1	99.985	1/2+	stable		
	H	2	0.015	1+			
	H	3		1/2+	12.33 y	β^-	100
2	He	3	0.000137	1/2+	stable		
	He	4	99.999863	0+			
	He	6		0+	806.7 ms	β^-	100
	He	8		0+	119.0 ms	β^-	100
3	Li	6	7.5	1+	stable		
	Li	7	92.5	3/2−	stable		
	Li	8		2+	838 ms	β^-	100
	Li	9		3/2−	178.3 ms	β^-	100
	Li	11		3/2−	8.5 ms	β^-	100
4	Be	7		3/2−	53.29 d	EC	100
	Be	9	100	3/2−	stable		
	Be	10		0+	1.51×10^6 y	β^-	100
	Be	11		1/2+	13.81 s	β^-	100

Nuclear Physics for Applications. Stanley G. Prussin
Copyright © 2007 WILEY-VCH Verlag GmbH & Co., Weinheim
ISBN: 978-3-527-40700-2

Z	Element	A	Abund (%)	J^π	$t_{1/2}$	Decay mode	Intensity (%)
	Be	12		0+	23.6 ms	β^-	100
	Be	14		0+	4.35 ms	β^-	100
5	B	8		2+	770 ms	EC	100
	B	10	19.9	3+	stable		
	B	11	80.1	3/2−	stable		
	B	12		1+	20.20 ms	β^-	100
	B	13		3/2−	17.36 ms	β^-	100
	B	14		2−	13.8 ms	β^-	100
	B	15			10.5 ms	β^-	100
	B	17		(3/2−)	5.08 ms	β^-	100
6	C	9		(3/2−)	126.5 ms	EC	100
	C	10		0+	19.255 s	EC	100
	C	11		3/2−	20.39 m	EC	100
	C	12	98.89	0+	stable		
	C	13	1.11	1/2−	stable		
	C	14		0+	5730 y	β^-	100
	C	15		1/2+	2.449 s	β^-	100
	C	16		0+	0.747 s	β^-	100
	C	17			193 ms	β^-	100
	C	18		0+	95 ms	β^-	100
	C	19			49 ms	β^-	
	C	20		0+	14 ms	β^-	100
7	N	12		1+	11.0 ms	EC	100
	N	13		1/2−	9.965 m	EC	100
	N	14	99.634	1+	stable		
	N	15	0.366	1/2−	stable		
	N	16		2−	7.13 s	β^-	100
	N	17		1/2−	4.173 s	β^-	100
	N	18		1−	624 ms	β^-	100
	N	19			290 ms	β^-	100
	N	20			142 ms	β^-	100
	N	21		(1/2−)	87 ms	β^-	100
	N	22			18 ms	β^-	100
8	O	13		(3/2−)	8.58 ms	EC	100
	O	14		0+	70.606 s	EC	100
	O	15		1/2−	122.24 s	EC	100
	O	16	99.762	0+	stable		
	O	17	0.038	5/2+	stable		
	O	18	0.2	0+	stable		
	O	19		5/2+	26.91 s	β^-	100
	O	20		0+	13.51 s	β^-	100
	O	21		(1/2, 3/2 52)+	3.42 s	β^-	100
	O	22		0+	2.25 s	β^-	100

Z	Element	A	Abund (%)	J^π	$t_{1/2}$	Decay mode	Intensity (%)
	O	23			82 ms	β^-	100
	O	24		0+	61 ms	β^-	100
9	F	17		5/2+	64.49 s	EC	100
	F	18		1+	109.77 ms	EC	100
	F	19	100	1/2+	stable		
	F	20		2+	11.163 s	β^-	100
	F	21		5/2+	4.158 s	β^-	100
	F	22		4+, (3+)	4.23 s	β^-	100
	F	23		(3/2, 5/2)+	2.23 s	β^-	100
	F	24		(1, 2, 3)+	0.34 s	β^-	100
	F	25		(5/2+)	87 ms	β^-	100
10	Ne	17		1/2−	109.2 ms	EC	100
	Ne	18		0+	1672 ms	EC	100
	Ne	19		1/2+	17.22 s	EC	100
	Ne	20	90.48	0+	stable		
	Ne	21	0.27	3/2+	stable		
	Ne	22	9.25	0+	stable		
	Ne	23		5/2+	37.24 s	β^-	100
	Ne	24		0+	3.38 m	β^-	100
	Ne	25		(1/2, 3/2)+	602 ms	β^-	100
	Ne	26		0+	0.23 s	β^-	100
	Ne	27			32 ms	β^-	
	Ne	28		0+	14 ms	β^-	
	Ne	29		(3/2+)	200 ms	β^-	
11	Na	20		2+	447.9 ms	EC	100
	Na	21		3/2+	22.49 s	EC	100
	Na	22		3+	2.6019 y	EC	100
	Na	23	100	3/2+	stable		
	Na	24		4+	14.9590 h	β^-	100
	Na	25		5/2+	59.1 s	β^-	100
	Na	26		3+	1.072 s	β^-	100
	Na	27		5/2+	301 ms	β^-	100
	Na	28		1+	30.5 ms	β^-	100
	Na	29			44.9 ms	β^-	100
	Na	30		2+	48 ms	β^-	100
	Na	31		3/2+	17.0 ms	β^-	100
	Na	32		(3−, 4−)	13.2 ms	β^-	100
	Na	33			8.2 ms	β^-	100
12	Mg	20		0+	90.8 ms	EC	100
	Mg	21		(3/2, 5/2)+	122 ms	EC	100
	Mg	22		0+	3.857 s	EC	100
	Mg	23		3/2+	11.317 s	EC	100
	Mg	24	78.99	0+	stable		
	Mg	25	10	5/2+	stable		

Z	Element	A	Abund (%)	J^π	$t_{1/2}$	Decay mode	Intensity (%)
	Mg	26	11.01	0+	stable		
	Mg	27		1/2+	9.458 m	β^-	100
	Mg	28		0+	20.91 h	β^-	100
	Mg	29		3/2+	1.30 s	β^-	100
	Mg	30		0+	335 ms	β^-	100
	Mg	31			230 ms	β^-	100
	Mg	32		0+	120 ms	β^-	100
	Mg	33			90 ms	β^-	100
	Mg	34		0+	20 ms	β^-	100
13	Al	23			0.47 s	EC	100
	Al	24		4+	2.053 s	EC	100
	Al	25		5/2+	7.183 s	EC	100
	Al	26		5+	7.17×10^5 y	EC	100
	Al	27	100	5/2+			
	Al	28		3+	2.2414 m	β^-	100
	Al	29		5/2+	6.56 m	β^-	100
	Al	30		3+	3.60 s	β^-	100
	Al	31		(3/2, 5/2)+	644 ms	β^-	100
	Al	32		1+	33 ms	β^-	100
	Al	34			60 ms	β^-	100
	Al	35			150 ms	β^-	100
14	Si	24		0+	102 ms	EC	100
	Si	25		5/2+	220 ms	EC	100
	Si	26		0+	2.234 s	EC	100
	Si	27		5/2+	4.16 s	EC	100
	Si	28	92.23	0+	stable		
	Si	29	4.67	1/2+	stable		
	Si	30	3.1	0+			
	Si	31		3/2+	157.3 m	β^-	100
	Si	32		0+	172 y	β^-	100
	Si	33			6.18 s	β^-	100
	Si	34		0+	2.77 s	β^-	100
	Si	35			0.78 s	β^-	100
	Si	36		0+	0.45 s	β^-	100
15	P	27		(1/2+)	260 ms	EC	100
	P	28		3+	270.3 ms	EC	100
	P	29		1/2+	4.142 s	EC	100
	P	30		1+	2.498 m	EC	100
	P	31	100	1/2+	stable		
	P	32		1+	14.262 d	β^-	100
	P	33		1/2+	25.34 d	β^-	100
	P	34		1+	12.43 s	β^-	100
	P	35		1/2+	47.3 s	β^-	100
	P	36			5.6 s	β^-	100
	P	37			2.31 s	β^-	100

Z	Element	A	Abund (%)	J^π	$t_{1/2}$	Decay mode	Intensity (%)
	P	38			0.64 s	β^-	100
	P	39			0.16 s	β^-	100
	P	40			260 ms	β^-	100
	P	41			120 ms	β^-	100
16	S	28		0+	125 ms	EC	100
	S	29		5/2+	187 ms	EC	100
	S	30		0+	1.178 s	EC	100
	S	31		1/2+	2.572 s	EC	100
	S	32	95.02	0+	stable		
	S	33	0.75	3/2+	stable		
	S	34	4.21	0+	stable		
	S	35		3/2+	87.51 d	β^-	100
	S	36	0.02	0+			
	S	37		7/2−	5.05 m	β^-	100
	S	38		0+	170.3 m	β^-	100
	S	39		(3/2 − 7/2)−	11.5 s	β^-	100
	S	40		0+	8.8 s	β^-	100
	S	41		(7/2−)	2.6 s	β^-	100
	S	42		0+	0.56 s	β^-	100
	S	43			220 ms	β^-	100
17	Cl	31			150 ms	EC	100
	Cl	32		1+	298 ms	EC	100
	Cl	33		3/2+	2.511 s	EC	100
	Cl	34		0+	1.5264 s	EC	100
	Cl	35	75.77	3/2+	stable		
	Cl	36		2+	3.01×10^5 y	β^-, EC	98.1, 1.9
	Cl	37	24.23	3/2+	stable		
	Cl	38		2−	37.24 m	β^-	100
	Cl	39		3/2+	55.6 m	β^-	100
	Cl	40		2−	1.35 m	β^-	100
	Cl	41		(1/2, 3/2)+	38.4 s	β^-	100
	Cl	42			6.8 s	β^-	100
	Cl	43			3.3 s	β^-	100
	Cl	44			434 ms	β^-	100
	Cl	45			400 ms	β^-	100
18	Ar	32		0+	98 ms	EC	100
	Ar	33		1/2+	173.0 ms	EC	100
	Ar	34		0+	844.5 ms	EC	100
	Ar	35		3/2+	1.775 s	EC	100
	Ar	36	0.3365	0+	stable		
	Ar	37		3/2+	35.04 d	EC	100
	Ar	38	0.0632	0+	stable		
	Ar	39		7/2−	269 y	β^-	100
	Ar	40	99.6003	0+	stable		
	Ar	41		7/2−	109.34 m	β^-	100

Z	Element	A	Abund (%)	J^π	$t_{1/2}$	Decay mode	Intensity (%)
	Ar	42		0+	32.9 y	β^-	100
	Ar	43		(3/2, 5/2)	5.37 m	β^-	100
	Ar	44		0+	11.87 m	β^-	100
	Ar	45			21.48 s	β^-	100
	Ar	46		0+	8.4 s	β^-	100
	Ar	47			~700 ms	β^-	100
19	K	35		3/2+	190 ms	EC	100
	K	36		2+	342 ms	EC	100
	K	37		3/2+	1.226 s	EC	100
	K	38		3+	7.636 m	EC	100
	K	39	93.2581	3/2+	stable		
	K	40	0.0117	4−	1.277×10^9 y	β^-, EC	89.28, 10.72
	K	41	6.7302	3/2+	stable		
	K	42		2−	12.360 h	β^-	100
	K	43		3/2+	22.3 h	β^-	100
	K	44		2−	22.13 m	β^-	100
	K	45		3/2+	17.3 m	β^-	100
	K	46		(2−)	105 s	β^-	100
	K	47		1/2+	17.50 s	β^-	100
	K	48		(2−)	6.8 s	β^-	100
	K	49		(3/2+)	1.26 s	β^-	100
	K	50		(0−,1, 2−)	472 ms	β^-	100
20	Ca	36		0+	102 ms	EC	100
	Ca	37		3/2+	181.1 ms	EC	100
	Ca	38		0+	440 ms	EC	100
	Ca	39		3/2+	859.6 ms	EC	100
	Ca	40	96.941	0+	stable		
	Ca	41		7/2−	1.03×10^5 y	EC	100
	Ca	42	0.647	0+	stable		
	Ca	43	0.135	7/2−	stable		
	Ca	44	2.086	0+	stable		
	Ca	45		7/2−	162.61 d	β^-	100
	Ca	46	0.004	0+	stable		
	Ca	47		7/2−	4.536 d	β^-	100
	Ca	48	0.187	0+	$> 6 \times 10^{18}$ y	β^-	
	Ca	49		3/2−	8.718 m	β^-	100
	Ca	50		0+	13.9 s	β^-	100
	Ca	51		(3/2−)	10.0 s	β^-	100
	Ca	52		0+	4.6 s	β^-	100
21	Sc	40		4−	182.3 ms	EC	100
	Sc	41		7/2−	596.3 ms	EC	100
	Sc	42		0+	681.3 ms	EC	100
	Sc	43		7/2−	3.891 h	EC	100
	Sc	44		2+	3.927 h	EC	100
	Sc	45	100	7/2−	stable		

Z	Element	A	Abund (%)	J^π	$t_{1/2}$	Decay mode	Intensity (%)
	Sc	46		4+	83.79 d	β^-	100
	Sc	47		7/2–	3.3492 d	β^-	100
	Sc	48		6+	43.67 h	β^-	100
	Sc	49		7/2–	57.2 m	β^-	100
	Sc	50		5+	102.5 s	β^-	100
	Sc	51		(7/2)–	12.4 s	β^-	100
	Sc	52		3+	8.2 s	β^-	100
	Sc	54			> 1 µs	β^-	100
22	Ti	40		0+	50 ms	EC	100
	Ti	42		0+	199 ms	EC	100
	Ti	43		7/2–	509 ms	EC	100
	Ti	44		0+	63 y	EC	100
	Ti	45		7/2–	184.8 m	EC	100
	Ti	46	8.25	0+	stable		
	Ti	47	7.44	5/2–	stable		
	Ti	48	73.72	0+	stable		
	Ti	49	5.41	7/2–	stable		
	Ti	50	5.18	0+	stable		
	Ti	51		3/2–	5.76 m	β^-	100
	Ti	52		0+	1.7 m	β^-	100
	Ti	53		(3/2)–	32.7 s	β^-	100
	Ti	54		0+	> 1 µs	β^-	100
	Ti	55		(3/2–)	570 ms	β^-	100
	Ti	56		0+	0.15 s	β^-	100
23	V	45		7/2–	547 ms	EC	100
	V	46		0+	422.37 ms	EC	100
	V	47		3/2–	32.6 m	EC	100
	V	48		4+	15.9735 d	EC	100
	V	49		7/2–	330 d	EC	100
	V	50	0.25	6+	1.4×10^{17} y	EC, β^-	83, 17
	V	51	99.75	7/2–	stable		
	V	52		3+	3.743 m	β^-	100
	V	53		7/2–	1.61 m	β^-	100
	V	54		3+	49.8 s	β^-	100
	V	55		(7/2–)	6.54 s	β^-	100
	V	56		3+	230 ms	β^-	100
	V	57		(7/2–)	0.32 s	β^-	100
	V	58			> 200 ns	β^-	100
	V	59			> 200 ns	β^-	100
	V	60		(3+)	220 ms	β^-	100
24	Cr	46		0+	0.26 s	EC	100
	Cr	47		3/2–	500 ms	EC	100
	Cr	48		0+	21.56 h	EC	100
	Cr	49		5/2–	42.3 m	EC	100
	Cr	50	4.345	0+	$> 1.8 \times 10^{17}$ y	2EC	

Z	Element	A	Abund (%)	J^π	$t_{1/2}$	Decay mode	Intensity (%)
	Cr	51		7/2–	27.7025 d	EC	100
	Cr	52	83.789	0+	stable		
	Cr	53	9.501	3/2–	stable		
	Cr	54	2.365	0+	stable		
	Cr	55		3/2–	3.497 m	β^-	100
	Cr	56		0+	5.94 m	β^-	100
	Cr	57		3/2– – 7/2–	21.1 s	β^-	100
	Cr	58		0+	7.0 s	β^-	100
	Cr	59			0.74 s	β^-	100
	Cr	60		0+	0.57 s	β^-	100
	Cr	61			> 200 ns	β^-	100
	Cr	62		0+	160 ms	β^-	100
25	Mn	49		5/2–	382 ms	EC	100
	Mn	50		0+	283.88 ms	EC	100
	Mn	51		5/2–	46.2 m	EC	100
	Mn	52		6+	5.591 d	EC	100
	Mn	53		7/2–	3.74×10^6 y	EC	100
	Mn	54		3+	312.3 d	EC, β^-	100, $< 2.9 \times 10^4$
	Mn	55	100	5/2–	stable		
	Mn	56		3+	2.5789 h	β^-	100
	Mn	57		5/2–	85.4 s	β^-	100
	Mn	58		1+	3.0 s	β^-	100
	Mn	59		3/2–, 5/2–	4.6 s	β^-	100
	Mn	60		0+	51 s	β^-	100
	Mn	61		(5/2)–	0.71 s	β^-	100
	Mn	62		(3+)	0.88 s	β^-	100
	Mn	63			0.25 s	β^-	100
26	Fe	50		0+	150 ms	EC	100
	Fe	51		5/2–	305 ms	EC	100
	Fe	52		0+	8.275 h	EC	100
	Fe	53		7/2–	8.51 m	EC	100
	Fe	54	5.845	0+	stable		
	Fe	55		3/2–	2.73 y	EC	100
	Fe	56	91.754	0+	stable		
	Fe	57	2.119	1/2–	stable		
	Fe	58	0.282	0+	stable		
	Fe	59		3/2–	44.503 d	β^-	100
	Fe	60		0+	1.5×10^5 y	β^-	100
	Fe	61		3/2–, 5/2–	5.98 m	β^-	100
	Fe	62		0+	68 s	β^-	100
	Fe	63		(5/2)–	6.1 s	β^-	100
	Fe	64		0+	2.0 s	β^-	100
	Fe	65			0.4 s	β^-	100
	Fe	66		0+	600 ms	β^-	100
	Fe	67			> 200 ns	β^-	100

Appendix 2 Nuclide Table

Z	Element	A	Abund (%)	J^π	$t_{1/2}$	Decay mode	Intensity (%)
27	Co	53		(7/2−)	240 ms	EC	100
	Co	54		0+	193.23 ms	EC	100
	Co	55		7/2−	17.53 h	EC	100
	Co	56		4+	77.233 d	EC	100
	Co	57		7/2−	271.74 d	EC	100
	Co	58		2+	70.86 d	EC	100
	Co	59	100	7/2−	stable		
	Co	60		5+	1925.1 d	β^-	100
	Co	61		7/2−	1.650 h	β^-	100
	Co	62		2+	1.50 m	β^-	100
	Co	63		(7/2)−	27.4 s	β^-	100
	Co	64		1+	0.30 s	β^-	100
	Co	65		(7/2)−	1.20 s	β^-	100
	Co	66		(3+)	0.233 s	β^-	100
	Co	67		(7/2−)	0.42 s	β^-	100
	Co	68			0.18 s	β^-	100
	Co	69			0.27 s	β^-	
28	Ni	55		7/2−	212.1 ms	EC	100
	Ni	56		0+	6.075 d	EC	100
	Ni	57		3/2−	35.60 h	EC	100
	Ni	58	68.077	0+	stable		
	Ni	59		3/2−	7.6×10^4 y	EC	100
	Ni	60	26.223	0+	stable		
	Ni	61	1.14	3/2−	stable		
	Ni	62	3.634	0+	stable		
	Ni	63		1/2−	100.1 y	β^-	100
	Ni	64	0.926	0+	stable		
	Ni	65		5/2−	2.5172 h	β^-	100
	Ni	66		0+	54.6 h	β^-	100
	Ni	67		(1/2−)	21 s	β^-	100
	Ni	68		0+	19 s	β^-	100
	Ni	69			11.4 s	β^-	100
	Ni	71			1.86 s	β^-	100
	Ni	72		0+	150 ms	β^-	100
29	Cu	57		3/2−	196.3 ms	EC	100
	Cu	58		1+	3.204 s	EC	100
	Cu	59		3/2−	81.5 s	EC	100
	Cu	60		2+	23.7 m	EC	100
	Cu	61		3/2−	3.333 h	EC	100
	Cu	62		1+	9.74 m	EC	100
	Cu	63	69.17	3/2−	stable		
	Cu	64		1+	12.700 h	EC, β^-	61, 39
	Cu	65	30.83	3/2−	stable		
	Cu	66		1+	5.120 m	β^-	100
	Cu	67		3/2−	61.83 h	β^-	100
	Cu	68		1+	31.1 s	β^-	100

Z	Element	A	Abund (%)	J^π	$t_{1/2}$	Decay mode	Intensity (%)
	Cu	69		3/2–	2.85 m	β^-	100
	Cu	70		1+	4.5 s	β^-	100
	Cu	71		(3/2–)	19.5 s	β^-	100
30	Zn	58		0+	86 ms	EC	100
	Zn	59		3/2–	182.0 ms	EC	100
	Zn	60		0+	2.38 m	EC	100
	Zn	61		3/2–	89.1 s	EC	100
	Zn	62		0+	9.186 h	EC	100
	Zn	63		3/2–	38.47 m	EC	100
	Zn	64	48.6	0+	stable		
	Zn	65		5/2–	244.26 d	EC	100
	Zn	66	27.9	0+	stable		
	Zn	67	4.1	5/2–	stable		
	Zn	68	18.8	0+	stable		
	Zn	69		1/2–	56.4 m	β^-	100
	Zn	70	0.6	0+	> 5E+14 y	$\beta^-\beta^-$	
	Zn	71		1/2–	2.45 m	β^-	100
	Zn	72		0+	46.5 h	β^-	100
	Zn	73		(1/2)–	23.5 s	β^-	100
	Zn	74		0+	95.6 s	β^-	100
	Zn	75		(7/2+)	10.2 s	β^-	100
	Zn	76		0+	5.7 s	β^-	100
	Zn	77		(7/2+)	2.08 s	β^-	100
	Zn	78		0+	1.47 s	β^-	100
	Zn	80		0+	0.545 s	β^-	100
31	Ga	62		0+	116.12 ms	EC	100
	Ga	63		3/2–, 5/2–	32.4 s	EC	100
	Ga	64		0+	2.627 m	EC	100
	Ga	65		3/2–	15.2 m	EC	100
	Ga	66		0+	9.49 h	EC	100
	Ga	67		3/2–	3.2612 d	EC	100
	Ga	68		1+	67.629 m	EC	100
	Ga	69	60.108	3/2–	stable		
	Ga	70		1+	21.14 m	β^-, EC	99.59, 0.49
	Ga	71	39.892	3/2–	stable		
	Ga	72		3–	14.10 h	β^-	100
	Ga	73		3/2–	4.86 h	β^-	100
	Ga	74		(3–)	8.12 m	β^-	100
	Ga	75		(3/2)–	126 s	β^-	100
	Ga	76		(2+, 3+)	32.6 s	β^-	100
	Ga	77		(3/2–)	13.2 s	β^-	100
	Ga	78		(3+)	5.09 s	β^-	100
	Ga	79		(3/2–)	2.847 s	β^-	100
	Ga	80		(3)	1.697 s	β^-	100
	Ga	81		(5/2–)	1.217 s	β^-	100

Z	Element	A	Abund (%)	J^π	$t_{1/2}$	Decay mode	Intensity (%)
32	Ge	64		0+	63.7 s	EC	100
	Ge	65		(3/2)−	30.9 s	EC	100
	Ge	66		0+	2.26 h	EC	100
	Ge	67		1/2−	18.9 m	EC	100
	Ge	68		0+	270.8 d	EC	100
	Ge	69		5/2−	39.05 h	EC	100
	Ge	70	21.23	0+	stable		
	Ge	71		1/2−	11.43 d	EC	100
	Ge	72	27.66	0+	stable		
	Ge	73	7.73	9/2+	stable		
	Ge	74	35.94	0+	stable		
	Ge	75		1/2−	82.78 m	β^-	100
	Ge	76	7.44	0+	stable		
	Ge	77		7/2+	11.30 h	β^-	100
	Ge	78		0+	88.0 m	β^-	100
	Ge	79		(1/2)−	18.98 s	β^-	100
	Ge	80		0+	29.5 s	β^-	100
	Ge	81		(9/2+)	7.6 s	β^-	100
	Ge	82		0+	4.60 s	β^-	100
33	As	67		(5/2−)	42.5 s	EC	100
	As	68		3+	151.6 s	EC	100
	As	69		5/2−	15.2 m	EC	100
	As	70		4(+)	52.6 m	EC	100
	As	71		5/2−	65.28 h	EC	100
	As	72		2−	26.0 h	EC	100
	As	73		3/2−	80.30 d	EC	100
	As	74		2−	17.77 d	EC, β^-	66, 34
	As	75	100	3/2−	stable		
	As	76		2−	1.0778 d	β^-	100
	As	77		3/2−	38.83 h	β^-	100
	As	78		2−	90.7 m	β^-	100
	As	79		3/2−	9.01 m	β^-	100
	As	80		1+	15.2 s	β^-	100
	As	81		3/2−	33.3 s	β^-	100
	As	82		(5−), (1+)	13.6 s, 19.1 s	β^-	100
	As	83		(5/2−, 3/2−)	13.4 s	β^-	100
34	Se	69		(3/2−)	27.4 s	EC	100
	Se	72		0+	8.40 d	EC	100
	Se	73		9/2+	7.15 h	EC	100
	Se	74	0.89	0+	stable		
	Se	75		5/2+	119.779 d	EC	100
	Se	76	9.36	0+	stable		
	Se	77	7.63	1/2−	stable		
	Se	78	23.78	0+	stable		
	Se	79		7/2+	1.1×10^6 y	β^-	100
	Se	80	49.61	0+	stable		

Z	Element	A	Abund (%)	J^π	$t_{1/2}$	Decay mode	Intensity (%)
	Se	81		1/2−	18.45 m	β^-	100
	Se	82	8.73	0+	1.08×10^{20} y	$2\beta^-$	100
	Se	83		9/2+	22.3 m	β^-	100
	Se	84		0+	3.10 m	β^-	100
	Se	85		(5/2+)	31.7 s	β^-	100
	Se	86		0+	15.3 s	β^-	100
	Se	87		(5/2+)	5.29 s	β^-	100
	Se	88		0+	1.53 s	β^-	100
35	Br	72		3+	78.6 s	EC	100
	Br	73		1/2−	3.4 m	EC	100
	Br	74		(0−)	25.4 m	EC	100
	Br	75		3/2−	96.7 m	EC	100
	Br	76		1−	16.2 h	EC	100
	Br	77		3/2−	57.036 h	EC	100
	Br	78		1+	6.46 m	EC, β^-	99.99, 0.01
	Br	79	50.69	3/2−	stable		
	Br	80		1+	17.68 m	β^-, EC	91.7, 8.3
	Br	81	49.31	3/2−	stable		
	Br	82		5−	35.30 h	β^-	100
	Br	83		3/2−	2.40 h	β^-	100
	Br	84		2−	31.80 m	β^-	100
	Br	85		3/2−	2.90 m	β^-	100
	Br	86		(2−)	55.1 s	β^-	100
	Br	87		3/2−	55.60 s	β^-	100
	Br	88		(1, 2−)	16.29 s	β^-	100
	Br	89		(3/2−, 5/2−)	4.40 s	β^-	100
	Br	90			1.91 s	β^-	100
	Br	91			0.541 s	β^-	100
	Br	92		(2−)	0.343 s	β^-	100
36	Kr	72		0+	17.2 s	EC	100
	Kr	73		5/2−	27.0 s	EC	100
	Kr	74		0+	11.50 m	EC	100
	Kr	75		5/2+	4.29 m	EC	100
	Kr	76		0+	14.8 h	EC	100
	Kr	77		5/2+	74.4 m	EC	100
	Kr	78	0.35	0+	stable		
	Kr	79		1/2−	35.04 h	EC	100
	Kr	80	2.25	0+			
	Kr	81		7/2+	2.29×10^5 y	EC	100
	Kr	82	11.6	0+	stable		
	Kr	83	11.5	9/2+	stable		
	Kr	84	57	0+	stable		
	Kr	85		9/2+	3934.4 d	β^-	100
	Kr	86	17.3	0+	stable		
	Kr	87		5/2+	76.3 m	β^-	100
	Kr	88		0+	2.84 h	β^-	100

Z	Element	A	Abund (%)	J^π	$t_{1/2}$	Decay mode	Intensity (%)
	Kr	89		3/2(+)	3.15 m	β^-	100
	Kr	90		0+	32.32 s	β^-	100
	Kr	91		5/2(+)	8.57 s	β^-	100
	Kr	92		0+	1.840 s	β^-	100
	Kr	93		1/2+	1.286 s	β^-	100
37	Rb	74		(0+)	64.9 ms	EC	100
	Rb	75		(3/2−)	19.0 s	EC	100
	Rb	76		1(−)	36.5 s	EC	100
	Rb	77		3/2−	3.77 m	EC	100
	Rb	78		0(+)	17.66 m	EC	100
	Rb	79		5/2+	22.9 m	EC	100
	Rb	80		1+	33.4 s	EC	100
	Rb	81		3/2−	4.576 h	EC	100
	Rb	82		1+	1.273 m	EC	100
	Rb	83		5/2−	86.2 d	EC	100
	Rb	84		2−	32.77 d	EC, β^-	96.2, 3.8
	Rb	85	72.165	5/2−	stable		
	Rb	86		2−	18.631 d	β^-, EC	99.99, 0.0052
	Rb	87	27.835	3/2−	4.75×10^{10} y	β^-	100
	Rb	88		2−	17.78 m	β^-	100
	Rb	89		3/2−	15.15 m	β^-	100
	Rb	90		0−	158 s	β^-	100
	Rb	91		3/2(−)	58.4 s	β^-	100
	Rb	92		0−	4.492 s	β^-	100
	Rb	93		5/2−	5.84 s	β^-	100
	Rb	94		3(−)	2.702 s	β^-	100
	Rb	95		5/2−	377.5 ms	β^-	100
	Rb	96		2+	202.8 ms	β^-	100
	Rb	97		3/2+	169.9 ms	β^-	100
	Rb	98		(1, 0)	114 ms	β^-	100
	Rb	99		(5/2+)	50.3 ms	β^-	100
	Rb	101		(3/2+)	32 ms	β^-	100
38	Sr	77		5/2+	9.0 s	EC	100
	Sr	78		0+	2.5 m	EC	100
	Sr	79		3/2(−)	2.25 m	EC	100
	Sr	80		0+	106.3 m	EC	100
	Sr	81		1/2−	22.3 m	EC	100
	Sr	82		0+	25.55 d	EC	100
	Sr	83		7/2+	32.41 h	EC	100
	Sr	84	0.56	0+	stable		
	Sr	85		9/2+	64.84 d	EC	100
	Sr	86	9.86	0+	stable		
	Sr	87	7.00	9/2+	stable		
	Sr	88	82.58	0+	stable		
	Sr	89		5/2+	50.53 d	β^-	100
	Sr	90		0+	28.79 y	β^-	100

Z	Element	A	Abund (%)	J^π	$t_{1/2}$	Decay mode	Intensity (%)
	Sr	91		5/2+	9.63 h	β^-	100
	Sr	92		0+	2.71 h	β^-	100
	Sr	93		5/2+	7.423 m	β^-	100
	Sr	94		0+	75.3 s	β^-	100
	Sr	95		1/2+	23.90 s	β^-	100
	Sr	96		0+	1.07 s	β^-	100
	Sr	97		1/2+	429 ms	β^-	100
	Sr	98		0+	0.653 s	β^-	100
	Sr	99		3/2+	0.269 s	β^-	100
	Sr	100		0+	202 ms	β^-	100
	Sr	101		(5/2−)	118 ms	β^-	100
	Sr	102		0+	69 ms	β^-	100
39	Y	79		(5/2+)	14.8 s	EC	100
	Y	81		(5/2+)	70.4 s	EC	100
	Y	82		1+	9.5 s	EC	100
	Y	83		(9/2+)	7.08 m	EC	100
	Y	84		1+, (5−)	4.6, 39.5	EC, EC	100
	Y	85		(1/2)−	2.68 h	EC	100
	Y	86		4−	14.74 h	EC	100
	Y	87		1/2−	79.8 h	EC	100
	Y	88		4−	106.65 d	EC	100
	Y	89	100	1/2−	stable		
	Y	90		2−	64.00 h	β^-	100
	Y	91		1/2−	58.51 d	β^-	100
	Y	92		2−	3.54 h	β^-	100
	Y	93		1/2−	10.18 h	β^-	100
	Y	94		2−	18.7 m	β^-	100
	Y	95		1/2−	10.3 m	β^-	100
	Y	96		0−, (8)+	5.34 s, 9.6 s	β^-, β^-	100
	Y	97		(1/2−)	3.75 s	β^-	100
	Y	98		(0)−	0.548 s	β^-	100
	Y	99		(5/2+)	1.470 s	$\beta-$	100
	Y	100		(3 − 5), 1−, 2−	0.94 s, 735 ms	β^-, β^-	100
	Y	101		(5/2+)	0.45 s	β^-	100
	Y	102			0.30, 0.36 s	β^-, β^-	100
40	Zr	81			15 s	EC	100
	Zr	82		0+	32 s	EC	100
	Zr	83		(1/2−)	44 s	EC	100
	Zr	85		7/2+	7.86 m	EC	100
	Zr	86		0+	16.5 h	EC	100
	Zr	87		(9/2)+	1.68 h	EC	100
	Zr	88		0+	83.4 d	EC	100
	Zr	89		9/2+	78.41 h	EC	100
	Zr	90	51.45	0+	stable		
	Zr	91	11.22	5/2+	stable		
	Zr	92	17.15	0+	stable		

Z	Element	A	Abund (%)	J^π	$t_{1/2}$	Decay mode	Intensity (%)
	Zr	93		5/2+	1.53×10^6 y	β^-	100
	Zr	94	17.38	0+	stable		
	Zr	95		5/2+	64.02 d	β^-	100
	Zr	96	2.8	0+	stable		
	Zr	97		1/2+	16.90 h	β^-	100
	Zr	98		0+	30.7 s	β^-	100
	Zr	99		(1/2+)	2.1 s	β^-	100
	Zr	100		0+	7.1 s	β^-	100
	Zr	101		(3/2+)	2.3 s	β^-	100
	Zr	102		0+	2.9 s	β^-	100
	Zr	103		(5/2−)	1.3 s	β^-	100
41	Nb	83		(5/2+)	4.1 s	EC	100
	Nb	85		(9/2+)	20.9 s	EC	100
	Nb	86		(5+)	88, 56 s	EC, EC	100
	Nb	87		(9/2+), (1/2−)	2.6, 3.7 m	EC, EC	100
	Nb	89		(9/2+)	2.03 h	EC	100
	Nb	90		8+	14.60 h	EC	100
	Nb	91		9/2+	680 y	EC	100
	Nb	92		(7)+	3.47×10^7 y	EC, β^-	100, < 0.05
	Nb	93	100	9/2+	stable		
	Nb	94		(6)+	2.03×10^4 y	β^-	100
	Nb	95		9/2+	34.975 d	β^-	100
	Nb	96		6+	23.35 h	β^-	100
	Nb	97		9/2+	72.1 m	β^-	100
	Nb	98		1+	2.86 s	β^-	100
	Nb	99		9/2+	15.0 s	β^-	100
	Nb	100		1+	1.5 s	β^-	100
	Nb	101		(5/2+)	7.1 s	β^-	100
	Nb	102		1+, ?	1.3 s, 4.3 s	β^-, β^-	
	Nb	103		(5/2+)	1.5 s	β^-	100
	Nb	104		(1+)	4.8 s	β^-	100
	Nb	105		(5/2+)	2.95 s	β^-	100
42	Mo	86		0+	19.6 s	EC	100
	Mo	87		(7/2+)	14.5 s	EC	100
	Mo	88		0+	8.0 m	EC	100
	Mo	89		(9/2+)	2.11 m	EC	100
	Mo	90		0+	5.56 h	EC	100
	Mo	91		9/2+	15.49 m	EC	100
	Mo	92	14.84	0+	stable		
	Mo	93		5/2+	4000 y	EC	100
	Mo	94	9.25	0+	stable		
	Mo	95	15.92	5/2+	stable		
	Mo	96	16.68	0+	stable		
	Mo	97	9.55	5/2+	stable		
	Mo	98	24.13	0+	stable		
	Mo	99		1/2+	65.94 h	β^-	100

Z	Element	A	Abund (%)	J^π	$t_{1/2}$	Decay mode	Intensity (%)
	Mo	100	9.63	0+	1.00×10^{19} y	$2\beta^-$	100
	Mo	101		1/2+	14.61 m	β^-	100
	Mo	102		0+	11.3 m	β^-	100
	Mo	103		(3/2+)	67.5 s	β^-	100
	Mo	104		0+	60 s	β^-	100
	Mo	105		(3/2+)	35.6 s	β^-	100
	Mo	106		0+	8.4 s	β^-	100
	Mo	107			3.5 s	β^-	100
43	Tc	89		(9/2+)	12.8 s	EC	100
	Tc	90		1+	8.7 s	EC	100
	Tc	91		(9/2)+	3.14 m	EC	100
	Tc	92		(8)+	4.23 m	EC	100
	Tc	93		9/2+	2.75 h	EC	100
	Tc	94		7+	293 m	EC	100
	Tc	95		9/2+	20.0 h	EC	100
	Tc	96		7+	4.28 d	EC	100
	Tc	97		9/2+	2.6×10^6 y	EC	100
	Tc	98		(6)+	4.2×10^6 y	β^-	100
	Tc	99		9/2+	2.111×10^5 y	β^-	100
	Tc	100		1+	15.8 s	β^-, EC	100, 0.0018
	Tc	101		9/2+	14.22 m	β^-	100
	Tc	102		1+	5.28 s	β^-	100
	Tc	103		5/2+	54.2 s	β^-	100
	Tc	104		(3+)	18.3 m	β^-	100
	Tc	105		(5/2+)	7.6 m	β^-	100
	Tc	106		(1, 2)	35.6 s	β^-	100
	Tc	107			21.2 s	β^-	100
	Tc	108		(2)+	5.17 s	β^-	100
44	Ru	91		(9/2+), (1/2−)	9 s, 7.6 s	EC, EC	100
	Ru	93		(9/2)+	59.7 s	EC	100
	Ru	94		0+	51.8 m	EC	100
	Ru	95		5/2+	1.643 h	EC	100
	Ru	96	5.52	0+	stable		
	Ru	97		5/2+	2.9 d	EC	100
	Ru	98	1.88	0+	stable		
	Ru	99	12.7	5/2+	stable		
	Ru	100	12.6	0+	stable		
	Ru	101	17	5/2+	stable		
	Ru	102	31.6	0+	stable		
	Ru	103		3/2+	39.26 d	β^-	100
	Ru	104	18.7	0+	stable		
	Ru	105		3/2+	4.44 h	β^-	100
	Ru	106		0+	373.59 d	β^-	100
	Ru	107		(5/2)+	3.75 m	β^-	100
	Ru	108		0+	4.55 m	β^-	100

Z	Element	A	Abund (%)	J^π	$t_{1/2}$	Decay mode	Intensity (%)
	Ru	109		(5/2+)	34.5 s	β^-	100
	Ru	110		0+	14.6 s	β^-	100
45	Rh	95		(9/2)+	5.02 m	EC	100
	Rh	96		(6+)	9.90 m	EC	100
	Rh	97		9/2+	30.7 m	EC	100
	Rh	98		(5+), (2)+	3.5, 8.7 m	EC, EC	> 0.00, 100
	Rh	99		1/2–	16.1 d	EC	100
	Rh	100		1–	20.8 h	EC	100
	Rh	101		1/2–	3.3 y	EC	100
	Rh	102		(1–, 2–)	207 d	EC, β^-	80, 20
	Rh	103	100	1/2–	stable		
	Rh	104		1+	42.3 s	β^-, EC	99.55, 0.45
	Rh	105		7/2+	35.36 h	β^-	100
	Rh	106		1+	29.80 s	β^-	100
	Rh	107		7/2+	21.7 m	β^-	100
	Rh	108		1+	16.8 s	β^-	100
	Rh	109		7/2+	80 s	β^-	100
	Rh	110		(> 4), 1+	28.5 s, 3.2 s	β^-, β^-	100
46	Pd	96		0+	122 s	EC	100
	Pd	97		(5/2+)	3.10 m	EC	100
	Pd	98		0+	17.7 m	EC	100
	Pd	99		(5/2)+	21.4 m	EC	100
	Pd	100		0+	3.63 d	EC	100
	Pd	101		5/2+	8.47 h	EC	100
	Pd	102	1.02	0+	stable		
	Pd	103		5/2+	16.991 d	EC	100
	Pd	104	11.14	0+	stable		
	Pd	105	22.33	5/2+	stable		
	Pd	106	27.33	0+	stable		
	Pd	107		5/2+	6.5×10^6 y	β^-	100
	Pd	108	26.46	0+	stable		
	Pd	109		5/2+	13.7012 h	β^-	100
	Pd	110	11.72	0+	stable		
	Pd	111		5/2+	23.4 m	β^-	100
	Pd	112		0+	21.03 h	β^-	100
	Pd	113		(5/2+)	93 s	β^-	100
	Pd	114		0+	2.42 m	β^-	100
	Pd	115		(5/2+)	25 s	β^-	100
	Pd	116		0+	11.8 s	β^-	100
	Pd	118		0+	1.9 s	β^-	100
47	Ag	98		(6+)	46.7 s	EC	100
	Ag	99		(9/2)+	124 s	EC	100
	Ag	100		(5)+	2.01 m	EC	100
	Ag	101		9/2+	11.1 m	EC	100
	Ag	102		5+	12.9 m	EC	100

Z	Element	A	Abund (%)	J^{π}	$t_{1/2}$	Decay mode	Intensity (%)
	Ag	103		7/2+	65.7 m	EC	100
	Ag	104		5+	69.2 m	EC	100
	Ag	105		1/2−	41.29 d	EC	100
	Ag	106		1+	23.96 m	EC, β^-	99.50, < 1
	Ag	107	51.839	1/2−	stable		
	Ag	108		1+	2.37 m	β^-	97.15, 2.85
	Ag	109	48.161	1/2−	stable		
	Ag	110		1+	24.6 s	β^-, EC	99.7, 0.03
	Ag	111		1/2−	7.45 d	β^-	100
	Ag	112		2(−)	3.130 h	β^-	100
	Ag	113		1/2−	5.37 h	β^-	100
	Ag	114		1+	4.6 s	β^-	100
	Ag	115		1/2−	20.0 m	β^-	100
	Ag	116		(2)−	2.68 m	β^-	100
	Ag	117		(1/2−)	72.8 s	β^-	~100
	Ag	118		1(−)	3.76 s	β^-	100
	Ag	119		(7/2+), (1/2−)	2.1 s, 6.0 s	β^-, β^-	100
	Ag	120		3+	1.23 s	β^-	100
	Ag	121		(7/2+)	0.78 s	β^-	100
48	Cd	100		0+	49.1 s	EC	100
	Cd	101		(5/2+)	1.36 m	EC	100
	Cd	102		0+	5.5 m	EC	100
	Cd	103		(5/2)+	7.3 m	EC	100
	Cd	104		0+	57.7 m	EC	100
	Cd	105		5/2+	55.5 m	EC	100
	Cd	106	1.25	0+	stable		
	Cd	107		5/2+	6.50 h	EC	100
	Cd	108	0.89	0+	stable		
	Cd	109		5/2+	461.4 d	EC	100
	Cd	110	12.49	0+	stable		
	Cd	111	12.8	1/2+	stable		
	Cd	112	24.13	0+	stable		
	Cd	113	12.22	1/2+	7.7×10^{15} y	β^-	100
	Cd	114	28.73	0+	stable		
	Cd	115		1/2+	53.46 h	β^-	100
	Cd	116	7.49	0+	stable		
	Cd	117		1/2+	2.49 h	β^-	100
	Cd	118		0+	50.3 m	β^-	100
	Cd	119		3/2+	2.69 m	β^-	100
	Cd	120		0+	50.80 s	β^-	100
	Cd	121		(3/2+)	13.5 s	β^-	100
	Cd	123		(3/2)+	2.10 s	β^-	100
	Cd	124		0+	1.25 s	β^-	100
	Cd	125		(3/2+)	0.65 s	β^-	100
	Cd	126		0+	0.506 s	β^-	100
	Cd	127		(3/2+)	0.37 s	β^-	100
	Cd	128		0+	0.34 s	β^-	100

Z	Element	A	Abund (%)	J^π	$t_{1/2}$	Decay mode	Intensity (%)
49	In	100			7.0 s	EC	100
	In	102		(6+)	22 s	EC	100
	In	103		(9/2)+	65 s	EC	100
	In	104		(6+)	1.80 m	EC	100
	In	105		(9/2+)	5.07 m	EC	100
	In	106		7+	6.2 m	EC	100
	In	107		9/2+	32.4 m	EC	100
	In	108		7+	58.0 m	EC	100
	In	109		9/2+	4.2 h	EC	100
	In	110		7+	4.9 h	EC	100
	In	111		9/2+	2.8047 d	EC	100
	In	112		1+	14.97 m	EC, β^-	56, 44
	In	113	4.29	9/2+	stable		
	In	114		1+	71.9 s	β^-, EC	99.5, 0.5
	In	115	95.71	9/2+	4.41×10^{14} y	β^-	100000
	In	116		1+	14.10 s	β^-, EC	99.97, < 0.06
	In	117		9/2+	43.2 m	β^-	100
	In	118		1+	5.0 s	β^-	100
	In	119		9/2+	2.4 m	β^-	100
	In	120		1+	3.08 s	β^-	100
	In	121		9/2+	23.1 s	β^-	100
	In	122		1+	1.5 s	β^-	100
	In	123		9/2+	5.98 s	β^-	100
	In	124		3+	3.11 s	β^-	100
	In	125		9/2+	2.36 s	β^-	100
	In	126		3(+)	1.60 s	β^-	100
	In	127		(9/2+)	1.09 s	β^-	100
	In	128		(3+)	0.84 s	β^-	100
	In	129		(9/2+)	0.61 s	β^-	100
	In	130		1(−)	0.32 s	β^-	100
	In	131		(9/2+)	0.28 s	β^-	100
	In	132		(7−)	0.201 s	β^-	100
50	Sn	104		0+	20.8 s	EC	100
	Sn	105			31 s	EC	100
	Sn	106		0+	115 s	EC	100
	Sn	107		(5/2+)	2.90 m	EC	100
	Sn	108		0+	10.30 m	EC	100
	Sn	109		5/2(+)	18.0 m	EC	100
	Sn	110		0+	4.11 h	EC	100
	Sn	111		7/2+	35.3 m	EC	100
	Sn	112	0.97	0+	stable		
	Sn	113		1/2+	115.09 d	EC	100
	Sn	114	0.65	0+	stable		
	Sn	115	0.34	1/2+	stable		
	Sn	116	14.54	0+	stable		
	Sn	117	7.68	1/2+	stable		
	Sn	118	24.22	0+	stable		

Z	Element	A	Abund (%)	J^π	$t_{1/2}$	Decay mode	Intensity (%)
	Sn	119	8.58	1/2+	stable		
	Sn	120	32.59	0+	stable		
	Sn	121		3/2+	27.06 h	β^-	100
	Sn	122	4.63	0+	stable		
	Sn	123		11/2−	129.2 d	β^-	100
	Sn	124	5.79	0+	stable		
	Sn	125		11/2−	9.64 d	β^-	100
	Sn	126		0+	$\sim 1 \times 10^5$ y	β^-	100
	Sn	127		(11/2−)	2.10 h	β^-	100
	Sn	128		0+	59.07 m	β^-	100
	Sn	129		(3/2+)	2.23 m	β^-	100
	Sn	130		0+	3.72 m	β^-	100
	Sn	131		(3/2+)	56.0 s	β^-	100
	Sn	132		0+	39.7 s	β^-	100
	Sn	133		(7/2−)	1.45 s	β^-	100
	Sn	134		0+	1.12 s	β^-	100
51	Sb	105			1.12 s	EC	
	Sb	109		(5/2+)	17.0 s	EC	100
	Sb	112		3+	51.4 s	EC	100
	Sb	113		5/2+	6.67 m	EC	100
	Sb	114		3+	3.49 m	EC	100
	Sb	115		5/2+	32.1 m	EC	100
	Sb	116		3+	15.8 m	EC	100
	Sb	117		5/2+	2.80 h	EC	100
	Sb	118		1+	3.6 m	EC	100
	Sb	119		5/2+	38.19 h	EC	100
	Sb	120		1+	15.89 m	EC	100
	Sb	121	57.21	5/2+	stable		
	Sb	122		2−	2.7238 d	β^-, EC	97.59, 2.41
	Sb	123	42.79	7/2+	stable		
	Sb	124		3−	60.20 d	β^-	100
	Sb	125		7/2+	2.75856 y	β^-	100
	Sb	126		(8)−	12.46 d	β^-	100
	Sb	127		7/2+	3.85 d	β^-	100
	Sb	128		8−	9.01 h	β^-	100
	Sb	129		7/2+	4.40 h	β^-	100
	Sb	130		(8−)	39.5 m	β^-	100
	Sb	131		(7/2+)	23.03 m	β^-	100
	Sb	132		(4+)	2.79 m	β^-	100
	Sb	133		(7/2+)	2.5 m	β^-	100
	Sb	134		(0−), (7−)	0.78 s, 10.22 s	β^-	100
	Sb	135		(7/2+)	1.68 s	β^-	100
52	Te	108		0+	2.1 s	EC, α	51, 49
	Te	109		(5/2+)	4.6 s	EC, α	96.1, 3.9
	Te	110		0+	18.6 s	EC, α	~100, ~0.003
	Te	111		(5/2+)	19.3 s	EC	100

Z	Element	A	Abund (%)	J^π	$t_{1/2}$	Decay mode	Intensity (%)
	Te	112		0+	2.0 m	EC	100
	Te	115		7/2+	5.8 m	EC	100
	Te	116		0+	2.49 h	EC	100
	Te	117		1/2+	62 m	EC	100
	Te	118		0+	6.00 d	EC	100
	Te	119		1/2+	16.03 h	EC	100
	Te	120	0.096	0+	stable		
	Te	121		1/2+	16.78 d	EC	100
	Te	122	2.603	0+	stable		
	Te	123	0.908	1/2+	$> 1 \times 10^{13}$ y	EC	100
	Te	124	4.816	0+	stable		
	Te	125	7.139	1/2+	stable		
	Te	126	18.952	0+	stable		
	Te	127		3/2+	9.35 h	β^-	100
	Te	128	31.687	0+	7.7×10^{24} y	$\beta^-\beta^-$	100
	Te	129		3/2+	69.6 m	β^-	100
	Te	130	33.799	0+	7.9×10^{20} y	β^-	100
	Te	131		3/2+	25.0 m	β^-	100
	Te	132		0+	3.204 d	β^-	100
	Te	133		(3/2+)	12.5 m	β^-	100
	Te	134		0+	41.8 m	β^-	100
	Te	135		(7/2−)	19.0 s	β^-	100
	Te	136		0+	17.5 s	β^-	100
	Te	137		(7/2−)	2.49 s	β^-	100
53	I	109		(5/2+)	100 μs	α	< 0.50
	I	113		(5/2+)	6.6 s	EC, α	100, 3.3×10^7
	I	116		1+	2.91 s	EC	100
	I	117		(5/2)+	2.22 m	EC	100
	I	118		2−	13.7 m	EC	100
	I	119		5/2+	19.1 m	EC	100
	I	120		2−	81.0 m	EC	100
	I	121		5/2+	2.12 h	EC	100
	I	122		1+	3.63 m	EC	100
	I	123		5/2+	13.27 h	EC	100
	I	124		2−	4.1760 d	EC	100
	I	125		5/2+	59.400 d	EC	100
	I	126		2−	13.11 d	EC, β^-	56.30, 43.7
	I	127	100	5/2+	stable		
	I	128		1+	24.99 m	β^-, EC	93.1, 6.9
	I	129		7/2+	157×10^7 y	β^-	100
	I	130		5+	12.36 h	β^-	100
	I	131		7/2+	8.02070 d	β^-	100
	I	132		4+	2.295 h	β^-	100
	I	133		7/2+	20.8 h	β^-	100
	I	134		(4)+	52.5 m	β^-	100
	I	135		7/2+	6.57 h	β^-	100
	I	136		(1−)	83.4 s	β^-	100

Z	Element	A	Abund (%)	J^π	$t_{1/2}$	Decay mode	Intensity (%)
	I	137		(7/2+)	24.5 s	β^-	100
	I	138		(2–)	6.49 s	β^-	100
	I	139		(7/2+)	2.280 s	β^-	100
54	Xe	112		0+	2.7 s	EC, α	99.16, 0.84
	Xe	113			2.74 s	EC, α	~100, ~0.01
	Xe	117		5/2(+)	61 s	EC	100
	Xe	118		0+	3.8 m	EC	100
	Xe	119		(5/2+)	5.8 m	EC	100
	Xe	120		0+	40 m	EC	100
	Xe	121		5/2(+)	40.1 m	EC	100
	Xe	122		0+	20.1 h	EC	100
	Xe	123		(1/2)+	2.08 h	EC	100
	Xe	124	0.1	0+	stable		
	Xe	125		1/2(+)	16.9 h	EC	100
	Xe	126	0.09	0+	stable		
	Xe	127		1/2+	36.4 d	EC	100
	Xe	128	1.91	0+	stable		
	Xe	129	26.4	1/2+	stable		
	Xe	130	4.1	0+	stable		
	Xe	131	21.2	3/2+	stable		
	Xe	132	26.9	0+	stable		
	Xe	133		3/2+	5.243 d	β^-	100
	Xe	134	10.4	0+	stable		
	Xe	135		3/2+	9.14 h	β^-	100
	Xe	136	8.9	0+	$> 9.3 \times 10^{19}$ y	$2\beta^-$	
	Xe	137		7/2–	3.818 m	β^-	100
	Xe	138		0+	14.08 m	β^-	100
	Xe	139		3/2–	39.68 s	β^-	100
	Xe	140		0+	13.60 s	β^-	100
	Xe	141		5/2(–)	1.73 s	β^-	100
	Xe	142		0+	1.22 s	β^-	100
55	Cs	113		(5/2+)	17 μs	EC	~0.03
	Cs	116		> 5+, (1+)	3.85 s, 0.70 s	EC	100
	Cs	117		(9/2+)	8.4 s	EC	100
	Cs	118		2	14 s	EC	100
	Cs	119		9/2+	43.0 s	EC	100
	Cs	120		high, 2	57 s, 64 s	EC, EC	100
	Cs	121		3/2(+)	128 s	EC	100
	Cs	123		1/2+	5.87 m	EC	100
	Cs	124		1+	30.8 s	EC	100
	Cs	125		1/2(+)	46.7 m	EC	100
	Cs	126		1+	1.63 m	EC	100
	Cs	127		1/2+	6.25 h	EC	100
	Cs	128		1+	3.66 m	EC	100
	Cs	129		1/2+	32.06 h	EC	100
	Cs	130		1+	29.21 m	EC, β^-	98.40, 1.6

Z	Element	A	Abund (%)	J^π	$t_{1/2}$	Decay mode	Intensity (%)
	Cs	131		5/2+	9.689 d	EC	100
	Cs	132		2+	6.479 d	EC, β^-	98.13, 1.87
	Cs	133	100	7/2+	stable		
	Cs	134		4+	2.0648 y	β^-, EC	100, 3×10^{-4}
	Cs	135		7/2+	2.3×10^6 y	β^-	100
	Cs	136		5+	13.16 d	β^-	100
	Cs	137		7/2+	30.07 y	β^-	100
	Cs	138		3–	33.41 m	β^-	100
	Cs	139		7/2+	9.27 m	β^-	100
	Cs	140		1–	63.7 s	β^-	100
	Cs	141		7/2+	24.94 s	β^-	100
	Cs	142		0–	1.70 s	β^-	100
	Cs	143		3/2+	1.78 s	β^-	100
	Cs	144		1	1.01 s	β^-	100
	Cs	145		3/2+	0.594 s	β^-	100
	Cs	146		1–	0.321 s	β^-	100
	Cs	147		(3/2+)	0.235 s	β^-	100
	Cs	148			140 ms	β^-	100
56	Ba	119		(5/2+)	5.4 s	EC	100
	Ba	120		0+	32 s	EC	100
	Ba	121		5/2(+)	29.5 s	EC	100
	Ba	124		0+	11.0 m	EC	100
	Ba	125		1/2(+)	3.5 m	EC	100
	Ba	126		0+	100 m	EC	100
	Ba	127		1/2+	12.7 m	EC	100
	Ba	128		0+	2.43 d	EC	100
	Ba	129		1/2+	2.23 h	EC	100
	Ba	130	0.106	0+	stable		
	Ba	131		1/2+	11.50 d	EC	100
	Ba	132	0.101	0+	stable		
	Ba	133		1/2+	10.51 y	EC	100
	Ba	134	2.417	0+	stable		
	Ba	135	6.592	3/2+	stable		
	Ba	136	7.854	0+	stable		
	Ba	137	11.23	3/2+	stable		
	Ba	138	71.7	0+	stable		
	Ba	139		7/2–	83.06 m	β^-	100
	Ba	140		0+	12.752 d	β^-	100
	Ba	141		3/2–	18.27 m	β^-	100
	Ba	142		0+	10.6 m	β^-	100
	Ba	143		5/2–	14.33 s	β^-	100
	Ba	144		0+	11.5 s	β^-	100
	Ba	145		5/2–	4.31 s	β^-	100
	Ba	146		0+	2.22 s	β^-	100
	Ba	147		(3/2+)	0.893 s	β^-	100
	Ba	148		0+	0.607 s	β^-	100

Z	Element	A	Abund (%)	J^π	$t_{1/2}$	Decay mode	Intensity (%)
57	La	128		4–, 5–	5.0 m	EC	100
	La	129		3/2+	11.6 m	EC	100
	La	131		3/2+	59 m	EC	100
	La	132		2–	4.8 h	EC	100
	La	133		5/2+	3.912 h	EC	100
	La	134		1+	6.45 m	EC	100
	La	135		5/2+	19.5 h	EC	100
	La	136		1+	9.87 m	EC	100
	La	137		7/2+	60000 y	EC	100
	La	138	0.0902	5+	1.05×10^{11} y	EC, β^-	66.4, 33.6
	La	139	99.9098	7/2+	stable		
	La	140		3–	1.6781 d	β^-	100
	La	141		(7/2+)	3.92 h	β^-	100
	La	142		2–	91.1 m	β^-	100
	La	143		(7/2)+	14.2 m	β^-	100
	La	144		(3–)	40.8 s	β^-	100
	La	145		(5/2+)	24.8 s	β^-	100
	La	146		2–	6.27 s	β^-	100
	La	147		(5/2+)	4.015 s	β^-	100
	La	148		(2–)	1.428 s	β^-	100
58	Ce	131		(7/2+)	10.2 m	EC	100
	Ce	134		0+	3.16 d	EC	100
	Ce	135		1/2(+)	17.7 h	EC	100
	Ce	136	0.19	0+	stable		
	Ce	137		3/2+	9.0 h	EC	100
	Ce	138	0.25	0+	stable		
	Ce	139		3/2+	137.640 d	EC	100
	Ce	140	88.48	0+	stable		
	Ce	141		7/2–	32.501 d	β^-	100
	Ce	142	11.08	0+	$> 5 \times 10^{16}$ y	$2\beta^-$	
	Ce	143		3/2–	33.039 h	β^-	100
	Ce	144		0+	284.893 d	β^-	100
	Ce	145		(3/2–)	3.01 m	β^-	100
	Ce	146		0+	13.52 m	β^-	100
	Ce	147		(5/2–)	56.4 s	β^-	100
	Ce	148		0+	56 s	β^-	100
	Ce	149		(3/2–)	5.3 s	β^-	100
	Ce	150		0+	4.0 s	β^-	100
59	Pr	131		(3/2+)	1.53 m	EC	100
	Pr	135		3/2(+)	24 m	EC	100
	Pr	136		2+	13.1 m	EC	100
	Pr	137		5/2+	1.28 h	EC	100
	Pr	138		1+	1.45 m	EC	100
	Pr	139		5/2+	4.41 h	EC	100
	Pr	140		1+	3.39 m	EC	100
	Pr	141	100	5/2+	stable		

Z	Element	A	Abund (%)	J^π	$t_{1/2}$	Decay mode	Intensity (%)
	Pr	142		2−	19.12 h	β^-, EC	99.98, 0.02
	Pr	143		7/2+	13.57 d	β^-	100
	Pr	144		0−	17.28 m	β^-	100
	Pr	145		7/2+	5.984 h	β^-	100
	Pr	146		(2)−	24.15 m	β^-	100
	Pr	147		(3/2+)	13.4 m	β^-	100
	Pr	148		1−	2.27 m	β^-	100
	Pr	149		(5/2+)	2.26 m	β^-	100
	Pr	150		(1)−	6.19 s	β^-	100
	Pr	151		(3/2−)	18.90 s	β^-	100
60	Nd	131		(5/2)	27 s	EC	100
	Nd	136		0+	50.65 m	EC	100
	Nd	137		1/2+	38.5 m	EC	100
	Nd	139		3/2+	29.7 m	EC	100
	Nd	140		0+	3.37 d	EC	100
	Nd	141		3/2+	2.49 h	EC	100
	Nd	142	27.13	0+	stable		
	Nd	143	12.18	7/2−	stable		
	Nd	144	23.8	0+	2.29×10^{15} y	α	
	Nd	145	8.3	7/2−	stable		
	Nd	146	17.19	0+	stable		
	Nd	147		5/2−	10.98 d	β^-	100
	Nd	148	5.76	0+	stable		
	Nd	149		5/2−	1.728 h	β^-	100
	Nd	150	5.64	0+	$> 1.1 \times 10^{19}$ y	$2\beta^-$	
	Nd	151		3/2+	12.44 m	β^-	100
	Nd	152		0+	11.4 m	β^-	100
	Nd	153		(3/2)−	31.6 s	β^-	100
	Nd	154		0+	25.9 s	β^-	100
	Nd	155			8.9 s	β^-	100
61	Pm	136		(2+), (5−)	47 s, 107 s	EC, EC	100
	Pm	139		(5/2)+	4.15 m	EC	100
	Pm	140		1+	9.2 s	EC	100
	Pm	141		5/2+	20.90 m	EC	100
	Pm	142		1+	40.5 s	EC	100
	Pm	143		5/2+	265 d	EC	100
	Pm	144		5−	363 d	EC	100
	Pm	145		5/2+	17.7 y	EC, α	100, 3×10^{-7}
	Pm	146		3−	5.53 y	EC, β^-	66, 34
	Pm	147		7/2+	2.6234 y	β^-	100
	Pm	148		1−	5.370 d	β^-	100
	Pm	149		7/2+	53.08 h	β^-	100
	Pm	150		(1−)	2.68 h	β^-	100
	Pm	151		5/2+	28.40 h	β^-	100
	Pm	152		1+	4.12 m	β^-	100
	Pm	153		5/2−	5.25 m	β^-	100

Z	Element	A	Abund (%)	J^π	$t_{1/2}$	Decay mode	Intensity (%)
	Pm	154		(0, 1)	1.73 m	β^-	100
	Pm	155		(5/2–)	41.5 s	β^-	100
	Pm	156		4(–)	26.70 s	β^-	100
62	Sm	137		(9/2–)	45 s	EC	100
	Sm	139		(1/2)+	2.57 m	EC	100
	Sm	140		0+	14.82 m	EC	100
	Sm	141		1/2+	10.2 m	EC	100
	Sm	142		0+	72.49 m	EC	100
	Sm	143		3/2+	8.83 m	EC	100
	Sm	144	3.1	0+	stable		
	Sm	145		7/2–	340 d	EC	100
	Sm	146		0+	1.03×10^8 y	α	100
	Sm	147	15	7/2–	1.06×10^{11} y	α	
	Sm	148	11.3	0+	8×10^{15} y	α	
	Sm	149	13.8	7/2–	$> 2 \times 10^{15}$ y	α	
	Sm	150	7.4	0+	stable		
	Sm	151		5/2–	90 y	β^-	100
	Sm	152	26.7	0+	stable		
	Sm	153		3/2+	46.284 h	β^-	100
	Sm	154	22.7	0+	stable		
	Sm	155		3/2–	22.3 m	β^-	100
	Sm	156		0+	9.4 h	β^-	100
	Sm	157		(3/2–)	482 s	β^-	100
	Sm	158		0+	5.30 m	β^-	100
63	Eu	140		1+	1.51 s	EC	100
	Eu	141		5/2+	40.7 s	EC	100
	Eu	142		1+	2.4 s	EC	100
	Eu	143		5/2+	2.59 m	EC	100
	Eu	144		1+	10.2 s	EC	100
	Eu	145		5/2+	5.93 d	EC	100
	Eu	146		4–	4.61 d	EC	100
	Eu	147		5/2+	24.1 d	EC, α	100, 2.2×10^{-3}
	Eu	148		5–	54.5 d	EC, α	100, 9.4×10^{-7}
	Eu	149		5/2+	93.1 d	EC	100
	Eu	150		5(–)	36.9 y	EC	100
	Eu	151	47.8	5/2+	stable		
	Eu	152		3–	13.537 y	EC, β^-	72.1, 27.9
	Eu	153	52.2	5/2+	stable		
	Eu	154		3–	8.593 y	β^-, EC	99.98, 0.02
	Eu	155		5/2+	4.7611 y	β^-	100
	Eu	156		0+	15.19 d	β^-	100
	Eu	157		5/2+	15.18 h	β^-	100
	Eu	158		(1–)	45.9 m	β^-	100
	Eu	159		5/2+	18.1 m	β^-	100

Z	Element	A	Abund (%)	J^π	$t_{1/2}$	Decay mode	Intensity (%)
64	Gd	143		(1/2)+	39 s	EC	100
	Gd	145		1/2+	23.0 m	EC	100
	Gd	146		0+	48.27 d	EC	100
	Gd	147		7/2−	38.06 h	EC	100
	Gd	148		0+	74.6 y	α	100
	Gd	149		7/2−	9.28 d	EC, α	100, 4.3×10^{-4}
	Gd	150		0+	1.79×10^6 y	α	100
	Gd	151		7/2−	124 d	EC, α	100, 8×10^{-7}
	Gd	152	0.2	0+	1.08×10^{14} y	α	
	Gd	153		3/2−	240.4 d	EC	100
	Gd	154	2.18	0+	stable		
	Gd	155	14.8	3/2−	stable		
	Gd	156	20.47	0+	stable		
	Gd	157	15.65	3/2−	stable		
	Gd	158	24.84	0+	stable		
	Gd	159		3/2−	18.479 h	β^-	100
	Gd	160	21.86	0+			
	Gd	161		5/2−	3.66 m	β^-	100
	Gd	162		0+	8.4 m	β^-	100
65	Tb	146		1+	8 s	EC	100
	Tb	147		(1/2+)	1.7 h	EC	100
	Tb	148		2−	60 m	EC	100
	Tb	149		1/2+	4.118 h	EC, α	83.3, 16.7
	Tb	150		(2−)	3.48 h	EC, α	100, < 0.05
	Tb	151		1/2(+)	17.609 h	EC, α	100, 9.5×10^{-3}
	Tb	152		2−	17.5 h	EC, α	100, $< 7 \times 10^{-7}$
	Tb	153		5/2+	2.34 d	EC	100
	Tb	154		0	21.5 h	EC, β^-	100, < 0.1
	Tb	155		3/2+	5.32 d	EC	100
	Tb	156		3−	5.35 d	EC	100
	Tb	157		3/2+	71 y	EC	100
	Tb	158		3−	180 y	EC, β^-	83.4, 16.6
	Tb	159	100	3/2+	stable		
	Tb	160		3−	72.3 d	β^-	100
	Tb	161		3/2+	6.88 d	β^-	100
	Tb	162		1−	7.60 m	β^-	100
	Tb	163		3/2+	19.5 m	β^-	100
	Tb	164		(5+)	3.0 m	β^-	100
66	Dy	146		0+	29 s	EC	100
	Dy	147		1/2+	40 s	EC	100
	Dy	148		0+	3.1 m	EC	100
	Dy	149		(7/2−)	4.20 m	EC	100
	Dy	150		0+	7.17 m	EC, α	64, 36
	Dy	151		7/2(−)	17.9 m	EC, α	94.4, 5.6
	Dy	152		0+	2.38 h	EC, α	99.9, 0.1
	Dy	153		7/2(−)	6.4 h	EC, α	99.99, 9.4×10^{-3}

Z	Element	A	Abund (%)	J^π	$t_{1/2}$	Decay mode	Intensity (%)
	Dy	154		0+	3.0×10^6 y	α	100
	Dy	155		3/2–	9.9 h	EC	100
	Dy	156	0.06	0+	stable		
	Dy	157		3/2–	8.14 h	EC	100
	Dy	158	0.1	0+	stable		
	Dy	159		3/2–	144.4 d	EC	100
	Dy	160	2.34	0+	stable		
	Dy	161	18.9	5/2+	stable		
	Dy	162	25.5	0+	stable		
	Dy	163	24.9	5/2–	stable		
	Dy	164	28.2	0+	stable		
	Dy	165		7/2+	2.334 h	β^-	100
	Dy	166		0+	81.6 h	β^-	100
	Dy	167		(1/2–)	6.20 m	β^-	100
	Dy	169		(5/2–)	39 s	β^-	100
67	Ho	149		(11/2–)	21.1 s	EC	100
	Ho	151		(11/2–)	35.2 s	EC, α	78, 22
	Ho	152		2–	161.8 s	EC, α	88, 12
	Ho	153		11/2–	2.01 m	EC, α	99.95, 0.05
	Ho	154		(2)–	11.76 m	EC, α	99.98, 0.02
	Ho	155		5/2+	48 m	EC	100
	Ho	157		7/2–	12.6 m	EC	100
	Ho	158		5+	11.3 m	EC	100
	Ho	159		7/2–	33.05 m	EC	100
	Ho	160		5+	25.6 m	EC	100
	Ho	161		7/2–	2.48 h	EC	100
	Ho	162		1+	15.0 m	EC	100
	Ho	163		7/2–	4570 y	EC	100
	Ho	164		1+	29 m	EC, β^-	60, 40
	Ho	165	100	7/2–	stable		
	Ho	166		0–	26.763 h	β^-	100
	Ho	167		7/2–	3.1 h	β^-	100
	Ho	168		3+	2.99 m	β^-	100
	Ho	169		7/2–	4.7 m	β^-	100
	Ho	170		(6+)	2.76 m	β^-	100
	Ho	171		(7/2–)	53 s	β^-	100
68	Er	152		0+	10.3 s	α, EC	90, 10
	Er	153		(7/2–)	37.1 s	α, EC	53, 47
	Er	154		0+	3.73 m	EC, α	99.53, 0.47
	Er	155		7/2–	5.3 m	EC, α	99.98, 0.02
	Er	156		0+	19.5 m	EC, α	100, 5×10^{-6}
	Er	157		3/2–	18.65 m	EC, α	~100, < 0.02
	Er	159		3/2–	36 m	EC	100
	Er	160		0+	28.58 h	EC	100
	Er	161		3/2–	3.21 h	EC	100
	Er	162	0.14	0+	stable		

Z	Element	A	Abund (%)	J^{π}	$t_{1/2}$	Decay mode	Intensity (%)
	Er	163		5/2−	75.0 m	EC	100
	Er	164	1.61	0+	stable		
	Er	165		5/2−	10.36 h	EC	100
	Er	166	33.6	0+	stable		
	Er	167	22.95	7/2+	stable		
	Er	168	26.8	0+	stable		
	Er	169		1/2−	9.40 d	β^-	100
	Er	170	14.9	0+	stable		
	Er	171		5/2−	7.516 h	β^-	100
	Er	172		0+	49.3 h	β^-	100
69	Tm	153		(11/2−)	1.48 s	α, EC	91, 9
	Tm	155		(11/2−)	21.6 s	EC, α	98.1, 1.9
	Tm	156		2−	83.8 s	EC, α	99.94, 0.06
	Tm	157		1/2+	3.63 m	EC	100
	Tm	159		5/2+	9.13 m	EC	100
	Tm	160		1−	9.4 m	EC	100
	Tm	161		7/2+	33 m	EC	100
	Tm	162		1−	21.70 m	EC	100
	Tm	163		1/2+	1.810 h	EC	100
	Tm	164		1+	2.0 m	EC	100
	Tm	165		1/2+	30.06 h	EC	100
	Tm	166		2+	7.70 h	EC	100
	Tm	167		1/2+	9.25 d	EC	100
	Tm	168		3+	93.1 d	EC, β^-	99.99, 0.01
	Tm	169	100	1/2+	stable		
	Tm	170		1−	128.6 d	β^-, EC	99.87, 0.13
	Tm	171		1/2+	1.92 y	β^-	100
	Tm	172		2−	63.6 h	β^-	100
	Tm	173		(1/2+)	8.24 h	β^-	100
	Tm	174		(4)−	5.4 m	β^-	100
	Tm	175		1/2+	15.2 m	β^-	100
	Tm	176		(4+)	1.9 m	β^-	100
70	Yb	156		0+	26.1 s	EC, α	90, 10
	Yb	157		7/2−	38.6 s	EC, α	99.5, 0.5
	Yb	158		0+	1.49 m	EC, α	100, $\sim 2.1 \times 10^{-3}$
	Yb	159		5/2(−)	1.58 m	EC	100
	Yb	163		3/2−	11.05 m	EC	100
	Yb	165		5/2−	9.9 m	EC	100
	Yb	166		0+	56.7 h	EC	100
	Yb	167		5/2−	17.5 m	EC	100
	Yb	168	0.13	0+	stable		
	Yb	169		7/2+	32.026 d	EC	100
	Yb	170	3.05	0+	stable		
	Yb	171	14.3	1/2−	stable		
	Yb	172	21.9	0+	stable		
	Yb	173	16.12	5/2−	stable		

Z	Element	A	Abund (%)	J^π	$t_{1/2}$	Decay mode	Intensity (%)
	Yb	174	31.8	0+	stable		
	Yb	175		7/2−	4.185 d	β^-	100
	Yb	176	12.7	0+	stable		
	Yb	177		(9/2+)	1.911 h	β^-	100
	Yb	178		0+	74 m	β^-	100
71	Lu	157		(1/2+, 3/2+)	6.8 s	α	> 0.00
	Lu	159			12.1 s	EC, α	100, 0.04
	Lu	163		(1/2−)	238 s	EC	100
	Lu	165		(7/2+)	10.74 m	EC	100
	Lu	166		(6−)	2.65 m	EC	100
	Lu	167		7/2+	51.5 m	EC	100
	Lu	168		(6−)	5.5 m	EC	100
	Lu	169		7/2+	34.06 h	EC	100
	Lu	170		0+	2.012 d	EC	100
	Lu	171		7/2+	8.24 d	EC	100
	Lu	172		4−	6.70 d	EC	100
	Lu	173		7/2+	1.37 y	EC	100
	Lu	174		(1)−	3.31 y	EC	100
	Lu	175	97.41	7/2+	stable		
	Lu	176	2.59	7−	4.00×10^{10} y	β^-	
	Lu	177		7/2+	6.734 d	β^-	100
	Lu	178		1(+)	28.4 m	β^-	100
	Lu	179		7/2(+)	4.59 h	β^-	100
	Lu	180		(3)+	5.7 m	β^-	100
72	Hf	160		0+	13.6 s	EC, α	99.3, 0.7
	Hf	161			18.2 s	EC	99.71, 0.29
	Hf	162		0+	39.4 s	EC, α	99.99, 8×10^{-3}
	Hf	169		(5/2)−	3.24 m	EC	100
	Hf	172		0+	1.87 y	EC	100
	Hf	174	0.162	0+	2.0×10^{15} y	α	
	Hf	175		5/2−	70 d	EC	100
	Hf	176	5.206	0+	stable		
	Hf	177	18.606	7/2−	stable		
	Hf	178	27.297	0+	stable		
	Hf	179	13.629	9/2+	stable		
	Hf	180	35.1	0+	stable		
	Hf	181		1/2−	42.39 d	β^-	100
	Hf	182		0+	9×10^6 y	β^-	100
	Hf	183		(3/2−)	1.067 h	β^-	100
	Hf	184		0+	4.12 h	β^-	100
73	Ta	161			2.89 s	EC, α	~95, 5
	Ta	163			11.0 s	EC, α	~99.8, ~0.20
	Ta	172		(3+)	36.8 m	EC	100
	Ta	174		3+	1.14 h	EC	100
	Ta	176		(1)−	8.09 h	EC	100

Z	Element	A	Abund (%)	J^π	$t_{1/2}$	Decay mode	Intensity (%)
	Ta	177		7/2+	56.56 h	EC	100
	Ta	178		1+	9.31 m	EC	100
	Ta	179		7/2+	1.82 y	EC	100
	Ta	180		1+	8.152 h	EC, β^-	86, 14
	Ta	181	99.988	7/2+	stable		
	Ta	182		3–	114.43 d	β^-	100
	Ta	183		7/2+	5.1 d	β^-	100
	Ta	184		(5–)	8.7 h	β^-	100
	Ta	185		(7/2+)	49.4 m	β^-	100
	Ta	186		(2–, 3–)	10.5 m	β^-	100
74	W	164		0+	6.0 s	EC, α	97.4, 2.6
	W	165			5.1 s	EC, α	100, < 0.20
	W	166		0+	18.8 s	EC, α	99.97, 0.04
	W	178		0+	21.6 d	EC	100
	W	179		(7/2)–	37.05n	EC	100
	W	180	0.12	0+	stable		
	W	181		9/2+	121.2 d	EC	100
	W	182	26.498	0+	stable		
	W	183	14.3		stable		
	W	184	30.642	0+	$> 3 \times 10^{17}$ y	α	
	W	185		3/2–	75.1 d	β^-	100
	W	186	28.426	0+	stable		
	W	187		3/2–	23.72 h	β^-	100
	W	188		0+	69.4 d	β^-	100
	W	189		(3/2–)	11.5 m	β^-	100
	W	190		0+	30.0 m	β^-	100
75	Re	165			2.4 s	EC, α	87, 13
	Re	178		(3+)	13.2 m	EC	100
	Re	179		(5/2)+	19.5 m	EC	100
	Re	180		(1)–	2.44 m	EC	100
	Re	181		5/2+	19.9 h	EC	100
	Re	182		7+	64.0 h	EC	100
	Re	183		5/2+	70.0 d	EC	100
	Re	184		3(–)	38.0 d	EC	100
	Re	185	37.4	5/2+	stable		
	Re	186		1–	3.7183 d	β^-, EC	92.53, 7.47
	Re	187	62.6	5/2+	4.35×10^{10} y	β^-, α	100, 1×10^{-4}
	Re	188		1–	17.005 h	β^-	100
	Re	189		5/2+	24.3 h	β^-	100
	Re	190		(2)–	3.1 m	β^-	100
	Re	191		(3/2+, 1/2+)	9.8 m	β^-	100
76	Os	168		0+	2.1 s	EC, α	51, 49
	Os	169			3.4 s	EC, α	89, 11
	Os	170		0+	7.3 s	EC, α	88, 12
	Os	178		0+	5.0 m	EC	100

Z	Element	A	Abund (%)	J^π	$t_{1/2}$	Decay mode	Intensity (%)
	Os	181		1/2−	105 m	EC	100
	Os	182		0+	22.10 h	EC	100
	Os	184	0.02	0+	$> 5.6 \times 10^{13}$ y	α	
	Os	185		1/2−	93.6 d	EC	100
	Os	186	1.58	0+	2.0×10^{15} y	α	
	Os	187	1.6	1/2−	stable		
	Os	188	13.3	0+	stable		
	Os	189	16.1	3/2−	stable		
	Os	190	26.4	0+	stable		
	Os	191		9/2−	15.4 d	β^-	100
	Os	192	41	0+	stable		
	Os	193		3/2−	30.11 h	β^-	100
	Os	194		0+	6.0 y	β^-	100
	Os	196		0+	34.9 m	β^-	100
77	Ir	169			0.4 s	α	~100
	Ir	181		(5/2)−	4.90 m	EC	100
	Ir	182		(5+)	15 m	EC	100
	Ir	184		5−	3.09 h	EC	100
	Ir	186		5+	16.64 h	EC	100
	Ir	187		3/2+	10.5 h	EC	100
	Ir	188		1−	41.5 h	EC	100
	Ir	189		3/2+	13.2 d	EC	100
	Ir	190		(4−)	11.78 d	EC	100
	Ir	191	37.3	3/2+	stable		
	Ir	192		4+	73.827 d	β^-, EC	95.13, 4.87
	Ir	193	62.7	3/2+	stable		
	Ir	194		1−	19.28 h	β^-	100
	Ir	195		3/2+	2.5 h	β^-	100
	Ir	196		(0−)	52 s	β^-	100
	Ir	197		3/2+	5.8 m	β^-	100
78	Pt	172		0+	0.096 s	α, EC	94, 6
	Pt	173			342 ms	α, EC	84, 16
	Pt	174		0+	0.889 s	α, EC	76, 24
	Pt	182		0+	3.0 m	EC, α	99.96, 0.04
	Pt	185		9/2+	70.9 m	EC	100
	Pt	186		0+	2.08 h	EC, α	100, ~1.4×10^{-4}
	Pt	188		0+	10.2 d	EC, α	100, 2.6×10^{-5}
	Pt	189		3/2−	10.87 h	EC	100
	Pt	190	0.01	0+	6.5×10^{11} y	α	
	Pt	191		3/2−	2.802 d	EC	100
	Pt	192	0.79	0+	stable		
	Pt	193		1/2−	50 y	EC	100
	Pt	194	32.9	0+	stable		
	Pt	195	33.8	1/2−	stable		
	Pt	196	25.3	0+	stable		
	Pt	197		1/2−	19.8915 h	β^-	100

Z	Element	A	Abund (%)	J^π	$t_{1/2}$	Decay mode	Intensity (%)
	Pt	198	7.2	0+	stable		
	Pt	199		5/2–	30.80 m	β^-	100
	Pt	200		0+	12.5 h	β^-	100
	Pt	201		(5/2–)	2.5 m	β^-	100
79	Au	173			59 ms	α	100
	Au	185		5/2–	4.25 m	EC, α	99.74, 0.26
	Au	186		3–	10.7 m	EC, α	100, 8×10^{-4}
	Au	190		1–	42.8 m	EC, α	100, $< 1.0 \times 10^{-6}$
	Au	191		3/2+	3.18 h	EC	100
	Au	192		1–	4.94 h	EC	100
	Au	193		3/2+	17.65 h	EC	100
	Au	194		1–	38.02 h	EC	100
	Au	195		3/2+	186.098 d	EC	100
	Au	196		2–	6.183 d	EC, β^-	92.8, 7.2
	Au	197	100	3/2+	stable		
	Au	198		2–	2.69517 d	β^-	100
	Au	199		3/2+	3.139 d	β^-	100
	Au	200		1(–)	48.4 m	β^-	100
	Au	201		3/2+	26 m	β^-	100
	Au	202		(1–)	28.8 s	β^-	100
	Au	203		3/2+	60 s	β^-	100
80	Hg	176		0+	34 ms	α	~100
	Hg	177			0.130 s	α, EC	85, 15
	Hg	178		0+	0.266 s	α, EC	~70, ~30
	Hg	186		0+	1.38 m	EC, α	99.98, 0.02
	Hg	191		(3/2–)	49 m	EC	100
	Hg	193		3/2–	3.80 h	EC	100
	Hg	194		0+	444 y	EC	100
	Hg	195		1/2–	9.9 h	EC	100
	Hg	196	0.15	0+	stable		
	Hg	197		1/2–	64.14 h	EC	100
	Hg	198	9.97	0+	stable		
	Hg	199	16.87	1/2–	stable		
	Hg	200	23.1	0+	stable		
	Hg	201	13.18	3/2–	stable		
	Hg	202	29.86	0+	stable		
	Hg	203		5/2–	46.612 d	β^-	100
	Hg	204	6.87	0+	stable		
	Hg	205		1/2–	5.2 m	β^-	100
	Hg	206		0+	8.15 m	β^-	100
	Hg	207		(9/2+)	2.9 m	β^-	100
81	Tl	197		1/2+	2.84 h	EC	100
	Tl	198		2–	5.3 h	EC	100
	Tl	199		1/2+	7.42 h	EC	100
	Tl	200		2–	26.1 h	EC	100

Z	Element	A	Abund (%)	J^π	$t_{1/2}$	Decay mode	Intensity (%)
	Tl	201		1/2+	72.912 h	EC	100
	Tl	202		2–	12.23 d	EC	100
	Tl	203	29.524	1/2+	stable		
	Tl	204		2–	3.78 y	β^-, EC	97.1, 2.9
	Tl	205	70.476	1/2+	stable		
	Tl	206		0–	4.199 m	β^-	100
	Tl	207		1/2+	4.77 m	β^-	100
	Tl	208		5(+)	3.053 m	β^-	100
	Tl	209		(1/2+)	2.20 m	β^-	100
	Tl	210		(5+)	1.30 m	β^-	100
82	Pb	182		0+	55 ms	α	100
	Pb	190		0+	1.2 m	EC, α	99.1, 0.9
	Pb	199		3/2–	90 m	EC	100
	Pb	200		0+	21.5 h	EC	100
	Pb	201		5/2–	9.33 h	EC	100
	Pb	202		0+	52500 y	EC, α	100, < 1
	Pb	203		5/2–	51.873 h	EC	100
	Pb	204	1.4	0+	$\geq 1.4 \times 10^{17}$ y	α	
	Pb	205		5/2–	1.53×10^7 y	EC	100
	Pb	206	24.1	0+	stable		
	Pb	207	22.1	1/2–	stable		
	Pb	208	52.4	0+	stable		
	Pb	209		9/2+	3.253 h	β^-	100
	Pb	210		0+	22.3 y	β^-, α	100, 1.9×10^{-6}
	Pb	211		9/2+	36.1 m	β^-	100
	Pb	212		0+	10.64 h	β^-	100
	Pb	214		0+	26.8 m	β^-	100
83	Bi	197		(9/2–)	9.33 m	EC, α	100, 1×10^{-4}
	Bi	198		(2+, 3+)	10.3 m	EC	100
	Bi	199		9/2–	27 m	EC	100
	Bi	200		7+	36.4 m	EC	100
	Bi	201		9/2–	108 m	EC, α	100, $< 1 \times 10^{-4}$
	Bi	202		5+	1.72 h	EC, α	100, $< 1.0 \times 10^{-5}$
	Bi	203		9/2–	11.76 h	EC, α	100, $\sim 1.0 \times 10^{-5}$
	Bi	204		6+	11.22 h	EC	100
	Bi	205		9/2–	15.31 d	EC	100
	Bi	206		6(+)	6.243 d	EC	100
	Bi	207		9/2–	31.55 y	EC	100
	Bi	208		(5)+	3.68×10^5 y	EC	100
	Bi	209	100	9/2–	stable		
	Bi	210		1–	5.013 d	β^-, α	100, 1.3×10^{-4}
	Bi	211		9/2–	2.14 m	α, β^-	99.72, 0.28
	Bi	212		1(–)	60.55 m	β^-, α	64.06, 35.94
	Bi	213		9/2–	45.59 m	β^-, α	97.91, 2.09
	Bi	214		1–	19.9 m	β^-, α	99.98, 0.02
	Bi	215			7.6 m	β^-	100

Z	Element	A	Abund (%)	J^{π}	$t_{1/2}$	Decay mode	Intensity (%)
84	Po	194		0+	0.392 s	α	100
	Po	203		5/2−	36.7 m	EC, α	99.89, 0.11
	Po	204		0+	3.53 h	EC, α	99.34, 0.66
	Po	205		5/2−	1.66 h	EC, α	99.96, 0.04
	Po	206		0+	8.8 d	EC, α	94.55, 5.45
	Po	207		5/2−	5.80 h	EC, α	99.98, 0.02
	Po	208		0+	2.898 y	α	100
	Po	209		1/2−	102 y	α, EC	99.52, 0.48
	Po	210		0+	138.376 d	α	100
	Po	211		9/2+	0.516 s	α	100
	Po	212		0+	0.299μs	α	100
	Po	213		9/2+	4.2μs	α	100
	Po	214		0+	164.3μs	α	100
	Po	215		9/2+	1.781 ms	α, β^-	100, 2.3×10^{-4}
	Po	216		0+	0.145 s	α	100
	Po	218		0+	3.10 m	α, β^-	99.98, 0.02
85	At	201		(9/2−)	89 s	α, EC	71, 29
	At	202		(2, 3)	184 s	EC, α	82, 18
	At	203		9/2−	7.4 m	EC, α	69, 31
	At	204		7+	9.2 m	EC, α	96.2, 3.8
	At	205		9/2−	26.2 m	EC, α	90, 10
	At	206		(5)+	30.0 m	EC, α	99.11, 0.89
	At	207		9/2−	1.80 h	EC, α	91.4, 8.6
	At	208		6+	1.63 h	EC, α	99.45, 0.55
	At	209		9/2−	5.41 h	EC, α	95.9, 4.1
	At	210		(5)+	8.1 h	EC, α	99.82, 0.18
	At	211		9/2−	7.214 h	EC, α	58.2, 41.8
	At	212		(1−)	0.314 s	α, EC, β^-	100.0, < 0.03, $< 2.0 \times 10^{-6}$
	At	213		9/2−	125 ns	α	100
	At	214		1−	558 ns	α	100
	At	215		9/2−	0.10 ms	α	100
	At	216		1−	0.30 ms	α, β^-, EC	100, $< 6 \times 10^{-3}$, $< 3 \times 10^{-7}$
	At	217		9/2−	32.3 ms	α, β^-	99.99, 0.01
	At	218			1.5 s	α, β^-	99.9, 0.1
	At	219			56 s	α, β^-	~97, ~3
86	Rn	198		0+	57 ms	α, EC	
	Rn	207		5/2−	9.25 m	EC, α	79, 21
	Rn	208		0+	24.35 m	α, EC	62, 38
	Rn	209		5/2−	28.5 m	EC, α	83, 17
	Rn	210		0+	2.4 h	α, EC	96, 4
	Rn	211		1/2−	14.6 h	EC, α	72.6, 27.4
	Rn	212		0+	23.9 m	α	100
	Rn	213		(9/2+)	25.0 ms	α	100

Z	Element	A	Abund (%)	J^π	$t_{1/2}$	Decay mode	Intensity (%)
	Rn	214		0+	0.27μs	α	100
	Rn	215		9/2+	2.30μs	α	100
	Rn	216		0+	45μs	α	100
	Rn	217		9/2+	0.54 ms	α	100
	Rn	218		0+	35 ms	α	100
	Rn	219		5/2+	3.96 s	α	100
	Rn	220		0+	55.6 s	α	100
	Rn	222		0+	3.8235 d	α	100
87	Fr	205		(9/2−)	3.85 s	α, EC	100, < 1.0
	Fr	206		(5+)	15.9 s	EC, α	
	Fr	207		9/2−	14.8 s	α, EC	95, 5
	Fr	208		7+	59.1 s	α, EC	90, 10
	Fr	209		9/2−	50.0 s	α, EC	89, 11
	Fr	210		6+	3.18 m	α, EC	60, 40
	Fr	211		9/2−	3.10 m	α, EC	> 80, < 20
	Fr	212		5+	20.0 m	EC, α	57, 43
	Fr	213		9/2−	34.6 s	α, EC	99.45, 0.55
	Fr	214		(1−)	5.0 ms	α	100
	Fr	215		9/2−	86 ms	α	100
	Fr	216		(1−)	0.70μs	α, EC	$100, < 2.0 \times 10^{-7}$
	Fr	217		9/2−	22μs	α	100
	Fr	218		1−	1.0 ms	α	100
	Fr	219		9/2−	20 ms	α	100
	Fr	220		1+	27.4 s	α, β−	99.65, 0.35
	Fr	221		5/2−	4.9 m	α, β−	100, < 0.10
	Fr	222		2−	14.2 m	β−	100
	Fr	223		3/2(−)	22.00 m	β−, α	$99.99, 6 \times 10^{-3}$
	Fr	224		1−	3.33 m	β−	100
	Fr	225		3/2−	4.0 m	β−	100
	Fr	226		1−	49 s	β−	100
	Fr	227		1/2+	2.47 m	β−	100
88	Ra	211		5/2(−)	13 s	α, EC	> 93, < 7
	Ra	212		0+	13.0 s	α, EC	~90, ~15
	Ra	213		1/2−	2.74 m	α, EC	80, 20
	Ra	214		0+	2.46 s	α, EC	99.94, 0.06
	Ra	215		(9/2+)	1.59 ms	α	100
	Ra	216		0+	182 ns	α, EC	$100, < 1.0 \times 10^{-8}$
	Ra	217		(9/2+)	1.6μs	α	100
	Ra	218		0+	25.6μs	α	100
	Ra	219		(7/2)+	10 ms	α	100
	Ra	220		0+	18 ms	α	100
	Ra	221		5/2+	28 s	α	100
	Ra	222		0+	38.0 s	α	100
	Ra	223		3/2+	11.435 d	α	100
	Ra	224		0+	3.66 d	α	100
	Ra	225		1/2+	14.9 d	β−	100

Z	Element	A	Abund (%)	J^π	$t_{1/2}$	Decay mode	Intensity (%)
	Ra	226		0+	1600 y	α	100
	Ra	227		3/2+	42.2 m	β^-	100
	Ra	228		0+	5.75 y	β^-	100
	Ra	229		5/2(+)	4.0 m	β^-	100
	Ra	230		0+	93 m	β^-	100
89	Ac	209		(9/2−)	0.10 s	α, EC	~99, ~1
	Ac	210			0.35 s	α, EC	~96, ~4
	Ac	211			0.25 s	α	~100
	Ac	212			0.93 s	α, EC	~97, ~3
	Ac	213			0.80 s	α	≤ 100
	Ac	214			8.2 s	α, EC	≥89, ≤ 11
	Ac	215		9/2−	0.17 s	α, EC	99.91, 0.09
	Ac	216		(1−)	~0.33 ms	α	100
	Ac	217		9/2−	69 ns	α, EC	100, ≤ 2
	Ac	218		(1−)	1.08 μs	α	100
	Ac	219		9/2−	11.8 μs	α	100
	Ac	220		(3−)	26.4 ms	α, EC	100, 5 × 10^{-4}
	Ac	221			52 ms	α	100
	Ac	222		1−	5.0 s	α, EC	99, 1
	Ac	223		(5/2−)	2.10 m	α, EC	99, 1
	Ac	224		0−	2.78 h	EC, α, β^-	90.9, 9.1, < 1.6
	Ac	225		(3/2−)	10.0 d	α	100
	Ac	226		(1)	29.37 h	β^-, EC, α	83, 17, 0.006
	Ac	227		3/2−	21.773 y	β^-, α	98.62, 1.38
	Ac	228		3+	6.15 h	β^-	100
	Ac	229		(3/2+)	62.7 m	β^-	100
	Ac	230		(1+)	122 s	β^-	100
	Ac	231		(1/2+)	7.5 m	β^-	100
	Ac	232		(1+)	119 s	β^-	100
90	Th	215		(1/2−)	1.2 s	α	100
	Th	216		0+	0.028 s	α, EC	100, ~0.01
	Th	217		(9/2+)	0.252 ms	α	100
	Th	218		0+	109 ns	α	100
	Th	219			1.05 μs	α	100
	Th	220		0+	9.7 μs	α, EC	100, 2.0 × 10^{-7}
	Th	221		(7/2+)	1.68 ms	α	100
	Th	222		0+	2.8 ms	α	100
	Th	223		(5/2)+	0.60 s	α	100
	Th	224		0+	1.05 s	α	100
	Th	225		(3/2)+	8.72 m	α, EC	~90, ~10
	Th	226		0+	30.57 m	α	100
	Th	227		(1/2+)	18.72 d	α	100
	Th	228		0+	1.9116 y	α	100
	Th	229		5/2+	7340 y	α	100
	Th	230		0+	7.538×10^4 y	α, SF	100, 5.x10^{-11}
	Th	231		5/2+	25.52 h	β^-, α	100, ~1.0 × 10^{-8}

Z	Element	A	Abund (%)	J^π	$t_{1/2}$	Decay mode	Intensity (%)
	Th	232	100	0+	1.405×10^{10} y	α, SF	$100, < 1.0 \times 10^{-9}$
	Th	233		1/2+	22.3 m	β^-	100
	Th	234		0+	24.10 d	β^-	100
	Th	235		(1/2+)	7.1 m	β^-	100
91	Pa	213		(9/2–)	5.3 ms	α	100
	Pa	214			17 ms	α	≤ 100
	Pa	215			15 ms	α	100
	Pa	216			0.20 s	α, EC	~98, ~2
	Pa	217			4.9 ms	α	100
	Pa	218			0.12 ms	α	100
	Pa	219		9/2–	53 ns	α	100
	Pa	220			0.78 µs	α, EC	$100, 3.0 \times 10^{-7}$
	Pa	221		9/2–	5.9 µs	α	100
	Pa	223			5 ms	α	100
	Pa	224			0.79 s	α	100
	Pa	225			1.7 s	α	100
	Pa	226			1.8 m	α, EC	74, 26
	Pa	227		(5/2–)	38.3 m	α, EC	85, 15
	Pa	228		3+	22 h	EC, α	98, 2
	Pa	229		(5/2+)	1.50 d	EC, α	99.52, 0.48
	Pa	230		(2–)	17.4 d	EC, β^-, α	$91.6, 8.4, 3 \times 10^{-3}$
	Pa	231		3/2–	32760 y	α, SF	$100, \leq 3\text{E}{-10}$
	Pa	232		(2–)	1.31 d	β^-, EC	100, 0.003
	Pa	233		3/2–	26.967 d	β^-	100
	Pa	234		4+	6.70 h	β^-	100
	Pa	235		(3/2–)	24.5 m	β^-	100
	Pa	236		1(–)	9.1 m	β^-	100
	Pa	237		(1/2+)	8.7 m	β^-	100
	Pa	238		(3–)	2.3 m	β^-, SF	$100, < 2.6 \times 10^{-6}$
92	U	219		(9/2+)	42 µs	α	100
	U	223		(7/2+)	18 µs	α	100
	U	224		0+	0.9 ms	α	100
	U	225			95 ms	α	100
	U	226		0+	0.35 s	α	100
	U	227		(3/2+)	1.1 m	α	100
	U	228		0+	9.1 m	α, EC	> 95, < 5
	U	229		(3/2+)	58 m	EC, α	~ 80, ~ 20
	U	230		0+	20.8 d	α	100
	U	231		(5/2–)	4.2 d	EC, α	100, ~ 0.004
	U	232		0+	68.9 y	α	100
	U	233		5/2+	1.592×10^5 y	α, SF	$100, < 6.0 \times 10^{-9}$
	U	234	5.5e–3	0+	2.455×10^5 y	α, SF	$100, 1.7 \times 10^{-9}$
	U	235	0.72	7/2–	7.038×10^8 y	α, SF	$100, 7.0 \times 10^{-9}$
	U	236		0+	2.342×10^7 y	α, SF	$100, 9.6 \times 10^{-8}$
	U	237		1/2+	6.75 d	β^-	100
	U	238	99.2745	0+	4.468×10^9 y	α, SF	$100, 5 \times 10^{-5}$

Z	Element	A	Abund (%)	J^π	$t_{1/2}$	Decay mode	Intensity (%)
	U	239		5/2+	23.45 m	β^-	100
	U	240		0+	14.1 h	β^-	100
93	Np	225		(9/2–)	> 2 µs	α	100
	Np	227			0.51 s	α	100
	Np	229			4.0 m	α, EC	> 50, < 50
	Np	230			4.6 m	EC, α	97, 3
	Np	231		(5/2)	48.8 m	EC, α	98, 2
	Np	233		(5/2+)	36.2 m	EC, α	100, < 0.001
	Np	234		(0+)	4.4 d	EC	100
	Np	235		5/2+	396.1 d	EC, α	100, 0.0026
	Np	236		(6–)	1.54×10^5 y	EC, β^-, α	87.3, 112.5, 0.16
	Np	237		5/2+	2.144×10^6 y	α, SF	100, 2×10^{-10}
	Np	238		2+	2.117 d	β^-	100
	Np	239		5/2+	2.3565 d	β^-	100
	Np	240		(5+)	61.9 m	β^-	100
	Np	241		(5/2+)	13.9 m	β^-	100
94	Pu	228		0+	~0.2 s	α	100
	Pu	229		(3/2+)	> 2 µs	α	100
	Pu	230		0+	~200 s	α	≤ 100
	Pu	232		0+	34.1 m	EC, α	80, 20
	Pu	233			20.9 m	EC, α	99.88, 0.12
	Pu	234		0+	8.8 h	EC, α	~94, ~6
	Pu	235		(5/2+)	25.3 m	EC, α	100, 0.0027
	Pu	236		0+	2.858 y	α, SF	100, 1.4×10^{-7}
	Pu	237		7/2–	45.2 d	EC, α	100, 0.0042
	Pu	238		0+	87.7 y	α, SF	100, 1.9×10^{-7}
	Pu	239		1/2+	24110 y	α, SF	100, 3×10^{-10}
	Pu	240		0+	6564 y	α, SF	100, 5.7×10^{-6}
	Pu	241		5/2+	14.35 y	β^-, α, SF	100, 2.5×10^{-3}, $< 2 \times 10^{-14}$
	Pu	242		0+	3.733×10^5 y	α, SF	100, 5.5×10^{-4}
	Pu	243		7/2+	4.956 h	β^-	100
	Pu	244		0+	8.08×10^7 y	α, SF	99.88, 0.12
	Pu	245		(9/2–)	10.5 h	β^-	100
	Pu	246		0+	10.84 d	β^-	100
95	Am	237		5/2(–)	73.0 m	EC, α	99.98, 0.03
	Am	238		1+	98 m	EC, α	> 99.99, 1×10^{-4}
	Am	239		(5/2)–	11.9 h	EC, α	99.99, 0.01
	Am	240		(3–)	50.8 h	EC, α	100, 1.9×10^{-4}
	Am	241		5/2–	432.2 y	α, SF	100, 4×10^{-10}
	Am	242		1–	16.02 h	β^-, EC	82.7, 17.3
	Am	243		5/2–	7370 y	α, SF	100, 3.7×10^{-9}
	Am	244		(6–)	10.1 h	β–	100
	Am	245		(5/2)+	2.05 h	β–	100
	Am	246		(7–)	39 m	β–	100

Z	Element	A	Abund (%)	J^π	$t_{1/2}$	Decay mode	Intensity (%)
96	Cm	238		0+	2.4 h	EC, α	90, 10
	Cm	240		0+	27 d	α, EC, SF	> 99.5, < 0.5, 3.9×10^6
	Cm	241		1/2+	32.8 d	EC, α	99, 1
	Cm	242		0+	162.8 d	α, SF	100, 6.2×10^{-6}
	Cm	243		5/2+	29.1 y	α, EC, SF	99.71, 0.29, 5.3E−9
	Cm	244		0+	18.10 y	α, SF	100, 1.3×10^{-4}
	Cm	245		7/2+	8500 y	α, SF	100, 6.1×10^{-7}
	Cm	246		0+	4760 y	α, SF	99.97, 0.03
	Cm	247		9/2−	1.56×10^7 y	α	100
	Cm	248		0+	3.48×10^5 y	α, SF	91.74, 8.26
	Cm	249		1/2(+)	64.15 m	β^-	100
	Cm	250		0+	~9700 y	SF, α, β^-	~80, ~11, ~9
	Cm	251		(1/2+)	16.8 m	β^-	100
97	Bk	243		(3/2−)	4.5 h	EC, α	~99.85, ~0.15
	Bk	244		(1−)	4.35 h	EC, α	99.994, 0.006
	Bk	245		3/2−	4.94 d	EC, α	99.88, 0.12
	Bk	246		2(−)	1.80 d	EC, α	100, < 0.20
	Bk	247		(3/2−)	1380 y	α	
	Bk	249		7/2+	320 d	β^-, α, SF	100, 1.4×10^{-3}, 4.7×10^{-8}
	Bk	250		2−	3.217 h	β^-	100
	Bk	251		(3/2−)	55.6 m	β^-, α	100, 1.0×10^{-5}
98	Cf	242		0+	3.7 m	α	~65
	Cf	244		0+	19.4 m	α	100
	Cf	246		0+	35.7 h	α, EC, SF	100, < 0.004, 2.5×10^{-3}
	Cf	247		(7/2+)	3.11 h	EC, α	99.97, 0.04
	Cf	248		0+	333.5 d	α, SF	100, 0.0029
	Cf	249		9/2−	351 y	α, SF	100, 5.2×10^{-7}
	Cf	250		0+	13.08 y	α, SF	99.92, 0.08
	Cf	251		1/2+	898 y	α	100
	Cf	252		0+	2.645 y	α, SF	96.91, 3.09
	Cf	253		(7/2+)	17.81 d	β^-, α	99.69, 0.31
	Cf	254		0+	60.5 d	SF, α	99.69, 0.31
99	Es	251		(3/2−)	33 h	EC, α	99.51, 0.49
	Es	252		(5−)	471.7 d	α, EC, β^-	76, 24, ~0.01
	Es	253		7/2+	20.47 d	α, SF	100, 8.7×10^{-6}
	Es	254		(7+)	275.7 d	α, EC, SF, β^-	100, < 1×10^{-4}, < 3×10^{-6}, 1.7×10^{-6}
	Es	255		(7/2+)	39.8 d	β^-, SF, α	92, 8, 0.0041

Z	Element	A	Abund (%)	J^π	$t_{1/2}$	Decay mode	Intensity (%)
100	Fm	246		0+	1.1 s	α, SF, EC	92, 8, ≤ 1
	Fm	248		0+	36 s	α, EC, SF	99, ~1.0, ~0.05
	Fm	250		0+	33 m	α, EC, SF	> 90, < 10, ~ .0006
	Fm	251		(9/2−)	5.30 h	EC, α	98.2, 1.8
	Fm	252		0+	25.39 h	α, SF	100, 0.00232
	Fm	253		1/2+	3.00 d	EC, α	88, 12
	Fm	254		0+	3.240 h	α, SF	99.94, 0.06
	Fm	255		7/2+	20.07 h	α, SF	100, 2.4×10^{-5}
	Fm	256		0+	157.6 m	SF, α	91.9, 8.1
	Fm	257		(9/2+)	100.5 d	α, SF	99.79, 0.21
101	Md	255		(7/2−)	27 m	EC, α, SF	92, 8, ≤ 30.15
	Md	256		(0−, 1−)	76 m	EC, α, SF	90.7, 9.3, < 3.0
	Md	257		(7/2−)	5.3 h	EC, α, SF	90, 10, < 4.0
	Md	258		(8−)	51.5 d	α, SF	100, ≤ 0.003
102	No	252		0+	2.30 s	α, SF	73.10, 26.9
	No	254		0+	54 s	α, EC, SF	90, 10, 0.25
	No	255		(1/2+)	3.1 m	α, EC	61.4, 38.6
	No	256		0+	2.91 s	α, SF	99.5, 0.5
	No	257		(7/2+)	25 s	α	~100
104	Rf	256		0+	6.7 ms	SF, α	98, ~2.2
106	Sg	260		0+	3.6 ms	α, SF	50, 50
108	Hs	264		0+	0.08 ms	α, SF	100, < 1.5

Appendix 3
Physical Constants

The table below contains a selected set of physical constants taken from the data contained in the 1986 CODATA recommended values of the fundamental physical constants. All values are given in S.I. units. For a number of the constants, such as masses and magnetic moments, several additional values are given in units especially useful for applications.

A complete authoritative list of physical constants can be obtained from the National Institute of Science and Technology at http://physics.nist.gov/PhysRefData/codata86/codata86.html.

Constant	Symbol	Value (S.I. units)
Speed of light in vacuum	c	299792458 m s^{-1}
Permeability of vacuum	μ_o	$1.25663706143592 \times 10^{-06}$ N A^{-2}
Permittivity of vacuum	ε_o	$8.854187817 \times 10^{-12}$ F m^{-1}
Newtonian constant of gravitation	G	$6.67259 \times 10^{-11} \pm 8.5 \times 10^{-15}$ m^3 kg^{-1} s^{-2}
Planck constant	h	$6.6260755 \times 10^{-34} \pm 4.0 \times 10^{-40}$ J s
		$4.1356692 \times 10^{-15} \pm 1.2 \times 10^{-21}$ eV s
	$\hbar$	$1.05457266 \times 10^{-34} \pm 6.3 \times 10^{-41}$ J s
		$6.582122 \times 10^{-16} \pm 2.0 \times 10^{-22}$ eV s
Elementary charge	e	$1.60217733 \times 10^{-19} \pm 4.9 \times 10^{-26}$ C
Bohr magneton	μ_B	$9.2740154 \times 10^{-24} \pm 3.1 \times 10^{-30}$ J T^{-1}
		$5.78838263 \times 10^{-05} \pm 5.2 \times 10^{-12}$ eV T^{-1}

Nuclear Physics for Applications. Stanley G. Prussin
Copyright © 2007 WILEY-VCH Verlag GmbH & Co., Weinheim
ISBN: 978-3-527-40700-2

Appendix 3 Physical Constants

Constant	Symbol	Value (S.I. units)
		13996241800 ± 4200 T^{-1} s^{-1}
Nuclear magneton	μ_N	$5.0507866 \times 10^{-27} \pm 1.7 \times 10^{-33}$ J T^{-1}
		$3.15245166 \times 10^{-08} \pm 2.8 \times 10^{-15}$ eV T^{-1}
		7622591.4 ± 2.3 T^{-1} s^{-1}
Fine structure constant	α	$0.00729735308 \pm 3.3 \times 10^{-10}$
Inverse fine structure constant	$1/\alpha$	$137.0359895 \pm 6.1 \times 10^{-06}$
Bohr radius	a_o	$5.29177249 \times 10^{-11} \pm 2.4 \times 10^{-18}$ m
Electron mass	m_e	$9.1093897 \times 10^{-31} \pm 5.4 \times 10^{-37}$ kg
		$0.000548579903 \pm 1.3 \times 10^{-11}$ u
		510999.06 ± 0.15 eV
Electron Compton wavelength	λ_C	$2.42631058 \times 10^{-12} \pm 2.2 \times 10^{-19}$ m
Electron Classical radius	r_e	$2.81794092 \times 10^{-15} \pm 3.8 \times 10^{-22}$ m
Electron Thompson Cross section	σ_e	$6.6524616 \times 10^{-29} \pm 1.8 \times 10^{-35}$ m^2
Electron magnetic moment		$9.2847701 \times 10^{-24} \pm 3.1 \times 10^{-30}$ J T^{-1}
		$1.001159652193 \pm 1.0 \times 10^{-11}$ Bohr magnetons
		$1838.282 \pm 3.7 \times 10^{-05}$ nuclear magnetons
Electron g-factor	g_e	$2.002319304386 \pm 2.0 \times 10^{-11}$
Muon mass		$1.8835327 \times 10^{-28} \pm 1.1 \times 10^{-34}$ kg
		$0.113428913 \pm 1.7 \times 10^{-08}$ u
		105658389 ± 34 eV
Proton mass	m_p	$1.6726231 \times 10^{-27} \pm 1.0 \times 10^{-33}$ kg
		$1.00727647 \pm 1.2 \times 10^{-08}$ u
		938272310 ± 280 eV
Proton magnetic moment		$1.41060761 \times 10^{-26} \pm 4.7 \times 10^{-33}$ J T^{-1}

Constant	Symbol	Value (S.I. units)
		$0.001521032202 \pm 1.5 \times 10^{-11}$ Bohr magnetons
		$2.792847386 \pm 6.3 \times 10^{-08}$ nuclear magnetons
Proton gyromagnetic ratio	γ_p	267522128 ± 81 T^{-1} s^{-1}
Neutron mass	m_n	$1.6749286 \times 10^{-27} \pm 1.0 \times 10^{-33}$ kg
		$1.008664904 \pm 1.4 \times 10^{-08}$ u
		939565630 ± 280 eV
Neutron magnetic moment		$9.6623707 \times 10^{-27} \pm 4.0 \times 10^{-33}$ J T^{-1}
		$0.00104187563 \pm 2.5 \times 10^{-10}$ Bohr magnetons
		$1.91304275 \pm 4.5 \times 10^{-07}$ nuclear magnetons
Deuteron mass	m_d	$3.343586 \times 10^{-27} \pm 2.0 \times 10^{-33}$ kg
		$2.013553214 \pm 2.4 \times 10^{-08}$ u
		1875613390 ± 570 eV
Avagadro constant	N_A	$6.0221367 \times 10^{23} \pm 3.6 \times 10^{17}$ mol^{-1}
Atomic mass unit	u	$1.6605402 \times 10^{-27} \pm 1.0 \times 10^{-33}$ kg
		931494320 ± 280 eV
Faraday constant	F	96485.309 ± 0.029 C mol^{-1}
Boltzmann constant	k	$1.380658 \times 10^{-23} \pm 1.2 \times 10^{-28}$ J K^{-1}
		$8.617385 \times 10^{-05} \pm 7.3 \times 10^{-10}$ eV K^{-1}
Molar volume (ideal gas), STP	V_m	$0.0224141 \pm 1.9 \times 10^{-07}$ m^3 mol^{-1}
Stefan–Boltzmann constant	σ	$5.67051 \times 10^{-08} \pm 1.9 \times 10^{-12}$ W m^{-2} K^{-4}
Electron volt	eV	$1.60217733 \times 10^{-19} \pm 4.9 \times 10^{-26}$ J

Appendix 4
First-Order Time-Dependent Perturbation Theory

The results of first-order perturbation theory have been used throughout this text. A complete discussion of the theory can be found in a number of books on quantum theory. The present discussion is somewhat abbreviated and is meant to provide a working knowledge of the fundamentals and the ideas contained therein.

The wave function of any quantized system that can undergo a spontaneous transformation is described by the time-dependent Schrödinger equation. The wave functions of such states, while not solutions of the time-independent equation, can nevertheless be expressed as an expansion in the complete set of eigenfunctions of a stationary Hamiltonian that contains the total potential to which the system is subject, with the exception of that part which gives rise to the decay itself. For most of our applications, such as β- and γ- decay or spontaneous fission, the lifetime of a state that can decay is very, very much longer than the characteristic nuclear time of 10^{-22}–10^{-21} s and thus the state is "almost" a stationary state. For example, the wave function of the ground state of a nucleus that can decay only by β– emission is quite well-described, up to the moment of decay, by a solution to the Schrödinger equation that contains the strong and electromagnetic interactions but completely neglects the weak interaction. Similarly, almost all of the properties of a low-lying excited state of a nucleus that can decay only by photon emission or internal conversion are well-described by the solution of the Schrödinger equation that contains the strong and electromagnetic interactions and completely neglects the fact that decay can occur. Differences between the real states and the stationary approximations do, of course, exist. In particular, a state that can decay will not have a "sharp" energy but will exist with appreciable probability over a range in energies. This implies that the decay energy associated with a transition will also vary in a range that reflects the energy distributions of the two states involved. But, as we have seen during the discussion of resonances in low-energy neutron reactions, the widths of levels are very, very small even for states with lifetimes as short as 10^{-17} s.

The fact that most of the states we consider are "almost" stationary means that their expansions in the eigenfunctions of the appropriate time-independent Hamiltonian will have but a single term whose amplitude will be very nearly unity. To be sure, all other eigenfunctions of the same angular momentum and parity will be present, but their amplitudes will be extremely small. The same can be said for the wave function of the state to which decay occurs, whether it is "stable" or also can

Nuclear Physics for Applications. Stanley G. Prussin
Copyright © 2007 WILEY-VCH Verlag GmbH & Co., Weinheim
ISBN: 978-3-527-40700-2

Appendix 4 First-Order Time-Dependent Perturbation Theory

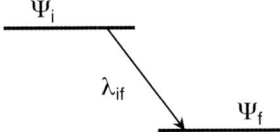

Fig. A4.1 Decay of a long-lived time-dependent state to a stable or long-lived state Ψ_f.

undergo decay. These observations lie at the heart of the treatment of decay probabilities in the limit of first-order time dependent perturbation theory. While rather mathematical, the development of the theory will be made much more tractable if the reader keeps these observations in mind.

We consider the simple decay shown schematically in Fig. A4.1. The initial time-dependent state Ψ_i undergoes decay to the stable or long-lived state Ψ_f with a decay constant λ_{if}. We assume that the total Hamiltonian H for the system can be approximated as

$$H = H_o + H' \tag{A4.1}$$

where H_o is a stationary Hamiltonian and H' is the perturbation potential that produces the transition. Both Ψ_i and Ψ_f are then eigenfunctions of H and are obtained by solution of the time-dependent Schrödinger equation

$$H\Psi = i\hbar \frac{\partial \Psi}{\partial t} = (H_o + H')\Psi \tag{A4.2}$$

and, of necessity, both H_o and H' are Hermitian. In keeping with the discussion above, we carry out an expansion of Ψ in the eigenfunctions of H_o that we write generally in the form ψ_n^o. Including the time-dependent part of the form $e^{-\frac{iE_n^o t}{\hbar}}$, the functions Ψ are then expanded as

$$\Psi = \sum_m a_m(t) \psi_m^o e^{-\frac{iE_m^o t}{\hbar}} \tag{A4.3}$$

where, if the state is to decay, the expansion coefficients $a_m(t)$ must be time-dependent. If Eq. (A4.3) is substituted directly into (A4.2), we have

$$\begin{aligned}H_o\Psi + H'\Psi &= i\hbar \sum_m \dot{a}_m(t) \psi_m^o e^{-\frac{iE_m^o t}{\hbar}} + i\hbar \sum_m a_m(t) \psi_m^o \left(\frac{-iE_m^o}{\hbar}\right) e^{-\frac{iE_m^o t}{\hbar}} \\ &= i\hbar \sum_m \dot{a}_m(t) \psi_m^o e^{-\frac{iE_m^o t}{\hbar}} + \sum_m a_m(t) E_m^o \psi_m^o e^{-\frac{iE_m^o t}{\hbar}}\end{aligned} \tag{A4.4}$$

Note that the second term in equation (A4.4) is nothing more than $H^o\Psi$, and we then must have

$$H'\Psi = i\hbar \sum_m \dot{a}_m(t)\psi_m^o e^{-\frac{iE_m^o t}{\hbar}} \quad (A4.5)$$

Now H' is the perturbation potential that transforms the initial state Ψ_i into the final state Ψ_f and we can then take $H'\Psi = H'\Psi_i$ for our specific problem. If we make this substitution in equation (A4.5), multiply from the left by Ψ_f^* and integrate over all spatial and spin coordinates represented by $d\tau$, we can then write

$$\int \Psi_f^* H'\Psi_i d\tau = i\hbar \int \Psi_f^* \sum_m \dot{a}_m(t)\psi_m^o e^{-\frac{iE_m^o t}{\hbar}} d\tau$$

$$= i\hbar \sum_m \dot{a}_m(t) e^{-\frac{iE_m^o t}{\hbar}} \int \Psi_f^* \psi_m^o d\tau \quad (A4.6)$$

Remember, we are only integrating over spatial and spin coordinates.

To proceed further, we can expand the final state Ψ_f in the complete set of orthonormal eigenfunctions of H_o and substitute this expansion into the integral in the second of Eq. (A4.6). While this is formally the correct step to take, we can now make use of the "almost stationary" natures of both the initial and final states of the system. Namely, because both are nearly stationary, each must be, to a good approximation, just one of the eigenstates of H_o. If we then take $\Psi_f^* \approx (\psi_f^o)^*$, the integrals that must be performed are all of the form

$$\int \Psi_f^* \psi_m^o d\tau \approx \int (\psi_f^o)^* \psi_m^o d\tau = \delta_{fm} \quad (A4.7)$$

the last equality simply reflecting the fact that the eigenfunctions of an Hermitian operator are orthogonal. Therefore, so long as we deal with nearly stationary initial and final states, all but one of the very many integrals in equation (A4.6) will vanish or very nearly so. If we set $f = m$, this approximation then leads to the result

$$\int \Psi_f^* H'\Psi_i d\tau \approx i\hbar \dot{a}_f(t) e^{-\frac{iE_f^o t}{\hbar}} \quad (A4.8)$$

Before we try to interpret this result, it is useful to write the integral on the right-hand side of (A4.8) by direct substitution of the expansion of the initial state in terms of the eigenfunctions of H_o (Eq. (A4.3)). That is,

$$\int \Psi_f^* H' \Psi_i d\tau = \int \Psi_f^* H' \sum_m a_m(t) \psi_m^o e^{\frac{iE_m^o t}{\hbar}} d\tau$$

$$= \sum_m a_m(t) e^{-\frac{iE_m^o t}{\hbar}} \int \Psi_f^* H' \psi_m^o d\tau \quad (A4.9)$$

If we now equate the right-hand sides of Eqs (A4.8) and (A4.9) and solve for $\dot{a}_f(t)$, we find

$$\dot{a}_f(t) = \frac{1}{i\hbar} \sum_m a_m(t) e^{\left[\frac{i}{\hbar}(E_f^o - E_m^o)t\right]} \int \Psi_f^* H' \psi_m^o d\tau \quad (A4.10)$$

where we have dropped the approximate sign and from here on will interpret our results as exact in the limit of our stationary approximation.

There are four changes that we can make to simplify our result somewhat. First, we recognize that we can use the approximation $\Psi_f^* \approx (\psi_f^o)^*$ in the integral. Second, it is customary to make the definition

$$\omega_m \equiv \frac{1}{\hbar}(E_f^o - E_m^o) \quad (A4.11)$$

Third, the integral in equation (A4.10) is usually written in terms of the Dirac notation used in the matrix formulation of quantum mechanics and we therefore write

$$\langle \psi_f^o | H' | \psi_m^o \rangle \equiv \int \Psi_f^* H' \psi_m^o d\tau \quad (A4.12)$$

where the left-hand side is referred to as the matrix element of the perturbation potential between the initial and final states. Finally, by assumption, the initial state is very nearly a single eigenstate of the stationary Hamiltonian. Thus, all of the integrals implied by the summation in (A4.10) must be negligibly small compared to that with the index m equal to i. With these changes, (A4.10) becomes

$$\dot{a}_f(t) = \frac{1}{i\hbar} a_i(t) e^{i\omega_i t} \langle \psi_f^o | H' | \psi_i^o \rangle \quad (A4.13)$$

The expression in equation (A4.13) must still be manipulated further to obtain the desired result, but there are a number of points that can now be made with respect to physical interpretation. The left-hand side represents the rate of change of the amplitude of the final state eigenfunction in the system. We know that the system initially is, for all intents and purposes, in the eigenstate ψ_i^o and thus $a_i(t = 0) \cong 1$ and $a_f(t = 0) \cong 0$. On the other hand, after the decay, $a_i(t \to \infty) \cong 0$ and $a_f(t \to \infty) \cong 1$. It is the perturbation potential that causes the transition and the matrix element or integral (Eq. (A4.13)) can be interpreted as the operation of the perturbation potential on the initial state to produce the final state. The rate at which the final state "grows" into the system clearly will depend upon the magnitude of the matrix element. If large, the amplitude of the final state wave function will increase rapidly and the transition from the initial to the final state will be "fast". If small, the transition will be "slow". Within our approximation that the initial and final states are very nearly stationary, we can interpret $\tilde{a}_f(t)$ as the "amplitude" of the transition, i.e., $\tilde{a}_f(t) = \tilde{a}_{i \to f}(t)$. At any time t, the amplitude of the transition is found by integrating Eq. (A4.13) over time and thus

$$a_{i \to f}(t) = \int \frac{1}{i\hbar} a_i(t') e^{i\omega_i t'} \langle \psi_f^o | H' | \psi_i^o \rangle dt' \tag{A4.14}$$

where the variable t' measures all time from the creation of the system, or the application of the perturbation potential, to the time t. For generality, we can take

$$H'(t') = \begin{cases} 0, & -\infty < t' < 0 \\ H'(t'), & t' \geq 0 \end{cases} \tag{A4.15}$$

Now we recognize that the transition actually takes place essentially instantly. That is, we do not see the transition take place over times long compared to the characteristic time for motion. Hence, so long as the transition has not occurred, $a_i(t') \approx 1$, and thus (A4.14) can be approximated by

$$a_{i \to f}(t) = \frac{1}{i\hbar} \int e^{i\omega_i t'} \langle \psi_f^o | H' | \psi_i^o \rangle dt' \tag{A4.16}$$

This expression is known as the Fermi Golden Rule #1. It relates the amplitude of a transition to the properties of the perturbation potential and the initial and final states of the system in the limit of first-order time-dependent perturbation theory.

In order to proceed further, the characteristics of the perturbation potential must be specified. The simplest case to consider is that where H' is just a constant and independent of time. This is what appears to be true in the case of β-decay. The weak interaction is a property of matter and is present from the moment of creation of a nuclide. If we make the assumption that

$$H'(t') = \begin{cases} 0, & t' < 0 \\ H', & t' \geq 0 \end{cases} \tag{A4.17}$$

the matrix element is not time dependent and can be removed from the integral. Thus

$$\begin{aligned} a_{i \to f}(t) &= \frac{1}{i\hbar} \langle \psi_f^o | H' | \psi_i^o \rangle \int_0^t e^{i\omega t'} dt' \\ &= -\frac{1}{\hbar \omega} \langle \psi_f^o | H' | \psi_i^o \rangle (e^{i\omega t} - 1) \end{aligned} \tag{A4.18}$$

In this equation, we have dropped the subscript i from ω as there should be no confusion with respect to its meaning.

The quantity $a_{i \to f}(t)$ is the amplitude of the final state in the system at the time t and therefore $|a_{i \to f}(t)|^2$ represents the probability for finding the system in the final state at the time t. Squaring the second of Eqs (A4.18) is straightforward but can be manipulated a bit to give a relatively simple result. First,

$$|e^{i\omega t} - 1|^2 = (e^{i\omega t} - 1)(e^{-i\omega t} - 1) = 2 - (e^{i\omega t} + e^{-i\omega t}) \qquad (A4.19)$$

Also,

$$\sin\theta = \frac{1}{2i}(e^{i\theta} - e^{-i\theta}) \qquad (A4.20)$$

and therefore

$$\sin^2\theta = \frac{1}{4}[2 - (e^{2i\theta} + e^{-2i\theta})] \qquad (A4.21)$$

We can then rewrite (A4.19) as

$$|e^{i\omega t} - 1|^2 = 4\sin^2\left(\frac{\omega t}{2}\right) \qquad (A4.22)$$

and $|a_{i \to f}(t)|^2$ can now be written as

$$|a_{i \to f}(t)|^2 = \frac{4}{(\hbar\omega)^2}|\langle \psi_f^o|H'|\psi_i^o\rangle|^2 \sin^2\left(\frac{\omega t}{2}\right) \qquad (A4.23)$$

This result is worth some study. It shows that the probability of a transition oscillates in time in proportion to $\sin^2(\omega t/2) = \sin^2[(t/2\hbar)(E_f^o - E_i^o)]$ and decreases with increasing difference in the total energies of the final and initial states as $(\hbar\omega)^{-2} = (E_f^o - E_i^o)^{-2}$. For any fixed time, the transition probability will be large only in the vicinity of $E_f^o - E_i^o = 0$. The overall result reflects two facts. First, as we have pointed out above, time-dependent states have a distribution in energy and thus real differences in the decay energies of individual transitions will be found. Second, the Heisenberg Uncertainty Principle in the form $\Delta E \Delta t \geq \hbar$ indicates that at very short times, energy need not be conserved. However, as time becomes large, conservation of energy must be upheld.

The total transition probability must account for the energy distributions of the time-dependent states and we therefore must perform an integration over the full range of $(E_f^o - E_i^o) = \hbar\omega$. That is,

$$\int_{-\infty}^{\infty} |a_{i \to f}(t)|^2 d(\hbar\omega) = 4|\langle\psi_f^o|H'|\psi_i^o\rangle|^2 \int_{-\infty}^{\infty} \frac{1}{(\hbar\omega)^2}\sin^2\left(\frac{\omega t}{2}\right) d(\hbar\omega) \qquad (A4.24)$$

The argument $\omega t/2$ is just $y = \hbar\omega(t/2\hbar)$. With a change in the variable of integration to y, the integral on the right-hand side of (A4.24) can be written as

$$\frac{t}{2\hbar}\int_{-\infty}^{\infty}\frac{\sin^2 y}{y^2}dy = \frac{t}{\hbar}\int_{0}^{\infty}\frac{\sin^2 y}{y^2}dy = \frac{\pi t}{2\hbar} \tag{A4.25}$$

and Eq. (A4.24) then gives the result

$$\int_{-\infty}^{\infty}|a_{i\to f}(t)|^2 dE_f^o = \frac{2\pi}{\hbar}t|\langle\psi_f^o|H'|\psi_i^o\rangle|^2 \tag{A4.26}$$

which expresses the probability for decay of the system in the limit of first-order perturbation theory over all possible energy differences between the initial and final states of the system. It indicates that the probability grows linearly in time and that the decay probability per unit time is just

$$\frac{1}{t}\int_{-\infty}^{\infty}|a_{i\to f}(t)|^2 dE_f^o = \frac{2\pi}{\hbar}|\langle\psi_f^o|H'|\psi_i^o\rangle|^2 \tag{A4.27}$$

Note that this expression has the dimensions of energy/time. Because of the differences in the decay energies, we must consider that there will be a number of different final states available for the system. As an example, remember that in β decay the leptons are emitted into a continuum of states. If $\rho(E_f)$ represents the density of final states, the number of final states per unit of energy at E_f, the expected decay probability will then be

$$w = \frac{2\pi}{\hbar}|\langle\psi_f^o|H'|\psi_i^o\rangle|^2\rho(E_f) \tag{A4.28}$$

This result is known as the Fermi Golden Rule #2 and it is the fundamental relation used to estimate transition probabilities in the limit of first-order perturbation theory. While we will not provide the derivation, it is no more difficult to demonstrate that the Golden Rule also applies to the case of perturbation potential that is oscillatory in nature, as is found in the case of photon emission.

Index

a

Å 27
α decay 36, 245, 247–248, 251–252, 256, 268, 272
  – angular momentum effect 272
  – barrier 256
  – decay constant 268
  – energetic 247
  – half-life 248, 252, 270
  – nuclear structure effect 252
  – odd, odd 252
  – Odd-A nuclide 274
  – probability 252
  – structure effect 274
  – total probability 251
α emission 260, 269
  – barrier width 260
  – decay probability 269
α emitter 124
α particle 248, 253, 255, 410
  – first-excited state 248
  – heavy nuclei 253
  – lectron 410
  – total potential 255
  – trajectory 410
α transition 275
  – angular momenta 275
  – parity 275
absolute decay intensity 364
activity 61
α-decay 250
  – decay energy 250
  – half-life 250
alkaline earth element 129
allowed β decay 300
  – angular momentum effect 300
allowed energy state 97
allowed photon transition 344
  – observable 344
allowed state 181
  – spectrum 181
allowed transition 301, 366
  – Fermi model 301
alpha decay 4
  – α-particles 4
$^{241}_{95}$Am 8, 37, 248, 274
^{241}Am 45
angular distribution 436, 458, 466
  – reciprocity theorem 436
angular distribution function 461–462
angular frequency 172, 214
angular momenta 155, 300
  – coupling 155
  – nonzero 300
  – orbital 300
angular momentum 22, 80–81, 136–137, 147, 149–150, 152, 154, 165, 167, 186, 215, 220–221
  – eigenfunction 221
  – intrinsic 80–81, 150
  – orbital 152, 167, 186
  – quantization 154
  – quantum number 186, 215, 221
  – rotational 215, 220
  – square 150, 167
  – total 154, 220
  – z-component 165
angular momentum vector 154, 168, 210
  – length 168
antineutrino 21, 31–32, 283–285, 287
areal density 535
asymmetry energy 100, 105
atomic form factor 511
atomic mass 19, 94, 545
atomic mass unit 20
atomic spectroscopy 308
  – K shell 308
  – L shell 309

attenuation 69–70, 523
 – beam 69
 – exponential 69
 – γ-ray 70
 – neutron 69
attractive potential 92
$^{197}_{79}$Au(n, γ)$^{198}_{79}$Au 454
 – resonance 429
Auger electron 313
 – ejection 313
Auger electron ejection 495
average nuclear interaction 83
average range 531
axe 216
 – semi-major 216
 – semi-minor 216

b

β decay 281, 303
 – angular momentum 302
 – characteristic time 298
 – conservation law 281
 – decay constant 299
 – lepton pair 302
 – selection rule 303
 – transition 303
β⁻ decay 30–31
β⁻ decay chain 386, 390
β⁻ emission 22
β⁻- radioactive nuclide
 – neutron capture 374
β spectra 292
β stability 47
β transition 306
 – classification 306
 – first-forbidden 307
 – forbidden 307
 – log$_{10}$ft 307
β⁻ transition 364
β⁺ decay
 – mass–energy relation 32
^{138}Ba 483
background 66
background radiation 62, 65
band head 221
barrier 260, 264–265, 273
 – arbitrary shape 265
 – height 273
 – rectangular 264
 – thickness 273
barrier penetration 245, 258, 261, 266, 383, 456
β-decay scheme 320

β-delayed particle emission 325
 – $^{85}_{35}$Br 325
 – $^{85}_{36}$Kr 325
β-delayed proton emission 326
$^{7}_{4}$Be 34, 357
 – electron capture decay 321
$^{8}_{4}$Be 215
^{8}Be 216, 218
beam
 – polarized 80
 – unpolarized 80
Becquerel 45
Bessel function 178–181
 – asymptotic form 179
 – spherical 178–179, 181
 – zero 180–181
Bessel's equation 178
beta decay 5
 – annihilation 5
 – antineutrino 5
 – β⁻ decay 5
 – β⁻ particle 5
 – lepton 5
 – $\bar{\nu}$ 5
$^{209}_{83}$Bi 200
binary fission 125, 376, 385
binding energy 28–29, 77, 91–92, 94, 100
 – average 91–92
 – electron 30
 – electronic 31
 – nuclear 129
 – odd/even effect 94
 – variation 77
binomial distribution 43, 54, 56
 – radioactive decay 55
Bohr orbit 316
boson 332
$^{85}_{35}$Br 323, 367
Breit–Wigner single-level cross section 478
Breit–Wigner single-level formula 451, 454, 475
bremsstrahlung 539

c

$^{12}_{6}$C(α, d)$^{14}_{7}$N 436
$^{14}_{6}$C 22, 32, 38
 – low-energy 39
 – n,p 39
$^{14}_{6}$C$_{8}$ 6, 9
^{40}Ca 481
$^{41}_{20}$Ca 197
$^{42}_{20}$Ca 364, 366
$^{49}_{20}$Ca 198

Index | 621

Cartesian component   150
Cartesian coordinate   136
$^{109}_{48}$Cd   34
$^{144}_{58}$Ce   48
^{144}Ce   51, 53
center of mass   407, 412–414
 – momentum   412, 414
 – motion   407
 – velocity   412–413
center of mass coordinate system
 – projectile velocity   413
 – target velocity   413
Central Limit Theorem   389
central potential   162
centrifugal barrier   426
centrifugal force   172–173
 – conservative   173
centrifugal potential   172, 180
$^{246}_{98}$Cf   35
$^{252}_{98}$Cf   8, 37, 368, 390
chain yield   386–388
characteristic nuclear time   442, 450
characteristic x-ray   310
charge density   85, 205
charge distribution   84, 87, 205, 208, 390
 – electric dipole moment   208
 – root-mean-square radius   84
charge exchange   534
charged particle   494, 525
 – heavy   525
 – light   525
classical electron radius   90, 500
classical liquid drop   110
 – total binding energy   110
closed shell   191, 198, 210
coherent scattering   495, 502
collective effect   205, 222
collective excitation   212
collective motion   212–213
 – rotation   212
 – vibration   212
collision   410–411
 – grazing   410
 – head-on   411
commutator   144–145, 149
comparative half-life   299
competitive decay   35
complex conjugate   139
compound nucleus   441, 443–445
 – decay   444
 – formation   444
 – reaction   445
 – total excitation energy   441

Compton edge   508
Compton electron   504, 508
 – energy spectra   508
 – kinetic energy   504
Compton scattering   495
conservation   283
 – angular momentum   148
 – energy   283
 – momentum   283
conservation law   22, 134, 138, 282
 – conservation of energy   134
conservation of lepton   283
conserved quantity   135, 144
constant
 – fine structure   78
 – Planck   78
constant density   85–86, 88
 – compressible component   86
 – incompressible component   86
continuity condition   472–473
 – combined   473
continuity equation   334
continuum   290
conversion coefficient   356
conversion electron   331, 352
 – spectrum   352
coordinate system   407
 – center of mass   407
 – center-of-mass   27
 – laboratory   14, 407
 – rest frame   13
Coriolis force   220
 – rotational state   220
cosmic ray   9
 – protons   9
Coulomb barrier   257
Coulomb energy   89–92, 377
 – stored   377
 – stored energy   89
Coulomb force   27, 78, 92
Coulomb interaction   193, 254
Coulomb potential   84, 91, 194, 254, 266, 268
 – uniformly charged sphere   194
coupling   155
cross section   66, 68, 71, 79, 82, 174, 424, 427,
 429, 440, 446, 453, 455, 457, 459, 464, 469,
 471, 499, 505, 509–511
 – (n,γ)   429
 – 1/v   457
 – 1/V behavior   455
 – barn   68
 – coherent scattering   511
 – definition   68, 424

- differential Compton collision 505
- differential Compton energy-scattering 509
- differential elastic scattering 459
- differential scattering 79
- direct reaction 446
- elastic scattering 440
- energy absorption 510
- first-order perturbation theory 424
- isolated resonance 453
- neutron absorption 427
- pair production 519
- partial 68
- reaction 68, 471
- reaction probability 66
- thermal 431
- Thompson scattering 499, 505
- total 68, 469
- total reaction 427

cross section.partial 464
Curie 45
current density 262, 469
- classical mechanic 262
- quantum mechanic 262
- quantum mechanical 469

d

d, p 446
daughter 33
De Broglie relation 440
decay
- $^{85}_{35}Br$ 323
- statistical nature 43

decay chain 47, 51, 53
- complex 53

decay constant 41, 267, 292, 450
- statistical nature 42

decay probability 296
- β^- and β^+ decay 296

decay scheme 51, 357
deformation 234, 376
- energy barrier 376
- shape 234

deformation parameter 378
deformed nuclei 205, 212, 215, 226, 229
- level structure 229
- nuclear shapes 205
- single-particle motion 212

deformed potential 229, 233
- Hamiltonian 229

degeneracy 186–187
degenerate 155
degenerate Fermi gas 101

delayed neutron precursor 326
density 89
- nuclear matter 89

density of states 425
deuterium 25
deuteron 24, 81, 446
δ-function 448
diatomic molecule 186, 213
- low-energy 186
- rotation 213
- vibration 186, 213

dimensionless energy 315
di-neutron 92
direct reaction 446
disintegration rate 49, 61
- decay rate 49
- secular equilibrium 49

distribution 89
- Fermi 89
- Saxon–Woods 89

double β-decay 24
double factorial 179

e

E1 transition 350
E2 and M1 transition 351
effective ionization constant 531
eigenfunction 136–138, 144, 146–147, 149–150, 164–165, 168, 172, 215
- angular part 172
- rotational 215
- single-valued 164
- stationary part 172

eigenvalue 136
EL 339
elastic scattering 10, 70, 79, 82, 405, 407, 412, 414, 426, 444, 458, 472, 476
- binary reaction 412
- center of mass coordinate system 412
- compound 444, 472, 476
- Coulomb 79
- neutrons on protons 426
- potential 472

elastic scattering of photons 496
- unbound electron 496

electric dipole emission 340
electric dipole moment 497
electric dipole radiation 339
electric field 19, 333, 498
- multipole moment 333
- nuclear 205
- oscillating electric dipole 498

electric potential 205, 207

electric radiation   343
 – decay constant   343
electromagnetic field   336, 339
 – multipole moment   339
electromagnetic interaction   79
electromagnetic wave   332
 – tranverse nature   332
electron   27, 287, 311, 528, 538
 – ejection   311
 – energy loss   538
 – energy transfer   528
 – ionization   538
electron binding energy   26
electron capture   34, 284, 308, 317–318
 – competition   318
 – detection   308
 – probability   317
 – relative probability   310
electron capture decay   307, 312, 364
 – measurement   312
electron mass   19
electron scattering   87
ellipse   376
 – eccentricity   376
ellipsoid   209
 – oblate   209
 – prolate   209
ellipsoid of revolution   219
 – prolate   219
endoergic reaction   418
energetic   13
energy barrier   25
energy degeneracy   99
energy level   26
energy level diagram   33–34
energy loss   525, 529
 – quantum mechanic   529
energy release   77
 – fission   77
 – fusion   77
energy state   162
 – allowed   162
energy unit   20
 – electron volt   20
 – gigaelectron volt   20
 – kiloelectron volt   20
 – megaelectron volt   20
error
 – propagation   62
evaluated   11
excitation   443
 – many-particle   443
excitation energy   226

 – 4^+ level   226
 – first 2^+ level   226
excited state   7–8, 189, 484
 – density   484
 – low-lying   189
 – metastable state   8
 – nuclear isomer   8
 – transformation   9
exoergic reaction   418
expectation value   139
experimental   11

f

$^{18}_{9}F$ 33, 39, 42
$^{18}_{9}F_9$ 6, 9
^{18}F 45–46
$^{55}_{26}Fe$ 7
^{55}Fe 312
$^{56}_{26}Fe$ 465
 – elastic scattering   465
Fermi   282, 288
 – allowed decay   282
Fermi energy   189–190
Fermi function   288, 292, 294, 297, 316
 – electron capture   316
 – integrated   297
Fermi gas model   189, 443
Fermi golden rule   315, 337, 353, 424
Fermi golden rule #1   615
Fermi golden rule #2   285–286, 617
Fermi operator   303
Fermi theory   291
 – allowed β^- decay   287, 291
Fermi transition   302, 327
 – nuclear matrix element   302
fermion   80–81, 93, 150, 152
fermion state
 – antiparallel intrinsic spin   171
final state   13, 30, 291, 315
 – density   291
 – neutrino   315
first-order perturbation   286
 – decay constant   286
First-order time-dependent perturbation theory   285
fission   375, 383, 385, 389, 399–400, 406
 – charge distribution   389
 – energetic   375
 – energy release   399
 – neutron-induced   383, 406
 – recoverable energy   400
 – shell structure   385
 – spontaneous   383

– total energy release   400
fission barrier   397, 401–402, 488
 – liquid-drop model   402
fission cross section   487
fission fragment   376, 385, 393
 – average kinetic energy   393
fission fragment energy   393
 – spectrum   393
fission isomer   385
fission parameter   381, 384, 401
 – critical value   382
fission probability   383–384, 401
 – nuclear-structure effect   384
 – spontaneous   401
fission process   396
 – shape change   396
fission product   375, 385
 – primary   385
fission yield   388
 – independent   388
fluorescence yield   312–313, 368
flux   469
 – scattered   469
$^{254}_{100}$Fm   390
$^{256}_{100}$Fm   37, 390–391
$^{258}_{100}$Fm   391
force   18, 77–78
 – relative strength   78
force in nuclei   75
Fourier transform   448
 – frequency function   448
fragment kinetic energy   392
 – average total   392
free neutron   374
 – source   374
free particle   458
function
 – even   146
 – mean value   64
 – odd   146
fusion   24

g
γ decay   331
γ radiation   274
 – penetrating   274
gamma decay   8
 – γ-ray emission   8
Gamow factor   267, 269
Gamow–Teller operator   305–306
Gamow–Teller transition   305
 – nuclear matrix element   305
Gauss's law   527

Gaussian unit   90
$^{148}_{64}$Gd   248
$^{154}_{64}$Gd   215
genetic relation   32
 – daughter   32
 – parent   32
geometric   217
 – mean radius   217
gravitational coupling constant   78
γ-ray   8, 333, 385, 392
 – prompt   385, 392
 – wavelength   333
grazing collision   416
ground state   7, 22, 25–26, 31, 190, 195
 – even, even   195
 – parity   195
 – spin   195
growth and decay   45, 47

h
$^{2}_{1}$H   24
$^{2}_{1}$H, p   446
^{2}H (^{2}H, n) ^{3}He   421
$^{3}_{1}$H   24, 31
$^{3}_{1}$H(p, n)$^{3}_{2}$He   437
$^{3}_{1}$H$_2$
 – tritium   6
^{3}H   283
half-life   24, 43, 46, 250, 345, 352, 357, 442
 – excited state   52
 – partial   52, 250
 – single-particle estimate   442
Hamiltonian   220
Hamiltonian operator   137
harmonic oscillator   186, 213, 234
 – anisotropic   234
 – isotropic   186
 – one-dimensional   186, 213
harmonic oscillator potential   176–177
 – isotropic   176–177
$^{3}_{2}$He   24, 32
$^{3}_{2}$He(n, p)$^{3}_{1}$H   431, 436, 479
^{3}He   283, 320
$^{4}_{2}$He   4
heavy charged particle   525
 – classical model   525
 – energy loss   525
 – stopping   525
heavy element   368
 – K-binding energy   368
heavy mass peak   390
Heisenberg Uncertainty Principle   135, 145, 448, 616

Hermitian operator   139
^{175}Hf   239
$^2_1$H(n, n)$^2_1$H   427
^{2}H (^{2}H, n) ^{3}H   423
hindered transition   350

i

identical nucleon   92
identical particle   92, 109
IHO   186
 – level spectrum   186
imaging
 – SPECT   50
impact parameter   526, 529
 – maximum   529
 – minimum value   529
impulse approximation   526
incident flux   460
incoherent scattering   495, 512
incoherent scattering function   513
independent fission yield   386
independent particle   143, 178
independent particle approximation   175
 – average central potential   175
independent particle model   144
indium isotope   347
inelastic scattering   70, 405, 444
 – compound   444
initial   30
integrated Fermi function   296, 299
interaction   27, 77–78, 183–184, 287
 – electromagnetic   27
 – gravitational   78
 – local   287
 – spin–orbit   183–184
 – strong   27
interference   463, 476
intermediate state   442
 – half-life   442
internal conversion   331, 351–353, 356, 368
 – $0^+ \to 0^+$ transition   351
 – elementary theory   353
 – perturbation potential   353
 – transition probability   353
internal conversion coefficient   352, 356
internal energy   21
internal transition   347
intrinsic angular momentum   148, 152
intrinsic spin   93, 152
 – quantum number   152
 – z-component   152
intrinsic state   221
 – rotational band   222

intristic state
 – $^{175}_{72}$Hf   221
invariant   16
inversion   168
ionization energy   128
ionization potential   26
ionizing radiation   493
isobar   387
 – β-stable   119
 – hypothetical most stable   119
 – most probable charge   387
isolated resonance   445
isomer
 – $^{42}_{21}$Sc   366
isomeric   53
isomeric state   346, 365
isomeric transition   347
isotropic harmonic oscillator   185, 191, 234
 – angular frequency   234
 – Saxon–Woods potential   185
isotropic harmonic oscillator potential   189
 – degeneracy   189

k

$^{42}_{19}$K   364, 366
Katz–Penfold relation   540
k_c 26
kinematic   502
 – Compton Scattering   502
 – recoil angle   502
 – scattering angle   502
kinetic energy   392
 – fission fragment   392
Klein–Nishina cross section   505, 507, 513
 – application   507
 – unpolarized radiation   505
$^{85}_{36}$Kr   323, 331, 367

l

laboratory coordinate system   408
 – elastic scattering   408
Legendre equation   166
Legendre polynomial   166, 207, 223, 461, 464
 – associated   166
 – orthogonality   464
lepton   300
level   450
 – total width   450
level density   239, 443
level diagram   187, 191, 193, 195, 197
 – comparison   187
 – empirical single-particle   191
 – IHO   187

- isotropic harmonic oscillator 191
- neutron 193
- proton 193
- spherical potential 187
level diagram for protons 192
- empirical 192
level energy 31
level scheme 188, 198
- single-particle 188
level sequence 182
- energy spacing 182
- single particle 182
$^{7}_{3}$Li 322
$^{6}_{3}$Li(n, $^{3}_{1}$H)α 479
$^{7}_{3}$Li(p, n)$^{7}_{4}$Be 423
linear attenuation coefficient 524
linear momentum 22, 136
liquid drop model 212
- rotation 212
- vibration 212
liquid-drop model 376, 402
local interaction 307
$\log_{10}$ 364
$\log_{10}$ft 299
long-wavelength approximation 288, 315–316
Lorentz transformation 15

m

$^{A}_{Z}$M or M (Z, A) 4
$^{A}_{Z}$M or M (Z, A)
- excited state 4
- ground state 4
M4 transition 348
macroscopic cross section 523
magic number 128, 187, 191, 236, 271
magnetic dipole radiation 334
magnetic field 18
magnetic moment 148, 150, 152
magnetic quadrupole radiation 339
magnetic radiation 343
- decay constant 343
major shell 198
- energy gap 198
many-particle system 158
- wave function 158
- total parity 158
mass 20, 29
- atomic 29
- daughter 30
- electron 18
- hydrogen atom 21
- neutral atom 30
- neutron 18, 21
- nuclide 29
- parent 30
- proton 18
- quark 18
mass attenuation coefficient 524
mass density 85, 214
mass excess 117
- isobar 117
mass number 92
mass peak 386
mass spectrometer 19
mass yield 386
mass–energy 20, 31
mass–energy balance 23, 36–37
- fission probability 37
mass-yield distribution 386–387
matrix element 286, 289, 354, 356
Maxwell equation 334
Maxwell–Boltzmann distribution 110, 386, 396, 432
mean free path 524
mean level lifetime 447
mean lifetime 44
mean radius 216
- geometric 216
metastable state 53, 346
$^{24}_{12}$Mg 215
microscopic reversibility 436, 445
mirror nuclei 197
mirror nuclide 120
ML 339
$^{100}_{42}$Mo 24
$^{99}_{42}$Mo 49
moment of inertia 214–216
momentum spectrum 292
- electron 292
most probable charge 389
most stable isobar 389
- hypothetical 389
motion 212
- correlated 212
multipole
- electric 207
multipole moment 208
multipole operator 339–340
- matrix element 339

n

$^{14}_{7}$N 22
$^{14}_{7}$N(d, α)$^{12}_{6}$C 436
$^{A}_{N}$N 4
N = 49 isotones 347, 349

n,n´ 432
$^{22}_{11}$Na 33
$^{22}_{10}$Ne 33
neutral atom 29
neutrino 32, 34, 282–284
- mass 283
neutron 3, 9, 19, 32, 320, 374, 385–386, 392, 405, 465
- configuration 320
- decay 32, 320
- discovery 374
- prompt 385, 392
- slowing down 405
- thermal 386
neutron absorption 71
neutron emission 386, 398
- delayed 386
- fission fragment 399
neutron excess 104
neutron flux 71
neutron mass 19
neutron scattering 432
- inelastic 432
neutron state 191
- single-particle 191
neutron width 456
neutron-capture 406
^{60}Ni 433
Nilsson 189
Nilsson diagram 237–238, 240, 384
Nilsson model 234, 236, 239, 257, 274, 277
- deformation parameter 236
- Hamiltonian 234–235
- Nilsson diagram 236
- rotational state 274
Nilsson state 242
$^{259}_{102}$No 8
nomenclatura or definitons
- barn 68
nonrelativistic nuclear reaction 418
- kinematic 418
nonrelativistic reaction 407
- kinematic 407
normalizable 138
normalization 165
notation 3
- atomic number 3
- isotope 3
- mass number 3
notation or definiton
- electron volt 20
- energy 20

- gigaelectron volt 20
- kiloelectron volt 20
- megaelectron volt 20
nuclear binding 94
- quantum effect 94
nuclear binding energy 75, 112
- average 75
- total 75
nuclear density 89, 212
nuclear fission 71, 373, 375, 385
- discovery 373
- induced 385
- liquid-drop model 375
- spontaneous 385
nuclear force 9, 82–84, 88, 92
- short range 82, 84
- spin dependence 80–81, 92
nuclear interaction 183
- spin-dependence 183
nuclear isomer 346, 367
nuclear mass 13, 113
- global fit 113
- local fit 113
nuclear mass surface 115
- isobar 116
nuclear matter 84
- constant density 84
nuclear physic
- low-energy 2
nuclear radii 83, 85, 90
- hard-sphere approximation 88
- proton 86
- radius 86
nuclear radius 89, 100
nuclear reaction 9, 13, 21, 38, 405–406, 424, 440, 468
- binary 9
- low-energy 405, 440
- mass–energy balance 38
- partial wave analysis 468
- projectile 9
- qualitative consideration 440
- shorthand notation 9, 406
- target 9
nuclear shell structure 129, 161
nuclear stability 24
nuclear stopping 494, 534
nuclei 161, 196–197, 212
- deformed 212
- doubly magic 197
- odd, odd 196
- odd-A 196
- structure 161

nucleon
- binding 28
- binding energy 75
nucleon motion 446
- characteristic time 446
nucleus 87–88, 94, 111
- charge distribution 88
- even, even 94, 111
- mass distribution 87–88
- odd, odd 111
- quantized structure 88
- radius 223
nuclide
- mass 75
nuclide table 565

o

$^{17}_{8}O$ 190, 196
- ground state 190
oblate shape 225
occupancy 189
operator 135, 137–138, 145, 147–148
- Cartesian component 148
- Hermitian 612
- parity 147
orbital angular momentum 148, 186, 232
- deformed potential 232
- quantization 186
- spherical potential 232
orthonormal function 141
oscillator frequency 234
oscillator quantum 235
outgoing wave 468
- amplitude change 468
- phase shift 468

p

pair production 494, 519
- electron field 520
pairing 92
pairing effect 109
pairing energy 92–94, 119, 190, 402
parent 33
parent–daughter relation 46
parity 146, 158
parity operator 146
partial decay constant 52
partial half-life 378
- spontaneous fission 378
partial wave analysis 457
partial width 451, 455
particle 101, 156–157, 297
- identical 156–157

- non-interacting 101
- total energy 297
- uncollided 70
particle current density 67, 262
particle in a box 230–231, 289
- energy level 231
- oblate 231
- prolate 231
particle in a box model 100
particle in a quantized box 94
Pauli 281
Pauli Exclusion Principle 4, 93, 101, 189
Pauli spin operator 305
$^{201}_{82}Pb$ 406
$^{206}_{82}Pb$ 245
$^{207}_{82}Pb$ 479
$^{208}_{82}Pb$ 427
$^{209}_{82}Pb$ 200
perturbation 83, 144
perturbation potential 426, 612–613
perturbation theory 282, 611
- first-order time-dependent 611
- time-dependent 282
phase factor 462
phase shift 463, 471
phonon 185
phonon state 226
photoelectric absorption 517
- K shell 517
photoelectric effect 494–495, 516
photoelectric interaction 516, 518
- edge 518
photon 311, 332, 494–495, 503, 522
- absorber 522
- angular momentum 311, 332
- attenuation coefficient 520
- collimator 522
- elastic scattering 494–495
- energy deposition 522
- inelastic scattering 494–495
- pair production 495
- photoelectric absorption 495
- reduced energy 503
- total cross section 520
- transport 522
photon detector 508
photon emission 334, 341, 353
- decay constant 342
- density of state 341
- matrix element 341
- parity 339
- theory 334
- transition probability 353

- transition rate 341
- Weisskopf single-particle estimate 344
pickup reaction 446
Planck's constant 96
plane monochromatic wave 496
$^{210}_{84}$Po 248
^{210}Po 254
Poisson distribution 43, 56–58, 62
  - binomial distribution 56
  - Gaussian approximation 66
  - Gaussian distribution 60
  - standard deviation 58
  - variance 58
Poisson statistic 65
positron emission 34, 284, 318
  - competition 318
  - Q_β 32
positron emitter 6
potential 28, 162–163, 175–176, 184, 188
  - harmonic oscillator 175
  - Saxon–Woods 176, 188
  - space-dependent 163
  - spherically symmetric 163
  - spin–orbit 184
potential barrier 382, 456
potential diagram 266
potential energy 26, 375, 384
  - Coulomb 375
  - deformation 384
  - equilibrium 384
potential energy surface 375
potential field
  - average 27
potential function 162
  - spin-dependent 162
potential scattering 440, 445, 475
potential scattering cross section 478
  - low energy 478
potential well 25, 28, 93, 258
  - finite 258
Poynting vector 496
$^{144}_{59}$Pr 48
^{144}Pr 51, 53
probability amplitude 138
probability density 138
probe 83
projectile 23, 38, 408–409, 419
  - property 409
  - Q value 38
projectile flux 424
prolate ellipsoid 376
prolate shape 225

prompt 8
- fission product 8
- γ-ray 8
- neutron 8
prompt γ-ray 399
- spectrum 399
prompt neutron 395–396
- energy spectrum 395
- evaporation 396
- kinetic energy 395
proton 3, 27, 320, 336, 530
- configuration 320
- deuteron 3
- intrinsic magnetic moment 336
- intrinsic spin vector 337
- isobar 4
- isotone 3
- magnetic moment 337
- nucleon 3
- nuclide 3
- spin g-factor 337
- stopping power 530
proton density 87
proton level 194
- ordering 194
$^{239}_{94}$Pu 8, 390, 400–401
^{239}Pu 405
$^{240}_{94}$Pu 401
$^{239}_{94}$Pu (n, f) 10

q

Q value 30–31, 418
Q_β 34
Q_β^- 32
Q_{EC} 34
quadrupole moment 208–210, 212
- electric 210, 212
- experimental 210
- nuclear 210
- single particle 210
quadrupole vibration 224–225
quantized state 182
- designation 182
- orbital angular momentum 182
quantized system 27
Quantum mechanic 133
quantum mechanical tunneling 261
quantum number 93, 165, 236
- asymptotic 236
quantum state 156
- discrete 156
quark 82

r

$^{226}_{88}$Ra   246, 374, 406
radial equation   172
radiation field   337
radioactive decay   13, 41, 45, 54
   – poisson statistic   54
   – spontaneous   13
   – statistical consideration   54
   – statistical fluctuation   60
radioactive decay series   245
radionuclide   71
   – production rate   71
radius parameter   86
Raleigh scattering   495
range-scaling law   532
rare gase   128
$^{85}_{37}$Rb   24
$^{87}_{37}$Rb   24
reaction   407
   – endoergic   407
   – exoergic   407
reaction cross section   425–426
   – energy dependence   426
reaction probability
   – center of mass coordinate system   412
reaction Q value   407
reaction rate   69
reciprocity theorem   435
   – forward reaction   435
   – reverse reaction   435
recoil angle   418
recoil energy   247
recommended mass   545
recurrence relation   179
reduced mass   174
relativity
   – special theory   13
resonance   427, 477
   – observable parameter   477
rest mass   16, 23, 29
   – energy equivalent   17
rigid body   214
   – rotation   214
rigid rotator   215
rigid rotor   218, 252
$^{222}_{86}$Rn   246
r_o 85–86
root-mean-squared radius   85
rotation   151, 214
   – frequency of rotation   151
   – Hamiltonian   214
   – kinetic energy   151, 214
   – momentum   214

rotational band   217–218, 252, 368
   – energy   218
rotational excitation   213, 219
   – even,even   213
   – Odd-A nuclei   219
rotational level   250
rotational motion   150, 173, 219
   – angular momentum   219
rotational spectrum   215
   – rigid body   215
rotational state   213, 215
   – angular momentum   215
   – energy spectrum   213
   – parity   215

s

saddle point   396
Saxon–Woods potential   175, 254
$^{41}_{21}$Sc   197
$^{42}_{21}$Sc   364, 366
$^{49}_{21}$Sc   198
scalar potential   334
scattered neutron   459
   – wave function   459
scattered photon   507
   – energy spectrum   507
scattered projectile   409
scattering   79, 84, 464, 510
   – coherent   510
   – electron   84
   – incoherent   510
   – s-wave   464
scattering angle   410, 417–418
   – maximum   410
scattering cross section   469
   – differential   469
scattering distribution   80
   – cylindrical symmetry   80
Schrödinger equation   96, 98, 134, 258, 611–612
   – time-dependent   611–612
$^{82}_{34}$Se   24
second-chance fission   489
secular equilibrium   247
selection rule   306
   – Gamow–Teller transition   306
semi-empirical mass formula   109, 114, 117, 123, 125, 128, 375–376, 382
   – α decay   123
   – asymmetry term   114
   – β decay   117
   – Coulomb term   114
   – discrepancy   128

– nuclear fission 125
– nuclear shell structure 123
– surface term 114
– volume term 114
separation energy 28, 130, 191, 253
– neutron 130
– proton 130
separation of variables 99, 162, 164, 172
shape isomer 385
sharp energy 611
shell gap 197
shell structure 4, 109, 148, 184, 249, 270
SI 20, 45
single particle level diagram 161
single-particle level 181, 366
single-particle model 143, 195
– experiment 195
single-particle states 191
skin thickness 89
$^{146,147}_{62}$Sm 248
$^{132}_{50}$Sn 391, 399
sources of nuclear data 11
– nuclear reaction 11
– nuclear structure 11
spectrum 156
spherical Bessel function 300, 333, 338, 340, 461
spherical harmonic 166–169, 223, 333, 338, 355, 461
– addition theorem 355
– even parity 168
– odd parity 168
– parity 168
spherical nuclei 109, 189
– single-particle level 189
spherical polar coordinate 136, 162
spherical potential 176
– finite 176
spherical potential well 178, 181, 184, 456, 472
– energy 181
– infinite 178, 184
– level scheme 184
– reaction 472
– state 181
– S-wave scattering 472
spherical shell model 161
spherical shell structure 390
spherical symmetry 208
spin-orbit 155
spin–orbit coupling 172
spontaneous 375
– fission 375

– neutron-induced 375
spontaneous decay 23
spontaneous fission 7, 37, 127, 378, 382–383, 391
– ^{234,235,238}U 127
– $^{252}_{95}$Cf 127
– $^{256}_{100}$Fm 127
– fission fragment 7
– half-life 383
– mass-yield distribution 391
– Q_f 37
– spontaneous fission 7
square integrable function 138
$^{87}_{38}$Sr 24
stability 25
stable nucleus 134
standard deviation 65
state 366
– bound 25–27
– discrete 26
– energy spectrum 366
– initial 13, 22
– unbound 25–27
stationary Hamiltonian 612
statistical factor 454, 477
statistical fluctuation 61–62
stopping 538
– electron 538
– positron 538
stopping medium 525, 532
– property 532
stopping power 530
straggling 535
stripping reaction 446
strong interaction 78–79, 84, 88, 148
– nuclear force 78
– probe 79
symmetric fission 375, 386, 391

t

target 23, 38, 69, 80, 408, 419
– polarized 80
– thick 69
– thin 69
Taylor serie 476
$^{99m}_{43}$Tc 49
^{99m}Tc 8
$^{230}_{90}$Th 246
$^{231}_{90}$Th 274, 276
$^{234}_{90}$Th 5
^{234}Th 251, 268–269
^{236}Th 253
thermal fission 390

thermal neutron   405
Thomas–Fermi
  – input page   30
  – statistical model   31
Thompson scattering   495–496, 500
  – total cross section   500
three-dimensional box   98
threshold   33–34, 421
threshold energy   422
time reversal   436
time-dependent   285
time-dependent state   447, 449
  – energy   447
  – energy dependence   449
  – probability distribution   449
time-independent solution   142
$^{201}_{81}$Tl   39, 406
total angular momentum   136
total decay constant   52
total energy   17, 21, 26, 29, 136
total level width   451
total reaction cross section   468
total wave disturbance   462
total wave function   459
total width   455
tracer experiment   374
transformation   22
  – spontaneous   21, 23
transition   306, 364, 366, 615
  – amplitude   615
  – first-forbidden   366
  – Gamow–Teller   306
  – hindered   364
  – isospin   307
  – selection rule   306
transition probability   616
  – total   616
transmission coefficient   263–266, 273, 276, 456–457
tritium   25, 320
  – decay   320
triton   24

u

$^{235, 238}_{92}$U   248
$^{235}_{92}$U   37, 274, 276, 378, 381, 386, 390, 392, 395, 399–401, 406, 421
$^{235}_{92}$U(n, f)   487
$^{235}_{92}$U   241, 405, 485
$^{236}_{92}$U   386, 392, 401, 406
$^{238}_{92}$U   5, 77, 215, 245
$^{238}_{92}$U(n, f)   487
^{238}U   216, 219, 251–253, 268–269, 484, 486

$^{235}_{92}$U (n, f)   10

v

valence nucleon   253
  – binding   253
valley of β stability   375
vector diagram   154
vector Helmholtz equation   338
vector model   152
  – angular momentum   152
vector potential   334, 336–337
vibration   222, 228
  – changes in shape   222
  – two-phonon   228
vibrational energy   222
  – quantum   222
vibrational excitation   222
vibrational quantum   224
vibrational spectra   228

w

wave equation   335
  – three-dimensional   335
wave function   134, 163, 165, 170, 172, 215, 285, 300–301, 447, 458
  – angular   163
  – antineutrino   300
  – asymptotic limit   458
  – neutrino   301
  – radial dependence   300
  – spherical nucleus   215
  – spin-dependent   172
  – symmetry   215
  – time dependent   447
weak interaction   284–285, 287, 298
  – potential   287
Weisskopf single-particle estimate   364
Weizsäcker semi-empirical mass formula   112

x

$^{135}_{54}$Xe   479
X-ray
  – naming   310
x-ray   8, 25, 495
X-ray emission   308
  – bound electron   308

y

^{89}Y   486

z

zero-point energy   185

Related Titles

Stacey, W. M.

Nuclear Reactor Physics

735 pages
2007
Hardcover
ISBN: 978-3-527-40679-1

Lilley, J. S.

Nuclear Physics

Principles and Applications

412 pages in 2 volumes
2001
Softcover
ISBN: 978-0-471-97936-4

Krane, K. S.

Introductory Nuclear Physics

862 pages in 2 volumes
1995
Hardcover
ISBN: 978-0-471-80553-3